Lecture Notes in Physics

Edited by J. Ehlers, München, K. Hepp, Zürich
R. Kippenhahn, München, H. A. Weidenmüller, Heidelberg
and J. Zittartz, Köln
Managing Editor: W. Beiglböck, Heidelberg

135

Group Theoretical Methods in Physics

Proceedings of the IX International Colloquium
Held at Cocoyoc, México, June 23–27, 1980

Edited by Kurt Bernardo Wolf

Springer-Verlag
Berlin Heidelberg New York 1980

Editor

Kurt Bernardo Wolf
Instituto de Investigaciones
en Matemáticas Aplicadas y en Sistemas (IIMAS)
Universidad Nacional Autónoma de México
México 20, D.F.

ISBN 3-540-10271-X Springer-Verlag Berlin Heidelberg New York
ISBN 0-387-10271-X Springer-Verlag New York Heidelberg Berlin

Printing and binding: Beltz Offsetdruck, Hemsbach/Bergstr.
2153/3140-543210

Preface

The importance which group theoretical methods have been acquiring in physics and the new mathematical developments in their track have made it increasingly necessary to have periodic conferences in this field. Thus in the early seventies, thanks to the efforts of members of the CNRS in Marseille and of the University of Nijmegen, the Colloquia on Group Theoretical Methods in Physics were started. Since then they have taken place regularly in '72 and '74 in Marseille, '73 and '75 in Nijmegen, '76 in Montréal, '77 in Tübingen, '78 in Austin, and '79 in Kiriat Anavim, Israel.

The IX International Colloquium on Group Theoretical Methods in Physics took place in Cocoyoc, an old hacienda converted into a resort and convention hotel, about 70 kilometers south of Mexico City, in the week of June 23 to 27, 1980. There were 121 participants from 22 countries who presented 15 invited talks and 93 contributed papers.

The local organizing committee consisted of:

Marcos Moshinsky (Chairman)

Clicerio Avilez	Laurence Jacobs
Charles P. Boyer	Jerzy Plebański
Elpidio Chacón	Kurt Bernardo Wolf.

Technical assistance was provided by Oscar Troncoso, with the cooperation of Beatriz Aizen, Armando Antillón, Enrique Henestroza, Gustavo López, and Socorro del Olmo.

The members of the Standing Committee of the Group Theory Colloquia are:

H. Bacry, L. C. Biedenharn, J. L. Birman, K. Bleuler, A. Bohm, L. L. Boyle, A. P. Cracknell, G. G. Emch, L. P. Horwitz, A. Janner, B. R. Judd, P. Kramer, V. I. Man'ko, L. Michel, M. Moshinsky, Y. Ne'eman, C. Piron, S. Sternberg, P. Winternitz, and J. A. Wolf.

The local organizing committee hopes that the atmosphere in Cocoyoc was conducive to strong interactions between the participants. In fact, it seemed that some of the more interesting discussions on physics and mathematics occurred, besides in the lecture rooms, in the outside bar and around the swimming pool.

The Colloquium was made possible thanks to the cooperation and financial assistance of the following institutions:

INTERNATIONAL UNION OF PURE AND APPLIED PHYSICS

UNIVERSIDAD NACIONAL AUTONOMA DE MEXICO

CONSEJO NACIONAL DE CIENCIA Y TECNOLOGIA

INSTITUTO NACIONAL DE INVESTIGACIONES NUCLEARES

CENTRO DE INVESTIGACION Y ESTUDIOS AVANZADOS, IPN

FOMENTO EDUCACIONAL A. C.

To all of them the local committee wishes to express its gratitude. We are also indebted to the participants for their cooperation and patience, and to the members of the Standing Committee who gave very helpful advice.

Marcos Moshinsky
(Chairman, Local
Organizing Committee)

The local organizing committee for the IX Colloquium made a call for papers from the scientific community working with the applications of group theoretical methods in physics, and invited some of the leading representatives of each field to deliver survey or important research talks. The program was set up giving one-hour plenary sessions to the invited speakers and half-hour simultaneous sessions to the contributed talks. No papers were rejected by the committee.

Simultaneous sessions were held in order to accomodate the number of presentations, dividing the papers into nine areas, in the way followed in these Proceedings. Care was taken so that a minimum of overlap should occur between talks delivered simultaneously. During the Conference, a Poster Session was created so as to accomodate for last-minute contributions, and for speakers who felt this was a better mode of interaction with their colleagues. In these Proceedings, all contributions -including invited speakers- are divided into the nine areas, and an author index is appended at the end.

A high point in the IX Colloquium was the 1980 Wigner Medal ceremony, honouring Professor Izrail Moiseevich Gel'fand. As Professor Gel'fand could not receive the medal in person, a congratulatory telegram was signed by most of the participants. A Section on the Wigner Medal award is included in these Proceedings.

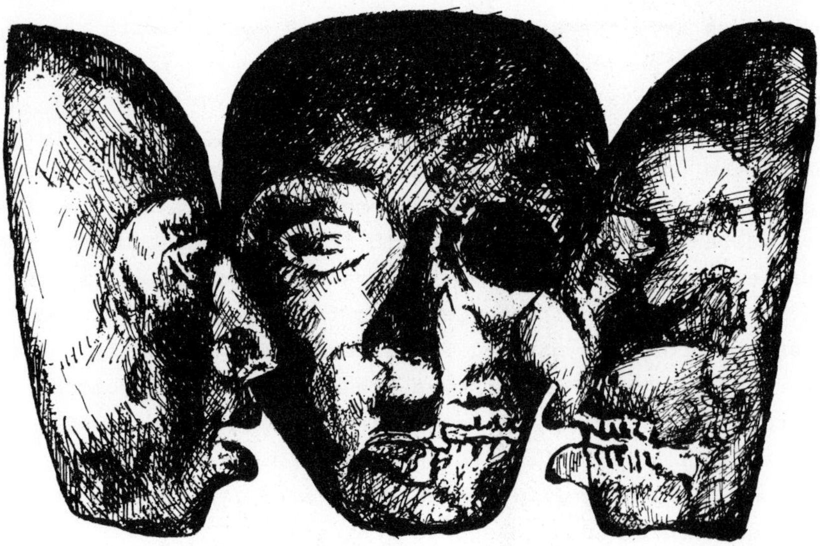

OUR LOGO

Early in the activities of the local organizing committee, we decided upon a logo which summarized a Mexican image in a group theoretical way: The duality between Life and Death as a broken symmetry. In the Totonac culture of the Atlantic coast there are a number of sculptures which depict a human face, one of whose halves shows the living person, and the other half his mask of death. Our logo was composed out of the two profiles and the front of one particularly beautiful clay figure, some 12 by 15 cm, belonging to Dr. Kurt Stavenhagen. His private collection of prehispanic art is among the finest in Mexico, and he graciously allowed us to use it for the Colloquium poster - printed in purple and Mexican pink.

The participants of the IX Colloquium, in Cocoyoc, Mexico.

LIST OF PARTICIPANTS

V. AMAR
Istituto di Fisica
Università di Parma
Parma, Italy

R.L. ANDERSON
Dept. of Physics & Astronomy
University of Georgia
Athens, Ga. 30602, USA

B. ANGERMANN
Institut für Theoretische Physik
der Universität Clausthal
Clausthal, West Germany

A. ANTILLON
Instituto de Física, UNAM
Apdo. Postal 20-364
México 20, D.F., México

J. AZCARRAGA
Departamento de Física Teórica
Facultad de Ciencias Físicas
Burjasot, Valencia, Spain

I. BARS
Physics Department
Yale University
New Haven, Conn. 06520, USA

A.O. BARUT
Physics Department
University of Colorado
Boulder, Colo. 80309, USA

M. BERRONDO
Instituto de Física, UNAM
Apdo. Postal 20-364
México 20, D.F., México

L.C. BIEDENHARN
Department of Physics
Duke University
Durham, N.C. 27706, USA

A.M. BINCER
Physics Department
University of Wisconsin
Madison, Wis. 53706, USA

A. BOHM
Center for Particle Theory
University of Texas
Austin, Texas 78712, USA

L.J. BOYA
Departamento de Física Teórica
Universidad de Salamanca
Salamanca, Spain

C.P. BOYER
IIMAS-UNAM
Apdo. Postal 20-726
México 20, D.F., México

A.J. BRACKEN
Dept. of Physics & Astrophysics
University of Colorado
Boulder, Colo. 80309, USA

J. BROECKHOVE
Dienst Theoretische eu
Wiskundige Natuurkunde
University of Antwerpen
B-2020 Antwerpen, Belgium

P.H. BUTLER
Department of Physics
University of Canterbury
Christchurch 1, New Zealand

J.F. CARIÑENA
Departamento de Física Teórica
Facultad de Ciencias
Universidad de Zaragoza
Zaragoza, Spain

O. CASTAÑOS
Centro de Estudios Nucleares, UNAM
Apdo. Postal 70-543
México 20, D.F., México

J.A. CASTILHO ALCARAS
Instituto de Física Teórica
Rua Pamplona, 145 - CEP 01405
Sao Paulo, Brasil

E. CHACON
Instituto de Física, UNAM
Apdo. Postal 20-364
México 20, D.F., México

J.-Q., CHEN
Department of Physics
Nanjing University
Nanjing, People's Rep. of China

M. COUTURE
Physics Department
McGill University
Montréal, Québec, Canada

A.P. CRACKNELL
Carnegie Laboratory of Physics
University of Dundee
Dundee DD1 4HN, Scotland, U.K.

B.J. DALTON
Department of Physics
Ames Laboratory
Iowa State University
Ames, Iowa 50011, USA

E. DEUMENS
Dienst Theoretische eu
Wiskundige Natuurkunde
University of Antwerpen
B-2020 Antwerpen, Belgium

A. DIN
C.E.R.N. - TH
CH-1211 Genève 23, Switzerland

R. DIRL
Institut für Theoretische Physik
Technische Universität Wien
A-1040 Wien, Austria

J.P. DRAAYER
Department of Physics
Louisiana State University
Baton Rouge, La. 70803, USA

H. ECKSTEIN
Centre de Physique Théorique II
F-13288 Marseille, Cedex 2, France

F.J. ERNST
Physics Department
Illinois Institute of Technology
Chicago, Ill. 60616, USA

D. FELDMAN
Department of Physics
Brown University
Providence, R.I. 02912, USA

F. GALLONE
Istituto di Fisica A. Pontremoli
Via Celoria 16
Milano 20133, Italy

A. GARCIA
CINVESTAV
Instituto Politécnico Nacional
México 14, D. F., México

J.P. GAZEAU
Laboratoire de Chimie Physique
Université Pierre et Marie Curie
11, rue Pierre et Marie Curie
75 231 Paris Cedex 05, France

G. GERMAN VELARDE
Instituto de Física, UNAM
Apdo. Postal 20-364
México 20, D.F., México

G.-C. GHIRARDI
Istituto di Fisica Teórica
Università di Trieste
34014 Miramare-Grignano, Italy

N. GIOVANNINI
Département de Physique Théorique
Université de Genève
121 Genève 4, Switzerland

S. GITLER
CINVESTAV
Instituto Politécnico Nacional
México 14, D. F., México

R. GLEISER
IMAF
Universidad de Córdoba
Córdoba, Argentina

J.C.H. GODDARD
Departamento de Matemáticas
Universidad Autónoma Metropolitana
México 14, D. F., México.

V. GORINI
Istituto di Fisica
Università di Milano
Milano, Italy

M. GOTAY
Department of Mathematics
University of Calgary
Calgary T2N 1N4, Canada

A. GROSSMANN
Centre de Physique Théorique II
F-13288 Marseille, Cedex 2, France

Z. HABA
Institute of Theoretical Physics
University of Wrocław
50-205 Wrocław, Poland

M. HAMERMESH
Department of Physics
University of Minnesota
Mineapolis, Minn. 55455, USA

J. HARNAD
CRMA
Université de Montréal
C.P. 6128
Montréal, Québec, Canada

I. HAUSER
Physics Department
Lewis College of Sciences
and Letters, IIT
Chicago, Ill. 60616, USA

G.C. HEGERFELDT
Department of Physics
Princeton University
Princeton, N.J. 08544, USA

E. HENESTROZA
Instituto de Física, UNAM
Apdo. Postal 20-364
México 20, D.F., México

A. HENRIQUEZ
Instituto de Física, UNAM
Apdo. Postal 20-364
México 20, D.F., México

P. HESS
Institut für Theoretische Physik
der Universität Frankfurt
Frankfurt, West Germany

Y. ILAMED
Soreq Nuclear Research Centre
Yavne 70600, Israel

G. IOMMI AMUNATEGUI
F.C.B.M.
Universidad Católica de Valparaíso
Valparaíso, Chile

M. IRAC
Laboratoire de Physique Théorique
Université de Paris VII,
Place Jussieu
75221 Paris Cedex 05, France

L. JACOBS
Instituto de Física, UNAM
Apdo. Postal 20-364
México 20, D.F., México

M.V. JARIĆ
Physics Department
University of California
Berkeley, Calif. 94720, USA

G. JOHN
Institut für Theoretische Physik
der Universität Tübingen
D-7400 Tübingen, West Germany

R. JUAREZ W.
Depto. de Física y Matemáticas
Instituto Politécnico Nacional
México 14, D. F., México

B.R. JUDD
Physics Department
John Hopkins University
Baltimore, Md. 21218, USA

C.K. JUE
Center for Particle Theory
University of Texas
Austin, Texas 78712, USA

P. KASPERKOVITZ
Institut für Theoretische Physik
Technische Universität Wien
A-1040 Wien, Austria

M. KIBLER
Institut de Physique Nucléaire
43, Bd. du 11 Novembre 1918
69621 Villeurbanne, France

Y.S. KIM
Dept. of Physics & Astronomy
University of Maryland
College Park, Md. 20742, USA

W. KINNERSELEY
Physics Department
Montana State University
Bozeman, Mont. 59717, USA

W.H. KLINK
Dept. of Physics & Astronomy
University of Iowa
Iowa City, Iowa 52242, USA

B. KOSTANT
Department of Mathematics
Massachusetts Institute of Technology
Cambridge, Mass. 02139, USA

P. KRAMER
Institut für Theoretische Physik
der Universität Tübingen
D-7400 Tübingen, West Germany

A.K. KWASNIEWSKI
Institute of Theoretical Physics
University of Wrocław
50-205 Wrocław, Poland

E. LACOMBA
Departamento de Matemáticas
Universidad Autónoma Metropolitana
Apdo. Postal 55-534
México 13, D. F., México

P.G.L. LEACH
Department of Mathematics
La Trobe University
Bundoora, Victoria 3083, Australia

D.B. LITVIN
Physics Department
Pennsylvania State University
Reading, Pa. 19607, USA

G. LOPEZ
Instituto de Física, UNAM
Apdo. Postal 20-364
México 20, D.F., México

G. LUGO
Department of Mathematics
University of Kansas
Lawrence, Kansas 66045, USA

J. LUKIERSKI
Institute of Theoretical Physics
University of Wrocław
50-205 Wrocław, Poland

K.T. MAHANTHAPPA
Physics Department
University of Colorado
Boulder, Colo. 80309, USA

V.I. MAN'KO
Lebedev Institute of Physics
USSR Academy of Sciences
Leninsky Prospekt 53
Moscow, Russian SFSR, USSR

W. MILLER Jr.
School of Mathematics
University of Minnesota
Minneapolis, Minn. 55455, USA

A. MONDRAGON
Instituto de Física, UNAM
Apdo. Postal 20-364
México 20, D.F., México

L.E. MORALES
CINVESTAV
Instituto Politécnico Nacional
México 14, D. F., México

M. MOSHINSKY
Instituto de Física, UNAM
Apdo. Postal 20-364
México 20, D.F., México

P. MOYLAN
Department of Physics, CPT
University of Texas
Austin, Texas 78712, USA

T. MURPHY
Physics Department
University of Wisconsin
Madison, Wis. 53706, USA

Y. NE'EMAN
Dept. of Physics & Astronomy
Tel-Aviv University
Ramat-Aviv, Israel

J. NIEDERLE
Physics Institute
Czechoslovak Academy of Sciences
180 40, Prague 8, Czechoslovakia

L. O'RAIFEARTAIGH
School of Theoretical Physics
Dublin Institute for
Advanced Studies
Dublin 4, Ireland

A. PALMA
Instituto Mexicano del Petróleo
Av. de los Cien Metros No. 174
México 14, D. F., México

J. PATERA
CRMA
Université de Montréal
C.P. 6128
Montréal, Québec, Canada

M. PAURI
Istituto Nazionale di
Fisica Nucleare
Sezione di Milano
Milano, Italy

M. PERROUD
Département de Mathematiques
Appliquées
Ecole Polytechnique
Montréal H3C 3A7, Canada

J. PLEBAŃSKI
CINVESTAV
Instituto Politécnico Nacional
México 14, D. F., México

J.F. POMMARET
Depart, of Mechanics
Ecole Nationale des
Ponts et Chaussées
28 rue des Saints Péres
Paris 7°, France

G. PRECIADO
IIMAS-UNAM
Apdo. Postal 20-726
México 20, D.F., México

J.F. PROVOST
Laboratoire de Physique Théorique
Université de Nice
06034 Nice Cedex, France

C. QUESNE
Phys. Théorique et Mathematique
Université Libre de Bruxelles
B-1050 Bruxelles, Belgium

F. REUSE
Departement de Physique Théorique
24, Quai Ernest-Ansermet
CH-1211, Genève 4, Switzerland

A. RIECKERS
Institut für Theoretische Physik
der Universität Tübingen
D-7400 Tübingen, West Germany

V. RITTENBERG
Physikalisches Institut der
Universität Bonn
5300 Bonn 1, West Germany

A. RONVEAUX
Facultes Universitaires
N.D. de la Paix
B-5000 Namur, Belgium

G. ROSENSTEEL
Physics Department
Tulane University
New Orleans, La. 70118, USA

D.J. ROWE
Physics Department
University of Toronto
Toronto, Ontario M5S 1A7, Canada

S. RUBINSTEIN
Instituto de Física, UNAM
Apdo. Postal 20-364
México 20, D.F., México

H. RUCK
Nuclear Science Division
Lawrence Berkeley Laboratory
Berkeley, Calif. 94720, USA

Y. SAINT-AUBIN
CRMA
Université de Montréal
C.P. 6128
Montréal, Québec, Canada

H. SALLER
Max-Planck Institut für
Physik und Astrophysik
Föhringer Ring 6
8000 München 40, West Germany

O.A. SANCHEZ VALENZUELA
IIMAS-UNAM
Apdo. Postal 20-726
México 20, D.F., México

M. SARACENO
Departamento de Física
Com. Nal. de Energía Atómica
Buenos Aires, Argentina

D. SEPULVEDA
Instituto de Física, UNAM
Apdo. Postal 20-364
México 20, D.F., México

R.T. SHARP
Department of Physics
McGill University
Montréal, Québec, Canada

M.C. SINGH
Dept. of Mechanical Engineering
University of Calgary
Calgary, Alberta, Canada

A.I. SOLOMON
Faculty of Mathematics
Open University
Milton Keynes, United Kingdom

P. SORBA
LAPP
Annecy-le-Vieux, France

S. STERNBERG
Mathematics Department
Harvard University
Cambridge, Mass., USA

J.J. SULLIVAN
Physics Department
University of New Orleans
New Orleans, La. 70122, USA

G. TORRES DEL CASTILLO
CINVESTAV
Instituto Politécnico Nacional
México 14, D. F., México

V. VANAGAS
Institute of Physics
Lithuanian Academy of Sciences
K. Pozelos 54
Vilnius, Lithuanian SSR, USSR

L. VINET
CRMA
Université de Montréal
C.P. 6128
Montréal, Québec, Canada

G. VITIELLO
Istituto di Fisica
Università di Salerno
Salerno, Italy

P. WINTERNITZ
CRMA
Université de Montréal
C.P. 6128
Montréal, Québec, Canada

K.B. WOLF
IIMAS-UNAM
Apdo. Postal 20-726
México 20, D.F., México

W. ZAHN
Institut für Theoretische Physik
der Universität Erlangen-Nürenberg
852 Erlangen, West Germany

A. ZEPEDA
CINVESTAV
Instituto Politécnico Nacional
México 14, D. F., México

THE COLLOQUIA ON GROUP THEORETICAL METHODS IN PHYSICS

I	1972	Centre de Physique Théorique de CNRS, Marseille, France.	Joint report of the University of Provence, the University at Aix-Marseille, and the CNRS.
II	1973	University of Nijmegen, Nijmegen, The Netherlands.	Printed by the Faculty of Science, University of Nijmegen.
III	1974	Centre de Physique Théorique, Marseille, France.	Printed by the Faculty of Science, University of Nijmegen.
IV	1975	University of Nijmegen, Nijmegen, The Netherlands.	Lecture Notes in Physics, No. 50, Springer 1976, ed. by A. Janner, T. Janssen and M. Boone.
V	1976	Université de Montréal, Montréal, Québec, Canada.	Academic Press 1977, ed. by R. T. Sharp and B. Kolman.
VI	1977	Universität Tübingen, Tübingen, West Germany.	Lecture Notes in Physics, No. 79, Springer 1978, ed. by P. Kramer and A. Rieckers.
VII	1978	University of Texas at Austin, Austin, Texas, U. S. A.	Lecture Notes in Physics, No. 94, Springer 1979, ed. by W. Beiglböck, A. Bohm and E. Takasugi.
VIII	1979	Kibbutz Kiriat Anavim, Israel.	A. Hilger and the Israel Physical Society, The American Institute of Physics 1980, ed. by L. P. Horowitz and Y. Ne'eman.

The 1980 Wigner Medal ceremony
honouring I. M. GEL'FAND

THE WIGNER MEDAL IS ADMINISTERED BY

THE GROUP THEORY AND FUNDAMENTAL PHYSICS FOUNDATION

PATRONS:

Hamad A. Alhendi
L. C. Biedenharn
Sharifa Bohm
H. B. G. Casimir
John D. Dollard
Lawrence Dresner
F. J. Dyson
Einstein Centennial Celebration Committee at Southern Illinois University
J. B. Garg
Marvin Goldberger
Morton Hamermesh
Richard Haymaker
Yehiel Ilamed
Erdal Inonu
Y. S. Kim
E. J. Konopinski
James D. Louck
W. M. MacDonald
Kazumi Maki
F. A. Matsen
Bernd T. Matthias
Esperanza de Moshinsky
Marcos Moshinsky
Tadao Nakano
Pedro L. Garcia Perez
T. O. Philips
Physical Chemistry Laboratory, ETH Zurich, Switzerland
Humitaka Sato
Marlan Scully
Frederick Seitz
The Solvay Institute, Brussels
SKM, Inc.
Edward Teller
Texas Nuclear
Tracor, Inc.
C. W. Ufford
Instituto de Fisica, Universidad Nacional Autonoma de Mexico
Janette Hegner Wheeler
John Archibald Wheeler

SPONSORS:

R. K. Adair
Norman Austern
V. Bargmann
Joseph Birman
D. Allan Bromley
Harvey Brooks
Philip B. Burt
E. G. D. Cohen
B. L. Davies
R. H. Dicke
G. and S. Domokos
Leonard Eisenbud
Arnold M. Feingold
Stig Flodmark
Stephen A. Fulling
Rudolph Haag
Christopher J. Henrich
Emmet J. Hudspeth
J. H. Irving
John R. Klauder
Wolfgang Ludwig
Louis Michel
R. Nottrot
Robert F. O'Connell
S. Okubo
L. C. Papaloucas
Constantin Piron
Luigi A. Radicati
Frederick Reines
Fritz Rohrlich
Paul Roman
Frank Stern
Marshall H. Stone
Eric W. Vogt
Steven Weinberg
V. F. Weisskopf
A. S. Wightman
Pavel Winternitz
Chen N. Yang

CONTRIBUTORS:

Julius Ashkin
A. Bohm
Burt V. Bronk
Morrel H. Cohen
Volker Enss
U. Fano
William Faris
Hans Frauenfelder
R. H. Gluckstern
Robert Hofstadter
A. Janner
B. R. Judd
P. Kielanowski
S. T. Kuroda
Francis J. Narcowich
Enrico Onofri
W. Opechowski
Herbert Piller
Arlan B. Ramsay
Martin G. Redlich
Peter Rejto
Yitzhak J. Sharon
L. C. Shepley
Eric Sheldon
A. Shimony
N. H. Zebouni

THE WIGNER MEDAL FOR 1980 - A TRIBUTE TO THE AWARDEE

INTRODUCTORY REMARKS BY A. BOHM

This is a very happy moment but it is also a very sad one. It is happy because we have come together to honor a great scientist. And it is sad, because the scientist we honor was unable to be with us.

I have written Gel'fand that he has received the 1980 Wigner Medal and have communicated with him a couple of times. Gel'fand felt deeply honored by the Wigner award and tried to come to our conference. We are unable to explain why he is not here.

The Wigner Medal is a truly international award and is not connected with any government or any national society. The recipient is chosen by an international selection committee whose members serve for four years, except Wigner who is a lifetime member. The selection committee for the 1980 award consisted of Morton Hamermesh, Valentine Bargmann, Luis Michel, Marcos Moshinsky and Eugene Wigner.

On behalf of the Group Theory Foundation I would like to express my gratitude to the committee members for their work and I would like to ask the chairman, Morton Hamermesh to tell us a little bit about the reasons for their choice of Israil Moisseevich Gel'fand.

Presiding the 1980 Wigner Medal ceremony: Professors M. Hamermesh, A. Bohm, L. C. Biedenharn, and M. Moshinsky.

Presentation of the Wigner Medal to

IZRAIL' MOISEEVICH GEL'FAND

by Morton Hamermesh

The Selection Committee has chosen I. M. Gel'fand to receive the Wigner Medal at this, the 9th International Colloquium on Group Theory Methods in Physics.

Professor Gel'fand's contributions to physics have been of prime importance, and have influenced many fields of research. For the past forty years he, with his colleagues and pupils, has touched with his mathematical genius many of the centrally interesting problems of physics:

1. Practical direct construction of representation of gl(n) and its subalgebras. This work of 1950, with Tsetlin, has had a continuing influence on calculations in physics and theoretical chemistry.

2. Representations of the Lorentz group and enumeration of Lorentz-invariant equations. A series of papers with Yaglom, published in 1947, discuss the classification of Lorentz-invariant equations and include an elegant generalization of the Pauli spin-statistics theorem to infinite-dimensional representations of the Lorentz group. A paper in the same year, with Naimark, gives the unitary representations of the Lorentz group. Later, with Minlos and Shapiro, Gel'fand published much of this material in textbook form.

3. Representations of the classical groups (1950). With Naimark, Gel'fand gave a detailed discussion of the method for solving this problem.

4. The Gel'fand-Levitan inversion method (1951). This result became the core of the inverse problem method and the treatment of evolution equations.

5. In a series of volumes with Shilov et al., Gel'fand covered a whole range of subjects - distribution theory, representation theory, rigged Hilbert spaces.

We have not cited the numerous papers of more purely mathematical interest with Kirillov and others. And we have not discussed Gel'fand's many important contributions to computational methods, to seismology, magnetohydrodynamics, biology and medicine. Professor Gel'fand's has truly been an all-embracing interest. But we must give special mention to the many monographs and texts that he has authored, books with a clarity that is refreshing, books that have enabled physicists to learn how to apply all these wonderful structures to the solution of their own particular problems.

I. M. Gel'fand is truly deserving of the Wigner Medal for his contributions to the development of group theory methods in physics. It is a pleasure and an honor for me, on behalf of the Selection Committee, to present it to him.

The Selection Committee: M. Hamermesh, Chairman
V. Bargmann, L. Michel
M. Moshinsky, E. P. Wigner

TEXT OF THE TELEGRAM SIGNED BY PARTICIPANTS OF THE IX COLLOQUIUM TO

PROFESSOR I. M. GEL'FAND

PROFESSOR I. M. GEL'FAND
FACULTY OF MATHEMATICS AND MECHANICS
MOSCOW STATE UNIVERSITY
MOSCOW, USSR

DEAR IZRAIL MOISEEVICH,

ON JUNE 24, 1980 AT COCOYOC, MEXICO, THE PARTICIPANTS IN THE 9TH INTERNATIONAL COLLOQUIUM ON GROUP THEORY IN PHYSICS ATTENDED THE AWARD CEREMONY OF THE WIGNER MEDAL.

ALL OF US ARE DELIGHTED TO HAVE TAKEN PART IN THIS PRESENTATION, BUT WE REGRET THAT YOU WERE UNABLE TO BE HERE WITH US TO RECEIVE THE MEDAL AND OUR INDIVIDUAL FELICITATIONS.

WE HOPE THAT AT SOME FUTURE DATE WE SHALL BE ABLE TO TRANSMIT TO YOU IN PERSON THE MEDAL AND OUR BEST WISHES.

R. L. ANDERSON, B. ANGERMANN, A. ANTILLON, J. DE AZCARRAGA, I. BARS, A. O. BARUT, M. BERRONDO, L. BIEDENHARN, A. M. BINCER, A. BOHM, L. BOYA, C. P. BOYER, A. J. BRACKEN, J. CARIÑENA, J. A. CASTILHO ALCARAS, E. CHACON, J.-Q. CHEN, M. COUTURE, F. DEUMENS, A. DIN, R. DIRL, J. P. DRAAYER, H. EKSTEIN, F. J. ERNST, D. FELDMAN, A. GARCIA, J. P. GAZEAU, S. GITLER, M. HAMERMESH, J. HARNAD, D. HAUSER, I. HAUSER, G. C. HEGERFELDT, E. HENESTROZA, S. HENRIQUEZ, P. O. HESS, M. IRAC, L. JACOBS, G. JOHN, R. JUAREZ, B. R. JUDD, B. KASPERKOVITZ, M. KIBLER, G. S. KIM, W. H. KLINK, P. KRAMER, E. LACOMBA, P. G. L. LEACH, G. LOPEZ, G. LUGO, K. T. MAHANTHAPPA, A. MONDRAGON, L. E. MORALES, M. MOSHINSKY, P. MOYLAN, T. MURPHY, Y. NE'EMAN, L. O'RAIFEARTAIGH, J. PATERA, M. PERROUD, C. QUESNE, A. RIECKERS, A. RONVEAUX, G. ROSENSTEEL, D. G. ROWE, H. RUCK, V. RUCK, Y. SAINT-AUBIN, H. SALLER, M. SARACENO, D. SEPULVEDA, R. T. SHARP, A. SOLOMON, P. SORBA, A. STERNBERG, S. STERNBERG, J. SULLIVAN, L. VINET, G. VITIELLO, P. WINTERNITZ, K. B. WOLF.

Contents

* Asterisks indicate invited papers

DIFFERENTIAL EQUATIONS

ELEMENTARY PARTICLES

GAUGE THEORIES

NUCLEAR PHYSICS

RELATIVITY

REPRESENTATION THEORY

CONTRIBUTIONS

GROUP THEORY OF THE MORSE POTENTIAL

Manuel Berrondo*
Instituto de Física, Univ. of Mexico,
Apdo. Postal 20-364, México 20, D. F.

and

Alejandro Palma,+
Instituto Mexicano del Petróleo,
Av. Cien Metros 152, México 14, D.F.

We map the problem of a Morse potential in one dimension into a two-dimensional harmonic oscillator. The symmetry group for this problem is U(2). Starting from the dynamical group Sp(4), we use two different chains of groups including SU(1,1) and U(2) respectively.

1. Introduction

The solution to the Morse potential in one dimension is known since the early days of quantum mechanics [1]. It is obtained by transforming the wave equation into the radial equation [2] of the Coulomb problem, from which one can actually find an algebraic formulation [3]. The virtues of the Morse potential, particularly in molecular theory, are well known: its spectrum includes anharmonic effects quadratic in the vibrational quantum number [1,2]; adding a centrifugal potential, it yields vibration-rotation interactions [2,4] thus obtaining a very good fit to diatomic molecular spectra [5]; it has a finite number of bound states and, contrary to the usual anharmonic potentials, it allows for dissociation of the molecule.

In dealing with collisions against a Morse potential [6,7], one needs however to compute matrix elements [8] between different Morse states (bound-bound for inelastic collisions or bound-continuum for collisional dissociation). The advantage of counting with an algebraic formulation is thus evident. It allows to simplify the calculations in the Born and genaralized Born (or Magnus) approximation [6,8], but it also helps in finding exact solutions to different model perturbing potentials [9], as we shall briefly illustrate at the end of this paper. The analogy

---

* Consultant at Instituto Mexicano del Petróleo, México.

+ Also at Escuela Superior de Física y Matemáticas (IPN)

with the Coulomb radial problem introduces in a natural way an SO(2, 1) algebra [3] for the Morse potential, whose ladder operators shift the potential's depth. By introducing an extra angular degree of freedom [10], we were however able to produce an SO(3) algebra wich avoids this difficulty.

In the present paper we introduce a different transformation which maps the Morse problem into the radial equation of a harmonic oscillator in two dimensions. In section 2 we extend the variable space to define two-dimensional boson operators. Section 3 introduces the dynamical group Sp(4) and the symmetry group U(2) of the transformed problem and its relation with the Morse problem. Finally in section 4 we conclude with some remarks.

2. Mapping to a harmonic oscillator.

The Morse potential is defined by:

$$V(u) = D\,(\exp(-2au) - 2\exp(-au)) \tag{2.1}$$

in terms of the two parameters a and D. It is chosen so that its minimum -D is at the origin u=o, and tends to zero asymptotically, $u\to\infty$. Writing the wave equation in natural units ($\hbar = a = 2m = 1$):

$$\frac{d^2\psi}{du^2} + (E - D\,(e^{-2u} - 2\,e^{-u}))\,\psi = 0 \tag{2.2}$$

we impose one-dimensional boundary condtions:

$$\psi(-\infty) = \psi(+\infty) = 0 \tag{2.3}$$

for the bound states.

In order to transform it to a simpler equation, let us introduce the new variable:

$$\rho = 2D^{1/4}\, e^{-u/2} \tag{2.4}$$

to obtain

$$\frac{1}{\rho}\,\frac{d}{d\rho}\,\rho\,\frac{d\psi}{d\rho} + \left(\frac{4E}{\rho^2} - \frac{1}{4}\,\rho^2 + 2\,D^{1/2}\right)\psi = 0. \tag{2.5}$$

This is the radial equation for a two-dimensional harmonic oscillator with unit frequency ($\omega = 1$), making the identifications:

$$4E \longleftrightarrow -m^2$$

$$2D^{1/2} \longleftrightarrow W = 2n + |m| + 1 \qquad (2.6)$$

where m is the angular quantum number of the equivalent oscillator, n the number of radial quanta, and W its eigenvalue. Introducing the angular variable ϕ[3,10], we obtain, in polar coordinates (ρ,ϕ):

$$\frac{1}{\rho}\frac{\partial}{\partial\rho}\rho\frac{\partial\Psi}{\partial\rho} + \left(\frac{1}{\rho^2}\frac{\partial^2}{\partial\phi^2} - \frac{1}{4}\rho^2 + 2D^{1/2}\right)\Psi = 0. \qquad (2.7)$$

Hence we can define two boson operators, which in cartesian coordinates (X_1, X_2) read:

$$a_i = \frac{1}{\sqrt{2}}\left(X_i - \frac{\partial}{\partial X_i}\right), \quad a_i^+ = \frac{1}{\sqrt{2}}\left(X_i + \frac{\partial}{\partial X_i}\right) \quad i = 1,2, \qquad (2.8)$$

and the corresponding generators of the symmetry group U(2);

$$C_{ij} = a_i a_j^+ \qquad i,j = 1,2. \qquad (2.9)$$

As a first consideration, we can find the spectrum of the Morse potential from the Casimir operator $C_{11} + C_{22}$, which corresponds to the number of quanta of the equivalent oscillator:

$$N = 2n + 2(-E)^{1/2} = 2D^{1/2} - 1, \quad n = 0,1,... \qquad (2.10)$$

using Eqs. (2.6). It follows then that

$$E_n = -D + 2D^{1/2}\left(n + \frac{1}{2}\right) - \left(n + \frac{1}{2}\right)^2 \qquad (2.11)$$

are the bound state energies of the Morse potential.

3. Two chains of groups.

The bilinear forms $\{E_k\}$ of the boson operatos (2.8) form the set of generators of the dynamical group for the oscillator [11], namely Sp(4):

$$\{E_k\} = \{C_{ij}, a_i^+ a_j^+, a_i a_j\} \qquad k = 1,...,12. \qquad (3.1)$$

Let us consider first the subgroups corresponding to the separation in polar coordinates, directly related to the Morse potential through Eq. (2.5). The angular part corresponds to the subgroup O(2) generated by the antisymmetric combination.

$$J_\phi = \frac{i}{2}(C_{12} - C_{21}). \qquad (3.2)$$

From the identification (2.6), the energy is given by the eigenvalue of $-J_\phi^2/4$. The subgroup associated to the radial equation (2.5) in the SU(1,1) group [11] generated by:

$$I_1 = \frac{1}{4}(p^2 - \rho^2),\; I_2 = \frac{1}{4}(\bar{p}.\bar{\rho} - \bar{\rho}.\bar{p}),\; I_3 = \frac{1}{4}(p^2+\rho^2), \quad (3.3)$$

with commutation relations:

$$[I_1, I_2] = -I_3,\; [I_3, I_1] = I_2,\; [I_2, I_3] = I_1, \quad (3.4)$$

so we have

$$Sp(4) \supset SU(1,1) \otimes O(2). \quad (3.5)$$

The ladder operators $I_\pm = I_1 \pm i\, I_2$ shift the number of radial quanta n, leaving m fixed. This implies changing the depth [3,12] of the Morse well, for a fixed energy E.

The second alternative is to include the symmetry group of the oscillator defined in section 2:

$$Sp(4) \supset U(2) \supset O(2). \quad (3.6)$$

In this case the ladder operators

$$J_\pm = \frac{1}{2}(C_{11} - C_{22}) \mp \frac{i}{2}(C_{12} + C_{21}) \quad (3.7)$$

change the value of m by one unit. Since they belong to the symmetry group U(2), they leave W, and hence the depth D, fixed. Recalling that $E = -m^2/4$, we see that we can shift the energy eigenvalues with the aid of $J_\pm$.

4. Concluding Remarks.

The symmetry group U(2) is very useful in the calculation of transition probabilities for the case of inelastic transitions between two states of the Morse potential. Expressing the transtion-inducing potentials in terms of the generators of U(2) allows for an exact solution, with the aid of the Baker-Hausdorff formulae [9]. For instance, if

$$V(t) = a(t)\, J_+ + a^*(t)\, J_- + b(t)\, J_\phi, \quad (4.1)$$

we can write down the transition probability between two Morse states m and m^1 as:

$$S_{mm^1} = \langle m\,|\exp(A\,J_+ + A^*\,J_- + B\,J_\phi)|\,m^1\rangle, \qquad (4.2)$$

which can be evaluated directly [9] in terms of the d-matrices for SU(2). We should be aware of the fact that the potential (4.1) in general is an operator in ρ and ϕ, and not merely a function of ρ. This can be regarded as an implicit time dependence, since ϕ is the canonical conjugate variable to $E^{1/2}$ in this description.

REFERENCES

[1] P.M. Morse, Phys. Rev. 34, 57 (1929).

[2] L. Pauling & E.B. Wilson, Introduction to Quantum Mechanics. (McGraw-Hill, New York, 1935).

[3] M. Berrondo & A. Palma, J. Phys. A13, 773 (1980).

[4] C.L. Pekeries, Phys. Rev. 45, 98 (1934).

[5] See e.g. J.N. Huffacker, J. Chem. Phys. 64, 3175 (1976).

[6] N. Cabrera, V. Celli, F.O. Goodman & R. Hanson, Surface Sci. 19, 67 (1970).

[7] A.P. Clark & A.S. Dickinson, J. Phys. B6, 164 (1973).

[8] N. Rosen, J. Chem. Phys. 1, 319 (1933).

[9] R.D. Levine & C.E. Wulfman, Chem. Phys. Lett. 60, 372 (1979).

[10] W. Miller, Lie Theory and Special Functions, (Academic, New York, 1968).

[11] C. Quesne & M. Moshinsky, J. Math. Phys. 12, 1780 (1971).

[12] J. N. Huffacker & P.H. Dwivedi, J. Math. Phys. 16, 862 (1975).

FOUR EUCLIDEAN CONFORMAL GROUP APPROACH TO THE MULTIPHOTON PROCESSES IN THE H-ATOM

J.P.Gazeau*
Centre de recherche de mathématiques appliquées
Université de Montréal, Montréal, Québec, Canada

The evaluation of transition probabilities for multiphoton processes requires drudgery calculations of matrix elements issued from the time dependent perturbation theory [1]. In the specific case of interaction of light with the H-atom, the latter arise in the general form (for a N-photon process)

$$I_{n\ell m\to n'\ell'm'} = (\psi_{n'\ell'm'}, \mathcal{O}\psi_{n\ell m})_{L^2(R^3)} \tag{1}$$

$\psi_{n\ell m}$ is a H-bound state wave function, $\mathcal{O}$ is the operator $[\prod_{i=1}^{N-1} A_i e^{i\vec{k}_i.\vec{r}} G(E_i)]A_N e^{i\vec{k}_N.\vec{r}}$ where $G(E)$ is the Coulomb resolvent $(\vec{p}^2/2m - E - g/r)^{-1}$ and $A_i = \vec{p}.\vec{\varepsilon}_i$ for all i (situation "P") or $A_i = \vec{\varepsilon}_i.\vec{r}$ for all i (situation "R"), with $\vec{\varepsilon}_i.\vec{k}_i = 0$. In this work, we make use of the Fock transformation [2] and some representation of group $g \to \mathcal{C}(g)$ [3][4] for obtaining a compact analytical expression for the matrix element (1).

1. Fock transformation and quaternionic conformal action

Let p_0 be a real positive momentum like parameter. In the momentum representation, a wave function will be viewed as a function of the four vector $x = (p_0,\vec{p}) = p_0\underline{1} + \Sigma_{i=1}^3 p_i\vec{e}_i$. We denote by $\underline{F}_{p_0,+1}$ the space of complex functions square integrable with respect to the scalar product $(\psi(p_0),\psi'(p_0))_{p_0,+1} = \int d\vec{p}\,\psi^*(x)\frac{|x|^2}{2p_0^2}\psi'(x)$, $|x|^2 \equiv p_0^2+\vec{p}^2$. The Fock transformation $\mathcal{F}_{p_0}$ is an isomorphism mapping $\underline{F}_{p_0,+1}$ onto $\underline{E} \equiv L^2(SU(2))$. It is defined in the following way:

$$\phi(\xi) = (\mathcal{F}_{p_0}\psi(p_0))(\xi) = 4p_0^{3/2}|\underline{1}+\xi|^{-4}\psi(s(p_0).\xi).$$

It should be recalled that the S^3-harmonic $Y_{n\ell m}$ is the $\mathcal{F}_{p_{0n}}$-transform, $p_{0n} = \sqrt{-2mE_n} = \frac{mg}{\hbar n}$, of the function $\psi_{n\ell m}$. $x = s(p_0).\xi$ is the stereographic projection of $\xi \in S^3 \approx SU(2)$. The projection $s(p_0)$ is conveniently described by introducing the quaternion field $\mathbb{H}$ $(\vec{e}_1\vec{e}_2 = \vec{e}_3$ etc...) and the conformal action of any 2×2 quaternionic matrix $g = \begin{pmatrix} a & b \\ c & d\end{pmatrix}$: $x \in \mathbb{H}$, $g.x \equiv (ax+b)(cx+d)^{-1}$. Explicitly, $s(p_0) = \frac{1}{\sqrt{2p_0}}\begin{pmatrix} 2\underline{p_0} & \underline{0} \\ \underline{1} & \underline{1}\end{pmatrix}$, $\underline{\lambda} \equiv (\lambda,\vec{0})$.

---

* Permanent address: Laboratoire de Chimie Physique, 11 rue Pierre et Marie Curie, F.75231, Paris Cedex 05, France.

2. Semigroup representation on $L^2(SU(2))$

The conformal group of $\mathbb{H}$ is defined by: $G = SU^*(4) \approx \overline{SO_+(1,5)} = \{g=\begin{pmatrix} a & b \\ c & d\end{pmatrix};$ $a,b,c,d \in \mathbb{H},\ g \in SL(4,\mathbb{C})\}$. Special consideration is given to two subsets of G, according to their conformal action on the unit ball in $\mathbb{H}$:

$$G_{\leq} = \{g \in G,\ |x| \leq 1 \Rightarrow |g.x| \leq 1\} \qquad \text{(semigroup)}$$

$$G_{<} = Sp(1,1) \approx \overline{SO_+(1,4)} = \{h \in G_{\leq},\ |x| = 1 \Rightarrow |h.x| = 1\}.$$

Then, we consider the irreducible representation $\mathcal{T}$ of $G_<$, equivalent to one complementary series unitary irreducible representation:

$$\phi \in \underline{E},\quad h \in G_<,\quad \mathcal{T}(h)\phi(\xi) = (\alpha(\xi,h))^{-2}\phi(h^{-1}.\xi),$$

where $\alpha(x,g) \equiv |cx+d|$ for $g^{-1} = \begin{pmatrix} a & b \\ c & d\end{pmatrix}$ and $x \in \mathbb{H}$. We extend $\mathcal{T}$ to $G_{\leq}^{-1}$ by means of the harmonic extension Φ of ϕ:

$$g \in G_{\leq}^{-1},\quad \mathcal{T}(g)\phi(\xi) = (\alpha(\xi,g))^{-2}\Phi(g^{-1}.\xi),$$

where

$$\Phi(x) = \frac{1}{2\pi^2}\int_{SU(2)} d\mu(\xi')\,\frac{1-|x|^2}{|\xi'-x|^4}\,\phi(\xi') \qquad |x| < 1$$

and

and $\Phi(\rho\xi) \underset{\rho\to 1.}{\to} \phi(\xi)$ almost everywhere.

3. Distributions with support in the semigroup G^{-1}.

Of particular importance are the following elements of $\mathcal{G}$, the Lie algebra of G:

$$D = \frac{1}{2}\begin{pmatrix} \underline{1} & 0 \\ \underline{0} & -\underline{1}\end{pmatrix},\quad L = \frac{1}{2}\begin{pmatrix} 0 & 1 \\ \underline{1} & \underline{0}\end{pmatrix},\quad \Omega = \frac{1}{2}\begin{pmatrix} 1 & 1 \\ -\underline{1} & -\underline{1}\end{pmatrix},\qquad L_i = L\vec{e}_i,\qquad \Omega_i = \Omega\vec{e}_i.$$

By exponentiating, we obtain

$\exp tD \equiv dn(t) \in G_{\leq}^{-1}$ for all $t > 0$ (conformal dilatation)

$\exp tL \equiv \ell(t) \in G_<$ for all t

$\exp t\Omega \equiv \omega(t) \in G_{\leq}^{-1}$ for all $t > 0$

$\exp \vec{x}.\vec{\Omega} \equiv \omega(\vec{x}) \in G_<$ for all $\vec{x}$.

Let X be an element of $\mathcal{G}$ such that $\exp tX \in G_{\leq}^{-1}$ for $0 \leq t < \varepsilon$. We can define the following operator acting on some dense subspace $\underline{E}_1$ of $\underline{E}$:

$$\mathcal{T}(T_X) \equiv \partial_{t=0^+}\mathcal{T}(\exp tX).$$

Of course, this shallow introduction of distributions T_X via the representation $\mathcal{T}$ has to be precised in the rigorous framework of "tempered vector distributions" T on a Lie group [3] [4] [5]. The action of these distributions T, with support in the semi-group $G_{\leq}^{-1}$, will be written formally,

$$\mathcal{T}(T) \equiv \int_{G_{\leq}^{-1}} dT(g)\,\mathcal{T}(g).$$

A convolution product can be introduced:

$$\mathcal{T}(T*T') = \mathcal{T}(T)\mathcal{T}(T') = \int_{G_{\leq}^{-1}} dT(g) \int_{G_{\leq}^{-1}} dT'(g')\mathcal{T}(gg').$$

We denote by T^{-1} the (eventually existing) *-inverse of T. For instance, $\mathcal{T}(T_D^{-1}) = -\int_0^{+\infty} dt\,\mathcal{T}(dn(t))$.

4. Calculation of the N-photon matrix element

We obtain the N-photon matrix element as "linear superposition" of matrix elements of the representation operators $\mathcal{T}(g)$, calculated with respect to the S^3 harmonics basis:

$$I_{n\ell m\to n'\ell' m'} = \int_{G_{\leq}^{-1}} dS(g)\,(\mathcal{T}(g))_{n'\ell' m',n\ell m}$$

The distribution S is given by:

$$S = c\int_0^{+\infty} dt_1 e^{\nu_1 t_1} \ldots \int_0^{+\infty} dt_{N-1} e^{\nu_{N-1} t_{N-1}} [\prod_{i=1}^{N}{}^{(*)} T_{X_i}]*\delta(g_n)*T_D^{-1}$$

with $X_i = \begin{cases} g_i\vec{\varepsilon}_i.\vec{L}g_i^{-1} & \text{(situation "P")} \\ g_i(\Omega+\vec{\varepsilon}_i.\vec{\Omega})g_i^{-1} & \text{(situation "R")} \end{cases}$, $g_1 = \ell(\lambda_{n'1})$, $g_i = g_{i-1}\omega(\hbar\frac{\vec{k}_{i-1}}{p_{0i}})dn(t_{i-1})\times$ $\times\ell(\lambda_{i-1\,i})$; $g_n = g_N\omega(\hbar\frac{\vec{k}_N}{p_{0n}})$, $p_{0i} = \sqrt{-2mE_i}$, $\nu_i = \frac{mg}{p_{0i}\hbar}$, $\lambda_{i-1\,i} = \text{Log}\,\frac{p_{0i-1}}{p_{0i}}$, $1 \leq i \leq N-1$

$\lambda_{01} = \lambda_{n'1}$, $\lambda_{N-1\,N} = \lambda_{N-1\,n}$, $\delta(g)$ is the Dirac distribution on G.

The general expressions of the matrix elements $\mathcal{T}(g)_{n\ell m,n'\ell' m'}$, $\mathcal{T}(T_X)_{n\ell m,n'\ell' m'}$ are known [3] [4]. The latter is zero except if $|n-n'|$, $|\ell-\ell'|$, $|m-m'| \leq 1$. Finally, we make the change of variables $z_i = e^{-t_i}$. Then, the matrix element $I_{n\ell m\to n'\ell' m'}$ appears as being a finite linear combination of "generalized Euler integrals"

$$\mathcal{J}_{N-1}(p_i,q_i;r,g_n) = \int_0^1 dz_1 \ldots \int_0^1 dz_{N-1} (\prod_{i=1}^{N-1} z_i^{p_i-1}(1-z_i)^{q_i-1})|d(g_n^{-1})|^{-2r}$$

where $g_n^{-1} = \begin{pmatrix} a & b \\ c & d(g_n^{-1}) \end{pmatrix} \in G_{\leq}$, $1 \leq r \leq n+n'+N$.

In the two photon case, we recover well known formulas with hypergeometric or Appell functions [1] [4]. It should be noted that all these expressions can be analytically continued to positive energies $E_i = -(p_{0i}^2/2m) > 0$ or equivalently to purely imaginary values of ν_i.

References

[1] P.Lambropoulos, "Topics on Multiphoton Processes in Atoms", in Advances in Atomic and Molecular Physics, ed. by D.R.Bates and B.Bederson, 12, 87 (Academic Press, New York, 1976) and references therein.

[2] W.Fock, Z.Phys. 98, 145 (1935).

[3] J.P.Gazeau, Lett.Math.Phys. 3, 285 (1979).

[4] J.P.Gazeau, J.Math.Phys. 19, 1048 (1978); preprint (1980).

[5] Harish-Chandra, Acta.Math. 116, 1 (1966).

LANDAU THEORY, SYMMETRY BREAKING AND THE CHAIN CRITERION

Marko V. Jarić

Department of Physics, University of California
Berkeley, California 94720, USA*

Abstract

A study of absolute minima of bounded below, real, G-invariant polynomials on R^m is initiated. The group Im(G), acting on R^m, is assumed orthogonal, irreducible and finite. Aforementioned polynomials are used for the free energy in a theory of phase transitions. Symmetry of an absolute minimum of such a polynomial is the broken symmetry. Several theorems on possible broken symmetries are proven.

Landau theory of continuous phase transitions has been around for some forty years now.[1] The main goal of this theory is determination of all low symmetry groups (broken symmetries) accessible from a high symmetry group (G) via a second order phase transition.[2] The transition is assumed to be driven by an m-dimensional order parameter (x) which spans an orthogonal, irreducible representation of G. As the temperature is changed passing the critical temperature the order parameter should change continuously from zero in one phase to a value different than zero in another phase. Value of the order parameter, x, is a vector (a linear form[3]) in R^m. The symmetry of this vector (i.e. little group of x) defines the symmetry of corresponding phase. Therefore, whatever this symmetry group is, it is necessarily amongst the little subgroups of G on R^m.[4] This assertion is the necessary symmetry condition contained in the Landau theory.[5]

In order to find the little groups it may be proceeded directly by determining the little group of each vector in R^m. However, I have developed[6] and applied[7,8] the Chain Criterion which lends itself to a more efficient use: characters of the representation determine all the little groups. The validity of the Chain Criterion has been proved.[5,9] Clearly, this criterion is not restricted to the Landau theory. It is valid in any theory where the symmetry is determined by a linear form of x; it is irrelevant whether this linear form is calculated using the Landau theory, Renormalization-group or something else.

The little groups are "allowed in principle" to occur in a phase transition. However, each little group need not correspond to an actual symmetry breaking--which ones do correspond, depends on the particular theory being used. For example, in order to determine which of the little groups actually occur in Landau theory, minimization of the Landau free energy (F) should be performed. This minimization is often tedious and intractable. I will devote the rest of this paper to some attempts to reduce this minimization.

I will slightly generalize the Landau theory: the free energy is a real, G-invariant, bounded below polynomial in x.[10] Unlike the Landau, the degree of the polynomial is not restricted to four, and the free energy is not necessarily conceived as an expansion in x. In this fashion multicritical and discontinuous, as well as non-mean-field transitions are included. Broken symmetry is still determined by the little group of an absolute minimum of F.

Our free energy can be written as:

$$F(x) = \sum_{j=0}^{n} J_j P_j \,(I_1,\ldots,I_m) \equiv F(I,J) \quad , \tag{1}$$

where $J_0 \equiv 1$; $J \equiv J_j(x)$ $(j = 1,\ldots,n < \infty)$ and $I \equiv I_i(x)(i = 1,\ldots,m)$

are G-invariant polynomials; (I,J) is an integrity basis for the ring of G-invariant polynomials; P_j $(j=1,\ldots,n)$ are polynomials; for groups generated by reflections n = 0, in Eq. 1. For invariants J_j there is a finite integer l_j such that

$$J_j^{l_j} - Q_j(I) = 0 \; (j = 1,\ldots,n) \quad , \tag{2}$$

where Q_j $(j = 1,\ldots,n)$ are polynomials. Therefore, the free energy is a polynomial in the integrity basis.

Before proceeding further I will review important notions of orbit, orbit space, stratum and generic stratum.

The little groups divide R^m into disjoined strata; stratum being a set of points (in R^m) whose little groups are conjugated. The generic stratum is associated with a minimal little group; the generic stratum is open dense. Strata, in turn, consist of disjoined orbits; orbit being a set of <u>all</u> points related by the action of G. The set of all orbits forms an orbit space. Orbit space can be represented by taking one point from each orbit. It is seen directly from these definitions that any G-invariant function is invariant on an orbit. Since the free energy and the integrity basis are G-invariant, they are invariant on any orbit. Therefore, they are in fact functions on the orbit space.

The values of (I,J) are from R^{m+n}. I will call this space an "extended order-parameter space". Therefore, the integrity basis defines a map, i, of the orbit space, order-parameter space, or R^m onto the extended order-parameter space, or R^{m+n}:

$$i : x \to i(x) \equiv [\, I(x), J(x) \,]. \tag{3}$$

Now I will prove that

the map i, Eq. 3, is one-to-one onto. That is

$$i(x) = i(y) \Rightarrow \exists g \in G,\ x = gy. \tag{4}$$

Proof:

From the assumption in Eq. (4) and the definition of the integrity basis it follows that for any G-invariant polynomial P:

$$P(x) = P(y). \tag{5}$$

Choose the polynomial P to be:

$$P(z) \equiv \prod_{g \in G} \| x - gz \|^2 , \tag{6}$$

where $\| \ \|$ denotes Euclidian norm on R^m. Clearly,

$$P(y) = \prod_{g \in G} \| x - gy \|^2 = P(x) = 0 \tag{7}$$

implies

$$\exists g \in G , \qquad x = gy . \tag{8}$$

Q.E.D.

The free energy is a polynomial on R^{m+n} whose domain d, the extended order-parameter space, is given by the map Eq. 3: $d = i(R^m)$. In fact, d is a piece of an m-dimensional surface in R^{m+n}. The equations of this surface are given explicitly by Eq. 2. Since the symmetry breaking is determined by the absolute minimum of F, it should be minimized with respect to (I, J) on d.[11] The theorem, Eq. 4, then ensures that the broken symmetry group will be determined uniquely (up to conjugation). The theorem further ensures that the disjoined strata will be mapped into disjoined sets (strata).

In particular, we can find any minimum on generic stratum by simple differentiation with respect to (I,J). However, constraints, Eq.2, must be included. This can be achieved by the use of Lagrange multipliers a_j ($j = 1,\ldots,n.$). Therefore, the effective free energy is defined:

$$F^*(I,J;a) \equiv F(I,J) + \sum_{j=1}^{n} a_j \, [J_j^{1_j} - Q_j(I)]. \qquad (9)$$

Hence, it is immediately clear:

If F^* contains only a linear term in any component of (I,J) then the symmetry cannot be broken into the minimal little group.

Thus, in order to eliminate symmetry breaking into minimal little group it is sufficient that F(I,J) contains a term which is a linear form in I, and that the constraints, Eq. 2, do not depend on those I_i which enter this particular linear form. For groups generated by reflections the situation is simpler (the second restriction is automatically satisfied). For example, let us consider a quartic (Landau) free energy. In this case there must exist at least one cubic or quartic I_i [otherwise, the effective G is O(m)]. This particular I_i must enter the free energy linearly. Therefore:

For orthogonal, irreducible, finite (image) groups generated by reflections the Landau theory predicts no symmetry breaking into the minimal little group.[12]

For the similar reason that there are no minima of F^* on the generic stratum, it is most likely that the symmetry will be broken into a maximal little group. This is in agreement with the conjecture of Ref. 12.

If we do not restrict the degree of the free energy, then we can certainly find a smooth one-parameter family of free energies for which the absolute minima will lie in any wanted stratum. However, in order for the high symmetry group to belong to the same family there are following alternatives: The phase transition is discontinuous or multicritical; or, the transition is simple continuous but some terms are omitted from the expansion of F in x. In the latter case a non-mean-field exponents are found, and the missing terms could be justified as a result of a "renormalization" procedure.

In conclusion, I would like to stress that this study is in its infancy. There are many obvious questions yet to be answered.

I benefited from a discussion with M. DePortu. I also acknowledge the Miller Fellowship at U.C. Berkeley.

References

*Present address: Freie Universität Berlin, FB20, Institut für Theorie der kondeusierten Materie (WE5), Arnimallee 3, 1000 Berlin 33.

1. L.D. Landau, Phys. Z. Sowj. Un. 11, 26 and 545 (1937). See also L.D. Landau and E.M. Lifshitz, Statisticheskaya Fizika (Nauka, Moskva 1976).

2. The theory was most successful in applications to structural phase transitions; see G. Ya. Lyubarskii, The Application of Group Theory in Physics (Pergamon, Oxford, 1960).

3. In the Landau theory this linear form is usually referred to as Landau's density function.

4. Note that the relevant group is the image of G defined by the representation. Thus, when we say G we will actually refer to Im(G).

5. M.V. Jarić, Phys. Rev. (To be published).

6. M.V. Jarić, Ph.D. Thesis (CUNY, New York, 1977).

7. M.V. Jarić, and J.L. Birman, Phys. Rev. B16, 2564 (1977).

8. M.V. Jarić, Phys. Rev. B (To be published).

9. Independently L. Michel proved the same, Rev. Mod. Physics (To be published).

10. I will only consider finite, orthogonal and irreducible groups Im(G).

11. Due to the theorem Eq. 4, even a local minimum (associated with a quasi-stable phase) can be obtained in the same fashion.

12. A similar theorem was proved by L. Michel, preprint Ref. TH.2716-CERN.

THE CONFIGURATION d^N IN CUBIC SYMMETRY : A SYMMETRY ADAPTED WEAK FIELD APPROACH

Maurice Kibler and Geneviève Grenet
Institut de Physique Nucléaire (et IN2P3)
Université Lyon-1, 69622 Villeurbanne Cedex, France

We propose a new Hamiltonian to describe the combined action of the interelectronic repulsion, the crystal-field interaction, and the spin-orbit interaction within a d^N manifold. The energy matrix of such an Hamiltonian in a symmetry adapted weak field basis is considerably easier to set up than the one of the conventional Hamiltonian in a strong field basis. The model arising from the use of the proposed Hamiltonian in conjunction with a symmetry adapted weak field basis yields results comparable to those of the well-known strong field model. The equivalence between the two models requires a relation between the ten strong field Coulomb integrals to be satisfied.

1. Introduction

This work deals with (atomic) d^N configurations acted upon by (crystalline) fields of cubic symmetry. These configurations in nonspherical symmetry turn out to be of interest in the study of magnetic and optical properties of transition-metal ions embedded in molecular, solid-state, or biological environments. The quantum mechanical treatment of such systems lies on the diagonalization, within a subspace of the relevant Hilbert space, of the Hamiltonian of each ion in its environment. The chosen subspace generally contains as many vectors than the atomic d^N manifold and the Hamiltonian to be used involves, at least, the Coulomb interaction H_C between the N electrons, the crystal-field interaction H_{cf} between the N electrons and the surrounding ions, and the spin-orbit interaction H_{so} among the N electrons. The models we may elaborate for describing each ion in its environment depend on the choices we make both for the vectors in the d^N manifold and the Hamiltonian [1-4]. It is the aim of this paper to report on an extended symmetry adapted weak field model (SAWFM) which is as easy to handle than the conventional SAWFM [4] and which gives rise to results comparable to the ones afforded by the extended strong field model (SFM) [1-3] without having the disavantages of the latter one. We shall sketch here only the underlying philosophy. A complete version will be the subject of a forthcoming paper [5].

2.   Two well-known models

We shall first briefly describe two existing models. Let us begin with the following (nonspontaneous) symmetry breaking. When passing from the rota-tional symmetry SO_3 of the free-ion to the octahedral symmetry O of the ion in its environment, the 5-dimensional irreducible representations class (IRC) $\ell = 2$ of SO_3 breaks down into the direct sum of the 3-dimensional IRC T_2 and the 2-dimensional IRC E of O. (We take the octahedral group O as the prototype of the five cubic groups. O is isomorphic with the permutation group S_4.) In other words, the atomic d shell splits into two subshells referred to as the t_2 and e crystalline shells and associated respectively with the IRC's T_2 ([31]) and E ([2^2]) of O (S_4) :

$$d \; - \; \textcircled{5} \; - \left\{ \begin{array}{l} \textcircled{2} \; - \; e \\[4pt] \textcircled{3} \; - \; t_2 \end{array} \right.$$

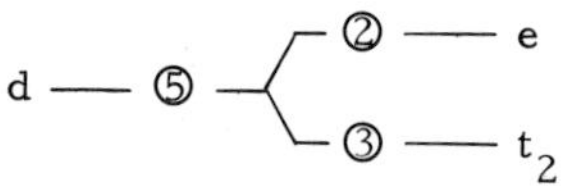

$$2 \;\; = \;\; T_2 \oplus E$$

$$SO_3 \; \supset \; O.$$

The next idea is to consider the t_2 and e shells as new entities, the vectors of which being not necessarily pure d vectors, and then to distribute the N electrons among these two shells by taking care of Pauli's principle. This leads to crystalline configurations of type $t_2^{N_1} e^{N-N_1}$. We need vectors for such configurations. We may take for example vectors of type

$$| S \, \Gamma \, \beta \Gamma^* M_\Gamma^* > \; \equiv \; | t_2^{N_1} (S_1 \Gamma_1) \, e^{N-N_1} (S_2 \Gamma_2) \, S\Gamma \beta \Gamma^* M_\Gamma^* > \, ,$$

where, among other things, the Γ's stand for IRC's of O, Γ^* results from the decomposition of $S \otimes \Gamma$ into IRC's of O^* (the double group of $O = O^*/Z_2$), and β is a decomposition multiplicity label. (Observe that $| S\Gamma \beta \, \Gamma^* M_\Gamma^* >$ resembles the Russell-Saunders coupling vector $| S \, L \, J \, M_J >$ used in atomic spectroscopy.) The diagonalization of $H_C + H_{cf} + H_{so}$ in a $| S\Gamma \beta \Gamma^* M_\Gamma^* >$ basis is at the root of the so-called SFM [1-3]. If the only assumption about the vectors relative to the t_2 and e shells is that they transform respectively according to the IRC's T_2 and E of O, the SFM (referred in that case to as the extended SFM) is a 13 parameter model : 10 for H_C (a, b, c, ..., j) , 1 for H_{cf} (Δ), and 2 for H_{so} (ζ_{so}, ζ'_{so}) [1-2]. We now examine two particular realizations of the (extended) SFM. First, <u>the covalent realization</u> : the vectors relative to the t_2 and e shells are molecular orbitals built from linear combinations of atomic orbitals of the central ion and its surrounding ions ; the 13 parameters may therefore be expressed in terms of many-center integrals. Second, <u>the ionic realization</u> : the vectors relative to the t_2 and e shells are simply built from pure d vectors ; the 13 parameters are not hence all independent and reduce to only 5 parameters : 3 for H_C (A, B, C), 1 for H_{cf} (10 Dq), and 1 for H_{so} (ζ). Indeed, the latter realization corresponds to the conventional SFM. An alternative to the conventional SFM is provided by the conventional SAWFM which also includes 5 parameters and theoretically leads to the same energy levels than the conventional SFM [4]. The conventional SAWFM requires the diagonalization of $H_C + H_{cf} + H_{so}$ in a basis of type

$$|J a \Gamma^* M_{\Gamma^*}> \equiv |d^N \alpha SLJa\Gamma^* M_{\Gamma^*}>,$$

where a is a branching multiplicity label to be used when the IRC J of SU_2 contains several times the IRC Γ^* of O^*.

3. <u>Towards a new model</u>

Needless to say that the calculations within the (extended) SFM are not so easy to conduct [1-3] since the angular momenta L and J do not appear in $|S\Gamma\beta\Gamma^* M_{\Gamma^*}>$. Therefore, the SFM calculations require a complete knowledge of various fractional parentage coefficients in the basis $|S\Gamma\beta\Gamma^* M_{\Gamma^*}>$ as well as of the Wigner-Racah algebra of O^*. In this respect, it should be noted that the $3\text{-}\Gamma^* M_{\Gamma^*}$, $6\text{-}\Gamma^*$, and $9\text{-}\Gamma^*$ symbols for the (not simply reducible) group O^* are not so well standardized than the corresponding quantities for the (simply reducible) group SU_2. On the contrary, the vectors $|Ja\Gamma^* M_{\Gamma^*}>$ exhibit the angular momenta S, L, and J so that it is possible to use the well standardized 6-j and 9-j symbols for SU_2 as well as fractional parentage coefficients and reduced matrix elements tabulated for atomic spectroscopy ; the only complication is that the magnetic quantum number M_J has to be replaced here by the label $a\Gamma^* M_{\Gamma^*}$. In other words, the calculations within the conventional SAWFM present the same characteristics than the ones in (nuclear or) atomic spectroscopy except that the $SU_2 \supset O^*$ Clebsch-Gordan coefficients have to be substituted to the ordinary $SU_2 \supset U_1$ Clebsch-Gordan coefficients [4]. Nevertheless, the crucial point is that the conventional SAWFM, a 5 parameter model, is far from being equivalent to the extended SFM, a 13 parameter model. Consequently, it appears highly desirable to ameliorate the conventional SAWFM by retaining its computational advantages. The best way to achieve this aim seems to keep intact the basis $|Ja\Gamma^* M_{\Gamma^*}>$ but to modify the Hamiltonian $H_C + H_{cf} + H_{so}$ in order to produce results covering those given by the conventional SAWFM and paralleling those given by the extended SFM.

We thus start from the spin- and orbit-dependent Hamiltonian

$$H = \sum_{\substack{i \neq j \\ k's \\ a_0}} \left\{ \left\{ u^{k_1}(i) \otimes u^{k_2}(j) \right\}^{k_S} \otimes \left\{ u^{k_3}(i) \otimes u^{k_4}(j) \right\}^{k_L} \right\}^{k}_{a_0 \Gamma_0} D\left[(k_1 k_2) k_S (k_3 k_4) k_L k a_0 \right] , \qquad (1)$$

where u^k denotes a Racah unit tensor, Γ_0 the identity IRC of O (O^*), and a_0 a branching multiplicity label. The coupled tensors $\{\ \}^{k_S}$ and $\{\ \}^{k_L}$ act respectively on the spin and orbit parts of $|Ja\,\Gamma^* M_{\Gamma}{}^* >$. The Hamiltonian H results from an hybridization of ideas more or less implicite in previous works (cf. [4] and [5] as well as references therein). Its matrix elements in a basis $|Ja\,\Gamma^* M_{\Gamma}{}^* >$ assume a very simple structure :

$$< ' | H | > = \delta(\Gamma^{*'}, \Gamma^*)\, \delta(M_{\Gamma}{}^{*'}, M_{\Gamma}{}^*) \sum (' \| \ \| \ )\ f\begin{pmatrix} J' & J & k \\ a'\Gamma^{*'} & a\Gamma^* & a_0\Gamma_0 \end{pmatrix} D[\ ], \qquad (2)$$

where f( ) stands for a coupling coefficient relative to the chain $SU_2 \supset O^*$ [4] and (' || ||) an atomic-like reduced matrix element. All the information on the group O (O^*) appears in the product f( ) D[ ].

The most general model underlying Eqs. (1) and (2) involves more than 13 parameters. These parameters can be seen to include : orbit-dependent parameters $\{k_3 k_4 k_L\} \equiv D[(00)\,0\,(k_3 k_4)\, k_L k_L a_0]$ and spin- and orbit-dependent parameters of the special type $[1 k_L k] \equiv D[(01)\,1\,(0 k_L)\, k_L k a_0]$. The conventional SAWFM arises as a limiting case of the most general model : A, B, and C are linear combinations of {000}, {220}, and {440} while 10Dq and ζ are respectively proportional to {044} and {110}. A nontrivial particular case of the most general model comes out from the parametrization spanned by the 10 orbit-dependent parameters {000}, {044}, {220}, {224}, {244}, {246}, {440}, {444}, {446}, {448} and the 2 spin- and orbit-dependent parameters [110], [134]. The latter parametrization corresponds to a 12 parameter model we refer to as the extended SAWFM.

The comparison of the extended SAWFM and the extended SFM leads to a 13-12 (compatible) linear system [5]. The equivalence between the two latter models requires the following relation

$$a + 2b - c/\sqrt{3} - d - 2e + 2f = 0 \qquad (3)$$

to be satisfied [5]. It is then possible to express the 12 parameters of the extended SAWFM in terms of 12 of the 13 parameters of the extended SFM (and vice versa) [5].

We close with some general comments about Eq. (3). To the best of our knowledge, this relation has never been mentioned in the literature. A pseudo-derivation of Eq. (3) can be directly achieved in the framework of Griffith's formalism [2] without passing through the above discussed equivalence between the extended SFM and SAWFM. We note that Eq. (3) is (is not) trivially satisfied when the extended SFM restricts to its ionic (covalent) realization. In the case of the covalent realization, it seems from the numerical values of a, b, c, ..., f scattered in numerous works that Eq. (3) becomes better and better satisfied as far as the involved molecular orbitals are more and more refined. Finally, it is perhaps worthwhile to stress that Eq. (3) can be rewritten as

$$3\,B(t_2^{\,2}) - B(t_2e) - 2\,B(e^2) = 0\,,$$

where $B(x)$ stands for the orbital barycenter of the configuration x; the latter relation compares with the identity

$$9\,B(t_2^{\,2}) + 12\,B(t_2e) + 4\,B(e^2) = 25\,B(t_2^{\,2} + t_2e + e^2)\,.$$

References

[1] S. Sugano, Y. Tanabe, and H. Kamimura, Multiplets of Transition Metals Ions in Crystals (Academic, New York, 1970).

[2] J.S. Griffith, The Theory of Transition-Metal Ions (Cambridge Univ. Press, 1961).

[3] D. T. Sviridov and Yu. F. Smirnov, Theory of Optical Spectra of Transition Metal Ions (Nauka, Moscow, 1977). (In Russian)

[4] M. Kibler, J. Molec. Spectrosc. 26, 111 (1968); Int. J. Quantum Chem. 3, 795 (1969).

[5] M. Kibler and G. Grenet, Preprint LYCEN 8034 (IPN de Lyon, 1980).

GROUP THEORY OF THE EFFECTIVE POTENTIAL*

W. Lang
Max Planck Institute, Föhringer Ring 6, München 40, Germany.

L. O'Raifeartaigh
Dublin Institute for Advanced Studies, Dublin 4, Ireland.

G. Parravicini
Institute of Theoretical Physics, University of Milan,
Via Celoria 16, Milan, Italy.

ABSTRACT

A simple group theoretical proof is given of the result that, to all orders of perturbation, the effective potential in scalar quantum field theory can be expressed as an expansion of the vacuum graphs. Although, this result has been established for some time we have felt it worthwhile to present our proof for a number of reasons. First, the result though not new, does not appear to be as well-known as it deserves. Second, the proof follows immediately from a very simple group-theoretical consideration and is much simpler than its forerunners. Third, the proof establishes a one-one correspondence between the convexity of the classical potential and the reality of the effective potential (thus clarifying the appearance of complex values for some effective potentials, notably in the case of spontaneous symmetry breakdown). Finally, a general relationship between the classical field and the quantum mechanical mean field is found as a bye-product.

1. INTRODUCTION

As mentioned in the abstract the purpose of this talk is to give a simple group-theoretical proof of the 'vacuum graph formula', which expresses the effective potential in scalar quantum field theory in terms of the vacuum graphs of the theory. More precisely, if

$$V(\phi) = \frac{1}{2} m_{ab} \phi_a\phi_b + \frac{1}{3!} f_{abc} \phi_a\phi_b\phi_c + \frac{1}{4!} g_{abcd} \phi_a\phi_b\phi_c\phi_d \quad (1.1)$$

denotes a renormalizable, convex, classical potential, and $\Sigma(m,f,g)$ denotes the sum of all (connected one-particle-irreducible) vacuum graphs in the corresponding quantized theory (depending only on the parameters $m = m_{ab}$, $f = f_{abc}$, $g = g_{abcd}$, as indicated), then the effective potential $U(\phi)$, defined as $V(\phi)$ plus its radiative corrections in the limit of constant fields, is given (to all orders

* Talk presented at the IXth International Conference on Group Theoretical Methods in Physics, Cocoyoc, Mexico, June 1980.

of perturbation) by the formula

$$U(\phi) = V(\phi) + \Sigma(\Delta^{\phi}m,\ \Delta^{\phi}f,\ g), \tag{1.2}$$

where

$$(\Delta^{\phi}m)_{ab} = m_{ab} + f_{abc}\,\phi_c + \tfrac{1}{2}g_{abcd}\,\phi_c\phi_d\ , \tag{1.3}$$

$$(\Delta^{\phi}f)_{abc} = f_{abc} + g_{abcd}\,\phi_d\ ,$$

The significance of the formula (1.3) will become clear later. The result (1.2) is very remarkable because it implies that to know all the Feynman graphs of the theory at zero external momentum (constant ϕ) one need know only the vacuum graphs. In other words, the graphs with any number of external legs (at zero momentum) can be obtained from those with no external legs by differentiation of (1.2) with respect to ϕ.

As the definition of the effective potential depends on the concept of the Legendre transforms, we begin by considering this concept separately. We then define the effective action, and potential, and show that they are determined by integro-differential equations whose kernels are simply related to the classical Lagrangian. The kernels are shown to have a simple group theoretical property which is discussed in some detail, and the expansion in terms of the vacuum graphs is shown to follow as a result of this group property. The role of convexity in the derivation is stressed, and as a by-product, the distinction between the classical field and the quantum mechanical mean field is made precise.

2. THE LEGENDRE TRANSFORM

A well-known example of a Legendre transform is the transform from the Lagrangian to the Hamiltonian formalism in classical mechanics

$$L(\dot{q},q) \rightarrow -H(p,q) = L(\dot{q},q) - p\dot{q}\ ,\ \text{where } p = \frac{\partial L}{\partial \dot{q}}\ . \tag{2.1}$$

In this case the variable q is a spectator, and the transformation takes place in $(\dot{q},p)$ space. Accordingly, the general definition of the Legendre transform for a function f(x) of one variable is defined to be

$$f(x) \rightarrow f^{\ell}(y) = f(x) - xy,\ \text{where } y = \frac{df}{dy}\ . \tag{2.2}$$

The geometrical interpretation of $f^{\ell}(y)$ is that it plots the point of intersection of the tangent to f(x) against the slope of the tangent. It follows immediately from (2.2) that the Legendre transform is involutive,

$$f^{\ell\ell}(x) = f(x)\ . \tag{2.3}$$

Note also that since from (2.2)

$$\frac{dy}{dx} = \frac{d^2f}{dx^2} , \qquad (2.4)$$

the transformation $x \to y$, and hence the Legendre transform is not well-defined at points of inflexion. In fact, from (2.4) it also follows that the second derivatives of $f^{\ell}(y)$ and $f(x)$ are inversely related

$$\frac{d^2f^{\ell}(y)}{dy^2} \cdot \frac{d^2f(x)}{dx^2} = -1 , \qquad (2.5)$$

a result which shows explicitly that the transform must break down when $f''(x)$ is zero.

For several variables x_a , $a = 1 \ldots\ldots n$, the generalization of the Legendre transform is

$$f^{\ell}(y\ ) = f(x\ ) - x_a y_a , \text{ where } y_a = \frac{\partial f(x)}{\partial x_a} , \qquad (2.6)$$

The geometrical interpretation is the same except that the tangent is to be replaced by the tangent-plane. It follows again that the transform is involutive,

$$f^{\ell\ell}(x_a) = f(x_a). \qquad (2.7)$$

Since the Jacobean of the transformation $x_a \to y_a$ is

$$\frac{\partial y_a}{\partial x_b} = \frac{\partial^2 f}{\partial x_a \partial x_b} , \qquad (2.8)$$

we see that the transform will be well-defined only for those ranges of x_a for which $\partial^2 f\ \partial x_a \partial x_b$ is not zero. These are just the ranges for which $f(x_a)$ is convex(or concave). From (2.8) we also see that the generalization of (2.5) is

$$\frac{\partial^2 f^{\ell}(y)}{\partial y_a \partial y_b} \cdot \frac{\partial^2 f(x)}{\partial x_b \partial x_c} = -\delta_{ac} . \qquad (2.9)$$

and this equation exhibits explicitly the breakdown when $f(x)$ fails to be convex(or concave).

For functionals $\phi_a(x)$, the Legendre transform is obtained by passing to the continuum limit, and we obtain

$$f^{\ell}(\pi_a(x)\ ) = \int dx \left[f(\phi_a(x)\ ) - \phi_a(x)\pi_a(x) \right] , \qquad (2.10)$$

where $\pi_a(x) = \frac{\delta}{\delta\phi_a(x)} f(\phi_a(x)\ )$, and $f(\phi_a(x))$ may contain any finite number of derivatives $\partial\phi_a(x)|\partial x_\mu$ of $\phi_a(x)$.

3. THE EFFECTIVE ACTION AND EFFECTIVE POTENTIAL

In order to make contact with standard concepts in defining the effective action and effective potential we start from the Schwinger[3] generating functional W(J) for the connected Greene's functions of the quantized theory. The functional W(J) can be written as a functional integral[4] and for scalar quantum field theory it then takes the form

$$e^{W(J,L)} = \int d\mu(\eta)\, e^{L(\eta) + \int J_a(x).\eta_a(x)d^4x},$$

$$L(\eta) = \int d^4x \quad (\eta(x))\,, \tag{3.1}$$

where $\eta_a(x)$ denotes any number of scalar fields, (η) is the Lagrangian density and $d\mu(\eta)$ the measure. The measure is assumed to be translationally-invariant in field space, and is normalized so that $W(o,L) = 0$.

In the classical limit the path-integral over fields is replaced by the single path which corresponds to the classical field $\eta^c(x)$. Since the classical field satisfies the Euler-Lagrange field equations

$$\frac{\delta L}{\delta\eta^c} \equiv \frac{\partial\mathcal{L}}{\partial\eta^c_a} - \partial_\mu \frac{\partial\mathcal{L}}{\partial\eta^c_{a,\mu}} = - J_a \,, \tag{3.2}$$

we see that in the classical limit W(J,L) reduces to the Legendre transform of L.

In both the classical and quantum cases the effective action $\Gamma(\phi,L)$ is defined to be the Legendre transform of W(J,L),

$$\Gamma(\phi,L) = W(J,L) - J\phi \quad \text{where } \phi = \frac{\delta W}{\delta J}\,. \tag{3.3}$$

In the classical limit Γ is therefore the double Legendre transform of L and hence reduces to the classical action L itself. In the quantized case, $\Gamma(\phi,L)$ becomes the generating functional for the connected one-particle irreducible (CIPI) Feynmann graphs[5]. (The reducible graphs are missing because these would become the tree-graphs in the classical limit and L contains no tree graphs except, of course, the original point interactions).

A most important special case of (3.3) is when $\phi = o$. Then the CIPI Feynmann graphs become just the vacuum graphs, and we have (up to a factor due to integration over the whole 4-space)

$$\Gamma(o,L) \equiv W(J,L)_{\frac{\partial W}{\partial J}=o} = \Sigma\,(m,f)\,, \tag{3.4}$$

where $\Sigma(m,f)$ denotes the sum of all vacuum graphs. Since the vacuum graphs have no external momenta or fields they can depend only on the mass and coupling constants parameters(denoted generally by m and f respectively) and it is to emphasize this point that m and f are exhibited explicitly in $\Sigma(m,f)$. Normally, of course, the fact that $\Sigma(m,f)$ depends only on m and f makes it uninteresting, but in the case if the effective action $\Sigma(m,f)$ becomes of crucial importance because, as we shall see, m and f become field-dependent.

The effective potential $U(\phi,L)$ is defined to be the effective action in the limit when the fields $\phi(x)$ become constant

$$U(\phi,L) = \operatorname{Lt} \Gamma(\phi(x), L) \quad , \text{ as } \phi(x) \to \text{constant}. \tag{3.5}$$

In this limit the functionals of this section reduce to functions (modulo a factor ℓ^4 corresponding to an integration over the total 4-volume). From the properties of the effective action it is clear that the effective potential reduces to the ordinary potential in the classical limit (modulo ℓ^4) and that in the quantum case it describes the CIPI Feynman graphs in the limit of zero external momentum. Thus the effective potential gives the leading terms in an expansion of the Feynman graphs in powers of the momentum.

4. THE CLASSICAL FIELD AND THE MEAN FIELD.

From the definition of the field $\phi(x)$ in (3.3) we have

$$\phi(x) = e^{-W(J,L)} \int du(n)n(x)\, e^{L(n) + \int J_a n_a} \quad , \tag{4.1}$$

from which we see that ϕ is the average field with respect to the integrand in $W(J,L)$. For this reason $\phi(x)$ is sometimes called the classical field, but to distinguish it from the truly classical field $n^c(x)$ which satisfies the classical Euler-Lagrange equation (3.2) we prefer to call $\phi(x)$ the mean field. The relationship between the classical and mean fields can be obtained by noting that the definition of $J_a(x)$ does not change in the classical limit. Hence we have, from the inverse Legendre transforms,

$$-J_a(x) = \frac{\delta L(n^c)}{\delta n_a^c} = \frac{\delta \Gamma(\phi)}{\delta \phi_a} \quad , \tag{4.2}$$

which gives the required relationship. Note that (4.2) has a simple intuitive interpretation: the classical field n^c satisfies the classical Euler-Lagrange equations for the classical action L while the mean field ϕ satisfies the classical Euler-Lagrange equations for the effective action Γ.

To illustrate that (4.2) does make a practical difference let us consider for example the one-loop contribution U_1 to U. Expanding (4.2) we obtain in the limit of constant fields

$$\overset{(1)}{\Delta\phi} \equiv (\phi - n^c) = -\frac{\partial U_1}{\partial \phi} \Big| \frac{\partial^2 L}{\partial \phi^2} \quad , \tag{4.3}$$

which gives the shift in the field to this order. In particular, at any extremum of V, (4.3) gives the shift in the field which is necessary to remain at the extremum of the potential, and agrees with the result that is obtained by calculating directly using tadpole graphs[6]. Note that the shift (4.3) vanishes if $v(\phi)$, and hence $u_1(\phi)$, is even at the extremum.

It should be remarked, however, that the shift (4.2) or (4.3) will occur only if we compare $u(\phi)$ with $V(\phi)$ (or $u(\eta^c)$ with $V(\eta^c)$). If we make the more natural comparison of $u(\phi)$ with $V(\eta^c)$ then, by definition, the extrema coincide and there is no shift in position. Thus what (4.2) and (4.3) really measure is the shift between the classical and the mean fields, not the shift in the position of the extremum. Similarly, since

$$-\frac{\partial J_a}{\partial \phi_b} = \frac{\partial^2 u}{\partial \phi_a \partial \phi_b} = \frac{\partial \eta_d}{\partial \phi_b} \cdot \frac{\partial^2 V}{\partial \eta\, \partial \eta_d} = \frac{\partial^2 V}{\partial \phi_a \partial \phi_b} + \text{rad. corr.} \tag{4.4}$$

we see that the character of the extremum (minimum, saddle-point, maximum) is the same for $u(\phi)$ as for $V(\eta_c)$. Thus the radiative corrections change neither the position nor the character of the local extrema. Of course, these results are strictly local since they assume both the Legendre transform and perturbation theory. Hence they do not preclude the emergence of new global extrema as in the Coleman-Weinberg example.

5. INTEGRO-DIFFERENTIAL EQUATIONS FOR THE EFFECTIVE POTENTIAL

By solving eq. (3.3) for J(x) we obtain

$$J(x) = -\frac{\delta\Gamma(\phi,L)}{\delta\phi(x)} , \tag{5.1}$$

and if we insert this result and (3.3) in (3.1) we find that $\Gamma(\phi,L)$ satisfies the integro-differential equation

$$e^{\Gamma(\phi,L)} = \int d\mu(\eta) e^{L(\eta) + \int(\phi-\eta)\cdot\Gamma'(\phi,L)} , \tag{5.2}$$

where prime denotes functional differentiation (and dot denotes inner product in the case of more than one field). This equation is, of course, quite convoluted since it contains $\Gamma'(\phi,L)$ on the right hand side, but it has the advantage that it no longer contains any reference to J or W(J) and therefore determines $\Gamma(\phi,L)$ directly as a functional of L. In fact, had we not wished to make the connection with the more standard formalism of sect. 2, the integro-differential equation (5.2) could have been taken as the starting point. Note that (5.2) contains also the boundary conditions for $\Gamma(\phi,L)$ since its derivative

$$\Gamma''(\phi,L) \int (\phi-\eta) d\mu(\eta)\, e^{L(\eta) + \int(\phi-\eta)\Gamma'(\phi,L)} = 0 , \tag{5.3}$$

determines $\Gamma'(\phi,L)$ implicitly in terms of ϕ. (If $\Gamma''(\phi,L)$ is zero, higher derivatives of (5.2) can be used). Of course, the solution of (5.2) may only exist in local domains of ϕ, where the implicit equation (5.3) has a(unique) solution.

For reasons which will be discussed in the next section, it is more convenient to work with the radiative corrections $\gamma(\phi,L)$ to $L(\phi)$ defined by

$$\Gamma(\phi,L) = L(\phi) + \gamma(\phi,L) \; , \tag{5.4}$$

than with $\Gamma(\phi,L)$ itself. Inserting (5.4) in (5.2) we find that $\gamma(\phi,L)$ satisfies the integro-differential equation

$$e^{\gamma(\phi,L)} = \int d\mu(\eta)\; e^{L(\eta)-L(\phi)-\int(\eta-\phi)\left[L'(\phi) + \gamma'(\phi,L)\right]} . \tag{5.5}$$

The right hand side of (5.5) appears complicated, but if we use the assumption that the measure is translationally invariant we can shift the field η to $\eta+\phi$ and (5.5) reduces to

$$e^{\gamma(\phi,L)} = \int d\mu(\eta)\; e^{\Delta^{\phi}L(\eta)-\int\eta\,\gamma'(\phi,L)} \; , \tag{5.6}$$

where $\Delta^{\phi}L(\eta)$ is the 'shifted and truncated' action

$$\Delta^{\phi}L(\eta) = L(\phi+\eta) - L(\phi) - \int\eta L'(\phi) \; . \tag{5.7}$$

Since (5.7) is equivalent to (5.2) it determines $\gamma(\phi,L)$ directly as a functional of L (at least in local patches of field-space). What distinguishes (5.7) from (5.2), however, is the occurence of the kernel $\Delta^{\phi}L(\eta)$, whose properties we shall now discuss.

6. PROPERTIES OF THE KERNEL $\Delta^{\phi}L(\eta)$

The first important property of $\Delta^{\phi}L(\eta)$ is that it is again an action. That is, it contains only quadratic and higher terms. Indeed, as we shall see below it is the same as the original action $L(\eta)$, except that the parameters become ϕ-dependent. It is for this reason that it is preferable to work with $\gamma(\phi,L)$ and the kernel $\Delta^{\phi}L(\eta)$, than with $\Gamma(\phi,L)$ and the untruncated kernel $L(\eta + \phi)$, which contains constant and linear terms.

In the case of a convex potential in particular, $\Delta^{\phi}L(\eta)$ has the property that if the absolute minimum of the original potential $V(\eta)$ is at $\eta = 0$ then the absolute minimum of the potential $\Delta^{\phi}V(\eta)$ also is at $\eta = 0$. To see this we note that since $\Delta^{\phi}V(\eta)$ has only terms which are quadratic and higher in η, it vanishes at $\eta = 0$ and hence $\eta = 0$ is an absolute minimum if, and only if, $\Delta^{\phi}V(\eta)$ is positive,

$$\Delta^{\phi}V(\eta) \equiv V(\eta+\phi) - V(\phi) - \eta V'(\phi) \geq 0. \tag{6.1}$$

But (6.1) is just the condition that $V(\eta)$ be convex. Thus for convex potentials, $\Delta^{\phi}V(\eta)$ is a 'true' potential for all real ϕ.

The second important property of $\Delta^{\phi}L(\eta)$ is that it satisfies the simple translational group relation

$$\Delta^{\phi}(\Delta^{\varepsilon}L(\eta)) = \Delta^{\phi+\varepsilon}(L(\eta)) \ . \tag{6.2}$$

If Δ^{ϕ} were only the operator of translation $L(\eta)\to L(\eta+\phi)$ the property (6.2) would be obvious, but the point is that it holds in spite of the truncation. The formal proof is as follows:

$$\begin{aligned}(\Delta^{\phi}(\Delta^{\varepsilon}L))\eta &= \Delta^{\phi}(L(\eta+\varepsilon)-L(\varepsilon)-\eta L'(\varepsilon)) \\ &= (L(\eta+\varepsilon+\phi)-L(\varepsilon)-(\eta+\phi)L'(\varepsilon))-(L(\phi+\varepsilon)-L(\varepsilon)-\phi L'(\varepsilon)) \\ &\qquad -\eta\,(L'(\phi+\varepsilon)-L'(\varepsilon)) \qquad (6.3) \\ &= L(\eta+\phi+\varepsilon)-L(\phi+\varepsilon)-\eta\, L'(\phi+\varepsilon) = \Delta^{\phi+\varepsilon}L(\eta)\ .\end{aligned}$$

However, the deeper reason for (6.2) is that although the translation group $f(\eta)\to f(\eta+\phi)$ does not leave invariant the space of truncated functions (they pick up constant and linear terms), it does leave invariant the complement, namely the space of linear functions. Hence the restriction of the group to the truncated functions, which is the quotient of all functions with respect to the linear functions, still forms a group representation. In other words,

$$\begin{pmatrix} a & b \\ 0 & c\end{pmatrix}\begin{pmatrix} a' & b' \\ 0 & c'\end{pmatrix} = \begin{pmatrix} A & B \\ 0 & C\end{pmatrix} \implies A = a.a'. \tag{6.4}$$

A third important property of $\Delta^{\phi}L(\eta)$ is that the effect of the operator Δ^{ϕ} may be absorbed by the parameters of $L(\eta)$, which then become ϕ-dependent. They also form a (triangular) linear representation of the translation group. For example, for the general renormalizable single-field action

$$L(\eta;\ m\ f\ g) = \int dx\ \ \tfrac{1}{2}(\partial\eta)^2 + \frac{m}{2!}\eta^2 + \frac{f}{3!}\eta^3 + \frac{g}{4!}\eta^4 \ , \tag{6.5}$$

one easily finds that

$$\Delta^{\phi}L(\eta) = \int dx\ \ \tfrac{1}{2}(\partial\eta)^2 + \frac{\Delta^{\phi}m}{2!}\eta^2 + \frac{\Delta^{\phi}f}{3!}\eta^3 + \frac{g}{4!}\eta^4 \ , \tag{6.6}$$

where

$$\begin{bmatrix}\Delta^{\phi}m \\ \Delta^{\phi}f \\ \Delta^{\phi}g\end{bmatrix} = \begin{bmatrix} 1 & \phi & \frac{\phi^2}{2} \\ 0 & 1 & \phi \\ 0 & 0 & 1\end{bmatrix}\begin{bmatrix} m \\ f \\ g\end{bmatrix} = \left[M(\phi)\right]\begin{bmatrix} m \\ f \\ g\end{bmatrix}, \tag{6.7}$$

and

$$M(\phi+\varepsilon) = M(\phi)M(\varepsilon) . \tag{6.8}$$

More generally, for a single field potential with Taylor expansion

$$V(\eta) = \sum_{n=2} \frac{c_n}{n!} \eta^n , \tag{6.9}$$

one finds that

$$\Delta^{\phi}V(\eta) = \sum_{n=2} \frac{c_n(\phi)}{n!} \eta^n = e^{\phi t} \sum_{n=2} \frac{c_n}{n!} \eta^n , \tag{6.10}$$

where t is the triangular operator defined by

$$tc_n = c_{n+1} . \tag{6.11}$$

and the group property of exp ϕt is evident.

Finally, for a renormalizable potential with n scalar fields as in (1.1) we obtain in the constant limit

$$\Delta^{\phi}V(\eta,m,f,g) = V(\eta,\Delta^{\phi}m,\Delta^{\phi}f,g) , \tag{6.12}$$

where $\Delta^{\phi}m$ and $\Delta^{\phi}f$ are the expressions given in (1.3).

7. PROOF OF THE VACUUM GRAPH FORMULA FOR CONVEX POTENTIALS

After this somewhat lengthy preparation the proof of the vacuum graph formula is relatively short. From equation (5.6) we see that the radiative part $\gamma(\phi L)$ of the effective action is generated by $\Delta^{\phi}L(\eta)$ through an integro-differential equation which contains no other functions. Thus

$$\gamma(\phi,L) = F(\Delta^{\phi}L) , \tag{7.1}$$

where F is a fixed functional (at least for local patches in ϕ-space). Then

$$\gamma(\phi+\varepsilon,L) = F(\Delta^{\phi+\varepsilon}L), \tag{7.2}$$

and

$$\gamma(\phi, \Delta^{\varepsilon}L) = F(\Delta^{\phi}(\Delta^{\varepsilon}L)) . \tag{7.3}$$

But from the group property (6.2) we have

$$\Delta^{\phi+\varepsilon}L = \Delta^{\phi}(\Delta^{\varepsilon}L). \tag{7.4}$$

Hence

$$\gamma(\phi+\varepsilon,L) = \gamma(\phi,\Delta^{\varepsilon}L) . \tag{7.5}$$

This is a general relation satisfied by the radiative part of the effective action. The vacuum graph formula is the special case obtained by setting $\phi = o$, in which case we obtain

$$\gamma(\varepsilon, L) = \gamma(o, \Delta^{\varepsilon}L) , \tag{7.6}$$

and hence

$$\Gamma(\varepsilon,L) = \Gamma(o,\Delta^{\varepsilon}L) + L(\varepsilon). \tag{7.7}$$

But from (3.4) we have

$$\Gamma(o,L) = \Sigma(m,f) , \tag{7.8}$$

and since for convex potential and constant ε, $\Delta^{\varepsilon}L(\eta)$ is just as valid a potential as $L(\eta)$ itself (with parameters $\Delta^{\varepsilon}m$, $\Delta^{\varepsilon}f$), we also have

$$\Gamma(o,\Delta^{\varepsilon}L) = \Sigma(\Delta^{\varepsilon}m,\Delta^{\varepsilon}f), \tag{7.9}$$

where $\Delta^{\varepsilon}m$ and $\Delta^{\varepsilon}f$ are the parameters of $\Delta^{\varepsilon}L$ from (1.3). It follows at once from (7.7) and (7.9) that

$$U(\varepsilon,L) \equiv \underset{\varepsilon=\text{constant}}{\text{Lt}\Gamma(\varepsilon(x),L)} = \Sigma(\Delta^{\varepsilon}m, \Delta^{\varepsilon}f) + V(\varepsilon) , \tag{7.10}$$

as required.

8. LIMITATIONS OF THE VACUUM GRAPH FORMULA

Apart from the obvious limitations of rigour, and the fact that the renormalization has not been taken explicitly into account, the formula (1.2) is limited in that it has been derived only for convex potentials. As we have seen in section 6, $\Delta^{\phi}V(\eta)$ is a 'true' potential, (and hence (7.9) is immediately valid) if, and only if, $V(\eta)$ is convex.

To see that the extension to non-convex potentials does raise problems, let us consider the one-loop contribution to the effective potential (1.2). The one-loop contribution is well-known [1,2,7] to be

$$U(\phi,L) = (64\pi^2)^{-1}\text{tr}\,(\Delta^{\phi}m)^2 \ln\left[(\Delta^{\phi}m)\,|\mu\right], \tag{8.1}$$

where μ is the renormalization point in momentum space. It is clear that the contribution (8.1) is real for all real ϕ if, and only if, $\Delta^{\phi}m$ is positive for all real ϕ. But since

$$(\Delta^{\phi}m)_{ab} \equiv \frac{\partial^2 V}{\partial\phi_a \partial\phi_b} , \tag{8.2}$$

such will be the case if, and only if, V is convex. Thus for non-convex V the effective potential becomes complex for some real values of ϕ. A detailed discussion of the non-convex case, and applications of the effective potential will be given in a subsequent paper.

REFERENCES

1. R. Jackiw, Phys. Rev. D9 1686 (1974)
2. See also J. Iliopoulos, C. Itzykson, A. Martin, Rev. Mod. Phys. 47, 165(1975); S.Y. Lee, A. Sciacculaga, Nucl. Phys. B96, 435(1975)
3. J. Schwinger, Proc. Nat. Acad. Sci. US 37, 452, 455 (1951)
4. D. Amit, Field Theory, Renormalization Group and Critical Phenomena. (McGraw-Hill, New York 1978)
5. G. Jona-Lasinio, Nuovo Cim. 34, 1790(1964)
6. S. Weinberg, Phys. Rev. D7, 2887(1973)
7. S. Coleman, E. Weinberg, Phys. Rev. D7 1888(1973)

BLOCH THEOREM FOR CRYSTALS WITH STRUCTURAL DISTORTIONS

Daniel B. Litvin
Department of Physics
The Pennsylvania State University
The Berks Campus, P.O. Box 2150
Reading, PA 19608 USA

ABSTRACT

A new Bloch Theorem, based on wreath group symmetry, is formulated for crystals with structural distortions. This new Bloch Theorem is applied to determine the form and corresponding charge density of one-electron eigenfunctions in the nearly free electron approximation for crystals with periodic structural distortions.

1. WREATH GROUPS

The use of wreath groups for the classification of color functions and spin functions defined on crystals was first introduced by Koptsik and Kotzev [(1-3)]. Recent reviews on the classification of color and spin functions have been given by Kotzev [(4)] and Opechowski [(5)]. Wreath groups have recently also been used in the classification of crystals with structural distortions by Koptsik [(6-8)] and Litvin [(9)]. It is an application of the theory of wreath groups to the derivation of a Bloch Theorem for crystals with structural distortions which is the topic of this contribution.

Let $E(3) = \{P, V_E(3)\}$ be a three-dimensional Euclidean point space consisting of the point space P and the Euclidean vector space $V_E(3)$. Let R denote the positions of atoms of a crystal, in P, whose symmetry group $\tilde{F}$ is one of the 230 three-dimensional space groups, and let $\vec{D}(R)$ denote a vector function defined on the crystal which maps points R into vectors of the vector space $V_E(3)$. $\vec{D}(R)$ represents the structural distortion of the position of the atom which in the undistorted crystal is at the position R.

The wreath group symmetry of a crystal with structural distortions $\vec{D}(R)$ is defined as the group of all operator pairs $(\vec{V}_F(R)|F)$, where $\vec{V}_F(R)$ is a vector function defined on the crystal which maps points R into vectors of $V_E(3)$, paired with an element F of the space group $\tilde{F}$ of the crystal and such that [(9,10)]:

$$(\vec{V}_F(R)|F)\vec{D}(R) = \vec{D}(F^{-1}R) + \vec{V}_F(R) = \vec{D}(R)$$

The product of two such operator pairs is:

$$(\vec{V}_{F_1}(R)|F_1)\ (\vec{V}_{F_2}(R)|F_2) = (\vec{V}_{F_1}(R) + \vec{V}_{F_2}(F_1{}^{-1}R)|F_1F_2)$$

The wreath group is a subgroup of the semi-direct product $\Omega_V\ s\Omega_F$, where

Ω_V is the group of all operators $(\vec{V}(R)|E)$ and Ω_F the group of all operators $(E|F)$. This semi-direct product is called the <u>wreath product</u> $V_E(3)\ w\Omega_F$ (11,12).

2. BLOCH THEOREM FOR CRYSTALS WITH STRUCTURAL DISTORTIONS

The Bloch Theorem for crystals whose symmetry group is one of the 230 three-dimensional space groups (13) is a statement concerning the structure of eigenfunction $\Psi(r)$ of the electronic Schrodinger equation $H\Psi(r) = E\Psi(r)$ of the crystal. The Bloch Theorem states that the structure of eigenfunctions $\Psi_k(r)$ are such that:

$$\Psi_k(r) = e^{-ik\cdot r}\ U_k(r) \quad ; \quad \{t\}U_k(r) = U_k(r)$$

where t is a translation of the space group $\tilde{F}$ of the crystal and k is a vector in the first Brillouin Zone.

We define operators $\{V(r)|F\}$ which act on the space of all scalar functions $U(r)$ defined on the point space P of E(3) in the following manner: First, we define the vector functions $\vec{W}(r)$ and $\vec{V}(r)$ defined on P which map <u>all</u> points r into vectors of the vector space $V_E(3)$. We define operators $(\vec{V}(r)|F)$ on the space of all vector functions $\vec{W}(r)$ such that:

$$(\vec{V}(r)|F)\vec{W}(r) = \vec{W}(F^{-1}r) + \vec{V}(r)$$

Second, we define the mappings $<\vec{V}(r)|F>$ of the point space P onto itself. A mapping $<\vec{V}(r)|F>$ maps the point r into the point denoted by $<\vec{V}(r)|F>r$ and defined by:

$$<\vec{V}(r)|F>r = Fr + \vec{V}(r)$$

Finally, we define the operators $\{V(r)|F\}$ on the space of all scalar functions $U(r)$ defined on P:

$$\{\vec{V}(r)|F\}\ U(r) = U(<\vec{V}(r)|F>^{-1}r)$$

Because of an isomorphism which exists between the operators $(V(r)|F)$, $<\vec{V}(r)|F>$, and $\{\vec{V}(r)|F\}$, and typographical reasons, we shall denote all these operators by $(\vec{V}(r)|F)$.

We assume the following relationship between the wreath group symmetry of the crystal with structural distortions and an invariance wreath group of the Hamiltonian: There exists an invariance wreath group of the Hamiltonian consisting of operator pairs $(\vec{V}_F(r)|F)$, one such pair for each element F of $\tilde{F}$, such that restricting the functions $\vec{V}_F(r)$ to $\vec{V}_F(R)$, the resulting operators $(\vec{V}_F(R)|F)$ constitute the wreath group of the crystal with structural distortions $\vec{D}(R)$. This invariance wreath group contains a subgroup $\tilde{T}_w$ of operator pairs $(\vec{V}_t(r)|t)$ one such pair for each translation t of the translational subgroup $\tilde{T}$ of $\tilde{F}$. $\tilde{T}_w$ is isomorphic to $\tilde{T}$ and the irreducible representation of $\tilde{T}_w$ are then $\Gamma^{\vec{k}}(\vec{V}_t(r)|t) = \exp(ik\cdot t)$.

We construct eigenfunctions $\Psi_k(r)$ of the Hamiltonian which are basis functions of the irreducible representations Γ^k of $\tilde{T}_w$ using a projection operator proceedure:

$$\Psi_k(r) = \sum_{t'} e^{-ik\cdot t'}(\vec{V}_{t'}(r)|t')\phi(r)$$

where $\phi(r)$ is an arbitrary eigenfunction of the Hamiltonian. The eigenfunctions $\Psi_k(r)$ are basis functions of irreducible representations Γ^k of $\tilde{T}_w$:

$$\begin{aligned}(\vec{V}_t(r)|t)\Psi_k(r) &= \sum_{t'} e^{-ik\cdot t'}(\vec{V}_t(r)|t)\ (\vec{V}_t{}'(r)|t')\phi(r) \\ &= \sum_{t'} e^{-ik\cdot t'}(\vec{V}_{t+t'}(r)|t+t')\phi(r) \\ &= e^{ik\cdot t}\sum_{t''} e^{-ik\cdot t''}(V_{t''}(r)|t'')\phi(r) \\ &= e^{ik\cdot t}\Psi_k(r)\end{aligned}$$

The Bloch Theorem for crystals with structural distortions follows from writing the eigenfunctions $\Psi_k(r)$ as:

$$\Psi_k(r) = e^{-ik\cdot r}U_k(r)$$

where, from the above, one finds:

$$(\vec{V}_t(r)|t)U_k(r) = e^{-ik\cdot\vec{V}_t(r)}U_k(r)$$

3. NEARLY FREE ELECTRON APPROXIMATION

In the nearly free electron approximation $U_k(r)$ is given by the fourier integral:

$$U_k(r) = \int A_k(p)e^{ip\cdot r}dp$$

The coefficients $A_k(p)$ are given by the inverse fourier integral:

$$A_k(p) = \int U_k(r)e^{-ip\cdot r}dr$$

Using the Bloch Theorem for crystals with structural distortions on the first of the two above equations, substituting into the second, and summing on all elements of the wreath group $\tilde{T}_w$, one derives an integral equation for the coefficients $A_k(p)$. Assuming that the structural distortions are of the form $\vec{D}(R) = \vec{D}\sin(Q\cdot R)$, summing over t and integrating over r, one obtains:

$$A_k(p) = (2\pi)^3 \sum_{m,m'} \int J_m(\vec{D}\cdot(k-p))J_{m'}(-\vec{D}\cdot(k-p))A_k(p')\times$$

$$\delta(K-p'-mQ)\ \delta(p'-p+(m+m')Q)dp'$$

where J_m are Bessel functions of the first kind, and K is a reciprocal lattice vector of the translational subgroup $\tilde{T}$ of $\tilde{F}$. $A_k(p)$ is equal to zero if the two delta functions in the above equation are not simultaneously satisfied. It follows that:

$$A_k(p) = 0 \quad \text{if} \quad p \neq K + mQ$$

Consequently, in the nearly-free electron approximation, the eigenfunctions of a one-electron Schrodinger equation for a crystal with periodic structural distortions $\vec{D}(R) = \vec{D}\sin(Q\cdot R)$ are given by:

$$\Psi_k(r) = e^{-ik\cdot r} \sum_{K,m} A_k(K,m)e^{i(K + mQ)\cdot r}$$

The charge density $\delta_k(r) = |\Psi_k(r)|^2$ corresponding to the kth eigenfunction $\Psi_k(r)$ can be calculated from the above. This charge density can be written in the form:

$$\rho_k(r) = \rho_k^o(r) + \sum_{m\neq o}\rho_k^m(r)\cos(mQ\cdot r)$$

where the functions $\rho_k^m(r)$, m=0,1,..., are functions invariant under translations t of $\tilde{T}$, i.e. $\rho_k^m(r + t) = \rho_k^m(r)$. Consequently, the charge density in the nearly-free electron approximation for a crystal with periodic structural distortions is a modulated charge density, consisting of charge densities having the translational periodicity of the undistorted crystal modulated by functions cos(mQ•r).

ACKNOWLEDGEMENTS

Financial support for this work by the Faculty Scholarship Support Fund - Phase IV of The Pennsylvania State University for a visit to the University of British Columbia, and by Dr. H. W. Perkins, Director, Berks Campus, The Pennsylvania State University, for travel funds to attend this colloquium, is greatfully acknowledged.

REFERENCES

1) V. A. Koptsik and J. N. Kotzev, Comm. JINR, P4-8067,P4-8068 (1974) Dubna.
2) V. A. Koptsik, Kristal un Technik 10 231 (1975).
3) J. N. Kotzev, Proceedings of the Neutron Diffraction Conference, Petten, The Netherlands (Reactor Center of Netherlands RCN-234, 1975) p. 126.
4) J. N. Kotzev, Proceedings International Symposium on Crystallographic Groups, Bielefeld, in press MATCH - Informal Communications in Mathematical Chemistry.
5) W. Opechowski, ibid. and in Group Theoretical Methods in Physics, Edited by R. T. Sharp and B. Kolman (Academic Press, NY 1977) p. 93.
6) V. A. Koptsik, Ferroelectrics 21 499 (1978).
7) V. A. Koptsik, Proceedings International Symposium on Crystallographic Groups, Bielefeld, in press MATCH - Informal Communications in Math-

ematical chemistry.

8) V. A. Koptsik, Doklady Akademii Nauk SSR 250 353 (1980).

9) D. B. Litvin, Phys. Rev. 21 3184 (1980).

10) D. B. Litvin, Annals of the Israel Physical Society, Vol. 3 Group Theoretical Methods in Physics, Editors L. Horwitz and Y. Ne'eman (APS, NY, 1980).

11) B. H. Neuman, Lectures on Topics in the Theory of Infinite Groups (Bombay, Tata Institute for Fundamental Research, 1961) Chapter V.

12) D. B. Litvin, Physica, in press (1980).

13) L. Jansen and M. Boon, Theory of Finite Groups, Applications in Physics (Amsterdam, North Holland 1967) p. 249.

GLOBAL SYMMETRIES OF SPIN SYSTEMS DEFINED ON ABELIAN MANIFOLDS

V. Rittenberg

Physikalisches Institut, Bonn, West-Germany

I would like to report on some work done in Bonn together with M. Marcu[1)] and related work done by some of my colleagues[2)].

We consider the generating functional for a statistical mechanical system defined on an abelian group A (called A-system):

$$W = \sum_{\alpha_P, \alpha_{P'}} \exp\left\{ -\beta S(\alpha_P, \alpha_{P'}, \dots) + \sum_{r,P} J_{r,P} \chi_r(\alpha_P) \right\} \tag{1}$$

Here $J_{r,P}$ are the sources, $\chi_r(\alpha_P)$ are the characters of A, $\alpha_P \in A$, P denotes the lattice site, $\beta = \frac{1}{kT}$, S is the action:

$$S = \sum_{\alpha_P, \alpha_{P'}, \dots} L(\alpha_P - \alpha_{P'}) \tag{2}$$

In Eq. (2) the summation is done on the appropriate lattice configuration (for example nearest neighbours on a two dimensional lattice). The Lagrangian L can be parametrized in two ways:

$$L(\alpha) = \sum_r a_r \chi_r(\alpha) \quad , \quad \alpha \in A \tag{3}$$

or

$$L(\alpha) = \sum_\beta b_\beta \delta(\alpha - \beta) \quad , \quad \alpha, \beta \in A \tag{4}$$

The parametrisation (3) is done in terms of characters while the parametrisation (4) is done in terms of orbits.

Models of this type are generalized Ising models ($A = Z_2$) and some of them have been recently intensively studied[3)]. The action S is by construction invariant under A:

$$\alpha'_P = \alpha_P + t \quad , \qquad t \in A \tag{5}$$

One is interested to know which are the larger global symmetries $g \in G$ which are obtained through special choices of the coupling constants a_n or b_β in Eqs. (3) and (4):

$$L(\alpha_1 - \alpha_2) = L(g(\alpha_1) - g(\alpha_2)) \tag{6}$$

We have been able to give the complete classification for the following groups: Z_p, Z_{p^2}, $Z_p \otimes Z_q$, $Z_p \otimes Z_p$ and $Z_2 \otimes Z_2 \otimes Z_2$. Here p and q are prime numbers. I will mention some of the results.

a) Z_p groups

Let Z_{p-1} be the automorphism group of Z_p, and Z_k one of its subgroups (k divides p-1). If the coupling constants in Eqs. (3) and (4) verify the conditions:

$$a_{un} = a_n \quad , \quad b_{u\beta} = b_\beta \tag{7}$$

The Lagrange function L is invariant under the transformations

$$\alpha' = u\alpha + t \quad , \qquad u \in Z_k \, , \, t \in Z_p \tag{8}$$

We denote this group by M_p^k which is a group of order k.p

If

$$a_1 = a_2 = \cdots = a_{p-1} \quad ; \quad b_1 = b_2 = \cdots = b_{p-1} \tag{9}$$

The symmetry group is S_p (the symmetric group of order p!).

Theorem: All the global symmetries of a Z_p-system are M_p^k and S_p.

b) Z_6 group

We give the relations and the corresponding symmetry group:

$$
\begin{array}{lll}
1) & a_1 = a_2 = a_4 = a_5 & (S_3 \wr Z_2) \\
2) & a_2 = a_4\,,\ a_1 = a_3 = a_5 & (Z_2 \wr S_3) \\
3) & a_1 = a_3 = a_5 & (Z_2 \wr Z_3) \\
4) & a_1 = a_4\,,\ a_2 = a_5 & (Z_3 \wr Z_2) \\
5) & a_1 = a_5\,,\ a_2 = a_4 & (S_3 \otimes Z_2) \\
6) & a_1 = a_2 = a_3 = a_4 = a_5 & (S_6)
\end{array}
\tag{10}
$$

In Eq. (10), $A \wr B$ denotes the wreath product of the permutation groups A and B. It is to our knowledge for the first time that wreath products of groups appear in physics.

A few comments on the M_p^k groups mentioned before. Those groups have only 1-dimensional and p-dimensional representations[2)] and they represent a class of metacyclic groups.

The metacyclic groups are defined through the multiplication law

$$g^{\alpha_1}_{\beta_1}\, g^{\alpha_2}_{\beta_2} = g^{\alpha_1+\alpha_2}_{\beta_1 + a^{\alpha_1}\beta_2} \tag{11}$$

where

$$
\begin{array}{l}
\alpha_1, \alpha_2 \in Z_n \ ; \ \beta_1, \beta_2 \in Z_m \\
a^n = 1 \ (\mathrm{mod}\ m) \ ; \ (a, m) = 1
\end{array}
\tag{12}
$$

For the metacyclic groups all irreducible representations and Clebsch-Gordan coefficients can be written in an analytic form[2)]. This is remarkable because those groups are not simply reducible. The knowledge of the Clebsch-Gordan coefficients, in particular for the M_p^k groups, allows the construction of many-body interaction terms which can be added to the Lagrangian (2).

A final comment on a purely group theoretical problem. We may ask the question: which is the smallest order group which has an irreducible representation of dimension N? If N = p - 1 the group is M_p^{p-1}.

References

1. M. Marcu, V. Rittenberg: The global symmetries of spin systems defined on abelian groups I, II, III, Nuclear Physics B (to be published)

2. A. Bovier, M. Lüling and D. Wyler: Representations and Clebsch Gordan coefficients of Z-metacyclic groups (J. Math. Phys. to be published)

3. R. Savit, Rev. Mod. Phys. 52, 453 (1980)

PHASES AND CONJUGACY CLASSES

A. I. Solomon
The Open University
Milton Keynes MK7 6AA
U.K.

Introduction

The relationship between symmetry groups and phase transitions is fairly well accepted; the transition of a system from one phase to another is usually accompanied by the spontaneous breaking of the symmetry group associated with the system. Examples abound - the transition from liquid to solid accompanied by the loss of translational symmetry, from electron gas to superconductor accompanied by the breakdown of phase-angle symmetry. However, there is as yet no corresponding analysis of the phase structure of a system in terms of the corresponding dynamical group, that is, the group which describes both the symmetry of the system and its spectrum. Following the orbit analysis of the spontaneous symmetry breaking case, one might expect a similar orbit analysis of the dynamical group to reveal details of the phase structure. In algebraic terms, restricting oneself to the spectrum-generating-algebra (SGA) of a given thermodynamic system, one might expect aspects of the phase structure to be revealed by a study of conjugacy classes of sub-algebras of the appropriate SGA. However, in previous models for which a SGA description has been given, the Lie algebra has usually been insufficiently complicated in structure to enable this speculation to be tested adequately. For example, the SGA for the BCS model of a superconductor has the form

$$g = \bigoplus_k g_k$$

where each g_k is isomorphic to so(3) - an algebra remarkably deficient in conjugacy classes! Nonetheless, in the light of the preceding remarks one might be tempted to say that the existence of a single non-trivial conjugacy class - that of the whole algebra so(3) itself - is indicative of the presence of a single superconducting phase. The case of superfluid Heluim Four is more suggestive. Here the appropriate SGA is so(2,1) and there it is known[(1)] that the two non-trivial conjugacy classes, those associated with a compact and non-compact generator respectively, do correspond to properties of the Helium interaction potential - repulsive and attractive - which give rise to superfluid and non-superfluid phases. But it is only with an algebra much richer in subalgebras than either of the two above examples that one can begin to test the relationship between conjugacy classes and phase structure. Such a model is afforded by that of an anisotropic fermi superfluid[(2)], which includes the BCS superconductor model and superfluid Helium Three. Details of this model will be presented elsewhere; for the present discussion it will suffice to outline the model and the derivation of its spectrum-generating-algebra. For the general fermi superfluid this turns out to be so(6), and for superfluid Helium Three so(5); the richness of the subalgebra structure of this latter algebra is reflected in the rich phase structure experimentally observed for superfluid Helium Three[3].

Conjugacy Classes

If we suppose that the SGA of our system is a semi-simple rank-ℓ Lie algebra g, a basis for g

$$\{h_1, h_2, \ldots, h_\ell; e_1, e_2, \ldots, e_{n-\ell}\}$$

may be chosen, where the ℓ elements h_i form a commuting basis of a Cartan subalgebra. As described in the Proceedings of the previous conference in this series[4], diagonalisation consists in finding an automorphism of the SGA which sends the hamiltonian (assumed to be an element of the SGA) to a sum of the mutually commuting h_i. As these are algebras over R, there will in general be several conjugacy classes of Cartan subalgebras; our hamiltonian will belong to just one such class. Further, it is the thesis of the present note that the algebraic structure of the SGA is reflected in the physical phase structure of the hamiltonian. However, in general there is an infinite number of subalgebras available - even in the case of so(6) and so(5) - so the question arises as to how to choose the subalgebras which in turn generate the conjugacy classes. Since the energy spectrum of the hamiltonian is ultimately to be expressed in terms of the h_i, it would seem reasonable to use the Cartan elements h_i to generate the conjugacy classes. An attempt to associate with the element h_i those elements of the algebra which map to h_i by some automorphism would not in general give a subalgebra, as can be seen by considering the case of so(2,1), with non-compact generators X,Y and compact generator Z. The elements X and Y are both conjugate to X, but $[X,Y] = -Z$ is conjugate to Z. We therefore consider the following construction - which always leads to a subalgebra. Let $A_i' = \text{centralizer of } h_i = \{x \in g : [x,h_i] = 0\}$. In general, the algebra A_i' will not be semi-simple, as the element $h_i \in A_i'$ generates an abelian ideal. However, define

$$A_i = A_i'/\text{centre } A_i$$

and then consider the conjugacy class $\bar{A}_i$ of subalgebras conjugate to A_i by automorphisms of g. It is these conjugacy classes A_i which we shall use to describe the phases associated with our given hamiltonian.

Anisotropic Superfluid Model

We consider a pairing model for an interacting system of fermions whose hamiltonian is

$$H = \sum_{k,\beta} E_k a^+_{k\alpha} a_{k\alpha} + \frac{1}{2} \sum_{k,k',\alpha,\beta} V_{kk'} a^+_{k\alpha} a^+_{-k\beta} a_{-k'\beta} a_{k'\beta}$$

where the fermi creation and annihilation operators obey

$$[a_{k\alpha}, a_{k'\beta}]_+ = \delta_{kk'} \delta_{\alpha\beta}$$

The indices k,k' refer to momentum, and α,β to spin (↑ or ↓). Hartree-Fock linearisation leads to a reduced hamiltonian

$$H^{red} = \sum_k H_k$$

where $H_k = E_k a^+_{k\alpha} a_{k\alpha} + V(k,\alpha,\beta) a^+_{k\alpha} a^+_{-k\beta} + V^+(k,\alpha,\beta) a_{-k\beta} a_{k\alpha}$

with $V(k,\alpha,\beta) = \sum \langle \frac{1}{2} V_{kk'} a_{k'\beta} a_{-k'\alpha} \rangle$.

If we define $(A_1,A_2,A_3,A_4) = (a_\uparrow, a_\downarrow, a^+_{-\downarrow}, a^+_{-\uparrow})$ we find that the hamiltonian can be expressed in terms of

$$J^{\mu,\nu} = A^+_i m^{\mu,\nu}_{ij} A_j \qquad (i,j = 1,2,3,4)$$

where the 4 × 4 matrices $m^{\mu,\nu}_{ij}$ are given by $\tau_\mu \times \tau_\nu$ $(\mu,\nu = 0,1,2,3)$ where τ_μ are the usual Pauli spinors (and $\tau_o \times \tau_o$ is absent). The 15 elements $J^{\mu,\nu}$ thus generate su(4) (~ so(6)); the generators may be represented by 3-vectors thus $\underline{E} = \frac{1}{2}\underline{\tau} \times \tau_o$, $\underline{T} = \frac{1}{2} \tau_1 \times \underline{\tau}$, $\underline{U} = \frac{1}{2} \tau_2 \times \underline{\tau}$, $\underline{S} = \frac{1}{2} \tau_o \times \underline{\tau}$, $\underline{W} = \frac{1}{2} \tau_3 \times \underline{\tau}$. Here $\underline{S}$ plays the role of a spin operator; restricting to the spin-0 case give the BCS so(3) hamiltonian, while restricting to the spin-1 case gives the Helium Three so(5) case, whose latter generators are E_3, $\underline{T}$, $\underline{U}$, $\underline{S}$. We select a basis for a Cartan subalgebra of so(6) as follows;

$$\{h_1, h_2, h_3\} = \{E_3, W_3, S_3\}.$$

The construction outlined in the previous section leads to three so(4) conjugacy classes - which we may refer to as the E, W and S phases respectively. The intersection of these classes with the Helium Three so(5) subalgebra of so(6) gives rise to various physically interpretable phases as follows: the pairwise intersection of the phases is in each case a "unitary state" of Helium Three and an so(3) subalgebra. The W phases are "equal-spin-pairing phases" - so the pairwise intersections (E,W) and (W,S) are unitary, equal-spin-pairing (ESP) states (A-states) while (E,S) is a unitary non-ESP state (B-state). The so(4) W-state may be written as a direct sum so(3) ⊕ so(3), where the so(3) subalgebras refer to A^{I} and A^{II} states, respectively (these are 'non-unitary' states).

Summary

The preceding section is a synopsis of the results obtained by a conjugacy class analysis of the SGA for an anisotropic fermi superfluid. I apologize for the jargon used in describing the various experimentally observed states; however, there is enough correspondence between subalgebras and phases to suggest that the relationship is not merely fortuitous. In each case the experimentally observed state reduces to an so(3) subalgebra; this is reasonable as, being of rank 1, such an algebra is associated with one observable (Casimir operator). This observable corresponds to the single energy gap associated with the state.

References

(1) A. I. Solomon, J. Math. Phys. 12, 390 (1971).
(2) A. I. Solomon, Annals N.Y. Academy of Sciences (to be published, 1980).
(3) A. J. Leggett, Rev. Mod. Phys. 47, 331 (1974).
(4) A. I. Solomon, Proceedings of the VIIIth International Colloquium on Group Theoretical Methods in Physics, Israel; page 357 (1979).

THE PROBABILITY DENSITY IN NON RELATIVISTIC QUANTUM MECHANICS AND THE (EXTENDED) GALILEI GROUP

V. Aldaya and J.A. de Azcárraga

Dpto. de Física Teórica,

Facultad de C. Físicas, Burjasot (Valencia), Spain

As is well known, every significant statement in Quantum Theory is a statement about rays and, because of that, the projective representations of the Galilei group G are the relevant ones in non-relativistic Quantum Mechanics. These projective representations come from a representation of a central extension of the Galilei group by the "phase" group Θ,

$$1 \to \Theta \to \tilde{G}_{(m)} \to G \to 1$$

which is usually denoted $\tilde{G}_{(m)}$ where m indicates the mass of the particle. On the Schrödinger wave function of a galilean free particle the representation of $\tilde{G}_{(m)}$ is given by (*)

$$[U(\theta,\tau,\vec{a},\vec{v},R)\psi](\vec{x}',t') = \exp im(\frac{1}{2}\vec{v}^2t + \vec{v}\cdot R\vec{x} + \frac{\theta}{m})\psi(\vec{x},t)$$

which fulfills the $\tilde{G}_{(m)}$ group law

$$(\theta';\tau',\vec{a}',\vec{v}',R')*(\theta;\tau,\vec{a},\vec{v},R) = (\theta+\theta+\xi_{(m)}(g',g);g'g)$$

(*) We shall take $\hbar=1$ throughout and the possibility of spin will be ignored as it is not relevant for the ensuing discussion. For a review of the Galilei groups, see Ref.1 .

where $\xi_{(m)}$ is the factor system of the extension and g'g is the Galilei group composition law, $g'g = (\tau+\tau', \vec{a}'+R'\vec{a}+\tau\vec{v}', \vec{v}'+R'\vec{v}, R'R)$

It is interesting to analyze the invariance of the Schrödinger Lagrangian under $\tilde{G}_{(m)}$ since, as it may be convenient to remark here, the galilean invariance of the quantum description of a particle is achieved through that group since the Schrödinger lagrangian is not invariant under the Galilei group (see below). Defining the trivial Schrödinger bundle $\eta_S = (E \xrightarrow{\pi} M)$ of coordinate system $(t,\vec{x};\psi,\psi^*)$, the Lagrangian density $\mathcal{L}_S$ is the following function on the bundle $J^1(E)$ of the 1-jets of E

$$\mathcal{L}_S = -\frac{1}{2m}\psi_i^*\psi^i + i\psi^*\psi_t$$

the coordinate system of $J^1(E)$ being given by $(t,x^i,\psi,\psi^*;\psi_t,\psi_t^*,\psi_i,\psi_i^*)$. The wave functions of the free particle are cross sections of E which satisfy the Schrödinger equation, which is obtained by applying the variational principle to the Hamilton functional defined by

$$I(\psi) = \int_{j^1(\psi)(M)} \mathcal{L}_S[j^1(\psi)]\,\omega \qquad \forall\psi\in\Gamma(E)$$

where $\Gamma(E)$ is the modulus of cross sections, ω the volume element on M and j^1 is the 1-jet prolongation.

To check the invariance of $\mathcal{L}_S$ under $\tilde{G}_{(m)}$ it is sufficient, once the expression of the vector fields X_a $(a=1,\dots,11)$ associated with the different parameters of the group has been obtained(*) to evaluate

$$L_{\bar{X}_a} \mathcal{L}_S \qquad (\bar{X}_a \equiv j^1(X_a))$$

to see that the Lie derivative is either zero or (on cross sections) a total derivative. Indeed, it may be verified that the Lie derivative is zero for all the vector fields of $\tilde{G}_{(m)}$ but for the "boosts", for which one finds a total derivative. (This is not the case for the galilean boosts of G, for which it is found that $L_{\bar{K}_{(i)}} \mathcal{L}_S = -i\psi_i\psi^*$ and thus there is no invariance as earlier mentioned.) Taking into account the $\tilde{G}_{(m)}$ invariance of $\mathcal{L}_S$, the Noether charges may now be evaluated. Although the 1-jet prolongations of the vector fields X_a are necessary to check the invariance of $\mathcal{L}_S$, only their components on E appear in the expression of the Noether currents, which are given by

$$j^{0,i} = -\mathcal{L}X^{0,i} + (X^t\psi_t + X^j\psi_j - X_\psi)\frac{\partial\mathcal{L}}{\partial\psi_{0,i}}$$

$$+ (\psi^*_t X^t + \psi^*_j X^j - X_{\psi^*})\frac{\partial\mathcal{L}}{\partial\psi^*_{0,i}}$$

(*) The X_a satisfy the same commutation relations of the Galilei group but for $[X_\theta$, any $X] = 0$, $[X_{K_i}, X_{P_j}] = m\,\delta_{ij}X_\theta$ were $X_\theta \equiv X_{11}$.

- where X^t, X^i, X^ψ indicate the different components of the vector field - plus an additional term $-\Delta^{0,i}$ when $L_{\tilde{X}}\mathcal{L}_S = \frac{D}{Dx^{0,i}} \Delta^{0,i}$. Applying this expression to the generator associated with the central element of the algebra of $\tilde{G}_{(m)}$, which is explicitly given by

$$X_\theta = -i\psi \frac{\partial}{\partial\psi} + i\psi^* \frac{\partial}{\partial\psi^*}$$

one obtains

$$j^0_{(\theta)} = -\psi^*\psi \equiv -\rho \qquad j^i_{(\theta)} = \frac{i}{2m}(\psi^*\psi^i - \psi^{i*}\psi) \quad .$$

Thus, the charge associated with θ is simply the nonrelativistic probability density, and the expression of conservation of the Noether current is the well known continuity equation of nonrelativistic quantum mechanics which thus appears as a consequence of the extended galilean invariance of the theory.

To conclude, let us mention that the above results are consistent with the customary galilean invariance of newtonian Classical Mechanics. The transition from $\tilde{G}_{(m)}$ to G is accomplished by averaging over charge densities i.e., going to the classical limit in Ehrenfest's sense. In this limit, the conserved magnitudes are the usual ones (the associated to the boosts gives the uniform motion of the center of mass) and the integrated probability density, one for a normalized wave function, becomes irrelevant in the classical limit. This is interesting because it is not possible to define a faithful action of $\tilde{G}_{(m)}$ on the configuration space $\{(t,q^i)\}$, which would have led to a $\tilde{G}_{(m)}$-invariant Classical Mechanics; indeed, the Schrödinger lagrangian density is the minimal lagrangian invariant under the action of $\tilde{G}_{(m)}$, group which appears as the quantum symmetry group of the nonrelativistic approximation. The relevance of $\tilde{G}_{(m)}$ is in agreement with other approaches: in the geometric quantization scheme, G is not quantizable, and $\tilde{G}_{(m)}$ appears as a quantizable extension of the Galilei group G (2)(*).

References

(1) J.M.Lévy-Leblond in Group Theory and its Applications, vol. II, E.M. Loebl Ed., Academic Press (1971)

(2) J.M.Souriau, Structure des Systèmes Dynamiques, Dunod, Paris (1970)

(3) M.Pauri and G. Prosperi, J. Math. Phys. 7, 366 (1966)

(*) There is, in fact, another reason to introduce $\tilde{G}_{(m)}$, namely, that the classical Poisson algebra of G does not close (3).

QUATERNIONIC QUANTUM MECHANICS AND ADLER'S CHROMOSTATICS

L.C. Biedenharn* and D. Sepunaru*+#
Physics Department, Duke University
Durham, North Carolina 27706 U.S.A.

and

L.P. Horwitz#
Department of Physics and Astronomy
Tel Aviv University, Ramat Aviv, Israel

I. *Introduction*

Quantum field theory involving non-abelian gauge fields has been highly developed during the past several years,but the lack of decisive results on the problem of confinement(a major objective for such theories)has led to the suggestion that achieving a semi-classical understanding of the dynamics of such systems would be a useful first step. It is difficult to define and study a semi-classical limit for a field theory of this type in three space and one time dimension, since there is no natural scale for achieving such a limit. A direction for the development of a semi-classical understanding of theories of this type was suggested by Khriplovich(1),Giles and McClerran(2) and Adler(3), and in a series of papers Adler(4) has worked out a systematic procedure for obtaining the dynamical equations describing field configurations,and the static potentials,in a semi-classical framework.

Adler's procedure imbeds the algebraic structures of the gauge field in a tensor product space constructed from the algebras generated by the charges carried by the sources. In a recent work(5), he has shown that this procedure can be applied to the construction from pre-quarks,of the eight composite systems of a quark-lepton generation thus providing--in the context of his model--a constructive theoretical basis for the model of Harari(6)and Shupe(7).

The correspondence between Adler's construction and the usual approach taken in quantum field theory has not yet been clarified.In this paper,we shall show that the algebraic structures associated by Adler's construction with an underlying U(2)gauge group can be obtained from quaternionic quantum theory.We shall first briefly review the construction given by Adler,and then summarize some of the special features of quaternionic quantum mechanics. We then show how Adler's construction is obtained in the framework of this latter theory.

*Supported in part by the National Science Foundation
+On leave of absence from Tel Aviv University,Ramat Aviv,Israel.
#Research supported in part by Binational Science Foundation (BSF), Jerusalem.

II. *Adler's Algebraic Chromostatics*[3,4,5]

Let us consider a Yang-Mills field for the SU(n) gauge group; the gauge potentials are denoted by $b^a_\mu(x)$, where $a=1,2\ldots,(n^2-1)$; $\mu=0\ldots3$ and x is a space-time point. Denoting the n x n generators of SU(n) by $\{\lambda^a\}$, we may write the potentials as a matrix:

$$B_\mu(x) \equiv \sum_{a=1}^{n^2-1} \tfrac{1}{2} b^a{}_\mu(x)\, \lambda^a \tag{1}$$

The field strength tensor can then be written as:

$$F_{\mu\nu}(x) = \partial_\nu B_\mu - \partial_\mu B_\nu - ig[B_\mu, B_\nu], \tag{2}$$

where:

$$F_{\mu\nu}(x) \equiv \sum_{a=1}^{n^2-1} \tfrac{1}{2} f^a_{\mu\nu}\, \lambda^a \quad . \tag{3}$$

In the classical version of the theory it is assumed that $b^a_\mu(x)$ is a c-number field; in the framework of field theory, it is a local quantized field.

Writing the covariant derivative as:

$$D_\mu W = \partial_\mu W + ig[B_\mu, W] \tag{4}$$

one finds that eqs. (1-4) imply

$$D_\lambda F_{\mu\nu} + D_\mu F_{\nu\lambda} + D_\nu F_{\lambda\mu} = 0. \tag{5}$$

The source equation is taken to be:

$$D_\nu F^{\mu\nu} = gJ^\mu, \tag{6}$$

with the (matrix) source current being:

$$J_\mu \equiv \sum_{a=1}^{n^2-1} \tfrac{1}{2} j^a_\mu(x) \lambda^a \tag{7}$$

That the sources are covariantly conserved follows from eq. (6); that is,

$$D_\mu J^\mu = 0 \ . \tag{8}$$

In the classical case, taking $j_\mu{}^a(x)$ to be a single point source, the solution to these equations is of Coulomb type, thus all non-linear structure disappears. To resolve this well-known difficulty, Khriplovich[1] suggested a solution for the case of *two* static point sources; here non-linearity is maintained in a non-trivial way. It was quickly pointed out[2,3,8] that for color singlet states (singlet state for the golbal gauge group constructed of the two source particles) this non-linearity again disappears. Giles and McClerran[2] and Adler[3] suggested that in the semi-classical limit of gauge theory, the functions $b^\mu_a(x)$ should not be assumed to be c-number fields, but that they should be algebra-valued as a consequence of assuming *non-commuting source charges.*

In particular, Adler assumes that, for N particles, one has the source charges*: $Q_1^{\ a}(x), Q_2^{\ a}(x), \ldots, Q_N^{\ a}(x)$--where now $a=0,1,\ldots,n^2-1$ --which obey the commutation rules:

$$[Q_i^a, Q_j^b] = i\delta_{ij} f^{abc} Q_\alpha^c, \quad (i,j=1\ldots N). \qquad (9)$$

The f^{abc} are the (totally anti-symmetric) structure constants of U(n), defining $f^{0bc}=0$. (The Lie algebra of U(n) carried by each of the charges is called by Adler *the underlying algebra.*)

The basic principle of Adler's construction is that the potentials, $b_\mu^{\ a}(x)$, and the field strengths, $f^a_{\mu\nu}(x)$, even though now algebra-valued and generally non-commuting, *must maintain their formal gauge-transformation properties under (algebraic-valued) gauge transformations*. This is actually accomplished easily, once one recognizes that the $\{\lambda^a\}$, with the unit tensor (λ^0) adjoined, form a matrix algebra:

$$\tfrac{1}{2}\lambda^a \ \tfrac{1}{2}\lambda^b = \tfrac{1}{2}q^{abc} \ \lambda^c, \qquad (10)$$

where: $q^{abc}=d^{abc}+ i f^{abc}$; with d^{abc} the usual, totally symmetric, D-operator [(9)] and $d^{0bc} \equiv (2/n)^{\frac{1}{2}} \delta^{bc}$.

Let us once again define matrix potentials by:

$$B_\mu(x) \equiv \sum_{a=0}^{n^2-1} \tfrac{1}{2} b_\mu^a(x) \ \lambda^a, \qquad (1')$$

and matrix field strength by:

$$F_{\mu\nu}(x) = \sum_{a=0}^{n^2-1} \tfrac{1}{2} f^a_{\mu\nu}(x) \ \lambda^a \qquad (3')$$

where now not only are $\{b^a_\mu\}$ and $\{f^a_{\mu\nu}\}$ algebra-valued but also components for a=0 have been adjoined. Note that the sources also have an adjoined a=0 component, so that:

$$J_\mu(x) = \sum_{a=0}^{n^2-1} \tfrac{1}{2} j^a_\mu \ \lambda^a, \qquad (7')$$

With these modifications, one now finds: <u>the algebra-valued fields, eqs. (1') and (3'), and sources, eq. (7'), obey the Yang-Mills equations (2), (4-7).</u>

[*Remark:* Adler expressed this basic result in a different way. Let $U = \sum_{a=0}^{n^2-1} \tfrac{1}{2} u^a (\lambda^a)$ and $V = \sum_{a=0}^{n^2-1} \tfrac{1}{2} v^a \lambda^a$. Then the commutator of U and V takes the form (using (10)):

*(For simplicity, we take static sources: $j_\mu^{\ a}(x)=\delta^0_\mu \, Q^a(x)$.)

$$[U,V] = \sum_{a=0}^{n^2-1} \frac{\lambda^a}{2} \{\sum_{b,c=0}^{n^2-1} q^{abc}(u^b v^c - v^b u^c)\} .$$

Adler called the (antisymmetric) term in brackets "the P-product" of $\{u^a\}$ with $\{v^a\}$. Note that the P-product has a complicated algebraic structure since it involves both commutators and anti-commutators of u and v. The use of the matrix commutator in place of P-product greatly simplifies the analysis.]

In order to gain an understanding of the solutions to these algebraic gauge field equations let us, following Adler, consider the (static) N particle case. One introduces for each source charge an independent set of n x n U(n) matrices: $\{\lambda^a_i\}$, $(a=0\ldots n^2-1;$ $i=1\ldots N)$ which satisfy eq. (9). (These matrices are to commute with the original set of $\{\lambda^a\}$.) Thus we have the N (matrix) charges:

$$Q_i \equiv \tfrac{1}{2} \sum_{a=0}^{n^2-1} \lambda^a_i \lambda^a , \qquad i=1,\ldots N. \tag{11}$$

The *charge algebra* is defined[1] to be the (Lie) algebra generated by the charges, Q_i, under commutation. The resulting charge algebra for N particles is particularly simple since the *N generating charges are ismorphic to transpositions belonging to the permutation group over N +1 objects.*[10-12] [This simplifying result is actually no surprise to nuclear structure physicists since (for U(2)) the charge Q_1 is easily seen to be Dirac's spin exchange operator; the generalization (the exchange operator for U(n)) is equally well-known in nuclear physics[13].]

The analysis of the charge algebra is further simplified if one observes that the charges, Q_1, which generate the algebra, are all invariant under the U(n) group generators: $2\Lambda^a \equiv \lambda^a + \sum_{i=1}^{N} \lambda^a_i$.

Consider, for example, the case N=2. Forming all possible scalars (under Λ^a) from the elements $\{\lambda^a\}$, $\{\lambda_1{}^a\}$ and $\{\lambda_2{}^a\}$ one finds the five quantities[10]: 1; $Q_1 = \lambda \cdot \lambda_1$; $Q_2 = \lambda \cdot \lambda_2$; $Q_{12} \equiv \lambda_1 \cdot \lambda_2$; and $Q_{012} = f^{abc} \lambda^a \lambda_1{}^b \lambda_2{}^c$. The charge algebra (using results from the permutation group S_3) splits into three ideals: two one-dimensional ideals and one three-dimensional ideal. Thus the charge algebra[10] is the Lie algebra of U(1) x U(2).

Similarly the N=3 charge algebra[10] is found to consist of 14 elements which split into ideals corresponding to the Lie algebra U(1) x U(2) x U(3).

To contrast with the Lie algebra generated by each of the charges $\{Q^a\}$--this is the *underlying algebra* (see eq.9)--the charge algebra, discussed above, is called the *overlying algebra*.

In order to understand the role of the overlying algebra let us consider the response of the algebraic gauge fields to an infinitesimal gauge transformation. By construction, the infinitesimal gauge transformation S:

$$S \equiv 1 + i\tfrac{1}{2}\sum_{a=0}^{n^2-1}\delta u^a \lambda^a = 1 + i\ \delta U \tag{12}$$

generates the gauge variations:

$$\delta B_\mu = -g^{-1}D_\mu(\delta U);\delta F_{\mu\nu}=i[\delta U,F_{\mu\nu}];\delta J_\mu=i[\delta U,J_\mu]. \quad (13a,b,c)$$

The last result in particular shows that the generators of the overlying algebra (the charges Q_i) undergo the gauge variation:

$$\delta Q_i=i[\delta U,Q_i]. \tag{14}$$

Using the fact that all other elements of the overlying algebra are generated by commutation, and using the result that $\delta Q_{12}=\delta[Q_1,Q_2]=[\delta U,[Q_1,Q_2]]$, we see that *all* elements of the overlying algebra obey eq. (14).

This has an important consequence: *If the gauge variations* δU *are assumed to be elements of the charge algebra* (and hence the algebra-valued fields, B_μ and $F_{\mu\nu}$, are spanned by the charge algebra), *then the gauge fields and the sources transform as the adjoint representation of the overlying gauge group.**

Remarks:

(1) For N=1, that is, for a single non-commuting charge source belonging to the U(n) *underlying* group, the charge algebra is trivial and the *overlying* group is U(1). Thus--contrary to what one might expect!--the two groups are distinct *even for one particle.*

(2) The net result of Adler's classical algebraic gauge field theory is to yield classical Yang-Mills fields belonging to the adjoint representation of the *overlying group*. This structure explicitly depends on the number of particles; in particular *the dimension of the gauge group increases with the number of particles.*

The dynamical aspects of these (classical) algebraic gauge fields have been developed quite far[4]; since, however, such results do not bear directly on our problem, we shall not review them here.

---

*This basic result is only implicit in ref. (1).

III. *Quaternion Quantum Mechanics*

In this section, we shall review the basic structure of quaternion quantum mechanics (QQM). A detailed study, including the construction of admissible tensor product spaces, has been given in ref.(14). The basic structure of a quantum theory is contained in the geometry of states, represented in a linear space by linear manifolds and their corresponding projection operators.[15] The usual complex quantum theory is concerned with the structure of complex or real linear manifolds; the range of applicability of the latter is very limited, since the reals are not algebraically closed, and the remarkable quantum mechanical effects of phase interference are not easily displayed in this context.[16] The reason for the central importance of the complex field in quantum theory is not completely understood; the necessity for hypercomplex systems such as quaternions will perhaps only become evident in the framework of phenomena concerning non-Abelian gauge theories. The manifestation of QQM is severely limited by the fact that tensor products, i.e., many-particle states, cannot be linear in the same sense that the one-particle states can, and the observables of many-body systems are not trivial copies of the possible observables of one-body systems. As we shall discuss below, only complex linearity can survive in a universal way the process of tensor product construction of spaces which can describe an increasing number of particles. The gauge group, associated with component transformations of the wave functions leaving scalar products invariant, admits operator valued transformations, reflecting the full quaternion structure of the space, of increasing complexity with the order of the tensor product. As we shall show, this structure contains that proposed (in the case of an underlying U(2) gauge theory) by Adler to obtain semi-classical approximations to the dynamics of non-Abelian gauge fields (confer Section **II**).

Let us consider a space, H_Q, of elements f, g, ..., which we shall call vectors, (right) linear over quaternions $q \in \mathbb{Q}$, where $\mathbb{Q}$ is the real algebra generated by the elements $1, e_1, e_2, e_3$ ($e_1^2 = e_2^2 = -1, e_1e_2 + e_2e_1 = 0, e_3 \equiv e_1e_2$) ; that is, if $f, g \in H_Q$, then $fq_1 + gq_2 \in H_Q$. A scalar product:

$$(f,g) = (g,f)^* \in \mathbb{Q} \tag{15}$$

is defined (where * is the quaternion involutory automorphism), for which the norm is:

$$\|f\|^2 = (f,f) \geq 0, \tag{16}$$

and the equality is valid **if** and only if **f=0**. The scalar product is linear over the quaternion algebra:

$$(f,gq) = (f,g)q \; . \tag{17}$$

We shall call a space of the type H_Q a module space.

In addition to the scalar product (15) it will be useful to define the complex scalar product:

$$(f,g)_C \equiv \mathrm{tr}(f,g) - e_1 \; \mathrm{tr}\,(f,g)e_1 \quad . \tag{18}$$

which is linear over the (complex) algebra $\mathbb{C}(1,e_1)$ generated by the elements $1,e_1$ over the reals. (The trace is defined intrinsically by $\mathrm{tr}q = \frac{1}{2}(q+q^*)$ with the normalization tr 1=1.) The norm defined by (18) coincides with the norm (16), so closure of the space (and linear manifolds) is equivalent in both cases.

We may also define left multiplication by quaternions. The left multiplying elements, which we shall also call $\{1,e_1,e_2,e_3\}$ over the reals, are not necessarily identical to right multiplying elements (they may correspond to a different realization), and the scalar products (15) and (18) are not linear with respect to left multiplication. Every element $f \in H_Q$ can be shown[14] to have the decomposition

$$f = f_0 + f_1 e_1 + f_2 e_2 + f_3 e_3, \tag{19}$$

where

$$e_i f_j = f_j e_i, \tag{20}$$

that is, the f_j, $j=0,1,2,3$, are real. An alternative representation is

$$f = \psi_0 + \psi_1 e_2, \tag{21}$$

where ψ_0,ψ_1 are complex ($\mathbb{C}(1,e_1)$-valued). Note that for $z \in \mathbb{C}(1,e_1)$,

$$fz = \psi_0 z + \psi_1 z^* e_2, \tag{21'}$$

i.e., ψ_0 is linear, and ψ_1 antilinear under multiplication by complex numbers.

The complex scalar product (18), in terms of the representation (21) is

$$(f,g)_C = (\psi_0,\chi_0) + (\psi_1,\chi_1)^*, \tag{21''}$$

where $g = \chi_0 + \chi_1 e_2$.

We shall define a *gauge group* as the set of transformations, generated by the left and right quaternion algebras, which leave the scalar products (15), respectively, (18), invariant. It can be proven[14] that:

$$(f,qg) = (q^* f,g), \tag{22}$$

where $q^*(qf) = f|q|^2$. Hence, under the transformation:

$$f \to qfz, \tag{23}$$

for $|q|^2=|z|^2=1, z\in ¢(1,e_1)$, the gauge group for the complex scalar product(18) is found to be U(2); under the transformation:

$$f \rightarrow qf \tag{24}$$

for $|q|^2=1$, the gauge group is SU(2) for the quaternion scalar product (15) .

These symmetries do not necessarily coincide with the automorphism group of the quaternion algebra. We may argue that different realizations of the quaternion algebra may be used equally well provided that the scalar product associated with measureable probability amplitudes is unaltered. With this restriction, the automorphisms which coincide with each of the above symmetries are said to correspond to the color gauge group[17]. These are: $f \rightarrow z^{-1}fz$ (U(1)) for the complex scalar product and only the trivial transformation for the quaternion scalar product. This definition of color is particularly appropriate if the systems described by the elements $f\varepsilon H_Q$ are considered for the description of pre-quarks, used as the building blocks for quarks and leptons[6,7].

One may show that self-adjoint operators (relative to the scalar products discussed above) have spectral resolutions in terms of projections that have the linearity of the corresponding scalar products.[18]

With the help of the spectral decomposition of the (quaternion linear) position operator x, we may define the wave functions $f(x)=\langle x|f)$.* It is usual to define the action of the translation group on wave functions as inducing a displacement on wave functions. Selecting** the complex field $¢(1,e_1)$, we may define the (quaternion linear) momentum as the generator of translations by means of the complex part $\langle x|f)_C$ of $\langle x|f)$:

$$\langle x|T(\delta x)f)_C=\langle x+\delta x|f)_C=\langle x|f)_C+\frac{1}{\hbar}\langle x|Pf)_C e_1, \tag{25}$$

so that

$$\langle x|Pf)_C = -e_1\hbar\frac{\partial}{\partial x}\langle x|f)_C. \tag{26}$$

Since (26) is valid for all f, and P is quaternion linear

$$\langle x|Pf)=\langle x|Pf)_C-\langle x|Pfe_2)_C e_2=-e_1\hbar\frac{\partial}{\partial x}\langle x|f)_C+e_1\hbar\langle x|fe_2)_C e_2=-e_1\hbar\frac{\partial}{\partial x}\langle x|f)_C. \tag{27}$$

Using the completeness of the spectral family $\{|x\rangle\langle x|dx\}$,

*This uses the notation of Jauch wherein $|f)$ denotes a proper ket vector and $\langle x|$ uses an improper (continuum) bra vector.
**See reference (14) for an investigation of alternative possibilities which lead to the conclusion reported here.

we obtain:

$$P = -\hbar\int dx\ |x>e_1\frac{\partial}{\partial x}<x|\ . \qquad (28)$$

The canonical commutator can then be evaluated in a straight-forward way:

$$[X,P] = \hbar I(e_1),\ \text{where}\ I(e_1) = \int dx|x>e_1<x|\ . \qquad (29a,b)$$

In analogy with states for which the x-component of the spin operator for a spin-½ particle has expectation value zero, states can be constructed in H_Q for which the expectation value of $I(e_1)$ vanishes. Moreover, there are even states for which *both* $[X,P]$ and $\{X,P\}$ vanish! Hence the usual procedure for establishing a lower bound for the uncertainty relation fails. One may nevertheless show, however, that $\Delta X\Delta P$ is minimized only if the wave function is complex-valued (with a possible constant quaternion coefficient), and that $\Delta X\Delta P \geqslant \hbar/2$, as in the usual complex theory[14].

We now turn to the structure of the tensor product, for the construction of many-body states. The basic result here is negative: *It is not possible to construct a tensor product of module spaces* H_Q *which maintains quaternion linearity*[19]. (One can see the problem in an intuitive way by observing that the scalar product of elements of such a tensor product would have to be constructed as products of scalar products from the constituent spaces. Since quaternion linear scalar products are quaternion valued, it would then not be possible to extract the quaternion coefficients due to lack of commutativity.)

Let us therefore study the possibility of constructing a tensor product linear over the complex sub-algebra $\mathbb{C}(1,e_1)$. Complex tensor products of the usual type between the components of $g=\chi_0+\chi_1e_2$ and $f=\psi_0+\psi_1e_2$, if restricted to $\psi_0\chi_0, \psi_0\chi_1^*, \psi_1^*\chi_0, \psi_1^*\chi_1^*$, are complex-linear under $g\to gz$ or $f\to fz$. Symmetrizing or antisymmetrizing for Bose-Einstein or Fermi-Dirac statistics, we define the functional $\Psi_{\alpha_1\alpha_2}(f,g)$ as

$$\Psi_{00}(f,g) = \frac{1}{\sqrt{2}}(\psi_0\chi_0 \pm \chi_0\psi_0) \qquad \Psi_{10}(f,g) = \frac{1}{\sqrt{2}}(\psi_1^*\chi_0 \pm \chi_1^*\psi_0)$$

$$\Psi_{01}(f,g) = \frac{1}{\sqrt{2}}(\psi_0\chi_1^* \pm \chi_0\psi_1^* \qquad \Psi_{11}(f,g) = \frac{1}{\sqrt{2}}(\psi_1^*\chi_1^* \pm \chi_1^*\psi_1^* \qquad (30)$$

Note that the symmetrization with respect to particle states does not interchange the 0,1 labels of the complex components of the quaternion functions. These indices, labelling the sectors of what is essentially a direct sum, are to be considered as part of the manifold over which the functions are defined (i.e. the functional $\Psi_{\alpha_1\alpha_2}(f,g)(x,y)$ may be thought of as $\Psi(f,g)(\alpha_1,x;\alpha_2,y)$), and not as labels for the physical states of the subsystems represented by f,g.

The norm of the tensor product vector is defined as

$$\|\Psi(f,g)\|^2 = \sum_{\alpha_1\alpha_2} \|\Psi_{\alpha_1\alpha_2}(f,g)\|^2 , \tag{31}$$

for which a simple calculation yields:

$$\|\Psi(f,g)\|^2 = \|f\|^2\|g\|^2 \pm |(f,g)_C|^2. \tag{32}$$

The scalar product, defined by

$$(\Psi(f,g),\ \Psi(f',g')) = \sum_{\alpha_1\alpha_2}(\Psi_{\alpha_1\alpha_2}(f,g),\Psi_{\alpha_1\alpha_2}(f',g')) \tag{33}$$

is then found to be

$$(\Psi(f,g),\ \Psi(f',g')) = (f,f')_C(g,g')_C \pm (f,g')_C(g,f')_C \tag{34}$$

These results are precisely what is required for the construction of annihilation-creation operators. (Except for the structure(21") for the complex scalar products appearing here, these results are parallel to those of the usual complex Hilbert space theory.)

The general form for N particle is

$$\Psi_{\alpha_1\alpha_2\cdots\alpha_N}(f_1,\ldots f_N) = \sum_P \frac{(\pm)^P P(\psi_{\alpha_1}^{r_1}\psi_{\alpha_2}^{r_2}\cdots\psi_{\alpha_N}^{r_N})}{\sqrt{N!}}, \tag{35}$$

where the permutations P act on the indices $r_1,\ldots r_N$, and for $\alpha_j=1$, the conjugate function must be used. Applied to one-particle states the corresponding functional is: $\Psi_0(f)=\psi_0, \Psi_1(f)=\psi_1^*$; the scalar product (33) is: $(\Psi(f),\Psi(g))=(\psi_0,\chi_0)+(\psi_1,\chi_1)^*$, as in (21").

We may define creation operators $a^\dagger(f_{N+1})$ as

$$(a^\dagger(f_{N+1})\Psi(f_N,\ldots f_1))_{\alpha_{N+1}\cdots\alpha_1} = \frac{1}{\sqrt{(N+1)!}}\sum_P(\pm)^P(\psi_{\alpha_{N+1}}^{r_{N+1}}\psi_{\alpha_N}^{r_N}\cdots\psi_1^{r_1}),$$

$$= \Psi_{\alpha_{N+1}\cdots\alpha_1}(f_{N+1},f_N,\ldots f_1), \tag{36}$$

and annihilation operators through the adjoint.

It then follows that

$$a(f_{N+1})\Psi(g_{N+1},\ldots g_1) = \sum_{j=0}^{N}(\pm)^j(f_{N+1},g_{N+1-j})_C\Psi(g_{N+1},\ldots\hat{g}_{N+1-j},\ldots g_1) \tag{37}$$

and

$$[a(f),\ a^\dagger(g)]_\pm = (f,g)_C. \tag{38}$$

IV. *Local Gauge Theory for Tensor Product Spaces*

In this section, we shall show how the structure assumed by Adler for algebraic gauge fields associated with several source particles (with an underlying U(2) gauge group) follows systematically from the properties of the tensor product spaces constructed in section III. The local gauge transformation on Ψ induced by transformations on the individual particle states is given by:

$$\Psi(qf_1z,qf_2z,\ldots qf_Nz)(x_1,\ldots x_N)\equiv\Psi(q\hat{z}f_1,q\hat{z}f_2\ldots q\hat{z}f_N)(x_1\ldots x_N)$$
$$= E_1(q(x_1)\hat{z}(x_1))E_2(q(x_2)\hat{z}(x_2))\ldots E_N(q(x_N)\hat{z}(x_N))\Psi(f_1,f_2\ldots f_N)(x_1,\ldots x_N) \qquad (39)$$

where $E_i(q\hat{z})$ (or $E_i(a)$, $a=0,1,2,3$) is a 2x2 matrix acting on the indices α_i of Ψ, and includes both the action of the left multiplication by $q(x_i)$ and the right multiplication by the (universal) complex-valued function $z(x_i)$. The matrices E_i are determined by the left action of quaternions on the one-particle functions. It follws from (21) that:

$$\Psi_0(e_1f)=e_1\psi_0, \qquad \Psi_1(e_1f)=-e_1\psi_1^{*};$$
$$\Psi_0(e_2f)=-\psi_1^{*} \qquad \Psi_1(e_2f)=\psi_0 ;$$
and
$$\Psi_0(e_3f)=-e_1\psi_1^{*} \qquad \Psi_1(e_3f)=-e_1\psi_0 . \qquad (40)$$

Multiplication by z on the right introduces a complex factor multiplying both ψ_0,ψ_1^{*} (complex linearity). Hence, the E_i have the form of matrices acting on the indices α_i given explicitly by:

$$E(1)=\begin{pmatrix} e_1 & 0 \\ 0 & -e_1 \end{pmatrix}, E(2)=\begin{pmatrix} 0 & -1 \\ 1 & 0 \end{pmatrix}, E(3)=\begin{pmatrix} 0 & -e_1 \\ -e_1 & 0 \end{pmatrix}, \text{ and } E(0)=\begin{pmatrix} e_1 & 0 \\ 0 & e_1 \end{pmatrix}, \qquad (41)$$

the representatives, respectively, of left multiplication by e_1,e_2,e_3 and right multiplication by e_1.

We may then write the general action of a local gauge field as:

$$B_\mu\Psi(f_1,f_2,\ldots f_N)(x_1\ldots x_N)=\sum_{j=1}^{N}(B_\mu^{a_1\ldots a_N}(x_j))\,E_1(a_1)\ldots E_N(a_N)$$
$$.\Psi(f_1,f_2\ldots f_N)\;(x_1,x_2\ldots x_N), \qquad (42)$$

where $B_\mu^{a_1\ldots a_N}$ is real-valued.

The operator B_μ lies in the tensor product space of the algebras associated with each particle, in agreement with the structure assumed by Adler.

Infinitesimal translation of the state (39) results in a sum of derivatives acting on each term in the product, and it might appear that a gauge field restricted to the direct sum of the algebras acting on each index would be adequate. Although the gauge field, which we shall call $B_\mu(x)$, acts locally at each point, its action may also alter the structure of the tensor product. It is this mechanism which opens the possibility of maintaining essential non-linearity in the Yang-Mills equations on a semi-classical level.

We define the covariant derivative as:

$$D_\mu \Psi(f_1 \ldots f_N)(x_1 \ldots x_N) = \sum_j \partial_{\mu j} \Psi(f_1, f_2 \ldots f_N)(x_1, x_2, \ldots x_j \ldots x_N)$$
$$+ e_1 g B_\mu^{\,a_1 \ldots a_N}(x_j) E_1(a_1) \ldots E_N(a_N) \Psi(f_1 \ldots f_N)(x_1 \ldots x_j \ldots x_N) \quad . \qquad (43)$$

Under the transformation (39), we see that the corresponding transformations (we represent the set of indices $a_1 \ldots a_N$ by a, and write q for $\overset{\otimes}{\underset{i}{\Pi}} E_i(a_i)$).

$$B'^{\,a}_\mu(x) = q(x) B_\mu^{\,a} q(x)^{-1} - ig^{-1} \partial^\mu q(x) q(x)^{-1} \delta a,o$$

and

$$E'_i(a_i) = q(x_i) E_i(a_i) q(x_i)^{-1}, \qquad (44)$$

ensure that $D^{\mu'} \Psi' = (D^\mu \Psi)'$. (The inhomogeneous gauge compensation terms only contribute when all of the tensor product cross-coupling indices are zero, since the derivative of (39) is of direct sum form.) Separating the left quaternion and right multiplication by z, the inhomogeneous term is of the form

$$\partial^\mu q\, q^{-1} f_j + f_j\, \partial^\mu z\, z^{-1}, \qquad (45)$$

i.e., it contains compensation for the full U(2) local gauge (the second term of eq. (45) transforms like the local electromagnetic potential).

It is straightforward to verify that the field strength tensor:

$$F_{\mu\nu}^{\;a}(x) \equiv \partial_\nu B_\mu^{\,a}(x) - \partial_\mu B_\nu^{\,a}(x) - ig\{B_\mu^{\,b}(x) B_\nu^{\,c}(x) - B_\nu^{\,b}(x) B_\mu^{\,c}(x)\} q^{abc}, \qquad (46)$$

where $q^{\bar{a}bc} = \prod_{i=1}^{N} q^{a_i b_i c_i}$,

transforms as

$$F_{\mu\nu}^{\;a'}(x) = q(x) F_{\mu\nu}^{\;a}(x) q(x)^{-1} \; . \qquad (47)$$

From the action (42) of B_μ, it is clear that the gauge field is an operator in the direct product space, and that there is a field (matrix-valued in the full direct product space) given by:

$$B_\mu(x) = B_\mu^{\,a}(x) E_1(e_1) \ldots E_N(e_N) \qquad (1'')$$

with which the action (43) may be expressed as:

$$B^\mu \Psi(f_1 \ldots f_N)(x_1 \ldots x_N) = \sum_j B^\mu(x_j) \Psi(f_1 \ldots f_N)(x_1 \ldots x_N). \qquad (48)$$

In terms of this field, defining

$$F_{\mu\nu}(x) = F^a_{\mu\nu}(x)E_1(e_1)\ldots E_N(e_N), \tag{3''}$$

equation (46) may be written as

$$F_{\mu\nu}(x) = \partial_\nu B_\mu(x)-\partial_\nu B_\mu(x)-ig[B_\mu(x),B_\nu(x)] \ . \tag{2''}$$

The covariant derivative on algebraic functions may be identified by the relation

$$\delta F_{\mu\nu}(x) = D_\mu \delta B_\nu - D_\nu B_\mu \ ,$$

yielding the usual form

$$D_\mu W = \partial_\mu W + ig[B_\mu, W] \ . \tag{4''}$$

Assuming the equations of motion (source equations)

$$D_\nu F^{\mu\nu} \quad J^\mu \ , \tag{6''}$$

where the currents J^μ are due to particle sources, it follows that

$$D_\nu J^\nu = 0 \ . \tag{8''}$$

It is also an identity(from the Jacobi identity) that

$$D_\mu F_{\nu\lambda} + D_\nu F_{\lambda\mu} + D_\lambda F_{\mu\nu} = 0 \ . \tag{5''}$$

We have thus recovered the algebraic structure discussed earlier in Section II. Let us note explicitly, however, that the gauge groups are much larger. For two particle states, the B_μ's carry the Lie algebra of U(4),and for three particle states, the Lie algebra of U(8). These structures are too large for application to presently known phenomena; subspaces corresponding to certain subalgebras, however, appear to be useful.

We are now in a position to understand Adler's construction in the light of this many body version of QQM. It is clear already that Adler's construction is a heuristic way to reduce the size of the $U(2^N)$ gauge group in QQM, but the real question is: How is this reduction actually accomplished? The answer is that *Adler's imposition of an algebraic constraint* (the "P-product") *is equivalent to the requirement that only those generators of* $U(2^N)$ *that transform as scalars and vectors* (under the U(2) group generated by $K^a = \frac{1}{2}\sum_{i=1}^{N} \lambda_i^a$) *are to be retained.* (To see this, recall that all elements of the charge algebra were scalars under $\Lambda^a = \frac{1}{2}\lambda^a + K^a$.)

The elements of the charge algebra are the scalars from this set of retained $U(2^N)$ generators plus the "scalars" (under Λ_a) obtained by using $\lambda^a(a=1,2,3)$ to saturate the vector index of the retained vector operators.

As an example, let us consider the N = 2 case once again. The U(4) algebra has 16 elements: $1, \vec{\lambda}_1, \vec{\lambda}_2, \vec{\lambda}_1 \otimes \vec{\lambda}_2$. The scalars and vectors (under $\vec{K} = \frac{1}{2}(\vec{\lambda}_1 + \vec{\lambda}_2)$) are the 11 elements: $1, \vec{\lambda}_1, \vec{\lambda}_2, \vec{\lambda}_1 \cdot \vec{\lambda}_2, \vec{\lambda}_1 \times \vec{\lambda}_2$. Using $\vec{\lambda}$ (Adler's "carrier matrices") to saturate the vector indices we get *precisely* the 5 elements of the N=2 charge algebra given earlier. It is interesting to note, that if we do not saturate the vector indices, the 11 retained generators do indeed generate a gauge group: U(1) x Sp(4).

As a second example consider N=3. The U(8) algebra for this case has 64 elements: $\lambda_1{}^a \otimes \lambda_2{}^b \otimes \lambda_3{}^c$, (a,b,c=0...3). Enumerating all scalars and vectors is lengthy, but the answer is: 5 scalars and 9 vectors. Saturating with $\vec{\lambda}$ yields the 14 elements of the n=3 charge algebra (as given by ref. 10). If we do not saturate the vector indices, the 32 elements of the retained set generate the gauge group U(4) x U(4).

This fully explains just how Adler achieved his reduction of the gauge group $U(2^N)$--which was our stated purpose. We must admit, however, that our result still leaves open the question as to just what his reduction means in physical terms.

Acknowledgements

We are grateful to Professors N. Kuiper and L. Michel for their hospitality during our visit at the Institut des Hautes Etudes Scientifiques, where much of the work reported here was done, and for their interest and encouragement. We wish to thank Professors J. Fröhlich and J. Tits for discussions at the I.H.E.S.

One of us (L.P.H.) wishes to thank S. Nussinov, H. Harari, M. Zieberg and U. Wolff and his other colleagues at Tel Aviv University and the Weizmann Institute of Science for helpful discussions.

References

1. I.B. Khriplovich, Sov.Phys.JETP 47,1(1978)[Zh.Eksp.Teor.Fiz. 74,37 (1978)]

2. R. Giles and L. McLerran, Phys.Lett. 79B,447(1978); Phys. Rev. D19, 3732(1979); Phys.Rev. D21,1672 (1980).

3. S.L. Adler, Phys. Rev. D17,3212 (1978).

4. S.L. Adler, Phys. Rev. D18, 411(1978); Phys.Lett. 86B,203(1979); Phys. Rev. D19,1168(1979); Phys. Rev. D20, 1386(1979).

5. S.L. Adler, "Quaternionic Chromodynamics as a Theory of Composite Quarks and Leptons", Inst. for Adv. Study preprint, December,1979.

6. H. Harari, Phys. Lett. 86B, 83 (1979).

7. M.A. Shupe, Phys. Lett. 86B, 87 (1979).

8. J.E. Mandula, Phys. Rev. D14, 3497 (1976).

9. L.C. Biedenharn, J. Math. Phys. 4, 436(1963).(D-operators for SU(n).)

10. V. Rittenberg and D. Wyler, Phys. Rev. D18, 4806(1978).

11. P. Cvitanovic, R.J. Gonsalves and D.E. Neville, Phys. Rev. D18, 3881 (1978).

12. S. C. Lee, Phys. Rev. D20, 1951 (1979).

13. L.C. Biedenharn, "Group Theoretical Approaches to Nuclear Spectroscopy",258-421,in"Lectures in Theoretical Physics",edited by W.E. Brittin, B.W. Downs and Joanne Downs, Vol. 5 (Interscience, New York) 1963.

14. L.P. Horwitz, D. Sepunaru, L.C. Biedenharn, "Quaternion Quantum Mechanics and Second Qunatization"(to be submitted to Comm. Math. Phys.)

15. C. Piron, Foundations of Quantum Physics, W.A. Benjamin, Reading Mass. (1976). See also L.P. Horwitz and L.C. Biedenharn, Helv. Phys. Acta, 38, 385 (1965); M. Jammer, The Conceptual Development of Quantum Mechanics, McGraw Hill, New York (1966).

16. E.C.G.Stueckelberg, Helv.Phys. Acta 33,727 (1960); E.C.G.Stueckelberg, and M. Guenin, Helv. Phys. Acta 34,621(1961); E.C.G.Stueckelberg, C. Piron and H. Ruegg, Helv. Phys. Acta 34, 675 (1961); E.C.G. Stueckelberg and M. Guenin, Helv.Phys.Acta 35, 673 (1962).

17. M. Günaydin and F. Gürsey, Lett. Nuovo Cimento 6, 401(1973); Jour. Math. Phys. 14, 1651 (1973); Phys. Rev. D9,3387(1974); F. Gürsey, in "Johns Hopkins University Workshop on Current Problems in High Energy Particle Theory", Baltimore, Md. (1974); M. Günaydin, Jour. Math. Phys. 17, 1875 (1976).

18. H.H. Goldstine and L.P. Horwitz, Math. Ann. 164,291 (1966).

19. If one has two vector spaces, one quaternion linear on the left and the other quaternion linear on the right, the juxtaposed

product is then quaternion linear on both the left and the right. However, this is not an acceptable tensor product since half of the quaternion action on each of the two vector spaces is lost. For more than two vector spaces even a tensor product of this type cannot be defined.

We wish to thank Professor Jacques Tits for discussing this subject with us.

COHERENT ANGULAR MOMENTUM STATES FOR THE ISOTROPIC OSCILLATOR

Anthony J. Bracken
Department of Physics
University of Colorado
Boulder, CO 80309, U.S.A.
(On leave from University of Queensland)

and

Howard I. Leemon
Department of Mathematics
University of Queensland
St. Lucia, Qld. 4067, Australia

For any 3-dimensional, isotropic, quantum system, the total angular momentum quantum number j takes values 0, 1, 2, ... or ½, 3/2, 5/2, ... Recalling the spectrum of the number operator for a simple harmonic oscillator, and the definition of coherent states for that system, we are led to consider the possibility of defining coherent angular momentum (CAM-) states for 3-dimensional systems by diagonalizing operators which lower the value of j by one unit. Such states would be quite distinct from the much-discussed coherent spin states, which are defined for a fixed value of j. Several authors [1] have constructed CAM-states, but always for systems such as the rigid rotor, which have only rotational degrees of freedom. Moreover, the methods used have been rather ad hoc, involving lowering operators unrelated to the dynamics of the system. For the 3-dimensional isotropic oscillator, we construct lowering operators from the usual dynamical variables, and are led to a set of CAM-states quite distinct from the usual Cartesian coherent (CC-) states for this system, but sharing many of their nice properties.

We start with the variables

$$H = \frac{1}{2M}\underline{p}^2 + \tfrac{1}{2}M\omega^2\underline{x}^2 = \hbar\omega(\underline{a}^\dagger \cdot \underline{a} + 3/2) = \hbar\omega(N + 3/2)$$

$$\underline{a} = (2M\hbar\omega)^{-\frac{1}{2}}(M\omega\underline{x} + i\underline{p}) , \quad \underline{a}^\dagger = (2M\hbar\omega)^{-\frac{1}{2}}(M\omega\underline{x} - i\underline{p})$$

$$\underline{L} = i\hbar\, \underline{a} \wedge \underline{a}^\dagger . \tag{1}$$

In an angular momentum basis, the eigenvalues of N appear in the form $(2k + \ell)$, where k is the "radial quantum number" and ℓ the total angular momentum quantum number. We want to introduce operators whose eigenvalues are the integers ℓ and k, so we define L by

$$\underline{L}^2 = \hbar^2 L(L+1), \; L^\dagger = L, \; L \geqslant 0, \; [L, \underline{L}] = \underline{0}, \tag{2}$$

and set $K = \tfrac{1}{2}(N - L)$ so that $N = 2K + L$. To identify lowering operators for k and ℓ, we note that the vector operator $\underline{a}$ can be resolved into operators which shift the value of ℓ up or down by one unit[2]:

$$\underline{a} = \underline{a}^{(+)} + \underline{a}^{(-)} , \qquad \underline{a}^{(\pm)} = [\,\underline{a}(L + \tfrac{1}{2} \pm \tfrac{1}{2}) \pm i\hbar^{-1}\underline{a} \wedge \underline{L}]\,[2L + 1]^{-1} . \tag{3}$$

Then, in particular $\underline{a}^{(-)}$ shifts both N and L down by one unit, and so commutes with K.

We also consider the operator $(\underline{a} \cdot \underline{a})$ which, being a scalar, commutes with L. Since it shifts N down by two units, it must shift K down by one. The commuting operators $\underline{a}^{(-)}$ and $(\underline{a} \cdot \underline{a})$ are a possible set of lowering operators for L and K to use for defining CAM-states, but nicer properties are obtained if one introduces instead

$$\underline{\lambda} = \underline{a}^{(-)} \left[(2L + 1)/(2K + 2L + 1)\right]^{\frac{1}{2}} , \quad \nu = (\underline{a} \cdot \underline{a})\left[1/2(2K + 2L + 1)\right]^{\frac{1}{2}} \quad (4)$$

and their conjugates $\underline{\lambda}^{\dagger}, \nu^{\dagger}$. We find

$$[L, \nu] = 0 , \quad L\underline{\lambda} = \underline{\lambda}(L - 1)$$

$$[K, \underline{\lambda}] = \underline{0} , \quad K\nu = \nu(K - 1),$$

$$[\nu, \nu^{\dagger}] = 1, \quad [\lambda_{\alpha}, \lambda_{\beta}] = 0, \quad [\nu, \lambda_{\alpha}] = 0 = [\nu^{\dagger}, \lambda_{\alpha}] \quad (5)$$

and the conjugate relations, as well as

$$\underline{\lambda} \cdot \underline{\lambda} = 0 = \underline{\lambda}^{\dagger} \cdot \underline{\lambda}^{\dagger} \quad (6)$$

$$(2L + 1)\, \lambda_{\alpha}\lambda_{\beta}^{\dagger} = (2L + 1)(\lambda_{\beta}^{\dagger}\lambda_{\alpha} + \delta_{\alpha\beta}) - 2\lambda_{\alpha}^{\dagger}\lambda_{\beta} \quad (7)$$

$$L = \underline{\lambda}^{\dagger} \cdot \underline{\lambda} , \quad K = \nu^{\dagger}\nu , \quad \underline{L} = -i\hbar\underline{\lambda}^{\dagger} \wedge \underline{\lambda}$$

$$H = \hbar\omega(2\nu^{\dagger}\nu + \underline{\lambda}^{\dagger} \cdot \underline{\lambda} + 3/2). \quad (8)$$

In terms of the operators $\underline{\lambda}$, $\underline{\lambda}^{\dagger}$, ν and $\nu^{\dagger}$ we can give a completely algebraic and representation-independent treatment of the oscillator in an an angular momentum basis [3]. We introduce a vacuum vector $|0\rangle$ such that

$$\nu|0\rangle = 0 = \lambda_{\alpha}|0\rangle \quad (\Rightarrow K|0\rangle = 0 = L|0\rangle) \quad (9)$$

and build up eigenvectors of K and L as

$$(\nu^{\dagger})^{k}\, \lambda_{\alpha}^{\dagger}\lambda_{\beta}^{\dagger} \cdots \lambda_{\gamma}^{\dagger}|0\rangle. \quad (10)$$

Because of Eqs. (5) and (6) this is completely symmetric and traceless in the vector subscripts: if these are ℓ in number, we have a tensor basis for the $(2\ell + 1)$-dimensional representation of SO(3), and L has the value ℓ. Alternatively, we can build up the (K, L, L_3)-basis vectors $|k\ell m\rangle$. The operators $\underline{\lambda}$ and $\underline{\lambda}^{\dagger}$ are not boson operators, but the relation (7) permits the calculation of the effect of applying an annihilation operator λ_{σ} to a vector like (10).

Now we define CAM-states $|z, \underline{\zeta}\rangle$ by

$$\nu|z,\underline{\zeta}\rangle = z|z,\underline{\zeta}\rangle, \quad \underline{\lambda}|z,\underline{\zeta}\rangle = \underline{\zeta}|z,\underline{\zeta}\rangle \quad (11)$$

where $\underline{\zeta} \cdot \underline{\zeta} = 0$ because of (6). We find the action of ν and $\underline{\lambda}$ on $|k\ell m\rangle$, look for $|z, \underline{\zeta}\rangle$ in the form

$$|z,\underline{\zeta}\rangle = \sum_{k,\ell,m} A_{k\ell m}(z, \underline{\zeta})\, |k\ell m\rangle \quad (12)$$

and solve (11) for the A_{klm}. We find (up to a phase) one normalized $|z,\underline{\zeta}\rangle$ for each $(z, \underline{\zeta}) \in C \times K_3$, where K_3 is the complex cone $(\underline{\zeta} \cdot \underline{\zeta} = 0)$. This vector has

$$A_{klm} = \exp\left[-\tfrac{1}{2}(|z|^2+|\underline{\zeta}|^2)\right] z^k (\zeta_3)^{\ell-|m|} (-\varepsilon\zeta_{\varepsilon})^{|m|} \left[(2\ell)!/k!\,2^{\ell}\,\ell!\,(\ell+m)!\,(\ell-m)!\right]^{\frac{1}{2}} \quad (13)$$

where ε is the sign of m and $\zeta_\pm = \zeta_1 \pm i\zeta_2$. These CAM-states have the following properties[4].

(1) Expectation values in the state $|z, \underline{\zeta}\rangle$:

$$\langle\underline{\lambda}\rangle = \underline{\zeta}, \quad \langle\nu\rangle = z, \quad \langle L\rangle = \underline{\zeta}^* \cdot \underline{\zeta}, \quad \langle K\rangle = z^*z,$$

$$\langle H\rangle = \hbar\omega(2z^*z + \underline{\zeta}^* \cdot \underline{\zeta} + 3/2), \quad \langle \underline{L}\rangle = -i\hbar\underline{\zeta}^* \wedge \underline{\zeta}. \qquad (14)$$

(2) The probabilities of obtaining particular values k and ℓ for K and L in a CAM-state follow <u>independent</u> Poisson distributions.

(3) If $\sqrt{2\hbar}\nu = \sigma + i\tau$, $\sqrt{2\hbar}\underline{\lambda} = \underline{\alpha} + i\underline{\beta}$, where σ, τ, $\underline{\alpha}$ and $\underline{\beta}$ are hermitian; and if $\Delta\sigma = [\langle\sigma^2\rangle - \langle\sigma\rangle^2]^{\frac{1}{2}}$, $\Delta\alpha = [\langle\underline{\alpha}\cdot\underline{\alpha}\rangle - \langle\underline{\alpha}\rangle \cdot \langle\underline{\alpha}\rangle]^{\frac{1}{2}}$, etc., then in a CAM-state

$$\Delta\sigma\Delta\tau = \tfrac{1}{2}\hbar, \qquad \Delta\alpha\Delta\beta = \hbar(1 + \tfrac{1}{2}\langle(2L+1)^{-1}\rangle). \qquad (15)$$

(absolute minimum) (minimum for prescribed $\langle(2L+1)^{-1}\rangle$)

(4) If the state-vector $|\psi(t)\rangle$ is a CAM-state at one time, it is a CAM-state at all times:

$$|\psi(0)\rangle = |z_o, \underline{\zeta}_o\rangle \rightarrow |\psi(t)\rangle = e^{-3i\omega t/2}\,|z(t), \underline{\zeta}(t)\rangle$$

$$z(t) = z_o e^{-2i\omega t}, \quad \underline{\zeta}(t) = \underline{\zeta}_o e^{-i\omega t}. \qquad (16)$$

The expectation values $z(t)$, $\underline{\zeta}(t)$ of ν and $\underline{\lambda}$ then follow the trajectory of the corresponding classical variables $(\hat{\nu}, \hat{\underline{\lambda}})$ in the "complex phase-space" $C \times K_3$. There is a minimal volume of uncertainty associated by quantum mechanics with the representative point in this space, as in Eq. (15). This volume remains constant along a trajectory.

(5) Let $(\Delta L)^2 = \langle \underline{L}^2\rangle - \langle\underline{L}\rangle \cdot \langle\underline{L}\rangle$. In the CAM-state $|z, \underline{\zeta}\rangle$, $(\Delta L)^2 = 2\hbar^2\underline{\zeta}^* \cdot \underline{\zeta}$, while in the CC-state $|\underline{z}\rangle$, $(\Delta L)^2 = 2\hbar^2\underline{z}^* \cdot \underline{z}$. Corresponding to the typical classical orbit with

$$\underline{x} = (A\cos\omega t, B\sin\omega t, 0) \qquad A \geqslant B \geqslant 0, \qquad (17)$$

we have

$$z = \tfrac{1}{2}[M\omega/\hbar]^{\frac{1}{2}}(A - B)e^{-2i\omega t}, \quad \underline{\zeta} = [M\omega AB/2\hbar]^{\frac{1}{2}}e^{-i\omega t}(1, i, 0)$$

$$\underline{z} = [M\omega/2\hbar]^{\frac{1}{2}}e^{-i\omega t}(A, iB, o) \qquad (18)$$

so we get for the CAM- and CC-states, respectively,

$$(\Delta L)^2 = 2M\omega\hbar AB, \qquad (\Delta L)^2 = M\omega\hbar(A^2 + B^2). \qquad (19)$$

The latter is greater, by an amount $M\omega\hbar(A - B)^2$, unless the orbit is circular (A = B), when the CAM- and CC-state in fact coincide. In this sense, the uncertainty in the angular momentum of the system is in general less in a CAM-state than in the corresponding CC-state.

(6) Non-Orthogonality: $|\langle z', \underline{\zeta}'|z, \underline{\zeta}\rangle|^2 = \exp[-\tfrac{1}{2}(|z'-z|^2 + |\underline{\zeta}'-\underline{\zeta}|^2)]$. (20)

(7) The vectors $|z, \underline{\zeta}\rangle$ are overcomplete. The most useful completeness relation is

$$\int d\mu\, |z, \underline{\zeta})(z, \underline{\zeta}| = I, \quad |z, \underline{\zeta}) = \exp[\tfrac{1}{2}(|z|^2 + |\underline{\zeta}|^2)]\, |z^*, \underline{\zeta}^*\rangle$$

$$d\mu = (2/\pi^3)d^2z\,d^6\zeta\,\delta(\underline{\zeta}\cdot\underline{\zeta})(2\underline{\zeta}^*\cdot\underline{\zeta} - 1)\exp[-(|z|^2 + |\underline{\zeta}|^2)]. \qquad (21)$$

Then, for arbitrary $|\psi\rangle$,

$$|\psi\rangle = \int d\mu\ \psi(z,\underline{\zeta})\,|z,\underline{\zeta}) \tag{22}$$

where $\psi(z,\underline{\zeta})$ $[= (z,\underline{\zeta}|\psi\rangle]$ is analytic on $C\ xK_3$, and

$$|\psi|^2 \leqslant |\langle\psi|\psi\rangle|^{\frac{1}{2}} \exp[\tfrac{1}{2}(|z|^2 + |\underline{\zeta}|^2)]\ ,\ \int d\mu|\psi|^2 < \infty \tag{23}$$

(8) We may set up a representation of the abstract algebraic structure in a Hilbert space of analytic functions on $C \times K_3$, with

$$\begin{aligned}
&|\psi\rangle \to \psi(z,\underline{\zeta})\\
&\langle\psi|\chi\rangle \to (\psi,\chi) = \int d\mu\psi^*\chi\\
&\nu^\dagger \to z,\ \nu \to \partial/\partial z,\ K \to z\partial/\partial z\\
&\underline{\lambda}^\dagger \to \underline{\zeta},\ \ L \to \underline{\zeta}\cdot\underline{\nabla}_\zeta\ ,\ \underline{L} \to -i\hbar\underline{\zeta}\Lambda\underline{\nabla}_\zeta\\
&(2L+1)\underline{\lambda} \to (2\underline{\zeta}\cdot\underline{\nabla}_\zeta + 1)\underline{\nabla}_\zeta - \underline{\zeta}\nabla_\zeta{}^2\\
&(\nu^\dagger)^k\lambda_\alpha{}^\dagger\lambda_\beta{}^\dagger \dots \lambda_\gamma{}^\dagger|0\rangle \to z^k\zeta_\alpha\zeta_\beta \dots \zeta_\gamma
\end{aligned} \tag{24}$$

This Hilbert Space has a reproducing kernel

$$K(z,\underline{\zeta};z',\underline{\zeta}') = \exp(z^*z' + \underline{\zeta}^*\cdot\underline{\zeta}'). \tag{25}$$

For $k = 0$ we have a simple space of functions $f(\underline{\zeta})$, analytic on K_3, carrying each representation of SO(3) just once. Thus $1,\ \zeta_\alpha,\ \zeta_\alpha\zeta_\beta,\ \dots$ correspond to $\ell = 0, 1, 2, \dots$. A similar space has been described in a different context by Bargmann and Todorov[5]; it is not to be confused with the Bargmann space for SU(2).

References:

1. P. Bonifacio, D. M. Kim and M. O. Scully, Phys. Rev. 187, 441 (1969); P. W. Atkins and J. C. Dobson, Proc. Roy. Soc. Lond. A321, 321 (1971); D. Bhaumik, T. Nag and B. Dutta-Roy, J. Phys. A: Math. Gen. 8, 1868 (1975); R. Delbourgo, ibid. 10, 1837 (1977); T. M. Makhviladze and L. A. Shelepin, in "Proceedings of the P. N. Lebedev Physics Institute, Vol. 70", ed. D. V. Skobeltsyn. (Consultants Bureau, New York, 1975); J. Mostowski, Phys. Letts. 56A, 369 (1976); T. S. Santhanam, preprint, Austral. Natnl. Univ., 1978.
2. A. J. Bracken and H, S. Green, J. Math. Phys. 12, 2099 (1971).
3. A. J. Bracken and H. I. Leemon, J. Math. Phys. (to appear).
4. A. J. Bracken and H. I. Leemon, J. Math. Phys. (to appear).
5. V. Bargmann and I. T. Todorov, J. Math. Phys. 18, 1141 (1977).

HOW TO MEASURE THE CANONICAL MOMENTUM p AND OPERATORS p + f(q)

H. EKSTEIN (*)

Centre de Physique Théorique, Section II
C.N.R.S. - Marseille

If the conventional equations for a particle in an external magnetic field $\vec{B}$

$$H = (\vec{p} - e\vec{A})^2 / 2m \tag{1}$$

or

$$H = \vec{\alpha}\cdot(\vec{p} - e\vec{A}) + \beta m \tag{2}$$

are to be verifiable by observations, (as they are for potentials), there must be an operational measurement procedure for the canonical momentum $\vec{p}$, and, of course, a unique choice for the vector potential $\vec{A}$. I propose to generalize the obvious method for measuring $\vec{p}$ in the potential case : switch off the field suddenly, and measure the momentum e.g. with a ballistic pendulum. It is shown in ref. 1 that the result is a possible choice for $\vec{p}$, if the radiation gauge $\nabla\cdot\vec{A}_0 = 0$ is used for the vector potential. The choice is unique, if one postulates that $\vec{p}$ be the mathematical image of the generator of translations, in the sense of presymmetry (2). That is, the outcome of the procedure is translation invariant, transforms like a vector under rotations, and undergoes a Galilei boost under Galilei transformations.

(*) Postal Address : C.N.R.S. - LUMINY - CASE 907
Centre de Physique théorique
F-13288 MARSEILLE CEDEX 2 (France)

Can one also give operational meaning to such operators as $\vec{p} + \vec{q}$? The question is of fundamental importance, because the determination of an initial state (and hence prediction of mean values of observables in the future) is possible only if one can (in principle) measure all observables (3). The fact that such procedures are known only for a few observables, is perhaps a more important deficiency of quantum mechanics than the much-discussed measurement theory.

On the basis of a remark by S.T. Epstein (4), a measurement procedure for observables of the kind

$$\vec{\pi} = \vec{p} + \nabla\varphi(\vec{q}) \tag{3}$$

(where $\vec{p}$ is the canonical momentum operationally defined as above, and φ is any differentiable function) is conceivable if one not only switches off existing field abruptly, but also generates charges and currents during the short switch-off time. Let $\vec{E}_0$ and $\vec{B}_0$ be the electric and magnetic external fields just before the switch-off. Then, the idealized magnetic switch-off can be written, for the switch-off period

$$\vec{B} = \vec{B}_0\left[1 - \Theta(t)\right], \tag{4}$$

where Θ is the unit step function. The equation for the electric field

$$\vec{E} = \vec{E}_0\left[1 - \Theta(t)\right] + (\vec{A}_0 + \nabla\varphi)\,\delta(t) \tag{5}$$

together with Eq.(4) satisfies Maxwell's homogeneous equations. The inhomogeneous equations generate these fields if the experiment provides, during the switch-off time, the charge ρ and current density $\vec{J}$

$$\rho = \rho_0\left[1 - \Theta(t)\right] + (4\pi)^{-1}\nabla^2\varphi\,\delta(t) \tag{6}$$

and

$$\vec{J} = \vec{J}_0\left[1 - \Theta(t)\right] + (4\pi)^{-1}\vec{E}_0\,\delta(t) - (4\pi)^{-1}(\vec{A}_0 + \nabla\varphi)\frac{\partial}{\partial t}\left[\delta(t)\right]. \tag{7}$$

In the simplest case, $\rho_0 = \nabla\varphi = 0$, and Newton's equation for the particle velocity $\vec{v}$, with the electric field (5), gives

$$m \vec{v}_F = m \vec{v}_0 + e \vec{A}_0 , \tag{8}$$

where $\vec{v}_0$ is the velocity before, and $\vec{v}_F$ that after the switch-off. Since

$$m \vec{v} = \vec{p} - e \vec{A}_0 , \tag{9}$$

a measurement of $m \vec{v}_F$ measures indeed the canonical momentum $\vec{p}$ before the switch-off, as stated above.

Consider now the procedure for measuring the operator

$$\vec{\pi}_3 = \vec{p} + \vec{q} \, |\vec{q}|^{-3} . \tag{10}$$

In Eqs.(5-7), we substitute, with $\varphi = - |q|^{-1}$

$$\nabla\varphi = \vec{q} \, |q|^{-3} , \tag{11}$$

so that the second term in Eq.(6) becomes

$$(4\pi)^{-1} \delta(t) \nabla^2 \varphi = \delta(\vec{q}) \, \delta(t). \tag{12}$$

An approximate realization of this charge generation is provided by 4 piezoelectric crystals disposed radially about the origin which are compressed and released instantly at the switch-off time $t = 0$ (Fig.1).

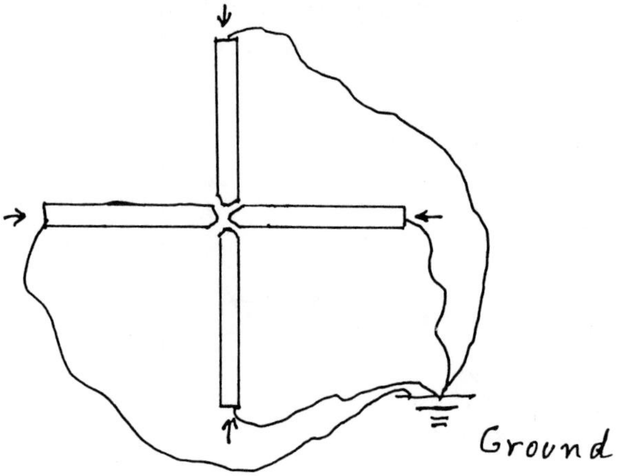

FIG. 1

If $\rho_0 = 0$, we have now

$$m \vec{v}_F = \vec{\pi}_3 \tag{12}$$

according to Eq.(9), and the measurement procedure for $\vec{\pi}_3$ is completely described.

REFERENCES

(1) H. EKSTEIN, Nuovo Cimento 54, 305 (1979).

(2) H. EKSTEIN, Phys. Rev. 184, 1315 (1969).

(3) W.C. DAVIDON and H. EKSTEIN, J. Math. Phys. 5, 1588 (1964) (N.Y.).

(4) S.T. EPSTEIN, Preprint, University of Wisconsin, Madison, Wis.

A UNIQUENESS RESULT FOR THE SEGAL QUANTIZATION OF A CLASSICAL SYSTEM WITH SYMMETRIES

Franco GALLONE and Antonio SPARZANI

Istituto di Scienze Fisiche dell'Università, via Celoria 16, I-20133 Milano, Italy.

Istituto Nazionale di Fisica Nucleare, Sezione di Milano.

Abstract

In the Segal approach, a classical system with symmetries can be quantized in a straightforward way when a complexification operator exists for both the symplectic space that describes the phase space and the symplectic transformations that represent the symmetry group of the system. If such group fulfils a real irreducibility condition, the complexification operator is unique. Two applications, to finite dimensional systems and to free Bose fields, are briefly described.

Introduction

In the Segal approach to the quantization of a linear classical system, a real symplectic space $(\mathcal{M}, B)$ is used to model the kinematical description of the system at a sharp time; to wit, $\mathcal{M}$ is the phase space of the system and B is related to the occurence of the dynamical variables in canonically conjugate pairs. Naturally enough, the group Aut $(\mathcal{M}, B)$ of the symplectic automorphisms of $(\mathcal{M}, B)$ is used to represent the symmetries of the system.

The space $(\mathcal{M}, B)$ determines uniquely, up to isomorphisms, a pair $(\mathcal{A}, w)$, which is called the Weyl algebra over $(\mathcal{M}, B)$. This is a first step toward the quantization of the classical kinematical description based on $(\mathcal{M}, B)$: $\mathcal{A}$ is a (non-abelian) C*-algebra and w a labeling of the elements of this algebra by the elements of $\mathcal{M}$; in fact w is a mapping from $\mathcal{M}$ into $\mathcal{A}$ that satisfies the Weyl form of the canonical commutation relations:

$$\forall m, m' \in \mathcal{M}, \quad w(m)\,w(m') = \exp[+(i/2)\,B(m,m')]\;w(m+m').$$

Moreover, the classical symmetries are represented by automorphisms of $\mathcal{A}$, since a unique homomorphism τ from Aut $(\mathcal{M},B)$ into the group of the c*-algebra automorphisms of $\mathcal{A}$ exists such that $\forall S \in \mathrm{Aut}(\mathcal{M},B)$, $\tau(S)w(m) = w(Sm)$. To be able to quantize the "canonical variables", however, we need a further step, i.e. we need to construct a regular state over the Weyl algebra or – equivalently – a Weyl system over $(\mathcal{M},B)$. In fact, a Weyl system over $(\mathcal{M},B)$ is a map W from $\mathcal{M}$ into the group $\mathcal{U}(\mathcal{H})$ of the unitary automorphisms of a complex separable Hilbert space $\mathcal{H}$ satisfying :

1) $\forall m, m' \in \mathcal{M}$, $W(m)\,W(m') = \exp[(i/2)\,B(m,m')]\,W(m+m')$,

2) the map $\mathbb{R} \ni t \longmapsto W(tm) \in \mathcal{U}(\mathcal{H})$ is weakly continuous, $\forall\, m \in \mathcal{M}$.

For any such system, $\forall m \in \mathcal{M}$, $t \longmapsto W(tm)$ is a continuous one-parameter group. The self-adjoint generators of these groups satisfy the Heisenberg canonical commutation relations and quantize the canonical variables (1).

Once a Weyl system is given, the Weyl algebra in realized in a Hilbert space. However, this does not entail that $\tau(S)$ is unitarily implemented. Now, if a classical symmetry Lie group is given over $(\mathcal{M},B)$, i.e. a homomorphism $g \longmapsto S_g$ of the group G into Aut $(\mathcal{M},B)$, it may be important to have a Weyl system W such that $\tau(S_g)$ is unitarily implementable by means of a continuous unitary representation of G in the Hilbert space in which W is defined. Indeed, if this is the case, we are able to represent the Lie algebra of G by means of self-adjoint operators acting in the Hilbert space where the quantum kinematical description is given (2). This can be achieved through the second quantization procedure.

Complexification operators.

To apply the second quantization to the classical description $(\mathcal{M},B,S_g)$ – where $g \longmapsto S_g$ is a classical symmetry group – one adopts the following strategy. First, one seeks a complexification operator for $(\mathcal{M},B,S_g)$.

Definition: a complexification operator for the classical description $(\mathcal{M},B,S_g)$ is an element J of Aut $(\mathcal{M},B)$ satisfying:

i) $J^2 = -\mathbb{1}$,

ii) $B(Jm,m) > 0 \quad if \quad m \neq 0$,

iii) $\forall g \in G \; , \; [S_g, J] = 0$.

Then, setting:

$$\alpha m := (\mathrm{Re}\,\alpha + (\mathrm{Im}\,\alpha)J)\, m \; ,$$

$$(m_1 \mid m_2) := B(Jm_1, m_2) - i\,B(m_1, m_2) \; ,$$

a complex inner product space structure is defined over the real space $\mathcal{M}$ in such a way that the real linear vector space structure of $\mathcal{M}$ is preserved and $B(\cdot,\cdot) = -\mathrm{Im}(\cdot\,|\,\cdot)$. Denoting by $\mathcal{M}_J$ the complex Hilbert space we get after completing the resulting inner product space, it is easy to note that there is a unitary representation $g \longmapsto \bar{S}_g$ of G in $\mathcal{U}(\mathcal{M}_J)$ such that $\bar{S}_g$ is an extension (indeed, the bounded extension) of S_g to $\mathcal{M}_J$.

Finally, through Segal quantization, a Weyl system over $(\mathcal{M},B)$ is constructed in the symmetric Fock space $\mathfrak{F}_s(\mathcal{M}_J)$ over $\mathcal{M}_J$; also, a unitary representation of G is defined in $\mathfrak{F}_s(\mathcal{M}_J)$ that unitarily implements $\tau(S_g)$. Indeed, let $\mathfrak{F}_s(\mathcal{M}_J)$ be the symmetric Fock space over $\mathcal{M}_J$, i.e. $\mathfrak{F}_s(\mathcal{M}_J) = \bigoplus_{n=0}^{\infty} \mathcal{M}_J^{(n)}$, where $\mathcal{M}_J^{(n)}$ is the n-fold symmetric tensor product of $\mathcal{M}_J$. Denoting by $a(m)$ and $a^+(m)$, with $m \in \mathcal{M}$, the annihilation and creation operators defined in the usual way in $\mathfrak{F}_s(\mathcal{M}_J)$, the operator $(1/\sqrt{2})(a(m) + a^+(m))$ is essentially self-adjoint; setting for $m \in \mathcal{M}$,

$$W_s(m) := \exp\left[(i/\sqrt{2})\,\overline{(a(m) + a^+(m))}\right]$$

(where the bar means closure), the mapping from $\mathcal{M}$ to the unitary automorphisms of $\mathfrak{F}_s(\mathcal{M}_J)$ given by $m \longmapsto W_s(m)$ is a Weyl system . Besides, if $\Gamma(\bar{S}_g)$ is the second quantization in $\mathfrak{F}_s(\mathcal{M}_J)$ of $\bar{S}_g$, $g \longmapsto \Gamma(\bar{S}_g)$ is a unitary representation of G in $\mathfrak{F}_s(\mathcal{M}_J)$ such that

$$\tau(S_g)\, W_s(m) = W_s(S_g m) = \Gamma(\bar{S}_g)\, W_s(m)\, \Gamma(\bar{S}_g)^{-1} \; ,$$

i.e. $\Gamma(\bar{S}_g)$ is a unitary operator which implements $\tau(S_g)$ (see, e.g. ,(3)).

Therefore, through a complexification operator we can actually construct a Weyl system and also unitarily implement the action of the classical symmetry group. Moreover, the program outlined at the end of the previous section is completely fulfilled if only a continuity condition is met. In fact, if G is a topological group and $\forall m, m' \in \mathcal{M}$, $G \ni g \longmapsto B(m, S_g m') \in \mathbb{R}$ is a continuous function, then the unitary representation $g \longmapsto \Gamma(\bar{S}_g)$ turns out to be continuous.

Now the question naturally arises as to the existence and uniqueness of complexification operators for a classical description $(\mathcal{M}, B, S_g)$. While no really general result is known about the existence problem, as far as uniqueness is concerned several results exist in the literature; we recently proved the following criterion (see (4) also for references to the other results).

Theorem: Suppose there exists a complexification operator for the classical description $(\mathcal{M}, B, S_g)$; then it is unique if the unitary representation $g \longmapsto \bar{S}_g$ of G in $\mathcal{U}(\mathcal{M}_J)$ is real irreducible in $\mathcal{M}_J$, i.e. no closed non-trivial real linear subspace exists in $\mathcal{M}_J$ that is left invariant by $\bar{S}_g$.

Examples.

As a first example of the situation described above, consider the very simple case $\mathcal{M} = \mathbb{R} \oplus \mathbb{R}$. On this linear space a unique (up to linear automorphisms) symplectic form B exists. If we want to study the classical (linear) motions of a classical system whose phase space is described by $(\mathcal{M}, B)$ we are led to consider the continuous homomorphisms of $\mathbb{R}$ into Aut $(\mathcal{M}, B)$. For every triple $(\gamma, \eta, \rho) \in \mathbb{R}^3$ the mapping

$$\mathbb{R} \ni t \longmapsto S_t^{(\gamma,\eta,\rho)} = \begin{bmatrix} \cos \omega t - 2\rho(\sin \omega t)/\omega \ , & 2\gamma(\sin \omega t)/\omega \\ -2\eta(\sin \omega t)/\omega \ , & \cos \omega t + 2\rho(\sin \omega t)/\omega \end{bmatrix},$$

where

$$\omega := \begin{cases} 2(\eta\gamma - \rho^2)^{1/2} & \text{if} \quad \eta\gamma - \rho^2 \geq 0, \\ 2i(\rho^2 - \eta\gamma)^{1/2} & \text{if} \quad \eta\gamma - \rho^2 < 0, \end{cases}$$

and $(\sin \omega t)/\omega$ means t when $\omega = 0$, is a continuous symplectic representation of the group $\mathbb{R}$ in this symplectic space, and there is no other continuous homomorphism of $\mathbb{R}$ into $\mathrm{Aut}(\mathfrak{M}, B)$. For the classical description $(\mathfrak{M}, B, S_t^{(\gamma,\eta,\rho)})$ there is one and only one complexification operator if $\eta\gamma - \rho^2 > 0$ and an infinity of them if $\gamma = \eta = \rho = 0$ (whereas there exists no complexification operator in any other case, see (5)). This is in agreement with the theorem since – as is easy to show – $t \mapsto S_t^{(\gamma,\eta,\rho)}$ is real irreducible if $\eta\gamma - \rho^2 > 0$, while $t \mapsto S_t^{(0,0,0)}$ is not ($S_t^{(0,0,0)}$ is just the identity operator). Also, this simple example shows that the condition of real irreducibility in the theorem cannot be weakened to complex irreducibility. Indeed, $t \mapsto S_t^{(0,0,0)}$ is complex irreducible in $\mathfrak{M}_J$ for any complexification operator J , since $\mathfrak{M}_J$ is a one-dimensional complex space.

As a further example, let $\mathfrak{M}$ be the set of the positive mass μ and zero spin real smooth solutions (see, e.g., (6)) of the Klein-Gordon equation $(\Box + \mu^2)\phi(\underline{x}, t) = 0$. There is a natural action of the restricted Poincaré group $P_+^\uparrow$ on $\mathfrak{M}$:

$$\phi(x) \longmapsto (S_{(a,\Lambda)}\phi)(x) := \phi(\Lambda^{-1}(x-a)) \,,\ \text{if } (a,\Lambda) \in P_+^\uparrow .$$

If we want to consider $\mathfrak{M}$ as the phase space of a system in which $P_+^\uparrow$ acts as a symmetry group through the action $(a,\Lambda) \longmapsto S_{(a,\Lambda)}$, we are led to look for a symplectic form B with respect to which $S_{(a,\Lambda)}$ is a symplectic transformation. This requirement determines B uniquely, up to a scalar factor, to be

$$B(\phi,\psi) = \int_{x_0=0} (\dot{\phi}\psi - \phi\dot{\psi})\, d\underline{x} \ .$$

A linear operator J is defined in $\mathfrak{M}$ by the mapping

$$(\dot{\phi}, \phi)_{x_0=0} \longmapsto (D\phi, -D^{-1}\dot{\phi})_{x_0=0}, \text{ where } D := (-\Delta + \mu^2)^{1/2},$$

since the initial data determine linearly a unique solution of the wave equation. Also, it can be easily proved that J is a complexification operator for the classical description $(\mathfrak{M}, B, S_{(a,\Lambda)})$. Therefore, through the second quantization defined by J , both the classical kinematical description based on $(\mathfrak{M}, B)$ and the classical symmetry group $(a,\Lambda) \longmapsto S_{(a,\Lambda)}$ can be quantized. Indeed, the quantum system that is thus obtained is the relativistic neutral free quantum field with zero spin and mass μ . The problem naturally arises whether this quantum system is the only one that can be obtained through second quantization from $(\mathfrak{M}, B, S_{(a,\Lambda)})$, i.e. whether the operator J defined above is the only complexification operator for $(\mathfrak{M}, B, S_{(a,\Lambda)})$. The theorem stated above proves that this is indeed the case, since the unitary representation $\overline{S_{(a,\Lambda)}}$ of $P_+^\uparrow$ in $\mathfrak{M}_J$ defined by $S_{(a,\Lambda)}$ and J is real irreducible. This can be either seen directly, or, using a result of Weinless (7), derived from the fact that $\overline{S_{(a,\Lambda)}}$ is complex irreducible and the self-adjoint generator of the time translations is strictly positive; indeed, $\overline{S_{(a,\Lambda)}}$ is (a realization of) the mass μ and zero spin continuous unitary irreducible representation of $P_+^\uparrow$.

It is worth mentioning that the uniqueness of the complexification operator for the classical description $(\mathfrak{M}, B, S_{(a,\Lambda)})$ described above appears to be a new result. Without using our theorem, from a result of Kay (8) the following weaker statement could have been obtained: the operator J defined above is the only complexification operator for $(\mathfrak{M}, B, S_{(a,\Lambda)})$ such that the unitary representation $\overline{S_{(a,\Lambda)}}$ of $\mathcal{P}_+^\uparrow$ contains a one-parameter subgroup having strictly positive self-adjoint generator.

References

(1) Segal I.E., Kgl. Danske Vidensk. Selsk. mat-fys. Medd. 31, 1 (1959).
- Canad. J. Math. 13, 1 (1961).
- Illinois J. Math. 6, 500 (1962).
- Mathematical Problems of Relativistic Physics (Providence, R.I.: Am. Math. Soc., 1963).

(2) Gårding L., Bull. Soc. Math. France, 88, 73 (1960) and Nelson E., Ann. Math., 70, 572 (1959).

(3) Reed M. and Simon B., Methods of Modern Mathematical Physics II. Fourier Analysis, Self-adjointness (New York: Academic, 1975).

(4) Gallone F. and Sparzani A., On the Uniqueness of the Segal Quantization, Preprint IFUM 241/FT Milano, June 1980.

(5) Gallone F. and Sparzani A., J. Math. Phys. 20, 1375 (1979).

(6) Reed M. and Simon B., Methods of Modern Mathematical Physics III. Scattering Theory (New York: Academic, 1979).

(7) Weinless M., J. Funct. Anal., 4, 350 (1969).

(8) Kay B. S., J. Math. Phys., 20, 1712 (1979).

ON THE SINGULAR POINTS OF THE MATRIX ELEMENTS OF THE RESOLVENT OPERATOR (*)

Gian Carlo Ghirardi

Istituto di Fisica dell'Università and ICTP P.O. Box 586, 34100 Trieste, Italy.

Vittorio Gorini and Guido Parravicini

Istituto di Fisica dell'Università, via Celoria 16, 20133 Milano, Italy.

As well known, the analytic properties of variuos quantities appearing in the quantum description of physical processes, have in many cases a direct physical meaning. Among these quantities, the matrix elements of the resolvent operator of the Hamiltonian play a particular role. Indeed, at least on the first sheet of the energy Riemann surface, their singular points are in a one-to-one correspondence with physically meaningful properties, such as the occurence of bound states and the opening of scattering or reaction channels. The correspondence mentioned above is actually one-to-one, provided one considers the set union of the singular points of the matrix elements evaluated on a dense set of states. Regarding the singular points of the analytic continuation of the matrix elements to the second sheet the situation is quite different. In fact, on the one hand, it is a very old idea to try to attach them a physical meaning by associating them to the unstable states of the theory, at least when they lie sufficiently close to the real axis. On the other hand, as we will discuss below, the occurrence and/or the location of singularities in the second sheet is crucially dependent on the set of states among which is one chooses to evaluate the matrix elements, even when one requires this set to be dense. Then, the problem naturally arises of properly characterizing a dense set of states such that the singular points of the corresponding matrix elements of the resolvent have a direct physical meaning.
To discuss this problem, let us consider a self-adjoint Hamiltonian H (which we assume for simplicity to posses only a simple Lebesgue spectrum from zero to $+\infty$) and the corresponding resolvent operator

$$G(z) = (z-H)^{-1} \tag{1}$$

As is well known [1], G(z) is an operator valued function of z, holomorphic for z in the resolvent set. G(z) cannot be continued analytically beyond the boundary of the resolvent set. However, if we consider states φ and ψ which are taken to vary in a properly chosen dense set of states X in the Hilbert space $\mathcal{H}$ of the system, the matrix elements $(\varphi, G(z)\psi)$ can be continued analytically across the spectrum from the upper half plane into the second sheet of the energy Riemann surface. We will study how the singularities of the continued matrix elements $(\varphi, G(z)\psi)^{II}$ depend upon the choice of the set X.

We introduce the following dense sets of states in $\mathcal{H}$:

1. The set A defined as $\varphi \in A$ iff $\varphi \in \mathcal{H}$ and $\langle \omega + | \varphi \rangle = \hat{\varphi}(\omega)$ is a function which can be analytically continued to the whole complex plane, the resulting continuation being an entire function. In the above definition the $|\omega +\rangle$ are the improper outgoing eigenstates of the operator H ($0 \leq \omega < +\infty$).

2. The set A_0 . It is defined exactly as the set A under 1., but by making reference to the improper eigenstates $|\omega\rangle$ of the free Hamiltonian H_0 (whose spectrum is assumed to coincide with the one of H), corresponding to a proper splitting $H = H_0 + H_I$.

With reference to the above definitions it is then easy to prove:

a). – The matrix elements $(\varphi, G(z)\psi)$ for $\varphi, \psi \in A$, can be analytically continued to the second sheet and are everywhere analytic there. This follows immediately from the expression for the analytic continuation

$$(\varphi, G(z)\psi)^{II} = \int_0^\infty \frac{\hat{\varphi}^*(\omega)\,\psi(\omega)}{z - \omega}\, d\omega - 2\pi i\, \hat{\varphi}^*(z^*)\,\hat{\psi}(z) \tag{2}$$

b). – The analytic properties of the continued matrix elements $(\varphi, G(z)\psi)^{II}$ for $\varphi, \psi \in A_0$, are determined essentially by the properties of the analytic continuation of the off-energy shell T-matrix elements. This also follows from eq. (2). In fact, by considering the improper eigenstates $|\omega\rangle$ of H_0 and setting $\tilde{\varphi}(\omega) = \langle \omega | \varphi \rangle$, we have

$$\begin{aligned} \hat{\varphi}(\omega) &= \int_0^\infty \langle \omega + | \omega' \rangle\, \tilde{\varphi}(\omega')\, d\omega' \\ &= \tilde{\varphi}(\omega) + \int_0^\infty \tilde{\varphi}(\omega')\, \frac{\langle \omega | T(\omega -) | \omega' \rangle}{\omega - i0 - \omega'}\, d\omega' \end{aligned} \tag{3}$$

where $T(z)$ is the T-matrix. Since $\varphi \in A_0$, $\tilde{\varphi}(\omega)$ is an entire function and therefore the properties of $(\varphi, G(z)\psi)^{II}$ are determined through eqs. (2) and (3) by the analytic structure of the function defined by the integral at the r.h.s. of eq. (3).

c). – By properly changing the dense set of states used to evaluate the matrix elements, the singular points of $(\varphi, G(z)\psi)^{II}$ can be made to occur in arbitrarly preassigned positions. In fact, starting e.g. from the set A, if one performs a unitary transformation **U on A**, one generates a new dense set A_U. For $\varphi_U \in A_U$ we then have

$$\hat{\varphi}_U(\omega) \equiv \langle \omega + | U | \varphi \rangle = \int_0^\infty U(\omega, \omega')\, \hat{\varphi}(\omega')\, d\omega' \tag{4}$$

with obvious meaning of the symbols. Since $\hat{\varphi}(\omega)$ is an entire function, equation (4) shows through eq. (2) that one can control the singularities of $(\varphi_U, G(z)\psi_U)^{II}$ by properly chosing the operator U. In ref. 2 this point has been discussed in great detail.

Concluding this analysis, we see that the analytic properties of the matrix elements of G(z) depend crucially on the set of states which is chosen to evaluate them, even when this set is required to be dense. Then , how can one identify an appropriate dense set of states for which the singular points of the continued matrix elements of the resolvent operator have a direct physical meaning?

The case discussed under b) is the one which is usually considered. In fact it is well known that, for a large class of problems whose analytic structure can be understood, the singular points identified according to the prescription under b), are correctly associated with the occurence of the unstable states of the system [3]. However,it has to be noted that the identification of the physically relevant singular points through the scheme outlined under b), makes explicit reference to the to some extent arbitrary splitting of H into a free and an interaction part.

It would then be interesting to find a different way of identifying a privileged set of states, without resorting to the above splitting, such that the singularities of the matrix elements $(\varphi, G(z)\psi)^{II}$ have a direct physical meaning. We have been able to prove [2] that, for a very large class of physical systems, i.e. for a particle subject to a rotationally invariant nonolocal interaction which is degenerate in all waves, or to a spherically symmetric local potential, such a privileged set of states can be identified by requiring that it consists of the spatially localized states. To be precise, with reference to a two-body system, we introduce the sets L_c and L_α by assuming that $\varphi \in L_c$ iff, in the relative co-ordinate representation, $\varphi(r)$ is a square summable function of compact support, while $\varphi \in L_\alpha$ iff $\varphi(r)$ decreases at infinity faster than $e^{-\alpha r}$. It is easy to prove that the set L_c has, for the above considered classes of physical systems, the desired property that the singularities occurring in the analytic continuation of the matrix elements of G(z) evaluated between states from L_c are physically significant, i.e. they can be correctly associated to unstable states of the theory (in the case of L_α the same is true in an appropriate strip of the lower momentum plane). The reader is referred to ref. 2 for a detailed discussion of the above statement. Here we will consider for illustrative purposes only the case of the ℓ-th wave part of a rotationally invariant local potential of compact support. In such a case, the expression of the matrix elements $(\varphi, G^{\ell}(k)\psi)$ for real momentum k in terms of the Jost function $f_\ell(p)$ is

$$(\varphi, G^\ell(k)\psi) = \frac{2}{\pi}\int_0^\infty dp \frac{p^{2\ell+2}}{f_\ell(p) f_\ell^*(p)} \times \frac{\check{\varphi}^*(p)\check{\psi}(p)}{k^2 - p^2} \tag{5}$$

where

$$\check{\varphi}(p) = \int_0^\infty \varphi_\ell(p,r)\varphi(r)\,dr \tag{6}$$

and $\varphi_\ell(p,r)$ is the regular solution. In the case under consideration, $\varphi_\ell(p,r)$ is an entire function of p , and $\check{\varphi}(p)$ is entire or, respectively, holomorphic in the strip $|\mathrm{Im}\ p| < \alpha$ according to whether $\varphi \in L_c$ or $\varphi \in L_\alpha$. Then, the singularities of $(\varphi, G^\ell(k)\psi)$ in the lower half of the complex momentum plane occur only at the points where the Jost function vanishes, and we know from other arguments that these points can be correctly associated to the unstable states of the system.

Summarizing, we have introduced a different way of identifying dense set of states for which the singularities of the continued matrix elements of the resolvent have a direct physical meaning. Even though, for the considered classes of physical systems, this way is essentially equivalent to the one based on scheme b), it is interesting to consider it, since it characterizes the dense set of states without making reference to the splitting of H into a free and an interaction part. It is also interesting to notice that the same requirement of spatial localization also plays a basic role in the preparation of an unstable system , as discussed elsewhere [4]

References

1. See e.g. M.Reed and B.Simon: Methods of Modern Mathematical Physics (Academic Press, New York, 1978), Vol. IV, Secs. XII.6 and XII.10.
2. G.C.Ghirardi , V.Gorini and G.Parravicini: Nuovo Cimento 57A , 1 (1980).
3. See e.g. R.G.Newton: Scattering Theory of Waves and Particles (Mc Graw Hill, N.Y.,1966)
4. L.Fonda and G.C.Ghirardi: Nuovo Cimento 67A , 257 (1970).

(*) Supported in part by INFN, MPI, NATO Research Grant No. 1380 and by CNR Research Contract No. 78.02740.63.

LIMITATIONS ON QUANTUM MEASURING PROCESSES

G.C. Ghirardi, T. Weber, Istituto di Fisica Teorica dell'Università, Trieste

F. Miglietta, A. Rimini, Istituto di Fisica Teorica dell'Università, Pavia

As well known[1-4], the existence of additive conserved quantities puts severe limitations on the quantum measurement process. In particular, any quantum mechanical operator which does not commute with the operator of an additive conserved quantity can be measured only approximately. The argument leading to this conclusion can be sketched, in the case in which the measured operator constitutes by itself a complete set of commuting observables, in the following simplified manner. Be M the operator, in the Hilbert space $\mathcal{H}^S$ of the observed system, corresponding to the measured quantity and

$$M|\mu_i> = \mu_i|\mu_i> \tag{1}$$

its eigenvalue equation. The ideal measurement scheme assumes then the existence of a normalized state $|A_0>$ in the Hilbert space $\mathcal{H}^A$ of the measuring apparatus and a unitary operator U onto $\mathcal{H}^S \times \mathcal{H}^A$, representing the effect of the system-apparatus interaction, such that,

$$U|\mu_i\ A_0> = |\mu_i\ A_i> \tag{2}$$

where the $|A_i>$'s are the final states of the apparatus which are eigenstates of an observable whose further detection yields the desired information about the result μ_i of the measurement of M. Consider then any additive conserved quantity $\Gamma = \gamma^S \times I^A + I^S \times \gamma^A$ of the system plus apparatus. From (2) one gets

$$<\mu_i\ A_0|\ \gamma^S + \gamma^A|\ \mu_j\ A_0> = <\mu_i|\gamma^S|\mu_j> + \delta_{ij}<A_0|\gamma^A|A_0> \tag{3}$$

Being $[U,\Gamma] = 0$, the l.h.s. of (3) can also be written

$$\begin{aligned} <\mu_i A_0 |\ U^+ (\gamma^S + \gamma^A)\ U|\mu_j\ A_0> &= <\mu_i A_i|(\gamma^S+\gamma^A)|\mu_j A_j> = \\ &= <\mu_i|\gamma^S|\mu_j>\ \delta_{ij} + <A_i|\gamma^A|A_j>\ \delta_{ij}. \end{aligned} \tag{4}$$

Comparison of (3) with (4) implies

$$[M, \gamma^S] = 0. \tag{5}$$

If (5) is not satisfied, eq. (2) cannot hold and M cannot be measured exactly. In place of (2) we must then resort to an approximate measure ment scheme. The above argument can be generalized to more complicated cases, e.g. to the one in which M exhibits degeneracy or to the one in which the apparatus states are described only by density operators ϱ_i belonging to orthogonal linear manifolds of $\mathcal{H}^A$. The above considerations have a great conceptual relevance and imply, among other things, that one cannot perform an ideal measurement process of the spin or orbital angular momentum component of a quantum system. In such cases one has to accept a measurement scheme according which there is a finite probability of an unsuccessful or wrong measurement. The problem then arises of understanding how small this probability can be made, and of defining an optimal measuring apparatus. It has been pointed out that the malfunctioning probability can be made very small provided $|A_o>$ contains a large amount of the conserved quantity.[1-2-4] In particular Yanase [5] has dealt with the problem of determining the minimum probability of an unsuccessful measurement consistent with a prescribed mean square value of the conserved quantity, in the case of the measurement of a spin component of a spin $\frac{1}{2}$ particle, obtaining the result that this minimum probability is inversely proportional to the mean square value of the conserved quantity and determining the proportionality constant. The derivation of ref.(5), however, suffers of some limitations. In particular it requires a very cumbersome playing with the angular momentum components of the apparatus states. Moreover the derivation involves a treatment of the angular momentum components values as a continuous variable, an assumption whose implications are not clear, since for physical reasons low angular momentum states must be present in the apparatus. Finally the formalism of ref.(5) does not allow simple generalizations to take into account the existence of the other additive conservation laws related to the other angular momentum components.

In this paper we want to reconsider this problem and show how invariance arguments allow to derive the Yanase bound in a very simple, rigorous and elegant way. The method works even when several conservation laws are simoultaneously taken into account. To get our result we start by denoting by $|\pm>$ the eigenvectors of σ_z

$$\sigma_z \,|\pm> \;=\pm\; |\pm> \,. \tag{6}$$

Since it is not possible to assume an ideal measurement scheme (2), we assume the existence of a unitary operator U of $\mathcal{H}^S \times \mathcal{H}^A$ and apparatus states $|A_\pm>$, $|\varepsilon_\pm>$, such that

$$U\,|\pm, A_o> = |\pm, A_\pm> + |\mp, \varepsilon_\mp> \quad , \tag{7}$$

the signs being correlated, with

$$< A_+ \,|\, A_- > = 0, \; < A_o \,|\, A_o > \; = 1. \tag{8}$$

The key problem is then to make as small as possible the error

$$\varepsilon^2 = <\varepsilon_+ \,|\, \varepsilon_+> + <\varepsilon_- \,|\, \varepsilon_-> \tag{9}$$

taking into account the conservation law for the total angular momentum $\underline{L} + \underline{S}$ of the system plus apparatus. With reference to eq.(7) we observe that U can be written

$$U = \tfrac{1}{2}\,(\, B_o + \underline{C} \cdot \underline{\sigma} \,) \quad , \tag{10}$$

where B_o and C_i (i=1,2,3) are operators of $\mathcal{H}^A$. Imposing $\left[U, \underline{L}+\underline{S}\right]=0$ one gets, as obvious

$$\left[B_o, L_i\right] = 0, \qquad \left[C_j \,,\, L_i\right] = i\;\varepsilon_{jik}\;C_k \,. \tag{11}$$

Unitarity of U implies

$$B_o^+ B_o + \sum C_i^+ C_i = 4 \quad , \tag{12}$$
$$B_o^+ C_k + C_k^+ B_o + i\;\varepsilon_{kji}\;C_j^+ C_i = 0$$

and the analogous relations coming from $UU^+ = 1$. Note that (12) implies that B_o and C_i are bounded operators, with bound equal to 4. The states $|A_\pm>$ and $|\varepsilon_\pm>$ are then expressed in terms of $|A_o>$, B_o and C_i by

$$|A_\pm> = \tfrac{1}{2}\,(\, B_o \pm C_3 \,)\,|A_o> \;, \qquad |\varepsilon_\pm> = \tfrac{1}{2}\,(\, C_1 \mp i C_2)\,|A_o> . \tag{13}$$

The orthogonality $<A_+ \,|\, A_-> = 0$ of the apparatus states implies

$$<A_o\,|\,B_o^+ B_o\,|\,A_o> \;= <A_o|\,C_3^+ C_3\,|\,A_o> \,, <A_o\,|\,C_3^+ B_o\,|\,A_o> = <A_o|\,B_o^+ C_3\,|\,A_o> \tag{14}$$

Getting the bound is now an elementary task. Infact from (11) one has

$$C_3^+ C_3 = i\;(\,L_2 C_1^+ C_3 - C_1^+ L_2 C_3) \,. \tag{15}$$

Use of the unitarity equations gives

$$C_3^+ C_3 = iL_2 C_3^+ C_1 - iC_1^+ L_2 C_3 + L_2 B_o^+ C_2 + C_2^+ L_2 B_o = iL_2 C_3^+ C_1 - iC_1^+ C_3 L_2 + C_1^+ C_1 + L_2 B_o^+ C_2 + C_2^+ B_o L_2 .$$

Taking the mean value on the state $|A_o>$, one has

$$<A_o|\,C_3^+ C_3\,|A_o> = i <A_o|L_2 C_3^+ C_1\,|A_o> - i<A_o|\,C_1^+ C_3 L_2\,|\,A_o> + \|C_1\,|A_o>\|^2 +$$
$$+ <A_o|\,L_2 B_o^+ C_2\,|A_o> + <A_o|\,C_2^+ B_o L_2\,|\,A_o> \,. \tag{16}$$

Using eq.(14) and the relation $\langle A_o | \{ B_o^+ B_o + \sum C_i^+ C_i \} | A_o \rangle = 4$ following from the unitary conditions one has

$$2 = i\langle A_o | L_2 C_3^+ C_1 | A_o\rangle - i\langle A_o | C_1^+ C_3 L_2 | A_o\rangle + \frac{3}{2} \| C_1 | A_o\rangle \|^2 + \frac{1}{2} \| C_2 | A_o\rangle \|^2 + \langle A_o | L_2 B_o^+ C_2 | A_o\rangle + \langle A_o | C_2^+ B_o L_2 | A_o\rangle \,, \tag{17}$$

i.e., taking the modulus,

$$2 \leqslant 2 |\langle A_o | L_2 C_3^+ C_1 | A_o\rangle| + 2 |\langle A_o | L_2 B_o^+ C_2 | A_o\rangle| + \frac{3}{2} \| C_1 | A_o\rangle \|^2 + \frac{1}{2} \| C_2 | A_o\rangle \|^2 \,. \tag{18}$$

Using the Schwarz inequality we have

$$2 \leqslant 2 \left\{ \| C_1 | A_o\rangle \| \| C_3 L_2 | A_o\rangle \| + \| C_2 | A_o\rangle \| \| B_o L_2 | A_o\rangle \| \right\} + \frac{1}{2} \| C_2 | A_o\rangle \|^2 + \frac{3}{2} \| C_1 | A_o\rangle \|^2 \,. \tag{19}$$

Considering the expression in curly brackets as a real scalar product, Schwarz inequality yields

$$2 \leqslant 2 \sqrt{\| C_1 | A_o\rangle \|^2 + \| C_2 | A_o\rangle \|^2} \cdot \sqrt{\| C_3 L_2 | A_o\rangle \|^2 + \| B_o L_2 | A_o\rangle \|^2} + \frac{3}{2} \| C_1 | A_o\rangle \|^2 + \frac{1}{2} \| C_2 | A_o\rangle \|^2 \,. \tag{20}$$

Now, the first square root is simply $\sqrt{2}\,\varepsilon$. Disregarding the last two terms which, for $\varepsilon \to 0$ are higher order infinitesimals, we have

$$\varepsilon^2 \geqslant 1 / 2 \left\{ \| C_3 L_2 | A_o\rangle \|^2 + \| B_o L_2 | A_o\rangle \|^2 \right\} . \tag{21}$$

From the unitarity equation, we have

$$\| C_1 L_2 | A_o\rangle \|^2 + \| C_2 L_2 | A_o\rangle \|^2 + \| C_3 L_2 | A_o\rangle \|^2 + \| B_o L_2 | A_o\rangle \|^2 = 4 \| L_2 | A_o\rangle \|^2 \tag{22}$$

so that the upper bound for the curly bracket in the denominator of eq. (21) is $4 \| L_2 | A_o\rangle \|^2$, yielding

$$\varepsilon^2 \geqslant 1 / 8 \| L_2 | A_o\rangle \|^2 \,, \tag{23}$$

which is the Yanase bound obtained in an elementary and rigorous way. One can observe that up to now we have actually used among the eqs. (11) only those guaranteeing the L_2 conservation. The method obviously applies also to the L_1 conservation equations. The corresponding procedure yields

$$\varepsilon^2 \geqslant 1 / 8 \| L_1 | A_o\rangle \|^2 \,. \tag{24}$$

We stress again the simplicity of the derivation. It should also be obvious that the method is particularly apt for generalizations. Finally, the procedure yields also a constructive method to build up the best functioning apparatus, since to get this one simply has to make the equality sign to be approximated as well as possible in all above inequalities. We stress that the limits (23, 24) cannot actually be reached exactly. Infact, to get the equality sign in (22) one needs to have, first of all

$$\| C_1 L_2 |A_o> \|^2 + \| C_2 L_2 |A_o> \|^2 = 0$$

which implies

$$C_1 L_2 |A_o> = C_2 L_2 |A_o> = 0 . \tag{25}$$

To have the equality sign also in the previous chain of inequalities one gets, from the condition that Schwarz inequality hold with the equal sign and that going from (17) to (18) no majorization is introduced, the following conditions

$$C_1 |A_o> = -ia^2 C_3 L_2 |A_o> , \quad C_2 |A_o> = a^2 B_o L_2 |A_o> , \tag{26}$$

a^2 being a real positive constant. Then, use of the unitarity condition

$$B_o^+ C_2 + C_2^+ B_o + iC_3^+ C_1 - iC_1^+ C_3 = 0 \tag{27}$$

with eqs. (25) and (26) yields immediately

$$\| C_1 |A_o> \|^2 + \| C_2 |A_o> \|^2 = 0 \tag{28}$$

which is absurd.

In a forthcoming paper we will discuss the possibility of indefinitely approaching the above derived limit.

References.

1. E.P. Wigner, Z. Physik, 131, 101 (1952).
2. H. Araki and M.M. Yanase, Phys. Rev., 120, 622 (1960).
3. H. Stein and A. Shimony, Varenna School, Course IL, pg. 56 (1970).
4. M.M. Yanase, Varenna School, Course IL, pg. 77, (1970).
5. M.M. Yanase, Phys. Rev., 123, 666 (1961).

CLASSICAL AND QUANTUM PHYSICS ON PHASE SPACE : THE STATE SPACES

Nicolao Giovannini
Département de Physique Théorique
Université de Genève
1211 Genève 4, Switzerland

1. Introduction

In classical as well as in quantum physics, the phase space can be seen as the set of possible values of a collection of observables (position, momentum) that correspond, independently of the evolution or of the interactions, to the physical system under consideration. The state of the physical system is on the other hand specified (in fact by definition) by the values, the possible outcomes, of the measurements corresponding to these observables. Our point of view is thus, in order to get information on the possible structure of the state spaces, to start from the space $\Gamma = \{\Gamma_A\}$, range of the possible values of an arbitrary but given set of observables $\{A\}$. On this space we also have, by usual physical equivalence postulates, a (kinematical) symmetry group G. Using this information and some elementary properties of a set of observables, we show that it is possible to construct a general representation framework. It turns out that the classical and the quantal single particle state spaces come out of this framework in a unified way, both in the relativistic and in the non-relativistic contexts. Moreover this framework is shown to allow to treat also problems where only part of the observables are of the classical type and play the role of, possibly continuous, superselection parameters.

In this note we shall (roughly and) briefly review some recent [1], new [2] and unpublished results concerning the group theoretical aspects and techniques involved in the above framework, getting rid as much as possible of technical details.

In order to fix ideas, let us first mention two examples.

2. Examples [1]

i) the single nonrelativistic particle : $\Gamma = \{\text{momentum } \vec{p},\ \text{position } \vec{q},\ \text{time } t\}$. As there exists no absolute zero for the momentum, the position or the time, as well as no privileged direction, the group G consists of elements $(\vec{w}, a^0, \vec{a}, \alpha) \in (\mathbb{R}^3, \mathbb{R}, \mathbb{R}^3, O(3))$ and its action on Γ is given by

$$\{\vec{p},\vec{q},t\} \longrightarrow \{\alpha(\vec{p}+\vec{w}),\ \alpha(\vec{q}+\vec{a}),\ t+a^0\} \qquad (1)$$

This group has been called the Newton group [1].

ii) *<u>the single relativistic particle</u>* : Γ = {4-momentum p, 4-position q}. *With no absolute zero for p and q and no absolute Lorentz frame, we generate a group, called the Einstein group [1] with elements $(w,a,\Lambda) \in (\mathbb{R}^4, \mathbb{R}^4, O(3,1))$ and whose action on Γ is given by*

$$\{p,a\} \longrightarrow \{\Lambda(p+w),\ \Lambda(q+a)\} \tag{2}$$

Note that the above time is purely kinematical and has a priori nothing to do with an evolution parameter.

3. Representation theory

The set {A} of observables is of course related to the properties of the physical system and the latter can be expressed in terms of lattices of propositions. It follows from the representation theorem of Piron [3] that (up to small technical restrictions) the most general such lattice $\mathcal{L}$ is isomorphic with the lattice of families of projectors on a family of Hilbert spaces

$$\mathcal{L} \cong V_S\ \mathcal{P}(\mathcal{H}_s),\ s \in S, \text{ some index set} \tag{3}$$

hence the most general state space K is given by a direct union of Hilbert spaces

$$K = V_S \mathcal{H}_s . \tag{4}$$

Correspondingly a symmetry of $\mathcal{L}$ (i.e. of the physical description) can be expressed in terms of automorphisms of K and the corresponding generalization of the theorem of Wigner reads :

<u>Theorem 1</u> [3]

Every symmetry of a proposition system defined by a family $V_S \mathcal{H}_s$ is given by a permutation π of the index set S and a family $U_{\pi(s)} : \mathcal{H}_s \longrightarrow \mathcal{H}_{\pi(s)}$ of (anti) unitary operators. Each $U_{\pi(s)}$ is defined uniquely up to a phase.

With our groups, we shall thus deal with K-representations, i.e. with mappings $S \times G \to Aut(K)$ satisfying

$$U_s(g_1)\quad U_{\pi_1^{-1}(s)}(g_2) = \omega_s(g_1,g_2)\ U_s(g_1 g_2) \tag{5}$$

where the families of phases $\Omega = \{\omega_s\}$ satisfy the generalized comultiplier equations

$$\Omega(g_1,g_2)\cdot\Omega(g_1g_2,g_3) = \Omega(g_1,g_2g_3)\cdot(\zeta(g_1)\Omega)(g_2,g_3) \tag{6}$$

where $\zeta(g)$ is a composition of the usual eventual complex conjugation $\eta(g)$ with a shift in the variable s :

$$(\zeta(g)\Omega)_s = \omega_{g^{-1}s}^{\eta(g)}, \text{ with } gs \equiv \pi(g)\cdot s. \tag{7}$$

A K-representation $\{U_s(g)\}$ will be called irreducible if (i) its restriction at some s to the stabilizer H of s is irreducible in the usual sense, and if (ii) the action of G on S is transitive (hence $S \cong G/H$). Remark that S is always a G-space in the sense of Mackey [4].

The phases of Theorem 1 define an equivalence and equivalence classes and the general solution of (6) is given by the following

Theorem 2 [2]

Let U be an irreducible (unitary) projective K-representation. Then the K-multiplier Ω is equivalent to Ω' with

$$(\Omega'(g_1,g_2))_s = \omega_{s_0}(\nu(s,g_1),\ \nu(g_1^{-1}s,g_2)) \tag{8}$$

with so arbitrary on S, H = Stab s_0 and $\nu(s,g)$ defined by the condition

$$\nu(s,g) \equiv k(s)\cdot g\cdot k^{-1}(g^{-1}s) \in H \tag{9}$$

with $k(s)$ some fixed coset representatives of G/H and $k^{-1}(s)s_0 = s$. In words : Ω is essentially determined by its restriction on any $\mathcal{K}_s \times H$, H = Stab s (and is in particular trivial if this restriction is trivial). Conversely, each $[\omega]\in H^2(H,U(1))$ gives rise to a unique solution $[\Omega]$.

One may also define classes of representations (modulo Aut(K)) and one then finds the following

Theorem 3

Each (unitary) irreducible projective K-representation U is equivalent with one of the form

$$U_s(g) = \iota_s^{s_0}\ L_{s_0}(\nu(s,g))\ \iota_s^{g^{-1}s} \tag{10}$$

with $L_{s_0}\in\mathcal{U}(\mathcal{K}_{s_0})$, the unitary group of $\mathcal{K}_{s_0}$, $s_0\in S$ and ι_s^{s}, the imbedding map of

$\mathcal{H}_s$ in $\mathcal{H}_{s'}$. In simple words each K-representation is equivalent with an "induced on S" one.

What about the observables in K ? They are defined by projections (like selfadjoint operators in Hilbert spaces are defined by spectral families). Obviously we demand that for each observable A in the given set, we have, on the Borel sets of Γ_A, $\mathcal{B}(\Gamma_A)$, a mapping in $\mathcal{P}(K)$ (the projections in K) i.e. in the families $\{P_s\}$ of projectors : $E \longrightarrow P_E \in \mathcal{P}(K)$, satisfying the commutativity of the diagrams

$$\begin{array}{ccc} \mathcal{B}(\Gamma_A) & \xrightarrow{\;P\;} & \mathcal{P}(K) \\ \big\downarrow \sigma(g) & & \big\downarrow U(g) \\ \mathcal{B}(\Gamma_A) & \xrightarrow{\;P\;} & \mathcal{P}(K) \end{array} \qquad (\sigma \text{ the defining representation}) \tag{11}$$

i.e. it is the natural extension of the notion of systems of imprimitivity (s.o.i.) of Mackey. The next result is now that as is already quite clear from the above theorems, every such mapping, on S, is essentially equivalent to a canonical one, i.e. to one defined by characteristic functions.

The general result is thus that, if we want to find all possible K-spaces, we first have to find each subset of {A} that gives rise to usual s.o.i. in Hilbert spaces for the corresponding subgroup H of G (use for that the imprimitivity theorem of Mackey) and then g-induce as above on G/H. Dropping the details, we now come back to our examples.

4. Applications

There are <u>two and only two</u> solutions for a (spinless) Newton particle :

i) <u>the classical one</u> : $S = \Gamma \cong G/H$, $H = \mathrm{Stab}\, s_0 \cong O(3)$, with $K = V_\Gamma \mathcal{H}_{\vec\gamma}$, $\mathcal{H}_\gamma$ one-dimensional. $(U(g)\psi)_\gamma = \psi_{g^{-1}\gamma}$ with the observables $\underline{\vec{p}}: \mathcal{B}(\mathbb{R}^3) \longrightarrow \Gamma$, $\underline{\vec{p}}(\Delta\vec{p}) = (\Delta\vec{p}, \vec{q}, t)$ and analogously for $\underline{\vec{q}}$ and $\underline{t}$.

ii) <u>the quantal one</u> : $S = G/H_t$, $H_t = \{(\vec{w}, o, \vec{a}, \alpha)\}$ and thus $K = V_{t\in\mathbb{R}}\mathcal{H}_t$, with $\mathcal{H}_t \cong \mathcal{L}^2(\mathbb{R}^3)$, $\forall t$. U is given by

$$((U(\vec{w}, a^0, \vec{a}, \alpha)\psi)(\vec{x}))_t = (\exp(-i\hbar^{-1}\vec{x}\vec{w}).\psi(\alpha^{-1}(\vec{x} - \vec{a})))_{t - a^0} \tag{12}$$

i.e. the time observable appears as a superselection variable. The observables : $(\underline{\vec{q}}\psi(\vec{x}))_t = (\vec{x}\psi(\vec{x}))_t$, $(\underline{\vec{p}}\psi(\vec{x}))_t = (-i\hbar\vec{\partial}\psi(\vec{x}))_t$, whereas the time observable is given by

$$(P_{\Delta_t}\psi(\vec{x}))_t = (\chi_{\Delta_t}(t)\psi(\vec{x}))_t\ , \quad \Delta_t \in \mathcal{B}(\mathbb{R}) \tag{13}$$

There are also two and only two (spinless) Einstein particles : a purely classical-one ($K = V_\Gamma \mathcal{H}_\gamma$. $\mathcal{H}_\gamma$ all 1-dim.) and a purely quantal one with $K = \mathfrak{L}^2(\mathbb{R}^4)$ and where in particular we have the following generalized Weyl commutation relations $U(w)\,U(a) = \exp{:}\hbar^{-1} g_{\mu\nu} w^\mu a^\nu\, U(a)\,U(w)$ whereas for the observables, we obtain $\underline{x}^\mu \psi(x) = x^\mu \psi(x)$ and $\underline{p}^\mu \psi(x) = -i\hbar\partial^\mu \psi(x)$. We refer to [1] for more details.

The previous models can be extended so to take spin into account [1]. We however would like to indicate here another line of generalization.

5. Other models, other examples

Our mathematical description and the interpretation of the physical objects is the same in all models : relativistic or not, classical or quantal. We can use this fact in order to understand perhaps better the well known difficulties of relativistic quantum dynamics. Namely we can then compare effectively the quantum relativistic model with its classical and nonrelativistic limits and counterparts, respectively, (see also [5]). For that purpose we need to enlarge the Γ of (1) with the energy observable E and the transformation $E \to E + w^o$ and the Galilean boosts $\vec{q} \to \vec{q} - \vec{v}t$, $E \to E - \vec{p}\vec{v}$. For this group, the above framework leads, apart from the classical solution, to only one class of solutions :

$$K \cong \mathfrak{L}^2(\mathbb{R}^4,\ \mathcal{H}(\Delta)) \tag{14}$$

with $(U(w)\phi)(x) = \exp(i\hbar^{-1}(\vec{w}\vec{x} - w_0 t)\phi(x)$, $U(a)\phi(x) = \phi(x - a)$, $(U(\alpha)\phi)(x) = \Delta(\vec{o},\alpha)\phi(t,\alpha^{-1}\vec{x})$, $(U(\vec{v})\phi)(x) = \Delta(\vec{v},1)\phi(t,\vec{x} + \vec{v}t)$, with Δ any irreducible representation of the homogeneous Galilei group, i.e. either an ∞-dim. on $\mathfrak{L}^2$ on the 2-dim. sphere, or of the spin type : $\Delta(\vec{v},\alpha| = D^j(\alpha|$. Obviously only the second one is satisfying (no ∞-dimensional internal degree of freedom). But the first one has some interesting analogies with the relativistic case with spin, if we do not impose a superselection variable in the Minkovski space [5]. The observables corresponding to (14) are given for $\vec{x}$ and $\vec{p}$ as before and for E and t by

$$\underline{E} = i\hbar\partial_t \qquad\qquad \underline{t} = t \tag{15}$$

References

[1] N. Giovannini and C. Piron, Helv. Phys. Acta 52 (1979), 518.
[2] N. Giovannini, Superselection Variables and Generalized Multipliers, preprint UGVA-DPT 1980/05-241.
[3] C. Piron, Foundations of Quantum Physics, Benjamin Read. Mass (1976).
[4] G.W. Mackey, The Theory of Unitary Group Representations, University of Chicago Press (1976).
[5] C. Piron, Particles, Dynamics and Covariance, preprint UGVA-DPT 1979/11-223.

QUANTUM DYNAMICAL SEMIGROUPS AND COMPLETE POSITIVITY. AN APPLICATION TO ISOTROPIC SPIN RELAXATION (*)

Vittorio Gorini

Istituto di Fisica dell'Università, via Celoria 16, 20133 Milano, Italy.

Maurizio Verri

Istituto di Matematica, Informatica e Sistemistica dell' Università, viale Ungheria 43, 33100 Udine, Italy.

E.C.G.Sudarshan

Department of Physics, CPT, The University of Texas at Austin, Austin, Texas 78712,USA.

1. Quantum Dynamical Semigroups.

Let S denote a quantum system with associated Hilbert space $\mathcal{H}$. As is well known, a state of S is described by a self-adjoint, non-negative, trace one, linear operator ρ on $\mathcal{H}$ called the statistical operator or, more commonly, the density matrix. The expectation value of an observable of S, represented by a linear self-adjoint operator A on $\mathcal{H}$, is given by $\langle A \rangle = \mathrm{tr}(\rho A)$ (whenever the expression at the r.h.s. exhists).

We consider a system S evolving irreversibly under the action of its surroundings R, which we think of as an unexhaustible energy reservoir for S. Whenever S and R are initially uncorrelated and the decay time of the reservoir's correlations is much smaller than the typical relaxation times of the system, the dynamical evolution of the state of S is described to a good approximation by a Markovian master equation of the form

$$\frac{d}{dt}\rho = L\rho, \tag{1.1}$$

where L is a linear transformation ("superoperator") acting on the space $T(\mathcal{H})$ of linear operators on $\mathcal{H}$ having finite trace. The integrated form of (1.1) writes

$$\rho = \rho(t) = e^{Lt}\rho(0) = T_t\,\rho(0),\ t \geq 0, \tag{1.2}$$

where T_t is expected to have the following properties:

(i) T_t is positive, namely $\rho \geq 0$ implies $T_t\rho \geq 0$;

(ii) T_t preserves the trace, namely $\mathrm{tr}(T_t\rho) = \mathrm{tr}(\rho)$ for all $\rho \in T(\mathcal{H})$;

(iii) $T_{t+s}\rho = T_t(T_s\rho)$, T_o = identity operator on $T(\mathcal{H})$;

(iv) $\mathrm{tr}\left[(T_t\rho)A\right]$ is a continuous function of t for all $\rho \in T(\mathcal{H})$ and for all $A \in B(\mathcal{H})$, where $B(\mathcal{H})$ is the space of linear bounded operators on $\mathcal{H}$.

Properties (i) and (ii) are demanded by the conservation of probability; (iii) is the semigroup property which formalizes the Markovian approximation; (iv) is a physical continuity requirement for expectation values of observables.

Conversely, if T_t, $t \geq 0$, is a one-parameter family of linear operators on $T(\mathcal{H})$ satisfying conditions (i) - (iv), there exists a (generally unbounded) linear operator L on $T(\mathcal{H})$, with dense domain of definition D(L), such that eq. (1.1) holds for all $\rho \in D(L)$ [1].

The family T_t gives the dynamics in the Schrödinger picture. By duality, we can define a dynamics T_t^* in the Heisenberg picture, acting on $B(\mathcal{H})$, as

$$\mathrm{tr}\left[\rho(T_t^*A)\right] = \mathrm{tr}\left[(T_t\rho)A\right], \qquad \rho \in T(\mathcal{H}),\ A \in B(\mathcal{H}). \tag{1.3}$$

Then T_t^* satisfies

(i') T_t^* is positive ;

(ii') $T_t^* \mathbb{1} = \mathbb{1}$;

(iii') $T_{t+s}^* = T_t^* T_s^*$, T_0^* = identity operator on $B(\mathcal{H})$,

as well as the continuity property following from (iv).

Actually, it turns out that the reduced dynamics T_t of the open system S must satisfy on physical grounds a considerably more stringent constraint than the positivity property (i) (or (i')). This requirement is called complete positivity and can best be expressed in the Heisenberg picture as follows. Let n be an arbitrary positive integer and, for any given n, let $\{\varphi_1, \ldots, \varphi_n\}$, $\varphi_i \in \mathcal{H}$, i = 1, ..., n, and $\{A_1, \ldots, A_n\}$, $A_i \in B(\mathcal{H})$, i=1, ..., n, be n arbitrary Hilbert space vectors and n arbitrary bounded operators. Then T_t^* must satisfy

(i'') $$\sum_{i,j=1}^{n} (\varphi_i, T_t^*(A_i^* A_j)\varphi_j) \geq 0, \qquad t \geq 0.$$

A linear map Φ on $B(\mathcal{H})$ satisfying (i'') is said to be completely positive. Taking n = 1 in (i'') we see that a completely positive map is positive. The converse is in general false.

Complete positivity is not an intuitive property of the reduced dynamics. On the other hand, it has a sound physical foundation. Indeed, it is a consequence of the assumption that the total dynamics of the system plus its surroundings, regarded globally as an isolated system, is Hamiltonian [2, 3]. Alternatively, it can be proved by an independent probability argument, even without making reference to the foregoing assumption [4]. In particular, a

Hamiltonian dynamics is completely positive.

A (quantum) dynamical semigroup is a one parameter family T_t , $t \geq 0$, of linear bounded operators on $T(\mathcal{H})$ satisfying conditions (ii), (iii), (iv) and (i"). From the above discussion, we conclude that the reduced dynamics of a quantum system is described in the Markovian limit by a dynamical semigroup. The operator L appearing in eq. (1.1) is called the (infinitesimal) generator of the semigroup. The general form of L was given in [4] for L bounded and independently in [3] for a finite- dimensional Hilbert space [5] .

2. Application to isotropic spin relaxation.

In the second part of this talk, we describe an application of the theory of dynamical semigroups to isotropic relaxation of two coupled spins, which is relevant in optical pumping phenomena [6, 7] . We find that complete positivity implies stringent restrictions on the reduced dynamics of the spins, in the form of inequalities among measurable parameters (such as relaxation rates of the irreducible spherical components of the density matrix). Our inequalities are stronger than those previously found by other authors [8] . For comparison between the conditions of complete positivity and of simple positivity, we examine as an illustration the simplest non trivial case of isotropic relaxation of a spin 1 magnetic moment. In this example, we exhibit explicitly the restrictions on the dipole and quadrupole relaxation rates imposed by positivity, and find that they are considerably weaker than those required by complete positivity. For similar comparisons in the case of axially symmetric spin 1/2 relaxation and of dynamical maps of two-level systems see [3, 9] . For a detailed discussion of the subject see our forthcoming papers [10, 11] .

For the applications that we have in mind, we can restrict our considerations to N-level systems. In this case, we can make the identifications $\mathcal{H} = \mathbb{C}^N$ and $B(\mathcal{H}) = T(\mathcal{H}) = M(N)$, the algebra of N x N complex matrices. The result of [3] can be stated in a slightly more general form as follows.

Theorem 2.1. [3] . Let $\{G_\alpha \in M(N) ;\ \alpha = 1, 2, \ldots, N^2\}$ be a complete orthonormal set (c.o.n.s.) in M(N), i.e. $\mathrm{tr}(G_\alpha^* G_\beta) = \delta_{\alpha\beta}$. Then, a linear transformation $L : M(N) \rightarrow M(N)$ is the generator of a dynamical semigroup iff it has the form

$$L: \rho \rightarrow L\rho = \frac{1}{2} \sum_{\alpha,\beta=1}^{N^2} d_{\alpha\beta} \left\{ [G_\alpha \rho, G_\beta^*] + [G_\alpha, \rho G_\beta^*] \right\} \tag{2.1}$$

for all $\rho \in M(N)$, where [12]

$$d_{\alpha\beta} = \overline{d_{\beta\alpha}} \tag{2.1a}$$

and

$$\sum_{\alpha,\beta=1}^{N^2} \overline{y_\alpha}\, d_{\alpha\beta}\, y_\beta \geq 0 \tag{2.1b}$$

for all vectors $\{y_\alpha\}_{\alpha=1,2,\ldots,N^2}$ such that

$$\sum_{\alpha=1}^{N^2} (\mathrm{tr}\, G_\alpha)\, y_\alpha = 0. \tag{2.1c}$$

Remark 2.1. Eq . (2.1) automatically incorporates the condition $\mathrm{tr}(L\rho) = 0$ and eq.(2.1a) ensures that $(L\rho)^* = L\rho^*$. The requirement of complete positivity is expressed by (2.1b) and (2.1c) .

Remark 2.2. Define

$$\hat{G}_\alpha = G_\alpha - [(\mathrm{tr}\, G_\alpha)/N]\mathbb{1} \tag{2.2}$$

and

$$H = H^* = \frac{i}{2N} \sum_{\alpha,\beta=1}^{N^2} \{(\mathrm{tr}\, G_\alpha^*)\overline{d_{\alpha\beta}}\, \hat{G}_\beta - (\mathrm{tr}\, G_\alpha) d_{\alpha\beta}\, \hat{G}_\beta^*\}. \tag{2.3}$$

Then (2.1) can be rewritten as

$$L: \rho \to L\rho = -i[H,\rho] + \frac{1}{2} \sum_{\alpha,\beta=1}^{N^2} d_{\alpha\beta} \{[\hat{G}_\alpha \rho, \hat{G}_\beta^*] + [\hat{G}_\alpha, \rho \hat{G}_\beta^*]\}. \tag{2.4}$$

The decomposition (2.4) of L into the sum of a Hamiltonian part $L_H = -i\,[H, .]$ plus a dissipative part $L_D = L - L_H$ is unique, namely it does not depend on the choice of the c.o.n.s. $\{G_\alpha\}$. In particular, choosing $G_\alpha = F_\alpha$, where $F_{N^2} = (1/\sqrt{N})\mathbb{1}$ (so that $\mathrm{tr}\, F_i = 0$, $i = 1, 2, \ldots, N^2-1$) we recover the form (2.3) of [3] , namely

$$L\rho = -i[H,\rho] + \frac{1}{2} \sum_{i,j=1}^{N^2-1} c_{ij} \{[F_i\rho, F_j^*] + [F_i, \rho F_j^*]\}, \tag{2.5}$$

where $\{c_{ij}\}$ is a self-adjoint non-negative matrix.

We consider the Markovian relaxation of two coupled spins $\vec{I}$ and $\vec{J}$. The situation that we have in mind is the relaxation, in an external magnetic field. among the Zeeman sublevels of an optically pumped atomic vapor with hyperfine structure [6, 7]. Here $\vec{J}$ stands for the electronic angular momentum and $\vec{I}$ for the nuclear spin. In typical experiments, mean free times between collisions are much smaller than spin relaxation times, so that the Markovian approximation is justified. Then the density matrix of $\vec{I} + \vec{J}$ satisfies a master equation of the form (1.1) and $N = (2I + 1)(2J + 1)$. We shall confine our considerations to the case when the external magnetic fields is sufficiently weak that the relaxation is to a good approximation isotropic. This situation has been frequently studied experimentally [6, 7]. The isotropy condition reads

$$L(\Delta(R)\rho\Delta(R)^*) = \Delta(R)(L\rho)\Delta(R)^*, \tag{2.6}$$

for all $\rho \in M(N)$ and for all $R \in SO(3)$, where Δ is the tensor product of the two irreducible representations $D^{(I)}$ and $D^{(J)}$ of SO(3) corresponding to spin I and J respectively. It is convenient to write (2.1) with the choice $G_\alpha \equiv T_{KQ}(FG)$ ($F, G = |I - J|, \ldots, (I+J)$; $K = |F - G|, \ldots, (F+G)$; $Q = -K, \ldots, K$), the standard basis of irreducible spherical tensors [13] :

$$\Delta(R)\, T_{KQ}(FG)\Delta(R)^* = \sum_M D^{(K|FG)}_{MQ}(R)\, T_{KM}(FG). \tag{2.7}$$

Then, using (2.6), we get

$$\begin{aligned}
\frac{d}{dt}\rho &= L\rho \\
&= \frac{1}{2}\sum_K \sum_{F,G,F',G'} \lambda_K(FG,F'G') \sum_Q \Big\{\big[T_{KQ}(FG)\rho,\ T_{KQ}(F'G')^*\big] \\
&\quad + \big[T_{KQ}(FG),\ \rho\, T_{KQ}(F'G')^*\big]\Big\} \\
&= -i[H,\rho] + \frac{1}{2}\sum_K \sum_{F,G,F',G'} \lambda_K(FG,F'G') \\
&\quad + \sum_Q \Big\{\big[\hat{T}_{KQ}(FG)\rho,\ \hat{T}_{KQ}(F'G')^*\big] + \big[\hat{T}_{KQ}(FG),\ \rho\,\hat{T}_{KQ}(F'G')^*\big]\Big\},
\end{aligned} \tag{2.8}$$

where

$$H = \sum_F h_F \mathbb{1}_F, \tag{2.9}$$

the h_F being arbitrary real constants (note that H is defined up to an additive multiple of the unit matrix $\mathbb{1}$). Define the matrices Λ_K (K = 1,...2(I+J)) by

$$(\Lambda_K)_{FG,F'G'} = \lambda_K(FG,F'G'). \tag{2.10}$$

Then, the complete positivity conditions (2.1b), (2.1c) are equivalent to

$$\Lambda_K \geq 0, \qquad K = 1,\ldots,2(I+J) \tag{2.11}$$

and

$$\sum_F \bar{y}_F \, \lambda_0(FF,GG)\, y_G \geq 0 \tag{2.12}$$

for all $\{y_F\}$ such that $\sum_F \sqrt{2F+1}\, y_F = 0$. Using the identity

$$\sum_Q T_{KQ}(FG)\, T_{LM}(F'G')\, T_{KQ}(F''G'')^* = (-)^{K+L+F''+G} \sqrt{2K+1}\, \delta_{F'G}\, \delta_{G'G''} \begin{Bmatrix} F'' & F & L \\ G & G'' & K \end{Bmatrix} T_{LM}(FF'') \tag{2.13}$$

and the relation $T^*_{LM}(FG) = (-)^{F-G+M}\, T_{L,-M}(G,F)$, eq. (2.8) gives

$$L\, T_{LM}(GG')^* = -\sum_{FF'} \gamma_L(FF';GG')\, T_{LM}(FF')^* \tag{2.14}$$

where

$$\gamma_L(FF';GG') = \sum_K (-)^{K+L+F'+G+1} \sqrt{2K+1} \begin{Bmatrix} F' & F & L \\ G & G' & K \end{Bmatrix} \lambda_K(F'G',FG). \tag{2.15}$$

The inverse of relation (2.15) is

$$\lambda_K(FG,F'G') = \sum_L (-)^{K+L+1-F-G'} \sqrt{2L+1} \begin{Bmatrix} G' & G & L \\ F & F' & K \end{Bmatrix} \gamma_L(F'F,G'G). \tag{2.16}$$

The master equation (2.8) can be written in terms of the expectation values of the $T_{KQ}(FF')$

(the standard irreducible components of the density matrix), $\langle T_{KQ}(FF')\rangle = \mathrm{tr}[\rho T_{KQ}(FF')]$ as [14]

$$\frac{d}{dt}\langle T_{KQ}(FF')\rangle = -\sum_{GG'} \gamma_K(FF';GG')\langle T_{KQ}(GG')\rangle. \qquad (2.17)$$

The coefficients $\gamma_K(FF, FF)$, $\gamma_K(FF, GG)$ $(F \neq G)$ and $\gamma_K(FF', FF')$ $(F \neq F')$ represent respectively the decay rates within the hyperfine multiplet F, the transfer rates between the multiplets F and G and the decay rates of " hyperfine coherences". The remaining γ's are usually expected to be small by the secular approximation if the hyperfine splittings are much greater than the natural widths. Inserting (2.16) into (2.11) and (2.12) we get the set of inequalities among the rates $\gamma_K(FF', GG')$ which follow from complete positivity and which therefore must be satisfied regardless of the interactions which are responsible for the relaxation. These inequalities are stronger than those previously reported by Omont [7] . Indeed, Omont's inequalities only amount to the condition of nonnegativity of the diagonal matrix elements $\lambda_K(FG, FG)$.

A particular class of master equations for the system $\vec{I} + \vec{J}$ having the form (2.8) can be obtained from a model in which the effect of the reservoir on the system is simulated by the action of a stationary isotropic fluctuating Hamiltonian

$$\tilde{H}(t) = \sum_{F,G,K,Q} T_{KQ}(FG)\, V_{KQ}^{FG}(t) \qquad (2.18)$$

with Gaussian correlations

$$\langle V_{KQ}^{FG}(t), \overline{V_{K'Q'}^{F'G'}(s)}\rangle$$
$$= \lambda_K(FG, F'G')\, \delta_{KK'}\, \delta_{QQ'}\, \frac{1}{\varepsilon\sqrt{\pi}}\, \exp\left[-(t-s)^2/\varepsilon^2\right]. \qquad (2.19)$$

The Markovian limit is obtained by letting $\varepsilon \downarrow 0$ (white noise) and one obtains a generator of the form (2.8), the $\lambda_K(FG, F'G')$ being those in (2.19) and H being the Hamiltonian of the isolated system $\vec{I} + \vec{J}$ (of course, the actual H is only approximatively of the form (2.9) since, besides the rotationally invariant coupling between $\vec{I}$ and $\vec{J}$, it also contains the small term due to the weak external magnetic field). The model based on (2.18) an (2.19) corresponds to a singular coupling to an "infinite temperature" bath. It is expected to give an accurate description of the spin dynamics in actual situations, since kT is in general much larger than the hyperfine splittings. In this model, the matrix $\{c_{ij}\}$ of (2.5) is real whenever we choose $F_i = F_i^*$ ($i = 1, \ldots, N^2-1$). This implies as expected that the central state $\rho_0 = \mathbb{1}/N$ (the equilibrium state at infinite temperature) is stationary and that the dynamics satisfies detailed balance with respect to ρ_0 [15, 9]. In general, for a dynamics of the form (2.8), if the stationary state is unique, by rotational invariance it will be of the form $\rho_0 = \sum_F \rho_F \mathbb{1}_F$, and one can prove that any initial state will approach ρ_0 as $t \to \infty$.

Next, consider the particular case $\vec{I} = 0$ (zero nuclear spin). In this circumstance, we have $\lambda_K(FG, F'G') = \lambda_K$ and $\gamma_K(FG, F'G') = \gamma_K$. Furthermore, conservation of probability implies $\gamma_0 = 0$. Complete positivity is equivalent to $\lambda_K \geqslant 0$ so that by (2.16) we have [16, 17]

$$\sum_{L=1}^{2J} (-)^{2J+1+L+K} (2L+1) \begin{Bmatrix} L & J & J \\ K & J & J \end{Bmatrix} \gamma_L \geqslant 0. \tag{2.20}$$

These are the inequalities which must be satisfied by the relaxation rates of the multipole components of the density matrix. As an example, we consider the case $J = 1$. There are two relaxation rates, the dipole rate γ_1 and the quadrupole rate γ_2, and (2.10) gives (see also [7])

$$\frac{3}{5}\gamma_1 \leqslant \gamma_2 \leqslant 3\gamma_1 . \tag{2.21}$$

The simple requirement of positivity, namely condition (i) of Sec. 1., is expressed by [11]

$$0 \leqslant \gamma_2 \leqslant 3\gamma_1 \tag{2.22}$$

which is weaker than (2.21) [18]. To our knowledge, all experimental data are consistent with (2.21) [6, 7].

FOOTNOTES AND REFERENCES.

1. Here the derivative at the l.h.s of (1.1) is defined as $\lim_{t\downarrow 0} \left\| \frac{d}{dt}\rho - t^{-1}(T_t\rho - \rho) \right\|_1 = 0$,

where $\|\sigma\|_1 = \mathrm{tr}[(\sigma^*\sigma)^{1/2}]$ is the trace norm on $T(\mathcal{H})$ (we denote by B^* the adjoint of an operator B). The domain D(L) is the set of all $\rho \in T(\mathcal{H})$ for which $d\rho/dt$ exists.

2. K.Kraus: Ann. Phys. (N.Y.) 64, 311 (1971).

3. V.Gorini, A.Kossakowski and E.C.G.Sudarshan: J.Math. Phys. 17, 821 (1976).

4. G.Lindblad: Commun. Math. Phys. 48, 119 (1976).

5. For a partial result when L is unbounded see E.B.Davies, Generators of dynamical semigroups, preprint (1977). For the classification of dynamical semigroups on arbitrary Von Neumann algebras and with bounded L see E. Christensen, Commun. Math. Phys. 62, 167 (1978).

6. W.Happer: Rev. Mod. Phys. 44, 169 (1972) and references contained therein.

7. A.Omont: Progr. Quantum Electronics 5, 69 (1977) and references contained therein.

8. See, e.g., Ref. 7 and J.F.Papp and F.A.Franz, Phys Rev. A5, 1763 (1972).

9. V.Gorini, A.Frigerio, M.Verri, A.Kossakowski and E.C.G.Sudarshan: Rep. Math. Phys. 13, 149 (1978).

10. M.Verri and V.Gorini: Quantum dynamical semigroups and isotropic relaxation of two coupled spins, in preparation.

11. V.Gorini, G.Parravicini, E.C.G.Sudarshan and M.Verri, Positive and completely positive SU(2) - invariant dynamical semigroups, in preparation.

12. A superscript bar denotes complex conjugation.

13. U.Fano and G.Racah: Irreducible tensorial sets, Academic Press, New York (1957).

14. A.Omont: J.Phys. 26, 26 (1965).

15. V.Gorini and A.Kossakowski: J.Math. Phys. 17, 1298 (1976).

16. W.Happer: Phys. Rev. B1, 2203 (1970).

17. M.Verri and V.Gorini: J.Math. Phys. 19, 1803 (1978)

18. The statement in [17] that for isotropic relaxation of a single spin positivity and complete positivity are equivalent is false. Actually, the argument given there allows only to prove that positivity implies $\lambda_{2J} \geqslant 0$.

(*) Partially supported by INFN, by NATO Research Grant No. 1380 and by CNR Research Contract No. 78.02740 .63.

ON GEOMETRIC QUANTIZATION OF THE RIGID BODY

G. John
Institut für Theoretische Physik
Universität Tübingen
D - 7400 Tübingen

A possible method of geometric quantization of a classical system is given by the following scheme /1/.

1) The phase space of the system is a symplectic manifold whose symplectic form σ_2 is integral (if we set Planck's constant equal to 1).

2) There exists a principal U(1)-bundle $(Y, U, \pi, U(1))$ over U with contact form α on Y so that $\Pi^* \sigma_2 = d\alpha$. The construction of Y is sometimes called prequantization.

3) A state space of sections over U in Y must be found which is invariant with respect to the lift to Y of a group of canonical transformations which acts transitively on U.

Here we shall only indicate how to do the first step for the rigid body. The second step turns out to be trivial but we shall supplement it by taking angular momentum into account. The third step has still to be investigated.

According to /1/ the symplectic form of the phase space of a system of n free particles is given by

$$\sigma_2^{\circ}(\delta x)(\delta' x) = \sum_i m_i \; \delta\vec{q} \cdot \delta\dot{\vec{q}} - \delta\dot{\vec{q}} \cdot \delta'\vec{q} \tag{1}$$

Here $\vec{q}_i$, $\dot{\vec{q}}_i$ are position and velocity coordinates (notation as in /1/), δx, $\delta' x$ are vector fields on the phase space and $\delta\vec{q}_i$, $\delta\dot{\vec{q}}_i$ are their components in the tangent space to the point space resp. velocity space of a single particle. If we restrict the movement to a rigid rotation $g \in SO(3,\mathbb{R})$ about a fixed point in space (for example the center of mass) we obtain

$$\vec{q}_i = g\,\vec{r}_i \;, \quad \dot{\vec{q}}_i = \dot{g}\,\vec{r}_i \;, \quad \delta\vec{q}_i = \delta g\,\vec{r}_i \;, \quad \delta\dot{\vec{q}}_i = \delta\dot{g}\,\vec{r}_i \tag{2}$$

where $\vec{r}_i$ are certain initial positions of the particles which together form the rigid body.

The classical state of the system is thus described by the pairs $(g, \dot{g})$ which are elements of the tangent bundle $TSO(3,\mathbb{R})$. Inserting (2) into (1) and writing $\delta = (\delta g, \delta\dot{g})$ we obtain

$$\sigma_2(\delta)(\delta') = tr\,\Theta\,\delta g^t\,\delta'\dot{g} - tr\,\Theta\,\delta\dot{g}^t\,\delta' g \tag{3}$$

Here we have introduced the symmetric matrix

$$\Theta = \sum m_i \vec{r}_i \, (\vec{r}_i)^t$$

g(g) and δg(g) can be related by left translations to elements of the tangent space at the neutral element of SO(3, ℝ). We write

$$\omega = g^{-1}\dot{g} \quad , \quad \delta_\lambda g = g^{-1}\delta g$$

The new coordinate **ω** is the matrix representation of the vector of angular velocity in body coordinates /1/. If we insert this into (3) and take into account that for example the following identity

$$\operatorname{tr} \Theta \delta_\lambda g \, \delta'\omega = \operatorname{tr} (\Theta \delta_\lambda g + \delta_\lambda g \, \Theta) \, \delta'\omega \tag{4}$$

holds, we obtain

$$\sigma(\delta)(\delta') = -\tfrac{1}{2} \operatorname{tr}(\Theta \delta_\lambda g + \delta_\lambda g \, \Theta) \delta'\omega - \tfrac{1}{2} \operatorname{tr}(\Theta \delta\omega + \delta\omega \, \Theta) \delta'_\lambda g - \tfrac{1}{2} \operatorname{tr}(\Theta\omega + \omega\Theta) [\delta_\lambda g, \delta'_\lambda g] \tag{5}$$

The identity (4) defines a metric on the manifold of SO(3, ℝ) if **Θ** fulfills certain conditions of non degeneracy

$$\langle \omega, \omega' \rangle = -\tfrac{1}{2} \operatorname{tr} (\Theta\omega + \omega\Theta) \, \omega' \tag{6}$$

If we choose the initial conditions in a way that Θ is a diagonal matrix we get

$$\langle \omega, \omega' \rangle = \sum_{i=1}^{3} I_i \, \omega_i \, \omega'_i \quad ; \quad I_1 = \Theta_2 + \Theta_3$$

The I_i are the moments of inertia of the rigid body. The form (5) coincides with the one given by Cushmann 1977 /1/. It is shown there that σ_2 is a symplectic form and, even more, that it is exact. This was not obvious from the construction given here, since a submanifold of a symplectic manifold is generally not symplectic.

The fact that σ_2 is exact implies that the phase space of the rigid body TSO(3, ℝ) can be prequantized in any case. This means that there is no restricting condition on the system in configuration space, i.e. any **Θ** is possible.

The symplectic form (5) is invariant with respect to left translations

$$\lambda : SO(3,\mathbb{R}) \times TSO(3,\mathbb{R}) \longrightarrow TSO(3,\mathbb{R}) \; ; \; \lambda(g', (g,\omega)) = (g'g, \omega)$$

Therefore, there exists a Ad^*-equivariant mapping for SO(3, ℝ) and we can construct a reduced phase space. This will be done using the theorems of Kirillov-Sourian-Kostant and Meyer /1/,/3/. For that purpose we need the momentum mapping which here can be most easily constructed if we use the 1-form σ_1 whose external derivative is σ_2:

$$\sigma_1(\delta) = -\tfrac{1}{2}\, tr\,(\Theta\omega + \omega\Theta)\,\delta_\lambda g$$

Let $\delta = (g^{-1}\eta\, g, 0)$ be the (left related) generator of a left translation on $SO(3, \mathbb{R})$ generated by the element η of the Lie algebra $so(3)$ of $SO(3, \mathbb{R})$. Then the momentum mapping

$$J : TSO(3,\mathbb{R}) \longrightarrow so(3)^*$$

(the dual of $so(3)$) is defined by

$$-\tfrac{1}{2}\, tr\, J(g,\omega)\,\eta = -\tfrac{1}{2}\, tr\,(\Theta\omega+\omega\Theta)g^{-1}\eta g = -\tfrac{1}{2}\, tr\, g(\Theta\omega+\omega\Theta)g^{-1}\eta$$

for any $\eta \in so(3)$. This implies

$$J(g,\omega) = g\,(\Theta\omega + \omega\Theta)\,g^{-1} = \mu$$

The reduced phase space is the set $P_\mu = J^{-1}(\mu)/G_\mu$ where G_μ is the isotropy group of the coadjoint representation of $SO(3, \mathbb{R})$ on $so(3)^*$ /2/,/3/. G_μ is isomorphic to $SO(2, \mathbb{R})$. For a fixed matrix μ $so(3)^*$ the set (g,ω), $TSO(3, \mathbb{R})$ for which $g(\Theta\omega+\omega\Theta)g^{-1} = \mu$ is determined by the vector field of constant angular momentum in space coordinates /2/. Thus $J^{-1}(\mu)$ is diffeomorphic to $SO(3, \mathbb{R})$. The reduced phase space is therefore symplectomorphic to the sphere S^2 provided with the symplectic form

$$\sigma_{red}(\mu)(J^*(g,\omega)\,(Ad_{g^{-1}}\eta))(J^*(g,\omega)(Ad_{g^{-1}}\xi)) = \langle \omega, [\xi,\eta]\rangle$$

Quantization of the sphere leads to discrete values of the parameters of the symplectic form. Since the parameters are given by the angular momentum we see that only discrete values for angular momentum are allowed. It has to be left to further investigation if we get the correct quantization of angular momentum by this method.

/1/ J.-M. Sourian, Structure des systèmes dynamiques, Dunod 1969

/2/ R. Abraham, J. Marsden, Foundations of mechanics[2], Benjamin 1978

/3/ G.-M. Marle, Hamiltonian Systems (in: Differential Geometry and Relativity, M. Cohen and M. Flato (Eds.)) 1976

A UNIFIED THEORY OF SIMPLE DYNAMICAL SYSTEMS

Peter Kasperkovitz and Josef Reisenberger
Institut für theoretische Physik, TU Wien, A-1040 Karlsplatz 13, Austria

1. This note deals with Hamiltonian systems where H is a polynomial of at most 2nd degree in the variables p,x or, stated differently, with systems whose evolution in time is described by a linear canonical transformation (LCT). We restrict the discussion to systems with one degree of freedom only ; everything can be generalized to f degrees ($f<\infty$). For these 'simple' systems a great number of relations between classical statistical mechanics (CL) and quantum mechanics (QM) is known. They all refer to the first two of the following topics :

(1) Transformations. The expectation values $\langle p\rangle,\langle x\rangle$ satisfy the classical equations of motion (Ehrenfest's theorem). The Wigner function, the quasi-classical substitute of the density operator, satisfies Liouville's equation ; if it is positive (which is not true in general) this property is conserved under all LCTs. These transformations can also be studied directly in QM since the unitary operators corresponding to LCTs have been given by Moshinsky and Quesne [1] (for propagators see also De Witt [2]).

(2.1) States. Quantum mechanical states of an oscillator allowing interpretation in terms of CL have already been discussed by Schrödinger [3] ; the free Gaussian wave packet is another example of this sort. Nowadays these states are known as 'coherent states'. In a classic paper Hudson [4] showed these states and their transforms under LCTs to be the only pure states for which the Wigner function is positive. Apart from convex linear combinations of these 'Hudson states' only one mixed state with positive Wigner function seems to be known : the canonical ensemble of oscillators.

(2.2) Observables. In CL observables are mostly unbounded functions, in QM unbounded operators (example : p,x). To ensure the existence of expectation values for all states of a given set and of the Weyl-transform and its inverse (relating operators to functions on phase space) restrictions have to be imposed on the set of observables. The least to be reqired for this set is that in CL it should contain the characteristic functions of bounded subsets of the phase space, and in QM that it should contain all projectors.

(3) Expectation value. A common feature of the expectation value in CL and QM is that it is a bilinear functional on pairs consisting of a state and an observable. This functional makes states and observables dual objects : The more restrictions are imposed on one set the less are needed for the other.

Restricting the set of admitted states both in CL and QM it is possible to define a unified dynamical theory of simple systems. In essence this theory is CL (systems evolve according to the classical equations of motion) with initial conditions compatible with QM (the uncertainity relation has to be satisfied). Four forms of the theory looking quite differently are outlined below. Their interrelations and the nature of their basic objects [items (1) to (3)] are indicated in the following scheme.

The four different forms of the theory :

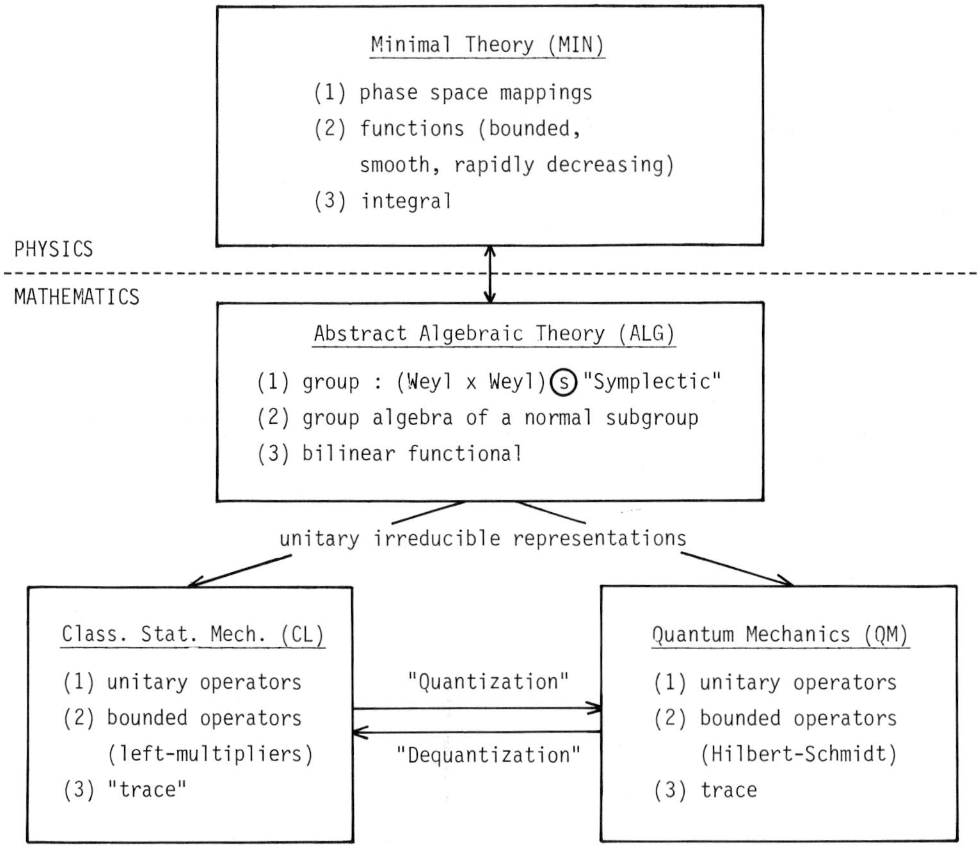

2. MIN. The term 'minimal' means that the results which can be interpreted are obtained with a minimum of mathematics. In this theory transformations of functions (LCTs) are induced by mappings of $\mathbb{R}^2$ (= phase space).

$$F \to F_{z_1,m} \quad ; \quad F_{z_1,m}(mz+z_1) = F(z) \quad ;$$

$$z_1 \quad \varepsilon \ \mathbb{R}^2 \ ; \quad m = \text{real } 2\text{x}2 \text{ matrix}, \quad \det m = 1 \Leftrightarrow ms\tilde{m} = s,\ s = \begin{pmatrix} 0 & 1 \\ -1 & 0 \end{pmatrix} \tag{1}$$

States are functions on $\mathbb{R}^2$ satisfying the following conditions :

$$\text{(i)} \quad W = W^* \qquad \text{(ii)} \quad 0 \leq W \leq 2/\hbar \qquad \text{(iii)} \quad < W,1 > := \int_{\mathbb{R}^2} dz\ W(z) \ = 1$$

$$\text{(iv)} \quad < W,W_{\Psi\Psi} > := \int_{\mathbb{R}^2} dz\ W(z)W_{\Psi\Psi}(z) \geq 0 \quad \text{ for all } \Psi \ \varepsilon \ L^2(\mathbb{R}) \tag{2}$$

The function $W_{\Psi\Psi}$ appearing in (iv) is the Weyl-transform of $|\Psi><\Psi|$, i.e.

$$\phi,\Psi \ \varepsilon \ L^2(\mathbb{R}) \ : \quad W_{\phi\Psi}(px) := \int_{\mathbb{R}} dv\ e^{(i/\hbar)pv}\ \phi(x-\tfrac{v}{2})\Psi^*(x+\tfrac{v}{2}) \quad . \tag{3}$$

(ii) and (iii) imply $W \ \varepsilon \ L^p(\mathbb{R}),\ 1 \leq p \leq \infty$. $\{W\}$ is a convex set, closed in the L^2-norm, and invariant under LCTs. $\{W\}$ contains the Wigner functions corresponding to coherent states.

$$W_{\phi z_1 \phi z_1}(z) = E_h(z-z_1) \ , \quad E_h(z) = (2/h)\ \exp(-z^2/\hbar) \tag{4}$$

More generally if P is a tempered distribution which can be interpreted as a classical distribution function,

$$\text{(i)} \quad P = P^* \qquad \text{(ii)} \quad 0 \leq P \qquad \text{(iii)} \quad < P,1 > = 1 \qquad \text{(iv)} \quad P \ \varepsilon \ S'(\mathbb{R}^2) \tag{5}$$

then

$$W_P(z) = P{*}E_h(z) := \int_{\mathbb{R}^2} dz_1\ P(z_1)E_h(z-z_1) \quad \varepsilon \ \{W\} \quad . \tag{6}$$

Whether $\{W_P\}$ is dense in $\{W\}$ or not is subject of current investigation. To meet the reqirements cited above observables are chosen to be rapidly decreasing functions, i.e. $A \ \varepsilon \ S(\mathbb{R}^2)$. The expectation value is defined by

$$E(W,A) = < W,A > \quad . \tag{7}$$

3. <u>ALG</u>. The definition of the phase space, the (possible) relations between p, x, d/dp, d/dx, and the LCTs (including their quantum mechanical representation), are all incorporated into the multiplication law of one single group G.

$$G = \{ \ (\rho,\tau|Z|\mu,M) \ \} \ \tilde{=} \ [W(1) \times W(1)] \ \textcircled{s} \ CSp(2,\mathbb{R}) \ ,$$

W(1) = Weyl group for one degree of freedom ;

CSp(2,$\mathbb{R}$) = central extension of the symplectic group Sp(2,$\mathbb{R}$) ;

$$\rho,\tau \ \varepsilon \ \mathbb{R} \ ; \quad Z = (z_0,z_1) \ \varepsilon \ \mathbb{R}^4 \ ; \quad \mu = \pm 1 \ ; \quad M = \begin{pmatrix} m & 0 \\ 0 & \tilde{m}^{-1} \end{pmatrix} , \quad m \text{ see (1) ;}$$

$$\rho_3 = \rho_1 + \rho_2 + (1/2)Z_1.M_1SZ_2 \qquad S = \begin{pmatrix} s & 0 \\ 0 & s \end{pmatrix} , \quad s \text{ see (1)}, \quad T = \begin{pmatrix} 0 & e \\ -e & 0 \end{pmatrix} ;$$

$$\tau_3 = \tau_1 + \tau_2 + (1/2)Z_1.M_1TZ_2$$

$$Z_3 = Z_1 + \tilde{M}_1^{-1}Z_2$$

$\mu_3 = \mu_1\mu_2\mu(M_1,M_2)$, $\mu(M,M')$ is a factor system of $Sp(2,\mathbb{R})$, see [5] ;

$$M_3 = M_1M_2 \quad . \tag{8}$$

The LCTs are contained in the subgroup $U = \{ (\rho,0|(0,z_1)|\mu,M) \}$. States and observables are represented by elements of the (extended) group algebra A(N), N being the normal subgroup $\{ (\rho,\tau|(z_0,0)|1,E) \}$. If $\bar{F}$ is the Fourier transform of F,

$$\bar{F}(z') = (\gamma/2\pi) \int_{\mathbb{R}^2} dz\, e^{-i\gamma z.z'} F(z) , \quad \gamma > 0 , \tag{9}$$

then states and observables are represented by the following expressions :

$$\begin{aligned}
&\underline{W}_{c,z_1,e} = \int_{\mathbb{R}^4} d\rho d\tau dz_0\, w_{c,z_1,e}(\rho,\tau,z_0)\ (\rho,\tau|(z_0,0)|1,E) , \\
&w_{c,z_1,e}(\rho,\tau,z_0) = \delta(\rho)\, \delta(\tau+z_0.[1-c]z_1)\, \bar{W}_{cz_1,e}(z_0) ; \\
&\underline{A} = \int_{\mathbb{R}^4} d\rho d\tau dz_0\, a(\rho,\tau,z_0)\ (\rho,\tau|(z_0,0)|1,E) , \\
&a(\rho,\tau,z_0) = \delta(\rho)\, \delta(\tau)\, \bar{A}(z_0) .
\end{aligned} \tag{10}$$

States are transformed according to $\underline{W} \to \underline{uWu}^{-1}$, $\underline{u} \in U$. If $w{*}a$ denotes the convolution corresponding to the product element $\underline{WA} \in A(N)$ the expectation value is

$$E_\gamma(\underline{W}_{c,z_1,e}\ \underline{A}) = \int_{\mathbb{R}^4} d\rho d\tau dz_0\, w_{c,z_1,e} * a\ (\rho,\tau,z_0)\ e^{i\gamma\tau} = < W_{z_1,e},A > . \tag{11}$$

(11) shows that the characterization of a state by the triple (W,z_1,c) is overcomplete ; an infinite class of elements (10) represents the same physical state.

Passing from G to unitary irreducible representations the theory changes its form from ALG to CL or QM. These representations are completely determined by the representation of the center of G consisting of the elements $(\rho,\tau|0|1,E)$. The elements of G and A(N) are both represented as integral operators with the kernels given below.

4. CL. The carrier space is $L^2(\mathbb{R}^2)$.

$$\begin{aligned}
&(\rho,\tau|(z_0,z_1)|\mu,M) \to <z|\,{}^{CL}\underline{U}(\rho,..,M)|z'> = \exp\{i\gamma[\tau-(1/2)z_0.z_1+z_0.z]\}\ \delta(z-z_1-mz') \\
&\underline{W}_{c,z_1,e} \to <z|\,{}^{CL}\underline{W}_{z_1,e}|z'> = (2\pi/\gamma)\ W_{z_1,e}(z)\ \delta(z-z') \\
&\underline{A} \to <z|\,{}^{CL}\underline{A}|z'> = (2\pi/\gamma)\ A(z)\ \delta(z-z') \\
&{}^{CL}E({}^{CL}\underline{W}_{z_1,e}\ {}^{CL}\underline{A}) = \text{"trace"}\ {}^{CL}\underline{W}_{z_1,e}\ {}^{CL}\underline{A} = < W_{z_1,e},A >
\end{aligned} \tag{12}$$

5. QM. The carrier space is $L^2(\mathbb{R})$.

$$(\rho,\tau|(x_0,p_0,p_1,x_1)|\mu,\begin{pmatrix} a & b \\ c & d \end{pmatrix}) \to <x|\,{}^{QM}\underline{U}(\rho,..,M)|x'> =$$

$$= (\gamma/4\pi|c|)^{1/2}\ (\text{sign } c - i)\ \exp\{i\gamma[\rho+\tau+(1/2)(x_0-x_1)(p_0+p_1)+(p_0+p_1)x+$$

$$+(1/2)(x+x_0-x_1)ac^{-1}(x+x_0-x_1)-(x+x_0-x_1)c^{-1}x'+(1/2)x'c^{-1}dx']\}$$

$$\underline{W}_{c,(p_1,x_1),e} \to \langle x|\,{}^{QM}\underline{W}(p_1,x_1)e|x'\rangle =$$

$$= \int_{\mathbb{R}^2} dx_0 dp_0\ \bar{\underline{W}}_{c(p_1,x_1),e}(x_0,p_0)\ \exp\{i\gamma[(c-1)(x_0p_1+p_0x_1)+(1/2)x_0p_0+p_0x]\}\ \delta(x+x_0-x')$$

$$\underline{A} \to \langle x|\,{}^{QM}\underline{A}|x'\rangle = \int_{\mathbb{R}^2} dx_0 dp_0\ \bar{A}(x_0,p_0)\ \exp\{i\gamma[(1/2)x_0p_0+p_0x]\}\ \delta(x+x_0-x')$$

$$^{QM}E(^{QM}\underline{W}_{z_1,e}\ {}^{QM}\underline{A}) = (\gamma/2\pi)\ \mathrm{trace}\ {}^{QM}\underline{W}_{z_1,e}\ {}^{QM}\underline{A}\ = \langle W_{z_1,e},A\rangle \tag{13}$$

[1] M.Moshinsky and C.Quesne, J.Math.Phys. 12(1971)1772

[2] B.S. De Witt, Rev.mod.Phys. 29(1957)377

[3] E.Schrödinger, Naturwissenschaften 14(1926)664

[4] R.L.Hudson, Rep.Math.Phys. 6(1974)249

[5] K.B.Wolf, Integral Transforms in Science and Engineering. Plenum Press, New York 1979.

GEOMETRY OF THE TIME-DEPENDENT VARIATIONAL PRINCIPLE IN QUANTUM MECHANICS

P. Kramer, Institut für Theoretische Physik, Universität Tübingen, Germany, and M. Saraceno, Dept. de Fisica, CNEA Buenos Aires, Argentina

1. Introduction

Various aspects of nuclear many-body systems have been studied by use of the time-dependent variational principle (TDVP). This principle allows one to introduce collective parameters into a many-body state and to investigate the dependence of these parameters on time. In many cases, as for example in the time-dependent Hartree-Fock method, the parameters are associated with a Lie group. The TDVP, its geometry and its relation to Lie groups are analyzed in the present contribution. More details are given in /KR 79/,/KR 80/. In section 2 the TDVP is shown to yield a generalization of classical mechanics on a phase space with a symplectic structure or generalized Poisson bracket. A standard form of this symplectic form is given for a complex parametrization. The relation between Lie algebras and symplectic structures which was explored by Kostant/KO 70/ is briefly sketched in section 3. In section 4 these symplectic structures are studied on group representation spaces. By use of an analytic parametrization of coset spaces one finds again a standard symplectic form. Applications are discussed in the remaining part.

2. The time-dependent variational principle

Consider the action functional

$$S = \int_{t_1}^{t_2} \mathcal{L}'(\psi,\bar{\psi})dt, \quad \mathcal{L}'(\psi,\bar{\psi}) = \langle \psi(t) \mid i\frac{\partial}{\partial t} - H \mid \psi(t)\rangle$$

From the requirement $\delta S=0$ for arbitrary variations of ψ one obtains the time-dependent Schrödinger equation. We prefer to use the Lagrangian

$$\mathcal{L}(\psi,\bar{\psi}) = \frac{1}{2} i(\langle \psi \mid \dot{\psi}\rangle - \langle \dot{\psi} \mid \psi\rangle)\langle \psi \mid \psi\rangle^{-1} - \langle \psi|H|\psi\rangle\langle \psi \mid \psi\rangle^{-1}$$

which is invariant under scale transformations of the state ψ. With $\mathcal{L}'$ replaced by $\mathcal{L}$, the principle $\delta S=0$ leads to an equation equivalent to the time-dependent Schrödinger equation. The typical applications of the TDVP arise by choosing time-dependent parameters for the description of the state ψ. The variational equations then lead to equations of motion for these parameters. In what follows it will be assumed that ψ depends on N complex parameters $z_1 z_2 \dots z_N$,

$$|\psi\rangle = |\psi(\bar{z}_1\bar{z}_2\dots\bar{z}_N)\rangle, \quad z_i=z_i(t), i=12\dots N.$$

Important quantities in the computations are the overlap

$$< \psi(z') | \psi(\bar{z}) > \; = \; < \psi(z_1' z_2' \ldots z_N') | \psi(\bar{z}_1 \bar{z}_2 \ldots \bar{z}_N) >$$

and the mean value of the Hamiltonian

$$\mathcal{H}(z,\bar{z}) = \; < \psi(z_1 z_2 \ldots z_N) | H | \psi(z_1 z_2 \ldots z_N) > \; < \psi | \psi >^{-1} .$$

For both quantities we assume that they are analytic in z' (or z) and in $\bar{z}$ respectively. An examination of the TDVP yields

2.1 Proposition. In the complex analytic parametrization, the Lagrangian is given by

$$\mathcal{L}(z,\bar{z},\dot{z},\dot{\bar{z}}) = \sum_{l} (\mathcal{Z}_l \dot{z}_l + \bar{\mathcal{Z}}_{\bar{l}} \dot{\bar{z}}_l) - \mathcal{H}(z,\bar{z})$$

where the N quantities $\mathcal{Z}_l$ are defined as

$$\mathcal{Z}_l = \frac{1}{2} i \frac{\partial}{\partial z'_l} \ln < \psi(z') | \psi(\bar{z}) > |_{z'=z} .$$

The equations of motion for the quantities $z_i, \bar{z}_i$ are given by

$$\sum_j C_{l\bar{j}} \dot{\bar{z}}_{\bar{j}} = -i \frac{\partial \mathcal{H}}{\partial z_l} , \qquad \sum_l C_{\bar{j}l} \dot{z}_l = i \frac{\partial \mathcal{H}}{\partial \bar{z}_j}$$

where the NxN matrix C is defined as

$$C_{l\bar{j}} = -i \left(\frac{\partial \mathcal{Z}_l}{\partial \bar{z}_{\bar{j}}} - \frac{\partial \bar{\mathcal{Z}}_{\bar{j}}}{\partial z_l} \right) = \frac{\partial^2}{\partial z_l' \, \partial \bar{z}_{\bar{j}}} \ln < \psi(z') | \psi(z) > |_{z'=z}$$

Assuming now that C be invertible one finds

2.2 Proposition. For C invertible, the equations of motion for the TDVP have the canonical form

$$\dot{z}_l = i \{ \mathcal{H}, z_l \} , \qquad \dot{\bar{z}}_{\bar{j}} = i \{ \mathcal{H}, \bar{z}_{\bar{j}} \}$$

with the symplectic form or generalized Poisson bracket $\{ , \}$, defined as

$$\{ \mathcal{F}, \mathcal{G} \} = \sum_{l\bar{j}} \left(\frac{\partial \mathcal{F}}{\partial z_l} g_{l\bar{j}} \frac{\partial \mathcal{G}}{\partial \bar{z}_{\bar{j}}} + \frac{\partial \mathcal{F}}{\partial \bar{z}_{\bar{j}}} g_{\bar{j}l} \frac{\partial \mathcal{G}}{\partial z_l} \right)$$

and the 2Nx2N matrix g given by

$$g = \begin{bmatrix} 0 & -{}^tC^{-1} \\ C^{-1} & 0 \end{bmatrix} .$$

From proposition 2.2, the TDVP appears as an approximation to the system in quantum mechanics where the time-dependence is given by canonical equations of motion for trajectories on the manifold whose coordinates are $z_1 z_2 \ldots z_N$. To these trajectories there corresponds an approximation of the time-evolution of the state ψ. The classical Hamiltonian is the expectation value of the hamiltonian operator.

3. Lie algebras and symplectic structures

Consider an r-dimensional Lie algebra $\mathcal{L}_G$ of a Lie group G,

$$\mathcal{L}_G = \{ A \mid A = \sum_i \alpha^i A_i \}$$

with basis $A_1 A_2 \ldots A_r$ and commutators

$$[A_i, A_j] = \sum_k c_{ij}^k A_k \quad .$$

On the dual of $\mathcal{L}_G$ which we denote as $\mathcal{L}_G^*$, any linear form B may be expressed in the dual basis as

$$B(A) = \sum_i \beta_i \alpha^i$$

For functions $\mathcal{F}, \mathcal{G}$ of the r coordinates we introduce

3.1 Definition. To any pair $\mathcal{F}, \mathcal{G}$ of functions on $\mathcal{L}_G^*$, we define the symplectic form by

$$\{ \mathcal{F}, \mathcal{G} \} = \sum \frac{\partial \mathcal{F}}{\partial \beta_i} g_{ij} \frac{\partial \mathcal{G}}{\partial \beta_j} \quad , \qquad g_{ij} = \sum c_{ij}^k \beta_k$$

3.2 Proposition. The functions $\mathcal{A}_i(\beta) = \beta_i$ yield a symplectic realization of $\mathcal{L}_G$, that is,

$$\{ \mathcal{A}_i, \mathcal{A}_j \} = \sum_k c_{ij}^k \mathcal{A}_k \quad .$$

Now let G act on $\mathcal{L}_G$ according to

$$G \ni g: A \to gAg^{-1} = \sum_{ij} \alpha^i A_j \, d_{ji}(g)$$

With respect to the components α^i, we define the adjoint representation d as

$$G \ni g: \alpha^i \to \sum_j \alpha^j d_{ij}(g)$$

This action may be transferred to $\mathcal{L}_G^*$ to define the coadjoint representation $\hat{d}$ by

$$G \ni g: \beta^j \leftrightarrow (\beta_g)_j = \sum_i \beta_i \hat{d}_{ji}(g)$$

where $\hat{d}(g)$ is the transposed inverse of $d(g)$. For functions $\mathcal{F}$ on $\mathcal{L}_G^*$ we now defince the action of G on $\mathcal{F}$ by

$$G \ni g: \mathcal{F}(\beta) \to (T_g \mathcal{F})(\beta) = \mathcal{F}(\beta_{g^{-1}})$$

Assuming the exponential parametrization $g = \exp \tau^j A_j$, one then finds

3.3 Proposition. The action of G on the function $\mathcal{F}(\beta)$ is given by the multiple Poisson bracket,

$$(T_g \mathcal{F})(\beta) = \sum_n (n!)^{-1} \quad \{ \tau^j \mathcal{A}_j, \mathcal{F} \}_{(n)} \quad ,$$

the functions $\mathcal{A}_i(\beta)$ are the symplectic generators of this action. The action of G on $\mathcal{L}^*_G$ in general is not transitive and we therefore must decompose $\mathcal{L}^*_G$ into its orbits under G. The determination of the orbits is simplified for semi-simple Lie groups:
The Killing form provides then a non-degenerate inner product on $\mathcal{L}^*_G$ which moreover is invariant under the adjoint representation. This inner product allows one to essentially identify the points on $\mathcal{L}_G$ and $\mathcal{L}^*_G$. The adjoint and coadjoint representation become equivalent, their orbits on $\mathcal{L}_G$ and $\mathcal{L}^*_G$ are in one-to-one correspondence. On $\mathcal{L}_G$, the determination of the orbits may be reduced to finding the classes of G and hence one has

3.4 Proposition. The symplectic orbits of a semi-simple Lie group G on $\mathcal{L}^*_G$ are in one-to-one correspondence to the classes of G. The implication for the symplectic form is that while $\{\ ,\ \}$ is degenerate on $\mathcal{L}^*_G$, it becomes non-degenerate when restricted to an orbit. This restriction is achieved by introducing on $\mathcal{L}^*_G$ a representative point for each orbit along with coordinates on the coset space which corresponds to the orbit.

4. Coherent states, analytic parametrization of orbits, and symplectic structures on group representation spaces.

Consider a complex semisimple group G with Lie algebra $\mathcal{L}_G$. Associated with its p lowering generators $E_{-\alpha}$, s weight generators H_i and p raising generators E_α there are three subgroups G_-, G_o, G_+ with elements

$$g_- = \exp\left(\sum_\alpha \bar{y}_\alpha E_{-\alpha}\right),\quad g_o = \exp\left(\sum_i v_i H_i\right),\quad g_+ = \exp\left(\sum_\beta z'_\beta E_\beta\right)$$

Any element $g \in G$ has a Gauss decomposition /BA 77/ p.100,

$g = g_- g_o g_+ = g_o g'_- g_+$

where we used $g_o^{-1} g_- g_o \in G_-$ in the second step. The coset space G/K where $K = G_o G_-$ is clearly in one-to-one correspondence to G_+, parametrized through p complex numbers z' . Consider now the maximal compact real subgroups $G^c < G$ and its subgroup $G^c_o = G^c \cap G_o$.

4.1 Proposition. The coset space G^c/G^c_o is in one-to-one correspondence to G_+, parametrized through p complex coordinates z' .

This proposition may be verified for the examples to be considered later. Consider now an irreducible representation of G^c with maximum weightstate $|w_{max}\rangle$.

4.2 Definition. The bra and ket coherent states $|z\rangle$ and $\langle z'|$ are defined as $|z\rangle = \exp\left(\sum_\alpha \bar{z}_\alpha E_{-\alpha}\right) |w_{max}\rangle$, $\langle z'| = \langle w_{max}| \exp\left(\sum_\beta z'_\beta E_\beta\right)$

with the overlap $\langle z' | z \rangle$.
For the irreducible representation we introduce

4.3 Definition. The stability group H associated with w_{max} is defined as

$$H = \{ h \,|\, \langle w_{max} | U(h)\, A_i\, U^{-1}(h) | w_{max} \rangle = \langle w_{max} | A_i | w_{max} \rangle \}$$

for $A_i = \{ H_i, E_\alpha, E_{-\alpha} \}$.
We call w_{max} of <u>regular type</u> if $H = G_o^c$. Now we connect the coherent states with the symplectic structures discussed in section 3.
From section 3 we know that for semi-simple groups the orbits correspond to the classes of G^c. Any representative of a class may be taken as the exponential from a maximal abelian subalgebra which we may identify with the algebra A^o. An element of A^o is called <u>regular</u>, /GI 74/p. 246 if its centralizer is generated by A^o. Hence for regular class representatives, the stability group G_o^c is the abelian subgroup generated by A^o.

4.4 Definition. The irreducible representation corresponding to the maximum weight is called of <u>regular type</u> if the stability group H according to definition 4.3 is the abelian group G_o^c.

4.5 Definition. The p x p matrix $C = (C_{i\bar{j}})$ is defined as

$$C_{i\bar{j}} = (\partial^2 / \partial z_i' \, \partial \bar{z}_{\bar{j}}) \ln \langle z' | z \rangle \big|_{z'=z}$$

4.6 Definition. The expectation values $\mathcal{H}_i$ $\mathcal{E}_{\pm\alpha}$ are defined as

$$\mathcal{H}_i(z,\bar{z}) = \langle z | H_i | z \rangle \langle z | z \rangle^{-1}$$
$$\mathcal{E}_{\pm\alpha}(z,\bar{z}) = \langle z | E_{\pm\alpha} | z \rangle \langle z | z \rangle^{-1}$$

We are now prepared to determine the symplectic structure of G^c on the orbit G^c/G_o^c.

4.7 Proposition. The restriction of the symplectic form to the orbits of intrinsic structure G^c/G_o^c has the form

$$\{ \mathcal{F}, \mathcal{G} \} = \sum_{ij} \left(\frac{\mathcal{F}}{z_i} g_{i\bar{j}} \frac{\mathcal{G}}{\bar{z}_{\bar{j}}} + \frac{\mathcal{F}}{\bar{z}_{\bar{j}}} g_{\bar{j}i} \frac{\mathcal{G}}{z_i} \right), \; g_{i\bar{j}} = -{}^tC^{-1}_{i\bar{j}}, \; g_{\bar{j}i} = C^{-1}_{\bar{j}i}$$

The expectation values $\mathcal{H}_i$, $\mathcal{E}_{\pm\alpha}$ generate the Lie algebra of G^c for this generalized Poisson bracket.
For the proof we refer to /KR 79/,/KR 80/ and prefer to give examples in the following sections.

5. The group SU(2).
The classes of SU(2) are characterized by the eigenvalues $\lambda_1, \lambda_2 = \bar{\lambda}_1$ of its elements. For $\lambda_2 \neq \lambda_1$ the orbits are of the type SU(2)/U(1). The complex coset parametrization of proposition 4.1 has the form

$$g_0' \, g_- \, g = \begin{bmatrix} a & 0 \\ 0 & b \end{bmatrix} \begin{bmatrix} 1 & 0 \\ \bar{y} & 1 \end{bmatrix} \begin{bmatrix} 1 & z \\ 0 & 1 \end{bmatrix}$$

where $a = b^{-1} = (1+z\bar{z})$, $y = -(1+z\bar{z})^{-1} z$

If $\beta\gamma$ are the usual polar coordinates of the sphere, one finds $z = \mathrm{ctg}\,\frac{1}{2}\beta \exp i\gamma$. The highest weight state $|JJ>$ for $J \neq 0$ has the stability group U(1) as required in definition 4.3, the coherent state is

$$|z> = \exp(\bar{z} J_-) \, |JJ>$$

and the overlap becomes

$$<z'|z> = (1 + z'\bar{z})^{2J}$$

The symplectic form is determined by

$$C_{1\bar{1}} = (\partial^2/\partial z' \partial \bar{z}) \ln(1 + z'\bar{z})^{2J} \Big|_{z'=z} = 2J(1 + z\bar{z})^{-2}$$

and given by

$$\{\mathcal{F}, \mathcal{G}\} = -(2J)^{-1}(1 + z\bar{z})^2 \left(\frac{\partial \mathcal{F}}{\partial z}\frac{\partial \mathcal{G}}{\partial \bar{z}} - \frac{\partial \mathcal{F}}{\partial \bar{z}}\frac{\partial \mathcal{G}}{\partial z}\right)$$

with the symplectic generators

$$\mathcal{J}_+ = \overline{\mathcal{J}_-} = 2J\,\bar{z}(1 + z\bar{z})^{-1} , \quad \mathcal{J}_0 = J(1 - z\bar{z})(1 + z\bar{z})^{-1}$$

At the point $z = \bar{z} = 0$ one has $\mathcal{J}_+ = \mathcal{J}_- = 0, \mathcal{J}_0 = J$ and hence the sphere on $\mathcal{L}^*$ has the radius J.

6. The Lipkin model

The Lipkin model/LI 65/ is a simple, though not trivial model that has been used as a testing ground for approximation schemes to many-fermion systems. Its basic simplicity rests with the fact that the hamiltonian has such symmetries that it can be written in terms of SU(2) generators.

The model consists of N fermions interacting via one- and two-body forces distributed in two levels each having an N-fold degeneracy. The fermion operators have two indices: The index σ labels the level and has values ± 1 and p labels the degenerate states within a level. The basic dimensionality of the state space is thus 2^N and the solution of a general hamiltonian eigenvalue problem would involve matrices of that dimension.

The hamiltonian is taken as

$$H = \frac{\varepsilon}{2} \sum_{\substack{p=1..N \\ \sigma=\pm 1}} \sigma \, a^+_{p\sigma} a_{p\sigma} - \frac{V}{2} \sum_{\substack{p,p'=1..N \\ \sigma=\pm 1}} a^+_{p\sigma} a^+_{p'\sigma} a_{p'-\sigma} a_{p-\sigma}$$

and its basic symmetry is that the potential with which two particles in one shell scatter to one another is independent of the "magnetic" quantum number p.

We introduce the quasi-spin operators

$$J_+ = \sum_{p=1}^{N} a^+_{p1}\, a_{p-1}$$

$$J_- = \sum_{p=1}^{N} a^+_{p-1}\, a_{p1} = (J_+)^+$$

$$J_o = \frac{1}{2} \sum_{\substack{p=1 \\ \sigma=\pm 1}}^{N} \sigma \; a^+_{p\sigma}\, a_{p\sigma} = J_3$$

which among themselves form a SU(2) Lie algebra. These operators allow us to rewrite the hamiltonian as

$$H = J_o - \frac{V}{2}(J_+^{\,2} + J_-^{\,2})$$

or alternatively as

$$H = J_o - V(J_1^{\,2} - J_2^{\,2})$$

where $J_{\pm} = J_1 \pm J_2$.

The hamiltonian commutes with J^2 and therefore the dimension of the matrices for an exact diagonalization is 2J+1. The relationship between N and J is obtained by noticing that J_o counts the difference between the number of pairs of particles in the upper and lower level. This difference is at most N/2 so that the maximum value taken by J_o and therefore the value of J is N/2. The exact solution of the eigenvalue problem for a given particle number N then involves at most matrices of dimension 2J+1 = N+1.

In accordance with the general methods developed in section 2 we parametrize the state as

$$|\psi(t)\rangle = U(g)\,|\psi_o\rangle$$

where U(g) is a unitary representation in Hilbert space of SU(2) and ψ_o is a reference state which we will take to be the Slater determinant

$$|\psi_o\rangle = \prod_{p=1}^{N} a^+_{p-1}\,|0\rangle$$

Physically this is the ground state of the non-interacting system (V=o). As the SU(2) generators are one-body operators, $\psi(t)$ will be a Slater determinant at all times, and therefore the parametrization imposes the TDHF restriction that the state be always a Slater determinant. Moreover U(g) also preserves the quantum number J thus adapting the parametrization to the symmetries of the hamiltonian.

The state ψ_o satisfies

$$J_-\,|\psi_o\rangle = 0, \quad J_o\,|\psi_o\rangle = -\frac{N}{2}\,|\psi_o\rangle$$

and therefore it is the minimum weight state of the representation J = N/2.

To follow the presentation in section 5 we will take instead the reference state $|JJ\rangle$,

$$|JJ\rangle = \prod_{p=1}^{N} a^{+}_{p1} |0\rangle .$$

The classical hamiltonian $\mathcal{H}$ can be written as

$$\mathcal{H} = \mathcal{H}(\mathcal{J}_1, \mathcal{J}_2, \mathcal{J}_0)$$

$$= \varepsilon \mathcal{J}_0 - \frac{\varepsilon \chi}{2J} (\mathcal{J}_1^{\,2} - \mathcal{J}_2^{\,2}) \quad , \quad \chi = \frac{V}{\varepsilon}(2J-1)$$

Notice that the classical hamiltonian, when expressed in terms of the symplectic generators, has almost the same form as the quantum hamiltonian but differs from it by terms of order $(2J)^{-1}$. This form of the hamiltonian is independent of the parametrization chosen for the orbit, but depends on its radius J.

The equations of motion for $\mathcal{J}_1$ are

$$\dot{\mathcal{J}}_1 = -i \{ \mathcal{J}_1, \mathcal{H} \}$$

and the bracket can be calculated directly. In this way no reference to the orbit parametrization is needed at all, the equations become

$$\dot{\mathcal{J}}_1 = \varepsilon(-\mathcal{J}_2 + \frac{\chi}{J}\mathcal{J}_2\mathcal{J}_3)$$

$$\dot{\mathcal{J}}_2 = \varepsilon(\mathcal{J}_1 + \frac{\chi}{J}\mathcal{J}_1\mathcal{J}_3)$$

$$\dot{\mathcal{J}}_3 = \varepsilon\frac{\chi}{J}(-2\mathcal{J}_1\mathcal{J}_2)$$

It is easy to see that these equations imply that

$$\mathcal{J}^2 = \mathcal{J}_1^{\,2} + \mathcal{J}_2^{\,2} + \mathcal{J}_3^{\,2} = \text{constant}$$

$$\mathcal{H} = \varepsilon(\mathcal{J}_3 - \frac{\chi}{2J}(\mathcal{J}_1^{\,2} - \mathcal{J}_2^{\,2})) = \text{constant}$$

Therefore the motion occurs in the intersection of these surfaces in the three-dimensional space $\mathcal{L}^*$.

The $\mathcal{H}$-surface is a parabolic hyperboloid shifted from the origin by ε along the 3-axis.

Let us now analyze qualitatively the possible intersections of the surfaces. There are two possible types of motion according to the value of the parameter χ,

(1) $\chi < 1$

The value of the curvature at the vertex of the parabola in the $\mathcal{J}_3\mathcal{J}_1$ and $\mathcal{J}_3\mathcal{J}_2$ planes is given by $\pm\chi/J$. For the sphere it is $1/J$ so that the curvature of the parabola is less than that of the sphere. The first intersection (as E/ε is varied) is obtained for $E/\varepsilon = J$ and there will be intersections until $E/\varepsilon = -J$. The main feature is that all trajectories turn by 2π around the $\mathcal{J}_3$ axis giving the picture shown in fig. 1 for several motions with different values of the energy and a fixed number of particles.

(2) $\chi > 1$

In this case the curvature of the parabola is greater than that of the sphere so that the intersection first occurs at a finite value of β. The value of β at which first contact occurs is given by

$$\cos \beta_c = \frac{1}{\chi}$$

and the energy is

$$\frac{E}{\varepsilon} = J\frac{\chi^2+1}{2\chi} > J$$

As the energy decreases the trajectory will describe a closed path around β_c with increasing amplitude until $E/\varepsilon = J$. At that point the vertex of the parabola touches the north pole and the trajectories start turning around the $\mathcal{J}_3$ axis as in the previous case. For negative values of the energy essentially the same situation arises, but the parabola is in the $\mathcal{J}_3\mathcal{J}_1$ plane and points upwards. The complete set of trajectories on the sphere is given in fig. 2.

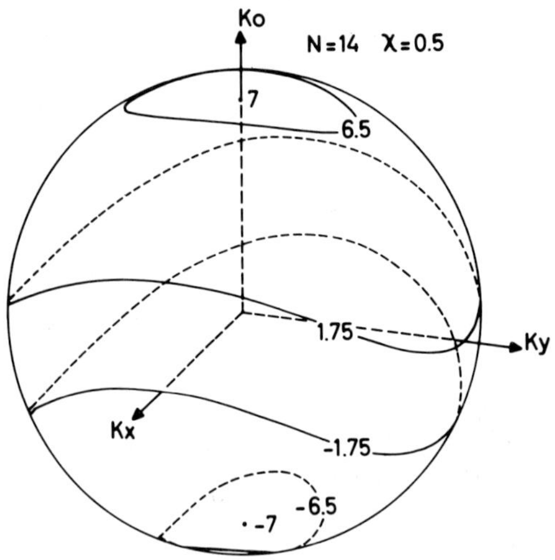

Fig.1 TDHF trajectories for the Lipkin model for N=2J=14 χ= o.5. The numbers labeling the trajectories are the energies and have been chosen for clarity of display Dots at the two poles are the static HF solutions.

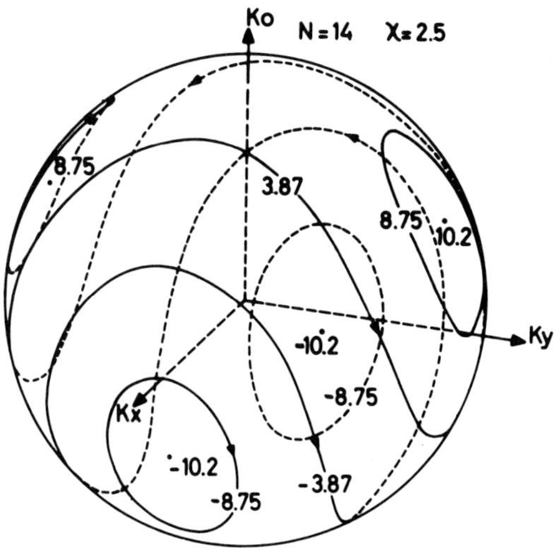

Fig. 2 TDHF trajectories for the Lipkin model for N=2J=14 and χ = 2.5. For details compare the caption of fig. 1.

Dequantizations for other groups and physical systems through the TDVP are treated in /KR 8o/. The dequantization with the usual coherent states is associated with the Weyl group. The time-dependent Hartree-Fock theory TDHF for N particles occupying M states is dequantized on the coset space $U(M)/U(N)\times U(M-N)$. Rotationally invariant two-body systems are dequantized on the coset space $SU(1,1)/U(1)$.

References

BA 77 A.O. Barut, R.Rączka, Theory of Group Representations and Applications, Warsaw 1977

GI 74 R. Gilmore, Lie groups, Lie algebras, and some of their applications, New York 1974

KO 70 B. Kostant, in: Lecture Notes in Math. vol. 170, Berlin 1970

KR 79 P. Kramer and M. Saraceno, Symplectic structures in many-body systems, Proc. Group Theor. Meth. Physics, Zvenigorod 1979

KR 80 P. Kramer and M. Saraceno, Geometry of the TDVP in quantum mechanics, preprint 1980

LI 65 H.J. Lipkin, N. Meshkov and A.J. Glick, Nuclear Physics 62 (1965) 188

VARIATIONAL CHARACTERIZATION OF HAMILTONIAN VECTORFIELDS AS LIE ALGEBRA OF CANONICAL DIFFEOMORPHISMS

Ernesto A. Lacomba *
Departamento de Matemáticas
Universidad Autónoma Metropolitana
Iztapalapa, Apdo. Postal 55-534
México 13, D.F. México.

Lucette Losco.
Laboratoire de Mécanique Théorique
Université de Besançon
25030 Besançon, France.

Reasoning by analogy to V.I. Arnold's characterization {1}, {3} of Euler equations in hydrodynamics by means of (infinite dimensional) groups of volume-preserving diffeomorphisms, we characterize from variational principles the Hamiltonian vectorfields in a compact manifold M with boundary, as Lie algebra of canonical diffeomorphisms on M. The momentum of the symmetry in this case is interpreted as the canonical one-form α defining the symplectic structure.

1. Variational principle in hydrodynamics.- We will begin by describing the variational principle which we will generalize later to apply to the Hamiltonian case.

Let M be a Riemannian manifold with boundary (usually a region in $\mathbb{R}^2$ or $\mathbb{R}^3$), which will be thought of as describing the initial positions of the fluid particles in an incompressible non viscous fluid at $t=0$.
Since the particles are confined to M, the position of the particles at any time t will be obtained by applying a volume-preserving diffeomorphism g_t: $M \to M$. (depending on t). Recall that a diffeomor-phism is a C^∞ one to one map, whose inverse is also C^∞. the term volume-preserving means that g_t locally preserves the volume element $d\mu$ associated to the Riemannian metric on M. Let G be the group of volume-preserving diffeomorphisms on M.

Then $x(t,x_0) = g_t(x_0)$ will be the trajectory actually followed by particle initially at x_0. Let $X(x,t)$ be the time-dependent vectorfield of velocities of the fluid particles on M. Such a

* Research partially supported by CONACYT (méxico), under grant PNCB170.

vectorfield is related to the curve of diffeomorphisms $g_t \in G$ by

$$\dot{g}_t(x) = \frac{d}{dt} g_t(x) = X(g_t(x),\ t),$$

or equivalently

$$X(x,t) = \dot{g}_t \circ g_t^{-1}(x).$$

We remark that this vectorfield must be tangent at ∂M, and have div $X = 0$ because of volume-preservation. The set $\mathcal{L}$ of vector fields satisfying these two conditions is actually the <u>Lie algebra</u> associated to the infinite dimensional group G.

The proposition below states the correspondence between the following two variational principles. It has already been exploited by Arnold {1},{2}, and later formalized by Ebin-Marsden {4} (see also Marsden{5, Sec. 4}). We denote by $|\ \ |$ the norm on tangent vectors, defined by the Riemannian metric.

The first variational principle is just a formulation of the principle of stationary action for the fluid flow on M. The second one would be its global form.

1) The trajectories $x(t, x_0)$ are extremals on M for the action functional

$$A = \int_{t_0}^{t_1} \left(\frac{|\dot{x}|^2}{2} - U(x,t)\right) dt,$$

where the potential U is the sum of an external potential V, plus the pressure P.

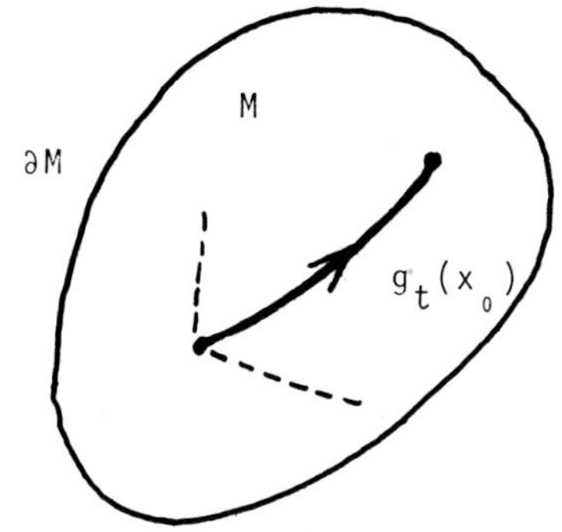

2) On G, the arc g_t describing the motion is an extremal of

$$\mathcal{A} = \int_{t_0}^{t_1} \int_M \frac{1}{2} \left| \dot{g}_t \circ g_t^{-1} \right|^2 d\mu \, dt.$$

The integral over M is the total kinetic energy of the fluid, so this is a global stationary action principle

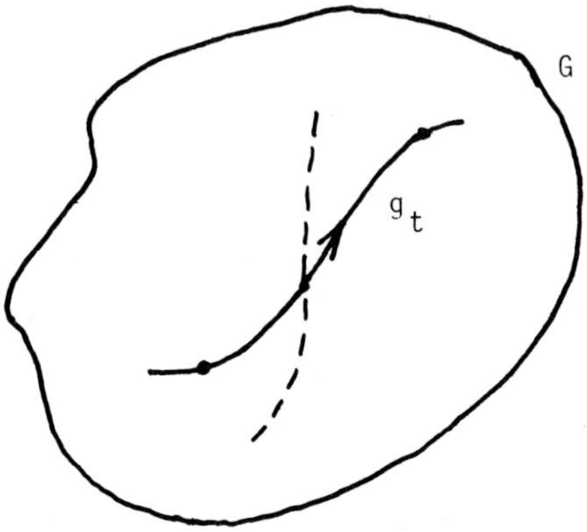

The arc g_t passes through the origin, but is not a one-parameter subgroup, unless X is time-independent.

Proposition (Arnold).- If 1) holds for the solutions of the vectorfield X, then 2) holds for its integral arc g_t. Conversely if 2) holds for g_t then there exists a function U(x,t) such that the solutions of $X = \dot{g}_t \circ g_t^{-1}$ on M are extremals of

$$A + \int_{t_0}^{t_1} U(x,t) \, dt.$$

2.- Generalization of our variational equivalence.

We generalize the proposition of §1 in the following theorem. From now on we use sumation convention for repeated indices.

Theorem 1.

A) If the solution curves of X are extremals on M of

$$A = \int_{t_0}^{t_1} f(x,x,t)\, dt,$$

then the integral curve $g_t \subset G$ of X is an extremal of

$$\mathcal{a} = \int_{t_0}^{t_1} \left(\int f(x, X(x,t), t)\, d\mu \right) dt.$$

B) Conversely, if $g_t \subset G$ is an extremal for $\mathcal{a}$, then there exists a function U: $M \times \mathbb{R} \to \mathbb{R}$ such that the solutions of $X = \dot{g}_t \circ g_t^{-1}$ are extremals on M of

$$A = \int_{t_0}^{t_1} (f(x,x,t) + U(x,t))\, dt.$$

In other words, the action A on M determines $\mathcal{a}$ on G, but the converse is true only modulo a function U(x,t). This corresponds to the fact that in the hydrodynamical case the fluid velocity at any given point may issue in different directions, corresponding to different choices of the extremizing arc $g_t \subset G$, that is, to different vector fields X.

The main idea in the proof of theorem 1 consists in computing the variation

$$\delta \mathcal{a} = \int_{t_0}^{t_1} \int_M \left(\frac{\partial f}{\partial x_i} - \frac{d}{dt} \frac{\partial f}{\partial \dot{x}_i} \right)_{\dot{x}=X} \delta g^i \, d\mu \, dt,$$

for a corresponding variation δg of g_t, together with the fact that any vector field Y = X + grad r where r is a function, and $X \in \mathfrak{g}$. We obtain the following result. Here i_X denotes interior product, and L_X is the Lie derivative associated to any vectorfield X {3}.

<u>Theorem 2</u>. The Euler - Lagrange equations for $\mathcal{a}$ are equivalent to the following 3 statements:

E1) $\beta = \left(\frac{\partial f}{\partial x_i} - \frac{d}{dt} \frac{\partial f}{\partial \dot{x}_i} \right)_{\dot{x} = X} dx_i$ is exact

E2) If $\alpha = \left. \frac{\partial f}{\partial \dot{x}_i} \right|_{\dot{x} = X} dx_i$, then $\frac{\partial \alpha}{\partial t} + i_X d\alpha$ is exact

E_3) If $\tilde{X} = X + \frac{\partial}{\partial t}$, then $L_{\tilde{X}}\,\alpha$ is exact.

For example, in the case of a fluid in $\mathbb{R}^3$, property E_2 is equivalent to Bernoulli equation,

$$\frac{\partial X}{\partial t} - X \times \text{rot } X = -\text{grad } h,$$

where $\alpha = X^i\, dx_i$, and h is the energy, $h(x,t) = \frac{X^i X^i}{2} + U$.

3. <u>Application to the canonical case</u>.- Let now M be a compact symplectic manifold with boundary with its Liouville volume element $d\mu = \omega^n$, where $\omega = d\alpha$ is its canonical 2-form. We consider the Lie algebra $\mathfrak{C}$ of Hamiltonian vectorfields X on M (therefore preserving $d\mu$) which are tangent to ∂M. We are now dealing with the subgroup $C \subset G$ of canonical diffeomorphisms of M, whose Lie algebra $\mathfrak{C}$ is a Lie subalgebra of $\mathfrak{g}$.

In canonical local coordinates we write $x = (q,p)$ and $\alpha = p^i dq_i$. To use the theory of § 2, we define $f(x,\dot{x}) = p^i\dot{q}_i$, so that

$f(x,X) = \alpha(X)$ for $X \in \mathfrak{C}$

Theorem 1 now says that the globally Hamiltonian vector fields on M are those whose integral arc is an extremal for

$$\mathfrak{a} = \int_{t_0}^{t_1} \int_M \alpha(X)\, d\mu\, dt,$$

while the corresponding trajectories on M are extremals of

$$A = \int_{t_0}^{t_1} (\alpha(\dot{x}) - H(x))\, dt.$$

But the lagrangian of A is $p^i\dot{q}_i - H(q,p)$, and we get just Hamilton equations for H (globally, $i_X\, d\alpha = -\,dH$):

$$\dot{q} = \frac{\partial H}{\partial p}, \quad \dot{p} = -\frac{\partial H}{\partial q}$$

The equivalent forms of Euler - Lagrange equations for the action in theorem 2 now read as following (the differential form α there, is in this case the above defined $\alpha = p^i dq$)

$E_1 : \beta = dH$

$E_2 : i_X d\alpha = -dH$ (since α is time-independent)

$E_3 : L_X\alpha = i_X d\alpha + di_X\alpha = d(\alpha(X) - H)$.

4. Momentum associated to the symmetry.- In this final section we consider again the general scheme of § 2, defining a momentum associated to the group G, and giving a corresponding Nöether's theorem

The idea comes from the situation for the motion a rigid body with a fixed point and no external forces {3}. If $\omega \in \mathbb{R}^3$ is the angular velocity in body coordinates, and a rotation $A \in SO(3)$ describes the position of the body, the kinetic energy T is defined by $2T = Q(\omega,\omega)$ where Q is a positive definite quadratic form (inertia tensor). Then the angular momentum in body coordinates is defined as $J = Q(\omega,\cdot) \in \mathbb{R}^{3*}$ naturally identified with a vector in $\mathbb{R}^3$. The angular momentum in space coordinates can be written as $J_e = AJ$, and Nöether's theorem states that $\dot{J}_e = 0$ along motions of the body.

Similarly, for fluid flow we just define

$$J_X(X') = \int_M \langle X, X' \rangle d\mu, \text{ for } X, X' \in \mathfrak{g}$$

where $\langle , \rangle$ denotes inner product in the Riemannian metric.

In general, we define the momentum J_X by.

$$J_X(X') = \int_M \alpha_X(X')d\mu, \text{ where } \alpha_X = \left.\frac{\partial f}{\partial \dot{x}_i}\right|_{\dot{x} = X} dx_i$$

Nöether's theorem goes as follows

Theorem 3.- For any X' invariant under the integral arc g_t of X, $J_X(X')$ is a first integral, that is, $\frac{d}{dt} J_{\dot{g}_t \circ g_t^{-1}}(X') = 0$.

For example, in the fluid case we need $\alpha_X = X^i dx_i$.

In the canonical case, we put

$$J_X(X') = \int_M \alpha(X')\, d\mu,$$

where $\alpha = \alpha_X = p^i\, dq_i$, so that J_X may be identified with α, for any X.

REFERENCES.

1. V.I.Arnold, Sur la géometriedifferentielle des groupes de Lie de dimension infinie et ses applications ā l'hydrodynamique des fluides parfaits, Ann, Inst.Founer, Grenoble 16 (1) (1966) 319-361.

2. V.I. Arnold, Sur un principe variationnel pour les écoulements stationnaires des liquides parfaits et ses applications aux problèmes de stabilité non lineáires, Journal de Mécanique 5(1)(1966) 29-43.

3. V.I. Arnold, Mathematical Methods of Classical Mechanics. Springer Verlag, New York, 1978.

4. D.G. Ebin & J. Marsden, Groups of diffeorphisms and the motion of an in compressible fluid, Ann. of Math. 92(1970) 102-163.

5. J. Marsden, Applications of Global Analysis in Mathematical Physics, Publishor Perish, Inc., Lecture Notes #3, Boston, Mass, 1974.

THE LIE THEORY OF EXTENDED GROUPS IN CLASSICAL MECHANICS - IS IT OF RELEVANCE TO QUANTUM MECHANICS?

P. G. L. Leach
Department of Applied Mathematics
La Trobe University, Bundoora 3083, Australia.

1. Time-Dependent Linear Transformations in Quantum Mechanics

Non-autonomous dynamical systems occur, for example, in the study of plasmas. The simplest of these is the time-dependent harmonic oscillator with Hamiltonian, in the usual notation,

$$H = \tfrac{1}{2} p^2 + \tfrac{1}{2} \omega^2(t) q^2 . \tag{1.1}$$

The germinal papers for the developments reported here were those of Lewis [1,2] and Lewis and Riesenfield [3]. They obtained an exact invariant for $H(1.1)$ and discussed the quantum mechanical aspects of the problem.

Lewis' approach was fairly sophisticated. A simpler method was to use the time-dependent linear canonical transformation

$$Q = \rho^{-1}(t)q, \qquad P = -\dot{\rho}(t)q + \rho(t)p, \tag{1.2}$$

where $\rho(t)$ is a solution of the auxiliary equation [4]

$$\ddot{\rho} + \omega^2\rho = \rho^{-3}, \tag{1.3}$$

to obtain the new Hamiltonian

$$\bar{H} = \tfrac{1}{2}\rho^{-2}(P^2 + Q^2). \tag{1.4}$$

The Schrödinger equation for the problem described by $\bar{H}(1.4)$ is readily solved and from it the solution to the equation corresponding to $H(1.1)$ is found to be

$$\psi_n(q,t) = |\rho|^{-\frac{1}{2}} \exp(\tfrac{1}{2} i\rho^{-1}\dot{\rho}\, q^2/\hbar)\, \psi_n'(\rho^{-1}q,T) , \quad T = \int^t \rho^{-2}(t')dt' \tag{1.5}$$

where $\psi_n'(Q,T)$ is the usual oscillator wave function and T is the new time measure. This type of result may be extended to more complicated linear transformations using the integral transforms described by Wolf [5,6].

The tool provided by time-dependent linear transformations was also useful in examining the symmetries of systems such as the multi-dimensional time-dependent oscillator [7] with which the group $SU(n)$ can be associated. Further work on quadratic Hamiltonians produced rather more symmetries, in terms of types of invariants, than appeared useful. It was also apparent that the type of time-dependent transformation which is permissible in quantum mechanics is restricted in its nature. This led to the work on classical systems which is described below. The first task was to examine classical symmetries. The second was to find out whether there is any sense in speaking of restrictions on the type of time-dependent linear transformations which should be permitted in classical mechanics.

2. The Lie Method of Extended Groups

There has been some recent work on the symmetries admitted by classical systems, especially linear systems [10, 11, 12]. The essential result was that such systems as the one-dimensional free particle and one-dimensional oscillator possessed eight generators of one-parameter point transformations forming the symmetry group $SL(3,\mathbb{R})$. Using a variety of methods, linear transformations [13], Lutzky's composite Noether's theorem/Lie method [14] and the Lie method of extended groups [15], the classical symmetry group of any one-dimensional linear system was found to be $SL(3,\mathbb{R})$.

In these analyses a preference for the Lie method emerged. It is best to define exactly what is meant by this method as different versions are used. The discussion is based on the Newtonian equations of motion. A parallel development has been given for Hamilton's equations of motion [16], but the former is used here as the inverse problem in mechanics is to be discussed below. Writing the generator of an infinitesimal transformation as

$$X(\underset{\sim}{q},t) = \xi(\underset{\sim}{q},t)\ \partial/\partial t + \underset{\sim}{\eta}(\underset{\sim}{q},t)^T\ \partial/\partial \underset{\sim}{q}\ , \tag{2.1}$$

the differential operator is a symmetry generator for the Newtonian system

$$\underset{\sim}{N}(\ddot{\underset{\sim}{q}},\ \dot{\underset{\sim}{q}},\ \underset{\sim}{q},\ t) = \underset{\sim}{0} \tag{2.2}$$

if, whenever (2.2) is satisfied,

$$X^{(2)}(\ddot{\underset{\sim}{q}},\dot{\underset{\sim}{q}},\underset{\sim}{q},t)\ N(\ddot{\underset{\sim}{q}},\dot{\underset{\sim}{q}},\underset{\sim}{q},t) = \underset{\sim}{0} \tag{2.3}$$

where $X^{(2)}$ is the second extension of X given by

$$X^{(2)} = X + (\dot{\underset{\sim}{\eta}} - \dot{\xi}\dot{\underset{\sim}{q}})^T\ \partial/\partial\dot{\underset{\sim}{q}} + (\ddot{\underset{\sim}{\eta}} - 2\dot{\xi}\ \ddot{\underset{\sim}{q}} - \ddot{\xi}\dot{\underset{\sim}{q}})^T\ \partial/\partial\ddot{\underset{\sim}{q}}\ . \tag{2.4}$$

The solution of the equations resulting from (2.3) with (2.2) taken into account is, when it exists, a linear combination of generators of one-parameter symmetry groups.

The preference for the Lie method of extended groups using point transformations only follows from several considerations. Firstly, in general the Lie method gives more generators than, say, Noether's theorem or, alternatively, allows increased freedom in the parameters of a particular problem, e.g. for the time-dependent anharmonic oscillator [17]. Secondly, because the transformations are point only, there is a finite number and the defining differential equations for their coefficients may be obtained by equating coefficients of powers of $\dot{q}$ to zero. On the other hand, if velocity-dependent transformations are admitted, there is an infinite number of generators and infinity is not really much better than zero. There has been some work in determining velocity-dependent transformations in a systematic way [18]. However, there does not seem to be any necessity to use them as all of the first integrals of the motion for a particular system may be obtained form point transformations. Furthermore velocity-dependent transformations generating first integrals can be derived very simply from the extended operators of the point transformations. This particular aspect of the theory is shortly to be reported by Prince [19]. This last point is the third consideration for preferring the version of the Lie method used here.

3. One-Dimensional Linear Systems

In §2 it was reported that all one-dimensional linear systems possess the symmetry group $SL(3,\mathbb{R})$. This group consists solely of generators of infinitesimal point transformations. Were velocity-dependent transformations to be taken into account, there would be an infinite number of generators all of which are related to the point transformations and commutation relations of which are determined by the symmetry group of the point transformations [19].

For a one-dimensional linear system the eight generators give rise to first integrals of the motion. Only two of these, the initial conditions, are functionally independent. A third integral, the ratio of the initial conditions, seems to play a rôle of equal importance to the initial condition integrals and in some sense is a "hidden symmetry" for the system. Starting from these three integrals, it is possible to work backwards to find the generators of the transformations under which they are invariant. To each integral there corresponds a triplet of generators. Of the total of nine generators only eight are linearly independent. Other constants of the motion are obtained by using generators which are suitable linear combinations of the nine. Each of the three triplets has the same set of commutation properties, viz.

$$[X_1,X_2] = X_2, \quad [X_1,X_3] = X_3, \quad [X_2,X_3] = 0. \tag{3.1}$$

The Lie algebra represented by the commutation relations in (3.1) is nonsemisimple since it contains an Abelian sub-algebra which is a proper ideal. The algebra does not appear to have any specific name, but is listed by Barut and Rączka[20] in the classes of solvable Lie algebras of dimension three. A feature of the nine operators taken together is that the commutation table is considerably simpler than the standard table for $SL(3,\mathbb{R})$. None of these triplets forms a proper ideal with respect to the nine generators [21]. An alternative approach to the relation of first integrals and their generators has been given by Prince [22].

4. Multi-dimensional Systems

In several recent papers devoted to multi-dimensional systems [14, 23, 24], the symmetry groups of infinitesimal point transformations have been shown to be $SL(n+2,\mathbb{R})$ for the isotropic oscillator (both autonomous and non-autonomous) and the free particle and $SO(4)$ for the Kepler problem. In the cases of the last problem, there is no further symmetry than that already known from other considerations. Indeed, often an autonomous non-linear system may possess only one generator [25] and a non-autonomous non-linear sytem none at all [7].

It follows that all n-dimensional quadratic Hamiltonians which are canonically related to the free particle will possess $SL(n+2,\mathbb{R})$ symmetry. There has been some doubt cast on whether this includes all quadratic Hamiltonians, but, for those which do, there is a fourth point to be considered. In general the linear canonical transformation from one quadratic Hamiltonian to another is not "point" but "velocity-dependent", e.g. the transformation between

$$H = \tfrac{1}{2}(p_1^2 + p_2^2) \tag{4.1}$$

$$\bar{H} = \tfrac{1}{2}(P_1^2 + Q_1^2 + P_2^2) \tag{4.2}$$

is given by

$$Q_1 = q_1 \cos t + p_1(\sin t - t\cos t),\; Q_2 = q_2 \tag{4.3}$$

$$P_1 = -q_1 \sin t + p_1 (\cos t - t\sin t),\; P_2 = p_2\,. \tag{4.4}$$

Using these transformations, the fifteen generators of an $SL(4,\mathbb{R})$ symmetry group for $\bar{H}(4.2)$ may be obtained from the generators for $H(4.1)$. However, eight of these generators are not generators of infinitesimal point transformations. Once velocity-dependent transformations are admitted, all possible of such transformations should be admitted with the effect objected to above. The point transformation symmetry group of $\bar{H}(4.2)$ contains only seven elements consisting of two sub-groups of dimensions four and three. The former corresponds to the initial conditions.

It should be noted that not all "velocity-dependent" linear canonical transformations have the effect of reducing the number of point transformation symmetries. A trivial example is found in the transformation from the free particle to the isotropic harmonic oscillator. However, in the example given above, it is apparent that the reduction in the number of point transformations is, in part at least, a reflection of a reduction in the geometric symmetry displayed by the problem.

5. The Inverse Problem in Classical Mechanics

The problem of the existence of a Lagrangian for a given Newtonian system has received some attention recently particularly in respect of linear systems [26]. Considering, by way of example, two-dimensional autonomous systems, in general there exist only seven generators of point symmetry transformations for the Newtonian system

$$\ddot{\underset{\sim}{x}} + A\underset{\sim}{x} = \underset{\sim}{0}. \tag{5.1}$$

This is the case whether (5.1) admits a Lagrangian or not. The group consists of two subgroups of dimensions four and three as noted above in connection with (4.2). Since the group structure is preserved under point transformation, it follows that systems which cannot be written in Lagrangian form either directly or after transformation will have a different group to those which can. Note that the Lie method does not distinguish between systems which are formally self-adjoint and systems which can be transformed into a formally self-adjoint form. A simple example is

$$A = \begin{bmatrix} 0 & 4 \\ 4 & 0 \end{bmatrix},\quad A = \begin{bmatrix} 0 & 1 \\ 16 & 0 \end{bmatrix},\quad \underset{\sim}{x} \to \underset{\sim}{x}' = \begin{bmatrix} 2 & 0 \\ 0 & \tfrac{1}{2} \end{bmatrix} \underset{\sim}{x}. \tag{5.2}$$

6. Conclusion

The question posed in the title of this paper may be answered with a qualified yes. The first qualification is that the Lie symmetries to be considered are only those arising from infinitesimal point transformations. The second qualification is that for autonomous systems, only those generators which commute with the time operator will produce first integral which will provide (possibly) useful quantum mechanical operators of similar utility to those already used to generate group such as *SU(3)* and *SO(4)*. A further use of the Lie method is to be found in identifying those Newtonian systems for which quantization via the Hamiltonian is possible.

References

[1] H. R. Lewis, Jr., Phys. Rev. Letters, 18, 510-512 (1967).
[2] H. R. Lewis, Jr., J. Math. Phys., 9, 1976-1986 (1968).
[3] H. R. Lewis Jr. and W. B. Riesenfeld, J. Math. Phys., 10, 1458-1473 (1969).
[4] C. J. Eliezer and A. Gray, SIAM J. Appl. Maths., 30, 463-468 (1976).
[5] K. B. Wolf, J. Math. Phys., 17, 601-613 (1976).
[6] K. B. Wolf, Integral Transforms in Science and Engineering (Plenum Press, New York, 1979), Part IV.
[7] N. J. Günther and P. G. L. Leach, J. Math. Phys., 18, 572-576 (1977).
[8] P. G. L. Leach, J. Math. Phys., 19, 446-451 (1978).
[9] P. G. L. Leach, J. Math. Phys., 21, 32-37 (1980).
[10] C. E. Wulfman and B. J. Wybourne, J. Phys. A, 9, 507-518 (1976).
[11] R. L. Anderson and S. M. Davison, J. Math. Anal. Applic., 48, 301-315 (1974).
[12] M. Lutzky, J. Phys. A, 11, 249-258 (1978).
[13] P. G. L. Leach, AM79:02 Research Report, Department of Applied Mathematics, La Trobe University.
[14] G. E. Prince and C. J. Eliezer, J. Phys. A, 13, 815-823 (1980).
[15] P. G. L. Leach, AM79:04 Research Report, Department of Applied Mathematics, La Trobe University.
[16] G. E. Prince and P. G. L. Leach, AM79:13 Research Report, Department of Applied Mathematics, La Trobe University.
[17] P. G. L. Leach, AM79:10 Research Report, Department of Applied Mathematics, La Trobe University.
[18] J. A. Kobussen, Research Report, Institut für Theoretische Physik der Universität Zürich, Schönberggasse 9, 8001, Zürich.
[19] G. E. Prince, La Trobe University, work in progress (private communication).
[20] A. O. Barut and R. Rączka, Theory of Group Representations and Applications. (Polish Scientific Publishers, Warsaw 1977), 49.
[21] P. G. L. Leach, AM80:02 Research Report, Department of Applied Mathematics, La Trobe University.
[22] G. E. Prince, AM80:03 Research Report, Department of Applied Mathematics, La Trobe University.
[23] G. E. Prince and C. J. Eliezer, AM79:06 Research Report, Department of Applied Mathematics, La Trobe University.
[24] P. G. L. Leach, AM80:01 Research Report, Department of Applied Mathematics, La Trobe University.
[25] P. G. L. Leach, AM79:12 Research Report, Department of Applied Mathematics, La Trobe University.
[26] L. Y. Bahar, Drexel University, and W. Sarlet, Rijksuniversiteit Gent, work in progress (private communication).

ON THE STRUCTURE OF PHASE SPACE

M. Moshinsky* and T. H. Seligman
Instituto de Física, UNAM
Apdo. Postal 20-364, México 20, D. F.

One of the fascinating aspects of mathematics and mathematical physics is the introduction of new concepts which initially have a simple formulation, but later become more and more subtle as their deeper meaning is understood better. Thus, for example, the concept of the complex plane envisaged at the beginning of the XIX century in a simple fashion, acquires through the work of Riemann and later Klein and Weyl a great subtlety. One can say that at the present time we have a complex manifold rather than plane, whose structure should allow us to carry out arbitrary conformal transformations in a bijective (i. e., one to one onto) fashion.

The above manifolds have been extensively discussed leading to what are now known as Riemann surfaces. Yet a similar problem in phase space -or more correctly in the phase plane for problems of one degree of freedom- has not received comparable attention. We can after all discuss in a somewhat parallel fashion[1-3)] the properties of the complex and phase planes as carriers, respectively, of conformal and canonical transformations. Thus the phase plane -and for more degrees of freedom the phase space- becomes a manifold with properties that resemble those of a Riemann surface[1-3)], and thus merits a more detailed analysis.

In previous references[1-3)] some of the analysis was carried out with the purpose of understanding better the representation of non-bijective canonical transformations in quantum mechanics. In the present note we are interested purely in the classical problem so as to focus more sharply on the properties of phase space. We stress in particular a duality in our characterization of observables in the phase plane: Under non-bijective canonical transformations we can either introduce many-sheeted structures and define the values of the observables in each sheet, or we can keep the idea of a single plane but give matrix form to the observables in such a way that they continue to form an algebra. The latter approach leads to the concepts of ambiguity group[1-3)] and ambiguity spin which are fundamental to our objectives.

The detailed discussion will be given in a paper under the same title, submitted for publication in the Journal of Mathematical Physics.

---

*Member of the Instituto Nacional de Investigaciones Nucleares and El Colegio Nacional.

1. P. Kramer, M. Moshinsky and T. H. Seligman, J. Math. Phys. 19, 683 (1978).
2. M. Moshinsky and T. H. Seligman, Ann. Phys. (N.Y.) 114, 243 (1978); 120, 402 (1979).
3. J. Deenen, M. Moshinsky and T. H. Seligman, Ann. Phys. (N.Y.) in press (1980).

FIBER BUNDLES IN NON-RELATIVISTIC QUANTUM MECHANICS: THE HYDROGEN ATOM

P. Moylan
The University of Texas at Austin
Austin, Texas 78712 U.S.A.

The problem of describing the Hydrogen atom by a fiber bundle is considered. The quantization of a fiber bundle is introduced. A fiber bundle for the Hydrogen atom is constructed. It is possible to do away with the potential by associating with the relative motion a fiber which is a curved space: the three sphere, S^3.

1. Introduction

Fiber bundles in physics are not new. The phase space of a system of n classical point particles in interaction is a fiber bundle over the configuration space of the center of mass of the system. The configuration space of a rigid body is a fiber bundle over the space of its translational degrees of freedom. Here we consider the possibility of constructing a fiber bundle for the hydrogen atom which will make clear the hidden O(4) symmetry of the problem [1], [2].

First we deal with necessary notions of fiber bundle theory. Then in section III, we consider the quantization of an arbitrary fiber bundle. In section IV, we give a fiber bundle description to the Hydrogen atom. Finally we present some concluding remarks.

2. Construction of a Fiber Bundle

In this paper, we are going to construct a fiber bundle over three-dimensional Euclidean space. All fiber bundles over contractible Euclidean space are trivializable and hence can be reduced to tensor products, a simplifying fact used throughout this discussion. For our purposes a fiber bundle is simply a triplet of objects: the total space E (a manifold), the base space B, and the fiber F. There is a group acting in the fiber called the structural group G. If $G \simeq F$ and G acts via left translation then the bundle is called a principle fiber bundle and denoted by P(B, G). If $F = G/G'$ then we can construct a fiber bundle associated to P(B, G) called the associated bundle and denoted by E(B, F, G, P) [3]. (G' is a subgroup of G.)

For the Hydrogen atom, the fiber bundle is $E = R^3 \times S^3$ with base space R^3 and fiber the three sphere, S^3. It can be chosen as an associated bundle with structural group SO(4).

3. Quantization of a Fiber Bundle

According to the axioms of quantum mechanics, a physical observable is represented by a linear operator which acts in a linear space (space of physical states) [4].

Unless the observable is to represent a nonclassical degree of freedom such as spin or isospin it must be an element of some extention algebra of the enveloping algebra [4] of the Heisenberg algebra or possibly a limit of such elements. For a fiber bundle we chose certain functions of 6n canonical variables $P_i^{(\alpha)}$ and $Q_i^{(\alpha)}$ for points in the fiber bundle, or functions of $-i\,\partial^{(\alpha)}/\partial x_i$ and $x_i^{(\alpha)}$ $(i = 1, 2, 3;\ \alpha = 1, 2, \ldots n)$ in the Schrödinger representation. In the case of the Hydrogen atom, the choice is dictated by our construction of a set of solutions to the Schrödinger equation for the problem. Points in the base space are chosen to be the x_i's of the center of mass, and points in the fiber are related to the momenta of the relative motion.

4. The Fiber Bundle Description of the Hydrogen Atom (Bound States)

We will associate with the Hydrogen atom a fiber bundle (only for the case of bound states). It will be a fiber bundle over the configuration space of the center of mass. Our analysis also applies to the scattering and zero energy states but we will not consider them here (see [2]).

Before we proceed with the fiber bundle construction, however, we find it necessary to describe a set of solutions to the bound state Schrödinger equation for the Hydrogen atom. This description goes back to Fock [1]. This particular set of solutions is also a set of eigenfunctions of the wave equation for a free particle moving in S^3.

The Schrödinger equation for the relative motion part of the Hydrogen atom is:

$$\left(\frac{\nabla^2}{2} - \frac{\alpha}{r}\right)\psi(\vec{r}) = E\psi(\vec{r}) \tag{1}$$

The wave equation for a free particle moving in S^3 is:

$$\Delta\psi(\alpha,\theta,\phi) = \left(\frac{\partial^2}{\partial\alpha^2} + 2\cot\alpha\,\frac{\partial}{\partial\alpha} + \frac{1}{\sin^2\alpha}\left\{\frac{\partial^2}{\partial\theta^2} + \cot\theta\,\frac{\partial}{\partial\theta} + \frac{1}{\sin^2\theta}\,\frac{\partial^2}{\partial\phi^2}\right\}\right)\psi(\alpha,\theta,\phi) = \lambda\,\psi(\alpha,\theta,\phi) \tag{2}$$

In order to demonstrate the unitary equivalence of a complete set of solutions of the above two equations we proceed as follows: first we write the Schrödinger equation in momentum space

$$\left(\frac{p^2}{2} - E\right)\hat{\phi}(\vec{p}) = \frac{\alpha}{2\pi^2}\int d^3p'\,\frac{\hat{\phi}(\vec{p}\,')}{|\vec{p}\,' - \vec{p}|^2} \tag{3}$$

[$\hat{\phi}(\vec{p})$ is the Fourier transform of $\psi(\vec{r})$]. Next let $p_0^2 = -2E$ so that Eqn. (3) becomes

$$(p^2 + p_0^2)\hat{\phi}(\vec{p}) = \frac{\alpha}{\pi^2}\int \frac{d^3p'\,\hat{\phi}(\vec{p}\,')}{|\vec{p}\,' - \vec{p}|^2} \qquad E < 0 \tag{4}$$

Now embed momentum space in R^4 as shown in Fig. (1). (Note that the momenta are scaled by a factor $\frac{1}{2}p_0$.) Let S^3 be the unit three sphere in R^4 centered at the origin. Next relate points in S^3 (minus the north pole) to points in momentum space by steriographic projection. Introducing

$$\hat{\phi}(\vec{\xi}) = (p_o)^{-\frac{1}{2}}\left[\frac{p_o^2 + p^2}{2p_o}\right]^2 \hat{\phi}(\vec{p}) \qquad , \quad \vec{\xi}\varepsilon S^3 \tag{5}$$

and

$$d\Omega_{S_3} = 2\delta(\xi^2 - 1)d^4\xi = \left(\frac{2p_o}{p_o^2 + p^2}\right)^3 d^3p \tag{6}$$

we have that Eqn. (4) becomes [2]

$$\hat{\phi}(\vec{\xi}) = \frac{\alpha}{2p_o\pi^2}\int \frac{\hat{\phi}(\vec{\xi}')}{|\vec{\xi} - \vec{\xi}'|^2}\, d\Omega'_{S_3} \qquad \xi,\ \xi' \ \varepsilon\ S_3 \tag{7}$$

This equation exhibits the O(4) symmetry of the problem.

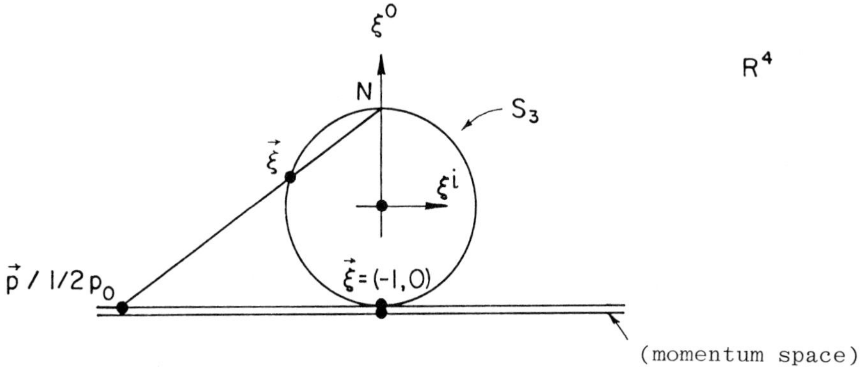

Fig. 1. The embedding of the fiber of the fiber bundle $E(R^3, S^3, SO(4))$ into R^4 and steriographic projection of the three-dimensional momentum space onto the unit sphere, $\vec{\xi}^2 = 1$, minus the north pole, N, in the Euclidean space, R^4.

Eqn. (7) should be compared with the equation satisfied by the 4-dim spherical harmonics [2]:

$$Y_{n\ell m}(\vec{\xi}) = \frac{n}{2\pi^2}\int \frac{d\Omega'_{S_3}}{|\vec{\xi} - \vec{\xi}'|^2}\, Y_{n\ell m}(\vec{\xi}') \tag{8}$$

A 4-dim spherical harmonic is defined to be $1/|\vec{r}|^{n-1}$ times a homogeneous polynomial $y_{n\ell m}(\vec{r})$ of degree n-1 in $\vec{r}\ \varepsilon\ R^4$, wherein also $y_{n\ell m}(\vec{r})$ is required to satisfy

$$\Delta^4 y_{n\ell m}(\vec{r}) = 0 \qquad (\Delta^4 \text{ is the 4-dim wave operator.}) \tag{9}$$

From Eqn. (9) it follows directly that the spherical harmonics of degree n-1 are eigenfunctions of the Laplace operator on S^3 (Eqn. (2)) with eigenvalue n^2-1:

$$\Delta Y_{n\ell m}(\vec{\xi}) = (n^2-1)Y_{n\ell m}(\vec{\xi}) \qquad n\varepsilon J^+ \quad .$$

This correspondence between solutions of the bound state Schrödinger equation and the eigenfunctions of the Laplacian on S^3 (given by the equations of steriographic projection and Eqn. (5)) induces a unitary isomorphism between $L^2(S^3)$ and the Hilbert space completion of the linear span of the bound state eigenfunctions of Eqn. (1) [1; Bander, Itzykson p. 332].

Now we are prepared to construct the fiber bundle for the Hydrogen atom. The Hydrogen atom is a two-particle system: two particles of masses m_1 and m_2 and corresponding momentum vectors $\vec{p}_1$ and $\vec{p}_2$ with position vectors $\vec{r}_1$ and $\vec{r}_2$. For the energy of the system we take

$$E = 2\left(\frac{(\vec{p}_1 + \vec{p}_2)^2}{2(m_1 + m_2)}\right) - \left(\frac{p_1{}^2}{2m_1} + \frac{p_2{}^2}{2m_2}\right) = \frac{P^2}{2M} - \frac{p^2}{2\mu} \tag{10}$$

($m = m_1 + m_2$ and $\mu = \dfrac{m_1 m_2}{m_1 + m_2}$).

The fiber bundle is

$$E(R^3,\ F = SO(4)/SO(3),\ G = SO(4),\ P) \simeq R^3 \times S^3 \tag{11}$$

We introduce a Riemannian metric on each fiber (the metric obtained by embedding S^3 as the unit three sphere in R^4). Then embed the fiber bundle in $R^3 \times R^4$: the base space being identified with R^3 and the fiber with the unit three sphere in R^4. Using the embedding we can introduce wave functions on the fiber bundle: a wave function on $R^3 \times S^3$ is the restriction to $R^3 \times S^3$ of a wave function on $R^3 \times R^4$.

Now we quantize the base space by identifying R^3 with the configuration space for the center of mass of the Hydrogen atom and letting $p_i \to -i\ \partial/\partial x_i = \bar{P}_i$ and $x_i \to x_i = \bar{Q}_i$. ($\bar{P}_i$ and $\bar{Q}_i$ are the operators of center of mass momentum and position.)

Next we require that the system satisfy the following two wave equations:

$$\frac{\nabla}{2M}\,\psi(\vec{x}, \vec{\xi}) = \frac{K}{2M}\,\psi(\vec{x}, \vec{\xi}) \qquad K \varepsilon R \tag{12}$$

$$\Delta\ \psi(\vec{x}, \vec{\xi}) = \lambda\ \psi(\vec{x}, \vec{\xi}) \qquad \lambda \varepsilon R \tag{13}$$

[$(\vec{x}, \vec{\xi})$ is a point in $R^3 \times S^3$, ∇ is the Laplacian on R^3 and Δ is the Laplacian on S^3.] The solutions of these equations that we take to describe the possible states of the Hydrogen atom are

$$\psi(\vec{x}, \vec{\xi}) = \phi_{k_1k_2k_3}(\vec{x})Y_{n\ell m}(\vec{\xi}) \tag{14}$$

$(k_1^2 + k_2^2 + k_3^2 \quad ; \quad \lambda = n^2-1)$.

Now we quantize the fiber. For a fixed n, let H_n be the subspace of $L_2(S^3)$ spanned by the spherical harmonics of degree n-1. Let $p_o = \alpha/n$ on H_n. Then embed the momentum space of the relative motion into R^4 as shown in Fig. 1. Relate coordinates on S^3 to momentum coordinates by steriographic projection as shown in Fig. 1. Further define wave functions on $L_2(R^3)$ in terms of wave functions on $L_2(S^3)$ as given by Eqn. (5). Under the mapping defined by steriographic projection and Eqn. (5), a $Y_{n\ell m}(\vec{\xi})$ corresponds unitarily to a solution $\hat{\phi}_{n\ell m}(\vec{p})$ of the Schrödinger equation in $L_2(R^3)$.

To calculate the energy of the Hydrogen atom, which is in the state $\phi_{k_1k_2k_3}(\vec{x})\, Y_{n\ell m}(\vec{\xi})$, we use the virial theorem. Let $\hat{\phi}_{n\ell m}(\vec{p})$ be the function in $L_2(R^3)$ which corresponds unitarily to $Y_{n\ell m}(\vec{\xi})$ under the above-mentioned isomorphism. For $\hat{\phi}_{n\ell m}(\vec{p})$ we have [1]

$$\int \hat{\phi}^*(\vec{p}) \left(\frac{\nabla^2}{2\mu} - \frac{\alpha}{r} \right) \hat{\phi}(\vec{p}) d^3p = - \int \frac{p^2}{2\mu} |\hat{\phi}(\vec{p})|^2 d^3p \tag{20}$$

Using this result and Eqn. (10), we obtain for the energy of the system

$$E = \frac{K^2}{2M} - \frac{\alpha^2}{2n^2} \mu c^2 \tag{21}$$

which is the usual non-relativistic expression for the energy of the nth bound state of the Hydrogen atom.

5. Concluding Remarks.

By associating a fiber bundle with the Hydrogen atom it was possible to eliminate the potential from the two-particle system. The method of constructing certain bases for irreducible representation spaces of SO(4) (the spherical harmonics) can be generalized to SO(n) and SO(n, 1). [Bander, Itzykson, Rev. Mod. Phys., 38, p. 330-358, (1966)].

References

[1] M. Bander, C. Itzykson, Rev. mod. Phys. 38, 330 (1966); J. Schwinger, J. Math Phys., 5, 1606 (1964); M. Engelfield, Group Theory and the Coulomb Problem, Wiley-Interscience, (1964); M. Bander, C. Itzykson, Rev. mod. Phys. 38, 346 (1966).

[2] P. Moylan, Forts. d. Phys. 28, 251-266 (1980).

[3] S. Kobayashi, K. Nomizu, Foundations of Differential Geometry, Vol. 1, Interscience (1963).

[4] A. Böhm, Quantum Mechanics, Springer-Verlag (1979).

QUANTIZATION AS MAPPING AND AS DEFORMATION

Jiří Niederle and Jiří Tolar
Institute of Physics,
Czechoslovak Academy of Sciences,
18040 Prague 8, Czechoslovakia.

Since the discussed topic already exists in a printed form (see [1],[2]) we present here only a brief summary.

In the paper, first, classes of quantizations as classes of maps from classical to quantum observables were discussed. More precisely the classes of all quantizations of polynomial observables for a system of one degree of freedom were classified. This is not too restrictive since the derived results can be generalized to several degrees of freedom in a straightforward way and since for an irreducible representation of Q and P by self-adjoint operators $\hat{Q}$ and $\hat{P}$ in $L^2(R)$ any self-adjoint operator in $L^2(R)$ can be approximated (in the sense of a strong resolvent convergence) by the self-adjoint closures of polynomials in $\hat{Q}$ and $\hat{P}$.

Secondly, the approach in which quantization was defined as deformation of the algebra of classical observables into the algebra of quantum observables was reviewed. Finally, a brief comparison of both quantization approaches was given. In particular, the *-products corresponding to the translationally invariant quantization maps were determined.

[1] Niederle J., in Proc. of the Colloquium on Mathematical Physics. Istanbul. 1979 (see also ICTP preprint IC/79/140, 1979).

[2] Niederle J., Tolar J., Czech. J. Phys. B29 (1979), 1358.

METRIC AND GROUP STRUCTURES FROM QUANTUM PHYSICS

J.G. Provost and G. Vallée
Physique Théorique, Université de Nice,
Parc Valrose, 06034 Nice Cedex, France.

1. Non gravitational examples of metric structures in physics.

It is a well established tradition in Physics to associate the concept of a metric with the theories of general relativity and gravitation. Apart from this field, the most favoured geometric structure is the symplectic one which plays an important role both in classical and in quantum mechanics. Yet, let us briefly recall that there exist some less known physical situations where a metric structure has been invoked from general arguments.

A first example comes from a remark by Sternberg [1] according to which the Euclidian structure of the ordinary space is connected with the dielectric properties of the vacuum. In this geometrical framework the counterparts of the electric vector field $\vec{E}$ and of the displacement vector field $\vec{D}$ are respectively a one-form E and a two-form D on R^3 and the usual equation $\vec{D} = \varepsilon_0 \vec{E}$ reads $D = \varepsilon_0 *E$. This later relation recuires the introduction of a $\star$-operation on forms and thus induces a metric structure on R^3.

As a second example we consider the geometrical description of classical mechanics where the kinetic energy $\frac{1}{2} m_{ik} \dot{q}_i \dot{q}_k$, written as $\frac{1}{2} \langle \dot{q},\dot{q}\rangle_q$ defines a metric structure on the tangent space $T_q(M)$ at the point q of the configuration space M. When M is a group G[2], it may happen that the metric be a left-invariant one; in that case $\langle \dot{g},\dot{g}\rangle_g = \langle g^{-1}\dot{g}, g^{-1}\dot{g}\rangle_e$ and, identifying $T_e(G)$ with the Lie algebra $\mathcal{G}$, one sees that the metric on G is entirely defined by the metric on $\mathcal{G}$. A well known illustration of this situation is given by the rotational motion of a solid body. The group G is SO(3) and $\omega = g^{-1}\dot{g} \in \mathcal{G}$ is the angular velocity vector. Then left-invariance means that the kinetic energy dependes only on the angular velocity ω. The metric is completely defined by the moment of inertia tensor which is, in this formalism, an operator from $\mathcal{G}$ to $\mathcal{G}^*$ such that $\langle \omega,\omega\rangle = (A\omega,\omega)$.

The last and (up to our knowledge) most recent example deals with equilibrium thermodynamics [3]. According to statistical mechanics, a system in contact with a fixed reservoir fluctuates from the most probable state parametrized by $s(s = s_1 \dots s_n)$ to a close state s'. The probability of this fluctuation is proportional to exp{S(s, s')} where S is the total change of the entropy of the system and of the reservoir; this probability may be interpreted as a measure of the two equilibrium states specified by s and s'. Expanding S up to second order:

$$S(s,s') = \left(\frac{\partial^2}{\partial s'_i \partial s'_j} S\right)_{s'=s} ds_i ds_j = g_{ij}(s) ds_i ds_j \qquad \left(\frac{\partial S}{\partial s'_j}\right|_{s'=s} = 0$$

from the second law of thermodynamics), one easily understands that the tensor components $g_{ij}(s)$ provide the manifold of equilibrium states with a physically reasonble Riemannian structure.

In this note, we want to show that a metric on a manifold M may be introduced in a natural and significant way in the case where M is a submanifold of some Hilbert space of quantum states [4]. If one takes the point of view that "classical" physics is a part of quantum physics, this case is not as peculiar as one could expect at first sight. More precisely the physical situations which we have in mind concern the classical collective behaviours of the macroscopic quantum systems; for example, "collective submanifolds" in Hilbert space have been introduced in nuclear physics [5] to extract the collective subdynamics from the manybody dynamics (treatment of nucleus deformations). In order to justify a little more our suggestion that the existence of a meaningful metric tensor may be invoked in the description of macroscopic systems, let us recell that, paradoxically, the distance between two quantum states has no physical interpretation in the ordinary quantum formalism; the physically relevant quantities are the transition probability amplitudes which are defined for any two states whatever be their relative distance. However for macroscopic systems these probability amplitudes may be thought to be zero as far as the macroscopic parameters specifying the two states are not infinitesimally related. As a consequence, we expect the metric properties to be essentially local properties of the manifold of macroscopic states.

Before we enter into some mathematical details, let us note that it is not a too drastic restriction to consider a manifold M equipped with a group action. Indeed, very often, the group structure is preserved when the procedure which leads from the many dimensional Hilbert space to the few-dimensional manifold M is performed. The group C may be the Heisenberg-Weyl group, the rotation group, the free motion group etc... Following Peremolov [6] we speak of generalized coherent states manifolds (hereafter noted G.C.S.M.) whenever the whole manifold is obtained form the action of G on some fixed collective state.

2. Geometrical structures on a manifold of quantum states.

Let $\{\psi(s)\}$ be a family of normalized state vectors of some Hilbert space which smoothly depend on an n-dimensional parameter $s = (s_1 \cdots s_n)$. The ordinary distance in Hilbert space induces a local metric on the family $\{\psi(s)\}$ through the relation

$$\|\psi(s+ds) - \psi(s)\|^2 = \left(\frac{\partial\psi}{\partial s_i}, \frac{\partial\psi}{\partial s_j}\right) ds_i ds_j = (\gamma_{ij} + i\sigma_{ij}) ds_i ds_j = \gamma_{ij} ds_i ds_j$$

we emphasize that the metric tensor γ so defined is meaningless as a metric tensor on a manifold of quantum states. Indeed, as far as the phase of a vector state is not an observable quantity [7], the physical states are represented by rays $\tilde{\psi}(s)$ of the Hilbert space. Identifying our manifold M with the family $\{\tilde{\psi}(s)\}$, one sees that the correct metric should be deduced from the distance D in the projective Hilbert space. One gets:

$$D^2(\psi(s+ds), \psi(s)) = g_{ij}(s)\, ds_i\, ds_j = dl^2$$

with $g_{ij} = \gamma_{ij} - \beta_i \beta_j$ ($\beta_i = \mathrm{Im}(\psi, \partial\psi/\partial s_i)$). In other words g is invariant under the "Gauge transformations" $\psi(s) \rightarrow e^{i\alpha(s)}\psi(s)$ on state vectors.

We recall that the usual symplectic structure is defined by the 2-form σ:

$$\sigma = \sigma_{ij}(s)\, ds_i \wedge ds_j \qquad (\sigma = d\beta\ ;\ \beta = \beta_i(s) ds_i)$$

In practice g_{ij} and σ_{ij} are easily obtained from the expression of the scalar product $(\psi(s), \psi(s'))$ of any two states of the family $\{\psi(s)\}$. The most trivial example is that of quantum states deduced from one another by translations in ordinary space. Let P_i ($i = 1, 2, 3$) be the generators of the translation group R^3, and let ψ_0 be some definite state, one obtains:

$$(\psi(x), \psi(x')) = (\psi_0, e^{-iP(x-x')}\psi_0)\ ;\ g_{ij} = (\psi_0, P_i P_j \psi_0) - (\psi_0, P_i \psi_0)(\psi_0, P_j \psi_0)$$

Some classical examples (coherent states, atomic coherent states, etc...) where the manifold is a G.C.S.M. can be found in reference [4]. Others such as the family of eigenvectors of some parameter dependent Hamiltonian H_λ or examples from field theory may be also considered. They all lead to interesting physical discussion.

3.- Curvature of G.C.S.M.

A G.C.S.M. may be defined as a set of rays $\{\widetilde{U(g)\psi_0}\}$ with U a unitary representation of a group G and ψ_0 a fixed ray in the representation space. We denote by small letters the elements of the Lie algebra $\mathcal{G}$ and by capital letters the corresponding operators in the representation. Let $K \subset G$, be the isotropy subgroup of $\tilde{\psi}_0$ and $\mathcal{K}$ be the corresponding subalgebra of $\mathcal{G}$. We introduce a decomposition $\mathcal{G} = \mathcal{K} \oplus \mathcal{X}$. The notation of section 2 is then recovered by identifying the manifold in the vicinity of ψ_0 with the family $\{\widetilde{\psi(s)}\}$:

$$\widetilde{\psi(s)} = \widetilde{e^{is_j x_j}\psi_0} \qquad (x_j \in \mathcal{X}\ ;\ j = 1 \cdots n)$$

From the group structure of the manifold, it is easily that the metric defined in section 2 is a left-invariant metric. At the origin $\tilde{\psi}_0$ ($s = 0$), the metric components are found to be:

$$g_{ij}(0) = \mathrm{Re}\,(\psi_0, x_i x_j \psi_0) - (\psi_0, x_i \psi_0)(\psi_0, x_j \psi_0)$$

This formula may be used to define a pseudometric on $\mathcal{G}$ in the following way. For any two elements a and $b \in \mathcal{G}$ we set:

$$\langle a, b \rangle = \mathrm{Re}\,(\psi_0, AB\,\psi_0) - (\psi_0, A\,\psi_0)(\psi_0, B\,\psi_0)$$

(In particular: $\langle x_i, x_j\rangle = g_{ij}(0)$). With respect to this scalar product $\langle\ ,\ \rangle$ the elements k of K are isotropic vectors in G ($\langle a, k\rangle = 0\ \forall\ a \in G$).

Arnold [2] has given an explicit formula for the curvature of a Lie group equipped with a left-invariant metric. This formula can be extended in a simple way to the case of homogeneous spaces [4]. The result is the following. Let $\{\hat{x}_i\}$ be an orthonormal basis in X ($\langle\hat{x}_i, \hat{x}_j\rangle = \delta_{ij}$) and let $[\hat{x}_i, \hat{x}_j] = i\ (\sum_l f^l_{ij}\ \hat{x}_l + K_{ij})$ be the commutator of any two basis vectors ($k_{ij} \in K$ and f^l_{ij} structural constants of G). The bisectional curvature in the 2-dimensional direction $\{\hat{x}_i, \hat{x}_j\}$ is

$$C_{\hat{x}_i\hat{x}_j} = Q_{ij}(f) - \frac{1}{2}\ (\langle i[\hat{x}_j, K_{ij}], \hat{x}_i\rangle - \langle i[\hat{x}_i, K_{ij}], \hat{x}_j\rangle)$$

where Q_{ij} is Arnold's quadratic function of the structural constants:

$$Q_{ij}(f) = \sum_l \left(\frac{1}{4}\ (f^i_{jl} + f^j_{il})^2 + \frac{1}{2}\ f_{ij}(f^i_{jl} - f^j_{il}) - \frac{3}{4}\ (f^l_{ij})^2 - f^i_{il}\ f^j_{jl}\right)$$

The scalar curvature, which is coordinate independent, is the sum of the n(n-1)/2 bisectional curvatures.

As non trivial examples we consider two G.C.S.M. which are three-dimensional. The algebras G are generated by the operators I (identity), x (Galilean transformation), p (space translation), H (time translation) with the two cases: $H = 1/2\ (x^2 + p^2)$ (M is an "harmonic oscillator manifold"); $H = \frac{p^2}{2}$ (M is a "free motion manifold"). Let $\hat{x}_1 = a_1\ x$; $\hat{x}_2 = a_2\ (p + \alpha_1\ x)$; $\hat{x}_3 = a_3\ (H + \alpha_2 x + \alpha_3 p)$ be an orthonormal basis in X. A straightforward calculation leads to the respective results for the scalar curvature of the harmonic oscillator and free motion manifolds:

$$C_{HO} = -\ a_3^2\ (\alpha_1^2 + \frac{1}{4}\ (\frac{a_2}{a_1}\ (1 + \alpha_1^2) - \frac{a_1}{a_2})^2)\ ;\ C_F = -\ \frac{1}{4}\ \frac{a_2^2\ a_3^2}{a_1^2}$$

An interesting remark concerns the possibility of getting a zero curvature. Paradoxically the free motion case (as long as one forbids unnormalized states) is not "free" of curvature whereas the interacting case provides a solution $\alpha_1 = 0$ and $\frac{a_1}{a_2} = 1$ which reads: $\mathrm{Re}\langle xp\rangle - \langle x\rangle\langle p\rangle = 0$ and $\langle x^2\rangle - \langle x\rangle^2 = \langle p^2\rangle - \langle p\rangle^2$. Classical text books tell us that these relations corresponds to the case where wave packets do not spread.

The above result reminds us that the metric on the manifold M has retained some information about the underlying quantum structure. As exemplified by the formula giving g_{ij}, this information concerns the quantum fluctuations (a situation to be compared with that encountered in thermodynamics): if Δx_i is the fluctuation of the quantum operator x_i which generates the variation Δs_i of the parameter s_i, the distance between two nearby states reads: $dl = (\Delta x_i)^{-1}\ ds_i$. Therefore the non spreading conditions considered above are clearly sufficient conditions for zero curvature. More generally, this suggests that the scalar curvature should be used to define a "collective" dispersion for a non commutative algebra of quantum operators.

References.

[1] In "Differential Geometrical Methods in Mathematical Physics II". Proceedings, Bonn 1977, Springer Verlag, Berlin-Heidelberg-New York, 1978.

[2] V.I. Arnold, Mathematical Methods of Classical Mechanics, Springer, New York (1978).

[3] G. Ruppeiner, Phys. Rev. A20, 1608-1613 (1979).

[4] Riemannian Structure on Manifolds of Quantum States, J.P. Provost and G. Vallée, to appear in Comm. Math. Phys.

[5] P.G. Reinhard and K. Goeke, Phys. Rev. C20, 1546-1559 (1979) and references therein. B.G. Giraund and D.J. Rowe, Journal de Physique, Lettre 8, 177-180 (1979).

[6] A.M. Peremolov, Comm. Math. Phys. 26, 222-236 (1972).

[7] J.P. Provost, F. Rocca and G. Vallée, Ann. Phys. (New York) 94 , 307-319 (1975).

SPONTANEOUS SYMMETRY BREAKING AND RESTRICTED SUPERPOSITION PRINCIPLE

A. Rieckers
Institut für Theoretische Physik
der Universität Tübingen
D - 7400 Tübingen, Germany

We want to discuss here some aspects of infinite many body systems and start - as is usual nowadays - by introducing the quasilocal C^*-algebra[1]. Let $\mathcal{R}$ be the abelian group $\mathbb{R}^3$ or $\mathbb{Z}^3$ and $\mathcal{L}$ be the set of all bounded open subsets of $\mathcal{R}$. For every $\Lambda \in \mathcal{L}$ the systemic algebra of observables $\mathcal{A}(\Lambda)$ is C^*-isomorphic to an algebra $\mathcal{B}(\mathcal{H}(\Lambda))$ where $\mathcal{H}(\Lambda)$ is a separable Hilbert space associated to Λ. Defining $\mathcal{A} := \overline{\bigcup_{\Lambda \in \mathcal{L}} \mathcal{A}(\Lambda)}^{\,\text{norm}}$ one can show that the quasilocal algebra $\mathcal{A}$ has a unit, is simple, and has a trivial center. The new features of infinite many body systems are intimately connected with the fact that $\mathcal{A}$ is a so-called antiliminal algebra, which has type II and III representations[2].

For every φ in the state space $\mathcal{S}(\mathcal{A})$ we introduce the GNS-triple $(\pi_\varphi, \mathcal{H}_\varphi, \Omega_\varphi)$. The universal envelopping von Neumann algebra is then $\mathcal{A}_u := (\Sigma_{\varphi \in \mathcal{S}} \pi_\varphi(\mathcal{A}))''$. The center $\mathcal{Z}(\mathcal{A}_u)$ is very large and contains the classical observables of the many body system. For every $C \in \mathcal{A}$ we introduce the state transformation $\varphi \to \varphi_C$ where

$$\langle \varphi_C ; A \rangle := \begin{cases} \langle \varphi ; C^* A C \rangle / \langle \varphi ; C^* C \rangle & \text{if defined} \\ \langle \varphi ; A \rangle & \text{otherwise} \end{cases} .$$

φ_C can be interpreted as a modification of the state φ induced by the local perturbation C.

Since every $\varphi \in \mathcal{S}(\mathcal{A})$ may be canonically identified with a normal state $\tilde{\varphi} \in \mathcal{S}_n(\mathcal{A}_u)$ [3] the following definition makes sense.

1. <u>Definition:</u> The support of $\varphi \in \mathcal{S}(\mathcal{A})$ is given by

$$S_\varphi := \inf \{ P \in \mathcal{A}_u \, ; \, P = P^2 = P^* , \, \tilde{\varphi}_P = \tilde{\varphi} \}$$

The central support of φ is defined as

$$C_\varphi := \inf \{ P \in \mathcal{Z}(\mathcal{A}_u) ; \, P = P^2 = P^* , \, \tilde{\varphi}_P = \tilde{\varphi} \}$$

A filter is usually a device to select states according to certain properties, especially according to the values which certain observables have in these states. S_φ may be viewed as the smallest filter which φ passes unchanged and C_φ as the smallest of those filters which can be constructed by means of classical observables. The states for which S_φ is minimal in $\mathcal{A}_u$ are just the pure states $\mathcal{S}_p(\mathcal{A})$ and the states for which C_φ is minimal in $Z(\mathcal{A}_u)$ are the factor states $\mathcal{S}_f(\mathcal{A})$, in which all classical observables are dispersion free. By means of the equivalence relation $\varphi \approx \psi$, given by $C_\varphi = C_\psi$, we decompose $\mathcal{S}_f(\mathcal{A})$ into classes $\mathcal{F}_\varphi$, φ being a representative of the class. If $\varphi \not\approx \psi$ then $C_\varphi \wedge C_\psi = 0$ ($\wedge$, $\vee$ denote the infimum resp. supremum of two projections).

2. Definition: A folium $\mathcal{F} \subset \mathcal{S}(\mathcal{A})$ is a norm-closed, convex subset of states, which is invariant under all mappings $\varphi \to \varphi_C$, $C \in \mathcal{A}$.

3. Proposition: $\mathcal{F}_\varphi$ is the smallest folium containing $\varphi \in \mathcal{S}_f(\mathcal{A})$. The type of $\pi_\varphi(\mathcal{A})''$ is an invariant of $\mathcal{F}_\varphi$. A many body system has over-countably many folia of type II and III. These folia contain neither pure nor extremal states.

Since our infinite quantum systems have no canonical Hilbert space representation we have beside other things to extract and reformulate the essence of Dirac's coherence principle[4] in a space-independent way. The relation that a state vector can be decomposed linearily into two other state vectors is a symmetric one in all three state vectors involved. If the state vectors lie in one and the same superselection sector, all three of them describe pure states. Since for infinite systems the pure states play no longer such a fundamental role we propose in accordance with [5] the following generalization of the coherence relation:

4. Definition: A triple of states $\varphi_1, \varphi_2, \varphi_3 \in \mathcal{S}(\mathcal{A})$ satisfies the symmetric coherence relation $K(\varphi_1, \varphi_2, \varphi_3)$, if

$$\left.\begin{aligned} S_{\varphi_i} \wedge S_{\varphi_j} &= 0 \\ S_{\varphi_i} \vee S_{\varphi_j} &= P \end{aligned}\right\} \quad i \neq j , \; 1 \leq i,j \leq 3$$

where P is some projection in $\mathcal{A}_u$.

The basic result for the discussion of the K-relation is the following criterium, the proof of which will be given in [6].

5. Theorem: For $\varphi_1, \varphi_2 \in \mathcal{S}(\mathcal{A})$ exists a $\varphi_3 \in \mathcal{S}(\mathcal{A})$ such that $K(\varphi_1, \varphi_2, \varphi_3)$ is valid, iff $S_{\varphi_1} \wedge S_{\varphi_2} = 0$ and $S_{\varphi_1} = U S_{\varphi_2} U^*$ for a unitary $U \in \mathcal{A}_U$.

Various aspects are combined in the following result.

6. Theorem: $K(\varphi_1, \varphi_2, \varphi_3)$ implies $C_{\varphi_i} = P$ for i = 1,2,3 and some $P \in \mathcal{A}_U$, and for factor states the K-relation can be fulfilled only within one folium. In a type III folium $S_{\varphi_1} \wedge S_{\varphi_2} = 0$ is already sufficient for $K(\varphi_1, \varphi_2, \varphi_3)$ with a certain φ_3.

A lot of more details will be given in [6]. Here we want to apply these considerations to the case of spontaneous symmetry breaking in thermodynamical equilibrium (≡ KMS-[1]) states. The typical situation is as follows: Assume that the net of local Gibbs states $\{\varphi_\beta(\Lambda); \Lambda \in \mathcal{L}\}$ belonging to the fixed temperature β converges in the w^*-topology to the state φ_β. Assume, furthermore, that there is a locally compact symmetry group G with the following properties: G acts bijectively in $\mathcal{L}$ such that $\Lambda \subset \Lambda'$ implies $\Lambda_g \subset \Lambda'_g, g \in G$. G acts bijectively in $\mathcal{A}$ by means of Jordan*- isomorphismus γ_g, $g \in G$, the restrictions of which to $\mathcal{A}(\Lambda)$ inducing the mappings

$$\gamma_g(\Lambda_g, \Lambda): \mathcal{A}(\Lambda) \xrightarrow{\text{onto}} \mathcal{A}(\Lambda_g).$$

If for all $\Lambda \in \mathcal{L}$ and all $A \in \mathcal{A}(\Lambda)$ we have

$$\langle \varphi_\beta(\Lambda_g); \gamma_g(\Lambda_g, \Lambda)[A] \rangle = \langle \varphi_\beta(\Lambda); A \rangle$$ then

$\varphi_\beta = \varphi_\beta \circ \gamma_g$; so that the limiting Gibbs state exhibits the full symmetry. For β being greater than some transition temperature β_0 it may happen, that φ_β is not factorial but has a non-trivial decomposition

$$\varphi_\beta = \int_{\boxminus} \varphi_{\beta,\lambda} \, d\mu(\lambda)$$

into factor states $\varphi_{\beta,\lambda}$, where λ is taken from some index set $\boxminus$ and μ is a probability measure on $\boxminus$. The extremal KMS-states $\varphi_{\beta,\lambda}$ describe the family of pure phases at the temperature β and in general transform as $\varphi_{\beta,\lambda} \circ \gamma_g = \varphi_{\beta,\lambda'}$, with $\lambda' =: g(\lambda) \neq \lambda$ if g is not taken from the reduced symmetry group $H \subset G$. For $\lambda \neq \lambda'$ one has for the corresponding central supports $C_{\beta\lambda} \wedge C_{\beta\lambda'} = 0$ and thus $\mathcal{F}_{\beta\lambda} \cap \mathcal{F}_{\beta\lambda'} = 0$ for the related folia, so that different phases are in fact macroscopically (classically) distinguishable. If G contains a gauge symmetry group which is broken for $\beta > \beta_0$, i.e.

not contained in H, then the ordered phases show new quasi-particles. The center $\mathcal{Z}(\pi_\beta(\mathcal{A})'')$ of the GNS representation corresponding to φ_β has just the $C_{\beta\lambda}$, $\lambda \in \Theta$, as atomic projections.* This illustrates the fact that spontaneous symmetry breaking leads to new classical observables and new restrictions for the superposition principle by the appearance of the disjoint folia $\mathcal{F}_{\beta,\lambda}$, $\lambda \in \Theta$. But even within one and the same folium $\mathcal{F}_{\beta\lambda}$ - which is of type III for finite β - the superposition principle has to be thoroughly reconsidered: If one adheres to the usual opinion that only pure states can be coherently superposed then the possibility of superposition breaks down completely in $\mathcal{F}_{\beta,\lambda}$ because of Proposition 3. If one finds arguments for a coherent superposition of local excitations in a pure phase or even of quasi-particle states then Definition 4 comes into play in its full generality and Theorem 6 opens up a wide area of coherence phenomena in many body physics.

1) G.G. Emch: "Algebraic Methods in Statistical Mechanics and Quantum Field Theory", J. Wiley & Sons, New York (1972)

2) G.K. Pedersen: "C*-Algebras and their Automorphism Groups", Academic Press, London (1979)

3) M. Takesaki: "Theory of Operator Algebras", Springer, New York (1979)

4) P.A.M. Dirac: "The Principles of Quantum Mechanics", Oxford at the Clarendon Press (1930)

5) E. Chen: J. Math. Phys. 14, 1462 (1973)

6) G. Raggio and A. Rieckers: "Coherence in Algebraic Quantum Mechanics", in preparation

* If the measure $d\mu(\lambda)$ is continuous then there are no atomic central projections in the strict sense but only in the sense that all central observables are integrals over the $C_{\beta\lambda}$.

MANY-BODY QUANTUM MECHANICS AS SYMPLECTIC DYNAMICS

David J. Rowe
Department of Physics
University of Toronto
Toronto, Canada M5S 1A7

1. Introduction

I present a formulation of many-body quantum mechanics as a Hamiltonian dynamics on a symplectic manifold developed in collaboration with G. Rosensteel and A. Ryman[1]. The motivation derives from the inadequacy of low dimensional truncated Hilbert spaces to describe some physical problems of interest and from the observation that much superior descriptions can emerge by taking slices of the Hilbert space in different ways. We have in mind large amplitude collective phenomena, like fission or heavy-ion scattering reactions, which one cannot hope to describe, for example, in the shell model but for which one achieves some limited success by restricting the dynamics to the space of Slater determinants. The latter space is undoubtedly a drastic suppression of the many-nucleon degrees of freedom and, as a consequence, has been subjected to much criticism. However, if one regards the constrained dynamics as a semi-classical approximation, one can subsequently hope to regain some of the suppressed quantal effects, e.g. barrier penetration by a fission fragment, by requantization. Indeed, we have recently shown that requantization of the manifold of Slater determinants (the Grassman manifold) leads to an exact coherent state representation of many-fermion quantum mechanics[2]. It seems worthwhile therefore to study in some detail the algebraic and geometric structures of such formulations of many-body quantum mechanics.

The conventional setting for many-body quantum mechanics is a Hilbert space $\mathbb{H}$, a complex linear space with a hermitian inner product $\langle\psi|\phi\rangle$. The dynamics is given by the Schrodinger equation, $H\psi = i\dot{\psi}$. On the other hand the fundamental space for the formulation as symplectic dynamics is the projective Hilbert space $P\mathbb{H}$, the space of all states in $\mathbb{H}$ of unit norm together with an equivalence relationship $\psi \equiv e^{i\delta}\psi$ which identifies states differing only in phase. This space is equipped with a symplectic form σ, an anti-symmetric bilinear form on the tangent vectors of $P\mathbb{H}$. σ is related to the hermitian inner product on $\mathbb{H}$ by

$$\sigma(X\psi, Y\psi) = -2\,\mathcal{I}m\langle X\psi|Y\psi\rangle, \tag{1}$$

where $X\psi$ and $Y\psi$ denote tangent vectors to $P\mathbb{H}$ expressible in terms of some coordinates (x^{α}) by

$$X\psi = X^{\alpha}\partial\psi/\partial x^{\alpha}. \tag{2}$$

We discuss the motivation for this definition shortly. The dynamics on $P\mathbb{H}$ is given by Hamilton equations of motion. Thus if F is a smooth function on $P\mathbb{H}$, representing some observable property of the system, and if $\mathcal{H}$ is the energy function, the time derivative of F is given by its Poisson bracket with $\mathcal{H}$,

$$\dot{F} = \{\mathcal{H},F\}, \tag{3}$$

where the Poisson bracket will be defined shortly in terms of the symplectic form σ.

Approximate Hilbert space dynamics are obtained by restricting the Schrodinger equation to a sub-Hilbert space $\mathbb{H}_0 \subset \mathbb{H}$, whereas approximate symplectic dynamics are obtained by restricting Hamilton's equations to a symplectic submanifold $M \subset P\mathbb{H}$. Now whereas there is always a symplectic manifold $P\mathbb{H}_0 \subset P\mathbb{H}$ associated with every $\mathbb{H}_0 \subset \mathbb{H}$, the converse is not true. There does not, in general, correspond a Hilbert subspace $\mathbb{H}_0 \subset \mathbb{H}$ to an arbitrary symplectic submanifold $M \subset P\mathbb{H}$. Thus the symplectic dynamical approach gives extra flexibility in formulating approximate quantal theories. The major merit of the latter, however, is that it has strong classical associations and is amenable to the exploitation of physical insight in the construction of physically reasonable manifolds.

2. Hamiltonian dynamics on $M \subset P\mathbb{H}$

The starting point for the formulation of constrained quantal dynamics is Dirac's extremal condition of an action integral

$$\delta\int_{t_1}^{t_2} L(\psi,\dot{\psi})dt = \delta\int_{t_1}^{t_2} \langle\psi(t)|\, i\frac{\partial}{\partial t} - H|\psi(t)\rangle dt = 0. \tag{4}$$

The corresponding Lagrange equation yields immediately

$$d\mathcal{H}(\psi) = -2\, Im\langle d\psi|\dot{\psi}\rangle \tag{5}$$

where $\mathcal{H}$ is the energy function on M

$$\mathcal{H}(\psi) = \langle\psi|H|\psi\rangle \tag{6}$$

and

$$\dot{\psi} = \frac{d\psi}{dt} \tag{7}$$

is tangent to M. In terms of a coordinate chart, $\dot{\psi} = \dot{x}^\nu \partial\psi/\partial x^\nu$ and eq.(5) becomes

$$\frac{\partial\mathcal{H}}{\partial x^\mu} = \sigma_{\mu\nu}\dot{x}^\nu \tag{8}$$

where

$$\sigma_{\mu\nu} = -2\ Im < \frac{\partial\psi}{\partial x^{\mu}} | \frac{\partial\psi}{\partial x^{\nu}} > = -\sigma_{\nu\mu} \ . \tag{9}$$

This expression for σ is seen to correspond to the coordinate independent expression (1). It can also be written explicitly as the two-form

$$\sigma = \frac{1}{2} \sigma_{\mu\nu}\ dx^{\mu} \wedge dx^{\nu} \tag{10}$$

which is seen to be the exterior derivative, $\sigma = d\omega$, of the one-form

$$\omega = Im\ \frac{\partial\psi}{\partial x^{\mu}} |\psi> \ . \tag{11}$$

Thus we observe that an arbitrary $M \subset PH$ has an exact, hence closed, two-form σ.

The question now arises as to whether or not eq.(8) defines dynamical paths $\psi(t)$ through any point $\psi \in M$. THe answer is evidently yes provided that the matrix $(\sigma_{\mu\nu})$ can be inverted in eq.(8) to give

$$\dot{x}^{\mu} = \sigma^{\mu\nu} \frac{\partial\mathcal{H}}{\partial x^{\nu}} , \tag{12}$$

where $\sigma^{\mu\nu} = (\sigma^{-1})_{\mu\nu}$. If $(\sigma_{\mu\nu})$ can be inverted, we say that the two-form σ is non-degenerate and, since it is closed, it endows M with the structure of a symplectic manifold. (Note that we shall assume throughout that our Hilbert spaces are of indefinitely large but nevertheless finite dimensions in order to avoid any pitfalls associated with infinite dimensional manifolds).

We conclude then that Dirac's action integral has extremal paths on $M \subset PH$, and hence defines a Hamilton dynamics on M, if and only if M is symplectic.

We now define the Poisson bracket, which gives a coordinate independent expression of Hamilton's equations of motion. If F is any smooth function on M, which may be thought of as some observable property of the system that evolves in time, $F(\psi(t))$, as the system evolves, then its time derivative is given by

$$\dot{F} = \frac{\partial F}{\partial x^{\mu}} \dot{x}^{\mu} = \frac{\partial F}{\partial x^{\mu}} \sigma^{\mu\nu} \frac{\partial\mathcal{H}}{\partial x^{\nu}} \ . \tag{13}$$

Thus if the Poisson bracket $\{F,G\}$ of any two smooth functions F and G is defined

$$\{F,G\} = \frac{\partial F}{\partial x^{\mu}} \sigma^{\nu\mu} \frac{\partial G}{\partial x^{\nu}} , \tag{14}$$

we obtain the familiar Poisson bracket expression of Hamilton's equations

$$\dot{F} = \{\mathcal{H}, F\}. \tag{15}$$

Finally, observe that the above equations can always be expressed to canonical form by an appropriate choice of coordinates. According to Darboux's theorem, there always exists canonical coordinates $(x^\nu) = (p_1, p_2, \ldots, q^1, q^2, \ldots)$ about any point of a symplectic manifold having the property that

$$\begin{aligned} \{q^i, q^j\} &= \{p_i, p_j\} = 0 \\ \{p_i, q^j\} &= \delta_i^j . \end{aligned} \tag{16}$$

It follows therefore, from the definition of the Poisson bracket, that in terms of such coordinates $(\sigma^{\mu\nu})$ takes the canonical form

$$(\sigma^{\mu\nu}) = \begin{pmatrix} 0 & -I \\ I & 0 \end{pmatrix} \tag{17}$$

and hence that Hamilton's equations become

$$\dot{q}^i = \frac{\partial \mathcal{H}}{\partial p_i}, \quad \dot{p}_i = -\frac{\partial \mathcal{H}}{\partial q^i}. \tag{18}$$

The Poisson bracket assumes the familiar form

$$\{F,G\} = \frac{\partial F}{\partial p_\alpha}\frac{\partial G}{\partial q^\alpha} - \frac{\partial F}{\partial q^\alpha}\frac{\partial G}{\partial p_\alpha}. \tag{19}$$

3. Practical applications

There are four primary uses to which the above formulation can be put.

(i) *The calculation of an approximate ground state*

A point $\psi \in M$ is evidently stationary with respect to the Hamiltonian dynamics if and only if $d\mathcal{H} = 0$; i.e. if the energy is stationary

$$\delta\langle\psi|H|\psi\rangle = 0 . \tag{20}$$

This is the standard variational equation for a stationary state which is known to be particularly good when the state not only makes $\mathcal{H}$ stationary but also a minimum. The state in question is then the variational approximation to the ground state. The familiar example is the Hartree-Fock approximation in which the energy is minimized on the Grassman manifold. The search for the minimum is particularly simple in this situation since the Hartree self-consistent field method applies. However, the variational approach is evidently much more general and minimum can be found using the Newton-Kantorovic method for finding the zeroes of $d\mathcal{H}^{(3)}$.

(ii) *A normal mode theory of excited states*

It is known that the small amplitude normal mode solutions of the time-dependent Hartree-Fock equations can be interpreted in a way known as the random phase approximation to give a theory of excitation energies and transition matrix elements. (References and a review of the RPA are given, for example, in my book[(4)]). We shall show in the following that the normal mode solutions on any symplectic manifold can be interpreted in a parallel manner.

(iii) *Large amplitude collective motion*

One of our prime objectives in investigating symplectic dynamics in the quantal context is to facilitate the description of large amplitude collective phenomena. The problem here is to reduce the dimensionality of the many-body problem to a size that is computationally tractable and at the same time physically plausible. For example, suppose one wanted to describe the scattering of a ^{40}Ca ion by a ^{208}Pb target. One might start with a state

$$\Psi(t = 0) = \Phi^{40}(r_i - R_1) \quad \Phi^{208}(r_j - R_2)$$

on the 248 particle Grassman manifold and let it evolve according to the Hamilton equations of motion. This is the time-dependent Hartree-Fock approximation which, in the previous section, we have shown can be generalized to an arbitrary symplectic manifold. (A useful recent review of the applications of TDHF methods in nuclear physics is given in the proceedings of the TDHF workshop of Paris 1979[(5)]).

(iv) *Extraction of a collective Hilbert space*

Considerable effort has been expended in recent years in examining the time-dependent Hartree-Fock equations to see if large amplitude collective motions can reasonably be constrained to much lower dimensional collective submanifolds than the Grassman manifold (6). In this way one hopes to be able to requantize the collective submanifold and regain a fully quantal Hilbert space formulation of large amplitude collective motion, as illustrated in the following diagram:

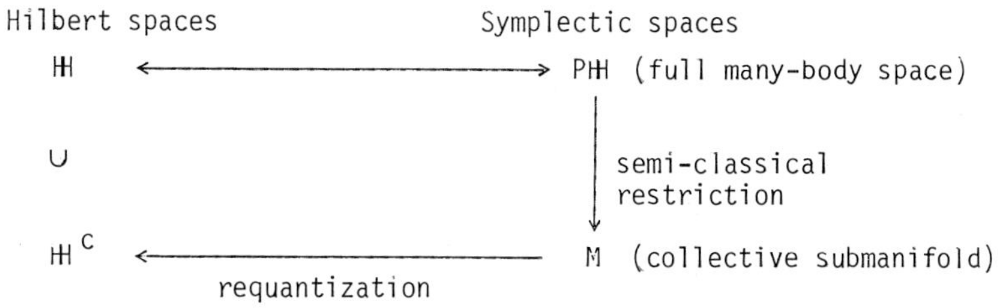

Our objective therefore is to obtain the most general and rigorous formulation of this procedure. Most of the effort to date by nuclear physicists has been to extract the collective submanifold M. The problem of requantizing an arbitrary symplectic manifold has been the subject of intense recent interest by mathematical physicists[7].

4. Normal mode theory of excited states

Let $\Psi_o \in ℍ$ be the ground state of the system with energy eigenvalue E_o and let $\Psi_\lambda \in ℍ$ be an excited state of energy E_λ. We can then construct the time-dependent state

$$\Psi(t) = \Psi_o e^{-iE_o t} + \delta\Psi_\lambda e^{-iE_\lambda t} \equiv \Psi_o + \delta\Psi_\lambda \, e^{-i(E_\lambda - E_o)t} \tag{21}$$

in Pℍ, where δ is a small parameter and a term of second order in δ is neglected. $\Psi(t)$ evidently represents a small amplitude normal mode vibration of the system about its ground state equilibrium configuration. For if F is any observable, its value on the time evolving path $\Psi(t)$ is given to leading order by

$$\begin{aligned} F(t) &= \langle\Psi(t)\,|F|\,\Psi(t)\rangle \\ &= F_o + \delta[F_{\lambda o}\, e^{i\omega_\lambda t} + \text{c.c.}] + \ldots \end{aligned} \tag{22}$$

where

$$\begin{aligned} F_{\lambda o} &= \langle\Psi_\lambda|F|\Psi_o\rangle \\ \omega_\lambda &= E_\lambda = E_o \,. \end{aligned} \tag{23}$$

On the other hand, the energy is constant in time and given by

$$\mathcal{H}(t) = \langle\Psi(t)\,|H|\,\Psi(t)\rangle = E_o + \delta^2\omega_\lambda \,. \tag{24}$$

Approximate theories of excited states are now obtained by looking at the normal mode solutions of Hamilton's equations on an arbitrary symplectic submanifold $M \subset PH$ and extracting transition matrix elements $F_{\lambda o}$ and excitation energies ω_λ from the expressions obtained for $F(t)$.

One proceeds as follows: Let $(p_1, p_2,\ldots,q^1, q^2,\ldots)$ be locally canonical coordinates about the energy minimum of a symplectic submanifold $M \subset PH$. Furthermore, for convenience of notation, choose the origin of the coordinate chart to be the energy minimum. The energy $\mathcal{H}$ can then be expressed as a function of these coordinates

$$\mathcal{H}(p,p) = \langle\Psi(p,q)|H|\Psi(p,q)\rangle \ . \tag{25}$$

We assume that $\mathcal{H}$ can be expanded and by a linear canonical transformation of coordinates brought to the normal form

$$\mathcal{H}(p,q) = \mathcal{H}_o + \frac{1}{2}\sum_\alpha (B_\alpha^{-1}\, p_\alpha p_\alpha + C_\alpha q^\alpha q^\alpha) + \ldots \ . \tag{26}$$

(When the Hessian $D^2\mathcal{H}$ is positive definite, as it is at an energy minimum, it is well-known that it always has this normal form[8]). The canonical coordinates which bring the Hessian to this normal form are called normal coordinates as in the standard elementary theory. They give rise to the familiar small amplitude solutions to Hamilton's equations

$$q^\alpha = \varepsilon_\alpha \cos \omega_\alpha t \ , \qquad p_\alpha = -\varepsilon_\alpha \omega_\alpha B_\alpha \sin \omega_\alpha t \tag{27}$$

with

$$\omega_\alpha^2 = C_\alpha / B_\alpha > 0 \ . \tag{28}$$

The normal mode solutions are those for which the infinitesimal amplitudes ε_α vanish for all but a single normal mode; i.e.

$$\varepsilon_\alpha = \varepsilon \delta_{\alpha\lambda} \ . \tag{29}$$

We then find, for such a mode,

$$\begin{aligned} F(t) &= F_o + \frac{\partial F}{\partial q^\alpha} q^\alpha(t) + \frac{\partial F}{\partial p_\alpha} p_\alpha(t) + \ldots \\ &= F_o + \frac{1}{2}\varepsilon\,[(\frac{\partial F}{\partial q^\lambda} + i\omega_\lambda B_\lambda \frac{\partial F}{\partial p_\lambda})e^{i\omega_\lambda t} + \text{c.c.}] + \ldots \end{aligned} \tag{30}$$

$$\mathcal{H}(t) = \mathcal{H}_o + \frac{1}{2}\varepsilon^2\omega_\lambda^2 B_\lambda . \tag{31}$$

Comparison with eqs. (22) and (24) respectively finally gives a generalization of the random phase approximation results; viz.

$$E_\lambda - E_o = \omega_\lambda \tag{32}$$

$$F_{\lambda o} = \frac{1}{\sqrt{2\omega_\lambda B_\lambda}} \left(\frac{\partial F}{\partial q^\lambda} + i\omega_\lambda B_\lambda \frac{\partial F}{\partial p_\lambda} \right). \tag{33}$$

It is of interest to note that this theory predicts the energy and ground state transition matrix elements for an excited state Ψ_λ without ever constructing the state. Indeed, one can readily ascertain by considering examples, that the excited state need not lie either on the manifold M or on one of its tangent spaces. Illustrative examples will be given in the following sections.

This theory of excited states generalizes and geometrizes an equations-of motion formalism proposed some years ago(9) .

5. Manifolds of coherent states

Of particular interest are the manifolds of coherent states. We follow the definitions of Perelomov(10) and Onofri(11) .

Let G be a Lie group acting unitarily on $\mathbb{H}$. Suppose that $\psi_o \in \mathbb{H}$ is a normalized highest (or lowest) weight state of this action. The G orbit in $P\mathbb{H}$ containing ψ_o is defined as the set of coherent states

$$O(\psi_o) = \{g\cdot\psi_o;\ g\in G\}. \tag{34}$$

The little group H of the orbit is the subgroup of elements of G that leave ψ_o fixed up to phase equivalence; i.e.

$$H = \{h\in G;\ h\cdot\psi_o \equiv \psi_o\}. \tag{35}$$

As Onofri has shown, $O(\psi_o) \sim G/H$ and $O(\psi_o)$ is symplectic.

(Note that an arbitrary group orbit is not in general symplectic. However, as S. Sternberg and B. Kostant have pointed out to us, those orbits which contain weight vectors other than those of highest or lowest weight are also symplectic.)

We now consider two examples and give a useful theorem.

(i) *Glauber coherent states*

Let $\mathbb{H} = L^2(\mathbb{R})$ be the Hilbert space of square integrable functions on the real line and G the Heisenberg-Weyl group with Lie algebra spanned by the basis $(b, b^\dagger, I)$ where

$$[b, b^\dagger] = I \tag{36}$$

etc. Let ψ_0 be the lowest weight state ($b\psi_0 = 0$). The Glauber coherent states[12] are then defined

$$\psi(z) = \exp(zb^\dagger - z^*b)\psi_0 \tag{37}$$

where z is any complex number

$$z = \frac{1}{\sqrt{2}}(q + ip). \tag{38}$$

One ascertains that (p,q) are canonical coordinates. Now for the harmonic oscillator Hamiltonian $H = \omega b^\dagger b$ one obtains

$$\mathcal{H}(p,q) = \frac{1}{2}\omega(p^2 + q^2) \tag{39}$$

and

$$F(p,q) = F_0 + \frac{1}{\sqrt{2}} q \langle\psi_0| F,(b^\dagger - b) |\psi_0\rangle$$

$$+ \frac{i}{\sqrt{2}} p \langle\psi_0| F,(b^\dagger + b) |\psi_0\rangle + \dots \quad . \tag{40}$$

Thus from the normal mode theory one obtains

$$E_1 - E_0 = \omega$$

$$F_{10} = \langle\psi_0|[b,F]|\psi_0\rangle \tag{41}$$

in full accord with the expected results for the harmonic oscillator.

From Hamilton's equations one also obtains the general results

$$\dot{q} = \frac{\partial\mathcal{H}}{\partial p} = \omega p \ , \quad \dot{p} = -\frac{\partial\mathcal{H}}{\partial q} = -\omega q \tag{42}$$

which imply

$$\dot{z} = -i\omega z \tag{43}$$

and hence the familiar Glauber result

$$\psi(t) \equiv \exp(ze^{-i\omega t}\, b^\dagger - z^* e^{i\omega t}\, b)\, \psi_o . \tag{44}$$

(ii) *A useful theorem*

If G is a Lie group with unitary action on $\mathbb{H}$, M is a symplectic orbit of G in $P\mathbb{H}$,

$$\underset{\sim}{g}_{\mathbb{C}} = \underset{\sim}{g} \oplus i\underset{\sim}{g} \tag{45}$$

is the complexified Lie algebra of G and $\mathcal{F}$ is a function defined on M for some $F \in \underset{\sim}{g}_{\mathbb{C}}$ by

$$\mathcal{F}(\psi) = \langle\psi|F|\psi\rangle, \tag{46}$$

then,

$$\dot{\mathcal{F}}(\psi) = i\langle\psi|[H,F]|\psi\rangle . \tag{47}$$

The proof of this theorem is given in ref. (1). It is obviously true for any observable F for the unconstrained dynamics on $P\mathbb{H}$ when it is simply the Heisenberg equation of motion. However, it is false for arbitrary F when M is a proper submanifold of $P\mathbb{H}$. The significance of this theory is that it shows the time derivative of any observable belonging to the complexified Lie algebra to be the same in the constrained as in the unconstrained dynamics. This is of particular import for Hartree-Fock theory where M is the orbit of the group of one-body unitary transformations whose complexified Lie algebra contains all one-body operators.

(iii) *The Grassman manifold (Hartree-Fock approximation)*

Let $\mathbb{H}^{(1)}$ denote the Hilbert space of a single fermion. The N-fermion Hilbert space $\mathbb{H}^{(N)}$ is then the exterior product of N copies of $\mathbb{H}^{(1)}$, i.e.

$$\mathbb{H}^{(N)} = \mathbb{H}^{(1)} \wedge \mathbb{H}^{(1)} \wedge \cdots \wedge \mathbb{H}^{(1)} \quad \text{(N copies)}. \tag{48}$$

An N-fermion Slater determinant is a simple state in $\mathbb{H}^{(N)}$ of the form

$$\Psi_{\alpha\beta\ldots} = \psi_\alpha \wedge \psi_\beta \wedge \cdots \quad . \tag{49}$$

The set of Slater determinants span $\mathbb{H}^{(N)}$, however not every state in $\mathbb{H}^{(N)}$ is a Slater determinant. In fact the set of all normalized Slater determinants, together with the usual phase equivalence relation, constitute a hypersurface in $\mathbb{H}^{(N)}$ known as the Grassman manifold and denoted $Gr_N(\mathbb{H}^{(1)})$.

Let $U(\mathbb{H}^{(1)})$ be the group of all unitary transformations of $\mathbb{H}^{(1)}$. Then $U(\mathbb{H}^{(1)})$ has a natural action on $Gr_N(\mathbb{H}^{(1)})$ given by

$$T(g)\Psi_{\alpha\beta\ldots} = (g\psi_\alpha)\ (g\psi_\beta)\ \ldots,\ g \in U(\mathbb{H}^{(1)}). \tag{50}$$

Furthermore, one can readily show that $U(\mathbb{H}^{(1)})$ acts transitively on $Gr_N(\mathbb{H}^{(1)})$, that $Gr_N(\mathbb{H}^{(1)})$ contains a lowest weight vector and hence that it is symplectic. Thus the above symplectic dynamical formulation of quantum mechanics applies.

The construction of coordinate charts for $Gr_N(\mathbb{H}^{(1)})$ and the detailed application of the formalism is discussed in ref. (1). Here we mention only some of the principal results. The stationary points of the energy function

$$d\mathcal{H} = 0 \tag{51}$$

give precisely the familiar Hartree-Fock equations. The normal mode theory reproduces the standard theory known as the 'random phase approximation' which has been widely and successfully used in the theory of collective and other excitations of many-body systems. The general Hamilton equations are the standard time-dependent Hartree-Fock equations[4]. These equations are expressed in a particularly convenient density matrix form by means of the above 'useful theorem'. The density matrix $(\rho_{\mu\nu})$ is defined for each $\Psi \in Gr_N(\mathbb{H}^{(1)})$ by

$$\rho_{\mu\nu} = \langle\Psi|\ X_{\nu\mu}\ |\Psi\rangle \tag{52}$$

where $(X_{\nu\mu})$ is a basis for $U(\mathbb{H}^{(1)})$. According to the theorem, it then follows that

$$\dot{\rho}_{\mu\nu} = i\ \langle\Psi|[H,X_{\nu\mu}]\,|\Psi\rangle\ . \tag{53}$$

This expression of the time-dependent Hartree-Fock dynamics is particularly useful because of a well-known diffeomorphism between the space of such densities and the Grassman manifold[1].

These expressions are of course all very familiar in the Hartree-Fock context. We mention them here primarily to illustrate that they are special cases of a more general formalism.

6. Pre-symplectic manifolds

The symplectic form (10) is evidently well defined on any submanifold $M \subset P\mathbb{H}$. However, in general, it is degenerate. Thus an arbitrary submanifold $M \subset P\mathbb{H}$ is pre-symplectic rather than symplectic and we have learned that Dirac's variational principle does not define a dynamics on a non-symplectic manifold. The question arises then as to whether or not one can find some way to regain a well-defined dynamics.

If M is a non-symplectic group orbit then M can be projected onto an associated co-adjoint orbit of the group which by a theorem of Kostant and Souriau [7] is known to be symplectic. The problem then remains to define the energy function on the co-adjoint orbit in order to define a dynamics on it. Such a dynamics in turn induces a dynamics on the pre-symplectic orbit M modulo the fibre of the projection. This approach and its possible application to physical problems is presented in the contribution to this conference by G. Rosensteel.

Another approach is to augment the pre-symplectic manifold M to a symplectic co-vector bundle. Consider, for example, the extreme situation in which the symplectic form σ is identically zero on M. M is then a 'configuration' space whereas its cotangent bundle is a symplectic 'phase' space[8]. In the general case, where σ is degenerate but not zero, it is sufficient to augment M with a subspace of the cotangent space at each point in order to make it symplectic. This approach has much in common with the Dirac procedure for handling singular Lagrangians[13]. In the event that M is a pre-symplectic group orbit, M can also be augmented to a symplectic manifold in a natural and closely related way simply by complexification of the group. (This observation was made to us by S. Sternberg and B. Kostant).

7. Vector bundle orbits

It should be recognized that the points of a symplectic manifold M do not have to be vectors in the Hilbert space $\mathbb{H}$ or in its associated projective space $P\mathbb{H}$. They might, for example, be Hilbert subspaces of $\mathbb{H}$. To be specific, recall that the nuclear shell model space V is a finite dimensional subspace of $\mathbb{H}^{(N)}$ and the question arises as to whether or not there is an optional way to choose V. The question is answered by considering M to be the set of possible shell model spaces.

Let $U(\mathbb{H}^{(1)})$ be the group of one-body unitary transformations which acts on $\mathbb{H}^{(N)}$ as detailed in section 5, and let $\Psi(g) = T(g)\Psi$ denote the transform of Ψ by $g \in U(\mathbb{H}^{(1)})$. Then if V is a Hilbert subspace $V \subset \mathbb{H}^{(N)}$, we define the transform $V(g)$ of V

$$V(g) = \{\Psi(g) \; ; \; \psi \in V\}. \tag{54}$$

In the shell model, V is defined by a partition of the one-body Hilbert space $\mathbb{H}^{(1)}$ into a direct sum of occupied, valence and unoccupied single-particle spaces; viz.

$$\mathbb{H}^{(1)} = \mathbb{H}^{(1)}_{occ} \oplus \mathbb{H}^{(1)}_{val} \oplus \mathbb{H}^{(1)}_{unocc}. \tag{55}$$

$\mathbb{H}^{(1)}_{occ}$ includes all single-particle states referred to as closed-shell states. If $\mathbb{H}^{(1)}_{occ}$ has dimension A, there is a unique way (up to phase equivalence) to construct a normalized A-fermion exterior product of A single-particle states in $\mathbb{H}^{(1)}_{occ}$. We call this state the closed-shell state Φ_{cs}. Now for an N-fermion system, where $0 < (N-A) < \dim \mathbb{H}^{(1)}_{val\ 1}$ we define the shell model space V as the span of all states of the form

$$\Psi = \Phi_{cs} \wedge \psi_1 \wedge \psi_2 \wedge \cdots \wedge \psi_{N-A} \tag{56}$$

where ψ_i, i=1,...,N-A are single-particle states in $\mathbb{H}^{(1)}_{val}$.

Now let M be the $U(\mathbb{H}^{(1)})$ orbit containing V; i.e.

$$M = \{V(g)\ ;\ g \in U(\mathbb{H}^{(1)})\}. \tag{57}$$

One observes by inspection that the little group is

$$H = U(\mathbb{H}^{(1)}_{occ}) \times U(\mathbb{H}^{(1)}_{val}) \times U(\mathbb{H}^{(1)}_{unocc}) \tag{58}$$

and hence that

$$M \sim U(\mathbb{H}^{(1)})/[U(\mathbb{H}^{(1)}_{occ}) \times U(\mathbb{H}^{(1)}_{val}) \times U(\mathbb{H}^{(1)}_{unocc})] \tag{59}$$

which one can ascertain is symplectic.

To define a Hamiltonian dynamics on M we first of all need an energy function. There are several possibilities. The simplest is to define $\mathcal{H}(V)$ to be the average energy of states $\psi \in V$; i.e.

$$\mathcal{H}(V) = \underset{\psi \in V}{\mathrm{Av}} \langle\psi|H|\psi\rangle . \tag{60}$$

Another is the minimum energy

$$\mathcal{H}(V) = \underset{\psi \in V}{\mathrm{Min}} \langle\psi|H|\psi\rangle. \tag{61}$$

For a finite temperature system, one might even consider taking a weighted average over a Boltzman distribution. However, once a definition has been made, the variational principle

$$d\mathcal{H} = 0, \tag{62}$$

and in particular, energy minimization, defines an optimal choice of the shell model space V. Thus the physical criteria for what one means by 'optimal' go into the definition of the energy function.

In addition to stationary states on M one may also consider the normal modes. With the minimum energy choice of the energy function, one obtains a theory of excited states known as the 'open shell random phase approximation'[14].

8. Dequantization and requantization

THe process of restricting the Hamiltonian dynamics to a symplectic submanifold $M \in P\mathbb{H}$ may be regarded as a process of dequantization. For example, the manifold of Glauber coherent states (cf. §5(i)) is known to be diffeomorphic to classical phase space. Thus we have a natural route from quantal to classical mechanics, in which the restriction of the dynamics is regarded as a suppression or partial suppression of some of the quantal dispersion effects. Recall that, for a simple harmonic oscillator, a Glauber coherent state evolves in time without any dispersion (i.e. spreading) of its wave packet and thus exhibits classical behaviour. An anharmonic oscillator similarly evolves without dispersion and exhibits classical behaviour if constrained to do so by restriction of its dynamics to the manifold of coherent states. In a similar way, one may regard the Hartree-Fock restriction of many-fermion dynamics to the Grassman manifold as a suppression of dispersion into non-determinantal modes.

The question arises as to whether or not the process of restriction (dequantization) can be reversed by requantization. I am not aware that a complete answer to this question is known although there is much interest in the problems of quantizing symplectic manifolds[(7)]. The best studied examples are the manifolds of coherent states.

In the case of a symplectic orbit of an irreducible unitary group action, one starts with a phase-space realization of the group defined by a transitive action on a symplectic manifold. The problem of requantization is now essentially the problem of reconstructing the corresponding unitary irreducible representation. For the manifold of Glauber coherent states, the unirrep. of the Heisenberg-Weyl group is carried by the Hilbert space of entire analytic functions (Bargmann-Segal space[(15)]). The corresponding quantum mechanics is known as the Bargmann-Segal coherent state representation. In a similar way, one can requantize the N-fermion Grassman manifold $Gr_N(\mathbb{H}^{(1)})$. The general construction follows that outlined by Onofri[(11)].

Since the set of Slater determinants span the many-fermion Hilbert space, one is not surprised to discover that requantization of the Grassman manifold regains the unconstrained many-fermion quantum mechanics. In general, however, requantization of an arbitrary symplectic submanifold $M \subset P\mathbb{H}$ gives back only a sub-Hilbert space quantum mechanics. Part of the hope then is that one can use one's classical and semi-classical intuition to construct low dimensional symplectic spaces of relevance for problems of collective motions such as fission and heavy-ion reactions which one can subsequently requantize to obtain a fully quantal but solvable Hilbert space theory of collective dynamics.

References

(1) D.J. Rowe, A. Ryman and G. Rosensteel, 'Many-body quantum mechanics as a symplectic dynamical system' (Physical Review, to be published).

(2) D.J. Rowe and A. Ryman, 'Coherent state representation of many-fermion quantum mechanics' (Physical Review Letters, to be published).

(3) D.J. Rowe, Nucl. Phys. A107 (1968) 99;
F. F. Catara, M. Di Toro, E. Pace and G. Schiffrer, Nuovo Cim. 11A (1972) 733;
G. Fonte, R. Mignani and G. Schiffrer, Commun. Math. Phys. 33 (1973) 293.

(4) D.J. Rowe, 'Nuclear Collective Motion' (Methuen, London, 1970).

(5) P. Bouche, B. Giraud and Ph. Quentin (eds), 'Time-Dependent Hartree Fock Method' (Orsay-Saclay, 1979).

(6) D.J. Rowe and R. Basserman, Nucl. Phys. A220 (1974) 404;
Can. Journ. Phys. 54 (1976) 1941;
G. Holzwarth and T. Yukawa, Nucl. Phys. A219 (1974) 125;
F. Villars, Nucl. Phys. A285 (1977) 269;
T. Marumori, Prog. Theor. Phys. 57 (1977) 112;
K. Goeke, P.-G. Reinhard, Ann. Phys. 112 (1978) 328;
M. Baranger and M. Veneroni, Ann. Phys. 114 (1978) 123.

(7) B. Kostant, 'Lecture notes in mathematics' (Springer, Berlin 1970);
J.-M. Souriau, 'Structure des systemes dynamiques' (Dunod, Paris, 1970);
D.J. Simms and N.M.J. Woodhouse, 'Lectures on geometric quantization' (Springer-Verlag, New York, 1976).

(8) R. Abraham and J.E. Marsden, 'Foundations of Mechanics' (Benjamin/Cummings, Reading, Mass., 1978).

(9) D.J. Rowe, Rev. Mod. Phys. 40 (1968) 153; Nucl. Phys. A107 (1968) 99.

(10) A.M. Perelomov, Commun. Math. Phys. 26 (1972) 222.

(11) E. Onofri, Journ. Math. Phys. 16 (1975) 1087.

(12) R.J. Glauber, Phys. Rev. 131 (1963) 2766.

(13) P.A.M. Dirac, Can. Journ. Math. 2 (1950) 129.

(14) D.J. Rowe and S.S.M. Wong, Nucl. Phys. A 153 (1970) 561.

(15) V. Bargmann, Commun. Pure and Appl. Math. 14 (1961) 187;
Proc. Natl. Acad. Sci. U.S. 48 (1962) 199;
I.E. Segal, Illinois J. Math. 6 (1962) 520.

A NONLINEAR SUPERPOSITION PRINCIPLE

FOR RICCATI EQUATIONS OF THE CONFORMAL TYPE

Robert L. Anderson
Department of Physics and Astronomy
University of Georgia
Athens, GA 30602/USA

Pavel Winternitz
Centre de Recherches
de Mathématiques Appliquées
Universite de Montréal
Montréal, Québec, H3C 3J7/CANADA

1. Introduction

In this communication, we sketch a proof that n-coupled Riccati equations of the conformal type admit a Vessiot-Guldberg-Lie nonlinear superposition principle. The method of proof yields directly an explicit expression for a nonlinear superposition law. By such a law we mean an expression for the general solution of these equations in terms of a finite number (here n+2) of particular solutions and n arbitrary constants, where the particular solutions are arbitrary up to certain independence conditions. The terminology Riccati equations of the conformal type will be defined in the next section.

The method of derivation of this nonlinear superposition law is of interest in its own right because it utilizes conformal group operations exclusively and hence exhibits the group morphology of the law.

2. Riccati Equations of the Conformal Type

Consider a real Minkowski space $M(p,q)$, $p+q=n$, then the conformal group $C(p,q)$ is realized as a nonlinear representation of $O(p+1, q+1)$ with $M(p,q)$ as the carrier space. The usual basis for its Lie algebra is [see e.g. 1]

$$\begin{aligned}
\hat{M}_{\mu\nu} &= -x_\mu \frac{\partial}{\partial x^\nu} + x_\nu \frac{\partial}{\partial x^\mu} \quad , \\
\hat{P}_\mu &= \frac{\partial}{\partial x^\mu} \quad , \\
\hat{D} &= x^\mu \frac{\partial}{\partial x^\mu} \quad , \\
\hat{C}_\nu &= (x\cdot x)\frac{\partial}{\partial x^\mu} - 2x_\mu x^\nu \frac{\partial}{\partial x^\nu} \quad ,
\end{aligned} \tag{1}$$

with scalar product $(x\cdot y) = x^\nu y_\nu = x^\mu g_{\mu\nu} y^\nu$ where $g = \mathrm{diag}\,(+\ldots+-\ldots-)$ with p pluses and q minuses. The corresponding Lie equations for a one-parameter subgroup are

$$\frac{dx^\mu}{dt} = a^\mu + \lambda x^\mu + \omega^\mu{}_\nu x^\nu + c^\mu (x\cdot x) - 2(c\cdot x)x^\mu \quad ,$$

$$\mu = 1,\ldots, p,p+1,\ldots\ p+q = n \quad , \tag{2}$$

with the $\frac{(n+1)(n+2)}{2}$ coefficients a^μ, λ, $\omega^\mu{}_\nu$, c^μ constant. The exponentiation of (2) gives the global action of a one-parameter subgroup of $C(p,q)$ on $M(p,q)$ and in particular can be written in the form [see e.g. 2]

$$x^\mu(t) = \frac{\rho(t)\Lambda^\mu{}_\nu(t)(x_o^\nu + C^\nu(t)(x_o\cdot x_o))}{1 + 2C(t)\cdot x_o + (C(t)\cdot C(t))(x_o\cdot x_o)} + A^\mu(t). \tag{3}$$

where $A^\mu(t)$, $\rho(t)$, $\Lambda^\mu{}_\nu(t)$, $C^\nu(t)$ are known functions of the constants a^μ, λ, $\omega^\mu{}_\nu$, c^ν and t, respectively, and $x_o = x(t)\big|_{t=0}$. Following Lie [3], we obtain the result that the solution for the nonautonomous version of (2) can still be written in the form (3), however in this case A^μ, ρ, $\Lambda^\mu{}_\nu$, C^ν are not known explicitly in terms of a^μ, λ, $\omega^\mu{}_\nu$, c^ν and time t. Thus for a nonautonomous system of the form (2) the exponentiated form (3) does not represent a one-parameter subgroup of $C(p,q)$ with group parameter t, however it does represent for each t the action of an element of $C(p,q)$ on x_o^μ. It is precisely this group structure that we shall exploit to derive a nonlinear superposition law for, in general, nonautonomous systems of equations of the type (2) which we call systems of Riccati equations of the conformal type.

3. System of Linear Equations Replacing Conformal Riccati Equations

The group $C(p,q)$ can also be realized as the linear action of $O(p+1,\ q+1)$ on the projective cone in a $(n+2)$-dimensional Minkowski space $M(p+1,\ q+1)$. The nonlinear representation discussed in Section 1 is a projection of this linear action onto $M(p,q)$. It is more illuminating as well as more convenient to work with the linear representation and then project these results. Therefore in this section we very briefly sketch this known connection in order to set the context of our results as well as establish notation.

Consider an $(n+2)$-dimensional Minkowski space $M(p+1,\ q+1)$ with the metric

$$g_c = \begin{pmatrix} g & & \\ & -1 & \\ & & 1 \end{pmatrix} \quad , \qquad g = \begin{pmatrix} I_p & \\ & -I_q \end{pmatrix} \quad .$$

A vector in this space is denoted by (ξ, ξ^a, ξ^b) with $\xi = (\xi_1, \ldots, \xi_n)$. Next, introduce the cone $K_{p+1,q+1}$ defined by:

$$\xi^2 - (\xi^a)^2 + (\xi^b)^2 = 0 \quad ,$$

and consider the rays on the cone. This space is invariant under the action of O(p+1, q+1). The Lie equations, in matrix notation, describing this action are:

$$\begin{pmatrix} \dot{\xi} \\ \dot{\xi}^a \\ \dot{\xi}^b \end{pmatrix} = \begin{pmatrix} \omega & a+c & a-c \\ (a^T+c^T)g & 0 & -\lambda \\ -(a^T-c^T)g & -\lambda & 0 \end{pmatrix} \begin{pmatrix} \xi \\ \xi^a \\ \xi^b \end{pmatrix} , \tag{4}$$

where a and c are n×1 matrices, the superscript T denotes the transpose, and ω is an n×n matrix satisfying $\omega g + g\omega^T = 0$.

The relation to conformal Riccati equations is then given by the projection

$$x = \xi/(\xi^a + \xi^b) \quad , \tag{5}$$

which implies equation (2). Thus the set of n Riccati equations (2) can be replaced by the n+2 linear equations (4).

4. A Nonlinear Superposition Law for Conformal Riccati Equations

If we know one particular solution x_a of (2), then we can homogenize (2) through a change of variables as follows: Construct a particular type of element of C(p,q), namely, a translation with x_a and then introduce new variables given by

$$\begin{pmatrix} \eta \\ \eta^a \\ \eta^b \end{pmatrix} = \begin{pmatrix} I_n & -x_a & -x_a \\ -x_a^T g & 1 + \frac{x_a^2}{2} & \frac{x_a^2}{2} \\ x_a^T g & -\frac{x_a^2}{2} & 1 - \frac{x_a^2}{2} \end{pmatrix} \begin{pmatrix} \xi \\ \xi^a \\ \xi^b \end{pmatrix} . \tag{6}$$

This implies through the projection

$$y = \eta/(\eta^a + \eta^b) = x - x_a \quad , \tag{7}$$

that

$$\dot{y} = [\lambda - 2(c\cdot x_a)]y + \omega y - 2(c\cdot y)x_a + 2(x_a\cdot y)c + c(y\cdot y) + 2(c\cdot y)y \quad . \tag{8}$$

Proceeding, if we know a second particular solution x_b such that $(x_a - x_b)^2 \neq 0$, then we can linearize (9) as follows: Introduce new variables by means of another type of element of C(p,q), namely, a special conformal transformation, i.e., let

$$\begin{pmatrix} \xi \\ \xi^a \\ \xi^b \end{pmatrix} = \begin{pmatrix} I_n & -b & b \\ -b^T g & 1 + \frac{b^2}{2} & -\frac{b^2}{2} \\ -b^T g & \frac{b^2}{2} & 1 - \frac{b^2}{2} \end{pmatrix} \begin{pmatrix} \eta \\ \eta^a \\ \eta^b \end{pmatrix} , \tag{9}$$

where b is an $n\times 1$ matrix. Then set

$$b = (x_b - x_a)/(x_a - x_b)^2 \quad . \tag{10}$$

This implies through the projection

$$z = \zeta/(\zeta^a + \zeta^b) \quad , \tag{11}$$

that z satisfies the following set of linear homogeneous equations

$$\dot{z} = \omega z - 2(c\cdot z)x_a + 2(x_a\cdot z)c + (-\lambda + 2(c\cdot x_a))z \quad . \tag{12}$$

It follows from the preceding changes of variables (6) and (9) that each $z \not\equiv 0$ is related to x_a, x_b and some other x by

$$z = \frac{(x_b - x_a)^2(x - x_a) - (x - x_a)^2(x_b - x_a)}{(x - x_b)^2} \tag{13}$$

Hence, if we know another n particular solutions $\{x_k\}_{k=1}^{n}$, subject only to the independence conditions

$$\det \begin{pmatrix} z_1^1 & \cdots & z_1^n \\ \vdots & & \\ z_n^1 & \cdots & z_n^n \end{pmatrix} \neq 0 \quad , \tag{14}$$

and

$$(x_k - x_b)^2 \neq 0, \quad (x_k - x_a)^2 \neq 0, \quad k = 1,\dots,n \quad (x_a - x_b)^2 \neq 0 \quad , \qquad (15)$$

then the general solution of (12) is given by

$$z = \sum_{k=1}^{n} \alpha_k z_k \quad , \qquad (16)$$

where the z_k are related to the x_k by (13) and the α_k are arbitrary constants.

Inversely, if we identify the pairs (z,x), $(z_1,x_1),\dots,(z_n,x_n)$ via (13), then (16) expressed in terms of the x's reads

$$\frac{(x_b-x_a)^2(x-x_a)-(x-x_a)^2(x_b-x_a)}{(x-x_b)^2} = \sum_{k=1}^{n} \alpha_k \frac{(x_b-x_a)^2(x_k-x_a)-(x_k-x_a)^2(x_b-x_a)}{(x_k-x_b)^2} . \qquad (17)$$

Equation (17) expresses, implicitly, the general solutions of (2) in terms of n+2 particular solutions, subject only to the conditions (14), (15), and n arbitrary constants, i.e., equation (17) is a Vessiot-Guldberg-Lie nonlinear superposition law for n-coupled Riccati equations of the conformal type.

For further details see [4] and for treatments of the projective group and historical surveys see [4] and [5].

REFERENCES

[1] J. Beckers, J. Harnad, M. Perroud, P. Winternitz, J. Math. Phys. 19, 2126 (1978).

[2] A. O. Barut, Helv. Phys. Acta. 46, 496 (1973).

[3] S. Lie, and G. Scheffers, Vorlesungen über Continuierliche Gruppen, Teubner, Leipzig, 1893.

[4] R. L. Anderson, J. Harnad, P. Winternitz, "Nonlinear Superposition Principles Admitted By Ordinary Differential Equations," (in preparation).

[5] R. L. Anderson, Lett. Math. Phys. 4, 1 (1980).

BÄCKLUND TRANSFORMATIONS, CONSERVATION LAWS AND LIE SYMMETRIES FOR THE GENERALIZED NONLINEAR SCHRÖDINGER EQUATION

J. Harnad and P. Winternitz
Centre de Recherche de Mathématiques Appliquées
Université de Montréal
Montréal, Québec, CANADA

1. INTRODUCTION

The purpose of this talk is to review some work recently performed at the CRMA in Montréal concerning the applications of group theory to the study of nonlinear phenomena, particularly those described by nonlinear partial differential equations.

We shall restrict ourselves to a discussion of the generalized nonlinear Schrödinger equation (GNLSE)

$$z_{xx} + iz_t = f(z,z^*) \tag{1}$$

where $z(x,t)$ is the dependent variable and $f(z,z^*)$ is some function of z and its complex conjugate z^*. This type of equation, with different specific functions $f(z,z^*)$ on the right hand side, occurs in many different applications (plasma physics, hydrodynamics, biophysics, etc.)

An earlier publication[1] was devoted to the Lie symmetries of this equation (or a somewhat more general equation, since the right hand side had the form $f(x,t,z,z^*)$). The Lie symmetries were studied in the context of symmetry breaking from the Schrödinger group (the invariance group of the free equation $z_{xx} + iz_t = 0$) to each of its subgroups and were used to generate invariant solutions of equation (1). The work to be discussed here[2] concerns a systematic search for Bäcklund transformations (BT) for the GNLSE and the relation between the BT and Lie symmetries. Use is made of the method of "pseudopotentials"[3-5] which provides a certain unity to a variety of different aspects of certain nonlinear partial differential equations (conservation laws, Bäcklund transformations, inverse scattering techniques).

2. SEARCH FOR PSEUDOPOTENTIALS AS A LIE ALGEBRAIC PROBLEM

A pseudopotential $y^\mu(x,t)$, $1 \le \mu \le n$ for equation (1) is introduced by a set of first order partial differential equations

$$\begin{aligned} y^\mu_x &= \psi^\mu(x,t,z,z_x,z_t,z^*,z^*_x,z^*_t,y,y^*) \\ y^\mu_t &= \phi^\mu(x,t,z,z_x,z_t,z^*,z^*_x,z^*_t,y,y^*) \end{aligned} \tag{2}$$

where the functions ψ^μ and ϕ^μ are such that the integrability conditions $y^\mu_{x,t} = y^\mu_{t,x}$ are satisfied if and only if $z(x,t)$ satisfies equation (1). If such functions ψ^μ and ϕ^μ exist then equations (2) will provide conservation laws if ψ^μ and ϕ^μ are independent of y. In other cases they will provide Bäcklund transformations from the function $z(x,t)$ satisfying (1), to the functions $y^{\mu}(x,t)$, themselves satisfying some partial differential equations, not involving z. If $n = 1$ $y(x,t)$ may again satisfy (1) and we have an inner Bäcklund transformation. If ψ^{μ} and ϕ^{μ} depend linearly on y then (2) provides the linear equations of the inverse scattering method.[6,7]

We shall here restrict ourselves to a special case when $n = 1$ (we drop the label μ) and when ψ and ϕ do not depend on x,t and y^*. We replace the GNLSE (1) by the closed Pfaffian system of two forms ω^1, ω^2, ω^{1*} and ω^{2*}:

$$\omega^1 = dz\wedge dt - z_x dx\wedge dt \qquad \omega^2 = i dz\wedge dt - dz_x\wedge dt + f dx\wedge dt \tag{3}$$

and introduce the pullback of the one form

$$\theta = dy - y_x dx - y_t dt \tag{4}$$

under the mapping $(x,t,y,z,z_x,z_t) \to (x,t,y,y_x,y_t)$ defined by (2). Following the Wahlquist and Estabrook method we require that the above forms generate a closed differential ideal, i.e. that

$$\begin{aligned} d\theta &= -d\psi\wedge dx - d\phi\wedge dt = \\ &= A\omega^1 + B\omega^2 + a\omega^{1*} + b\omega^{2*} + (Cdy+Ddx+Edt)\wedge(dy-\psi dx-\phi dt) \end{aligned} \tag{5}$$

we $A,B,\ldots$ are functions of x and t.

Equation (5) implies a set of first order p.d.e. for ψ and ϕ:

$$\frac{\partial\psi}{\partial z_x} = \frac{\partial\psi}{\partial z_x^*} = \frac{\partial\psi}{\partial z_t} = \frac{\partial\psi}{\partial z_t^*} = \frac{\partial\phi}{\partial z_t} = \frac{\partial\phi}{\partial z_t^*} = 0 \tag{6a}$$

$$\frac{\partial\psi}{\partial z} + i\frac{\partial\phi}{\partial z_x} = 0 \qquad \frac{\partial\psi}{\partial z^*} - i\frac{\partial\phi}{\partial z_x^*} = 0 \tag{6b}$$

$$\phi\frac{\partial\psi}{\partial y} - \psi\frac{\partial\phi}{\partial y} - \frac{\partial\phi}{\partial z} z_x - \frac{\partial\phi}{\partial z^*} z_x^* = \frac{\partial\phi}{\partial z_x} f + \frac{\partial\phi}{\partial z_x^*} f^* \tag{6c}$$

Equations (6a) and (6b) can easily be solved directly to obtain

$$\begin{aligned} \psi &= i[|z|^2 Q + zP - z^*R + U] \\ \phi &= (zz_x^* - z^*z_x)Q - z_x P - z_x^* R + i[|z|^2 V - zX - z^*Y + S] \end{aligned} \tag{7}$$

Here $Q,P,\ldots,S$ are functions of y alone and (6c) imposes further conditions on them. They can be written in algebraic form by introducing the vector fields $\hat{Q} = Q(y)\frac{d}{dy}$, ..., $\hat{S} = S(y)\frac{d}{dy}$. Equations (6) are then equivalent to (7) together with the commutation relations

$$\begin{aligned} &\hat{V} = [\hat{P},\hat{R}] \quad , \quad \hat{X} = [\hat{P},\hat{U}] \quad , \quad \hat{Y} = [\hat{R},\hat{U}] \\ &[\hat{Q},\hat{P}] = 0 \quad , \quad [\hat{Q},\hat{R}] = 0 \quad , \quad [\hat{Q},\hat{U}] = 0 \\ &[|z|^2\hat{V} - z\hat{X} - z^*\hat{Y} + \hat{S},\ |z|^2\hat{Q} + z\hat{P} - z^*\hat{R} + \hat{U}] = (z^*\hat{Q}+\hat{P})f + (-z\hat{Q}+\hat{R})f^*. \end{aligned} \tag{8}$$

We now make the additional assumption that the 8 operators $\hat{Q},\ldots,\hat{S}$ generate a finite dimensional Lie algebra. The only Lie algebras that can be realized in terms of holomorphic vector fields in one complex variable are $s\ell(2,C)$ and its subalgebras. Hence, to find a general solution of the commutation relations (8) we introduce a basis τ_+,τ_- and τ_3 for $s\ell(2,C)$, satisfying

$$[\tau_3,\tau_\pm] = \mp i\tau_\pm \quad , \quad [\tau_+,\tau_-] = -2i\tau_3 \tag{9}$$

write each of the operators in (8) in the form $A = a_+\tau_+ + a_-\tau_- + a_3\tau_3$ (a_μ are complex constants) and determine the constants so as to satisfy (8). Each set of operators $\hat{Q},\dots,\hat{S}$ thus determined will provide us, via (7) with the functions ψ and ϕ figuring in (2). To obtain y_x and y_t explicitly, we need a specific realization of τ_i as vector fields, e.g.

$$\tau_3 = iy\frac{d}{dy} \quad , \quad \tau_+ = \frac{d}{dy} \quad , \quad \tau_- = y^2\frac{d}{dy} \tag{10}$$

3. THE BÄCKLUND TRANSFORMATIONS AND CONSERVATION LAWS

The existence of a nonzero solution of (8) imposes constraints on the "interaction" $f(z,z^*)$ in (1). We have shown[2] that solutions exist for 8 classes of functions, 5 of them leading to conservation laws, 3 to Bäcklund transformations. Here we can only give a list of these classes of interactions and discuss two examples.

a. Interactions involving an arbitrary real function $g(z,z^*) = g^*(z,z^*)$:

$$\begin{aligned} f_1 &= \{g(z,z^*) + \frac{i}{2}a(1+\frac{b}{|z|^2})\}z \\ f_2 &= g(z,z^*) \quad , \quad f_3 = g(z,z^*) + ia(z+z^*) + b(z-z^*) + ic \end{aligned} \tag{11}$$

b. Interactions involving arbitrary constants

$$\begin{aligned} f_4 &= \frac{i}{z+z^*}\{\alpha(|z|^2 + z-z^* + \beta)(z+1) - \alpha^*(|z|^2 - z+z^* + \beta^*)(z-1)\} \\ f_5 &= \alpha z + \beta z^* + \gamma \quad , \quad f_6 = |z|^2(-\varepsilon z+z^*) + \varepsilon\alpha z - \alpha^* z^* \\ f_7 &= (\varepsilon|z|^2+a)(z+bz^*) \quad , \quad f_8 = \frac{\varepsilon}{2}|z|^2 + az \end{aligned} \tag{12}$$

In (11) and (12) we have $\varepsilon = \pm 1$, $a,b,c, \in \mathbb{R}$, $\alpha,\beta,\gamma \in C$.

Equations (2) for $f_1,\dots,f_5$ lead imply to conservation laws. For example in the case of $f_1(z,z^*)$ equations (2) reduce to :

$$y_x = i(|z|^2+b) \qquad y_t = (-z_x z^* + z z^*_x) + ay \tag{13}$$

implying the following conservation law and conserved quantity

$$\begin{aligned} i\frac{\partial}{\partial t}(|z|^2+b)e^{-at} + \frac{\partial}{\partial x}(z_x z^* - z z^*_x)e^{-at} &= 0 \\ I = \int_{-\infty}^{\infty}(|z|^2+b)e^{-at}\,dx \end{aligned} \tag{14}$$

It is interesting to note that equation (1) with interaction $f_1(z,z^*)$ has no linear Lie symmetries except space and time translations and these are not the source of the conservation law (14).

Interactions f_6, f_7 and f_8 have genuine Bäcklund transformations. Consider the usual nonlinear Schrödinger equation, i.e. the interaction f_8 with $b = 0$. Equations (2) reduce to

$$\begin{aligned} y_x &= -\frac{1}{2}(zy^2 - \varepsilon z^* + 2\alpha y) \\ y_t &= -\frac{i\alpha}{2}(zy^2 - \varepsilon z^* + 2\alpha y) + \frac{i}{2}(-z_x y^2 - \varepsilon z_x^* + \varepsilon |z|^2 y) \end{aligned} \tag{15}$$

(this is an outer BT which already exists in the literature.[3,4])

The Lie symmetries of equation (1) in this case are

$$\begin{aligned} H &= \frac{\partial}{\partial t} \quad , \quad P = \frac{\partial}{\partial x} \quad , \quad E = i \\ B &= -t\frac{\partial}{\partial x} + \frac{ix}{2} \quad , \quad D = 2t\frac{\partial}{\partial t} + x\frac{\partial}{\partial x} + 1 \end{aligned} \tag{16}$$

corresponding to time and space translations, multiplication by a phase factor, Galilei boosts and dilations. An important feature of the BT (15) is the presence of the arbitrary complex constant α. This is due precisely to the dilation and Galilei invariance of the NLSE, since the transformation

$$\tilde{z}(x,t) = [e^{aD} e^{bB} z](x,t) = a e^{i\frac{b}{2}(x - \frac{b}{2}t)} z[a(x-bt), a^2 t]$$

leaves (1) in this case invariant but generates the complex parameter $\alpha = (2-ib)/a$ in (15).

This interplay between Lie symmetries and BT, generating free parameters in the BT, is very important[8], since the free parameters are necessary to obtain permutability thoerems, multisoliton solutions, infinite numbers of conservation laws, etc. For further considerations we refer to our article to be published elsewhere[2].

REFERENCES

1. C.P. Boyer, R.T. Sharp and P. Winternitz, J. Math. Phys. 17, 1439 (1976).
2. J. Harnad and P. Winternitz, Preprint CRMA-952, Montréal, 1980.
3. H.D. Wahlquist and F.B. Estabrook, J. Math. Phys. 16, 1 (1975), 17, 1293 (1976).
4. J. Corones, J. Math. Phys. 17, 756 (1976).
5. F.A.E. Pirani, D.C. Robinson and W.F. Shadwick, "Local Jet Bundle Formulation of Bäcklund Transformations", D. Reidel, Dordrecht, 1979.
6. C.S. Gardner, J.M. Greene, M.D. Kruskal and R.M. Miura, Phys. Rev. Lett. 19, 1095 (1967).
7. M.J. Ablowitz, D.J. Kaup, A.C. Newell and H. Segur, Studies Appl. Math. 53, 249 (1974).
8. R. Sasaki, Nucl. Phys. B154, 343 (1979).

KILLING TENSORS AND VARIABLE SEPARATION FOR HAMILTON-JACOBI EQUATIONS *

Willard Miller, Jr.
School of Mathematics
University of Minnesota
Minneapolis, Minnesota 55455

1. Introduction.

In this paper we describe precisely the relation between separable coordinates for the Hamilton-Jacobi equation

$$g^{ij}\partial_{x^i}W\partial_{x^j}W = E \tag{1.1}$$

and Killing tensors (constants of the motion) for an n-dimensional pseudo-Riemannian manifold V_n. (Here g_{ij} is the metric tensor expressed in terms of local coordinates x^i and $g^{i\ell}g_{\ell j} = \delta^i_j$.) All of the principal results announced here were obtained in collaboration with Ernie Kalnins, [1], [2].

Every Hamiltonian system for which (1.1) separates in some coordinates is completely integrable [3]. However, in the past ten years numerous examples of completely integrable Hamiltonian systems have been found where variables do not separate in (1.1). Here we will characterize those constants of the motion that correspond to variable separation and show how to compute these separable coordinates from the Killing tensors.

Although we will not treat it here, the case where a scalar potential is added to the left-hand side of (1.1) can easily be handled by our methods. Moreover, since every (multiplicative) separable system for the Helmholtz equation

$$\frac{1}{\sqrt{g}}\partial_{x^i}(\sqrt{g}\ g^{ij}\partial_{x^j}\Psi) = E\Psi\ ,\ g = \det(g_{ij}) \tag{1.2}$$

is an (additively) separable system for (1.1), our treatment has direct applicability to the Helmholtz equation and the important families of special functions that arise as the separable solutions of this equation. (See [4] for a discussion of the relationship between these two equations together with additional references.) See [5] for a review of the theory of variable separation for the H-J equation prior to 1978.

2. Nonorthogonal Separation.

A general definition of separation of variables for the H-J equation (where the separable coordinates need not be orthogonal) can be found in Ref. [5]. A complete

* Research partially supported by NSF Grant MCS 78-26216

integral of (1.1) is a solution $W(\underset{\sim}{x},\underset{\sim}{\alpha})$ depending on n parameters $\alpha_1,\dots,\alpha_{n-1}$, $\alpha_n = E$ such that $\det(\partial_{x^i \alpha^j} W) \neq 0$. A local coordinate system $\{y^j\}$ in V_n is separable if (1.1) admits a complete integral of the form $W(y,\alpha) = \Sigma_{j=1}^{n} W^{(j)}(y^j, \underset{\sim}{\alpha})$. Unfortunately, this intuitive definition does not directly provide a practical means for constructing and characterizing separable coordinate systems. Furthermore, the definition permits inclusion of many pathological types of separable systems which are trivially equivalent to much simpler separable systems. Accordingly, Boyer, Kalnins and Miller [4], [7], [8], introduced a constructive definition of separation which picks out only the "canonical" separation types. Independently, Benenti [9] arrived at this same definition and also showed explicitly that every coordinate system separable in the sense of Levi-Civita is equivalent to one of the "canonical" systems presented here. (It is worth remarking that the Stäckel separable coordinate systems are always orthogonal and are only special cases of the following results.)

Our definition of separation of variables for the H-J equation (1.1) is based on a division of the separable coordinates into three classes: ignorable, essential of type 1, and essential of type 2. Let $\{x^1,\dots,x^n\}$ be a coordinate system on the manifold with metric (g^{ij}) such that the n_1 coordinates x^a, $1 \le a \le n_1$, are essential of type 1, the n_2 coordinates x^r, $n_1+1 \le r \le n_1+n_2$, are essential of type 2, and the n_3 coordinates x^α, $n_1+n_2+1 \le \alpha \le n_1+n_2+n_3 = n$, are ignorable. (In the following, indices a, b, c range from 1 to n_1, indices r, s, t range from n_1+1 to n_1+n_2, indices α,β,γ range from n_1+n_2+1 to n, and indices i, j, k range from 1 to n.) This means that the metric (g^{ij}) expressed in terms of coordinates $\{x^k\}$, is independent of the x^α and that the separation equations take the form

$$W_a^2 + \sum_{\alpha,\beta=n_1+n_2+1}^{n} A_a^{\alpha,\beta}(x^a)\, W_\alpha W_\beta = \sum_{\ell=1}^{n_1+n_2} E_\ell\, \theta_{a\ell}(x^a) \equiv \Phi_a , \tag{2.1}$$

$$\sum_{\alpha=n_1+n_2+1}^{n} 2B_r^\alpha(x^r)\, W_r W_\alpha + \sum_{\alpha,\beta=n_1+n_2+1}^{n} C_r^{\alpha,\beta}(x^r)\, W_\alpha W_\beta \tag{2.2}$$

$$= \sum_{\ell=1}^{n_1+n_2} E_\ell\, \theta_{r\ell}(x^r) \equiv \Phi_r ,$$

$$n_1+1 \le r \le n_1+n_2 .$$

$$W_\alpha = E_\alpha , \quad n_1+n_2+1 \leqq \alpha \le n . \tag{2.3}$$

Here $A_a^{\alpha,\beta}(=A_a^{\beta,\alpha})$, B_r^α, $C_r^{\alpha,b}(=C_r^{\beta,\alpha})$, and θ_{jk} are defined and analytic in a neighborhood $N \times S \subseteq C^{n_1+n_2} \times C^{n_1+n_2}$ where N is a neighborhood of $(x_0^1,\dots,x_0^{n_1+n_2})$ and S is a neighborhood of $(0,\dots,0)$ in the Euclidean space with coordinates $E_1,\dots,E_{n_1+n_2}$ $(E_1 = E)$. The parameters E_α are arbitrary.

We say that the coordinates $\{x^j\}$ are <u>separable</u> for the H-J equation if there exist analytic functions A, B, C, θ above and functions $U_a(x^i)$, $V_r(x^i)$, analytic in N, such that the H-J equation

$$\Sigma g^{ij} \partial_i W \partial_j W = E \tag{2.4}$$

can be written in the form

$$\sum_a U_a(x^i) \Phi_a + \sum_r V_r(x^i) \Phi_r = E \tag{2.5}$$

(identically in the parameters $E_1 = E$, $E_2, \ldots, E_{n_1+n_2}$) where $W = \sum_{j=1}^{n} W^{(j)}(x^j)$
$W_i = \partial_i W = \partial_i W^{(i)}$.

Comparison of (2.4) and (2.5) determines the functions U_a, V_r uniquely. Furthermore, differentiating (2.5) with respect to E_ℓ we have

$$\sum_a U_a \frac{\partial \Phi_a}{\partial \lambda_\ell} + \sum_r V_r \frac{\partial \Phi_r}{\partial \lambda_\ell} = \delta_{1\ell}$$

and this leads to the usual Stäckel form

$$U_a(x^i) = \theta^{a1}/\theta \ , \quad V_r(x^i) = \theta^{r1}/\theta \tag{2.6}$$

where $\theta^{\ell m}$ is the (ℓm)-cofactor of the matrix (θ_{ij}). The nonzero components of the contravariant metric tensor are thus

$$\begin{aligned} g^{ab} &= (\theta^{a1}/\theta)\,\delta^{ab} \ , \quad g^{r\alpha} = g^{\alpha r} = (\theta^{r1}/\theta)\, B_r^{\alpha}(x^r) \ , \\ \tfrac{1}{2} g^{\alpha\beta} &= \sum_a A_a^{\alpha,\beta}(x^a)\, \theta^{a1}/\theta + \sum_r C_r^{\alpha,\beta}(x^r)\, \theta^{r1}/\theta \ , \quad \alpha \neq \beta \ , \\ g^{\alpha\alpha} &= \sum_a A_a^{\alpha,\alpha}\, \theta^{a1}/\theta + \sum_r C_r^{\alpha,\alpha}\, \theta^{r1}/\theta \ . \end{aligned} \tag{2.7}$$

These expressions are the master equations for separation of variables in the Hamilton-Jacobi equation (1.1).

<u>Remarks</u>:

1) Since the metric tensor (g^{ij}) is nonsingular, $n_3 \geq n_2$.

2) $\displaystyle\sum_{\ell=1}^{n_1+n_2} \frac{\theta^{\ell m}}{\theta} \Phi_\ell = E_m \ , \quad m = 1,\ldots,n_1+n_2 \ .$

Thus,

$$\begin{aligned} A_m(\underset{\sim}{x},\underset{\sim}{p}) &= E_m \ , \quad m = 1,\ldots,n_1+n_2 \\ L_\alpha(x,p) &= E_\alpha \ , \quad \alpha = n_1+n_2+1,\ldots,n \end{aligned} \tag{2.8}$$

where

$$A_m(\underset{\sim}{x},\underset{\sim}{p}) = \sum_{i,j=1}^{n} a_{(m)}^{i,j}\, p_i\, p_j \ , \quad L_\alpha(\underset{\sim}{x},\underset{\sim}{p}) = p_\alpha \tag{2.9}$$

$$p_i = \partial_{x^i} W$$

and the nonzero terms of the symmetric quadratic form $(a^{ij}_{(m)})$ are given by

$$a^{ab}_{(m)} = (\theta^{am}/\theta)\,\delta^{ab}\ , \quad a^{r\alpha}_{(m)} = (\theta^{rm}/\theta)\,B^{\alpha}_{r} \tag{2.10}$$

$$\frac{1}{2}a^{\alpha\beta}_{(m)} = \sum_{c} A^{\alpha,\beta}_{c}\,\theta^{cm}/\theta + \sum_{r} C^{\alpha,\beta}_{r}\,\theta^{rm}/\theta\ , \quad \alpha \neq \beta \tag{2.10}$$

$$a^{\alpha\alpha}_{(m)} = \sum_{c} A^{\alpha,\alpha}_{c}\,\theta^{cm}/\theta + \sum_{r} C^{\alpha,\alpha}_{r}\,\theta^{rm}/\theta\ .$$

(Note that $A_1 = E_1$ is the original Hamilton-Jacobi equation.)

3) By definition the quadratic form $H = \Sigma_{\ell=1}^{n_1+n_2} H_{\ell}^{-2} P_{\ell}^{2}$ is in Stäckel form if $H_{\ell}^{2} = \theta/\theta^{\ell 1}$ where

$$\Theta = (\theta_{\ell m}(x^m))$$

is a Stäckel matrix, $\theta = \det \Theta$ and $\theta^{\ell 1}$ is the $(\ell,1)$ minor of Θ. It is well known [10], Appendix 13, that necessary and sufficient conditions that H be in Stäckel form are

$$\partial^2_{x^j x^k} \ln H_i^2 - \partial_{x^j} \ln H_i^2\, \partial_{x^k} \ln H_i^2 + \partial_{x^j} \ln H_i^2\, \partial_{x^k} \ln H_j^2 \tag{2.11}$$

$$+ \partial_{x^k} \ln H_i^2\, \partial_{x^j} \ln H_k^2 = 0\ , \quad j \neq k\ .$$

4) If H is in Stäckel form as in 3), the expressions $\theta^{\ell m}/\theta \equiv \rho_{\ell}^{(m)}\theta^{\ell 1}/\theta = \rho_{\ell}^{(m)} H_{\ell}^{-2}$ are characterized by the equations

$$\partial_{x^k} \rho_{\ell} = (\rho_k - \rho_{\ell})\, \partial_{x^k} (\ln H_{\ell}^{-2})\ , \quad k \neq \ell \tag{2.12}$$

$$\partial_{x^{\ell}} \rho_{\ell} = 0\ ,$$

see [10]. In particular, equations (2.11) constitute the integrability conditions for the system (2.12) and this system admits an $(n_1 + n_2)$ - dimensional space of vector-valued solutions $(\rho_1, \ldots, \rho_{n_1+n_2})$. To any basis of solutions $(\rho_j^{(m)})$ with $\rho_j^{(1)} = 1$ there corresponds a Stäckel matrix with $\theta^{\ell m}/\theta = \rho_{\ell}^{(m)} H_{\ell}^{-2}$.

5) To understand the significance of the quadratic forms A_m and linear forms L_{α}, (2.9), it is useful to utilize the natural symplectic structure on the cotangent bundle $\tilde{V}_n$ of the Riemannian manifold V_n. Corresponding to local coordinates $\{x^j\}$ on V_n we have coordinates $\{x^j, p_j\}$ on $\tilde{V}_n$. If $\{\hat{x}^k(x^j)\}$ is another local coordinate system on V_n then it corresponds to $\{\hat{x}^k, \hat{p}_k\}$ where $\hat{p}_k = p_{\ell}\, \partial x^{\ell}/\partial \hat{x}^k$. The <u>Poisson bracket</u> of two functions $F(x^j, p_j)$, $G(x^j, p_j)$ on $\tilde{V}_n$ is the function

$$[F,G] = \partial_{x^i} F \partial_{p_i} G - \partial_{p_i} F \partial_{x^i} G\ . \tag{2.13}$$

(We are employing the summation convention for variables that range from 1 to n .)

It is straightforward, though tedious to verify the relations

$$[A_\ell , A_m] = 0 \ , \quad [L_\alpha , A_\ell] = 0 \qquad (2.14)$$

$$[L_\alpha , L_\beta] = 0 \ .$$

Thus, the A_m for $m \geq 2$ are second order Killing tensors and the L_α are Killing vectors (first order Killing tensors) for the manifold V_n. Moreover, the family of $n-1$ Killing tensors $\{A_m (m \geq 2), L_\alpha\}$ is in involution. (A Killing tensor is a polynomial in the p_j whose Poisson bracket with A_1 vanishes.)

The relations (2.14) associating separable coordinates on V_n with an involutive family of Killing tensors are not difficult to prove. Much more difficult is the characterization of exactly those involutive families of Killing tensors that define variable separation and the development of a constructive procedure to determine the coordinates from a knowledge of the Killing tensors.

3. Generalized Stäckel Form.

Here, we are given a Riemannian manifold V_n and the contravariant metric tensor g^{ij} on V_n, expressed in terms of the local coordinates $x^1,\ldots,x^n$. We wish to determine necessary and sufficient conditions on the g^{ij} in order that (1.1) permits separation in these local coordinates.

If $g^{ij} = H_i^{-2}\delta^{ij}$, i.e., if the coordinates $\{x^k\}$ are orthogonal, then the necessary and sufficient condition for separation is that $H = g^{ij}p_i p_j$ be in Stäckel form, Ref [10], Appendix 13. In other words, the relations (2.11) must be satisfied.

For non-orthogonal coordinates the conditions are somewhat more complicated. To derive these conditions we need some preliminary lemmas related to Stäckel form. Let $ds^2 = h_i^2(dx^i)^2 = g_{ij}\,dy^i dy^j$ be a metric that is in Stäckel form with respect to the local coordinates $y^1,\ldots,y^N$, i.e., there exists an $N \times N$ Stäckel matrix Θ such that $h_i^2 = \theta/\theta^{i1}$ where $\theta = \det \Theta$ and θ^{i1} is the $(i1)$ minor of Θ. A scalar valued function $f(y)$ is a Stäckel multiplier (for ds^2) if the metric $d\hat{s}^2 = f ds^2 = f h_i^2 (dy^i)^2$ is also in Stäckel form.

Lemma 1: f is a Stäckel multiplier for ds^2 if and only if it satisfies the relations

$$\partial_{y^j y^k} f + \partial_{y^j} f \, \partial_{y^k} \ln h_j^2 + \partial_{y^k} f \, \partial_{y^j} \ln h_k^2 = 0 \ , \quad j \neq k \ . \qquad (3.1)$$

Proof: These relations follow directly from the fact that equations (2.11) must hold for $H_i^2 = h_i^2$ and also for $H_i^2 = f h_i^2$ if f is a Stäckel multiplier.

Lemma 2: f is a Stäckel multiplier for ds^2 if and only if there exist local analytic functions $\varphi_\ell = \varphi_\ell(y^\ell)$ such that

$$f(y) = \sum_{\ell=1}^{N} \varphi_\ell(y^\ell)\, h_\ell^{-2} \quad . \tag{3.2}$$

Proof: See Ref. [1].

Let (g^{ij}) be a given contravariant metric in the coordinates $x^1,\ldots,x^n$. We wish to determine if these coordinates permit separation for the Hamilton-Jacobi equation. It is convenient to reorder the coordinates in a standard form. Let n_3 be the number of ignorable variables x^α. (Recall that x^α is ignorable if $\partial_{x^\alpha} g^{ij} = 0$ for all i,j.) Of the remaining $n - n_3$ variables suppose n_2 variables x^r have the property $g^{rr} = 0$ and the remaining n_1, variables x^a satisfy $g^{aa} \neq 0$. We relabel the variables so that $1 \le a \le n_1$, $n_1 + 1 \le r \le n_1 + n_2$, and $n_1 + n_2 + 1 \le \alpha \le n_1 + n_2 + n_3 = n$.

Theorem 1: Suppose (g^{ij}) is in standard from with respect to the variables $\{x^i\}$. The Hamilton-Jacobi equation (1.1) is separable for this system if and only if

1) The contravariant metric assumes the form

$$(g^{ij}) = \begin{array}{c} \begin{array}{ccc} \quad n_1 \quad & \quad n_2 \quad & \quad n_3 \quad \end{array} \\ \left(\begin{array}{c|c|c} \delta^{ab} H_a^{-2} & 0 & 0 \\ \hline 0 & 0 & H_r^{-2} B_r^{\alpha} \\ \hline 0 & H_r^{-2} B_r^{\alpha} & g^{\alpha\beta} \end{array}\right) \end{array} \begin{array}{c} \\ n_2 \\ n_2 \\ n_3 \end{array}$$

where $B_r^\alpha = B_r^\alpha(x^r)$.

2) The metric $d\tilde{s}^2 = \sum_{a=1}^{n_1} H_a^2 (dx^a)^2 + \sum_{r=n_1+1}^{n_1+n_2} H_r^2 (dx^r)^2$ is in Stäckel form, i.e., relations (2.11) hold for $1 \le i, j, k \le n_1 + n_2$.

3) Each $g^{\alpha\beta}(\underset{\sim}{x})$ is a Stäckel multiplier for the metric $d\tilde{s}^2$.

Proof: The theorem follows immediately from expressions (2.7) and Lemmas 1 and 2.

Note that Theorem 1 reduces the problem of determining whether the Hamilton-Jacobi equation is separable in given coordinates to the verification of two systems of partial differential equations. If the coordinates are orthogonal, then $n_2 = 0$ and the separation requirement is simply that the metric be in Stäckel form.

Let $A = a^{ij}(\underset{\sim}{x})\, p_i p_j$, $B = b^{ij}(\underset{\sim}{x})\, p_i p_j$ be symmetric quadratic functions on $\tilde{V}_n$. It follows from (2.13) that these functions are in involution with respect to the Poisson bracket if and only if

$$a^{[i,j}\,\partial_j\, b^{\boxed{k\ell}} = b^{[i,j}\,\partial_j\, a^{\boxed{k\ell}} \quad , \quad 1 \le i, k, \ell \le n \tag{3.3}$$

where

$$a^{[i,j}\partial_j b^{\overline{k\ell]}} = a^{ij}\partial_j b^{k\ell} + a^{\ell j}\partial_j b^{ik} + a^{kj}\partial_j b^{\ell i} .$$

A scalar valued function $p(\underset{\sim}{x})$ is a <u>root</u> of the form $a^{ij}(x)$ if

$$\det(a^{ij}(\underset{\sim}{x}) - \rho(\underset{\sim}{x})\, g^{ij}(\underset{\sim}{x})) = 0 \tag{3.4}$$

in a coordinate neighborhood, where (g^{ij}) is the metric on V_n. A form $\psi = \lambda_j(\underset{\sim}{x})dx^j$ such that

$$(a^{ij} - \rho g^{ij})\lambda_j = 0$$

in the same coordinate neighborhood is an <u>eigenform</u> corresponding to the root ρ.

<u>Theorem 2</u>: Let (g^{ij}) be the contravariant metric tensor on V_n in the coordinates $\{x^i\}$. If the Hamilton-Jacobi equation is separable in these coordinates then there exists a Q dimensional vector space $\mathcal{G}$ of second order Killing tensors on V_n such that

1) $[A,B] = 0$ for each $A,B \in \mathcal{G}$. (3.5)
2) For each of the n_1 essential coordinates of type 2, x^a, the form dx^a is a simultaneous eigenform for every $A \in \mathcal{G}$, with root ρ^A_a.
3) For each of the n_2 essential coordinates of type 1, x^r, the form dx^r is a simultaneous eigenform for every $A \in \mathcal{G}$, with root ρ^A_r. The root ρ^A_r has multiplicity 2 but corresponds to only one eigenform.
4) $\partial_i(a^{\alpha\beta}) - \rho^A_i \partial_i g^{\alpha\beta} = 0$, $i = n_1+1, \ldots, n_1+n_2$ for all $A \in \mathcal{G}$, and all n_3 ignorable variables $\alpha,\beta = n_1+n_2+1, \ldots, n$.
5) $[A,L_\alpha] = 0$ for each $A \in \mathcal{G}$ and $L_\alpha = p_\alpha$, $\alpha = n_1+n_2+1, \ldots, n$.
6) $Q = n + n_3(n_3-1)/2$.

This theorem is easily obtained from the proof of the following deeper result. Let $\{x^i\}$ be a coordinate system on V_n with coordinates divided into three classes containing n_1, n_2, and n_3 variables, respectively $(n = n_1+n_2+n_3)$. (We will call them essential variables of type 1, essential varaibles of type 2 and ignorable variables, respectively, even though at this point they have nothing to do with separation.) Let $H = g^{ij} p_i p_j$.

<u>Theorem 3</u>: Suppose there exists a Q dimensional vector space $\mathcal{G}$ of second order Killing tensors on V_n such that $H \in \mathcal{G}$ and conditions 1) - 6), (3.5), are satisfied. Furthermore, suppose $g^{ab} = 0$ if $1 \le a < b \le n_1$, and $g^{ar} = g^{a\alpha} = g^{rs} = 0$ for $1 \le a \le n_1$, $n_1+1 \le r$, $s \le n_1+n_2$, $n_1+n_2+1 \le \alpha \le n$. Then the Hamilton-Jacobi equation (1.1) is separable in the coordinates $\{x^i\}$. The Killing tensors A_m, $m = 1,\ldots,n_1+n_2$, (2.9), and $L_\alpha L_\beta = p_\alpha p_\beta$, $n_1+n_2+1 \le \alpha \le \beta \le n$, form a basis for $\mathcal{G}$.

The proof is simple in principle though intricate; see Ref. [2] for the details. One merely exploits the relations $[A_i, A_j] = 0$.

4. The Main Result.

We come now to the fundamental question: given an involutive family of $n-1$ Killing tensors, how do we determine if this family corresponds to a separable coordinate system for the Hamilton-Jacobi equation?

Let $\{x^j\}$ be a local coordinate system on the Riemannian manifold V_n and let $\theta_{(j)} = \lambda_{i(j)} dx^i$, $1 \le j \le n$, be a local basis of one-forms on V_n. The dual basis of vector fields is $X^{(h)} = \Lambda^{i(h)} \partial_{x^i}$, $1 \le h \le n$, where $\Lambda^{i(h)} \lambda_{i(j)} = \delta^{(h)}_{(j)}$. We say that the forms $\{\theta_{(j)}\}$ are <u>normalizable</u> if there exist local analytic functions $g_{(j)}$, y^j such that $\theta_{(j)} = g_{(j)} dy^j$, (no sum). It is classical that the forms are normalizable if and only if the coefficient of $X^{(\ell)}$ is zero in the expansion of $[X^{(h)}, X^{(k)}]$ in terms of the $\{X^{(j)}\}$ basis, whenever $h,k \neq \ell$, see Ref. [10], Section 35.

<u>Lemma 3</u>: The one-forms $\{\theta_{(j)}\}$ are normalizable if and only if

$$(\partial_{x^j} \lambda_{i(\ell)} - \partial_{x^i} \lambda_{j(\ell)}) \Lambda^{i(h)} \Lambda^{j(k)} = 0 , \quad h, k \neq \ell . \tag{4.1}$$

<u>Theorem 4</u>: Suppose there exists a Q dimensional vector space of second order Killing tensors on V_n such that $H \in \mathfrak{a}$ and

1) $[A,B] = 0$ for each $A,B \in \mathfrak{a}$.
2) There is a basis of one forms $\theta_{(h)} = \lambda_{i(h)} dx^i$, $1 \le h \le n$ such that
   a) the n_1 forms $\theta_{(a)}$, $1 \le a \le n_1$ are simultaneous eigenforms for every $A \in \mathfrak{a}$ with root ρ^A_a:
   $$(a^{ij} - \rho^A_a g^{ij}) \lambda_{j(a)} = 0 . \tag{4.5}$$
   b) the n_2 forms $\theta_{(r)}$, $n_1 + 1 \le r \le n_1 + n_2$, are simultaneous eigenforms for every $A \in \mathfrak{a}$ with root ρ^A_r
   $$(a^{ij} - \rho^A_r g^{ij}) \lambda_{j(r)} = 0 .$$
   The root ρ^A_r has multiplicity 2 but corresponds to only one eigenform.
3) $X^{(h)}(\lambda_{i(\alpha)} a^{ij} \lambda_{j(\beta)}) = \rho^A_h X^{(h)}(\lambda_{i(\alpha)} g^{ij} \lambda_{j(\beta)})$, $h = n_1+1,\dots,n_1+n_2$ for all $A \in \mathfrak{a}$ and all $\alpha,\beta = n_1+n_2+1,\dots,n$.

4) $[L_\alpha , L_\beta] = 0$ where $L_\alpha = \Lambda^{i(\alpha)} p_i$.

5) $[A, L_\alpha] = 0$ for each $A \in \mathfrak{a}$.

6) $Q = \frac{1}{2}(2n + n_3^2 - n_3)$ where $n_3 = n - n_1 - n_2$.

7) $G_{(ab)} = 0$ if $1 \leq a < b \leq n_1$, and $G_{(ar)} = G_{(a\alpha)} = G_{(rs)} = 0$ for $1 \leq a \leq n_1$, $n_1 + 1 \leq r$, $s \leq n_1 + n_2$, $n_1 + n_2 + 1 \leq \alpha \leq n$.

Then there exist local coordinates $\{y^j\}$ for V_n such that $\theta_{(j)} = f^{(j)}(\underset{\sim}{y})\, dy^j$ for suitably chosen functions $f^{(j)}$, and the Hamilton-Jacobi equation is separable in these coordinates. Conversely, to every separable coordinate system $\{y^j\}$ for the Hamilton-Jacobi equation there corresponds a family $\mathfrak{a}$ of second order Killing tensors on V_n with properties 1) - 7).

The proof involves use of the commutation relations $[A_i , A_j] = 0$ to show that the eigenvector fields are normalizable. For the tedious details see [2].

Note the drastic simplification in these conditions when $n_2 = 1$. In this case conditions 2b) and 3) become vacuous. If further $n_3 = 0$ then the conditions for separation are merely that the n_1 - dimensional involutive family of Killing tensors be simultaneously diagonalizable. (This is the classical Stäckel case.) The eigenvector fields are then guaranteed to be normalizáble and they determine the orthogonal separable coordinates.

5. An Example.

To show how Theorem 4 can be employed in practice we treat a single non-trivial exampli in some detail. (For some simpler examples see [1].) The real Hamilton-Jacobi equation

$$W_t^2 - W_x^2 - W_y^2 = E \tag{5.1}$$

admits the pseudo-Euclidean algebra e(2,1) as its symmetry algebra of Killing vectors. A basis for the symmetry algebra is

$$\begin{gathered} K_1 = xp_t + tp_x \ , \ K_2 = yp_t + tp_y \ , \ L_3 = yp_x - xp_y \\ P_0 = p_t \ , \ P_1 = p_x \ , \ P_2 = p_y \ . \end{gathered} \tag{5.2}$$

As is well known, e.g. [1], the space of second order Killing tensors for the pseudo-Riemmanian manifold with (5.1) as its associated equation is spanned by products of Killing vectors (5.2). Thus, it is easy to display the second order Killing tensors for this manifold.

Recall that two separable coordinate systems for a Hamilton-Jacobi equation are considered as equivalent if the defining symmetry operators for the two systems are equivalent under the adjoint action of the local Lie symmetry group of the equation [4], [7]. Thus, if we are looking for all separable coordinate systems with one ignorable variable we can limit our search to those cases where the Killing vector X_α corresponding to this variable is an explicitly chosen representative of one of the conjugacy classes of one-dimensional subalgebras of $e(2,1)$. We consider the particularly interesting case where $L_\alpha = P_0 + P_2$. (As shown in Ref. [4], all nonorthogonal separable corrdinates for (5.1) correspond to this case. Moreover, it is easily shown that any coordinate system with $P_0 + P_2$ as a generator for an ignorable coordinate must necessarily be nonorthogonal [8].) For such a system the Killing tensor A must commute with L_α. Thus A can be chosen from the real vector space of homogeneous second order polynomials in the symmetries (5.2). Furthermore, we can identify two Killing tensors that lie on the same orbit under the adjoint action of the normalizer for $P_0 + P_2$. The normalizer has basis $\{K_2, L_3 - K_1, P_0 + P_2, P_1, P_0 - P_2\}$. (See Ref. [11] for a more detailed discussion of this problem.) One family of orbit representatives is

$$(L_3 - K_1)^2 + 4P_1^2 + a(L_3 - K_1)(P_0 - P_2) + b(P_0 - P_2)^2$$
$$+ c(P_0 + P_2)^2 + dP_1(P_0 - P_2) . \tag{5.3}$$

(That is, two such representatives lie on the same orbit if and only if they are identical. We could, of course, easily compute all possible families of orbit representatives and apply the following considerations to each such family.) Group theory can take us no further than this point. We still have to determine which, if any, of the Killing tensors (5.3) actually correspond to separable coordinates.

In the following it is convenient to choose new coordinates $\{x,\tau,w\}$ such that $\tau = \frac{1}{2}(y+t)$, $w = \frac{1}{2}(y-t)$, so $p_\tau = p_y + p_t$, $p_w = p_y - p_t$. In terms of these coordinates

$$A - \rho H = \begin{pmatrix} 4w^2 + 4 + \rho & -2xw & -aw - d/2 \\ -2xw & x^2 + c & (ax+\rho)/2 \\ -aw - d/2 & (ax+\rho)/2 & b \end{pmatrix} . \tag{5.4}$$

Since $n = 3$, $n_3 = 1$ and the coordinates associated with $P_0 + P_2$ are necessarily nonorthogonal, we must have $n_1 = n_2 = 1$ for any separable coordinates. Thus A must have a single root ρ_1 and a distinct root ρ_2 of multiplicity 2 which has only one eigenform. The characteristic equation $f(\rho) = \det(A - \rho H) = 0$ reads

$$\frac{\rho^3}{4} + \rho^2\left(w^2 + 1 + \frac{ax}{2}\right) - \rho\Big(bx^2 - \frac{a^2}{4}x^2 - 2ax + cb - axw^2$$
$$+ dxw\Big) + \Big(dacw - c(4b - a^2)w^2 + \big(-4b + a^2 + \frac{d^2}{4}\big)x^2 \tag{5.5}$$
$$+ \frac{d^2c}{4} - 4cb\Big) = 0 .$$

Since ρ_2 is a double root we must have $f'(\rho_2) = 0$. Also, $f(\rho) = \frac{1}{4}(\rho - \rho_2)^2 (\rho - \rho_1)$. It is straightforward, though tedious, to verify that these conditions on ρ_1, ρ_2 are inconsistant unless $a = b = c = d = 0$, in which case

$$\rho_1 = -4(w^2 + 1), \quad \rho_2 = 0. \tag{5.6}$$

Thus,

$$A = \begin{pmatrix} 4w^2 + 4 & -2xw & 0 \\ -2xw & x^2 & 0 \\ 0 & 0 & 0 \end{pmatrix} \tag{5.7}$$

and $\theta_{(1)} = (2w^2 + 2)\,dx - 2xwdw$, $\theta_{(2)} = dw$. To satisfy conditions 5) and 7) of Theorem 4 we must require $\theta_{(3)} = xw(x^2 + 1)^{-1}dx + d\tau + f\,dw$. We choose f such that $\theta_{(3)}$ is a perfect differential and obtain

$$(\lambda_{j(k)}) = \begin{pmatrix} 2w^2+2 & 0 & \frac{xw}{(w^2+1)} \\ 0 & 0 & 1 \\ -2xw & 1 & \frac{x^2(1-w^2)}{2(1+w^2)^2} \end{pmatrix}, \quad (\Lambda^{(\ell)j}) = \begin{pmatrix} \frac{1}{2w^2+2} & \frac{-xw}{2(w^2+1)^2} & 0 \\ \frac{xw}{w^2+1} & \frac{-x^2}{2(1+w^2)} & 1 \\ 0 & 1 & 0 \end{pmatrix} \tag{5.8}$$

Condition 3) can be verified directly.
Finally, $Q = 3$ and $\mathfrak{G}$ has the basis $\{A, H, X^{(3)}\}$.

We conclude that among the operators (5.3), only $A = (L_3 - K_1)^2 + 4P_1^2$ corresponds to a separable coordinate system. Furthermore, in this case it is now straightforward to derive the separable coordinates. They are $\{x^1, x^2, x^3\}$ where

$$x = x^1[1 + (x^2)^2]^{1/2}, \quad \tau = [x^3 - (x^1)^2 x^2]/2, \quad w = x^2 \tag{5.9}$$

Indeed,

$$X^{(1)} = \frac{1}{2}(1 + w^2)^{-3/2}\partial_1, \quad X^{(2)} = \partial_2, \quad X^{(3)} = 2\partial_1. \tag{5.10}$$

REFERENCES

[1] Kalnins, E.G. and Miller, W. Jr., Killing tensors and variable separation for Hamilton-Jacobi and Helmholtz equations, SIAM J. Math. Anal. (to appear.)

[2] Kalnins, E.G. and Miller, W. Jr., Killing tensors and nonorthogonal variable separation for Hamilton-Jacobi and Helmholtz equations, SIAM J. Math. Anal. (to appear).

[3] Arnold, V.I., "Mathematical Methods of Classical Mechanics", Springer Verlag, New York, 1978.

[4] Boyer, C., Kalnins, E.G. and Miller, W. Jr., Commun. Math. Phys. 59, 285-302 (1978).

[5] Benenti, S. and Francaviglia, M., The theory of separability of the Hamilton-Jacobi equation and its application to general relativity, in "General Relativity and Gravitation, Vol. 1", A Held, ed., Plenum, New York, 1980.

[6] Levi-Civita, T., Math. Ann. 59, 383-397 (1904).

[7] Kalnins, E.G. and Miller, W. Jr., Separable coordinates for three-dimensional complex Riemannian spaces, J. Diff. Geometry (1979).

[8] Kalnins, E.G. and Miller, W. Jr., J. Phys. A: Math. Gen., 12, 1129-1147 (1979).

[9] Benenti , S., Separability structures on Riemannian manifolds, Proceedings of "Conference on Differential Geometrical Methods in Mathematical Physics", Salamanca 1979, Springer-Verlag (to appear).

[10] Eisenhart, L.P., "Riemannian Geometry", Princeton University Press, Princeton (2nd printing), 1949.

[11] Miller, W. Jr., Patera, G. and Winternitz, P., Subgroups of Lie groups and separation of variables, J. Math. Phys., (to appear).

LIE PSEUDOGROUPS AND GAUGE THEORY

Jean-François POMMARET
Département de mécanique
Ecole Nationale des Ponts et Chaussées
28 rue des Saints-Pères,75007 Paris ,France

As a matter of fact,unified gauge theories have,up to now,a common differential geometric framework,namely that of connections on principal bundles combined with differential identities induced by the exterior derivative,such as the Maurer-Cartan structure equations and the Bianchi identities.

The purpose of our communication at this colloquium is to criticise these methods while giving a new general mechanism that works not only for Lie groups as before,but also for Lie pseudogroups.(We recall that a Lie pseudogroup is any group of transformations defined by a system of partial differential equations).As the formal methods involved are very far from the standard known ones,we only sketch our program thereafter in the form of a syllogism and refer the reader to our forthcoming book "Differential Galois theory" (4).

1)The differential identities existing in gauge theories are just particular cases of the first and second non-linear Spencer sequences (2).It does not seem that any physicist has ever noticed that fact, though both sequences become isomorphic when the Lie pseudogroup involved corresponds to the action of a Lie group.Therefore,if one does want to achieve the development of classical gauge theory in the same spirit,this must necessarily be done along these lines.However,the way of dealing with systems of infinitesimal Lie equations instead of finite dimensional Lie algebras is quite different and brings a new light on the classical framework.

2)The use of connections and exterior calculus hides completely the preceding argument.In particular,Maurer-Cartan equations for Lie pseudogroups are absolutely not the proper generalizations of the Maurer-Cartan equations for Lie groups.As a byproduct,the most surprising aspect is that both the first and second Spencer sequences are certainly not the proper tools for studying Lie pseudogroups,natural (geometric) objects and related differential identities.The key piece of machinery is another sequence called Janet sequence (after the work of M.Janet on systems of p.d.e. in 1920 (1)) that we link with the Spencer sequences.

3)As a result,if one does want to use the later framework which is the only convenient one in practice,the whole phylosophy of gauge theory must be revisited within this new scheme. The reader must also notice that only this second picture may be used in the Galois theory of systems of p.d.e..

To summarize, we face a <u>dilemna</u> which is of a purely mathematical nature and therefore <u>must</u> be solved by future field theory <u>as it cannot be avoided</u>.

We sketch thereafter the basic arguments on a picture:

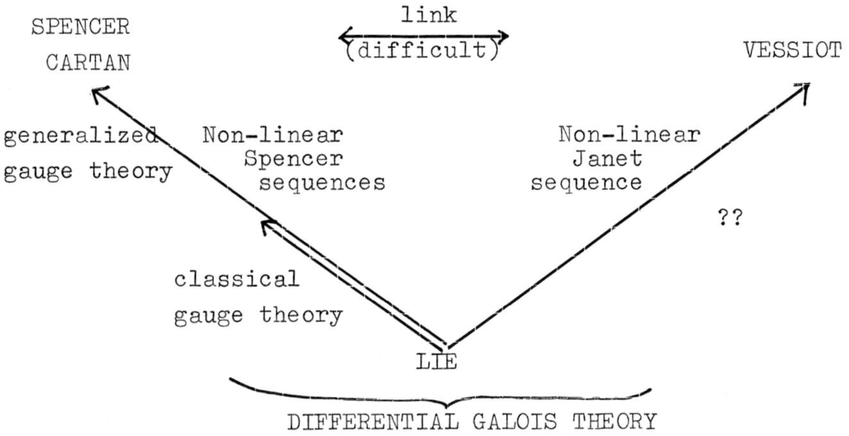

I / <u>CLASSICAL GAUGE THEORY</u>:

Let X be the <u>base manifold</u> and $\mathcal{P} \rightarrow X$ be a <u>principal bundle</u> over X with <u>structure group</u> $\mathcal{G}$. The reader must notice that $\mathcal{G}$ is in general a <u>linear algebraic group</u> and this provides a first link with the differential Galois theory. Now, there may be many ways to use $\mathcal{P}$ and we refer the reader to (5) for a very good account of that fact. However, the only one that can be generalized is to construct the following short exact sequence pulled back on $\mathcal{P}$ that defines the <u>vertical bundle</u> $V(\mathcal{P})$ of $\mathcal{P}$:

$$0 \longrightarrow V(\mathcal{P}) \longrightarrow T(\mathcal{P}) \longrightarrow T(X) \longrightarrow 0$$

Quotienting by the induced action of $\mathcal{G}$ we define a <u>connection</u> χ as a splitting of the sequence:

$$0 \longrightarrow V(\mathcal{P})/\mathcal{G} \longrightarrow T(\mathcal{P})/\mathcal{G} \overset{\chi}{\rightleftarrows} T(X) \longrightarrow 0$$

that now contains only <u>vector bundles</u> over X. In particular, $V(\mathcal{P})/\mathcal{G}$ is a <u>bundle of Lie algebras</u> over X.

We may obtain from χ a vertical 1-form ω on $\mathcal{P}$ with value in the Lie algebra $\mathfrak{g}$ of $\mathcal{G}$ with bracket $[\ ]$. Introducing the <u>horizontal exterior derivative</u>, we obtain the <u>Maurer-Cartan structure equations</u>:

$$d_h \omega = d\omega + \frac{1}{2}[\omega, \omega]$$

and the <u>Bianchi identities</u>:

$$d_h d_h \omega = 0$$

The rough idea is finally to use a differential sequence as follows:

$$\underset{\text{potential}}{\overset{\text{Yang-Mills}}{Y}} \xrightarrow[\text{M.C. equations}]{d_h} \underset{\text{field}}{\overset{\text{Yang-Mills}}{M}} \xrightarrow[\text{Y.M. equations}]{d_h} \overset{\text{matter}}{J} \xrightarrow[\text{B. identities}]{d_h} \cdots$$

II / SPENCER SEQUENCES:

Let $\Pi_q(X,X)$ be the jet bundle of order q of invertible maps from X to X and consider an involutive system of finite transitive Lie equations $\mathcal{R}_q \subset \Pi_q(X,X)$ whose solution sheaf Γ is a Lie pseudogroup.

EXAMPLE: $\frac{y_{xxx}}{y_x} - \frac{3}{2}\left(\frac{y_{xx}}{y_x}\right)=0$ admits the explicit solutions $y = \frac{ax+b}{cx+d}$ that define homographic transformations. However, $\frac{\partial(y^1,\dots,y^n)}{\partial(x^1,\dots,x^n)} = 1$ cannot be integrated though its solutions behave like a group by composition.

If $id_q = j_q(id)$ is the q-jet of the identity map $id = id_X$ we may introduce the involutive system $R_q = id_q^{-1}(V(\mathcal{R}_q)) \subset J_q(T)$ of infinitesimal transitive Lie equations with solution sheaf Θ .

For any given point $x_0 \in X$ we may introduce the principal bundle $\mathcal{P}_q = \mathcal{R}_q(X,x_0)$ over X of jets with fixed target x_0 ,with structure group $\mathcal{G}_q = \mathcal{R}_q(x_0,x_0)$ of order q .

The analogy with part I is to introduce a splitting χ_q of the short exact sequence:

$$0 \longrightarrow R_q^0 \longrightarrow R_q \underset{\chi_q}{\overset{\longleftarrow}{\longrightarrow}} T \longrightarrow 0$$

where R_q^0 is a bundle of Lie algebras over X for any q and R^0_{q,x_0} is the Lie algebra of $\mathcal{G}_q$.

Using the above objects,we may construct the first non-linear Spencer sequence:.

$$0 \longrightarrow aut(X) \xrightarrow{j_{q+1}} \Pi_{q+1}(X,X) \xrightarrow{\bar{D}} T^*\otimes J_q(T) \xrightarrow{\bar{D}'} \Lambda^2T^*\otimes J_{q-1}(T)$$

and its restriction:

$$0 \longrightarrow \Gamma \xrightarrow{j_{q+1}} \mathcal{R}_{q+1} \xrightarrow{\bar{D}} T^*\otimes R_q \xrightarrow{\bar{D}'} \Lambda^2T^*\otimes J_{q-1}(T)$$

Linearizing,we obtain the first linear Spencer sequence:

$$0 \longrightarrow T \xrightarrow{j_{q+1}} J_{q+1}(T) \xrightarrow{D} T^*\otimes J_q(T) \xrightarrow{D} \Lambda^2T^*\otimes J_{q-1}(T)$$

and its restriction:

$$0 \longrightarrow \Theta \xrightarrow{j_{q+1}} R_{q+1} \xrightarrow{D} T^*\otimes R_q \xrightarrow{D} \Lambda^2T^*\otimes J_{q-1}(T)$$

However,these sequences have bad formal properties because $\bar{D}'$ does not involve all the differential identities that could be obtained from $\bar{D}$ .Therefore we must introduce new vector bundles:

$$\begin{cases} C_r(T) = \Lambda^r T^* \otimes J_q(T) / \delta(\Lambda^{r-1}T^*\otimes S_{q+1}T^*\otimes T) \\ C_r = \Lambda^r T^* \otimes R_q / \delta(\Lambda^{r-1}T^*\otimes G_{q+1}) \end{cases}$$

where $\wedge^r T^*$ is the r-exterior product of T^* , $S_q T^*$ is the q-symmetric product of T^* , δ is the algebraic Spencer operator and G_{q+1} is the first prolongation of the symbol G_q of R_q .

We obtain the <u>second Spencer sequences</u>:

$$0 \longrightarrow aut(X) \xrightarrow{\dot{j}_q} \Pi_q(X,X) \xrightarrow{\bar{D}_1} C_1(T) \xrightarrow{\bar{D}_2} C_2(T) \longrightarrow ?$$

$$0 \longrightarrow \Gamma \xrightarrow{\dot{j}_q} \mathcal{R}_q \xrightarrow{\bar{D}_1} C_1 \xrightarrow{\bar{D}_2} C_2 \longrightarrow ?$$

$$0 \longrightarrow T \xrightarrow{\dot{j}_q} C_0(T) \xrightarrow{D_1} C_1(T) \xrightarrow{D_2} \cdots \xrightarrow{D_n} C_n(T) \longrightarrow 0$$

$$0 \longrightarrow \Theta \xrightarrow{\dot{j}_q} C_0 = R_q \xrightarrow{D_1} C_1 \xrightarrow{D_2} \cdots \xrightarrow{D_n} C_n \longrightarrow 0$$

The use of some tricks <u>simplifies greatly</u> the exposition of (2) that could not be applied otherwise to concrete cases.

III / <u>JANET SEQUENCES</u>:

<u>The part II is unuseful in practice as it never brings any known material even in the case of classical examples</u> and we must turn to another point of view (3).

First we notice that there is a correspondence between a system $\mathcal{R}_q$ of finite Lie equations and a section ω of the natural bundle $\mathcal{F} = \Pi_q(X,x_0)/\mathcal{G}_q$ over X where $\mathcal{G}_q$ acts on the target. In fact, using the sectional point of view, we have:

$$\mathcal{R}_q = \{ f_q \in \Pi_q(X,X) \mid f_q(\omega) = \omega \} \quad , \quad R_q = \{ \xi_q \in J_q(T) \mid L(\xi_q)\,\omega = 0 \}$$

where we have generalized the concept of Lie derivative. We may then introduce the epimorphism $\phi : \Pi_q(X,X) \to \mathcal{F} : f_q \to f_q^{-1}(\omega)$ and the source projection $\alpha : X \times X \to X$.

The <u>structure equations</u> come from <u>integrability conditions</u> and may be expressed by a non-linear system:

$$\mathcal{B}_1(c) \subset J_1(\mathcal{F}) \qquad\qquad I(j_1(\omega)) = c(\omega)$$

depending on the <u>structure constants</u> c . In fact the symbol $\mathcal{N}_1$ of $\mathcal{B}_1(c)$ is a natural vector bundle over $\mathcal{F}$ that only depends on $\mathcal{G}_q$ and we may define the following natural vector bundles over $\mathcal{F}$:

$$\mathcal{F}_r = \wedge^r T^* \otimes \mathcal{F}_0 / \delta(\wedge^{r-1} T^* \otimes \mathcal{N}_1) \qquad \text{with} \quad \mathcal{F}_0 = V(\mathcal{F})$$

Then c is just a <u>natural section</u> of $\mathcal{F}_1$ over $\mathcal{F}$ and we have the <u>non-linear Janet sequence</u>:

$$0 \longrightarrow \Gamma \longrightarrow aut(X) \underset{\omega \,\circ\, \alpha}{\overset{\phi \,\circ\, \dot{j}_q}{\rightrightarrows}} \mathcal{F} \underset{c}{\overset{I \,\circ\, j_1}{\rightrightarrows}} \mathcal{F}_1 \longrightarrow ?$$

<u>EXAMPLE</u>: There is no conceptual difference between the Maurer-Cartan equations for the Maurer-Cartan forms in the case of a Lie group acting simply transitively and the constant riemannian curvature for a metric in the case of a group of isometries. It follows that the analogy with the riemannian case in classical gauge theory (5) is

a pure coincidence.

Linearizing and setting $F_n = \omega^{-1}(\mathcal{F}_n)$ we obtain the linear Janet sequence:

$$0 \longrightarrow \Theta \longrightarrow T \xrightarrow{\mathcal{D}} F_0 \xrightarrow{\mathcal{D}_1} F_1 \xrightarrow{\mathcal{D}_2} \cdots \xrightarrow{\mathcal{D}_n} F_n \longrightarrow 0$$

where $\mathcal{D}$ is a Lie operator.

EXAMPLE: $\mathcal{D}\xi = \mathcal{L}(\xi)\,\omega$ for any tensor field ω.

The next diagram (3) explains in the linear case why our structure equations must be used in place of the Yang-Mills equations because the middle row is exact and there is an isomorphism between the cohomology at C_{r+1} of the top second Spencer sequence and the cohomology at F_r of the bottom Janet sequence:

$$\begin{array}{ccccccccccccccccc}
 & & & & & & 0 & & 0 & & 0 & & & & 0 & & \\
 & & & & & & \downarrow & & \downarrow & & \downarrow & & & & \downarrow & & \\
 & & & & & & C_0 & \xrightarrow{D_1} & C_1 & \xrightarrow{D_2} & C_2 & \xrightarrow{D_3} & \cdots & \xrightarrow{D_n} & C_n & \longrightarrow & 0 \\
 & & & & & & \downarrow & & \downarrow & & \downarrow & & & & \downarrow & & \\
 & & 0 & \longrightarrow & T & \xrightarrow{j_q} & C_0(T) & \xrightarrow{D_1} & C_1(T) & \xrightarrow{D_2} & C_2(T) & \xrightarrow{D_3} & \cdots & \longrightarrow & C_n(T) & \longrightarrow & 0 \\
 & & & & \| & & \downarrow & & \downarrow & & \downarrow & & & & \downarrow & & \\
0 & \longrightarrow & \Theta & \longrightarrow & T & \xrightarrow{\mathcal{D}} & F_0 & \xrightarrow{\mathcal{D}_1} & F_1 & \xrightarrow{\mathcal{D}_2} & F_2 & \xrightarrow{\mathcal{D}_3} & \cdots & \xrightarrow{\mathcal{D}_n} & F_n & \longrightarrow & 0 \\
 & & & & & & \downarrow & & \downarrow & & \downarrow & & & & \downarrow & & \\
 & & & & & & 0 & & 0 & & 0 & & & & 0 & &
\end{array}$$

The final argument is to use the differential Janet sequence as follows:

$$\underset{\text{potential}}{E} \xrightarrow{\mathcal{D}} \underset{\text{field}}{F_0} \xrightarrow[\text{field equations}]{\mathcal{D}_1} \underset{\text{matter}}{F_1} \xrightarrow[\text{conservation laws}]{\mathcal{D}_2} F_2 \rightarrow ?$$

but the physical interpretation has to be created!

We hope that the use of Lie pseudogroups in unifying the formalism of gauge theory will give the same impulse to theoretical physics as the one brought by Lie groups.

BIBLIOGRAPHY

1)M.JANET:J.Math.pures et appl.,t 3,1920,p 65

2)A.KUMPERA,D.C.SPENCER:Lie equations,Ann.of math. studies 73 Princeton University Press,1972

3)J.F.POMMARET:Systems of partial differential equations and Lie pseudogroups,Gordon and Breach,1978

4)J.F.POMMARET:Differential Galois theory (to appear in 1981)

5)W.DRECHSLER,M.E.MAYER:Fibre bundle techniques in gauge theory Springer lecture notes in physics 67,1977

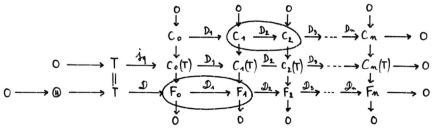

CONFORMAL INVARIANCE AND SYMMETRY BREAKING FOR THE ONE-DIMENSIONAL WAVE EQUATION

Luc Vinet
Centre de recherche de mathématiques appliquées
Université de Montréal
Montréal, Québec, Canada H3C 3J7

1. The purpose of this note is to present an outline of the results of a systematic application of Lie theory to linear hyperbolic equations (L.H.E.) in two variables. A detailed version can be found in reference [1]. The most general form of homogeneous, second order L.H.E. in two variables is

$$a\partial_{xx}u + 2b\partial_{xy}u + c\partial_{yy}u + d\partial_{x}u + e\partial_{y}u + fu = 0 \qquad (1)$$

where $a,b,\ldots$ are functions of the two independent variables x and y satisfying

$$ac - b^2 < 0 \qquad (2)$$

Substituting for x and y the characteristic curves ξ and η, eq. (1) with condition (2) satisfied can always be cast in the following form [2]:

$$\Delta_H u = \partial_{\xi\eta}u - F(\xi,\eta)u - G(\xi,\eta)\partial_\xi u - H(\xi,\eta)\partial_\eta u = 0 \qquad (3)$$

The problem we now address ourselves to, is to find and classify all equations $\Delta_H u = 0$ which have symmetries.

By a symmetry of an equation $\Delta u = 0$, we mean a differential operator transforming solutions into solutions, that is satisfying

$$[\Delta,Z] = \lambda\Delta \qquad (4)$$

where λ is an arbitrary operator (of order less than Z). The commutator of two symmetries is also a symmetry and the set of such operators thus realizes an abstract Lie algebra called the symmetry algebra of the equation. Given the realization $\{Z_i, i=1,2,\ldots\}$ of some basis for a symmetry algebra, the following set of invariance conditions must be verified:

$$[\Delta,Z_i] = \lambda_i\Delta \qquad i = 1,2,\ldots \qquad (5)$$

Our problem is consequently to solve that system for Δ_H. This amounts first to finding all possible realizations of symmetry algebras for L.H.E. and then to determine the invariant L.H.E. associated to each of these.

The motivations for such a study come both from physics and mathematics. A physical system is often described by P.D.E.'s. The symmetries of the system are then reflected in the invariance properties of these equations. It is interesting to know how the description is modified when the system is perturbed. (See [3] for a more detailed discussion). Consider for example the one-dimensional wave equation

$$\partial_{\xi\eta} u = 0. \qquad (6)$$

This equation is invariant under all conformal transformations of the pseudo-Euclidean plane, that is under all coordinate maps of the form

$$\xi' = r(\xi) \qquad \eta' = s(\eta) \qquad (7)$$

where r and s are "arbitrary" functions of their respective argument. The wave equation thus has an infinite dimensional symmetry algebra†. The addition on the right hand side of (6) of the interactions Fu, $G\partial_\xi u$, $H\partial_\eta u$ will lower that symmetry. Consequently the solution of the proposed problem will result in a classification of all possible breakings of the conformal symmetry of the one-dimensional wave equation, hence the title of this paper.

On the mathematical side, we know the intimate relation between separation of variables as a method for solving a given P.D.E. and the symmetry properties of that equation [4]. We have not undertaken a systematic investigation of all separable coordinate systems for the invariant L.H.E. we have found. Nevertheless we have considered in each case the separation of variables in all coordinate systems with at least one negligible variable. More precisely, let Z be a symmetry of some equation $\Delta u = 0$, to that equation we can add the following eigenvalue condition

$$Zu = \mu u \qquad (8)$$

(since compatibility is assured). Note that this auxiliary condition can be acted upon by the symmetry transformations of the equation $\Delta u = 0$. This enables one to classify these conditions into equivalence classes under the group of inner automorphisms of the symmetry algebra. It is hence possible to simultaneously diagonalize operators realizing a basis for an abelian subalgebra of the symmetry algebra. It turns out that these supplementary conditions lead to the separation of variables. Hence, mutually nonconjugate maximal abelian subalgebras of the symmetry algebra will be associated to inequivalent systems of separable coordinates. We have completely investigated such coordinate systems for each invariant L.H.E.

Let us now summarize our results. First we will discuss how we arrive at the solution of the problem. Eight classes of invariant L.H.E. are shown to exist. Two classes are associated with symmetry algebras of dimension one, three classes with symmetry algebras of dimension two (one corresponding to the Abelian 2-dimensional algebra while the other two, to realizations of the unique solvable algebra of that dimension). For 3-dimensional symmetry algebras, there is the class of the Klein-

---

† The situation is in this case exceptional in that the symmetry algebras are usually finite. For instance we know that the wave equation in p+q dimensions for $p+q > 2$, has $o(p+1,q+1)$ as its maximal symmetry algebra.

Gordon equation with $E(1,1)$ as invariance algebra and another class with $SU(1,1)$ as its symmetry algebra. The remaining class is that of the wave equation which is the only invariant L.H.E. to possess an infinite dimensional symmetry algebra. Except for those equations admitting only one symmetry involving arbitrary functions, the functional form of the invariant L.H.E., is completely determined. We have provided explicit solutions for each of these equations by separating variables in every co-ordinate system with at least one negligible variable. After having presented the invariant L.H.E.'s and their symmetries we will only present one example of how solutions to these equations are obtained by separation of variables.

2. We now put the problem into a more tractable form. We take the symmetries to be of the following linear form:

$$Z = a(\xi,\eta)\partial_\xi + h(\xi,\eta)\partial_\eta + c(\xi,\eta). \tag{9}$$

Recall that $\Delta_H = \partial_{\xi\eta} - F - G\partial_\xi - H_\eta$. The substitution of these two operators into the invariance condition $[\Delta_H, Z] = \lambda\Delta_H$ gives the following system of equations:

$$a_\eta = 0, \quad b_\xi = 0, \quad \lambda = a_\xi + b_\eta \tag{10}$$

$$aG_\xi + bG_\eta + c_\eta = -b_\eta G \tag{11}$$

$$aH_\xi + bH_\eta + c_\xi = -a_\xi H \tag{12}$$

$$aF_\xi + bF_\eta - c_\xi G - c_\eta H + c_{\xi\eta} = -(a_\xi + b_\eta)F. \tag{13}$$

We first observe from eq. (10) that the functions a and b only depend respectively on ξ and η. Further if a and b are identically zero, c must be a constant and the symmetry Z is trivial since it then only reflects the linearity of the equation. In conclusion, all non-trivial symmetries of the L.H.E. must be of the form

$$Z = a(\xi)\partial_\xi + b(\eta)\partial_\eta + c(\xi,\eta) \tag{14}$$

where a and b are not identically zero simultaneously. Also, the eqs.(11) and (12) are compatible if and only if the expression $G_\xi - H_\eta$ takes one of the following three forms:

i) $\dfrac{1}{a(\xi)b(\eta)}T(q_1)$ with $q_1 = \int^\xi \dfrac{d\zeta}{a(\zeta)} - \int^\eta \dfrac{d\zeta}{b(\zeta)}$

ii) $\dfrac{1}{a(\xi)}T(\eta)$

iii) $\dfrac{1}{b(\eta)}T(\xi)$

where T is an arbitrary function of its argument. This is thus a necessary (but not sufficient) condition for some L.H.E. to admit a symmetry. Finally if $F \equiv G \equiv H \equiv 0$, that is if $\Delta_H = \partial_{\xi\eta}$, we see that the wave equation possesses as symmetries all the operators $Z = f(\xi)\partial_\xi + g(\eta)\partial_\eta$ with arbitrary f and g. This set of operators constitutes the (infinite) conformal symmetry algebra of the wave equation. Following this discussion, it is clear that to obtain all L.H.E. with one or more symmetries, we must first find all realizations of algebras by operators of the form (14) and subsequently, given a basis $\{Z_i\}$ for each of these realizations, we then have to determine the corresponding invariant L.H.E. by solving for F,G,H the systems of eqs. given by the conditions $[\Delta_H, Z_i] = \lambda_i \Delta_H$.

We simplify the problem by introducing an equivalence relation. We will say that two L.H.E.'s $\Delta_H u = 0$ and $\Delta'_H r = 0$ are *convertible* [5,6] if there exists a function p such that the relation

$$u = pr \tag{15}$$

establishes a one-to-one correspondance between solutions of $\Delta_H u = 0$ and solutions of $\Delta'_H r = 0$. This notion is appropriate to the study of the symmetries of the L.H.E.'s since it is easy to prove [6] that the symmetry algebras of two convertible equations are isomorphic with their realizations related by the transformation

$$Z' = \frac{1}{p} Zp . \tag{16}$$

It suffices then to obtain a representative in each convertibility class of the L.H.E.'s with at least one symmetry.

So let $\Delta_H u = 0$ be some general L.H.E. with at least one symmetry. It is always possible to use the equivalence transformation (15) to convert $\Delta_H u = 0$ into an equation of the form

$$\partial_{\xi\eta} v = Fv + H\partial_\eta v \tag{17}$$

(i.e. to make $G \to 0$) while eliminating *at the same time* the functional part $C(\xi,\eta)$ of *one* arbitrary symmetry. Such will be the representatives we will look for in each convertibility class of invariant L.H.E.'s. Now note that because of eq.(11) when $G \equiv 0$, $c_\eta = 0$ so that all equations of the type (17) have symmetries for which the functional part only depends on ξ, $c = c(\xi)$.

Our initial problem has thus been reduced to the following one:

i) Find all realizations of algebras by operators of the form

$$Z_1 = a_1(\xi)\partial_\xi + b_1(\eta)\partial_\eta$$

$$Z_i = a_i(\xi)\partial_\xi + b_i(\eta)\partial_\eta + c_i(\xi) \qquad i=2,\ldots$$

(The equivalence transformation has been used to set one c equal to zero.)

ii) For each of these realizations find the most general equation

$$\partial_{\xi\eta} u = Fu + H\partial_\eta u$$

which admits these symmetries.

3. To find all the classes of L.H.E. symmetry algebra realizations, we proceed as follows. First find all the possible realizations of algebras by derivatives of the form

$$\overline{Z}_i = a_i(\xi)\partial_\xi + b_i(\eta)\partial_\eta.$$

For each of these, obtain the compatible system of functions $\{c_i(\xi),\ i=2,\ldots\}$. Such a set of functions is compatible with a given realization if it satisfies the following conditions

$$\overline{Z}_i(c_j(\xi)) - \overline{Z}_j(c_i(\xi)) = \sum_k D^k_{ij} c_k(\xi)$$

where

$$[\overline{Z}_i,\overline{Z}_j] = \sum_k D^k_{ij}\,\overline{Z}_k.$$

This method is exhaustive since the $\{\overline{Z}_i\}$ already realize effectively all the possible symmetry algebras. (Recall that we cannot have symmetries of the form $Z = c(\xi,\eta)$ other than constants.)

First consider derivatives in one real variable. It was shown by Lie (see [5]) that SU(1,1) is the maximal finite algebra that can be realized by operators $X_i = f_i(\xi)\partial_\xi$. If we take for SU(1,1) a basis $\{e_i,\ i=1,2,3\}$ such that $[e_1,e_2] = e_3$, $[e_2,e_3] = -e_1$, $[e_3,e_1] = -e_2$, we have for it a realization given by the operators $X_1 = f(\xi)\partial_\xi$, $X_2 = f(\xi)\mathrm{sh}\Xi\partial_\xi$ and $X_3 = f(\xi)\mathrm{ch}\Xi\partial_\xi$ with $\Xi = \int^\xi \frac{d\zeta}{f(\zeta)}$. The occurrence of an arbitrary function $f(\xi)$ allows for all changes of variable. Using derivatives in one variable we can thus construct only three different algebras, i.e. su(1,1) and its subalgebras $A_2: [e_1,e_2+e_3] = e_2+e_3$ and A_1 realized as $X_1 = f(\xi)\partial_\xi$.

Now we come to realizations by derivatives in two real variables of the type $\overline{Z} = f(\xi)\partial_\xi + g(\eta)\partial_\eta$. We have shown [1] that $o(2,2) \simeq su(1,1) \oplus su(1,1)$ is the maximal finite algebra that can be realized by such operators. All algebras that can be realized by such operators are ine one-to-one correspondence with the subalgebras of o(2,2) classified with respect to the conformal transformations of the pseudo-Euclidean plane. Patera et al [7] have given such a list of conjugacy classes of o(2,2) subalgebras under the action of the group O(2,2). The list of representatives of these o(2,2) subalgebra realizations is given in Table 1. It agrees with the list of Ref.7 if we take into account that different classes will collapse into one when conjugacy is defined with respect to the pseudogroup of conformal transformations in place of O(2,2).

Now let us obtain the functions $\{c_i(\xi),\ i=2,\ldots\}$. A simple analysis demonstrates that the only admissible systems of that kind are those that arise in realizations of algebras by operators in one variable $X_i = a_i(\xi)\partial_\xi + c_i(\xi)$. Consequently the compatible system of functions for a given realization by operators $\overline{Z}_i = a_i(\xi)\partial_\xi + b_i(\eta)\partial_\eta$ is determined by identifying the algebra realized by the ξ-dependent part of $\overline{X}_i = a_i(\xi)\partial_\xi$ of the operators $\overline{Z}_0$. It is the system of $\{c_i(\xi)\}$ corresponding to that subrealization in one variable that is to be adjoined to the $\overline{X}_i$. Finally we find that the only realization in one variable with a non-trivial system of $\{c_i\}$ is that of su(1,1) given by $\overline{X}_1 = f(\xi)\partial_\xi$, $\overline{X}_2 - \lambda \mathrm{ch}\Xi = f(\xi)\mathrm{sh}\Xi\partial_\xi - \lambda\mathrm{ch}\Xi$, $\overline{X}_3 - \lambda\mathrm{sh}\Xi = f(\xi)\mathrm{ch}\Xi\partial_\xi - \lambda\mathrm{sh}\Xi$.

4. The list of representatives for each class of L.H.E. symmetry algebra realizations is given in Table 1, together with the functions $F(\xi,\eta)$ and $H(\xi,\eta)$ that define the corresponding invariant equations. These functions were determined by solving for each realization the equations $[\Delta_H, Z_i] = \lambda_i\Delta_H$. The following notation is used in the table:

$$\Xi(\xi) = \int^\xi \frac{d\zeta}{f(\zeta)} \qquad \Theta = \int^\eta \frac{d\zeta}{g(\zeta)}$$

$$\overline{X}_1 = f(\xi)\partial_\xi \qquad \overline{Y}_1 = g(\eta)\partial_\eta$$

$$\overline{X}_2 = f(\xi)\mathrm{sh}\Xi\partial_\xi \qquad \overline{Y}_2 = g(\eta)\mathrm{sh}\Theta\partial_\eta$$

$$\overline{X}_3 = f(\xi)\mathrm{ch}\Xi\partial_\xi \qquad \overline{Y}_3 = g(\eta)\mathrm{ch}\Theta\partial_\eta$$

R and T are arbitrary functions of their argument, c, e and λ arbitrary real constants and $\delta = \pm 1$.

5. For each equation in the list we have determined the separable coordinate systems with at least one negligible variable and for every equation with symmetry algebra of dimension greater than one we have obtained complete solutions by separation of variables in each of these coordinate systems. We now show in an example how this is done. Consider the equation

$$\partial_{\xi\eta}u = \frac{c}{f(\xi)g(\eta)}\cdot\frac{\exp(-\Xi+\Theta)}{[1-\delta\exp(-\Xi+\Theta)]^2}\,u + \frac{e}{f(\xi)}\cdot\frac{\exp(-\Xi+\Theta)}{[1-\delta\exp(-\Xi+\Theta)]}\,\partial_\eta u \quad (\delta=\pm 1) \tag{18}$$

It admits A_2 realized by $Z_1 = f(\xi)\partial_\xi + g(\eta)\partial_\eta$ and $Z_2 = f(\xi)\exp\Xi\partial_\xi + \delta g(\eta)\exp\Theta\partial_\eta$ as symmetry algebra. Now A_2 has two mutually non-conjugate one-dimensional subalgebras realized by $\{Z_1\}$ and $\{Z_2\}$. This will enable us to separate variables in two different systems of coordinates.

a) Let us first use the condition

$$Z_1 u = f\partial_\xi u + g\partial_\eta u = \mu u \tag{19}$$

It yields the subsidiary equations

$$\frac{d\xi}{f} = \frac{d\eta}{g} = \frac{du}{\mu u} \tag{20}$$

from which we obtain

$$u = \phi(q_1)\exp\{\mu\int^{\xi}\frac{d\zeta}{f(\zeta)}\} \tag{21}$$

with

$$q_1 = \int^{\xi}\frac{d\zeta}{f(\zeta)} - \int^{\eta}\frac{d\zeta}{g(\zeta)} = \Xi - \Theta \tag{22}$$

In terms of this last variable the equation reads

$$\partial_{\xi\eta}u = \frac{c}{f(\xi)g(\eta)}\frac{e^{-q_1}}{(1-\delta e^{-q_1})^2}u + \frac{e}{f(\xi)}\frac{e^{-q_1}}{(1-\delta e^{-q_1})}\partial_{\eta}u \tag{23}$$

and substitution of eq. (21) into eq. (23) gives the following O.D.E. for $\phi(q_1)$:

$$\frac{d^2\phi}{dq_1^2} + [\mu - \frac{ee^{-q_1}}{(1-\delta e^{-q_1})}]\frac{d\phi}{dq_1} + \frac{ce^{-q_1}}{(1-\delta e^{-q_1})^2}\phi = 0 \tag{24}$$

The general solution of eq.(24) may be expressed in terms of Riemann's P symbol as

$$\phi(q_1) = P\begin{Bmatrix} 0 & 1 & \infty & \\ 0 & 0 & \alpha_1 & \dfrac{1}{1-\delta e^{-q_1}} \\ -\mu-\delta e & \mu & \alpha_2 & \end{Bmatrix} \tag{25}$$

with

$$\alpha_{\substack{1\\2}} = \frac{1}{2}[(1+\delta e) \mp \sqrt{(1+\delta e)^2-4\delta c}\ ] \tag{26}$$

b) Define $k(\xi) = f(\xi)e^{\Xi}$ and $\ell(\eta) = g(\eta)e^{\Theta}$. If in place of (19) we take

$$Z_2u = k\partial_{\xi}u + \ell\partial_{\eta}u = \nu u \tag{27}$$

we get

$$u = \phi(q_2)\exp\{\nu\int^{\xi}\frac{d\zeta}{k(\zeta)}\} \tag{28}$$

with

$$q_2 = \int^{\xi}\frac{d\zeta}{k(\zeta)} - \int^{\eta}\frac{d\zeta}{\ell(\zeta)}\ . \tag{29}$$

It is possible to show that

$$q_2 = -(\frac{f}{k} - \delta\frac{g}{\ell}) = -e^{-\Xi} + \delta e^{-\Theta} \tag{30}$$

In terms of this variable eq.(18) takes the form

$$\partial_{\xi\eta}u = \frac{c}{k(\xi)\ell(\eta)}\cdot\frac{1}{q_2^2}u + \frac{e\delta}{k(\xi)}\cdot\frac{1}{q_2}\partial_{\eta}u \tag{31}$$

Again substituting (28) into (31) we find for $\phi(q_2)$ an O.D.E., in this case

$$\frac{d^2\phi}{dq_2^2} + [\nu - \frac{e}{q_2}]\frac{d\phi}{dq_2} + \frac{c}{q_2^2}\phi = 0 \tag{32}$$

The solution of (32) depends on whether or not $\nu = 0$.

1° <u>if $\nu \neq 0$</u>

$$\phi(q_2) = q_2^{\varkappa} e^{-\frac{1}{2}\chi} w(\varkappa,m,\chi) \tag{33}$$

where $\chi = \nu q_2$, $\varkappa = e/2$, $m = \frac{1}{2}[(1+e)^2-4c]^{1/2}$ and $w(\varkappa,m,\chi)$ is the general solution to Whittaker's equation.

2° <u>if $\nu = 0$</u>

$$\phi(q_2) = A q_2^{m_1} + B q_2^{m_2} \tag{34}$$

with

$$m_{\substack{1\\2}} = \frac{1}{2}[(1+e) \mp \sqrt{(1+e)^2-4c}] \ .$$

In conclusion we would like to point out some directions for future work. The elliptic equations should obviously lend themselves to considerations similar to the ones presented here. Also a more thorough investigation of separation of variables in the invariant equations that were found could prove interesting. Finally these methods should be applicable to nonlinear equations.

This work was supported in part by the Ministère de l'Education du Gouvernement du Québec.

TABLE 1. Symmetry Algebra Realizations and Invariant Equations

REALIZATIONS		INTERACTIONS	
GENERATORS	TYPE OF ALGEBRA	$F(\xi,\eta)$	$H(\xi,\eta)$
$\overline{X}_1$	A_1	$\frac{1}{f(\xi)} R(\eta)$	$\frac{1}{f(\xi)} T(\eta)$
$\overline{X}_1+\overline{Y}_1$	A_1	$\frac{1}{f(\xi)g(\eta)} R(\Xi-\Theta)$	$\frac{1}{f(\xi)} T(\Xi-\Theta)$
$\overline{X}_1,\ \overline{Y}_1$	$A_1 \oplus A_1$	$\frac{c}{f(\xi)g(\eta)}$	$\frac{e}{f(\xi)}$
$\overline{X}_1,\ \overline{X}_2+\overline{X}_3$	A_2	0	0
$\overline{X}_1+\overline{Y}_1,\ \overline{X}_2+\overline{X}_3$	A_2	$\frac{c}{f(\xi)g(\eta)} \cdot \exp(-\Xi+\Theta)$	$\frac{e}{f(\xi)} \cdot \exp(-\Xi+\Theta)$

TO BE CONTINUED

TABLE 1 (continued)

$\bar{X}_1+\bar{Y}_1$, $\bar{X}_2+\bar{X}_3+\delta(\bar{Y}_2+\bar{Y}_3)$	A_2	$\frac{c}{f(\xi)g(\eta)}\cdot\frac{\exp(-\Xi+\Theta)}{[1-\delta\exp(-\Xi+\Theta)]^2}$	$\frac{e}{f(\xi)}\cdot\frac{\exp(-\Xi+\Theta)}{[1-\delta\exp(-\Xi+\Theta)]}$
$\bar{X}_1$, $\bar{Y}_1$ $\bar{X}_2+\bar{X}_3$	$A_2 \oplus A_1$	0	0
$\bar{X}_1+\bar{Y}_1$, $\bar{X}_2+\bar{X}_3$ $\bar{Y}_2+\bar{Y}_3$	E(2)	0	0
$\bar{X}_1-\bar{Y}_1$, $\bar{X}_2+\bar{X}_3$ $\bar{Y}_2+\bar{Y}_3$	E(1,1)	$\frac{c}{f(\xi)g(\eta)}\cdot\exp[-(\Xi+\Theta)]$	0
$\bar{X}_1+a\bar{Y}_1$, $\bar{X}_2+\bar{X}_3$ $\bar{Y}_2+\bar{Y}_3$		0	0
$\bar{X}_1$, $\bar{X}_2-\lambda\mathrm{ch}\Xi$ $\bar{X}_3-\lambda\mathrm{sh}\Xi$	**su**(1,1)	- 0	- $\lambda \neq 0$ 0 $\lambda = 0$
$\bar{X}_1+\bar{Y}_1$, $\bar{X}_2+\delta\bar{Y}_2-\lambda\mathrm{ch}\Xi$ $\bar{X}_3+\delta\bar{Y}_3-\lambda\mathrm{sh}\Xi$	su(1,1)	$\frac{c}{f(\xi)g(\eta)}\cdot\frac{\exp(-\Xi+\Theta)}{[1-\delta\exp(-\Xi+\Theta)]^2}$	$\frac{\lambda}{f(\xi)}\ \frac{1+\delta\exp(-\Xi+\Theta)}{1-\delta\exp(-\Xi+\Theta)}$
$\bar{X}_1$, $\bar{X}_2-\lambda\mathrm{ch}\Xi$ $\bar{X}_3-\lambda\mathrm{ch}\Xi$, $\bar{Y}_1$	su(1,1)$\oplus A_1$	- 0	- $\lambda \neq 0$ 0 $\lambda = 0$
$\bar{X}_1$, $\bar{X}_2+\bar{X}_3$ $\bar{Y}_1$, $\bar{Y}_2+\bar{Y}_3$	$A_2 \oplus A_2$	0	0
$\bar{X}_1$, $\bar{X}_2-\lambda\mathrm{ch}\Xi$ $\bar{X}_3-\lambda\mathrm{sh}\Xi$, $\bar{Y}_1,\bar{Y}_2+\bar{Y}_3$	su(1,1)$\oplus A_2$	- 0	- $\lambda \neq 0$ 0 $\lambda = 0$
$\bar{X}_1$, $\bar{X}_2-\lambda\mathrm{ch}\Xi$ $\bar{X}_3-\lambda\mathrm{sh}\Xi$, $\bar{Y}_1,\bar{Y}_2,\bar{Y}_3$	o(2,2)	- 0	- $\lambda \neq 0$ 0 $\lambda = 0$

References

[1] L.Vinet, *Les symétries des équations hyperboliques linéaires à deux variables*, Thèse de Doctorat, 3ième cycle, Université Pierre et Marie Curie, Paris 6 (1979).

[2] A.D. Forsyth, *Theory of differential equations*, part IV, vol.6, Dover Publications Inc. New York (1959).

[3] C.Boyer, R.T.Sharp and P.Winternitz, J.Math.Phys. *17*, 1439 (1976).

[4] W.Miller Jr., *Symmetry and Separation of Variables*, Addison Wesley, Reading, Mass. (1977).

[5] W.Miller Jr., *Lie Theory and Special Functions*, Academic Press, New York (1968).

[6] U,Niederer, Helv.Phys.Acta *46*, 191 (1973); Proceedings of the IInd International Colloquium on Group Theoretical Methods in Physics, Nijmegen (1973).

[7] J.Patera, R.T.Sharp, P.Winternitz and H.Zassenhaus, J.Math.Phys. *18*, 2259 (1977).

THE FLAVOUR SEQUENCE AND SUPERSELECTION RULES

Asim O. Barut

Department of Physics, University of Colorado, Boulder, Co. 80308

Luigi Mercaldo, Ileana Rabuffo and Giuseppe Vitiello

Istituto di Fisica, Università di Salerno, 84100 Salerno, Italia

In the lepton (flavour) sequence e , μ , τ , ... the extra mass of the muon and of the other heavy leptons may be interpreted as coming from a non-zero vacuum expectation value of the electromagnetic field [1] (similar to the mass derivation by Higgs mechanism from a vacuum expectation value of a scalar field). In this way a simple derivation of a generalized renormalizable QED [2] for the electron-muon system is obteined . In ref. [2] the two subspaces of states describing the electron and muon, respectively, are separated by a superselection rule in such a way that only pairs from one of the two sectors, e.g. $(\mu^-\mu^+)$, can be connected to the other one by electromagnetic process, or viceversa. The energy is definite positive and in each sector the ordinary QED holds. A single fermionic field ψ is introduced and the theory accounts for the equal charges of the fermions. Since the muon and the other heavy leptons are viewed as different states (excitations) of the field ψ, they should already be contained in QED and the renormalization procedure should be formulated in such a way as to incorporate also the muon sector, the τ sector, etc. We will see later that the present procedure to obtain this generalized QED can be useful in the formulation of such a new renormalization procedure.

We further discuss the indefinite metric problem and show that it can be dealt with by giving a structure to the vacuum of the theory; from this a superselection rule will emerge in a natural way.

We start with the standard electromagnetic theory

$$(-i\gamma^{\mu}\partial_{\mu}-m-e\gamma^{\mu}A_{\mu}(x))\psi(x)=0;\quad \Box A_{\mu}(x)=e\bar{\psi}\gamma_{\mu}\psi \qquad (1)$$

where m is the bare mass of the electron, the lowest state of <u>leptonic matter</u> . When the theory is renormalized it should contain also other leptons besides the physical electron and the physical photon. For this purpose, we assume that

$$A_\mu(x)=\langle A_\mu\rangle+\tilde{A}_\mu \tag{2}$$

with $\langle A_\mu\rangle$ such that

$$i(\partial_\mu-e\tilde{A}_\mu)C\psi=e\langle A_\mu\rangle\psi \tag{3}$$

where $C=C(\gamma^\mu\tilde{D}_\mu)$, $\tilde{D}_\mu=(\partial_\mu-e\tilde{A}_\mu)$. Usually C is assumed to be zero. Here we must determine it. In order that the ψ field admits an excitation with mass $k\neq m$, it must also satisfy the equation

$$(-i\gamma^\mu\tilde{D}_\mu-k)\psi=0 \tag{4}$$

Operating with C on (4) we have

$$-iC\gamma^\mu\tilde{D}_\mu\psi=kC\psi \tag{5}$$

By inserting (2) and (3) into (1) we have also

$$(-i\gamma^\mu\tilde{D}_\mu-m)\psi=i\gamma^\mu\tilde{D}_\mu C\psi \tag{6}$$

Thus, from (5) and (6),

$$C\psi=-\frac{1}{k}iC\gamma^\mu\tilde{D}_\mu\psi=-\frac{1}{k}(-i\gamma^\mu\tilde{D}_\mu-m)\psi \tag{7}$$

where the commutator of C with $\gamma^\mu\tilde{D}_\mu$ has been dropped since it gives a Pauli term which does not affect the mass spectrum [2] . By using (7), it is now possible to write (6) as

$$(-i\gamma^\mu\tilde{D}_\mu-m)(-i\gamma^\mu\tilde{D}_\mu-k)\psi=0 \tag{8}$$

which is the generalized Dirac equation proposed in the generalized renormalizable QED to study the electron-muon system [2] . The procedure can be continued successively to the excitation of the muon and so on.

It is well known that the "higher order" equation (8) leads to the difficulties of indefinite metric and of physical interpretation. To face such problems we introduce [1] the physical vacuum $|0_e\rangle$ for the electron sector as

$$|0_e\rangle=\prod_{p,\sigma}(\mu^-_{p\sigma})^\dagger(\mu^+_{p\sigma})^\dagger\ |0\rangle \tag{9}$$

which contains a "sea" of $\mu^-\mu^+$ pairs. In this sector , physical states satisfy the condition

$$[(\mu^-)^\dagger(\mu^-) - (\mu^+)^\dagger(\mu^+)]\,|phys.> = 0 \qquad (10)$$

wich is analogous to the Gupta-Bleuler condition. The condition (10) express the fact that the physical states are built ,in the electron sector , on the vacuum $|0_e>$. Due to (10) only pairs of $\mu^-\mu^+$ can be excited to the electron sector: a superselection rule is thus naturally implied by this construction . From (9) and (10) is also evident that since the vacuum carries always pairs of "ghost" muons, physical states have positive norm.

The esplicit construction of the vacuum state (9) can be performed by a GNS-type construction and has been used in the discussion of the indefinite metric in the non relativistic Lee-model [3].The procedure is as follows. Assume the volume V to be finite. Let $G^\dagger_{\bar{k}}$ $(G_{\bar{k}})$ be the creation (annihilation) fermionic operator with <u>anomalous</u> anticommutations

$$[G_{\bar{k}}, G^\dagger_{\bar{k}'}]_+ = -\delta_{\bar{k}\bar{k}'} \;\; ; \;\; [G_{\bar{k}}, G_{\bar{k}'}]_+ = [G^\dagger_{\bar{k}}, G^\dagger_{\bar{k}'}]_+ = 0 \qquad (11)$$

Let $|0>$ denote the vacuum: $G_{\bar{k}}|0>=0$,$<0|0>=1$. We want to construct a state $|0(\phi)>$,with ϕ a function of certain relevant parameters of the theory ,e.g. the coupling constant g in the Lee model, such that

$$<0(\phi)|G^\dagger_{\bar{k}}G_{\bar{k}}|0(\phi)> < 0 \qquad (12)$$

but

$$<0(\phi)|\Theta_c|0(\phi)> > 0 \qquad (13)$$

Here Θ_c is a product of operators with <u>canonical</u> (anti)commutators.Introduce then a fermionic operator $\tilde{G}^\dagger_{\bar{k}}(\tilde{G}_{\bar{k}})$:

$$[\tilde{G}_{\bar{k}}, \tilde{G}^\dagger_{\bar{k}'}]_+ = -\delta_{\bar{k}\bar{k}'} \;\; ; \;\; [\tilde{G}_{\bar{k}}, \tilde{G}_{\bar{k}'}]_+ = [\tilde{G}^\dagger_{\bar{k}}, \tilde{G}^\dagger_{\bar{k}'}]_+ = [\tilde{G}_{\bar{k}}, G_{\bar{k}}]_+ = 0 \qquad (14)$$

and a tilde-vacuum $|\tilde{0}>$: $\tilde{G}_{\bar{k}}|\tilde{0}>=0$; $<\tilde{0}|\tilde{0}>=1$. Denote further $|0>\otimes|\tilde{0}>$ by $|0>$. The state $|0(\phi)>$ is then given by

$$|0(\phi)\rangle = exp\,[-iF]\,|0\rangle = \prod_{\bar{k}} (cos\phi_{\bar{k}} - sin\phi_{\bar{k}} G^{\dagger}_{\bar{k}} \tilde{G}^{\dagger}_{\bar{k}})\,|0\rangle \tag{15}$$

$$F = -i\sum_{\bar{k}} \phi_{\bar{k}} (G_{\bar{k}} \tilde{G}_{\bar{k}} - \tilde{G}^{\dagger}_{\bar{k}} G^{\dagger}_{\bar{k}}) \tag{16}$$

F is the generator of the Bogoliubov transformations

$$G_{\bar{k}}(\phi) = G_{\bar{k}} cos\phi_{\bar{k}} - \tilde{G}^{\dagger}_{\bar{k}} sin\phi_{\bar{k}}\ ;\ \tilde{G}_{\bar{k}}(\phi) = \tilde{G}_{\bar{k}} cos\phi_{\bar{k}} + G^{\dagger}_{\bar{k}} sin\phi_{\bar{k}} \tag{17}$$

and h.c.. Note that $G_{\bar{k}}(\phi)\,|0(\phi)\rangle = 0$ and due to the presence of pairs $G^{\dagger}_{\bar{k}} \tilde{G}^{\dagger}_{\bar{k}}$ in $|0(\phi)\rangle$ is

$$\langle 0(\phi)\,|0(\phi)\rangle = 1 \tag{18}$$

Also note that equations (12) and (13) are satisfied. Our prescription now is that physical states must be constructed by cyclic operation of operators with canonical (anti)commutators on the "vacuum" $|0(\phi)\rangle$; this is equivalent to the condition (cf.eq.(10))

$$[G^{\dagger}_{\bar{k}} G_{\bar{k}} - \tilde{G}^{\dagger}_{\bar{k}} \tilde{G}_{\bar{k}}]\,|phys\rangle = 0. \qquad \forall\ \bar{k} \tag{19}$$

Thus

$$\langle phys\,|phys\rangle \geq 0 \tag{20}$$

In the Lee model there is a ghost state for $g > g_{crit}$; then we put [3] $\phi = \phi(g)$, $\forall\ \bar{k}$ and

$$\begin{aligned} sin\phi(g) &= 0 \qquad for \quad g < g_{crit} \\ sin\phi(g) &= 1 \qquad for \quad g > g_{crit} \end{aligned} \tag{21}$$

When (21) is used in (15) it can be shown that the ghost is excluded from the physical sector, the expectation of the Hamiltonian in the physical states is real [3], the unitarity of the S-matrix can be explicitely shown and the "higher order" field equation guarantees the renormalizability of the theory [4]. It should be noticed that in the infinite volume-limit is

$$\langle 0\,|0(\phi)\rangle = \prod_{\bar{k}} cos\phi_{\bar{k}} = exp\,[\sum_{\bar{k}} log\ cos\phi_{\bar{k}}] \xrightarrow[V \to \infty]{} 0 \tag{22}$$

unless $\phi_{\bar{k}} = 0$, $\forall \bar{k}$, since $\sum_{\bar{k}} \longrightarrow \frac{V}{(2\pi)^3} \int d\bar{k}$. In this limit the generator F became formal and the representations $\{ |0(\phi)> \}$ for different ϕ's are unitarily inequivalent representations of the canonical commutation relations (we observe that the Bogoliubov transformations preserve the canoical commutation relations). In this connection we note that an interesting aspect of indefinite-metric QFT examples is that in some way one can reduce to a definite metric theory by paying a convenient price: in QED, for example, indefinite metric is avoided by choosing a convenient gauge; the price is the loss of covariance of the theory. The gauge invariance ensures the physical equivalence of different formulations. In other examples, say the Lee model, one can avoid indefinite metric by choosing the renormalized coupling constant g smaller than a certain critical value g_{crit} : in such a case one obtains a theory of different physical content . Thus the occourrence of indefinite metric is related to a certain degree of freedom in the choice of the framework of the theory. In the present approach we have that the ghost state can be excluded from the physical sector by a convenient choice of the representation $\{ |0(\phi)> \}$.

In conclusion a feature which emerges from this work is that the ghost states in the generalized QED, as well in the Lee model and in the Pauli-Villars regularization, could be associated with a physical particle, say $|\mu>$ in a second sector , in such a way that only pairs $(\mu^+\mu^-)$ belong to the electron sector and a superselection rule separates the two sectors.

[1] A.O.Barut, I.Rabuffo and G.Vitiello; Lett.Nuovo Cimento, 26, 253, (1979)

[2] A.O.Barut and J.Crawford; Phys.Letters, B82, 233, (1979)

[3] I.Rabuffo and G.Vitiello; Nuovo Cimento, A44, 401, (1978)

[4] L.Mercaldo, I.Rabuffo and G.Vitiello; in preparation.

THE ZITTERBEWEGUNG OF THE ELECTRON AND ITS COMPACT PHASE SPACE

Asim O. Barut and Anthony J. Bracken
Department of Physics
University of Colorado
Boulder, CO 80309, U.S.A.

The recent discovery of new leptons raises the question of a possible sub-structure of these particles, and suggests in particular a fresh look at the nature of the Zitterbewegung of the Dirac electron [1]. Schrödinger [2] discovered this highly oscillatory, microscopic motion when he examined the evolution in time of the coordinate operator $\vec{x}$ associated with Dirac's equation. He attempted to identify macroscopic (or center of mass) variables and microscopic (or Zitterbewegung) variables for the electron, in association with a resolution of the motion described by $\vec{x}(t)$ into a macroscopic motion on which is superposed a microscopic motion-the Zitterbewegung. Our outlook is essentially the same as Schrödinger's, but we make a different identification of some of the basic "microscopic" variables and are then able to characterize the Zitterbewegung, in the rest frame of the electron, as a finite quantum system having a compact phase space associated with the compact symplectic group Sp (4) [$\sim$SO(5).].

The description of the Zitterbewegung involves the use of so-called "odd" operators which mix positive and negative energy states of the system, and for this reason the Zitterbewegung is sometimes dismissed as an unphysical curiosity. But it has observable effects-for example, they will appear if one calculates the matrix elements of $x_i x_j$ between positive energy states. We work in the vector space spanned by positive and negative energy states, in order to examine the geometrical and dynamical structure of the Zitterbewegung.

The hermitian dynamical variables for the Dirac electron are $\vec{x}$, a canonically conjugate momentum variable $\vec{p}$, and the rather mysterious variables α_1, α_2, α_3 and β, which commute with $\vec{x}$ and $\vec{p}$, anti-commute with each other, and have unit squares. The Hamiltonian is

$$H = c\vec{\alpha} \cdot \vec{p} + mc^2\beta \tag{1}$$

and the Heisenberg equations of motion (obtained using $i\hbar\dot{A} = [A,H]$) are then

$$\dot{\vec{x}} = c\vec{\alpha}, \quad i\hbar\dot{\vec{\alpha}} = 2c\vec{p} - 2H\vec{\alpha}, \quad \dot{\vec{p}} = \vec{0}. \tag{2}$$

It follows that $\vec{x}$ does not satisfy the second-order equation $\ddot{\vec{x}} = \vec{0}$, but rather the third-order equation

$$i\hbar\dddot{\vec{x}} = -2H\ddot{\vec{x}}. \tag{3}$$

Following Schrödinger, we find the solution

$$\vec{x}(t) = \vec{x}_A(t) + \vec{\xi}(t) \tag{4}$$

where

$$\vec{x}_A(t) = \vec{a} + c^2H^{-1}\vec{p}t$$

$$\vec{a} = \vec{x}(0) - \tfrac{1}{2}i\hbar c\vec{\alpha}(0)H^{-1} + \tfrac{1}{2}i\hbar c^2H^{-2}\vec{p}$$

$$\vec{\xi}(t) = \tfrac{1}{2}i\hbar c[\vec{\alpha}(0) - cH^{-1}\vec{p}]H^{-1}\exp(-2iHt/\hbar). \tag{5}$$

We can regard $\vec{x}_A$ as a center of mass coordinate - its time dependence is what we would expect classically - and $\vec{\xi}$ as the coordinate of the charge relative to this center of mass. Note that because H is of the order of mc^2, $|\vec{\xi}|$ is of the order of $\hbar/2mc$ ($\approx 10^{-11}$cm), while the angular frequency of $\vec{\xi}$ is of the order of $2mc^2/\hbar$ ($\approx 10^{21}\text{sec}^{-1}$).

We consider the relative motion in the rest frame of the center of mass, where $\vec{p} = \vec{0}$. Then $\vec{\xi}(t)$ reduces to what we call $\vec{Q}(t)$,

$$\vec{Q}(t) = \tfrac{1}{2}i(\hbar/mc)\vec{\alpha}(0)\beta\exp(-2imc^2\beta t/\hbar) = \tfrac{1}{2}i(\hbar/mc)\vec{\alpha}(t)\beta, \tag{6}$$

because H reduces to $mc^2\beta$ and H^{-1} to β/mc^2. The variable β is a constant in this frame. Now $\dot{\vec{x}}_A = \vec{0}$ in this frame, but the velocity of the charge is not zero:

$$\dot{\vec{x}}(t) = \dot{\vec{\xi}}(t) = \dot{\vec{Q}}(t) = c\vec{\alpha}(t), \tag{7}$$

and we are led to define a relative momentum vector $\vec{P}(t)$ (to be thought of as the momentum of the charge in this frame) by

$$\vec{P}(t) = mc\vec{\alpha}(t). \tag{8}$$

Here we differ from Schrödinger, who identified as the "microscopic" momentum an anti-hermitian variable which in this frame takes the value $mc\vec{\alpha}(t)\beta$, equal to $-2i(m^2c^2/\hbar)\vec{Q}(t)$.

In terms of $H(= mc^2\beta)$, $\vec{Q}$ and $\vec{P}$ the relative motion in this frame is described as a harmonic oscillator, for we have

$$[\vec{Q},H] = i(\hbar/m)\vec{P}, \quad [\vec{P},H] = -4i(m^3c^4/\hbar)\vec{Q} \tag{9}$$

implying

$$\ddot{\vec{Q}} + \omega^2\vec{Q} = \vec{0}, \quad \omega = 2mc^2/\hbar. \tag{10}$$

But the phase space of this oscillator is unusual: we have the commutation relations

$$[Q_i, P_j] = -i\hbar\delta_{ij}\beta \tag{11}$$

where $\vec{S} = -\tfrac{1}{4}i\hbar\vec{\alpha}\times\vec{\alpha}$ is the spin angular momentum (a constant in this frame). Since β commutes with $\vec{S}$, and

$$[\beta,\vec{Q}] = -i(\hbar/m^2c^2)\vec{P}, \quad [\beta,\vec{P}] = -4i(m^2c^2/\hbar)\vec{Q} \tag{13}$$

$$[Q_i, S_j] = i\hbar\,\varepsilon_{ijk}Q_k, \quad [P_i, S_j] = i\hbar\,\varepsilon_{ijk}P_k \tag{14}$$

$$[S_i, S_j] = i\hbar\,\varepsilon_{ijk}S_k, \tag{15}$$

the Lie algebra generated by $\vec{Q}$ and $\vec{P}$ closes on the algebra spanned by the ten hermitian operators β, $\vec{Q}$, $\vec{P}$ and $\vec{S}$. This is the Lie algebra of compact Sp(4), or SO(5). It is to be compared with the algebra spanned by I, $\vec{Q}$, $\vec{P}$ and $\vec{L}$ (= $\vec{Q} \times \vec{P}$) in the case of the usual non-compact, flat phase-space and Heisenberg relations $[Q_i, P_j]=i\hbar\delta_{ij}I$. In fact there is a contraction from the first algebra to the second. The Sp(4) or SO(5) algebra appearing here is the one well-known to be associated with the Dirac algebra, but now it appears with a dynamical interpretation, as indeed do the Dirac matrices. The representation of Sp(4) involved is of course the 4-dimensional one, so the Zitterbewegung appears in the rest frame of the center of mass as a finite quantum system associated with a compact symplectic structure.

Weyl [3] suggested the possible interest of finite-dimensional analogues of the Weyl-Heisenberg group in quantum mechanics, and Santhanam [4] has recently examined analogues of the canonical commutation relations in finite-dimensional Hilbert spaces, for one Q and one P, following Weyl's ideas. In this connection we note that the operators

$$A_j = i\,\exp(i\pi mcQ_j/\hbar) = i\beta\alpha_j$$

$$B_j = i\,\exp(i\pi P_j/2mc) = -\alpha_j \tag{16}$$

generate under multiplication a 2-valued representation of a finite Abelian group, in accordance with Weyl's general proposals.

Let us try to understand the structure of the finite quantum system, and then of Dirac's equation, from a more basic point of view. Suppose we wish to describe a point charge as a finite quantum system in three (spatial) dimensions, this system to represent the internal structure of a relativistic "particle" in its rest frame. Then we may suppose the necessity of hermitian operators $\vec{Q}$ and $\vec{P}$ acting in the finite-dimensional space. These cannot satisfy the canonical commutation relations, as is well-known. The possible relations closest to the canonical ones in form are as in (11), where β is hermitian and traceless, and $\beta^2 = I$. Then β can be thought of as diagonal, with an equal number of (+1)'s and (-1)'s on the diagonal. If the system is to be isotropic, we require also the existence of hermitian angular momentum operators $\vec{S}$, commuting with β and satisfying (14) and (15). The commutator $[Q_i, Q_j]$ cannot vanish, for then $\vec{Q}$ and $\vec{S}$ would generate a unitary finite-dimensional representation of E(3), and none such exist. The next simplest possibility is

$$[Q_i, Q_j] = +i(\lambda^2/\hbar)\varepsilon_{ijk}S_k \tag{17}$$

where λ is a constant with dimensions of length. The appearance of such a constant is inevitable for a finite quantum system, where for example each Q_i must have discrete dimensional eigenvalues. Note that the plus sign in the right-hand side of

(17) is essential: if it were a minus, then $\vec{Q}$ and $\vec{S}$ would generate a unitary finite-dimensional representation of SL(2,C), and none such exist. By a similar argument, the simplest form for the commutator $[P_i, P_j]$ is

$$[P_i, P_j] = +i(b^2\hbar/\lambda^2)\varepsilon_{ijk}S_k \tag{18}$$

where b is a positive dimensionless constant. Consistency with the Jacobi identity of the relations imposed so far requires that

$$[\beta, \vec{Q}] = -i(\lambda^2/\hbar)\vec{P}, \quad [\beta, \vec{P}] = i(b^2\hbar/\lambda^2)\vec{Q} \tag{19}$$

and we arrive again at the Lie algebra of Sp(4). It is not hard to show that the only irreducible representation of Sp(4) in which $\beta^2 = I$ can hold is the 4-dimensional one, and then only if $b = 2$. We see in this way that the structure of the phase-space algebra of the Zitterbewegung is the simplest possible under the conditions imposed. There is also little freedom in the choice of the dynamics of this finite quantum system. We know from our familiarity with the Dirac matrices that the only 3-scalar hermitian Hamiltonians one can construct from β, $\vec{Q}$, $\vec{P}$ and $\vec{S}$ have the general form

$$H_r = (\hbar c/\lambda)(u\beta + vI) \tag{20}$$

where u (which can be assumed non-negative without loss of generality) and v are real, dimensionless constants. (We exclude pseudo-scalars like $(c/\lambda^2)\vec{Q} \times \vec{Q} \cdot \vec{P}$.) Since we then have

$$[\vec{Q}, H_r] = iuc\lambda\vec{P}, \quad [\vec{P}, H_r] = -4iu(c\hbar^2/\lambda^3)\vec{Q}, \tag{21}$$

the only possible non-trivial dynamics is that of an harmonic oscillator, with angular frequency $\omega = 2cu/\lambda$.

Now we supposed that this finite quantum system represents the internal dynamics in the rest frame of a relativistic "particle". If this particle has energy-momentum 4-vector $(H/c, \vec{p})$, we have $H = H_r$ when $\vec{p} = \vec{0}$, and the rest-mass operator M of the particle satisfies

$$M^2c^4 = H_r^2 = (\hbar^2/\lambda^2c^2)[(u^2 + v^2)I + 2uv\beta] \tag{22}$$

in that frame. But if $v \neq 0$, it is impossible to boost the rest-frame states and operators in such a way that the operator $[(u^2 + v^2)I + 2uv\beta]$ does represent the form of an invariant in the rest-frame, while $(\hbar c/\lambda)(u\beta + vI)$ represents the form of the fourth component of a 4-vector (energy-momentum) there. We must take $v = 0$, and then we have

$$H_r = mc^2\beta, \quad M^2 = m^2I, \quad m = (u\hbar/\lambda c). \tag{23}$$

Note that we do not assign any mass to the point charge at $\vec{Q}$. Rather we prescribe the constant λ, which determines the curvature of the internal phase space, and the rest-mass energy of the particle appears as simply the energy of the internal motion in the rest frame. Likewise the spin in this frame is simply the angular momentum of the internal motion.

In order to be able to describe the dynamics in an arbitrary Lorentz frame, we need to identify a suitable dynamical group for the finite system, containing a Lorentz subgroup with generators $\vec{M}$ satisfying

$$[M_i, M_j] = -i\hbar\,\varepsilon_{ijk}S_k, \quad [M_i, S_j] = i\hbar\varepsilon_{ijk}M_k. \tag{24}$$

Acting with hindsight, we take $\vec{M} = (1/uc)H_r\vec{Q} = -\frac{1}{2}i\lambda\vec{P}$. The dynamical algebra should also contain H_r and its symmetry generators $\vec{S}$; and the algebra generated by H_r, $\vec{S}$ and $\vec{M}$ is that of SO(3,2) or Sp(4,R), with basis H_r, $\vec{S}$, $\vec{M}$ and $i\vec{Q}$. It is important that this SO(3,2) algebra contains a 4-vector $\gamma^\mu = [\beta,(2i/\lambda)\vec{Q}]$. That enables us essentially to follow the procedure used elsewhere [5] to provide a description of the H-atom as a relativistic "particle". The SO(3,2) group of the Zitterbewegung can also be extended to SO(4,2), by including as generators the pseudo-scalars $\vec{P}\times\vec{P}\cdot\vec{Q}$ and $\vec{Q}\times\vec{Q}\cdot\vec{P}$, and the pseudo-vector $\vec{Q}\times\vec{P}$.

Now consider a rest-frame state $\chi_{r\pm}(\vec{p})$ with $\pm$ve energy, in a $\vec{p}$-representation. It has the general form

$$\chi_{r\pm}(\vec{p}) = \chi_\pm \cdot \delta(\vec{p}), \quad \text{where} \quad \beta\chi_\pm = \pm\chi_\pm. \tag{25}$$

Boosting this state in the usual way to the frame where $\vec{p} = \hbar\vec{k}$:

$$\chi_{r\pm} \to \chi_{\vec{k}\pm} = B^\pm(\vec{\theta})\chi_\pm \cdot \delta(\vec{p} - \hbar\vec{k})$$

$$B^\pm(\vec{\theta}) = \exp[\pm(i/\hbar)\vec{\theta}\cdot\vec{M}]$$

$$\vec{\theta} = (\vec{k}/k)\ \text{arctanh}\ [\hbar k/(m^2c^2 + \hbar^2k^2)^{\frac{1}{2}}]$$

$$k = |\vec{k}|, \tag{26}$$

we find

$$(c\hbar\gamma^0\vec{\gamma}\cdot\vec{k} + mc^2\gamma^0)\chi_{\vec{k}\pm} = \pm(m^2c^2 + \hbar^2k^2)^{\frac{1}{2}}\chi_{\vec{k}\pm}. \tag{27}$$

This implies that the Hamiltonian of the particle is given by

$$H = c\gamma^0\vec{\gamma}\cdot\vec{p} + mc^2\gamma^0 = c\vec{\alpha}\cdot\vec{p} + mc^2\beta. \tag{28}$$

It is now possible to go to the usual configuration representation via a Fourier transform, and then go to the Schrödinger picture to obtain the familiar Dirac wave functions $\psi(\vec{x},t)$ satisfying Dirac's equation. Although the coordinate vector $\vec{x}$ introduced in this way is mathematically conjugate to $\vec{p}$, it is not the coordinate of the center of mass, for if we now couple the particle to an external electromagnetic field via the usual minimal coupling prescription, we are implying that the charge has coodinate $\vec{x}$ at time t. But in the rest-frame, according to (6), the relative coordinate of the charge has the value

$$\vec{\xi} = \tfrac{1}{2}i(\hbar/mc)\vec{\alpha}\beta = \tfrac{1}{2}i(\lambda/u)\vec{\alpha}\beta = (1/u)\vec{Q}. \tag{29}$$

Since we began by supposing that the relative coordinate of the charge is $\vec{Q}$ in this frame, we must for consistency new set $u = 1$, so that

$$H_r = (\hbar c/\lambda)\beta = mc^2\beta, \quad m = \hbar/\lambda c \ . \tag{30}$$

We conclude that the only dynamics of the finite, charged quantum system which is consistent with relativity, is the Zitterbewegung, and Dirac's equation is the relativistic wave equation appropriate to the description of this system in an arbitrary moving frame of reference.

References:

1. A.O. Barut and A. J. Bracken, "The Zitterbewegung and the Internal Geometry of the Electron," submitted for publication.
2. E. Schrödinger, Sitzungsb. Preuss. Akad. Wiss. Phys. - Math. Kl. 24, 418 (1930).
3. H. Weyl, "The Theory of Groups and Quantum Mechanics," (Dover, New York, 1950), pp. 272-280.
4. T. S. Santhanam, Foundns. Phys. 7, 121 (1977).
5. A. O. Barut and W. Rasmussen, J. Phys. B 6, 1695, 1713 (1973).

NEW EXPERIMENTAL EVIDENCE FOR A NEW INTERPRETATION OF FLAVOR SU(3)

A. Bohm
The University of Texas at Austin
Austin, Texas 78712 U.S.A.

Abstract: Present experimental data require a refinement of the Cabibbo model in which SU(3) is interpreted as spectrum generating group.

The emphasis in this talk is on physics and experimental numbers and it may, therefore, differ from the usual talk at this conference. The connection between group theoretical methods and experiment has been the subject of the spectrum generating group (S.G.) approach which has been successfully applied to a variety of physical problems including the decay processes $P \to \ell\nu$, $P \to P'\ell\nu$, $V \to e\bar{e}$, $V \to P_\gamma$ and $B \to B'\ell\nu$ where P, V, B is any pseudoscalar meson, vector meson and baryon. The subject has been reviewed in the 1978 Group Theory Colloquium by Teese, Garcia and Kielanowski.[1)]

There are three reasons for which I have chosen the process $B \to B'\ell\nu$ for my talk at this conference in Mexico. The first reason is that this is the subject of an international collaboration[2)] in which my Mexican friend Augusto Garcia played the most significant part. When I met him seven years ago he had investigated the phenomenology of the process $B \to B'\ell\nu$ and concluded that the prevailing belief that the Cabibbo model with SU(3) symmetry is the ultimate answer for these processes can really not be justified by a fit of the model to the experimental data. The custom at that time--and even still today--is to fit the theory not to the experimental data (which are the decay rates $\Gamma(B\to B')$ and the correlation coefficients and asymmetries $\alpha_{e\nu}^{B\to B'}$, $\alpha_{e}^{B\to B'}$, $\alpha_{\nu}^{B\to B'}$, $\alpha_{B'}^{B\to B'}$) but to some other quantities (the "experimental" g_A/g_V-rations) which were extracted from the experimental data under a theoretical assumption that was part of the model to be tested, namely that the g_2-terms (second class contributions) are zero. A fit to these hybrid data for all known decays of the baryon octet is indeed very good[3)] but hardly the right way to test the model.

The matrix elements of the SU(3) octet currents V_μ^β, A_μ^β, $\beta = \pm1, \pm2, \pm3, 0, 8$ for the process $B \to B'\ell\nu$ are usually taken between momentum eigenvectors $|p\sigma;B\rangle$ and written

$$\langle B',\sigma'p'|V_\mu^\beta + A_\mu^\beta|p\sigma B\rangle = \bar{u}_{B'}(p'\sigma')\Big\{f_1^{B'\beta B}(q^2)\gamma_\mu + f_2^{B'\beta B}(q^2)i\sigma_{\mu\nu}q^\nu + f_3^{B'\beta B}(q^2)q_\mu + \tag{1}$$
$$+ g_1^{B'\beta B}(q^2)\gamma_\mu\gamma_5 + g_2^{B'\beta B}(q^2)i\sigma_{\mu\nu}q^\nu\gamma_5 + g_3^{B'\beta B}(q^2)\gamma_5 q_\mu\Big\}u_B(p\sigma)$$

The $f_1^{B'\beta B}$, $f_2^{B'\beta B}$, $f_3^{B'\beta B}$, $g_1^{B'\beta B}$, $g_2^{B'\beta B}$, $g_3^{B'\beta B}$ are the unknown form factors which enter the theoretical expressions for rates and asymmetries in a complicated way[5)]: $\Gamma(B\to B') = \Gamma(f_1^{B'\beta B},\cdots g_3^{B'\beta B})$; $\alpha^{B\to B'} = \alpha(f_1^{B'\beta B},\cdots g_1^{B'\beta B}, g_2^{B'\beta B}, g_3^{\cdots})$. If one determines

f_2/f_1 by CVC from the magnetic moments, ignores g_3 and f_3 (because it enters with a small factor m_e/m_B) and puts $g_2 = 0$ (as demanded by the Cabibbo model) then the experimental value for $\alpha^{B\to B'}$ determines, for each process $B \to B'\ell\nu$ separately, a value for $g_1^{B'\beta B}/f_1^{B'\beta B}$ which is the so-called "experimental" $(g_A/g_V)^{B\to B'}$ ratio.

One aspect of the Cabbibo model is that

$$(2)\qquad \begin{aligned} f_i^{B'\beta B}(q^2) &= \sum_{\gamma=1,2} C(\gamma;B\beta B')f_i^{\gamma}(q^2) \\ g_i^{B'\beta B}(q^2) &= \sum_{\gamma=1,2} C(\gamma;B\beta B')g_i^{\gamma}(q^2) \end{aligned}$$

where $C(\gamma;B\beta B')$ are the Clebsch-Gordon coefficients (see Table I) and $f_i^1\ g_i^1$ are the F-type and $f_i^2\ g_i^2$ are the D-type reduced matrix elements. (2) follows from the Cabibbo hypothesis that V_μ^β, A_μ^β are octet operators with respect to an exact SU(3)-symmetry group i.e. that $[P_\mu, SU(3)] = 0$, a hypothesis which the S.G. model will replace as discussed below. (The other aspect of the Cabibbo model is given by eq. (3) below.) If one then obtains the remaining reduced matrix elements g_1^1, g_1^2 (and the angle θ in (3)) from a least square fit to the hybrid data $\Gamma(B \to B')$ and $(g_A/g_V)^{B\to B'}$ for all the processes $B \to B'$ ($n \to p$, $\Sigma^+ \to \Lambda$, $\Sigma^- \to \Lambda$, $\Lambda \to p$, $\Sigma^- \to n$, $\Xi \to \Lambda$, $\Xi \to \Sigma^0$), then the fit ("Cabibbo fit") is very good.[3)]

However Garcia noticed--first for the decay $\Lambda \to p$[4)] and then for all available hyperon decay data[6)] (see Table I) that a fit to the real experimental quantities $\Gamma(B \to B')$ and $\alpha^{B\to B'}$ under the above described assumptions of the Cabibbo model (including the assumption $g_2 = 0$) was really very poor and that an acceptable fit required a large g_2-term, much larger than that estimated by conventional SU(3) breaking calculations. This work of Garcia was the starting point for the application of the S.G. approach to the baryon decay.

My second reason for this talk is a new experimental value for $\alpha_e^{\Sigma\to n}$ by the Ohio State-Chicago-Argonne group.[7)] The predictions of the Cabibbo model and the S.G. model differ most in the value for $\alpha_e^{\Sigma\to n}$, with ≈ -0.7 for the Cabibbo model and ≈ 0 for the S.G. model. (The larger the mass differences between B and B' the larger are the deviations of the S.G. predictions from the Cabibbo predictions.) The old experimental average value was $\alpha_e^{\Sigma\to n} = 0.04 \pm 0.27$, the new--not yet official--value is $\alpha_e^{\Sigma\to n} = 0.17 \pm 0.26$ so that the discrepancy between experiment and Cabibbo model predictions is now sufficient to take revisions of this model more seriously.

The third reason for my choice of this subject is a new fit with a new ansatz for the weak hadronic current which was recently done (before the preliminary values of the experiment in reference 7 became available) by P. Kielanowski.[8)] One assumption of the Cabibbo model, which was also used in the S.G. model of reference 2, is that the weak current operator is given by

$$(3)\qquad J_\mu^{had} = \cos\theta\left(V_\mu^{-1} + A_\mu^{-1}\right) + \sin\theta\left(V_\mu^{-2} + A_\mu^{-2}\right) + \text{h.c.}$$

where $V_\mu^{\pm 1}$ are the strangeness-non-changing components of an octet operator and $V_\mu^{\pm 2}$ are the strangeness changing components of an octet operator and θ is the Cabibbo angle. It has so far been considered essential that the suppression of the strangeness changing decay should be given by θ with $|\sin\theta| \approx 1/4$.

However Kielanowski suggested now in place of the Cabibbo universality (3) the following universality assumption for the hadronic weak current

(4a) $$J_\mu^{had} = j_\mu^S + A_1[M,[M,j_\mu^S]] + A_2[M[M[M[M,j_\mu^S]]]]$$

with $A_1 = (\delta^2/2!)$, $A_2 = (\delta^4/4!)$ and $\delta \approx 0.007\ (\mathrm{MeV})^{-1}$ universal constants. M is the mass operator and j_μ^S is the universal SU(3) octet current

(4b) $$j_\mu^S = V_\mu^{-1} + A_\mu^{-1} + V_\mu^{-2} + A_\mu^{-2} + \text{h.c.}$$

(4) exhibits a higher universality than the Cabibbo universality and reduces the difference in the value of the coupling constant for decays between different states of the baryon octet to their mass differences. It explains the suppression as a symmetry breaking effect expressed in terms of suppression factors $\phi_{B'B}$ of the universal coupling constant. According to (4) the suppression factors are the following functions of the masses:

(5) $$\phi_{B'B} = (1 + A_1(m_B - m_{B'})^2 + A_2(m_B - m_{B'})^4)$$

whereas in case (3) $\phi_{B'B}$ is given by:

$$\phi_{B'B} = \begin{cases} \cos\theta & \text{for } B \to B' \text{ having } \Delta Y = 0 \\ \sin\theta & \text{for } B \to B' \text{ having } \Delta Y = 1 \end{cases} .$$

(4) can be conjectured from some general requirements[8)] but basically it is a phenomenological ansatz like (3) that can ultimately only be justified by the agreement with experimental data.

We shall see in the fits that the S.G. model with (4) gives a fit better than the S.G. model of reference 1 (using (3)) and we shall see that these fits are better than the fit to the experimental data of the conventional Cabibbo model. But first I shall give a brief review of the S.G. model for SU(3).

Relations (1) with (2) is the result of the Wigner-Eckart theorem for the SU(3)-octet operators V_μ^β, A_μ^β for the case that SU(3) is an exact symmetry group i.e. if $[P_\mu, SU(3)] = 0$. As P_μ is not an invariant of the particle classifying SU(3) -- because if it were then all masses in an octet would have to be the same, $m_B = m_{B'} = m_{B''} =$ etc., and no decays were possible--the formfactors $f_i^\gamma(q^2)$ and $g_i^\gamma(q^2)$ are not SU(3) invariant reduced matrix elements (and (1) with (2) is not the Wigner-Eckart theorem) but depend through the masses m_B and $m_{B'}$ upon the SU(3) charges $B(= I_B, I_{3B}, Y_B \cdots)$ and B'. For this theoretical reason the Cabibbo model can only be an approximation valid for the case $m_B = m_{B'} =$ etc., and Garcia's phenomenological results of references

4 and 6 were really no surprise in view of the large mass differences.

In the S.G. approach this is overcome by assuming that the SU(3) which classifies the particles is not an approximate symmetry group but a spectrum generating group[9] that fulfills the Werle relation[10]

(6) $$[\hat{P}_\mu, SU(3)] = 0$$

where $\hat{P}_\mu = P_\mu M^{-1}$ is the four-velocity operator. Again assuming that V^β_μ, A^β_μ are octet operators with respect to this spectrum generating SU(3) one now calculates their matrix elements not between momentum eigenvectors but between eigenvectors of $\hat{P}_\mu$, $|\hat{p}\sigma B\rangle$, and obtains from the Wigner-Eckart theorem

(7)
$$\langle B'\sigma'\hat{p}'|V^\beta_\mu + A^\beta_\mu|\hat{p}\sigma B\rangle = (m_B m_{B'})^{3/2}\,\bar{u}_B(\hat{p}',\sigma')\Big\{F_1^{B'\beta B}(\hat{q}^2)\gamma_\mu + F_2^{B'\beta B}(\hat{q}^2)i\sigma_{\mu\nu}\hat{q}^\nu + F_3^{B'\beta B}(\hat{q}^2)\hat{q}_\mu + G_1^{B'\beta B}(q^2)\gamma_\mu\gamma_5 + G_2^{B'\beta B}(\hat{q}^2)i\sigma_{\mu\nu}\hat{q}^\nu\gamma_5 + G_3^{B'\beta B}(\hat{q}^2)\gamma_5\hat{q}_\mu\Big\}u_B(\hat{p},\sigma)$$

with

(8)
$$F_i^{B'\beta B} = \sum_{\gamma=1,2} C(\gamma;B\beta B')F_i^\gamma(\hat{q}^2)$$
$$G_i^{B'\beta B} = \sum_{\gamma=1,2} C(\gamma;B\beta B')G_i^\gamma(\hat{q}^2)$$
$$\hat{q} = \hat{p}' - \hat{p} = [(p'/m_{B'}) - (p/m_B)]$$

Because of (6) the $F_i^\gamma(\hat{q}^2)$ and $G_i^\gamma(\hat{q}^2)$ are now genuine SU(3)-invariant reduced matrix elements which are functions of the SU(3)-invariant velocity transfer $\hat{q}^2$. One can now express the ordinary form factors of (1) in terms of the SU(3)-invariant formfactors F_i^γ, G_i^γ and obtain:

(9)
$$f_1^{B'\beta B} = \phi_{B'B}\sum_{\gamma=1,2} C(\gamma;B\beta B')\Big[F_1^\gamma + \Big(2 - \frac{(m_B+m_{B'})^2}{2m_Bm_{B'}}\Big)F_2^\gamma + \frac{m_B^2 - m_{B'}^2}{2m_Bm_{B'}}F_3^\gamma\Big]$$
$$f_2^{B'\beta B} = \frac{\phi_{B'B}}{2m_Bm_{B'}}\sum_{\gamma=1,2} C(\gamma;B\beta B')[(m_B+m_{B'})F_2^\gamma - (m_B-m_{B'})F_3^\gamma]$$
$$f_3^{B'\beta B} = \frac{\phi_{B'B}}{2m_Bm_{B'}}\sum_{\gamma=1,2} C(\gamma;B\beta B')[-(m_B-m_{B'})F_2^\gamma + (m_B+m_{B'})F_3^\gamma]$$
$$g_1^{B'\beta B} = \phi_{B'B}\sum_{\gamma=1,2} C(\gamma;B\beta B')\Big[G_1^\gamma + \frac{m_B^2 - m_{B'}^2}{2m_Bm_{B'}}G_2^\gamma - \frac{(m_B-m_{B'})^2}{2m_Bm_{B'}}G_3^\gamma\Big]$$
$$g_2^{B'\beta B} = \frac{\phi_{B'B}}{2m_Bm_{B'}}\sum_{\gamma=1,2} C(\gamma;B\beta B')[(m_B+m_{B'})G_2^\gamma - (m_B-m_{B'})G_3^\gamma]$$
$$g_3^{B'\beta B} = \frac{\phi_{B'B}}{2m_Bm_{B'}}\sum_{\gamma=1,2} C(\gamma;B\beta B')[-(m_B-m_{B'})G_2^\gamma + (m_B+m_{B'})G_3^\gamma]$$

With the usual assumptions one obtains

(10) $F_1^1 = \sqrt{6}$ (normalization) $\qquad F_1^2 = 0$

(11) $F_2^1 = (\frac{\mu_p}{2} + \frac{\mu_n}{4})\sqrt{6}$ $\qquad F_2^2 = \frac{\mu_n}{4}\sqrt{30}$

} from CVC

(12) $F_3^{1,2} = 0$ $G_2^{1,2} = 0$ from absence of genuine second class current.

(11) has become very doubtful recently because the new experimental values for the hyperon magnetic moments indicate that the Gell-Mann-Nishijima form of the electromagnetic current operator, upon which (11) is based, may have to be amended.[11)] The first class condition for the axial vector current, $G_2^{1,2} = 0$, $F_3^{1,2} = 0$, means that the current operators have a definite time inversion transformation property and a definity hermiticity property.[12)]

Assuming (10)-(12) one is left with the four free parameters G_1^1, G_1^2, G_3^1, G_3^2 (functions of velocity transfer) in addition to the Cabibbo angle θ (if one uses (3)) or the universal constants A_1, A_2 (if one uses Kielanowski's ansatz (4)) with which one can fit the experimental data. The essential difference between the conventional Cabibbo model and the S.G. models can be seen from g_2 of eq. (9): Though one starts with a first class axial vector current, $G_2^{1,2} = 0$, one can still obtain large induced second class contributions $g_2^{B'\beta B}$ for the individual processes $B \to B'$, in particular when the mass difference $(m_B - m_{B'})$ is large.

The results of the fits for the conventional Cabibbo model, the S.G. model with Cabibbo current (3) and the S.G. model with Kielanowski's current (4) are shown in Table II, Table III gives the values of the fitted parameters. These fits have been made by P. Kielanowski before the new value for $\alpha_e^{\Sigma n}$ was available and the old world average for $\alpha_e^{\Sigma n}$ has been used. As already mentioned, the conventional Cabibbo model gives a very poor fit and the new value for $\alpha_e^{\Sigma n}$ will make it still worse. The S.G. model with Cabibbo current has a 10% probability. The main reason for this improvement is the large induced g_2 term. From (9) one calculates:

$$m_\Lambda g_2^{p\beta\Lambda}/g_1^{p\beta\Lambda} = -0.55 \qquad m_\Sigma g_2^{n\beta\Sigma^-}/g_1^{n\beta\Sigma} = -9.43 \qquad m_\Sigma g_2^{\Lambda\beta\Sigma^-}/g_1^{\Lambda\beta\Sigma} = -0.87$$

as compared with $g_2^{B'\beta B} = 0$ by first class assumption for the conventional Cabibbo model. This result was not surprising because the main feature of the S.G. model is to provide induced pseudotensor terms that are proportional to $m_B - m_{B'}$, the quantity which expresses the symmetry breaking. The surprising result was that the Kielanowski current gave a still better fit (60% confidence level but with still larger values for g_2). Even if this cannot be interpreted as a phenomenological proof of Kielanowski's ansatz (4), it gives some support to an old hope that the suppression angle may be an expression of symmetry breaking.[13)]

The main purpose of all these fits was to find the real meaning of the flavor SU(3) (or general SU(n)) and to test the Werle relation $[\hat{P}_\mu, S.G.] = 0$. This relation I had learned from J. Werle in 1965 who suggested it to circumvent the O'Raifeartaigh theorem. His paper was never published, but starting around 1967 a relativistic quantum mechanics was developed in the framework of the spectrum generating groups that made it possible to apply and test this relation.[10)2)1)] The predictions for most experimental quantities do not only depend upon the Werle relation but also upon other assumptions (e.g. the ansatz for the current operator). The g_2-terms give--besides the formfactor ratios $\xi = f^-/f^+$ in $K_{\ell 3}$ decay--the cleanest test of the Werle relation (if one assumes the absence of genuine second class contributions; $m_n g_2^{p\beta n}/g_1^{p\beta n} = -0.01$ for our fit). The above described results show that SU(3) symmetry is no longer an adequate approximation but that the SU(3) flavor group still leads to good agreement if interpreted as spectrum generating group fulfilling the Werle relation.

References and Footnotes

1) Proceedings of the 7th International Group Theory Colloquium, pp. 301-324, Springer Lecture Notes in Physics, Vol. 94 (1979).
2) A. Bohm, R. B. Teese, A. Garcia, J. S. Nilsson, Phys. Rev. D15, 689 (1977); A. Bohm, Phys. Rev. D13, 2110 (1976); A. Bohm, J. Werle, Nucl. Phys. B106, 165 (1976).
3) K. Kleinknecht, Proceedings of the XVIII International Conference on High Energy Physics, London (1974); R. C. Shrock, L. L. Wang, Phys. Rev. Letters 41, 692 (1978).
4) A. Garcia, Phys. Rev. D3, 2638 (1971).
5) Formulas (9)···(13) of reference 4.
6) A. Garcia, Phys. Rev. D9, 177 (1974).
7) T. A. Romanowski, R. Winston, Ohio State-Chicago Argonne group; private communication (1980).
8) P. Kielanowski, J. Werle, private communications (1980).
9) A. O. Barut, A. Bohm, Phys. Rev. 139B, 1107 (1965); Y. Dothan, M. Gell-Mann, Y. Ne'eman, Phys. Rev. Letters 17, 145 (1965).
10) J. Werle, ICTP report, Trieste (1965), unpublished; A. Bohm, Phys. Rev. 158, 1408 (1967); D7, 2701 (1973); A. Bohm, E. C. G. Sudarshan, Phys. Rev. 178, 2264 (1969); H. van Dam, L. C. Biedenharn, Phys. Rev. D14, 405 (1976).
11) R. B. Teese, R. Settles, Phys. Lett. 87B, 111 (1979); A. Bohm, Phys. Rev. D17, 3127 (1978); D18, 2547 (1978);
12) E.g., A. Bohm, R. B. Teese, J. Math. Phys. 18, 1434 (1977).
13) R. Oehme, Ann. Phys. (N.Y.) 33, 108 (1965); M. Gell-Mann, P.R. 125, 1067 (1962).

Table I

SU(3) Clebsch Gordon coefficients in the normalization used here.

B	B'	C(γ=1;BβB')	C(γ=2;BβB')
n	p	$1/\sqrt{6}$	$-\sqrt{3/10}$
$\Sigma^{\pm}$	Λ	0	$-1/\sqrt{5}$
Λ	p	$-1/2$	$1/\sqrt{20}$
Σ^-	n	$-1/\sqrt{6}$	$-\sqrt{3/10}$
Ξ^-	Λ	$1/2$	$1/\sqrt{20}$
Ξ^-	Σ	$1/2\sqrt{3}$	$-1/2\sqrt{3/5}$

Table II gives the fit of the three models to the experimental data. (All transition rates are in 10^6 sec^{-1} except for neutron decay, which is in 10^{-9} sec^{-1}.) Table III gives the values of the fitted parameters. A_1 and A_2 are in MeV^{-2} and MeV^{-4} respectively.

Table II

Process	Conventional Cabibbo model		S.G.-model with Cabibbo current (3)		S.G.-model with Kielanowski current (4)		Experimental Value
	Predicted value	Contribution to χ^2	Predicted value	Contribution to χ^2	Predicted value	Contribution to χ^2	
$n \to pe\nu$ (rate)	1.064	2.175	1.061	2.634	1.090	0.005	1.089 ± 0.017
$\Sigma^+ \to \Lambda e\nu$(rate)	0.301	0.694	0.306	0.840	0.263	0.038	0.252 ± 0.059
$\Sigma^- \to \Lambda e\nu$(rate)	0.499	5.230	0.503	5.687	0.398	0.032	0.405 ± 0.041
$\Lambda \to pe\nu$(rate)	3.107	0.141	3.096	0.077	3.082	0.021	3.066 ± 0.109
$\Sigma^- \to ne\nu$(rate)	7.134	0.284	7.129	0.304	7.292	0.001	7.283 ± 0.280
$\Xi^- \to \Lambda e\nu$(rate)	2.720	0.606	3.006	0.527	0.257	1.524	6.928 ± 5.404
$\Xi^- \to \Lambda e\nu$(rate), $\Xi^- \to \Sigma^0 e\nu$(rate)	3.258	0.704	3.554	0.321	3.173	0.840	4.172 ± 1.090
$\Lambda \to p\mu\nu$(rate)	0.597	0	0.510	0.426	0.520	0.332	0.597 ± 0.133
$\Sigma^- \to n\mu\nu$(rate)	3.091	0.044	3.192	0.341	3.013	0.006	3.034 ± 0.271
$n \to pe\nu(\alpha_{e\nu})$	-0.081	0.244	-0.081	0.235	-0.076	0.452	-0.095 ± 0.028
$n \to pe\nu(\alpha_e)$	-0.091	1.446	-0.091	1.613	-0.085	0.002	-0.085 ± 0.005
$n \to pe\nu(\alpha_\nu)$	0.986	0.148	0.986	0.150	0.988	0.115	1.001 ± 0.038
$\Sigma^\pm \to \Lambda e\nu(\alpha_{e\nu})$	-0.400	0	-0.453	0.007	-0.491	0.254	-0.4 ± 0.18
$\Sigma^- \to ne\nu(\alpha_{e\nu})$	0.311	0.447	0.304	0.230	0.277	0.031	0.284 ± 0.041
$\Sigma^- \to ne\nu(\alpha_e)$	-0.714	7.798	-0.011	0.036	0.237	0.534	0.04 ± 0.27
$\Lambda \to pe\nu(\alpha_e)$	0.015	3.431	0.003	4.186	0.002	4.251	0.134 ± 0.064
$\Lambda \to pe\nu(\alpha_{e\nu})$	0.014	0.035	-0.008	0.167	0.027	0.290	0.007 ± 0.037
$\Lambda \to pe\nu(\alpha_\nu)$	0.978	4.699	0.964	3.833	0.917	1.466	0.839 ± 0.064
$\Lambda \to pe\nu(\alpha_p)$	-0.579	0.564	-0.563	0.284	-0.530	0.003	-0.526 ± 0.07
Total χ^2/n_D		28.69/14		21.98/14		10.20/13	
Confidence level		1%		8%		68%	

Table III

Conventional Cabibbo	S.G.-model with Cabibbo current (3)	S.G.-model with Kielanowski current (4)
$\theta = 0.230$	$\theta = 0.238$	$A_1 = -0.28494 \cdot 10^{-4}$ $A_2 = 0.154661 \cdot 10^{-9}$
$g_1^{\gamma=1} = 1.065$	$G_1^{\gamma=1} = 1.007$	$G_1^{\gamma=1} = 0.789$
$g_1^{\gamma=2} = -1.551$	$G_1^{\gamma=2} = -1.596$	$G_1^{\gamma=2} = -1.714$
$g_3^{\gamma=1} = 205.6$	$G_3^{\gamma=1} = -8.744$	$G_3^{\gamma=1} = -11.823$
$g_3^{\gamma=2} = -106.2$	$G_3^{\gamma=2} = -42.156$	$G_3^{\gamma=2} = -76.309$

BARYON MAGNETIC MOMENTS IN BROKEN SU(6)

M. Böhm
Physikalisches Institut der Universität Würzburg
Würzburg, FRG

R. Huerta* and A. Zepeda
Centro de Investigación y de Estudios Avanzados del IPN
Apdo. Postal 14-740, México 14, D.F.

Abstract

The magnetic moments of the baryons belonging to the low lying SU(3) octet are computed using second order perturbation theory taking into account the effect of the spin dependent interactions that are expected to arise in QCD. The unperturbed eigenfunctions of the confining Hamiltonian are approximated by harmonic oscillator wave functions and the flavor x spin parts belong, in the limit of equal quark masses, to irreducible representations of SU(6). In this basis we then calculate the mixing of the $(56, 0^+)$ ground state wave function with the orbital and radial excitations labeled by $(56, 0^+_R)$, $(70, 0^+)$, $(20, 1^+)$ and $(70, 2^+)$. This mixing arises not only from the spin dependent interactions but also from the differences among quark masses. Finally we comment on other contributions to the magnetic moments not taken into account in this work.

Introduction

The magnetic moments of the baryons in the lowest lying octet are all related [1] to one of them by (nonrelativistic) SU(6). These relations include those obtained in the (flavor) SU(3) limit [2]. On the other hand the symmetric quark model, where it is assumed that the quarks in a baryon are in a S wave and that when two quarks are identical they are in a spin 1 state, predicts [3] the following 9 relations between the baryon magnetic moments and the quark magnetic moments:

$$\mu_p = (4\mu_u-\mu_d)/3, \quad \mu_n = (4\mu_d-\mu_u)/3, \quad \mu_{\Sigma^+} = (4\mu_u-\mu_s)/3$$

$$\mu_{\Sigma^-} = (4\mu_d-\mu_s)/3, \quad \mu_{\Xi^0} = (4\mu_s-\mu_u)/3, \quad \mu_{\Xi^-} = (4\mu_s-\mu_d)/3$$

$$\mu_{\Sigma^0} = (2\mu_u+2\mu_d-\mu_s)/3, \quad \mu_\Lambda = \mu_s, \quad \mu_{\Sigma\Lambda} = (\mu_d-\mu_u)/\sqrt{3} \tag{1}$$

These relations correspond to SU(6) broken by quark mass differences and thus the SU(3) limit coincides with the SU(6) limit and implies $\mu_u = -2\mu_d = -2\mu_s$.

With the advent [4] of Quantum Chromodynamics (QCD) it soon became clear [5] that the interchange of colored gluons between quarks gives rise to a spin dependent interaction which in the nonrelativistic approximation resembles the Fermi-Breit one. The application [5,6] of these ideas to the spectroscopy of hadrons and to decay amplitudes has met a remarkable success. These spin dependent interaction breaks both SU(3) and SU(6) and should modify the relations (1). Here we want to report our calculation of these corrections. We have approached the problem perturbatively and have approximated the spin independent Hamiltonian by that of a harmonic oscillator. The effects we find are of the order of 10% and when compared with the experimental data they improve the predictions given in eqs. (1).

Dynamics

The Hamiltonian that describes in the nonrelativistic approximation the structure of the baryons is of the form [5,6]

$$H = \sum_{i=1}^{3} (m_i + \vec{p}_i^{\,2}/2m_i) + H_{conf} + H_s \tag{2}$$

where m_i and $\vec{p}_i$ are the mass and momentum of quark i, H_{conf} is the flavor and spin independent interaction responsible for confinement, and H_s arises from the interaction of the color magnetic moments of the quarks, As we have already mentioned, we will approximate H_{conf} by a harmonic oscillator potential so that $H = H_o^{ho} + H_s$, where

$$H_o^{ho} = \sum_i (m_i + \vec{p}_i^{\,2}/2m_i) + \frac{\kappa}{2} \sum_{i<j} r_{ij}^2 \quad . \tag{3}$$

κ is the oscillator constant and $\vec{r}_{ij} = \vec{x}_i - \vec{x}_j$ is the separation between quarks. H_s will be specified below.

We define two relative coordinates, $\vec{\rho}$ and $\vec{\lambda}$, which together with the center of mass coordinate, $\vec{X}/\sqrt{3}$, are given by

$$\vec{X}_i = M_{ij}\, \vec{x}_j \quad , \tag{4}$$

$$M = \begin{pmatrix} \sqrt{3}\,\frac{m_1}{M}, & \sqrt{3}\,\frac{m_2}{M}, & \sqrt{3}\,\frac{m_3}{M} \\ 1/\sqrt{6}, & 1/\sqrt{6}, & -2/\sqrt{6} \\ 1/\sqrt{2}, & -1/\sqrt{2}, & 0 \end{pmatrix} \tag{5}$$

where $\vec{X}_i = (\vec{X}, \vec{\lambda}, \vec{\rho})$ and $M = m_1+m_2+m_3$. In terms of the relative coordinates H_o^{ho} becomes

$$H_o^{ho} = M + \vec{p}_\lambda^{\,2}/2m_\lambda + \vec{p}_\rho^{\,2}/2m_\rho + 3\kappa(\rho^2+\lambda^2)/2 \tag{6}$$

while the orbital part of the magnetic moment operator

$$\vec{\mu}_L = \sum_{i=1} \mu_{q_i} \vec{L}_i \qquad \text{becomes}$$

$$\vec{\mu}_L = \vec{L}_\rho\left[(m_2 + \tfrac{1}{2} m_3)\mu_{q_1} + (m_1 + \tfrac{1}{2} m_3)\mu_{q_2}\right] / M$$
$$+ \vec{L}_\lambda\left[\tfrac{1}{2} m_3 (\mu_{q_1} + \mu_{q_2}) + (m_1 + m_2)\mu_{q_3}\right] / M$$
$$+ \vec{\lambda} \times \vec{p}_\rho \; 3m_3(\mu_{q_1} - \mu_{q_2})/M\sqrt{12}$$
$$+ \vec{\rho} \times \vec{p}_\lambda\left[\mu_{q_1}(2m_2+m_3) - \mu_{q_2}(2m_1+m_3) + 2\mu_{q_3}(m_1-m_2)\right] / \sqrt{12}\, M \qquad (7)$$

where $\vec{L}_\rho = \vec{\rho} \times \vec{p}_\rho$, $\vec{L}_\lambda = \vec{\lambda} \times \vec{p}_\lambda$ are related to the total angular momentum by $\vec{L} = \vec{L}_\rho + \vec{L}_\lambda$. We recall that the total magnetic moment is given by $\vec{\mu} = \vec{\mu}_s + \vec{\mu}_L$ where

$$\vec{\mu}_s = \mu_{q_1}\vec{\sigma}_1 + \mu_{q_2}\vec{\sigma}_2 + \mu_{q_3}\vec{\sigma}_3 \qquad (8)$$

We are going to make a perturbative expansion around $\overline{H}_o^{ho}$ defined as H_o^{ho} evaluated at $m_1 = m_2 = m_3$. Let us introduce quark mass shifts by $m_i = m + \Delta_i$, $\sum_i \Delta_i = 0$ (m varies from baryon to baryon). Then

$$H = \overline{H}_o^{ho} + H_1 + H_2 + \ldots \qquad (9)$$

where

$$\overline{H}_o^{ho} = M + \vec{p}_\lambda^{\,2}/2m + \vec{p}_\rho^{\,2}/2m + 3\kappa(\rho^2 + \lambda^2)/2 \qquad (10)$$

$$H_1 = H_\Delta + \overline{H}_s \qquad (11)$$

$$H_\Delta = -\sum_i p_i^2 \Delta_i / 2m^2 \qquad (12)$$

$$\overline{H}_s = H_s(m_i = m) \qquad (13)$$

$$H_2 = \sum_i p_i^2 \Delta_i / 2m^3 + \sum_i \left.\frac{\partial H_s}{\partial m_i}\right|_{m_j = m} \Delta_i \qquad (14)$$

Since the mass differences inside the octet are of the same order as those between the octet and the decuplet, H_1 is a first order perturbative term and H_2 is of second order.

Let $a = m\omega$, $\omega = (3\kappa/m)^{1/2}$, then the ground state eigenfunction of $\overline{H}_o^{ho}$ and the eigenfunctions of the second excited state are (the first excited state is of negative parity)

$$56:\ (0_o^+)_{123} = \left(\frac{a}{\pi}\right)^{3/2} \exp\{-a(\rho^2 + \lambda^2)/2\} \qquad (15)$$

$$56:\ (0_R^+)_{123} = (0_o^+)_{123}\,(3 - a\rho^2 - a\lambda^2)/\sqrt{3}\ ,$$

$$70:\ (0_2^+)_{12} = (0_o^+)_{123}\,a(\rho^2 - \lambda^2)/\sqrt{3}\ ,$$

$$70:\ [0_2^+]_{12} = (0_o^+)_{123}\ a2\vec{\rho}\cdot\vec{\Lambda}/\sqrt{3}\ ,$$

$$20:\ [1_2^+]_{123} = (0_o^+)_{123}\ (8\pi/3)a\rho\lambda Y_1(\rho)Y_1(\lambda)(11|1)\ ,$$

$$56:\ (2_2^+)_{123} = (0_o^+)_{123}\ (8\pi/15)^{1/2}a(\rho^2Y_2(\rho) + \lambda^2Y_2(\lambda))\ ,$$

$$70:\ (2_2^+)_{12} = (0_o^+)_{123}\ (8\pi/15)^{1/2}\ a(\rho^2Y_2(\rho) - \lambda^2Y_2(\lambda))\ ,$$

$$70:\ [2_2^+]_{123} = (0_o^+)_{123}\ (8\pi/3)a\rho\lambda Y_1(\rho)Y_1(\lambda)(11|2), \tag{16}$$

The number in front of each wave function indicates the SU(6) representation to which it is associated, in the limit $m_u = m_d = m_s$, according to its symmetry properties under the permutation of space variables indicated by $(\ )_{123}$ = completely symmetric, $[\ ]_{123}$ = completely antisymmetric, $(\ )_{12}$ and $[\ ]_{12}$ of mixed symmetry. The total wave function should be completely symmetric under the simultaneous interchange of space, flavor, and spin variables and thus they are, writteh in the order space x flavor x spin,

$$|56,\ L,\ 10_4\rangle = (L)_{123}(10)_{123}(4)_{123},$$

$$|56,\ L,\ 8_2\rangle = \frac{1}{\sqrt{2}}\ (L)_{123}\{(8)_{12}(2)_{12}+[8]_{12}[2]_{12}\}; \tag{17a}$$

$$|20,\ L,\ 8_2\rangle = \frac{1}{\sqrt{2}}\ [L]_{123}\{(8)_{12}[2]_{12}-[8]_{12}(2)_{12}\},$$

$$|20,\ L,\ 1_4\rangle = [L]_{123}[1]_{123}(4)_{123}; \tag{17b}$$

$$|70,\ L,\ 8_4\rangle = \frac{1}{\sqrt{2}}\ \{(L)_{12}(8)_{12}+[L]_{12}[8]_{12}\}(4)_{123},$$

$$|70,\ L,\ 10_2\rangle = \frac{1}{\sqrt{2}}\ \{(L)_{12}(10)_{123}(2)_{12}+[L]_{12}(10)_{123}[2]_{12}\},$$

$$|70,\ L,\ 8_2\rangle = \frac{1}{2}\{[L]_{12}(8)_{12}[2]_{12}+[L]_{12}[8]_{12}(2)_{12} + (L)_{12}[8]_{12}[2]_{12}-(L)_{12}(8)_{12}(2)_{12}\}\ ,$$

$$|70,\ L,\ 1_2\rangle = \frac{1}{\sqrt{2}}\ \{(L)_{12}[1]_{123}[2]_{12}-[L]_{12}[1]_{123}(2)_{12}\}; \tag{17c}$$

where L is the orbital angular momentum and where Δ, D, and d in $|\Delta,\ L,\ D_d\rangle$ denote, respectively, the dimension of the irreducible representation of SU(6), (flavor) SU(3), and (spin) SU(2).

The eigenvalue of $\bar{H}_o^{ho}$ in the state $(0_o^+)_{123}$ is $M+3\omega$ while that in the second lavel states is $M + 5\omega$.

In the space of completely symmetric wave functions and in the limit of equal masses, $m_1=m_2=m_3=m$ (but not necessarily $m_u=m_d=m_s$), H_s is given by

$$\bar{H}_s = \bar{H}_{ss} + \bar{H}_{so} + \bar{H}_T \tag{18}$$

$$\bar{H}_{ss} = (2\pi\alpha_s/3m^2\sqrt{2})\ \vec{\sigma}_1\cdot\vec{\sigma}_2\delta^{(3)}(\hat{\rho}) \tag{19a}$$

$$\bar{H}_{so} = (3/2m^2)\{(\vec{s}_1+\vec{s}_2)\cdot\vec{L}_\rho[(\alpha_s/\rho^3\sqrt{2}) - \kappa]$$
$$- (\vec{s}_1-\vec{s}_2)\cdot\vec{\rho}\times\vec{p}_\lambda[(\alpha_s/\rho^3 3\sqrt{2}) + \kappa]/\sqrt{3}\} \tag{19b}$$

$$\bar{H}_T = (\alpha_s/m^2\sqrt{2})(3\vec{s}_1\cdot\hat{\rho}\ \vec{s}_2\cdot\hat{\rho} - \vec{s}_1\cdot\vec{s}_2)/\rho^3 \tag{19c}$$

where $\vec{s}_i$ is the spin of quark i-th and α_s is the strong fine structure constant.

If we expand the magnetic moment operator around the limit $m_1=m_2=m_3=m$, $\vec{\mu} = \vec{\mu}_o + \vec{\mu}_1 + \vec{\mu}_2 + \ldots$, then the expansion of the ground state matrix elements of $\vec{\mu}$ up to second order is

$$\langle\psi|\vec{\mu}|\psi\rangle = \langle 0|\vec{\mu}|0\rangle\ (1 - \sum_i \frac{|\langle 0|H_1|i\rangle|^2}{(\varepsilon_o-\varepsilon_i)^2}) + 2\mathrm{Re}\sum_i \frac{\langle 0|\vec{\mu}_1|i\rangle\langle i|H_1|0\rangle}{\varepsilon_o-\varepsilon_i}$$
$$+ \sum_{i,j} \frac{\langle 0|H_1|i\rangle\langle i|\mu_o|j\rangle\langle j|H_1|0\rangle}{(\varepsilon_o-\varepsilon_i)(\varepsilon_o-\varepsilon_j)}\ . \tag{20}$$

where $|0\rangle = |56,\ 0_o^+,\ 8_2,\ 1/2\rangle$, $|i\rangle\varepsilon\{|56,\ 0_R^+,\ 8_2,\ 1/2\rangle,\ |70,\ 0_2^+,\ 10_2,\ 1/2\rangle,$ $|70,\ 0_2^+,\ 8_2,\ 1/2\rangle,\ |70,\ 0_2^+,\ 1_2,\ 1/2\rangle,\ |20,\ 1_2^+,\ 1_4,\ 1/2\rangle,\ |20,\ 1_2^+,\ 8_2,\ 1/2\rangle,$ $|56,\ 2_2^+,\ 10_4,\ 1/2\rangle,\ |70,\ 2_2^+,\ 8_4,\ 1/2\rangle\}$. The last number in each ket $|i\rangle$ is the eigenvalue of the total angular momentum. To arrive to eq. (22) we used the relation $\langle 0|\vec{\mu}_o|i\rangle = 0$ which is easy to prove.

Since $\langle 0|\vec{\mu}|0\rangle$ gives rise to relations (1) and these are not modified when SU(6) is broken by quark mass differences, the effect of H_Δ in eq. (20) cancels. As for $\bar{H}_s$, using the conservation of orbital angular momentum and the symmetry properties of $\bar{H}_{ss}$, $\bar{H}_{so}$, and $\bar{H}_T$ one obtains that the only nonvanishing matrix elements are $\langle 0|\bar{H}_{ss}|56,\ 0_R^+,\ 8_2\rangle$, $\langle 0|\bar{H}_{ss}|70,\ 0_2^+,\ 8_2\rangle$, $\langle 0|\bar{H}_{so}|20,\ 1_2^+,\ 8_2\rangle$ and $\langle 0|\bar{H}_T|70,\ 2_2^+,\ 8_2\rangle$. However, since $\langle 56,\ 0_R^+,\ 8_2|\vec{\mu}|56,\ 0_R^+,\ 8_2\rangle = \langle 0|\vec{\mu}|0\rangle$ and $\langle 0|\vec{\mu}_1|56,\ 0_R^+,\ 8_2\rangle \equiv 0$, the terms with $\langle 0|\bar{H}_{ss}|56,\ 0_R^+\rangle$ in eq. (20) cancel among themselves. The surviving matrix elements are

$$\langle 70,\ 0_2^+,\ 8_2,\ 1/2|\bar{H}_{ss}|56,\ 0_o^+,\ 8_2,\ 1/2\rangle = \alpha_s\omega(\omega/3\pi m)^{1/2} \tag{21a}$$

$$\langle 20,\ 1_2^+,\ 8_2,1/2|\bar{H}_{so}|56,\ 0_o^+,\ 8_2,\ 1/2\rangle = -\omega[\alpha_s(\omega/9\pi m)^{1/2}+\omega/2m]/\sqrt{6} \tag{21b}$$

$$\langle 70,\ 2_2^+,\ 8_4,\ 1/2|\bar{H}_T|56,\ 0_o^+,\ 8_2,\ 1/2\rangle = -\ \alpha_s\omega(\omega/30\pi m)^{1/2} \tag{21c}$$

The last items needed to evaluate the r.h.s. of eq. (20) are the magnetic moments in the excited states. The calculation of the matrix elements of the spin part of $\vec{\mu}$ is straightforward and the nonzero results, for states with maximum projection of the total angular momentum are

$$<70,\ 0_2^+,\ 8_2,\ 1/2,\ B|\mu_s^3|70,\ 0_1^+,\ 8_2,\ 1/2,\ B>$$

$$= -<70,\ 2_2^+,\ 8_4,\ 1/2,\ B|\mu_s^3|70,\ 2_2^+,\ 8_4,\ 1/2,\ B> = (\mu_a + \mu_b + \mu_c)/3 \quad , \qquad (22)$$

$$<20,\ 1_2^+,\ 8_2,\ 1/2,\ p|\mu_s^3|20,\ 1_2^+,\ 8_2,\ 1/2,\ p> = -\mu_d/3, \qquad (23a)$$

$$<20,\ 1_2^+,\ 8_2,\ 1/2,\ \Sigma^o|\mu_s^3|20,\ 1_2^+,\ 8_2,\ 1/2,\ \Sigma^o> = -\mu_s/3, \qquad (23b)$$

$$<20,\ 1_2^+,\ 8_2,\ 1/2,\ \Lambda|\mu_s^3|20,\ 1_2^+,\ 8_2,\ 1/2,\ \Lambda> = -(2\mu_u+2\mu_d-\mu_s)/9, \qquad (23c)$$

$$<20,\ 1_2^+,\ 8_2,\ 1/2,\ \Sigma^o|\mu_s^3|20,\ 1_2^+,\ 8_2,\ 1/2,\ \Lambda> = (g_d-g_u)/3\sqrt{3}, \qquad (23d)$$

where a, b, and c are the quark flavors in baryon B. From (23a) we obtain the expressions for the other outer states (n, Σ^+, Σ^-, Ξ^o, Ξ^-) substituting d, the less abundant flavor, by the corresponding flavor index.

We now discuss the matrix elements of the orbital part of $\vec{\mu}$. For the outer states we arrange the labels 1 and 2 in such a way that they correspond to the equal flavors and then $\vec{\mu}_L$ becomes

$$\vec{\mu}_L = (m_3\mu_{q_1} + 2m_1\mu_{q_3})\vec{L}_\lambda/M + \mu_{q_1}\ \vec{L}_\rho \qquad (24)$$

from which we conclude that $\vec{\mu}_L|0> = 0$. The calculation of the nonzero matrix elements involves only the multiple use Clebsch Gordan-coefficients and the results, for the equal mass case, are

$$<20,\ 1_2^+,\ 8,\ 1/2|\mu_L^3|70,\ 0_2^+,\ 8,\ 1/2> = (\mu_1-\mu_3)2/9 \qquad (25a)$$

$$<20,\ 1_2^+,\ 8,\ 1/2|\mu_L^3|20,\ 1_2^+,\ 8_2,\ 1/2> = (2\mu_1+\mu_3)2/9 \qquad (25b)$$

$$<70,\ 2_2^+,\ 8_4,\ 1/2|\mu_L^3|70,\ 2_2^+,\ 8_4,\ 1/2> = (2\mu_1+\mu_3)/3 \qquad (25c)$$

For the states with three different flavors, Σ^o and Λ, one has to consider also the three particle operators in the r.h.s. of eq. (7). Obviously $\vec{\mu}_L$ can mix the ground state only with the $|20,\ 1_2^+>$ state and one finds that

$$<20,1_2^+,8_2,1/2,\Lambda|\vec{\mu}_L|56,0_o^+,8_2,1/2,\Sigma^o> = -<20,1_2^+,8_2,1/2,\Sigma^o|\vec{\mu}_L|56,0_o^+,8_2,\Lambda> \qquad (26)$$

which implies that the term containing $<0|\mu_1|i>$ in eq. (20) cancels. For the rest of the matrix elements of $\vec{\mu}_L$ we need only the equal mass limit and the results are

$$<20,\ 1_2^+,\ 8_2,\ 1/2,\ \Lambda|\mu_L^3|70,\ 0_2^+,\ 8_2,\ 1/2,\ \Lambda>$$

$$=-<20,\ 1_2^+,\ 9_2,\ 1/2,\ \Sigma^o|\mu_L^3|70,\ 0_2^+,\ 8_2,\ 1/2,\ \Sigma^o> = -(\mu_u+\mu_d-2\mu_s)/9 \qquad (27a)$$

$$<20,\ 1^+_2,\ 8_2,\ 1/2,\ \Lambda|\mu^3_L|20,\ 1^+_2,\ 8_2,\ 1/2,\ \Lambda>$$

$$= <20,\ 1^+_2,\ 8_2,\ 1/2,\ \Sigma^o|\mu^3_L|20,\ 1^+_2,\ 8_2,\ 1/2,\ \Sigma^o> = 2(\mu_u+\mu_d+\mu_s)/9 \qquad (27b)$$

$$<70,\ 2^+_2,\ 8_4,\ 1/2,\ \Lambda|\mu^3_L|70,\ 2^+_2,\ 8_4,\ 1/2,\Lambda> =$$

$$= <70,\ 2^+_2,\ 8_4,\ 1/2,\ \Sigma^o|\mu^3_L|70,\ 2^+_2,\ 8_4,\ 1/2,\ \Sigma^o> = (\mu_u+\mu_d+\mu_s)/3 \qquad (27c)$$

Results and Conclusions

To obtain the corrected expressions for the magnetic moments it is a matter of substituting in eq. (20) the results obtained in eqs. (21), (22), (23), (25), (26) and (27). To compare these theoretical predictions with the experimental values we have set $\alpha_s = 1$ and ω_p (the oscillator frequency for the proton) = 500 MeV. Then we have assumed that the quark magnetic moments are Dirac moments and we have used the experimental values of μ_p, μ_n, and μ_Λ as inputs to fix the values of the quark masses and we obtain m_u = 311 MeV, m_d = 280 MeV, m_s = 469 MeV. These allows us to predict the values of the magnetic moments for the rest of the baryons in the lowest lying octet and the results are shown in table 1. For comparison we also show the results obtained with eq. (1) which ignores the spin forces. We obtain corrections, with respect to the case with no spin forces, of the order of 2 to 10% and, in three out of five cases, in the correct direction. However the improvement is, if any, far from sufficient to claim an understanding of the baryon magnetic moments. On the basis of the chiral perturbation theory analysis of the magnetic moments done in Ref. [7] we suspect that the most important effects not taken into account in our calculations are the effects of exchange currents. While we were finishing this work we became aware of a related work by N. Isgur and G. Karl. We thank G. Karl for letting us know their results prior to publication.

References

*Work supported by CONACyT

1. M.A.B. Bég, B.W. Lee, and A. Pais, Phys. Rev. Lett. 13 (1964) 514.
2. S. Coleman and S.L. Glashow, Phys. Rev. Lett. 6 (1961) 423.
3. J. Franklin, Phys. Rev. 172 (1968) 1807.
4. H. Fritzsch, M. Gell-Mann and H. Leutwyler, Phys. Lett. 47B (1973) 365.
5. A. De Rújula, H. Georgi, and S.L. Glashow, Phys. Rev. D12 (1975) 147.
6. M. Böhm, Z. Physik C 3 (1980) 321; N. Isgur, G. Karl, and R. Koniuk, Phys. Rev. Lett. 41 (1978) 1269.
7. D.G. Caldi and H. Pagels, Phys. Rev. D10 (1974) 3739.

Table 1 Baryon Magnetic Moments (in units of $e_p/2Mp$)

	P	n	Λ	Σ^+	Σ^-	Ξ^o	Ξ^-	$(\Sigma^o\Lambda)$
Exp.	2.793	-1.913	-.6138±.0047	2.33±.13	-1.4±.25	-1.236±.014	-0.75±.07	$-1.82\pm^{.25}_{.18}$
No Spin Forces	input	input	input	2.67	-1.09	-1.43	-.49	-1.63
With Spin Forces	input	input	input	2.74	-1.21	-1.46	-.52	-1.68

The Racah-Wigner Algebra and the Masses of Exotic Hadrons

Philip H. Butler
Physics Department, University of Canterbury,
Christchurch, New Zealand

Various attempts have been made to calculate masses of exotic hadrons using the MIT bag model. In order to exploit the $SU_6 > SU_2 \times SU_3$ structure of the problem one needs a generalization of the tensor operator techniques familiar to atomic and nuclear physics. The use of different masses for different quarks leads to a need to couple irreducible tensor operators and therefore requires a knowledge of the 6j and 9j symbols of all the groups in addition to the $SU_6 - SU_2 \times SU_3$ - 3jm coupling factors.

We discuss various aspects of such a generalization: the reality of the 6j symbols, orientation phases, multiplicity separations, and progress with systematic computer calculations. The algebra is compared with that of $SO_3 > SO_2$ and that of the point groups.

The MIT bag model shows some promise of giving a phenomenological description of those multiquark systems which are more general than the S-wave q^3 and $q\bar{q}$ baryons and mesons. Various alternatives have been studied. Our interest is in the so-called exotic hadrons, such as mesons formed as $q^2\bar{q}^2$ [1,2] and baryons formed as $q^4\bar{q}$ and $q^5\bar{q}^2$ [3]. One assumes three quark flavours and may or may not assume equal quark masses. The bag model leads to a Hamiltonian which is expressed in terms the quark masses and other parameters, and which is to be minimized with respect to the bag radius. The colour-spin interactions, involving the SU_6^{cs} structure of the states, is of particular interest to us. Except for the simplest models with a small number of quarks of a single mass, it is important to have access to the modern techniques of the Racah algebra of the unitary groups. Such techniques allow the matrix elements of the Hamiltonian to be evaluated without further approximations. It is important to have also a means of checking the accuracy of the Hamiltonian matrices produced [4].

When using the physical assumptions of refs [1-4], one is faced with basis states of the form,

$$|n_a\ n_b\ \bar{n}_a\ \bar{n}_b\ \alpha\ \lambda^{2s+1}\ \mu\ i> \qquad (1)$$

where n_a and n_b specify the number of quarks of masses m_a and m_b, $\bar{n}_a$ and $\bar{n}_b$ specific the number of antiquarks of the same masses, α denotes parentage, λ gives the transformation properties under the colour-spin group SU_6^{cs}, s is the spin, μ is the colour, and i gives additional quantum numbers. Every one of the labels shown is the irrep label of some group or groups, with the occasional exception of the occurence of multiplicity labels. The Pauli exclusion principle indicates that the n_a quarks form a basis for an irrep labelled by the Young diagram $\{1^{n_a}\}$ of the U_{12}^{fcsa} group of changes of flavour, colour, and spin of the quarks of mass m_a--it is assumed here that two flavours have mass m_a, one flavour has mass m_b. The direct product of the four groups of such transformations

$$U_{12}^{fcsa} \times U_6^{fcsb} \times U_{12}^{fcs\bar{a}} \times U_6^{fcs\bar{b}} \qquad (2)$$

has many subgroups. We require a chain of subgroups which includes SU_6^{cs}, the group of colour-spin tranformations which does not descriminate mass, flavour or quark-antiquark distinctions. The irrep labels of such intermediate groups are collectively the parentage label α.

Ideally one should choose parentage groups so that the Hamiltonian is as diagonal as possible, but the avaliability of tables of 3jm, 6j, and 9j symbols is a constraint. Different schemes will mean that the states of eq. (1) will be different combinations of single quark states. When the generalized tensor operator techniques are applied, different combinations of reduced matrix elements, 3jm factors, 6j symbols, and 9j symbols occur. To use the full power of the Racah-Wigner algebra, Jaffe's Hamiltonian [1], must be written in terms of tensor operators whose transformation properties are known for each of the groups of the fractional parentage chain [4]. Most terms within the Hamiltonian transform as the adjoint irrep of SU_6, labelled $\{21^4\}$.

We must stress that the operator-techniques long familiar to atomic physicists [5], are fully applicable to non-simply reducible groups [6-7]. However, these techniques depend, for their power, on symmetrized coupling coefficients we call 3jm (rather than 3j or 3Γ), 6j and 9j symbols. For many years it was held that it would be difficult to generalize Wigner's treatment of simply reducible groups but Derome and Sharp [8] showed that this was not so, and a proof of the Wigner-Eckart theorem for all compact groups is also available, see [9]. The coupling coefficient tables of So and Strottman [10] do not have the symmetries needed for a ready use of the many reduction theorems, and as they note, they treat irreps such as $\{0\}$ and $\{1^6\}$ differently, that is they present tables for U_6 rather than SU_6.

Previously we published partial tables required for the present calculation, namely, SU_6-6j symbols and SU_6-$SU_2 \times SU_3$-3jm factors because they appeared in an E_7 model of elementary particles [11]. Currently the programs used in the production of molecular and crystallographic point group 3jm and 6j tables [6] are being generalized to handle any compact group. The modifications are a matter of minor changes to the recusion relations, as suggested earlier [12], and preliminary tables are avaliable.

Various group theoretic results come from our calculations. Given the usual choice of symmetries of 6j symbols under complex conjugation--a unit choice for the Derome-Sharp matrix A [8]--some SU_n-6j symbols are complex. An example is known for the unitary groups ($SU_6 > SU_2 \times SU_3$) where an orientation phase occurs--such phases have been explored for the point groups [12]. As with the point groups [6,13], various multiplicity separations may be made and the corresponding vanishing 6js explored.

Our preliminary calculations with the MIT bag model [4], would reinforce the conclusions of earlier authors, that the exotic hadrons have properties that are not much different from the simple hadrons.

REFERENCES

[1] Jaffe, R.L. Phys. Rev. D. 15, 26T (1977).
[2] Jaffe, R.L. Phys. Rev. D. 15, 281 (1977).
[3] Strottman, D. Phys. Rev. D. 20, 748 (1979).
[4] Bickerstaff, P. and Wybourne, B.G. preprint, University of Canterbury (1980).
[5] Judd, B.R. Operator Techniques in Atomic Spectroscopy, New York: McGraw Hill (1963).
[6] Butler, P.H. Point Group Symmetry Applications: Methods and Tables, New York: Plenum (1980).
[7] Butler, P.H. "The Wigner-Racah algebra for finite and continuous compact groups", in Symmetries in Science, Eds. Gruber, B. and Millman, R., New York: Plenum (1980)(in press).
[8] Derome, J.R. and Sharp, W.T. J. Math. Phys. 6, 1584 (1965).
[9] Butler, P.H. Trans. Roy. Soc. (London) 277, 545 (1975).
[10] So, S.I. and Strottman, D. J. Math. Phys. 20, 153 (1979).
[11] Butler, P.H. Haase, R.W. and Wybourne, B.G. Austral J. Phys. 32, 137 (1979).
[12] Reid, M.F. and Butler, P.H. J. Phys. A. 13 (1980)(in press).
[13] Butler, P.H. and Ford, A.M. J. Phys. 12, 1357 (1980).

SIMPLE CONSTRUCTION OF A RELATIVISTIC WAVE EQUATION WITH ARBITRARY MASS AND SPIN AND BELINFANTE'S 1/S-CONJECTURE

Jean Pierre Gazeau
CRMA, Université de Montréal, Montréal H3C 3J7, Canada.
Marcel Perroud
Dép. Math. Appl., Ecole Polytechnique, Montréal H3C 3A7, Canada.

1. Introduction

We present a new relativistic wave equation for a particle with arbitrary mass $m > 0$ and spin s [1] which does not lead to the various inconsistencies shared by many wave equations proposed for $s > 1$ [2]. This equation is invariant with respect to a unitary representation of the orthochronous Poincaré group and, with minimal coupling, it satisfies to the Belinfante's conjecture $g = 1/s$ concerning the Landé spin factor [3].

The construction is a direct generalization of the spin $\frac{1}{2}$ case. By introducing the notations

$$\underline{P} = (\underline{E}/c, \vec{\underline{P}}),\ \underline{E} = i\hbar \frac{\partial}{\partial t},\ \vec{\underline{P}} = -i\hbar\, \vec{\nabla},$$

the Klein-Gordon equation for a particle of mass m and spin s is given by

$$\underline{P}^2 \psi(x) = (\underline{E}^2/c^2 - \vec{\underline{P}}^2)\psi(x) = m^2c^2\psi(x) \tag{1}$$

where $\psi(x)$ is a $(2s+1)$-components spinor function and $x = (ct,\vec{x})$. In the case where $s = \frac{1}{2}$, this equation can be written under the form

$$\sigma(\pi\underline{P})\ \sigma(\underline{P})\ \psi(x) = m^2c^2\psi(x) \tag{2}$$

where

$$\sigma(p) = \sigma(p)^{\dagger} = p_0 I_2 + \sigma(\vec{p}) = p_0 I_2 + p_i\sigma_i \qquad (\sigma_i = \text{Pauli matrices})$$

and $\pi p = \pi(p_0,\vec{p}) = (p_0,-\vec{p})$ $(p_0 = E/c)$. By putting

$$\psi_0(x) = \psi(x)\ ,\ \psi_1(x) = (mc)^{-1}\ \sigma(\underline{P})\,\psi(x)$$

the equation (2) is equivalent to the Dirac equation (in the Weyl representation)

$$\begin{pmatrix} -mc & \sigma(\underline{P}) \\ \sigma(\pi\underline{P}) & -mc \end{pmatrix}\begin{pmatrix} \psi_1(x) \\ \psi_0(x) \end{pmatrix} = 0. \tag{3}$$

Let us recall furthermore that the Ansatz of minimal coupling consists of replacing in equation (2) $\underline{P}$ by $\underline{P} - \frac{e}{c}A(x)$ $(A(x) = (A_0(x), \vec{A}(x)))$. After some computations we obtain

$$\left(\frac{1}{c^2}(\underline{E}-eA_0(x))^2 - (\vec{\underline{P}} - \frac{e}{c}\vec{A}(x))^2 + \frac{2e}{c}\vec{S}\cdot(\vec{B}(x)+i\vec{E}(x))\right)\psi(x) = m^2c^2\psi(x) \tag{4}$$

where

$$\vec{S}\cdot\vec{p} = \hbar/2\ \sigma(\vec{p}) \tag{5}$$

is the spin $\frac{1}{2}$ operator (in the $\vec{p}$ direction). The non relativistic limit of this equation is the well known Pauli equation

$$(\underline{E}-eA_0(x) - \frac{1}{2m}(\vec{\underline{P}} - \frac{e}{c}\vec{A}(x))^2 + \frac{2e}{2mc}\vec{S}\cdot\vec{B}(x))\,\underset{\sim}{\psi}(t,\vec{x}) = 0$$

which exhibits the Landé spin factor $g = 2$.

The generalization to the case of arbitrary spin is formally obvious: we have to find a linear matrix valued function $p \mapsto w(p)$ (by relaxing the hermiticity condition, it is a generalization of the Pauli matrices $p \mapsto \sigma(p)$) such that equation (1) is equivalent to

$$w(\pi \underline{P})^{\dagger}\, w(\underline{P})\ \psi(x) = m^2 c^2 \psi(x) \tag{6}$$

and such that, via minimal coupling, we obtain the generalization of Eq. (4):

$$\left(\frac{1}{c^2}\,(\underline{E}-eA_0(x))^2 - (\vec{\underline{P}} - \frac{e}{c}\vec{A}(x))^2 + \frac{e}{cs}\,\vec{S}^s \cdot (\vec{B}(x)+i\vec{E}(x))\right)\psi(x) = m^2c^2\psi(x) \tag{7}$$

where $\vec{S}^s \cdot \vec{p}$ is the spin s operator. Then it follows directly that the non relativistic limit of this equation exhibits the Landé spin factor $g = 1/s$ according to the Belinfante's conjecture. It is easy to show that the conditions to impose on the matrices $w(p)$ in order to obtain Eq(6) and Eq(7) are

$$w(\pi p)^{\dagger}\, w(p') = p\cdot p'\, I_{2s+1} + \sigma^S(p_0\vec{p}' - p_0'\vec{p} - i\,\vec{p}\times\vec{p}') \tag{8}$$

with

$$\sigma^S(p) = p_0 I_{2s+1} + \sigma^S(\vec{p}) = p_0 I_{2s+1} + s\hbar\,\vec{S}^s\cdot\vec{p}. \tag{9}$$

This is exactly the generalization of the well known property of the Pauli matrices

$$\sigma(\pi p)^{\dagger}\,\sigma(p) = p\cdot p'\, I_2 + \sigma(p_0\vec{p}' - p_0'\vec{p} - i\,\vec{p}\times\vec{p}').$$

As usual, we suppose that the positive (negative) energy solutions of Eq(1) are of the form

$$\psi^{\pm}(x) = (2\pi\hbar)^{-3/2}\int_{\mathbb{R}^3} d^3p\ \exp\{i/\hbar(\vec{p}\cdot\vec{x} \pm |E|t\}\ \phi^{\pm}(\vec{p})$$

where $\phi^{\pm} \in L^2(\mathbb{R}^3, d^3p; \mathbb{C}^{2s+1})$. Then it follows from the foregoing that a unitary action of the "Aristotle" group $\mathbb{R}^4 \square SU(2)$ on $L^2(\mathbb{R}^3, d^3p; \mathbb{C}^{2s+1})$ has to be

$$(U(a,u)\ \phi)(\vec{p}) = \exp\{(i/\hbar)a\cdot p\}\ D^S(u)\ \phi\ (R(u)^{-1}\vec{p}) \tag{10}$$

where D^S is a spin s representation of $SU(2)$. It will remain to examine the extension of this action to the Poincaré group.

2. Dirac equation

Let $J : V \to T$ be an isometry $(J^{\dagger}J = I_V,\ JJ^{\dagger} = I_T)$

$$J(\phi_0, \phi_R) = w_0\phi_0 + A\phi_R \tag{11}$$

where $V = \mathbb{C}^{2s+1} \times \mathbb{C}^R$, $R = 2^{2s} - (2s+1)$ and $T = (\mathbb{C}^2)^{\otimes 2s} = T_{sym} \oplus T_{nsym}$, such that

$$\text{Range}(w_0) = T_{sym}\ ,\ \text{Range}(A) = T_{nsym} \tag{12}$$

(An explicit construction of J is given in Ref [1]).

By defining

$$w(p) = \Sigma(p)\ w_0 = p_0 w_0 + w(\vec{p}) \tag{13}$$

where

$$\Sigma(p):\ v_1 \otimes v_2 \otimes \ldots \otimes v_{2s} \mapsto (\sigma(p)v_1) \otimes v_2 \otimes \ldots \otimes v_{2s},$$

it can be shown, by using the isometry conditions for J, that these matrices $w(p)$ satisfy the conditions (8) and (9). (Simply recall that the restriction to T_{sym} of the 2s-tensor product of the spin $\frac{1}{2}$ representation of the Lie algebra $su(2)$ is a spin s representation).

Now let $H_W = H_W^+ \oplus H_W^- \subset L^2(\mathbb{R}^3, d^3p; \mathbb{C}^{2s+1} \times (\mathbb{C}^2)^{\otimes 2s})$ be the Hilbert space with

$$H_W^{\pm} = \left\{ \begin{pmatrix} \phi_1 \\ \phi_0 \end{pmatrix} \middle| \begin{pmatrix} -mc & w(p_\pm) \\ w(\pi p_\pm)^\dagger & -mc \end{pmatrix} \begin{pmatrix} \phi_1(\vec{p}) \\ \phi_0(\vec{p}) \end{pmatrix} = 0 \right\}$$

where $p_\pm = (\pm|E|/c, \vec{p})$, $p_\pm^2 = m^2c^2$, $|E| = c\sqrt{m^2c^2 + \vec{p}^2}$. H_W^+ and H_W^- are two orthogonal subspaces with associated projectors

$$\pi_W^{\pm} \begin{pmatrix} \phi_1^{\pm} \\ \phi_0^{\pm} \end{pmatrix} (\vec{p}) = \frac{1}{2|E|/c} \begin{pmatrix} w(p_\pm)(\sigma^s(\pm p_\pm))^{-1} w(p_\pm)^\dagger & mc\, w(p_\pm)(\sigma^s(\pm p_\pm))^{-1} \\ mc(\sigma^s(\pm p_\pm))^{-1} w(p_\pm)^\dagger & m^2c^2(\sigma^s(\pm p_\pm))^{-1} \end{pmatrix} \begin{pmatrix} \phi_1^{\pm}(\vec{p}) \\ \phi_0^{\pm}(\vec{p}) \end{pmatrix}.$$

For any $\Phi_W \in H_W$, put $\Phi_W^{\pm} = \pi_W^{\pm} \Phi_W$; then the function

$$\Psi_W(x) = \begin{pmatrix} \psi_1(x) \\ \psi_0(x) \end{pmatrix} = (F\Phi_W)(x) = \int_{\mathbb{R}^3} d^3p \left(e^{(i/\hbar)p_- \cdot x} \phi_W^+(\vec{p}) + e^{(i/\hbar)p_+ \cdot x} \phi_W^-(\vec{p})\right)$$

satisfies the generalized Dirac equation (in Weyl representation)

$$\begin{pmatrix} -mc & w(\underline{P}) \\ w(\pi \underline{P})^\dagger & -mc \end{pmatrix} \begin{pmatrix} \psi_1(x) \\ \psi_0(x) \end{pmatrix} = 0 .$$

In the Dirac representation, the matrix $\begin{pmatrix} 0 & w_0 \\ w_0^\dagger & 0 \end{pmatrix}$ has to be diagonal. This can be achieved by considering the isometry

$$\underset{\sim}{B}: L^2(\mathbb{R}^3, d^3p; \mathbb{C}^{2s+1} \times \mathbb{C}^{2s+1} \times \mathbb{C}^R) \to L^2(\mathbb{R}^3, d^3p; \mathbb{C}^{2s+1} \times (\mathbb{C}^2)^{\otimes 2s})$$

defined by (see (11) and (12))

$$(\underset{\sim}{B}\Phi)(\vec{p}) = B\Phi(\vec{p}) \quad \text{with} \quad B = \frac{1}{\sqrt{2}} \begin{pmatrix} \sqrt{2}A & w_0 & w_0 \\ 0 & I_{2s+1} & -I_{2s+1} \end{pmatrix}.$$

Then the function $\Psi_D(x) \equiv \begin{pmatrix} \psi_R(x) \\ \psi_L(x) \\ \psi_S(x) \end{pmatrix} = (F\underset{\sim}{B}^\dagger F^{-1}\Psi_W)(x)$ satisfies the equation

$$\begin{pmatrix} -mc & \frac{1}{\sqrt{2}} C(\underline{\vec{P}}) & -\frac{1}{\sqrt{2}} C(\underline{\vec{P}}) \\ -\frac{1}{\sqrt{2}} C(\underline{\vec{P}})^\dagger & -mc + \frac{1}{c}\underline{E} & -\sigma^s(\underline{\vec{P}}) \\ \frac{1}{\sqrt{2}} C(\underline{\vec{P}})^\dagger & \sigma^s(\underline{\vec{P}}) & -mc - \frac{1}{c}\underline{E} \end{pmatrix} \begin{pmatrix} \psi_R(x) \\ \psi_L(x) \\ \psi_S(x) \end{pmatrix} = 0$$

where $C(\vec{p}) = A^\dagger w(\vec{p})$. The "redundant" component ψ_R can be eliminated (for $m \neq 0$) and we obtain

$$\begin{pmatrix} -mc + \frac{1}{c}\underline{E} - Q(\underline{\vec{P}}) & -\sigma^s(\underline{\vec{P}}) + Q(\underline{\vec{P}}) \\ \sigma^s(\underline{\vec{P}}) + Q(\underline{\vec{P}}) & -mc - \frac{1}{c}\underline{E} - Q(\underline{\vec{P}}) \end{pmatrix} \begin{pmatrix} \psi_L(x) \\ \psi_S(x) \end{pmatrix} = 0$$

where $Q(\vec{p}) = \frac{1}{2mc}(\vec{p}^2 - (\sigma^s(\vec{p}))^2)$ is quadratic in $\vec{p}$. By taking the non relativistic limit, ψ_L and ψ_S appear respectively as a "large" and a "small" component.

3. Relativistic invariance

The Wigner realization of the U.R. $U \sim U_{m,s,+} \oplus U_{m,s,-}$ of the Poincaré group $P(3,1) = \mathbb{R}^4 \,\square\, SL(2,C)$ is given by [4]

$$U(a,g)\begin{pmatrix} f^+ \\ f^- \end{pmatrix}(\vec{p}) = \begin{pmatrix} \exp\{(i/\hbar)p_+\cdot a\}\, D^s(h(p_+)^{-1}\, g\, h(L(g)^{-1}p_+))\, f^+(L_+(g)^{-1}\vec{p}) \\ \exp\{(i/\hbar)p_-\cdot a\}\, D^s(h(p_-)^{-1}\, g\, h(L(g)^{-1}p_-))\, f^-(L_-(g)^{-1}\vec{p}) \end{pmatrix}$$

where $\begin{pmatrix} f^+ \\ f^- \end{pmatrix} \in L^2(\mathbb{R}^3, \frac{d^3p}{|E|/c}; \mathbb{C}^{2s+1} \times \mathbb{C}^{2s+1})$, $h(p_\pm)$: $(\pm m, \vec{0}) \mapsto p_\pm$ is a Wigner boost and $L_\pm(g)$ denotes the usual non linear action of $SL(2,\mathbb{C})$ on $\mathbb{R}^3$.

This unitary representation of $P(3,1)$ can be mapped on the space H_W by means of the isometry

$$T_W\begin{pmatrix} f^+ \\ f^- \end{pmatrix}(\vec{p}) = (T_W^+ f^+)(\vec{p}) + (T_W^- f^-)(\vec{p})$$

where

$$(T_W^\pm f^\pm)(\vec{p}) = \frac{1}{\sqrt{2|E|/c}}\begin{pmatrix} w(p_\pm)(\sigma^s(\pm p_\pm))^{-\frac{1}{2}} f^\pm(\vec{p}) \\ mc(\sigma^s(\pm p_\pm))^{-\frac{1}{2}} f^\pm(\vec{p}) \end{pmatrix}.$$

Note that for $p_0 > 0$, the spectrum of $\sigma^s(p)$ lies in the interval $[p_0 - |\vec{p}|, p_0 + |\vec{p}|] \subset R^+$. The representation $U_W(a,g) = T_W U(a,g) T_W^{-1}$ is given by

$$(U_W(a,g)\ \phi_W)(\vec{p}) = (U_W^+(a,g)\ \phi_W^+)(\vec{p}) + (U_W^-(a,g)\ \phi_W^-)(\vec{p})$$

where

$$\begin{cases} (U_W^\pm(a,g)\ \phi_W^\pm)(\vec{p}) = \exp\{(i/\hbar)p_\pm\cdot a\} \left|\dfrac{(L(g)^{-1}p_\pm)_0}{p_0}\right| R(p_\pm,g)\ \phi_W^\pm(L_\pm(g)^{-1}\vec{p}) \\ R(p_\pm,g) = \begin{pmatrix} \frac{1}{m^2c^2} w(p_\pm)\ \Delta(p_\pm,g)\ w(\pi L(g)^{-1}p_\pm)^\dagger & 0 \\ 0 & \Delta(p_\pm,g) \end{pmatrix} \\ \Delta(p_\pm,g) = (\sigma^s(\pm p_\pm))^{-\frac{1}{2}}\, D^s(h(p_\pm)^{-1}\, g\, h(L(g)^{-1}p_\pm))(\sigma^s(\pm L(g)^{-1}p_\pm))^{\frac{1}{2}}. \end{cases}$$

When $g \in SU(2)$, $\Delta(p_\pm,g)$ reduces to $D^s(g)$ and the requirement (10) is fulfilled. In the case $s = \frac{1}{2}$ we recover the familiar results because the boosts $h(p_\pm)$ can always be chosen so that $(\sigma(\pm p_\pm))^{\frac{1}{2}} = \sqrt{mc}\ h(p_\pm)$.

The parity operator is obviously defined on H_W through $P_W = T_W P T_W^{-1}$, where P denotes the parity operator on $L^2(\mathbb{R}^3, \frac{d^3p}{|E|/c}; \mathbb{C}^{2s+1} \times \mathbb{C}^{2s+1})$.

Similar results hold of course for the "Dirac representation space" $H_D = \underset{\sim}{B}^\dagger H_W$; actually on this space the Poincaré group operates via the unitary representation $U_D(a,g) = T_D U(a,g) T_D^{-1}$ with $T_D = B^\dagger T_W$. Finally let us mention that these mappings T_W and T_D are two generalizations of the Foldy-Wouthuysen transformations.

References

[1] J.P. Gazeau, Preprint CRMA-946, submitted to Jour. of Phys. (1980).

[2] A.S. Wightman, in "Invariant Wave Equations", Proceedings, Erice, 1977, Lectures Notes in Physics 73 (Springer, Berlin, 1978).

[3] F.J. Belinfante, Phys. Rev. 92, 997 (1953).

[4] E.P. Wigner, Ann. Math. 40, 149 (1939).

POINCARÉ GROUP, SYMPLECTIC GROUP, AND RELATIVISTIC EXTENDED HADRONS CONSISTING OF SPIN 1/2 QUARKS

Paul E. Hussar and Y. S. Kim
Center for Theoretical Physics, Department of Physics and Astronomy
University of Maryland, College Park, Maryland 20742

Marilyn E. Noz
Department of Radiology, New York Univeristy, New York
New York 10016

In order to construct a quantum mechanics of relativistic extended hadrons consisting of quarks, we follow the prescription given by Dirac. In his classic paper[1] published in the special issue of the Reviews of Modern Physics in honor of Einstein's 70th Birthday, Dirac noted that the procedure of constructing relativistic quantum mechanics is equivalent to finding spacetime representations of the Poincaré group satisfying one of the subsidiary conditions which reduces the four-dimensional Minkowskian space into a three-dimensional Euclidian space in which nonrelativistic quantum mechanics is valid.

It has been shown that the covariant harmonic oscillator formalism provides such a representation.[2] In order to see this point, let us consider an extended hadron consisting of two quarks bound together by a harmonic oscillator of unit strength. If x_1 and x_2 represent spacetime coordinates for the two constituent quarks, it is a standard practice to define the hadronic coordinate X and relative quark separation x:

$$X = (x_1 + x_2)/2 \quad , \qquad x = (x_1 - x_2)/2\sqrt{2} \quad . \tag{1}$$

In terms of these coordinate variables, the generators of the Poincare group take the form

$$P_\mu = i\,\partial/\partial X^\mu \text{ , and } \quad M_{\mu\nu} = L^*_{\mu\nu} + L_{\mu\nu} \quad , \tag{2}$$

where

$$L^*_{\mu\nu} = i[X_\mu \partial/\partial X^\nu - X_\nu \partial/\partial X^\mu] \text{ , } \quad L_{\mu\nu} = i[x_\mu \partial/\partial x^\nu - x_\nu \partial/\partial x^\mu] \quad .$$

The Casimir operators then are[3]

$$P^\mu P_\mu \text{ and } \quad W^\mu W_\mu, \quad \text{where } W_\mu = (1/2)\, \varepsilon_{\mu\nu\alpha\beta}\, P^\nu M^{\alpha\beta} \quad . \tag{3}$$

For the hadron moving along the z direction with velocity parameter β, it has been shown[2] that the solutions of the above differential equation which are diagonal in the Casimir operators take the form

$$\phi(X, x) = e^{-iP\cdot X}\, \psi_\beta^{n\ell m}(x) \quad . \tag{4}$$

with

$$\psi_\beta^{n\ell m}(x) = (1/\pi)^{1/4}[\exp(-\, t'^2)]\, R_{n\ell}(r')\, Y_\ell^m(\theta', \phi') \quad ,$$

where

$$t' = (t - \beta z)/(1 - \beta^2)^{1/2} \quad , \quad z' = (z - \beta t)/(1 - \beta^2)^{1/2} \quad , \tag{5}$$

and r', θ', ϕ' are polar coordinate variables in the three-dimensional Euclidian space spanned by the x, y and z' variables.

The operator P^2 is constrained to take the discrete eigenvalues of the harmonic oscillator operator

$$P^2 = (1/2)[(\partial/\partial x^\mu)^2 - (x^\mu)^2] + m_0{}^2 \quad . \tag{6}$$

The wave function $\psi_\beta^{n\ell m}(x)$ of Eq.(4) is an eigenfunction of the above operator subject to the subsidiary condition

$$P^\mu a_\mu^+\, \psi_\beta^{n\ell m}(x) = 0 \ , \quad \text{where} \quad a_\mu^+ = \partial/\partial x^\mu + x_\mu \ . \tag{7}$$

This subsidiary condition allows a ground-state uncertainty along the t' direction but does not allow any excitations. This is precisely the quantum mechanical representation of Dirac's instant form constraint which forbids motions along the t' axis.[4]

The wave function of Eq.(4) is basically a polynomial in the Cartesian variables x, y, z', t' multiplied by a Gaussian function, and the only factor affected by a boost along the z direction is

$$\psi_\beta^n(z, t) = (\pi 2^n n!)^{-1/2} H_n(z') \exp[-\, (z'^2 + t'^2)] \quad . \tag{8}$$

Since this wave function is well localized in the zt plane, its Lorentz deformation property is dictated by the geometry of Lorentz transformation. The Lorentz transformation in Eq.(5) can be written as

$$(t' + z') = [(1 - \beta)/(1 + \beta)]^{1/2} (t + z) ,$$
$$(t' - z') = [(1 + \beta)/(1 - \beta)]^{1/2} (t - z) . \tag{9}$$

Thus the transformation elongates one light-cone axis and contracts the other in such a way that the quantity

$$(t + z)(t - z) = (t^2 - z^2) = (t'^2 - z'^2) = (t' + z')(t' - z') \tag{10}$$

is invariant.[5] This means that the area of the rectangle whose sides are parallel to the light cones is a Lorentz-invariant quantity. This indeed is a symplectic transformation.[6] The localized spacetime distribution of the wave function given in Eq.(4) undergoes this symplectic transformation, while preserving the total probability.

The important point is that this symplectic property manifests itself in basic high-energy hadronic phenomena including the mass spectra, form factors, parton model, and the jet phenomenon.[4] It explains how a fast-moving hadron consisting of a finite number of quarks appears as a collection of an infinite number of free independent partons.[7]

Let us next consider the relativistic extended hadron consisting of spin 1/2 quarks. The spacetime translation operator in this case remains the same as before, but the generators of the Lorentz transformation is modified to

$$M_{\mu\nu} = L^*_{\mu\nu} + L_{\mu\nu} + S_{\mu\nu} , \tag{11}$$

where $S_{\mu\nu}$ is the sum of the spin operators for the two quarks, and can take the form

$$S_{\mu\nu} = (1/4i)([\gamma^1_\mu, \gamma^1_\nu] + [\gamma^2_\mu, \gamma^2_\nu]) , \tag{12}$$

where the superscripts 1 and 2 specify the first and second constituent quarks respectively.

As before, the Casimir operators will be P^2 and W^2, where W^2 is to be constructed according to Eq.(3) with the new $M_{\mu\nu}$ given in Eq.(11). The mass spectrum condition of Eq.(6), and the subsidiary condition of Eq.(7) should also be modified in order to take into account the effect of the quark spin.[8] There are at present several approaches to this problem in the literature.[9,10,11] For the purpose of illustration, we shall discuss here the model proposed by Takabayashi.[10] According to his model, the mass spectrum condition becomes

$$P^2 = (1/2)[(\partial/\partial x^\mu)^2 + (x^\mu)^2 - \gamma^1_\mu \gamma^{2\mu} + m_0{}^2] \ . \tag{13}$$

Takabayashi then linearizes the subsidiary condition of Eq.(7) as

$$[\gamma^1{}_\mu p_1{}^\mu + \gamma^2{}_\mu p_2{}^\mu + (i/2)(\gamma^1{}_\mu - \gamma^2{}_\mu)(x_1{}^\mu - x_2{}^\mu)] \ \phi(x_1, x_2) = 0 \ . \tag{14}$$

The key question here is whether the above two conditions are consistent with each other. More specifically, the operators in these two equations should commute with each other in the function space which satisfies the condition of Eq.(14) and which is diagonal in the operator P^2.

We can solve the above equations by constructing first the spacetime solution of the equation

$$(1/2)[(\partial/\partial x^\mu)^2 - (x^\mu)^2 + 2\ m_0{}^2] \ f(x) = \lambda \ f(x) \quad , \tag{15}$$

and the spin wave function χ which satisfies

$$- \gamma^{1\mu}\gamma^2_\mu \ \chi = \lambda' \ \chi \quad . \tag{16}$$

Then the eigenvalue of P^2 will be

$$P^2 = M^2 = (\lambda + \lambda') \quad , \tag{17}$$

and the wave function in general will take the form

$$\phi(X, x) = e^{-iP\cdot X} \ [\chi_1 \ f_1(x) + \ \chi_2 \ f_2(x) + \text{---}] \quad , \tag{18}$$

where each term has the same $(\lambda + \lambda')$ eigenvalue. The spacetime solution f(x) has already been extensively discussed.[2] The construction of the spin states is not difficult. There are four singlet wave functions and four triplets. There are therefore altogether 16 spin states.

The next criterion in determining the correct wave function is that $\phi(X, x)$ should also be an eigenstate of the Casimir operator W^2. This procedure is not different from the usual angular momentum addition in the frame in which the hadron is at rest. Diagonalizing the complete set of eigenstates of P^2 developed in Eqs. (15)-(18) under the action of W^2 now gives us a complete set of eigenstates of the Casimir operators of the Poincaré group.

The wave functions in this model are required to obey the subsidiary condition given in Eq.(14). These solutions can be represented as degenerate linear combinations of the Casimir opertor eigenstates just constructed. The final step in determining the physical wave function is to choose those with the correct parity and charge conjugation.[9] We have used here Takabayshi's model only as an example of Dirac's prescription of constructing representations of the Poincaré group subject to a covariant subsidiary condition. The method outlined here can be used for other models of relativistic extended hadrons consisting of spin 1/2 quarks.

REFERENCES

1. P. A. M. Dirac, Rev. Mod. Phys. 21, 392 (1949).
2. Y. S. Kim, M. E. Noz, and S. H. Oh, J. Math. Phys. 20, 1341 (1979); 21, 1228 (1980).
3. E. P. Wigner, Ann. Math. 40, 149 (1939).
4. Y. S. Kim and M. E. Noz, Found. of Phys. 9, 375 (1979); Y. S. Kim, M. E. Noz, and S. H. Oh, Found. of Phys. 9, 947 (1979).
5. D. Han and Y. S. Kim, Bull. Am. Phys. Soc. 25, 492 (1980).
6. Y. S. Kim and M. E. Noz, Univ. of Maryland CTP Tech. Rep. # 80-094 (1980).
7. Y. S. Kim and M. E. Noz, Phys. Rev. D 15, 335 (1977).
8. P. E. Hussar, Bull. Am. Phys. Soc. 25, 491 (1980).
9. T. J. Karr, Univ. Pierre et Marie Curie Preprint PAR-LPTHE 78/1 (1978).
10. T. Takabayashi, Nagoya Univ. Preprint, Nov., 1978 [Excerpt from the talk given at the Japan Physical Society Meeting, Shinshu Univ., 1978].
11. H. Leutwyler and J. Stern, Phys. Lett. 73B, 75 (1978).

AN OPERATOR S MATRIX THEORY OF THE STRONGLY INTERACTING PARTICLES

William H. Klink
The University of Iowa
Iowa City, Iowa 52242 USA

I. Introduction

Although S matrix theory is not a fashionable theory of elementary particles today, it continues to be worthy of investigation at the very least as a means of comparison with the scattering amplitudes that are calculated from gauge field theories. One of the reasons that analytic S matrix theory does not play the prominent role that it did ten years ago is because of multiparticle reactions. It has not been possible to exhibit a representation for amplitudes of multiparticle reactions analogous to the Mandelstam representation for two-particle to two-particle reactions.

However, it is not necessary that an S matrix theory necessarily be an analytic S matrix theory. In Heisenberg's original papers [1], the emphasis was on the operator properties rather than the analyticity properties of the scattering matrix. The goal of this paper is to show what is meant by an operator S matrix theory, and in particular to show the degree to which group theoretical considerations can be used to express physical requirements in an operator S matrix language. Some of the requirements that any sensible relativistic scattering amplitude for the strongly interacting particles must have include unitarity, cluster properties, invariance under discrete symmetries such as parity and time reversal, correct connection between spin and statistics, and finally, appropriate crossing and causality properties. Because crossing properties seem rather ill suited to an operator language, they will be discussed in Section III, after the other physical requirements listed above, have been discussed in Section II.

The aim of any S matrix theory is to exhibit a representation of scattering amplitudes that automatically satisfies the above (and possibly more) physical requirements. An operator S matrix theory uses operator techniques to find such scattering amplitudes. After showing how properties such as cluster properties may be expressed in an operator language, it will be shown that these properties depend on coupling schemes for coupling together free relativistic particles to form initial or final states for multiparticle scattering amplitudes. To keep the discussion as brief and simple as possible, we will consider a model world consisting of only spinless particles of mass m; the goal will be to find a representation for the multiparticle scattering amplitudes describing reactions of the form

$$\underbrace{\pi + \pi + \ldots + \pi}_{A} \rightarrow \underbrace{\pi + \pi + \ldots + \pi}_{B} \quad ,$$

where A denotes the initial cluster, and B the final cluster, respectively.

II. Unitarity and Cluster Properties of Multiparticle Scattering Operators

Let $S_{B,A}$ denote a scattering operator from an A particle partial wave Hilbert space $\mathcal{H}_A$ to a B particle partial wave Hilbert space $\mathcal{H}_B$. An n particle Hilbert space is defined to be the n-fold tensor product of one-particle Hilbert spaces; these one-particle Hilbert spaces arise as irreducible representation spaces of the Poincaré group, for particles of zero spin and mass m. An n particle partial wave space is an n particle Hilbert space in which the total momentum $\vec{P}$, invariant mass $\sqrt{s}$ $(s = (p_1 + \ldots + p_n)^2)$, and total angular momentum and angular momentum projection J,σ have been extracted. Since an n particle Hilbert space is a direct integral of n particle partial wave Hilbert spaces, these partial wave spaces should be denoted by $\mathcal{H}_{\vec{P}sJ\sigma;n}$; however, because we are dealing exclusively with partial wave spaces in this paper, such spaces will be written simply as $\mathcal{H}_n$, with the variables $\vec{P}sJ\sigma$ fixed, but suppressed.

Partial wave Hilbert spaces are of interest because both unitarity and cluster properties of scattering operators can be simply expressed in terms of them. Let Λ_n be the projection operator from Fock space to an n particle partial wave space. Then the unitarity equations $S^\dagger S = S S^\dagger = I$ on Fock space become, for an initial cluster A and final cluster B,

$$\sum_n S^\dagger_{B,n} S_{n,A} = 0 \quad , \quad B \neq A \quad ; \quad = I \in \mathcal{H}_B \quad , \quad A = B$$

$$\sum_n S_{B,n} S^\dagger_{n,A} = 0 \qquad\qquad = I \in \mathcal{H}_B \tag{II.1}$$

where the channel scattering operators are $S_{B,A} \equiv \Lambda_B S \Lambda_A$, and the sum on n is over all intermediate clusters allowed by the fixed invariant mass $\sqrt{s}$. It should be noted that the $2 \to n$ or $n \to 2$ reactions $(n > 2)$ are described by scattering operators $S_{n,2}$ and $S_{2,n}$, respectively, whose kernels are actually elements in $\mathcal{H}_n$, since a two-particle partial wave space is one dimensional.

The solutions to the projected unitarity equations can be expressed in terms of a spectral representation of $S_{B,A}$ [2]:

$$S_{B,A} = \int d\mu(\gamma)\lambda_\gamma \, e^B_\gamma \otimes \bar{e}^{A\dagger}_\gamma \tag{II.2}$$

where λ_γ are (generalized) eigenvalues equal to one except for one eigenvalue (denoted by $\gamma = 1$) whose value is η, the inelasticity parameter of the $2 \to 2$ reaction. The projected eigenvectors e^n_γ span $\mathcal{H}_n$ (but are, in general, not orthogonal in $\mathcal{H}_n$); e^n_1 has direct physical significance in that

$$\mathcal{A}_{n\to 2} = e^{i(\frac{\pi}{2}+\delta)} \sqrt{1-\eta^2}\, e^n_1 \quad , \tag{II.3}$$

where $\mathcal{A}_{n\to 2}$ is the partial wave amplitude for the $n \to 2$ reaction and δ is the phase shift for the $2 \to 2$ reaction. $\bar{e}^n_\gamma$ is the complex congugate of e^n_γ. Thus, the content of unitarity is to generate a spectral representation for multiparticle scattering operators, in which one of the eigenvectors ($\gamma = 1$) is proportional to the production partial wave amplitudes.

By a cluster property is meant that property of a channel scattering operator such that when a subcluster of initial particles is moved far from the remaining particles in the cluster, the channel scattering operator splits into a tensor product of subcluster channel scattering operators. Let $U_{\vec{a}}$ be the unitary operator that translates by an amount $\vec{a}$. Then the cluster property can be expressed in operator language as

$$\underset{|\vec{a}| \to \infty}{\text{strong limit}}\; U_{\vec{a}}\, S_{B,A}\, U_{\vec{a}}^{\dagger} = S_{B_1A_1} \otimes S_{B_2A_2} \tag{II.4}$$

Reference [3] shows that this limit becomes an angular momentum limit in partial wave spaces; that is, if all the particles in cluster A_1 are coupled together, all the particles in cluster A_2 coupled together, and then these two subclusters coupled together to form the overall multiparticle, then the limit as $|\vec{a}| \to \infty$ in Eq. (II.4) becomes a limit as $J \to \infty$. Thus, the nature of the limit as J gets large depends on the properties of the Racah coefficients of the Poincaré group; these coefficients can be viewed as unitary operators on the partial wave Hilbert spaces, changing one coupling scheme to another.

III. Crossing Properties of Scattering Operators

While it may seem natural that physical requirements such as unitarity or time reversal invariance can be expressed as operator requirements, the same is not true of crossing, for crossing involves the analytic continuation of scattering amplitudes into unphysical regions to reach the physical region of the crossed channel reaction. By crossing we shall mean the crossing of one particle, so that the crossed reaction of $A \to B + c$ is $A + \bar{c} \to B$, where particle c has been crossed to become the antiparticle $\bar{c}$. Now the partial wave amplitudes associated with these reactions will, in general, have variables that behave in a very complicated way under analytic continuation. Reference [4] shows that if all the particles in cluster B are coupled together, and the resulting multiparticle coupled to particle c, the resulting variables (Racah coefficient variables) have the property (except for s_A and s_B) that under analytic continuation they remain in their physical region (with a possibly new physical meaning). Here $\sqrt{s_A}$ and $\sqrt{s_B}$ are the invariant mass of clusters A and B, respectively, and from kinematical considerations must satisfy $\sqrt{s_A} \geq \sqrt{s_B} + m$ for the $A \to B + c$ channel, and $\sqrt{s_A} \geq \sqrt{s_B} + m$ for the $A + \bar{c} \to B$ channel. Thus, analytic continuation in many variables is reduced to analytic continuation in two standard variables.

Now consider a definite coupling scheme in which all of the B cluster particles are coupled together in a stepwise fashion, particle 1 to 2, (12) to 3 and so forth, until all the particles in the cluster have been coupled together. The resulting variables become the invariant mass, angular momentum, and angular momentum projection of each subcluster. In these variables the projected eigenvectors may be written as [2]

$$\sqrt{\lambda_\gamma}\, e^B_\gamma(s_B J_B \sigma_{B-1} s_{B-1} J_{B-1} \cdots \sigma_2 s_2 J_2) = \int d\mu(\gamma_1) \cdots \int d\mu(\gamma_{B-3})$$
$$\times f_{\gamma\gamma_1}(s_B J_B \sigma_{B-1} s_{B-1} J_{B-1}) \cdots f_{\gamma_{B-4}\gamma_{B-3}}(s_4 J_4 \sigma_3 s_3 J_3)\, e^3_{\gamma_{B-3}}(s_3 J_3 \sigma_2 s_2 J_2) \qquad \text{(III.1)}$$

where the $f_{\gamma\gamma'}$ are functions of the appropriate stepwise coupled variables and e^3_γ is a projected three-particle eigenvector. For example, if B is a 5 particle cluster, then

$$\sqrt{\lambda_\gamma(s_5 J_5)}\; e^5_\gamma\,(s_5 J_5 \sigma_4 s_4 J_4 \sigma_3 s_3 J_3 \sigma_2 s_2 J_2) = \int d\mu(\gamma_1) d\mu(\gamma_2)$$
$$\times f_{\gamma\gamma_1}(s_5 J_5 \sigma_4 s_4 J_4) f_{\gamma_1\gamma_2}(s_4 J_4 \sigma_3 s_3 J_3) e^3_{\gamma_2}(s_3 J_3 \sigma_2 s_2 J_2) \quad . \qquad \text{(III.2)}$$

The $f_{\gamma\gamma'}$ are the basic building blocks of the operator S matrix theory and must have the following properties: (1) $\int d\mu(\bar{\gamma})(f_{\gamma\bar{\gamma}}, f_{\gamma'\bar{\gamma}}) + (e^3_\gamma, e^3_{\gamma'}) = \delta_{\gamma\gamma'}$ (orthogonality properties coming from unitarity, see Reference [2], Eq. (25)). (2) The $f_{\gamma\gamma'}$ should satisfy crossing relations, see Reference [5]; in particular, these crossing relations imply that e^3_γ is related to $f_{1\gamma}$. (3) The $f_{\gamma\gamma'}$ should satisfy certain functional relations coming from the interchange of identical particles. (4) The cluster properties of Section II can be translated into requirements on the $f_{\gamma\gamma'}$ (see Reference [3]). (5) Macrocausality [6] is related to rescattering diagrams, which will impose requirements on the spectral measure $d\mu(\gamma)$; however, this remains to be worked out.

For $f_{\gamma\gamma'}$ satisfying the above conditions, the scattering amplitude (i.e., the kernel of the scattering operator $S_{B,A}$) becomes

$$\mathcal{a}_{B,A} = \int d\mu(\gamma_1) \cdots \int d\mu(\gamma_1') e^3_{\gamma_1} f_{\gamma_1\gamma_2} f_{\gamma_2\gamma_3} \cdots f_{\gamma_2'\gamma_1'} e_{\gamma_1'} \qquad \text{(III.3)}$$

(in appropriate stepwise coupled variables) and satisfies the physical requirements listed in the introduction.

References

[1] W. Heisenberg, Z. Physik 120 (1942) 513; 120 (1943) 6; the acausal behavior of Heisenberg's scattering amplitudes was noted by E. C. G. Stueckelberg in Helv. Phys. Act. 27 (1954) 667.

[2] W. H. Klink, J. Math. Phys. 20 (1979) 2514; and a paper to be submitted for publication to Phys. Rev. D.

[3] W. H. Klink, Phys. Rev. A 20 (1979) 1864.

[4] W. H. Klink, Nucl. Phys. B86 (1975) 175.

[5] W. H. Klink, J. Math. Phys. 20 (1979) 2511.

[6] D. Iagolnitzer, The S Matrix, North-Holland Publishing Company, Amsterdam, 1978, ch. II.

GENERALIZED GALILEI INVARIANT PARTIAL WAVE EXPANSIONS OF THE SCATTERING AMPLITUDE FOR COLLISIONS BETWEEN TWO PARTICLES WITH ARBITRARY SPIN*

Alfonso Mondragón, Daniel Sepúlveda

Instituto de Física,
Universidad Nacional Autónoma de México
Apdo. Postal 20-364, México 20, D.F.
MEXICO

Abstract: The invariant operators of the Euclidean group $E(3)$ and its chains of subgroups $E(3) \supset O(3) \supset O(2)$ and $E(3) \supset E(2) \times T_{\perp} \supset O(2) \times T_{\perp}$ provide bases of eigenfunctions for the construction of generalized Galilei invariant partial wave expansions of the scattering amplitude for non-relativistic collisions between particles of arbitrary spin. These expansions are generalizations of those obtained by Kalnins et.al.[1] for spinless particles. The first chain of groups produces a spherical expansion which is a generalization of the well known helicity formalism. The second chain of groups gives rise to two different cylindrical representations of the scattering amplitude, each one related to one of the two symmetry axes in the collision. The cylindrical expansion associated to the total momentum axis of symmetry is a generalization of the impact parameter eikonal expansion supplemented with an additional expansion in the remaining kinematical variable. Associated to the momentum transfer axis, there is another cylindrical expansion of the scattering amplitude which coincides with the non-relativistic limit of the crossed channel expansion of the relativistic amplitude as shown by Cocho and Mondragón[2]. In every case, the scattering amplitude and the partial wave amplitude are integral transforms one of the other. The kernels of these transforms are expressed in terms of matrix elements of the group operators appropriate to each case.

I. INTRODUCTION

The elastic collision of two particles is described in non-relativistic quantum mechanics by means of the scattering amplitude A, which is the matrix element of the transition operator $\mathbb{A} = \mathbb{S} - \mathbb{1}$, between the initial or incoming and the final or outgoing states of the system

*Work supported partially by Instituto Nacional de Investigación Nuclear de México.

$$A(p_1 s_1 \lambda_1, p_2 s_2 \lambda_2; p_1' s_1' \lambda_1', p_2' s_2' \lambda_2') = \langle p_1' s_1' \lambda_1', p_2' s_2' \lambda_2' | \mathbb{A} | p_1 s_1 \lambda_1, p_2 s_2 \lambda_2 \rangle \tag{1}$$

For an elastic scattering process, $\mathbb{A}$ is defined on the Hilbert space $\mathcal{H}_1 \otimes \mathcal{H}_2$ of two non-interacting particles. Invariance of the system under the Galilei group makes it possible to expand the scattering amplitude in such a way that its dependence on all kinematical parameters, such as energies, angles, momentum transfers, etc., is displayed explicity. The kinematical parameters appear as arguments in known special functions while the dynamics specific to each process is thus isolated in the expansion coefficients or generalized partial wave amplitudes. The introduction of the kinematical assumptions through the matrix elements of the group of kinematical transformations —the Galilei group in this case— and its chains of subgroups provides a natural and convenient way of choosing the set of functions for the generalized partial wave expansions. The basis functions are then eigenfunctions of a complete set of commuting operators, all of which are either Casimir operators of the Galilei group itself or operators of one of the subgroups in the reduction chains. To each form of the expansions corresponds also a choice of a definite frame of reference in which the symmetries implicit in the reduction chain are most clearly exhibited.

In what follows we derive two such generalized expansions for the following two chains of groups

i) The spherical expansion, which is the generalization of the ordinary partial wave expansion and is made according to the chain

$$E(3) \supset O(3) \supset O(2)$$

ii) The non-relativistic crossed channel expansion made according to the chain

$$E(3) \supset E(2) \times T_\perp \supset O(2) \times T_\perp$$

This is a cylindrical or generalized eikonal expansion. The axis of symmetry is chosen parallel to the momentum transfer. This expansion coincides with the non-relativistic limit of the Regge-Jost $O(2,1)$ expansion[3)] for relativistic collisions.

There is another cylindrical representation of the scattering amplitude. It corresponds to a choice of the symmetry axis parallel to the total momentum of the system and it leads to a generalization of the familiar impact parameter representations. The analysis leading to the two cylindrical expansions is very similar, and for this reason we will treat here only the less well-known case of the non-relativistic, crossed channel generalized cylindrical expansion.

II. SPHERICAL EXPANSION

The in and out states are eigenstates of translations so that the total momentum $\vec{P}$ is already diagonal. In the center of mass frame of reference $\vec{P}=0$ and the in and out states are labelled with the eigenvalues of the momentum $\vec{p}$ of the relative motion and the helicity which is the projection of the spin of the states on the momentum $\vec{p}$. Coupling the particle spin s_1, s_2 to s, and s_1', s_2' to s' we get

$$\langle \vec{p}_1' s_1' \lambda_1', \vec{p}_2' s_2' \lambda_2' | A | \vec{p}_1 s_1 \lambda_1, \vec{p}_2 s_2 \lambda_2 \rangle =$$
$$= \sum_{\lambda_1 \lambda_2} \sum_{\lambda_1' \lambda_2'} \{ (s_1 \lambda_1, s_2 \lambda_2 | s \lambda)(s_1' \lambda_1' s_2' \lambda_2' | s' \lambda') \times$$
$$\times \langle \vec{p}' s' \lambda' | A | \vec{p} s \lambda \rangle \tag{1}$$

Making a rotation on $|\vec{p}'\rangle$ so as to align it with $|\vec{p}\rangle$ we obtain the familiar helicity expansion

$$\langle \vec{p}' s' \lambda' | A | \vec{p} s \lambda \rangle = \sum_j \{ (2j+1) \mathcal{D}^{j*}_{\lambda' \lambda}(\varphi, \theta, -\varphi) \times$$
$$\times \langle p s' \lambda' | A^j | p s \lambda \rangle \} \tag{2}$$

In (2), the rotation matrices $\mathcal{D}^{j*}_{\lambda'\lambda}(\varphi, \theta, -\varphi)$ display the angular dependence of the scattering amplitude, they are eigenfunctions of the Casimir operator and the generator of the subgroup of $G(3)$ that leaves $|\vec{P}|$ invariant, this is of course the ordinary group of rotations $O(3)$. In the same way, the helicity amplitudes $A^j(p s' \lambda', p s \lambda)$ will be now expanded so as to exhibit the invariance with respect to translations.

The initial and final states can be generated by a translation acting on a zero momentum state

$$|p s \lambda\rangle = T(0 0 p) |0 s \lambda\rangle \tag{3}$$

inserting this expression in (1), and making a decomposition in eigenstates of the total angular momentum and helicity, amplitude A^j becomes

$$\langle p s' \lambda' | \mathbb{A}^{j} | p s \lambda \rangle =$$

$$= \sum_{j m_j} \sum_{j' m'_j} \{ \langle 0 s \lambda' \lambda | T^{+}(0 0 p) | j' m'_j \lambda \rangle \times$$

$$\times \langle j' m'_j \lambda' | \mathbb{A}^{j} | j'' m''_j \lambda \rangle \langle j'' m''_j \lambda | T(0,0,p) | 0 s \lambda \lambda \rangle \} \quad (4)$$

The matrix elements of pure translations between eigenstates of the total angular momentum and helicity appearing in (4) are obtained from the matrix elements of the projective unitary irreducible representations of the Euclid group $E(3)$ defined by A.S. Wigthman[4]. Explicit expressions for the PUIR of $E(3)$ have been given by W. Miller Jr.[5]

$$\langle h s m | U(\bar{r}, A) | j h m \rangle =$$

$$= \sum_{\nu=-j}^{j} \langle h s m | T(0,0,p) | j \nu m \rangle \mathcal{D}^{j}_{\nu m}(A)$$

where

$$\langle j m_j \nu | T(0,0,p) | s \lambda' \lambda \rangle = \langle j m_j \nu | e^{i p.r} | s \lambda' \lambda \rangle =$$

$$\frac{1}{4\pi} \int_0^{\pi} \int_0^{2\pi} \mathcal{D}^{j*}_{\nu m_j}(-\varphi, \theta, \varphi) e^{i p.r} \mathcal{D}^{s}_{\lambda' \lambda}(-\varphi, \theta, \varphi) \times$$

$$\times \sin\theta \, d\theta \, d\varphi \quad (5)$$

where (θ, φ) define the direction of $\vec{p}$ with respect to some fixed system of coordinates.

Now, keeping j fixed and letting m_j vary we define a spinor

$$\chi^{(p,\lambda)}_{j m_j \lambda'}(\vec{r}) \equiv \langle j m_j \lambda | e^{i \vec{p}.\vec{r}} | s \lambda' \lambda \rangle \quad (6)$$

in terms of which the helicity amplitude can be written as

$$\langle p s' \lambda' | \mathbb{A} | p s \lambda \rangle =$$

$$= \sum_{m_j} \sum_{m'_j} \int d^3 r \int d^3 r' \, \chi^{(p,\lambda)+}_{j m'_j \lambda'}(\vec{r}') \mathbb{A}^{j}(\vec{r}' \vec{r}) \chi^{(p,\lambda)}_{j m_j \lambda}(\vec{r}) \quad (7)$$

in going from (4) to (7) we have made use of the fact that the collision matrix $\mathbb{S}$ and the transition operator $\mathbb{A}=\mathbb{S}-\mathbb{1}$ commute with the total angular momentum $\dot{\mathbf{J}}$.

The spinor $\chi^{(p\lambda)}_{j m_j s \lambda}(\vec{r})$ satisfy the equations

$$\hat{p}^2 \chi^{(p,\lambda)}_{j m_j s \lambda} = p^2 \chi^{(p,\lambda)}_{j m_j s \lambda} , \tag{8}$$

$$\hat{J}^2 \chi^{(p,\lambda)}_{j m_j s \lambda} = j(j+1) \chi^{(p,\lambda)}_{j m_j s \lambda} , \tag{9}$$

$$\hat{J}\cdot\hat{p} \chi^{(p,\lambda)}_{j m_j s \lambda} = \lambda p \chi^{(p,\lambda)}_{j m_j s \lambda} , \tag{10}$$

$$\hat{J}_3 \chi^{(p,\lambda)}_{j m_j s \lambda} = m_j \chi^{(p,\lambda)}_{j m_j s \lambda} , \tag{11}$$

where

$$\hat{p}_j = \frac{1}{i}\frac{\partial}{\partial x_j} ,$$

and

$$\hat{J}_i = \frac{1}{i} \varepsilon_{ijk} x_j \frac{\partial}{\partial x_k} + S_i$$

and S_i is the ith component of the spin operator, showing that the $\chi^{(p\lambda)}_{j m_j s \lambda}(\vec{r})$'s are labelled with a complete set of labels of an irreducible representation of $E(3)$

The orthogonality and completeness relations satisfied by the spinors $\chi^{(p\lambda)}_{j m_j s \lambda}(\vec{r})$ are[5)]

$$\iiint_{-\infty}^{+\infty} d^3\vec{r}\, \chi^{(p',\lambda')\dagger}_{j'm'_j s'\nu'}(\vec{r})\, \chi^{(p,\lambda)}_{jm_j s\nu}(\vec{r}) = \tag{12}$$

$$= 4\pi\, \delta_{jj'}\, \delta_{m_j m'_j}\, \delta_{ss'}\, \delta_{\nu\nu'}\, \frac{\delta(p-p')}{p^2}$$

and

$$\frac{1}{2J+1}\int_0^\infty \chi^{(\rho,n)}_{jm_j JM}(\vec{r})\, \chi^{(\rho,n)\dagger}_{jm_j JM'}(\vec{r}')\rho^2 d\rho = \tag{13}$$

$$= \delta^{(3)}(\vec{r}-\vec{r}')\,\delta_{MM'}$$

Explicit expressions for the spinors $\chi^{(p\lambda)}_{jm_j s\nu}(\vec{r})$ are given by Miller[5)]

$$\chi^{(p\lambda)}_{jm_j s\lambda}(\vec{r}) = \sqrt{4\pi}\sum_{l=0}^{\infty}\left\{\sqrt{\frac{(2l+1)(2j+1)}{(2s+1)}}\, e^{i\frac{\pi}{2}l}\, j_l(pr)\,\times\right. \tag{14}$$

$$\times\, Y_{l\lambda-m_j}(\theta_r\varphi_r)(l j 0 m_j|s m_j)(l j \lambda\text{-}m\, m|s\lambda)$$

Finally, inserting (14) in (7) we obtain the generalized Galilei spherical expansion

$$\langle p's'\lambda'|A|ps\lambda\rangle = \sum_{j=|\lambda|}^{\infty} \mathcal{D}^{j*}_{\lambda'\lambda}(\varphi,\theta,-\varphi)\,\times$$

$$\times\sum_{l=0}^{\infty}\sum_{l'=0}^{\infty}\sqrt{\frac{(2l+1)(2l'+1)}{(2s+1)(2s'+1)}}\; e^{i\frac{\pi}{2}(l-l')}\int_0^\infty r^2dr\int_0^\infty r'^2dr\;\times$$

$$j_{l'}(pr')\, j_l(pr)(l j 0\lambda|s\lambda)(l' j 0\lambda'|s\lambda')\, b^{j}_{l's',ls}(r',r) \tag{15}$$

The $j_l(pr)$ and $Y_{lm}(\theta,\varphi)$ are spherical Bessel functions and spherical harmonics respectively.

The coefficients of the expansion $b^{j}_{\ell' S'; \ell S}(r' r)$ are the Galilei spherical partial wave amplitudes for elastic collisions between particles with spins S' and S.

The inverse relation is readily obtained by use of the completeness relations, eq. (13)

$$b^{j}_{\ell' S'; \ell S}(r', r) = \sum_{\mu} \sum_{\mu'} \Big\{ (2j+1)(\ell' 0 j \mu' | S' \mu')(\ell 0 j \mu | S \mu) \times$$

$$\times \sum_{L=0} \sum_{L'=0} \sqrt{\frac{(2L+1)(2L'+1)}{(2S+1)(2S'+1)}}\, e^{i \frac{\pi}{2}(L'-L)} \quad (L 0 j \mu | S \mu)(L j \mu\text{-}\nu\, \nu | S \mu) \times$$

$$\times (L' j 0 \mu' | S' \mu')(L' j m' \mu'\text{-}m' | S' \mu') \int_0^{\infty} k^2 dk\, j_{L'}(k r') \times$$

$$\times \; A^{j}_{S' \mu', S \mu}(k)\, j_{L}(k r) \Big\}$$

(16)

The usual partial wave expansions of the helicity amplitude[5) is recovered from the generalized spherical expansion, equations (15) and (16), writing

$$a^{j}_{\ell' S', \ell S}(k) = \int_0^{\infty} r^2 dr \int_0^{\infty} r'^2 dr'\, j_{\ell'}(k r')\, b^{j}_{\ell' S', \ell S}(r', r)\, j_{\ell}(k r)$$

(17)

and identifying the $a^{j}_{\ell' S', \ell S}(k)$'s with the energy dependent partial wave amplitudes of the usual helicity expansion of the scattering amplitude.

The generalized spherical Galilei expansion we have just discussed was derived in the helicity formalism, it can easily be translated into a vector spherical harmonics formalism by means of standard methods[6).

III. THE NON-RELATIVISTIC CROSSED CHANNEL EXPANSION OR GENERALIZED GALILEI CYLINDRICAL EXPANSION OF THE MOMENTUM TRANSFER

Now, we want to make an expansion of the scattering amplitude $\langle out | \mathbb{A} | in \rangle$ using the basis of functions provided by the eigenfunctions of a complete set of commuting operators of the chain of groups $E(3) \supset E(2) \times T_{\perp} \supset O(2) \times T_{\perp}$. The plane in which the Euclidean group in two dimensions is defined can be chosen in two natural ways: perpendicular to the total momentum $\vec{P}$, or perpendicular to the momentum $\vec{q}$ transferred in the collision. The analysis of the problem is the same for both cases. Here, we shall consider the second choice, that is, $E(2)$ is defined in a plane perpendicular to the momentum transfer $\vec{q}$.

The symmetry of the chain of groups is exhibited in the brick wall frame of reference, defined by the equations

$$\left.\begin{aligned} \vec{p}_1 &= \tfrac{1}{2}\vec{q} \\ \vec{p}_2 &= -\tfrac{1}{2}\vec{q} + \vec{k} \end{aligned}\right\} \quad \text{incoming momenta}$$

$$\left.\begin{aligned} \vec{p}_1{}' &= -\tfrac{1}{2}\vec{q} \\ \vec{p}_2{}' &= \tfrac{1}{2}\vec{q} + \vec{k} \end{aligned}\right\} \quad \text{outgoing momenta}$$

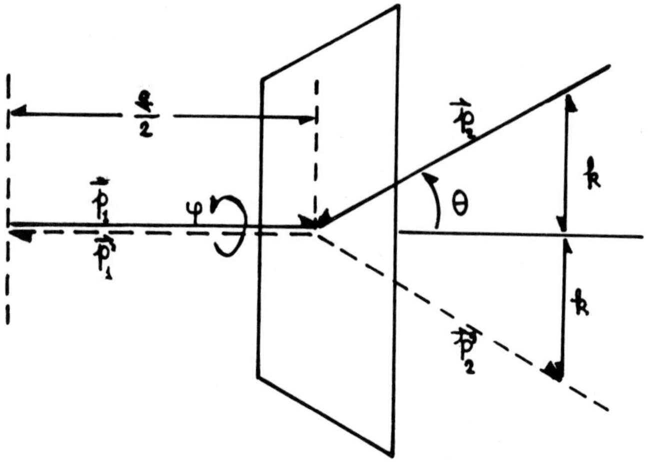

Two-particle collision in the brick wall frame of reference

In the discussion of the generalized spherical expansion the scattering amplitude was written as a matrix element of the transition operator labelled in the usual way, that is, $p_1\ \lambda_1$ and $p_2\ \lambda_2$ are paired to form a column index, while $p'_1\ \lambda'_1$ and $p'_2\ \lambda'_2$ are paired to form a row index

$$A(p_1\lambda_1, p_2\lambda_2; p'_1\lambda'_1, p'_2\lambda'_2) = \mathbb{A}^{(p_1\lambda_1, p_2\lambda_2)}_{(p'_1\lambda'_1, p'_2\lambda'_2)} =$$

$$= \langle p'_1 s'_1 \lambda'_1, p'_2 s'_2 \lambda'_2 | \mathbb{A} | p_1 s_1 \lambda_1, p_2 s_2 \lambda_2 \rangle \tag{18}$$

To exhibit the cylindrical symmetry, it is convenient to use a crossed labelling of the $\mathbb{A}$ matrix instead of the usual labelling

$$A(p_1\lambda_1, p_2\lambda_2; p'_1\lambda'_1, p'_2\lambda'_2) = \mathbb{A}^{(p_2\lambda_2, p'_2\lambda'_2)}_{(p_1\lambda_1, p'_1\lambda'_1)} \tag{19}$$

Now, the rows and columns of $\mathbb{A}$ are labelled with left and right pseudostates $|p_1\lambda_1\rangle\langle p'_1\lambda'_1|$ and $|p_2\lambda_2\rangle\langle p'_2\lambda'_2|$. The pseudostates are not considered as operators but as elements of the Hilbert space $\mathcal{H}\otimes\mathcal{H}'^{+}$ where $|p_1\lambda_1\rangle \in \mathcal{H}$ and $|p'_1\lambda'_1\rangle \in \mathcal{H}'$, and similarly for $|p_2\lambda_2\rangle$ and $|p'_2\lambda'_2\rangle$. Introducing the notation

$$|p_1\lambda_1; \tilde{p}'_1\tilde{\lambda}'_1\rangle \equiv |p_1\lambda_1\rangle\langle p'_1\lambda'_1|$$

$$|p_2\lambda_2; \tilde{p}'_2\tilde{\lambda}'_2\rangle \equiv |p_2\lambda_2\rangle\langle p'_2\lambda'_2|$$

equation (19) reads

$$A(p_1\lambda_1, p_2\lambda_2; p'_1\lambda'_1, p'_2\lambda'_2) =$$

$$= \langle \tilde{p}_2\tilde{\lambda}_2; p'_2\lambda'_2 | \mathbb{A} | p_1\lambda_1; \tilde{p}'_1\tilde{\lambda}'_1 \rangle \tag{20}$$

Note that (20) is the same matrix element of eq. (18) and that the change in the way the row and column indices are defined does not imply any crossing relations in non-relativistic quantum mechanics.

As before, we can diagonalize P_i

$$\mathrm{P}_i |p_1\lambda_1; \tilde{p}'_1\tilde{\lambda}'_1\rangle = q_i |p_1\lambda_1; \tilde{p}'_1\tilde{\lambda}'_1\rangle \tag{21}$$

where

$$\vec{q} = \vec{p}_1 - \vec{p}'_1$$

For non-forward elastic scattering, the momentum transfer $\vec{q}$ is invariant under rotations and translations in the two-dimensional k-plane, perpendicular to $\vec{q}$. The generators of such an $E(2)$ group are ξ_1, ξ_2 and L_{12}, and obey the commutation rules

$$[\xi_i, \xi_j] = 0 \tag{23}$$

$$[L_{12}, \xi_i] = i(\delta_{i1}\xi_2 - \delta_{i2}\xi_1) \tag{24}$$

They are realized as

$$\xi_i = i\frac{\partial}{\partial k_i} \; ; \quad i = 1, 2 \tag{25}$$

and

$$L_{12} = -i\left(k_1 \frac{\partial}{\partial k_2} - k_2 \frac{\partial}{\partial k_1}\right) \tag{26}$$

In general, the only invariant operator is

$$\xi_1^2 + \xi_2^2 = -\nabla_k^2 = -\frac{1}{k}\frac{\partial}{\partial k} k \frac{\partial}{\partial k} - \frac{1}{k^2}\frac{\partial^2}{\partial \varphi^2}$$
$$k^2 = k_1^2 + k_2^2, \quad \varphi = \tan^{-1}\frac{k_2}{k_1} \tag{27}$$

In the subspace where this operator vanishes there is one additional invariant operator, namely L_{12}. In this subspace, a complete set of labels for the vectors in an irreducible representation of $E(2)$ is provided by the eigenvalues of L_{12}.

The simultaneous eigenfunctions of (26) and (27) will be denoted by $h_{\nu m}(k, \varphi)$.

$$(\xi_1^2 + \xi_2^2)\, h_{\nu m} = \nu^2 h_{\nu m}, \quad 0 \leq \nu < \infty \tag{28}$$

$$L_{12}\, h_{\nu m} = m\, h_{\nu m}, \quad m = 0, \pm 1, \pm 2, \ldots \tag{29}$$

The ranges that have been indicated for ν and m are those that correspond to unitary representations of $E(2)$.

The regular solutions of (28) and (29) are

$$h_{\nu m} = J_m(\nu k)\, e^{im\varphi} \tag{30}$$

The normalization is such that

$$\frac{1}{2\pi}\int d^2k\, h_{\nu m}(k\varphi)\, h_{\nu' m'}(k,\varphi) = \frac{1}{\nu}\delta(\nu-\nu')\,\delta_{mm'} \tag{31}$$

$$\int_0^\infty \nu\, d\nu \sum_m h_{\nu m}(k,\varphi)\, h^*_{\nu m}(k,\varphi) = \delta^2(\vec{k}-\vec{k}') =$$

$$= \frac{\delta(k-k')}{k}\,\delta(\varphi-\varphi') \tag{32}$$

A complete set of labels of $E(2)\times T_\perp$ is provided by (ν, m) and the eigenvalue $\mathcal{J}$ of $\xi_3 = i\frac{\partial}{\partial k_3}$. In order to have a complete set of labels of $E(3)$, we supplement $(\nu, m, \mathcal{J})$ with the eigenvalues of the invariant operators $\xi_1^2 + \xi_2^2 + \xi_3^2$ and $\vec{\xi}\cdot\vec{J}$.

In the brick wall frame of reference, the cross labelled scattering matrix, eq. (20), can be written as

$$\langle \tfrac{1}{2}\vec{q}+\vec{k}, \lambda_2; -\tfrac{1}{2}\vec{q}+\vec{k}, \lambda_2' |A| -\tfrac{1}{2}\vec{q}_1 \lambda_1; \tfrac{1}{2}\vec{q}_1 \lambda_1' \rangle =$$

$$= \sum_n \Big\{ \int_0^\infty \nu\, d\nu\; a^{(n)}_{\lambda_2\lambda_2';\lambda_1\lambda_1'}(\nu, q) \;\times$$

$$\times \langle \lambda_2, \lambda_2' | T_\nu^{(m)}(k00) | \lambda_1 \lambda_1' \rangle e^{in\varphi} \Big\} \tag{33}$$

where $\langle \lambda_2 \lambda_2' | \mathbf{T}_\nu^{(m)}(k00) | \lambda_1 \lambda_1' \rangle e^{im\varphi}$ is a matrix element of the projective unitary irreducible representation of $E(2)$, $\langle \lambda_2 \lambda_2' | T_\nu^{(m)} | \lambda_1 \lambda_1' \rangle$ is a matrix element of a pure translation along $\vec{k}$ in the representation labelled with (ν, m). This matrix is readily evaluated to give

$$\langle \lambda_2 \lambda_2' | T_\nu^{(m)} | \lambda_1 \lambda_1' \rangle = \frac{1}{2\pi} \int_0^\infty e^{-i(\lambda_2 - \lambda_2')\varphi} e^{ipr\cos\varphi} e^{i(\lambda_1 - \lambda_1')\varphi} d\varphi =$$

$$= i^m J_m (k\nu) \; ; \quad m = (\lambda_2' - \lambda_2) - (\lambda_1 - \lambda_1') \tag{34}$$

The physical meaning of the quantum number m is clear, it is the number of units of angular momentum exchanged between the particles during the collision. This result allows us to write

$$A(p_1\lambda_1, p_1'\lambda_1'; p_2\lambda_2, p_2'\lambda_2') =$$

$$= \sum_m \int_0^\infty \nu d\nu \, i^m J_m(\nu k) \, e^{im\varphi} a^{(m)}_{\lambda_1\lambda_1'\lambda_2\lambda_2'}(\nu, q) \tag{35}$$

The partial wave amplitude $a^{(m)}_{\lambda_1\lambda_1', \lambda_2\lambda_2'}(\nu, q)$ is now further expanded in terms of the eigenfunctions e^{-iqy} of ξ_3, to give

$$A(p_1\lambda_1, p_2\lambda_2; p_1'\lambda_1', p_2'\lambda_2') =$$

$$= \sum_m \int_0^\infty \nu d\nu \int_{-\infty}^{\infty} dq \, e^{im\varphi} J_m(\nu q) \, e^{-iqy} \tilde{a}^{(m)}_{\lambda_1\lambda_1'; \lambda_2\lambda_2'}(\nu y) \tag{36}$$

In this equation all the spins are quantized along the direction of the momentum transfer $\vec{q}$. If we want to obtain the expression for the helicity amplitude in the brick wall frame of reference we need to invert the helicity of particle 1 and rotate the spins of particles 2 and 2'. We obtain

$$A(p_1\lambda_1, p_1'\lambda_1'; p_2\lambda_2, p_2'\lambda_2') = \sum_\mu \sum_\sigma \{ d^{s_2'}_{\lambda_2'\mu}(\theta) \, d^{s_2}_{\lambda_2\sigma}(\pi - \theta) \times$$

$$\sum_m \int_0^\infty \nu d\nu \int_{-\infty}^{\infty} dq \, i^m e^{im\varphi} J_m(\nu k) \, e^{-iqy} a^{(m)}_{\mu, \sigma, -\lambda_1\lambda_1'}(\nu, y) \} \tag{37}$$

in this equation the $d^{s}_{\lambda\mu}(\theta)$ are rotation matrices and

$$\theta = \tan^{-1} \frac{2k}{q}$$

$$\varphi = \tan^{-1} \frac{k_2}{k_1}$$

Equation (37) gives the non-relativistic crossed channel expansion or generalized Galilei transfer momentum cylindrical expansion of the scattering amplitude for elastic collisions between two particles with arbitrary spins. The inverse relation is obtained by means of the completeness and orthogonality relations eq. (31) and (32).

$$\tilde{a}^{(m)}_{\lambda_2\lambda_2';\lambda_1\lambda_1'}(\nu,\tau) = 2\pi\int_0^\infty k\,dk\int_{-\infty}^{\infty} dq \int_0^{2\pi} d\varphi \times$$

$$\times\, e^{-im\varphi}\Big\{\sum_\mu\sum_\sigma J_{\mu-\sigma-(\lambda_1+\lambda_2)}(k\nu)\, e^{im\varphi} \times$$

$$\times\, e^{iq\tau}\, d^{s_2'}_{\lambda_2\mu}\Big(\tan^{-1}\frac{2k}{q}\Big)\, d^{s_1'}_{\lambda_1\sigma}\Big(\pi-\tan^{-1}\frac{2k}{q}\Big) \times$$

$$\times\, A(p_1\lambda_1, p_2\lambda_2; p_1'\lambda_1', p_2'\lambda_2')\Big\}$$

$$m = (\lambda_2' - \lambda_2) - (\lambda_1' - \lambda_1) \qquad (38)$$

The coefficients $\tilde{a}^{(m)}_{\lambda_2\lambda_2';\lambda_1\lambda_1'}(\nu,\tau)$ are the non-relativistic crossed channel partial wave amplitudes. The physical meaning of the parameter that labels the non-relativistic crossed-partial wave expansion coefficients has been discussed by Cocho and Mondragón[7,8]. In their work, it is shown that the values of this parameter may be related to a Wigner time delay for a cylindrical wave packet.

IV. FORWARD SCATTERING AND E(3) EXPANSION

In the special case of forward scattering the momentum transfer $\vec{q}$ is equal to zero and the little group is $E(3)$, which is the group of motion in ordinary three-dimensional space. The generators of this group are $\xi_i = i\frac{\partial}{\partial k_i}$, $i = 1, 2, 3$ and L_{12}, L_{23}, L_{31}, they satisfy the commutation rules

$$[\xi_i, \xi_j] = 0$$

$$[L_{ij}, \xi_k] = (\delta_{ik}\xi_j - \delta_{jk}\xi_i)$$

$$[L_{ij}, L_{jk}] = i L_{ki}$$

If we select the chain $E(3) \supset O(3) \supset O(2)$, the basis for an irreducible representation will be formed by the eigenfunctions of the operators $\xi_1^2 + \xi_2^2 + \xi_3^2$, $L^2 = L_{12}^2 + L_{23}^2 + L_{31}^2$ and L_{12}. We shall denote such eigenfunctions by $f_{\mu l m}(k, \theta, \varphi)$, then we have

$$(\xi_1^2 + \xi_2^2 + \xi_3^2) f_{\mu l m} = \mu^2 f_{\mu l m} \tag{39}$$

with $0 \leq \mu < \infty$

$$L^2 f_{\mu l m} = l(l+1) f_{\mu l m} \tag{40}$$

with $l = 0, 1, 2$

and $m = -l, -l+1 \ldots (l-1), l$;
$$L_{12} f_{\mu l m} = m f_{\mu l m} \tag{41}$$

For $\mu = 0$, every value of L^2 defines an irreducible representation.

The solutions of equations (39) to (41) are

$$f_{\mu l m} = j_l(k\mu) Y_{lm}(\theta\varphi) \tag{42}$$

Let us now expand the forward scattering amplitude using as basis the eigenfunctions of eq. (42). In the brick wall frame of reference $\vec{q}=0$ and $\vec{p}_1 = \vec{p}_{1'} = 0$. The matrix element of the scattering amplitude is

$$A(0\lambda_1, 0\lambda_{1'}; p_2\lambda_2, p_{2'}\lambda_{2'}) =$$

$$\langle k\lambda_{2'}, k\lambda_{2'} | A | 0\lambda_1, 0\lambda_{1'} \rangle =$$

$$= \int_0^\infty \mu^2 d\mu \, a_{\lambda_2\lambda_{2'};\lambda_1\lambda_{1'}}(\mu) \langle s_{2'}\lambda_{2'} s_2\lambda_2 | T_\mu(k) | s_1\lambda_1 s_{1'}\lambda_{1'} \rangle \quad (43)$$

where $T_\mu(k)$ is the generator for a pure translation along $\vec{k}$ in the representation μ . If we choose as quantization axis the direction of $\vec{k}$, we obtain

$$\langle s_{2'}\lambda_{2'}, s_2\lambda_2 | T_\mu(k) | s_{1'}\lambda_{1'} s_1\lambda_1 \rangle =$$

$$= \sum_S \sum_{S'} (s_{2'} s_2 \lambda_{2'} \lambda_2 | S'M')(s_{1'} \lambda_{1'} s_1 \lambda_1 | SM) \times$$

$$\times (S'SM'M | \ell 0) \, \dot{j}_\ell(k\mu) \quad (44)$$

a sum of products of Clebsch-Gordan coefficients and spherical Bessel functions. The physical meaning of the parameter μ may also be related to a Wigner time delay[7,8].

The non-relativistic crossed channel expansion of the scattering amplitude is a decomposition of the scattering amplitude into partial amplitudes, each one of these corresponding to the "exchange" of a set of pseudostates transforming irreducibly under the Galilei group. A small range of the expansion parameters ν or μ gives the small curvature components of the cross-section, making the expansion especially appropriate for the elastic scattering of particles with arbitrary spin for the non-forward and forward scattering, where the little groups are $E(3)$ and $E(2)$ respectively. The concept of Euclidean poles and trajectories can be introduced in close analogy with the Regge poles and Regge trajectories. An application of this formalism to the scattering of nucleons by alpha particles is given in Cocho and Mondragón[2].

BIBLIOGRAPHY

1. Kalnins, E.G., Patera, J., Sharp, R.T. and Winternitz, P.
   Phys. Rev. D8, (8), 2552, (1973)
2. Cocho, G. and Mondragón, A.
   Nucl. Phys. A25, 417, (1969)
3. Joos, H.
   "Complex angular momentum and the representations of the Poincaré group with space-like momentum" in Lectures in Theoretical Physics (1964).
   University of Colorado Press, Boulder, Colorado 1964
4. Wightman, A.S.
   Rev. Mod. Phys. 34, 845, (1962)
5. Miller Jr., W.
   Comm. in Pure and Applied Mathematics XVII, 527, (1964)
6. Newton, R.G.
   "Scattering Theory of Waves and Particles"
   McGraw-Hill Book Co., New York (1966) p. 451
7. Cocho, G., Mondragón, A. and Colón-Vela, M.
   Nucl. Phys. A25, 417-424 (1969)
8. Cocho, G. and Mondragón, A.
   Rev. Mex. Fis. XVII (1) 59-67 (1968)

TERNARY ALGEBRAS AS THE BASIS OF A DYNAMICAL THEORY OF SUBCONSTITUENTS * †

Itzhak Bars **
Yale University, Physics Department
J.W. Gibbs Laboratory
New Haven, Connecticut 06520

The SU(3) x SU(2) x U(1) gauge model works well at low energies (10^{-15} - 10^{-16} cm). Because it contains large numbers of degrees of freedom and parameters, it is considered by many theorists as a good phenomenological model which should be derivable from a more complete and fundamental theory. Thus, we ask the question: What is the correct theoretical description of physics at much shorter distances than the present accelerator energies? Two possibilities have emerged (i) Grand unification schemes in which a large number of fields are taken as elementary (ii) Composite schemes of quarks, leptons, gauge bosons and higgs bosons whose goal is to make a viable theory in terms of few subconstituents.

Grand unification is based on the idea that the gauge principle which is successful at low energies can be extrapolated all the way to 10^{16} - 10^{19} GeV. Using symmetry schemes one could unify the many degrees of freedom within few representations of a unifying group. In a model based on the exceptional group E_8, recently proposed in collaboration with M. Günaydin(1), the maximum such unification has been achieved: Only one and the smallest possible representation for each spin are used, namely, the 248 adjoint representation for gauge bosons and fermions and the 3875 for higgs bosons. Note that the "successful" grand unification groups, including E_6, are all members of the E-series in the classification of Lie algebras(2) (i.e. SU(5) is isomorphic to E_4, SO(10) is isomorphic to E_5). The exceptional group E_8 is the last member of the series and has very special properties. The major prediction of the model, independent of detailed calculations, is that the next three SU(5) families to be discovered below 1 TeV are of V+A type with respect to weak interactions. If E_8 or some other special scheme of this type proves to be successful we may believe that it is fundamental. Otherwise, in my opinion, grand unification remains as a useful but probably not fundamental approach.

* Research supported by DOE Contract No. EY-76-C-02-3075

† Talk delivered at (i) IXth International Colloq. on Group Theoretical Methods in Physics, Cocoyoc, Mexico, June 1980; (ii) XXth International Conference on High Energy Physics, Madison, Wisconsin, July 1980.

**Alfred P. Sloan Foundation Fellow

The idea of subconstituents is at its infancy. Several schemes have been proposed and the field is growing. The common factor in all the schemes is the goal of constructing a theory with few degrees of freedom. But what kind of dynamics should the subconstituents satisfy in order to produce the dynamics of a gauge theory (SU(3) x SU(2) x U(1)) at low energies, including the phenomenologically successful quarks, leptons, gauge bosons and higgs bosons as effective degrees of freedom? I will describe some ideas developed since 1978 in collaboration with M. Günaydin[3,4] which are based on ternary algebras[3,5].

In the ternary algebraic approach our basic idea is to try to give physical meaning to the mathematical fact that ternary algebras are building blocks of all Lie algebras and Lie superalgebras. Given that gauge theories are based on Lie algebras, it seems natural to explore ternary algebras as a basis for the dynamics of the fundamental subconstituents.

Ternary algebras close under triple products (abc). An example of a triple product is

$$(abc) = a.(\bar{b}.c) + c.(\bar{b}.a) - b.(\bar{a}.c)$$

where a,b,c belong to some vector space and the product (.) could be associative as well as nonassociative. The relation of the ternary algebra to the Lie algebra can be seen by a grading of the form

$$\begin{array}{ccccccccc} . & . & . & -1 & 0 & +1 & . & . & . \\ . & . & . & \tilde{U}_b & S_{ab} & U_a & . & . & . \end{array}$$

where the Lie algebra generators U_a, S_{ab}, $\tilde{U}_b$ (which are multiplied by infinitesimal parameters) obey the commutation rules

$$[U_a, \tilde{U}_b] = S_{ab}$$

$$[S_{ab}, U_c] = U_{(abc)}$$

$$[S_{ab}, S_{cd}] = S_{(abc)d} - S_{c(bad)}$$

$$\vdots$$

This is explained fully in refs. (3,5).

There is a one-to-one correspondence between ternary algebras and (symmetric) coset spaces. Thus any physical application based on ternary algebras can be reproduced via coset spaces and vice versa. But ternary algebras provide an unusual way of looking at coset spaces and suggest mathematical structures useful for physical applications which are not available with the usual methods. For example a ternary algebra formed by the direct product of n quarternions $a = H_1 \otimes H_2 \otimes \ldots \otimes H_n$ leads to $SO(3x2^n)$ for n = even and $sp(3x2^n)$ for n = odd. Thus, it can be related to the coset spaces $SO(3x2^n)/SO(2^n) \times SO(2^{n+1})$ and $Sp(3x2^n)/Sp(2^n) \times Sp(2^{n+1})$. From the usual coset space methods the above quaternionic structure of this space is not at all obvious while in contrast, it is an explicit input in the ternary algebraic method. Similarly, octonionic ternary algebras associated with the coset spaces $E_6/SO(10) \times U(1)$, $E_7/SU(8)$, $E_8/SO(16)$, etc., provide the octonionic properties as an input. These structures do not emerge in standard coset space approaches. Vast classes of associative (non-octonionic) ternary algebras with such properties can be found in refs. (3,5). You may construct your own brand new ternary algebra, since no complete classification is available. Thus, ternary algebras provide a really novel tool for physical applications.

In our approach the fundamental fields are associated with the elements of the ternary algebra. We have called them ternons. Gauge bosons, quarks, leptons, etc., are taken as composites of ternons. As described below, at this stage it appears quite likely, as shown in ref. (4), that the dynamics of a low energy gauge theory will emerge from a ternon theory in four dimensions. Furthermore, a convergence of ideas seems to be developing between the ternary algebra point of view and Harari's rishon scheme[(6)], if one uses the ternary algebra of a complex octonion[(7)] from which the SO(10) and E_6 groups can be constructed. Also, the recent ideas of Ellis, Gaillard and Zumino[(8)] which developed from SO(8) supergravity[(9)] are in agreement with our approach since their SU(8) composite gauge potential, just like our gauge potentials, are constructed from a coset space. Their coset space $E_7/SU(8)$ is associated with the ternary algebra formed by the direct product of two octonions $O_1 \otimes O_2$, where the first octonion is purely imaginary while the second one contains only the five directions $(e_0, e_7, e_4, e_5, e_6)$. Both of the ternary algebras mentioned above leading to E_6 (Rishons) and E_7 (supergravity) are special cases of the ternary algebra of two arbitrary octonions[(3,5)] $O_1 \otimes O_2$ which leads to E_8.

The general procedure for constructing a composite gauge potential from any ternary algebra was given in ref. (4), where supersymmetric schemes including composite fermions were also described. The simplest case of a rectangular M x N complex ternon field $\phi(x)$, including the quantum theory, has been investigated more thoroughly in collaboration with M. Günaydin. In this case our approach is related to the Grassmannian generalization of the CP_N type[(10)] composite gauge potential, except that in 4 dimensions we consider a more general model[(11)].

The U(M) gauge potential is given by $A_\mu(\phi) = iW^\dagger \partial_\mu W$, where the $(M+N) \times M$ matrix W is constructed from the $N \times M$ matrix $\phi(x)$ as follows

$$W(\phi) = \begin{pmatrix} \phi(1 + \phi^\dagger\phi)^{-1/2} \\ (1 + \phi^\dagger\phi)^{-1/2} \end{pmatrix}$$

and it satisfies automatically $W^\dagger W = 1_M$. If $\phi(x)$ is subjected to a global U(M+N) transformation

$$\phi \to \phi' = (\alpha\phi+\beta)(\gamma\phi+\delta)^{-1} \quad ; \quad U_{M+N} = \begin{pmatrix} \alpha & \beta \\ \gamma & \delta \end{pmatrix} ,$$

then $A_\mu(\phi)$ transforms like a local U(M) gauge field

$$A_\mu(\phi') = U_M^\dagger(x)(A_\mu(\phi) + i\partial_\mu)U_M(x)$$

where $U_M(x)$ depends on $\alpha,\beta,\gamma,\delta$ as well as $\phi(x)$.

Recall that $\phi(x)$ belongs to the coset space $U(N+M)/U(N) \times U(M)$. The above construction of the gauge field A reminds us of the gauge formulation of gravity, where the elementary field, the vierbein, belongs to the coset space of the (Poincaré group)/(Lorentz group), and the connection ω_μ^{ab} is a composite gauge potential for the Lorentz subgroup.

The ternon dynamics will be described by the Lagrangians

$$L_4 = -\frac{1}{4g^2} \operatorname{Tr}[F_{\mu\nu}(A(\phi))]^2$$

$$L_2 = \frac{1}{2\lambda^2} \operatorname{Tr}[(D_\mu W)^\dagger (D_\mu W)]$$

in 4 and 2 dimensions respectively. It is interesting to note that according to a theorem due to Narasimhan and Ramanan[12] all the information of a classical U(M) gauge field is contained in the construction above, provided N is larger than a certain number. Our aim is, of course, to show that the quantum ternon theory leads to the dynamics of a quantum gauge theory at low energies only. Thus, we will not be bound, a priori, by the classical limits on the minimum dimension of $\phi(x)$.

The composite gauge potential has the same cubic and quartic couplings of a gauge field. If its propagator behaves at low energies like the propagator of an elementary gauge field then the low energy physics will reduce to a gauge theory. This is already known to occur in the CP_N model[10] and in its Grassmannian generalizations[4,13]. To investigate this problem we need to calculate the generating function in the quantum theory of ternons

$$Z(J) = \int d\mu(\phi)\, e^{-S(\phi) + \int Tr(A_\mu(\phi) J^\mu)}$$

The measure $d\mu(\phi)$ is given in ref. (4). The path integral can be manipulated by introducing auxiliary fields and integrating out the original ternon fields $\phi(x)$. Then, we arrive at a new form[4b]

$$Z(J) = \int [dA_\mu]\, e^{-\int \frac{1}{4g^2} Tr(F_{\mu\nu}(A))^2 + \text{corrections} + \int A.J}$$

where the path integral is done over an <u>elementary</u> gauge field. Thus, we learn that the effective action is the Yang-Mills action plus corrections which are a function of only the covariant derivative D_μ. These corrections can depend only on $F_{\mu\nu}$ and its derivatives as well as possible non-local terms associated with $(D_\mu)^{-2}$ etc. Thus, provided we can show that the corrections are absent or neglible at low energies we will obtain a low energy gauge theory. This point remains to be proven. The large N limit could be used as a tool to investigate this problem as discussed in ref. (4b).

Fermions can be included easily in the theory of ternons. A supersymmetric scheme based on superfield techniques was included in our original paper[4], but supersymmetry is not a requirement for the inclusion of fermions. More intricate supersymmetric models can probably be constructed via superternary algebras[3].

We started out with two goals: (i) To obtain the dynamics of a gauge theory at low energies from the dynamics of ternons (ii) To arrive at the correct phenomenological variables, namely, quarks, leptons, etc., as composites of ternons. We have shown that the first goal is likely to be achieved in our framework and we are now ready to investigate the second one. We are encouraged that other attempts such as the Rishon scheme[6] and the supergravity scheme based on $E_7/Su(8)$[8] are compatible with our ternary algebra approach as already mentioned above. Given the rich structures provided by ternary algebras we have no doubt that the model building stage will be rewarding.

References

(1) I. Bars and M. Günaydin, Yale preprint YTP80-09, to be published in Phys. Rev. Lett. and manuscript in preparation.

(2) This remark is due to F. Gürsey.

(3) I. Bars and M. Günaydin, J. of Math. Phys. 20, 1977 (1979) and lectures by I. Bars and M. Günaydin in the Proc. of the 8th Int. Colloq. on Group Theor. Methods in Physics March 1979, Kiriat Anavim, Israel.

(4a) I. Bars and M. Günaydin, Yale preprint YTP79-05, to be published in the Phys. Rev. D.

(4b) I. Bars and M. Günaydin, Yale preprint YTP80-14, to be published Phys. Lett.

(5) J. L. Kantor, Sov. Math. Dokl. 44, 254 (1973) and Trudy Sem. Vector Anal. 16, 407 (1972) (Russian); B. N. Allison, Am. J. Math 98, 285 (1976) and Trans. Am. Math. Soc. 114, 75 (1976).

(6) H. Harari, Phys. Lett. 86B, 83 (1979).

(7) This observation followed from a conversation with H. Harari. This ternary algebra is described by Kantor in ref. (5) and also in ref. (3).

(8) J. Ellis, M. K. Gaillard, B. Zumino, CERN preprint, April 1980.

(9) E. Cremmer and B. Julia, Phys. Lett. 80B, 48 (1978).

(10) A. D'Adda, M. Luscher, P. DiVecchia, Nucl. Phys. B146, 63 (1978).

(11) This model was independently constructed for different physical applications by a number of authors:
C.K. Chao, T.T. Sheng, Y.T. Nan, Scientiae Sinica 22, 34 (1979);
A.P. Balachandran, A. Stern, G. Trahern, Syracuse preprint SU-4213-126 (1978);
I. Bars in ref. (3); F. Gürsey and C. Tze, Yale preprint, Aug. (1979)
L. Lukierski, Summer Inst. Elem. Part. Phys., Kaiserlautern, Germany, Aug. 1979.

(12) M. S. Narasimhan and S. Ramanan, Am. J. Math. 83, 563 (1961) and 85, 223 (1963).

(13) E. Brezin, S. Hikami and J. Zinn-Justin, Saclay preprint 1980.

(14) J. Lukierski and B. Milewski, Phys. Lett. B93, 91 (1980).

SOLITONS AS SECTIONS IN NON TRIVIAL BUNDLES

Luis J. Boya and Juan Mateos
Departamento de Física Teórica,
Facultad de Ciencias Universidad de Salamanca,
Salamanca, Spain.

We interpret soliton solutions to classical wave equations as sections in some vector bundles; the peculiarity of soliton behaviour comes from the non-trivial character of the fibration, which also includes the "broken symmetry" aspect. We review in this light the most well-known cases: Kink, sine-Gordon soliton, vortex, monopole and instanton, and make contact with the former homotopy classification through the construction of bundles over spheres.

I. Motivation.

1. Peculiar solutions to classical field equations, called *solitons*, have been intensively studied in the last few years in particle physics ([1], [2], [3]); they arise most naturally in some theories exhibiting spontaneous symmetry breakdown ([3], [4]). In this paper we would like to remark the nature of solitons as particular functions ("sections" in non-trivial associated fibre bundles); the broken symmetry aspect is expected to be also a *consequence* of this fibre product: the ultimate reason of both seems to be the twisted "wedding" of the gauge (or discrete) symmetry group with the topology of the manifold representing spacetime.

Our aim in this paper is only to exhibit the fibre-bundle properties of the standard solitons in various dimensions. In § - 2 we establish briefly the mathematical setting, emphasizing the fundamental difference between "functions with n components" and "sections in a n dimensional vector bundle". § 3 will apply these constructs to the case of a discrete group, as exemplified by the kink and the sine-Gordon soliton; in § 4 a short reminder is given of electromagnetism as connection in a U(1)-bundle, contemplating in this light the Nielsen - Olesen vortex and the Dirac monopole. § 5 considers briefly non-abelian gauge groups, focusing in particular in Polyakov "instanton".

We acknowlegde our debt with the work of Yang and Wu [5], who first did this sort of analysis for the Dirac monopole.

II. Fibre bundle theory.

2. This is not the place to review the whole mathematical theory of principal and associated (in particular vector) bundles; as standard mathematical treatises we refer to Steenrod [6] or Milnor [7]. Introductory expositions for physicists are in Trautman [8], Cho [9], etc. That fibre bundles are the natural framework for gauge theories is becoming more and more apparent; see e.g. the "dictionary"

of Wu - Yang ([5], [15]) and also the recent book by Mayer [10].

We denote by β : P (V, G) a principal fibre bundle of total space P, base manifold V, and structure group G; our notation will be

$$\beta : G \circ\!\longrightarrow P \xrightarrow{\ \pi\ } V \tag{2.1}$$

$\pi^{-1}(x)$ is the fibre over $x \in V$; β is always, by definition, *locally trivial*, namely there is a covering of V by open sets, $V = \bigcup_\alpha U_\alpha$ such that $\pi^{-1}(U_\alpha)$, i.e. the *portion of* P *over* U_α is like a direct product:

$$\psi_\alpha : U_\alpha \times G \xrightarrow{\ \sim\ } \pi^{-1}(U_\alpha) \tag{2.2}$$

and provides *local sections* $s_\alpha : U_\alpha \longrightarrow \pi^{-1}(U_\alpha)$ given by

$$s_\alpha(x) = \psi_\alpha(x, e) \ , \ e = 1_G . \tag{2.3}$$

In the overlapping of two regions there should be *transition functions* $g_{\alpha\beta} : U_\alpha \cap U_\beta \to G$ to "match" the trivialization functions s_α, s_β of (2.2) and (2.3):

$$s_\beta(x) = s_\alpha(x)\, g_{\alpha\beta}(x) \ , \quad x \in U_\alpha \cap U_\beta \tag{2.4}$$

$g_{\alpha\beta}(x)$ is unique because G operates freely in the fibres.

The transition functions characterize the bundle completely. For any pair (V, G), the set of (equivalence classes of) principal bundles is a "cohomology set", denoted $H^1(V, G)$, where $H = \check{H}^*$ means Čech cohomology. Only for abelian G is H^1 an (abelian) group; see Koszul [11] or Hirzebruch [12]; the distinguished element of the cohomology set is the *direct product* β_o : P = $V \times$ G, or trivial bundle; for it (and only for it) there are (global) sections $s : V \longrightarrow P$ (e.g. $s_o(x) = (x, e)$).

If the base manifold V is euclidean (e.g. $\mathbb{R}^n$), any G-bundle over it is trivial, there are (global) sections, and there is no reason for transition functions; hence sections $S : V \longrightarrow P$ (with $\pi \circ s(x) = x$) and functions $\sigma : V \longrightarrow G$ can be identified. There are at least three reasons why we are interested, at this level, in non-trivial base spaces:

i) There might be *singular points* in V (such as sources, sinks, monopoles, etc.), which one would like to *remove*; this apparently innocent operation changes the topology radically, e.g. taking away just a point in $\mathbb{R}^n$,

$$\mathbb{R}^n - \{0\} \cong \mathbb{R} \times S^{n-1} \tag{2.5}$$

where S^n is the n-sphere.

ii) One might wish to *add* points to V, either directly, e.g. to "compactify" spacetime and make inversion and conformal transformations possible, or indirectly by considering functions which "behave well at infinity", thus prolongating naturally to functions from some completed spacetime.

iii) Finally, one should keep an open mind to generalize gauge theories to general relativity, whose spacetime manifold is definitely non-euclidean.

3. A few words about associated bundles are in order. Let F be a space acted upon by G (e.g. F is a linear space and $\rho : G \longrightarrow$ Linear group of F, is a representation). The *associated bundle* is defined by the quotient

$$E = P \times_G F \tag{3.1}$$

of the direct product $P \times F$ by the equivalence relation due to G : $(u, y) \simeq (u', y') = (ug, \rho(p)^{-1} y)$, $g \in G$. Our standard notation for the associated bundle is

$$\begin{array}{lccccc} \beta : & G & \circ\!\!\longrightarrow & P & \xrightarrow{\ \pi\ } & V \\ & \rho \downarrow & & & & \| \\ \zeta : & F & \circ\!\!\longrightarrow & E & \xrightarrow{\ \pi_E\ } & V \end{array} \quad \text{or just} \quad \begin{array}{lcccc} & G & & & \\ & \rho \downarrow & & & \\ \zeta : & F & \longrightarrow & E & \xrightarrow{\ \pi_E\ } V \end{array} \tag{3.2}$$

The most important case, also for physics, is when F is a linear space, i.e. *vector bundles*.

Sections are maps $\psi : x \in V \longrightarrow$ point in $\pi_E^{-1}(x)$, which is the fibre over x; note that for vector bundles a section is *not* a "vector value function" $\tilde{\psi}(x) = (\phi_1(x), \phi_2(x), \ldots \phi_n(x))$, $\phi_i(x) \in \mathbb{R}$ or $\mathbb{C}$; but there is a relation, crucial for what we are going to describe, between the mathematically intrinsic concept of section, and the (pedestrian) physicist's concept of "field with n components", provided by the local sections s_α in the principal bundle, that we introduced before:

Let $\psi : V \longrightarrow E$ be a section and define $h_\psi : P \longrightarrow F$ by $h_\psi(u) = y \in F$ such $E \ni [u, y] = \psi(\pi(u))$. Then the diagram

$$\begin{array}{lccccc} \beta : & G & \circ\!\!\longrightarrow & P & \overset{s_\alpha}{\underset{\pi}{\rightleftarrows}} & V = \bigcup_\alpha U \\ & \rho \downarrow & h_\psi \swarrow & & \psi & \| \\ \zeta : & F & \circ\!\!\longrightarrow & E & \underset{\pi_E}{\rightleftarrows} & V \\ & & \tilde{\psi}_\alpha & & & \end{array} \tag{3.3}$$

shows how one can define *locally* the vector-valued functions

$$\tilde{\psi}_\alpha : U_\alpha \subset V \longrightarrow F \ ; \ \psi(x) = h_\psi \circ s_\alpha(x) \tag{3.4}$$

For points in overlapping rerions, $x \in U_\alpha \cap U_\beta$ one has the transformation law

$$\tilde{\psi}_\beta(x) = h_\psi \circ s_\beta(x) = h_\psi \circ s_\alpha(x)\, g_{\alpha\beta}(x) = \rho\, [g_{\alpha\beta}(x)^{-1}]\, \tilde{\psi}_\alpha(x) \ ; \tag{3.5}$$

(3.5) is the "local gauge transformations of the fields", in physical terms.

4. Finally we recall very briefly how the cohomology set of principal bundles $H^1(V, G)$ is constructed from the pair (V, G): for each G, there is a "universal bundle" $\beta : E_G (B_G, G)$ such that any bundle $\beta \in H^1(V, G)$ is *induced* by a map

$f : V \longrightarrow B_G$:

$$\begin{array}{ccccc} & & V \\ & & \downarrow f \\ \hat{\beta} : G \circ\!\!\longrightarrow E_G & \xrightarrow{\hat{\pi}} & B_G \end{array} \quad \Rightarrow \quad \begin{array}{lccccc} f^*_{\hat{\beta}} = \beta : & G & \circ\!\!\longrightarrow & P & \longrightarrow & V \\ & \| & & & & \downarrow f \\ \hat{\beta} : & G & \circ\!\!\longrightarrow & E_G & \xrightarrow{\hat{\pi}} & B_G \end{array} \tag{4.1}$$

(Recall that P is that part of $E_G \times V$, say $\{(\bar{u}, x)\}$, such that $\hat{\pi}\ (\bar{u}) = f\ (x)$. Two maps $f, f' : V \longrightarrow B_G$ induce the same bundle β (in the sense of equivalence) if and only if they are *homotopic* (i.e. there is a map $h : V \times [0, 1] \to B_G$ with $f(x) = h\ (x, 0)$ and $f'\ (x) = h\ (x, 1)$).

For G-bundles over spheres there is a simpler construction which we sketch because it is the case we are mainly interested in; we follow [6], § 18 closely. Let the sphere S^n and group G be given; then $H^1(S^n, G)$ is calculated constructively as follows: in S^n draw the equator and the tropics; let U_1 be the part of S^n north of the tropic of Capricornius, U_2 be south of the other tropic; then

$$U_1, U_2 \approx \mathbb{R}^n \ ; \ U_1 \cup U_2 = S^n \ ; \ U_1 \cap U_2 \cong S^{n-1} \times I \tag{4.2}$$

where $I = [0, 1]$; then any function $g_{12} : U_1 \cap U_2 \to G$ is already a 1-cocycle, because there is no $U_3 \neq \emptyset$; the equivalence mod a 0-coboundary is easily seen to be an homotopy; and changing U_1 and U_2 just means to change "the angle of the ecliptic": the direct limit of the covarings is just the closed manifold of the equator, S^{n-1}, therefore the bundles are homotopy classes of maps $S^{n-1} \longrightarrow G$ or

$$H^1\ (S^n, G) = \pi_{n-1}\ (G) \tag{4.3}$$

We shall use this result repeatedly in the sequel.

III. Discrete groups.

5. *Covering spaces.* The fibre bundles with discrete structural group over a connected manifold V can be obtained by the following recipe (Steenrod [6] § 13): let $\pi_1\ (V)$ be the fundamental group of V (first homotopy, Poincaré group); it leads to an universal covering bundle

$$\tilde{\beta} : \pi_1\ (V) \circ\!\!\longrightarrow \tilde{V} \longrightarrow V \tag{5.1}$$

with $\tilde{V}$ connected and simply connected; then any (V, K) bundle with K discrete is obtained by the morphisms $\pi_1\ (V) \xrightarrow{\varphi} K$; we have the diagram

$$\begin{array}{lccccc} & H & = & H & & \\ & \downarrow & & \downarrow & & \\ \tilde{\beta} : & \pi_1(V) & \longrightarrow & \tilde{V} & \longrightarrow & V \\ & \varphi \downarrow & & \downarrow & & \| \\ \beta : & K & \longrightarrow & P & \longrightarrow & V \end{array} \tag{5.2}$$

where $H = \ker \varphi$; note the vertical fibration with base P. For most of spaces used in physics $\pi_1\ (V)$ as well as $\widetilde{V}$ are easy to visualize; as an example, for $V =$ = compactified Minkowski space $\approx S^1 \times S^3$, we have $\pi_1 = \mathbb{Z}$ and universal covering $\approx \mathbb{R} \times S^3$.

Discrete groups are the pertinent ones, for soliton solutions, in two dimensional spacetime (reasons are to be found in [4]). If we take $\mathbb{R}^2$ as spacetime, retain only the space part $\mathbb{R}^1$ (we look here only for *static* solitons) and compactify, $\mathbb{R}^1 \longrightarrow \dot{R}^1 = S^1$, we have to calculate $K \longrightarrow ? \longrightarrow S^1$ for discrete K; but bundles over the sphere S^n are classified by the homotopy $\pi_{n-1}(K)$: this is why the search for solitons in fully broken symmetry theories coincides with the search for non-trivial bundles: both reduce to the finding the same homotopy group, namely $\pi_{n-1}\ (G)$!

For the circle S^1, $\pi_o\ (K)$ just counts the elements of K (for any set X , $\pi_o(X)$ counts the number of arcwise connected components, see [4], III-2).

The fundamental group of the circle is $\mathbb{Z}$; any morphism $\mathbb{Z} \longrightarrow K$ has kernel $n\,\mathbb{Z}$ for some n. The two outstanding examples are described next:

6.i) Look at the diagram

$$\begin{array}{cccccc} \widetilde{\beta}: & \mathbb{Z} & \longrightarrow & \mathbb{R}^1 & \longrightarrow & S^1 \\ & \downarrow & & & & \| \\ \beta: & \mathbb{R}^1 & \longrightarrow & E & \longrightarrow & S^1 \end{array} \tag{6.1}$$

where the group action in the lower line is defined by $n : x \longrightarrow x + n$; the fibre bundle is non-trivial; if we mimick sections as functions $\varphi : S^1 \longrightarrow \mathbb{R}^1$, they should "coincide" as angles "at the infinity", i, e, $\varphi\ (+\infty) \equiv \varphi\ (-\infty) \bmod 2\pi$, or

$$\varphi\ (+\infty) - \varphi\ (-\infty) = 2\pi n \tag{6.2}$$

This is the situation that happens in the soliton of the sine-Gordon equation: for $n = \pm 1$ these are

$$\varphi_{S,A}(x) = \pm 4 \text{ arctg } \exp x \tag{6.3}$$

ii) *The kink.* Take now $K = Z_2$ and $\mathbb{Z} \longrightarrow Z_2$ given by $n \longrightarrow 2n$; associate the line bundle due to the action $x \longrightarrow \pm x$ in $\mathbb{R}$; then

$$\begin{array}{cccccc} \beta: & Z_2 & \hookrightarrow & P = S^1 & \longrightarrow & S^1 = \mathbb{R}P^1 \\ & \downarrow & & & & \| \\ \gamma_1^1: & \mathbb{R} & \longrightarrow & E & \longrightarrow & S^1 \end{array} \tag{6.4}$$

is still a more important fibre bundle, because, γ_1^1 is the prototype bundle for Z_2 -characteristic classes [7]: the β bundle is the twofold covering of the circle, and γ_1^1 the associated (vector) line bundle; $\mathbb{R}P^n$ is the real projective space associated to $\mathbb{R}^{n+1}$. Here any section has to have zeroes, and as functions verify

$$\varphi\,(+\infty) = -\,\varphi\,(-\infty)\ , \tag{6.5}$$

which is, of course, what happens with the kink solution to $(\varphi^4)_2$ theory [2], $\varphi_k\,(x) = \tanh x$. This example is essentially the (infinite) Möbius band, the simplest example of non-trivial bundle ([6], § 1; [7], 2, Fig. 2).

Of course, the detailed analytical form of both the s - G soliton or the kink are not implied by our topological analysis: on the contrary, they realize, via a particular Lagrangian or differential equation, the topological possibility.

IV. U(1)-Bundles.

7. For higher dimensional spacetime one needs continous groups (if searching for solitons), and then the identification of fibres with the standard fibre is no longer path-independent: one has to put a *connection* on the bundle (here a standard source is Ch. II of Nomizu [13]).

For 2 + 1 spacetime we take S^2 as base manifold, so for any group G the bundles are one-to-one with π_1 (G), as explained before. A possible candidate for G is U(1), because $\pi_1(U(1)) = \mathbb{Z}$: the n = 1 generating class is given by

$$\beta\ :\ U(1) = S^1 \longrightarrow S^3 \longrightarrow S^2 = \mathbb{C}\,P^1 \tag{7.1}$$

which is the famous Hopf fibering, a seminal discovery for homotopy, fibre bundles and characteristic classes (H. Hopf, 1931); the associated (complex) line bundles is

$$\begin{array}{cccc} & U\,(1) & & \\ & \downarrow & & \\ (\gamma_1^1)_G & \mathbb{C}^1 \;\circ\!\!\longrightarrow & E \longrightarrow & S^2 \end{array} \tag{7.2}$$

which is again fundamental for the so-called *Chern classes* of complex bundles ([7], § 13). As U(1) is the gauge group of electromagnetism (H. Weyl, 1929), the physical relevance is hard to overemphasize. Here we have

$$H^1\,(V,\,S^1) = H^2\,(V,\,\mathbb{Z}) \tag{7.3}$$

and hence any U(1)-bundle defines and is defined by a *entire* cohomology class of V, called the (first) Chern class. The proof of (7.3) is based on the "exact cohomology sequence" applied to $0 \longrightarrow \mathbb{Z} \longrightarrow \mathbb{R}^1 \longrightarrow S^1 \longrightarrow 0$, see Kostant [14]. For U(n)-bundles, $n > 1$, even the n Chern classes do not classify.

There is a remarkable connection between U(1)-bundles and quantization, discovered by Kostant and Souriau [14]. The deepest reasons for this, however, escapes us.

8. *Electromagnetism.* Normal electromagnetism obtains with the trivial bundle $0 \in H^2(V, \mathbb{Z})$; electromagnetism with monopoles is equivalent to "connections in a non-trivial U(1)-bundle", as emphasized so remarkably by Wu and Yang [15].

For our purposes it will suffice to show the bundle nature of two U(1)-solitons:

i) *Nielsen - Olesen vortex* [16]. The diagram

$$\begin{array}{llccccc} \beta : & \theta) & U(1) & \longrightarrow & S^3 & \longrightarrow & S^2 \\ & & \downarrow & & & & \| \\ \zeta : & \nabla) & C^1 & \longrightarrow & E & \longrightarrow & S^2 \end{array} \qquad (8.1)$$

which is the same as (7.1), (7.2), realizes the N-O vortex with a connection θ in the upper row, which produces a covariant derivative ∇ in the lower line bundle. The sections are the Higgs fields, which have to have zeroes, and in the vortex solutions their behaviour "at infinity" just indicates the broken symmetry of the solution: they are transformed by the gauge group; the connection form θ produces the background potential vector $A = A_\mu$ via the following:

$$\text{Connection } \theta \xrightarrow{D} \text{Curvature } \Omega \xrightarrow{w} \text{Field strength } F \qquad (8.2)$$

with D = exterior covariant derivative and w = Weil homomorphism ([13], Ch. XIII); $D\Omega = 0$ implies $dF = 0$ (d is the ordinary exterior derivative); but non-triviality and (7.3) imply *inexactness* of F, hence there is $A = A_\mu$ *locally, but not globally,* with $dA = F$. The integer character of the cohomology class can be recovered by the Gauss - Bonnet theorem ([13], Ap. 21; [7], Ap. C): if $c_1(\zeta)$ is the first Chern class of ζ,

$$\frac{1}{2\pi}\, c_1(\zeta) = \frac{1}{2\pi} \int_{S^2} F = 2 = \chi(S^2) \qquad (8.3)$$

with $\chi(V)$ the so-called Euler - Poincaré characteristic of the manifold V; for spheres $\chi(S^{2n}) = 2$, $\chi(S^{2n+1}) = 0$.

ii) *Dirac monopole* [17]. This has been formerly considered in this light by Stora [18], Mayer [19], and of course Yang and Wu [15], [20]; the diagram is

$$\begin{array}{ccccc} U(1) & \longrightarrow & P & \longrightarrow & V = \mathbb{R}^3 - \{0\} \\ \downarrow & & & & \| \\ C^1 & \longrightarrow & E & \longrightarrow & V \end{array} \qquad (8.4)$$

i.e. in $0 \in \mathbb{R}^3$ we just put a magnetic monopole; in $\mathbb{R}^3 - \{0\} \cong \mathbb{R} \times S^2$ we have a non-trivial Chern class, possible because $H^2(\mathbb{R} \times S^2) = H^2(S^2) = \mathbb{Z}$ and this implies the analysis of [15]: $A = A_\mu$ has to be given in two overlapping regions by *different* expressions; one has to solve the Schrödinger equation in both, and match the solutions through the overlap. The section in the lower row are just these solutions to the (double) quantum-mechnical wave equation, which can be solved in terms of "monopole harmonics".

To end, let us add that a single *electric* charge in the same situation *does* have a global (scalar) potential, because div $\underset{\sim}{E} = 0$ in $\mathbb{R}^3 - \{0\}$ implies the existence of a global φ with $\underset{\sim}{E} = -$ grad φ, because $H^1(\mathbb{R}^3 - \{0\}) = H^1(S^2) = \pi_1(S^2) = 0$!; of course $\varphi \propto \Omega^{-1}$.

V . Yang - Mills fields.

9. For realistic, 3-D spacelike space, even compactified to $R^3 = S^3$, we come to a surprise, because for any group G, the G - bundles over S^3 are classified by $\pi_2(G)$, which is zero for most groups! (in fact, for all discrete or compact; see [4]). Of course "surgery", will work, i.e. extirpate points of S^3 or $\mathbb{R}^3$; but this will more or less reproduce the case of the Dirac monopole.

Instead we turn to *Euclidean* space, which when compactified, $\mathbb{R}^4 \longrightarrow S^4$ produces a plethora of non-trivial bundles, as $\pi_3(G) \neq 0$ for many non-abelian Lie groups; in fact, $\pi_3(G) = \mathbb{Z}$ for simple compact Lie groups ([4]), $\pi_3(O(4)) = \mathbb{Z} \oplus \mathbb{Z}$; $\pi_3(U(1)) = 0$.

Here the pure Euclidean Yang - Mills theory with group SU(2) already contains solitons. These "solitons" are of course of paramount importance in providing a "tunneling" mechanism, for possible quark confinement (see Polyakov [21]). For the Polyakov instanton, our standard diagram is

$$\begin{array}{llcccc} \beta: & \theta)\ S^3 = SU(2) & \hookrightarrow & S^7 = P & \longrightarrow & S^4 = \mathbb{H}P^1 \\ & \qquad\downarrow & & & & \| \\ (\gamma_1^1)_H: & \nabla)\ \mathbb{R}^4 = \mathbb{H}^1 & \hookrightarrow & E & \longrightarrow & S^4 \end{array}$$

in which again β is the standard principal bundle with the symplectic group Sp(1) = SU(2) and coincides with another important Hopf fibering; the lower row has been taken as the line bundle over quaternions. Of course, the instanton displays many beautiful properties not covered by our "boundary conditions" study; see e.g. Jackiw Rebbi [22] or Nielsen - Schrör [23].

Let us end by writing the whole Hopf diagram of spheres

$$\begin{array}{lccccccc} \alpha: & S^0 \longrightarrow & S^1 & \longrightarrow S^1 & & & & \\ & & \| & & & & & \\ \beta: & & S^1 & \longrightarrow & S^3 & \longrightarrow S^2 & & \\ & & & & \| & & & \\ \delta: & & & & S^3 & \longrightarrow & S^7 & \longrightarrow S^4 \\ & & & & & & \| & \\ \zeta: & & & & & & S^7 & \longrightarrow S^{15} \longrightarrow S^8 \end{array}$$

because α, β and δ provide the first three examples of solitons (namely kink, vortex, and instanton); one can legitimally ask: is there any physics in fibration φ, over a superspace S^8?

Bibliography.

[1] Scott A.C. *et al* Proc. I.E.E. 61, 1443 (1973).

[2] Rajaraman, R. Phys. Lett 21C, 5 (1975) (Physics Reports).

[3] Coleman, S. Erice Lectures 1975 (Zichichi ed., Academic Press 1977).

[4] Boya L.J., Cariñena J.F. and Mateos J.: To appear (1978) in "Fortschritte der Physik".

[5] Wu T.T. and Yang, C.N.: Phys. Rev. D12, 3845, (1975).

[6] Steenrod, N: The topology of fibre bundles. Princeton U.P. (1951).

[7] Milnor J.: Characteristic classes. American Math. Soc. Pub. 76, (1974).

[8] Trautman, A.: Rep. Math. Phys. (Poland) 1, 29 (1970).

[9] Cho Y.: J. Math. Phys. 16, 2029 (1975).

[10] Mayer M.E. and Dreschler W.: Lectures on Fibre bundles and gauge theories (Lecture Notes in Physics Nº 67, 1977).

[11] Koszul, J.L.: Differential Geometry. Tata Institute (India), 1967.

[12] Hirzebruch, F.: Topological Methods in Algebraic Geometry. Springer, 1966.

[13] Kobayashi, I. and Nomizu K.: Foundations of Differential Geometry John Wiley; I, 1963; II, 1969; quoted as "Nomizu".

[14] Kostant B.: Quantization and group representations: in "Lectures in analysis and Applications III" (Lecture Notes in Mathematics, 170; Springer, 1970).- García, P.L.: Cuantificación Geométrica (Salamanca, 1976).

[15] Wu T.T. and Yang C.N.: Nucl. Phys. B107, 365 (1976).

[16] Nielsen H.B. and Olesen P.: Nucl. Phys. B61, 45 (1973).

[17] Dirac P.A.M.: Phys. Rev. 74, 817 (1948).

[18] Stora R.: "Gauge Theories" (Cargèse Lectures 1976; Marseille preprint).

[19] Mayer, M.E.: Lecture Notes in Math. 570 (Bonn Symposium) (pág. 307); Springer 1976.

[20] Wu T.T. and Yang C.N.: Phys. Rev. D16, 1018, (1977).

[21] Polyakov, A.M. - Phys. B120, 429 (1977).

[22] Jackiw, R. and Rebbi, C. - Phys. Rev. D16, 1052 (1977) and previous work.

[23] Nielsen, H.B. and Schroer, B. - Nucl. Phys. B127, 493 (1977).

EFFECTS OF HEAVY COLORED HIGGS SCALARS IN GRAND UNIFIED THEORIES

G. P. Cook, K. T. Mahanthappa and M. A. Sher
Department of Physics
University of Colorado
Boulder, Colorado
80309, U.S.A.

There have been grand unified models of strong, electromagnetic and weak interactions based on SU(n)($n \geq 5$), SO(4n+2)($n \geq 2$) and some exceptional groups. Especially various versions of SU(5), SO(10) and E(6) have been studied in more detail. In all these models the symmetry is broken down to SU(3)xSU(2)xU(1) which in turn is broken down to SU(3)xU(1). These symmetry breakdowns are achieved by using Higgs scalars. The last stage of the symmetry breaking generates masses of fermions. Usually it is assumed that the colored physical Higgs scalars are very heavy having masses of the order of the unification mass so that their effect on predictions become negligible. The purpose of my talk is to point out[1] that the uncertainties in the masses of the colored Higgs scalars lead to least uncertainties in the predictions of SU(5) with only 24 and 5 representations of Higgs scalars, whereas if one uses 45 instead of 5 or uses a bigger group which requires higher dimensional representations of Higgs scalars, the predictions like τ_p, proton life time, have large uncertainties. I confine myself to SU(5) throughout and comment on others at the end.

The SU(5) symmetry[2] is broken down to SU(3)xSU(2)xU(1) by the introduction of a 24 of Higgs field ϕ. The potential is given by

$$V(\phi)=-\mu^2 Tr(\phi^2)+\tfrac{1}{4}a[Tr(\phi^2)]^2+\tfrac{1}{2}bTr(\phi^4)$$

where $\mu^2>0$ and $a+(7/15)b>0$; ϕ acquires a vacuum expectation value given by $\langle\phi\rangle = diag(v,v,v,-(3/2)v,-(3/2)v)$. The $(mass)^2$ of vector bosons X and Y are of $O(g^2v^2)$. Out of 24 Higgs scalars, 12 remain physical and have $(mass)^2$ of $O(av^2,bv^2)$. Arguments[3] can be made to suggest a and b are of $O(g^2)$ so that the masses of these Higgs scalars are close to or greater than M_X and thus do not affect the renormalization group equations (RGEs). The Higgs multiplets which can break the SU(3) xSU(2)xU(1)symmetry down to SU(3)xU(1) and also give mass to the fermions are the 5 and the 45. Consider first the breaking by the 5. The potential is

$$V=-\tfrac{1}{2}\nu^2H^+H+\tfrac{1}{4}\lambda(H^+H)^2+\alpha\; H^+HTr\phi^2+\beta H^+\phi^2H$$

Minimization of this potential leads to the vacuum expectation value $\langle H\rangle = (0,0,0,0,v_0/\sqrt{2})$. This leads to $(mass)^2$ values of $O(10^{-1}g^2v_0{}^2)$ for $W^\pm$ and Z^0. The familiar SU(2) doublet contained in the 5 of Higgs scalars leads to a physical Higgs of $(mass)^2$ of $\lambda v_0^2/2$, while the color triplet gets a large $(mass)^2$ of $-5\beta v^2/4$ due to mixing. Usually the effects of the doublet are included in detailed calculations, but those of the triplet are ignored[4]. In order to know the effects of this, we have to know β. Even though we do not know β precisely, we know it must be smaller than a,b, or λ in order that the SU(3)xSU(2)xU(1) minimum be the absolute minimum. If β is extremely small, the renormalization effects will produce an effective β of $O(\alpha^2)$ where $\alpha=g^2/4\pi=0.024$. Thus we can conclude that $O(\alpha^2)\lesssim\beta\lesssim O(g^2)$ and hence

$$\alpha M_X \lesssim M_H \lesssim M_X$$

where M_H is the mass of Higgs color triplet. Since $M_H \leq M_X$, there will be an effect on RGEs, which neglecting thresholds, occurs for $q^2>4M_H^2$. Because the closer M_H is to M_X, the smaller the effect on RGEs, we will set $M_H=\alpha M_X$ in order to get the largest possible effect. The difference in the effects between $M_H=M_X$ and $M_H=\alpha M_X$ will yield the uncertainties in the calculations of $\sin^2\theta$, M_X (and thus τ_p) and m_b caused by our lack of knowledge of β, and thus of M_H.

The calculation is simple. We simply modify the beta-functions to include an additional scalar color triplet for $q^2>4\alpha^2 M_X{}^2$. For definiteness, we set $\alpha_s(M_W^2)=.137$ where α_s is the QCD coupling constant. The results are given in Table 1. From this table we see that M_X is uncertain by about 5%, $\sin^2\theta$ by 0.5% and m_b, the mass of the b-quark is unchanged. Since the proton life-time τ_p is proportional to M_X^4, it is uncertain by about 20%. The uncertainties are much smaller than the previously estimated uncertainties, and hence the neglect of the Higgs color triplet is justified.

Table 1. **Values of** $\sin^2\theta$,M_X and m_b for different values of the mass of the colored triplet in the <u>5</u> of Higgs, M_H.

M_H	$\sin^2\theta(M_W)$	M_X(GeV)	m_b(GeV)
$M_H=M_X$(no effect)	.206	3.5×10^{14}	5.2
$M_H=\alpha M_X$	.205	3.7×10^{14}	5.2

Now consider having the <u>45</u> of Higgs scalars. The <u>45</u> decomposes into seven SU(3)xSU(2) multiplets of Higgs given by $(8,2),(6^*,1),(3,3),(3^*,2),(3,1)$ and $(1,2)$. The (1,2) is the usual Higgs doublet and acquires a light mass and the other six multiplets acquire heavy masses. Call these masses M_i(i=1-6)in the order they occur. Their effects on the RGEs can be easily found. The full potential is given in Ref.5. In this case there are 5 β_i's which mix the <u>45</u> and <u>24</u> of Higgs multiplets. As before the m_i's are of $O(\beta_j M_X{}^2)$ and cannot be determined precisely. However, as before, we can conclude

$$\alpha M_X \leq M_i \leq M_X$$

In this instance, the largest change in $\sin^2\theta$ and M_X occur when <u>some</u> of the Higgs multiplets have masses αM_X and others have masses M_X. We have examined all possible combinations of masses consistent with the Higgs potential.

Table 2. Values of $\sin^2\theta$ and M_X for various values of M_i(i=1-6)

M_i/M_X	$\sin^2\theta(M_W)$	M_X(GeV)
α(i=1,2,3,4,5,6)	.205	$3.7\text{x}10^4$
α(i=3,4,5,6) 1(i=1,2)	.209	$1.3\text{x}10^{14}$
α(i=1,2,6) 1(i=3,4,5)	.201	$10.3\text{x}10^{14}$
α(i=3) 1(i=1,2,4,5,6)	.214	$2.8\text{x}10^{14}$
α(i=2,5,6) 1(i=1,3,4)	.199	$4.3\text{x}10^{14}$

The results for some choices of M_i are given in Table 2, which includes extreme values. Since we have no knowledge of the actual masses, any of the results of $\sin^2\theta$ and M_X are possible. It can be seen that the uncertainty in $\sin^2\theta$ is 4% and in M_X is a factor of 2.8. Thus, if a 45 is involved in the Higgs structure of SU(5), then the uncertainty due to our lack of knowledge of the colored scalar masses is 4% in $\sin^2\theta$ and a factor of $(2.8)^4$=60 in τ_p. Thus combined with earlier uncertainties, they are increased. In the proton life time the uncertainty becomes almost a factor of 10^4. The uncertainty in m_b, in the models which contain a 45, is very model dependent but turns out to be several hundred MeV.

In other grand unified theories, there are a large number of colored Higgs scalars which require heavy masses. For example in SO(10), we need 120 to give masses to fermions. SU(5) decomposition is given by

$$120=45+45^*+10+10^*+5+5^*$$

Similarly in E(6) we have Higgs representation like 351,351 . Thus, in other than the minimal SU(5), i.e. with only 24 and 45 of Higgs, the precise predictive power of the model is reduced as there are a large number of colored Higgs which give rise to large uncertainties in predictions as we are ignorant of the precise value of their masses.

References

1. G. P. Cook, K. T. Mahanthappa and M. A. Sher, Phys. Lett. 90B, 398 (1980)

2. H. Georgi and S. L. Glashow, Phys. Rev. Lett. 32, 438 (1975)

3. J. Ellis, M. K. Gaillard, A. Peterman and C. T. Sachrahda, Nucl. Phys. B164, 253 (1980)

4. D. V. Nanopoulos and D. A. Ross, Nucl. Phys. B157, 273 (1979)
   K. T. Mahanthappa and M. A. Sher, Phys. Rev. 21, 2036 (1980)
   T. Goldman and D. A. Ross, Phys. Lett. 84B, 208 (1979)
   W. Marciano, Phys. Rev. D20, 274 (1979)

5. P. H. Frampton, S. Nandi and J. J. G. Scanio, Phys. Lett. 85B, 255 (1979)

CATEGORIES OF NONLINEAR GROUP REALIZATIONS:

A POSSIBLE EXPLANATION FOR THE MULTIPLE STATES OF CHARGE

Bill J. Dalton
Ames Laboratory, US Department of Energy
Iowa State University
Ames, Iowa 50011/USA

Electromagnetic charge must be addressed as one of the key fundamental concepts in physics. Since first observed by early man, this intriguing structure of nature has over the years evaded attempted explanations [1-4] to arrive with us today, still a major challenge. Here I put forth and discuss the conjecture that the three observed states of charge (i.e. +q, -q, 0) correspond to different categories of nonlinear (including the linear) realizations of a group. The support for this conjecture is the recently [5] demonstrated mathematical existence of multiple categories of nonlinear realizations of a continuous group. Here I will outline the essence of this development and, for the Lorentz group (or the SL(2,C) covering group), present a family of nonlinear realizations. I will then consider nonlinear realizations of the Lorentz group acting on the four-velocities of particles and use "superconducting" [6] solutions to explain a) why the nonlinearity can produce non-zero forces, and b) why the forces differ in each category.

To describe these nonlinear realizations, I consider two spaces H_1 and H_2 on which there is a simultaneous realization of a continuous group G such that the transformation on H_2 has a nonlinear dependence on H_1; that is, for $g(\alpha)\varepsilon G$, $g(\alpha)$: $\Phi \to \Phi'(\alpha,\Phi), \Psi \to \Psi'(\alpha,\Psi,\Phi), \Phi\varepsilon H_1, \Psi\varepsilon H_2$, where α is a group parameter. I will follow the notation of [5] to which the reader may refer for more detail. The implied sums on repeated indices range over the number of generators. The discussion here is limited to infinitesimal realizations $\Psi' = \Psi + \alpha_i [T_i,\Psi]$ where the action $[T_i,\Psi]$ is restricted to the form $[T_i,\Psi] = q_{ij}(\Phi)t_j\Psi$ with $[t_i,t_j] = C_{ijk}t_k$. The C_{ijk} are the structure constants of the Lie Algebra of G so that the operators t_i themselves generate a representation. Requiring the action $[T_i,\Psi]$ to satisfy the Jacobi identities, $[T_i,[T_j,\Psi]] - [T_j,[T_i,\Psi]] = C_{ijk}[T_k,\Psi]$, gives the following equation for the functions $q_{ij}(\Phi)$.

$$[T_i,q_{jk}] - [T_j,q_{ik}] - q_{im}q_{jn}C_{mnk} - C_{ijn}q_{nk} = 0 \qquad (1)$$

Because of the nonlinearity of these equations we expect to find more than one solution and that the sum of two solutions is not necessarily a solution. The solutions of these equations do not depend upon the particular representations $\{t_i\}$.

For the group SL(2,C) there are six generators, and we use the vector basis [7,8] $T_i = (J_i + iK_i)/2$, $Z_{\hat{i}} = (J_{\hat{i}} - iK_{\hat{i}})/2$ where $i \varepsilon \{1,2,3\}$ and $\hat{i} \varepsilon \{4,5,6\}$. For this group there are 36 functions q_{ij} which must solve Eqn. (1). To demonstrate the existence of multiple categories of realizations we consider H_1 to be an arbitrary 4-vector, $H_1 = \{\Phi_1,\Phi_2,\Phi_3,\Phi_4\}$ i.e., a carrier space for the $(\frac{1}{2},\frac{1}{2})$ representation of SL(2,C). Either q_{ij} or $q_{\hat{i}\hat{j}}$ independently have among others, the following solutions:

$$q_{ij}^{(1)} = 0, \quad q_{ij}^{(3)\pm} = -\delta_{ij} \mp \frac{\Phi_4}{\Phi^2}\varepsilon_{ijk}\Phi_k + \frac{1}{\Phi^2}\Phi_i\Phi_j$$

$$q_{ij}^{(4)\pm} = \mp\frac{\Phi_4}{\Phi^2}\varepsilon_{ijk}\Phi_k - \frac{1}{\Phi^2}\Phi_i\Phi_j, \quad q_{ij}^{(2)} = -\delta_{ij}$$

$$q_{ij}^{(5)\pm} = -\frac{1}{2}\delta_{ij} + \left(\frac{\pm\Phi_4+D}{2\Phi^2}\right)\varepsilon_{ijk}\Phi_k, \quad q_{ij}^{(6)\pm} = -\frac{1}{2}\delta_{ij} + \left(\frac{\pm\Phi_4-D}{2\Phi^2}\right)\varepsilon_{ijk}\Phi_k$$

$$q_{ij}^{(7)\pm} = \frac{1}{2}\left(-1 \pm \left(\frac{D^2}{\Phi^2-D^2}\right)^{\frac{1}{2}}\right)\delta_{ij} - \frac{1}{2\Phi_4}\varepsilon_{ijk}\Phi_k$$

$$q_{ij}^{(8)\pm} = \frac{1}{2}\left(-1 \pm \left(\frac{D^2}{\Phi^2-D^2}\right)^{\frac{1}{2}}\right)\delta_{ij} + \frac{1}{2\Phi_4}\varepsilon_{ijk}\Phi_k \tag{2}$$

In these equations $D^2 = \Phi_\mu\Phi_\mu = \Phi_4^2 + \Phi^2$ and because we are dealing with the $(\frac{1}{2},\frac{1}{2})$ representation the sign on Φ_4 must be changed in going from q_{ij} to $q_{\hat{i}\hat{j}}$; and $\varepsilon_{\hat{i}\hat{j}\hat{k}}$ is cyclic on 4,5,6. Because of the particular Lie structure of this group, the equations for the cross functions $q_{\hat{i}j}$ and $q_{i\hat{j}}$ are linear but involve the functions q_{ij} and $q_{\hat{i}\hat{j}}$. One solution is $q_{\hat{i}j} = q_{i\hat{j}} = 0$. From the above list it is obvious that there are several different nonlinear realizations for a single space H_1. Each space H_2 must be labeled by the category under which it transforms. Although the sum of two or more solutions is generally not a solution, there are exceptions. From (2) we have for instance $q_{ij}^{(7)\pm} + q_{ij}^{(8)\mp} = -\delta_{ij} = q_{ij}^{(5)\pm} + q_{ij}^{(6)\mp}$. It should be clear that these solutions cannot be scaled, that is, if q_{ij} is a solution $2q_{ij}$ is <u>not</u> a solution.

We will first consider the force equations for classical electrodynamics. We use $\dot{x}^T = (\dot{x}_1,\dot{x}_2,\dot{x}_3,\dot{x}_4)$ with x_4 = ict and s for the invariant line element of a particle. The experimental fact that the action of the electromagnetic field leaves the quantity $\dot{x}_u\dot{x}_u$ invariant (not all physical forces have this property) suggests that electrodynamics could correspond to a local Lorentz gauge acting on the four-velocities [9]. For contrast with the nonlinear theory described below we first discuss this gauge picture. Consider the gauge transformation $\dot{x}' = U\dot{x}$ where U is an orthogonal 4 x 4 matrix with parameters which may be two point functions; that is, they may depend on two points x(1) and x(2) which lie on the world line of the particle. Since $\ddot{x}$ does not transform like $\dot{x}$ we introduce the covariant acceleration D (here D plays the role of the covariant derivative) with the standard form

$$D = \ddot{x} + A\dot{x}, \quad A' = UAU^{-1} + U\frac{d}{ds}U^{-1} \tag{3}$$

where A is the gauge potential (Weyl connection). For the covariant "superconducting" case D = 0 [6] we have the relations

$$\ddot{x}_\alpha = \frac{q}{mc^2}F_{\alpha\rho}\dot{x}_\rho, \quad \frac{qF_{\alpha\rho}}{mc^2} = -A_{\alpha\rho} \tag{4}$$

which are the Lorentz force equations. We see that the gauge potential A represents the product of the traditional charge and field. Although this gauge picture is consistent with the Lorentz Force equation, it gives us little insight into the three different states of experimental charge.

In contrast to the above local Lorentz gauge picture we now consider a picture of electrodynamics which involves nonlinear dependent realizations of the Lorentz group. As in the above gauge theory we are considering the Lorentz group because under this group the quadratic form $\dot{x}_u\dot{x}_u$ is left invariant. In the language discussed earlier, the space H_2 is the four-velocity space of a particle. We have $\dot{x}' = U\dot{x}$ where

$$U = \exp(B_j t_j), \quad B_j \equiv \alpha_i q_{ij}(\Phi), \tag{5}$$

the indices range from 1 to 6 and the t_j generate a four-vector representation. For the present we consider the parameters α_i to be global constants. Since the q_{ij} may depend upon space-time via $\Phi(x)$, the acceleration $\ddot{x}$ does not transform like $\dot{x}$ (except in the linear case) so that we introduce a covariant acceleration D and Weyl connection as in (3). Again, considering the "superconducting" solution D = 0, we obtain the Lorentz force equation (4). However, in this nonlinear theory we have, for the same group parameter α, a different charge-field product for each different nonlinear category.

To illustrate how the three states of charge could be described in this theory we consider the particular solution $A = U \frac{d}{ds} U^{-1}$ for the Weyl connection. From [10] we have the following formal expansion for this expression

$$A = U \frac{d}{ds} U^{-1} = [(1 - e^{-K})/K] \frac{d}{ds} (\alpha_i q_{ij}(\Phi) t_j) \tag{6}$$

where we define K by $KX = - [\alpha_i q_{ij} t_j, X]$. From this expression (with $\dot{\alpha} = 0$) we see that the linear case $q_{ij} = - \delta_{ij}$ gives a zero force. Thus, within this theory, we identify the linear case with elementary particles with zero charge. To describe the other two states of charge, consider two solutions of (1) which have the form $q^{\pm}_{ij} = - \frac{1}{2} \delta_{ij} \pm \eta_{ij}(\Phi)$. For instance, the solutions $q^{(5)+}$ and $q^{(6)-}$ in (2) have this form. For a small parameter α we have from (6) the relation

$$A^{\pm} = - \alpha_i \frac{dq^{\pm}_{ij}}{ds} (\Phi) t_j = \mp \alpha_i \frac{d\eta_{ij}}{ds} (\Phi) t_j \tag{7}$$

so that with (4) these two categories give Lorentz forces that are equal but opposite in sign.

This example illustrates how it is possible to explain the experimental states of charge in terms of different categories of nonlinear realizations. In contrast to the Lorentz gauge picture, only one parameter space is needed for the three different states of charge. Since the parameters α_i are global constants, one could argue that by transforming to a different Lorentz coordinate frame the above theory would lose content. If we had only linear realizations this would be true. However, with the nonlinear categories the group parameters α_i determine the strength of the force. We stress here that this dependence on the group parameter arises because of the particular solution for A discussed above. In general we have solutions for A that do not depend on the group parameters.

We may introduce the force equations as extremum conditions on an invariant form. Towards this end we consider a matrix space M on which we define a nonlinear realization of the Lorentz group;

$$\Lambda' = U_\ell \Lambda U^{-1} \simeq \Lambda - \alpha_i \, [a t_i \Lambda + \Lambda q_{ij} t_j] , \tag{8}$$

$\Lambda \epsilon M$ is a 4×4 matrix, U is given by (5) and $U_\ell = \exp(a\alpha_i t_i)$ where we set $a = 1$ for the regular linear realization and $a = 0$ for the null case. One can show that the Jacobi identities acting on Λ reduce to Eqn. (1) so that no new conditions are needed. Here Λ plays the role of H_2 in the earlier discussion. It is easy to show that $\Lambda^{-1} \frac{d\Lambda}{ds}$ transforms like the Weyl connection so that with D = 0 we have the force equations

$$\ddot{x} = - \Lambda^{-1} \dot{\Lambda} \dot{x} \, , \quad \dot{\Lambda} = \frac{d\Lambda}{ds} \, . \tag{9}$$

To obtain force equations via variation we consider a vector z, $z^T = (z_1, z_2, z_3, z_4)$ which transforms linearly, i.e., $z' = U_\ell z$ and the following invariant Lagrangian

$$L = \int (\frac{mc\dot{x}_\mu \dot{x}_\mu}{2} + \lambda \, \Lambda_{\mu\beta} \, \dot{x}_\beta \, z_\mu) \; ds \, . \tag{10}$$

The extremum equations for this Lagrangian have the form

$$\ddot{x}_\alpha = \frac{\lambda}{mc} \, [\frac{\partial}{\partial x_\alpha} \, (\Lambda_{\mu\rho} z_\mu) - \frac{\partial}{\partial x_\rho} \, (\Lambda_{\mu\alpha} z_\mu)] \dot{x}_\rho \tag{11}$$

From this form we see that the electromagnetic vector potential would be proportional to $(\Lambda_{\mu\rho} z_\mu)$ so that we have a different vector potential for each different nonlinear category. With the classical canonical momentum given by

$$\pi_\alpha = \frac{\partial L}{\partial \dot{x}_\alpha} = mc\dot{x}_\alpha + \lambda \, \Lambda_{\mu\alpha} z_\mu \tag{12}$$

one can easily take the quantum mechanical correspondence to obtain

$$D_\alpha \Psi = (-i\hbar \partial_\alpha + \lambda \, \Lambda_{\mu\alpha} z_\mu) \; \Psi \tag{13}$$

where Ψ is the wavefunction. We emphasize here however that this particular choice for the potential $(\Lambda_{\mu\alpha} z_\mu)$ is determined by the above choice of Lagrangian. It is not clear at this point that a suitable Lagrangian can be found such that the extremium equations are identical to those obtained from the covariant superconducting condition D = 0.

In the above discussion we have given some plausible arguments for the conjecture that the states of charge correspond to different categories of nonlinear realizations. At this stage in development the supporting arguments are not conclusive. However, if this conjecture is true, there are definite consequences, two of which we may ascertain from our above discussion: 1) If q_{ij} is a solution of (1), an arbitrary integer multiple of q_{ij} is not a solution. Within this theory this would suggest that for elementary particles (non-composite ones) integer multiples of the basic charge unit are not mathematically possible! This is in direct contradiction with a traditional theory [3,4]. However, from physical experiments, how many elementary particles have been observed with say 5 or 8 units of the electron's charge? Answer: None! One may also ask a similar question about the experimental observation of magnetic monopoles. 2) If we can identify certain solutions of (1) with the three observed charge states +q, -q and 0 then we must consider the physical meaning of other solutions. For instance, in example (2) there are solutions which add to $-\delta_{ij}$ but which are not symmetric in composition. From the number and type of solutions in (2) it is clear that if we have here the beginnings of a viable theory of electrodynamics, then we also have other related possibilities within this theory. In addition to a variety of possible solutions we also have rules for combining solutions to obtain other solutions.

Within the content of this discussion it is worthwhile to consider a combination of nonlinear and gauge realizations. We first consider a "linear" gauge transformation $U_g = \exp(\alpha_i(x) t_i)$, $A'_g = U_g A_g U_g^{-1} + U_g \frac{d}{ds} U_g^{-1}$. For the combined gauge and nonlinear realizations we have $\dot{x}' = U\dot{x}$ where $U = \exp(\alpha_i(x)\, q_{ij}(\Phi) t_j)$, $A = UAU^{-1} + U \frac{d}{ds} U^{-1}$. Now in place of (8) we consider a realization on a matrix space of form $\Lambda' = U_g \Lambda U^{-1} \simeq \Lambda + \alpha_i(x)\ [-t_i\Lambda - \Lambda q_{ij} t_j]$ which differs from (8) only in that the parameters α_i depends upon space-time. With the above equations it is easy to obtain the following expression for the covariant acceleration D. $D = \ddot{x} + \Lambda^{-1}\dot{\Lambda}\dot{x} + \Lambda^{-1} A_g \Lambda\dot{x}$. From this expression we see that for a given gauge potential A_g we have a family of potentials, one for each nonlinear realization. One term of the potential $\Lambda^{-1}\dot{\Lambda}\dot{x}$ depends only on the nonlinear part and the second is a transformed potential.

ACKNOWLEDGMENT

This work was supported by the U.S. Department of Energy, Contract No. W-7405-Eng-82, Office of Basic Science, Nuclear Physics Division.

REFERENCES

1. "Vier Abhandlumgen über de Elektricität und Den Magnetismus von Couloumb," by Walter König, Leipzig (1890).
2. H. Weyl, Z. Phys. 56 330, (1929).
3. P. A. M. Dirac, Phys. Rev. 74 817 (1948).
4. C. N. Yang, Ann. N. Y. Acad. Sci. 294 86 (1977). Many references to earlier work are given in this article.
5. B. J. Dalton. To be published.
6. Y. Nambu, Ann. N. Y. Acad. Sci 294 74 (1977).
7. B. J. Dalton, J. Math. Phys. 20 7 (1979).
8. B. J. Dalton, J. Math. Phys. 19 1335 (1978).
9. B. Kursunoglu, p. 24, in "Fundamental interactions in physics, Edited by B. Kursunoglu and A. Perlmutter, (Plenum Press, New York) 1973.
10. Curtus G. Callan, Jr., Sidney Coleman, J. Wess and Bruno Zumino, Phys. Rev. 177 2247 (1969).

CLASSICAL SOLUTIONS OF NON-LINEAR σ-MODELS AND THEIR QUANTUM FLUCTUATIONS

Allan M. Din

Laboratoire d'Annecy de Physique des Particules
B.P. 909 - F-74019 Annecy-le-Vieux, France

A B S T R A C T

I study the properties of $O(N)$ and CP^{n-1} non-linear σ-models in the two dimensional Euclidean space. All classical solutions of the equations of motion can be characterized and in the CP^{n-1} model they can be expressed in a simple and explicit way in terms of holomorphic vectors. The topological winding number and the action of the general CP^{n-1} solution can be evaluated and the latter turns out always to be an integer multiple of 2π. I further discuss the stability of the solutions and the problem of one-loop calculations of quantum fluctuations around classical solutions.

1. INTRODUCTION

Field theory models in two space-time dimensions have several attractive properties from both a mathematical and a physical point of view. Mathematically the classical structure of the theory can often be investigated completely using analytical, topological and other standard techniques. The quantum structure of the models may also to a large extent be treated using rather mathematically rigorous methods.

Physically the two-dimensional models have some direct analogies in solid state physics, but from the point of view of particle physics, the crucial question is, to what extent one can draw parallels with four-dimensional non-abelian gauge theories. The models thus selected should have properties like conformal invariance, asymptotic freedom non-trivial topological structure, etc.

The discussion below will be limited to the socalled $O(N)$ and CP^{n-1} models in two dimensional Euclidean space time. I will first define these models and then describe some well-known classical and quantum properties.

The $O(N)$ model[1)] is defined in terms of an N-dimensional real vector field $q(x)$

$$q(x) = (q_1(x), \ldots, q_N(x)) \tag{1.1}$$

where x belongs to the two-dimensional Euclidean plane E^2. Actually for most applications one considers a compactified E^2, i.e. $x \varepsilon S^2$.

The q is taken to fulfil the constraint

$$q^2 = 1 \tag{1.2}$$

and as a Lagrangian (energy functional) one takes

$$\mathcal{L}(q) = (\partial_\mu q)^2 \tag{1.3}$$

The theory is thus invariant under global $O(N)$ transformations.

In general one might be interested in finding the harmonic maps[2)] of the theory, i.e. solutions of the equation of motion

$$\partial^2 q + \mathcal{L}(q)\, q = 0 \tag{1.4}$$

with finite action

$$S = \int d^2x\, \mathcal{L}(q) \tag{1.5}$$

This problem was studied in Refs 3 and 4 and we will return to it in the more general framework of the CP^{n-1} model in the next section.

The O(N) model has from a physical point of view the drawback, that there is only a non-trivial topological structure for $N = 3$. In this case it is easy to establish that the only solutions of (1.4) are the socalled instantons and anti-instantons (fulfilling the simpler self-duality equations) which can be described by a complex field w related to q by

$$w = \frac{q_1 + iq_2}{1 + q_3} \tag{1.6}$$

and with w being just a rational function of x_+ or x_- (for instantons and anti-instantons respectively). Here

$$x_\pm = x_1 \pm ix_2 \tag{1.7}$$

In contrast the CP^{n-1} model[5] has a non-trivial topological structure for any n. The model is defined in terms of a complex n-dimensional vector field z

$$z(x) = (z_1(x),\dots,z_n(x)) \tag{1.8}$$

fulfilling

$$|z|^2 = 1 \tag{1.9}$$

and such tnat one identifies z's which only differ by an overall phase factor.

The Lagrangian is taken to be

$$\mathcal{L}(z) = \overline{D_\mu z}.D_\mu z \tag{1.10}$$

where the covariant derivative D_μ is

$$D_\mu = \partial_\mu - \bar{z}\partial_\mu z \tag{1.11}$$

The theory thus has a global $U(n)$ invariance and a local (abelian) $U(1)$ gauge invariance

$$z \to e^{i\theta} z \tag{1.12}$$

The equations of motion are

$$D^2 z + \mathcal{L}(z) z = 0 \tag{1.13}$$

The self-duality equations are in this model simply

$$D_\mp z = 0 \tag{1.14}$$

(differentiating with respect to the $x_\mp$ of (1.7)) where $-$ and $+$ refer to instantons and anti-instantons respectively. Thus the general instanton solution (of course also fulfilling (1.13)) is given by

$$z_\alpha = \frac{p_\alpha(x_+)}{|p(x_+)|} \tag{1.15}$$

where $p_\alpha(x_+)$ are polynomials in x_+ of degree k.
The degree k is precisely the topological winding number $\tilde{Q}$ of the solution

$$\tilde{Q} = \frac{1}{2\pi} \int d^2x \, Q(x) \tag{1.16}$$

where $Q(x)$ is the topological charge density

$$Q(x) = 2\left[|D_+z|^2 - |D_-z|^2\right]. \tag{1.17}$$

The instantons and anti-instantons are only special solutions of the equations of motion (1.13) and in the next section I will show how to construct the general solution. A k-instanton solution corresponds to a local minimum of the action, the value of the action at the point being

$$S_o = 2\pi k \tag{1.18}$$

The quantum fluctuations around such local minima have been considered by various authors[6),7)]. The starting point in such studies is the functional integral

$$Z = \int \mathcal{D}z \, \mathcal{D}\bar{z} \quad e^{-S(z)} \tag{1.19}$$

This integral is then approximated by a sum over the various topological sectors (instanton number) taking account only quadratic terms in the fluctuations ϕ around a given instanton solution z. For small fluctuations ϕ

$$z' = z\sqrt{1-|\phi|^2} + \phi \tag{1.20}$$

where

$$\bar{z}.\phi = 0 \tag{1.21}$$

and one finds

$$S(z') = 2\pi k + \int d^2x \left[|D_-\phi|^2 - |\bar{z}.D_-\phi|^2\right] \tag{1.22}$$

Thus one finds

$$Z \sim \sum_k e^{-2\pi k} \int \mathcal{D}\phi \, \mathcal{D}\bar{\phi} \; e^{-\int d^2x\left[|D_-\phi|^2 - |\bar{z}.D_-\phi|^2\right]} \tag{1.23}$$

The problem is therefore reduced to finding the determinant of a certain second order differential operator corresponding to a specific instanton background. This in general is quite complicated to evaluate because of problems with zero modes and regularization. I will not go into further technical details here but just remark that the calculation can be done in an explicit way for the CP^1 model and the interesting result comes out that the summation over the different instanton sectors eliminates the infrared (large x) divergences present for each k.

Physically however the above calculation is not quite relevant since it is not possible to do an exact calculation taking into account both instantons and anti-instantons. It is therefore natural to try to find all stationary points of the action, i.e. all solutions of (1.13), and subsequently investigate their contribution to the functional integral (1.19).

2. GENERAL CLASSICAL CP^{n-1} SOLUTIONS

It is a simple observation[8] that embedding $O(n)$ solutions in the CP^{n-1} model one gets CP^{n-1} solutions which are neither instantons nor anti-instantons. For the $O(N)$ model it is possible to characterize all solutions although only in an implicit way[3],[4]. It turns out, as will be described below (for technical details see Ref. 9), that it is possible to characterize CP^{n-1} solutions in a similar way, but surprisingly in this case one gets completely explicit expressions for the solutions.

The basic idea is to start with an arbitrary finite action solution z and then, in terms of z, define a vector f which has the simple property of being holomorphic $f = f(x_+)$ (or equivalently anti-holomorphic).

For a solution z which is not a trivial embedding from a lower dimensional space we will in general have spaces

$$H_\ell = \{D_-z,\ldots, D_-^\ell z\} \tag{2.1}$$

and

$$H'_{\ell'} = \{D_+z,\ldots, D_+^{\ell'} z\} \tag{2.2}$$

with dimensions ℓ and ℓ' respectively, such that the n-dimensional complex vector space is precisely spanned by $\{z, H_\ell, H_{\ell'}\}$ i.e.

$$\ell + \ell' = n - 1 \tag{2.3}$$

The spaces $z, H_\ell, H_{\ell'}$ can be proven to be mutually orthogonal. The relation (2.3) may in fact not hold at a finite number of points but we will not consider this complication here. If we now define the vector $f \in \{z, H_\ell\}$ to fulfil

$$\bar{f}.D_-^i z = \omega\, \delta^{i\ell} \qquad i = 0,\ldots,\ell \tag{2.4}$$

where ω is taken to be a solution of

$$\partial_+\omega - \bar{z}\partial_+ z\, \omega = 0 \tag{2.5}$$

then it is possible to show that f is holomorphic

$$\partial_- f = 0 \tag{2.6}$$

i.e. $f = f(x_+)$.

It is easy to see that a gauge transformation $z \to e^{i\theta} z$ corresponds to a transformation $f \to \lambda(x_+)f$.

Taking derivatives of f with respect to x_+ one can span the space $\{z, H_\ell\}$ and one is thus able to express z in terms of f. The formula turns out to be

$$z = \frac{\hat{z}^{(\ell)}}{|\hat{z}^{(\ell)}|} \tag{2.7}$$

where

$$\hat{z}^{(\ell)} = \partial_+^\ell f - \sum_{j,i=o}^{\ell-1} \partial_+^i f \, M_{i,j}^{(\ell)^{-1}} \partial_+ M_{j,\ell-1}^{(\ell)} \tag{2.8}$$

and the positive definite matrix $M^{(\ell)}$ is defined by

$$M_{ij}^{(\ell)} = \overline{\partial_+^i f} . \partial_+^j f \qquad i,j = 0,\ldots, \ell-1 \tag{2.9}$$

To get the dimensions ℓ and ℓ' right we should demand that $f,\ldots,\partial_+^{n-1} f$ be linearly independent. The finiteness of the action is guaranteed by having a rational f.

On the other hand defining z in terms of f via (2.7), (2.8) and (2.9) it is possible to show that the equation of motion (1.13) is fulfilled.

It has thus been established (for further details see Ref. 9) that there is a one-to-one correspondence between solutions z defined up to a phase factor and holomorphic vectors $f(x_+)$ defined up to a factor $\lambda(x_+)$. The formulas giving z in terms of f are completely explicit. This is in contrast to the situation in the O(N) model[3),4)] (N odd = 2N' + 1) where a similar characterization of solutions exists but with an additional constraint on f requiring it to be a totally isotropic holomorphic curve. This means that one has to solve the constraint equations

$$\partial_+^i f \; \partial_+^j f = \delta^{iN'} \; \delta^{jN'} . \tag{2.10}$$

Let us investigated some properties of the general CP^{n-1} solutions[10)]. It is clear that when $\ell = 0$ in (2.8) then $D_- z = 0$ (from 2.1) and we have an instanton solution. If $\ell = n - 1$ we must have $D_+ z = 0$, i.e. an anti-instanton.

It is possible to evaluate the winding number and action for a general solution (2.7). To see this one starts by observing that the action can be written

$$S_\ell = 2\pi \, \tilde{Q}_\ell + 4I_\ell \tag{2.11}$$

where $\tilde{Q}_\ell$ is (an integer winding number) given by (1.16) and

$$I_\ell = \int d^2x \; |D_- z|^2 \tag{2.12}$$

It is not difficult to prove that

$$|D_-z|^2 = \partial_-(M^{(\ell)-1}\,\partial_+\,M^{(\ell)})_{\ell-1,\ell-1} \tag{2.13}$$

Since this is a total divergence it is clear that I_ℓ will be an integer multiple of 2π. The same will therefore be true for the action S.

Introducing the determinant of $M^{(\ell)}$

$$|M^{(\ell)}| = \det\,(M^{(\ell)}_{ij}) \tag{2.14}$$

it is further possible to prove that

$$I_\ell = \int d^2x\;\partial_-\partial_+\,\log\,|M^{(\ell)}|. \tag{2.15}$$

The topological charge density can be seen to be

$$Q = 2\,\partial_+\partial_-\,\log\,|\hat{z}^{(\ell)}|^2 \tag{2.16}$$

and after some algebra one shows

$$|\hat{z}^{(\ell)}|^2 = [M^{(\ell)}\,\partial_-\,(M^{(\ell)-1}\partial_+\,M^{(\ell)})]_{\ell-1,\ell-1} \tag{2.17}$$

$$= \frac{|M^{(\ell+1)}|}{|M^{(\ell)}|}$$

Thus

$$2\pi\,\tilde{Q}_\ell = 2\int d^2x\;\partial_+\partial_-\,\log\frac{|M^{(\ell+1)}|}{|M^{(\ell)}|} \tag{2.18}$$

or

$$\tilde{Q}_\ell = \frac{1}{\pi}\,[I_{\ell+1} - I_\ell] \tag{2.19}$$

and

$$S_\ell = 2\,[I_{\ell+1} + I_\ell] \tag{2.20}$$

The actual value of I_ℓ as given by (2.15) depends on the behaviour of $|M^{(\ell)}|$ at infinity and near the singular points of f. For physical applications it is only necessary to discuss this behaviour for the case of a rational $f(x_+)$ with no special relation between the parameters of f (i.e. excluding certain hyperplanes of the parameter space). For the case where f is simply a polynomial of degree $\alpha(\geq n-1)$, the degree in x_+ of $M^{(\ell)}$ will be

$$\alpha^{(\ell)} = \ell(\alpha - \ell + 1) \tag{2.21}$$

so that

$$\tilde{Q}_\ell = \alpha^{(\ell+1)} - \alpha^{(\ell)} = \alpha - 2\ell \tag{2.23}$$

$$S_\ell = 2\pi(\alpha^{(\ell+1)} + \alpha^{(\ell)}) = 2\pi((2\ell+1)\alpha - 2\ell^2) \tag{2.24}$$

Similar formulas can be derived in the case f has singularities. Formulas valid for all values of the parameters can be derived using the techniques of Ref. 3 but do not appear to be very transparent.

3. DISCUSSION AND OUTLOOK

The question arises of what effect the stationary points of the action will have when evaluating the functional integral (1.19). Considering again a small fluctuation around ϕ a general solution z as in (1.20) one finds to second order in ϕ[9)]

$$S = S_o + \int d^2x\, V(\phi) \tag{3.1}$$

where

$$V(\phi) = |D_-\phi|^2 - |\phi|^2|D_-z|^2 - |\bar{z}.D_-\phi + \bar{\phi}.D_-z|^2 \tag{3.2}$$

For the case of an instanton $D_-z = 0$ and $V(\phi) = |D_-\phi|^2 - |\bar{z}.D_-\phi|^2$ as in (1.22), which is non-negative. If z is a general solution it might be possible to have a negative contribution from the integral over $V(\phi)$. As a matter of fact the choice

$$\phi = \varepsilon\, D_+z \tag{3.3}$$

can be seen to give rise to such a negative contribution if z is neither an instanton nor an anti-instanton. Thus we have seen that in general the CP^{n-1} solutions are unstable under small fluctuations.

In evaluating the functional integral this means that one in general has to find the determinant of an operator which may have negative eigenvalues. Properly speaking the Gaussian functional integral is not well defined and one would have to devise a prescription for how to evaluate it. Even if this can be done, the problem remains whether one can calculate the determinant explicitly as was the case for a pure instanton background.

A different approach to evaluation the functional integral might be to find (suitably defined) complex solutions of the equations of motion[11),12)], which can be done using the technique described in section 2, and subsequently saturate the integral by the contribution of the complex saddle points. This approach might conceivably also shed some light on the instanton gas problem in the CP^1 model, where there are no othe "real" solutions than the instantons and anti-instantons.

Let me finally mention some other problems connected with understanding the structure of the CP^{n-1} model. It has been found that the $1/n$ expansion method as applied to the CP^{n-1} model[13),14)] gives physically interesting results like dynamical mass generation and confining longe range forces. Superficially there seems to be a discrepancy between this approach and the instanton gas method[6),14)]. However taking account of the natural regularization imposed by considering the theory at a finite temperature T it has been found[15)] that there is no contradiction in a region where the two approximation schemes are supposed to be valid. It would be interesting to study further the relation between these two schemes taking account of the complete classical structure of the model.

Related to this point, it has recently been shown[16] the $1/n$ expansion of σ-models in two-dimensional Minkowski space can be directly related to the problem of finding solutions of the classical equations of motion fulfilling a certain boundary condition. It may very well turn out that this feature is also present in four-dimensional non-abelian gauge theories. For this, as well as other reasons, it should be interesting to try to generalize the technique developed in section 2 for finding classical solutions to the case of four-dimensional quaternionic and non-abelian gauge theories, both in Euclidean and Minkowski space.

REFERENCES

1) J. Eells and J.H. Sampson, Amer. Journ. Math. 86 (1964) 109.

2) J. Eells and L. Lemaire, Bull. London Math. Soc. 10 (1978) 1.

3) J. Barbosa, Trans. Am. Math. Soc., vol. 210 (1975) 75.

4) M.J. Borchers and W.D. Garber, Comm. Math. Phys. 72 (1980) 77.

5) H. Eichenherr, Nucl. Phys. B146 (1978) 215;
E. Cremmer and J. Scherk, Phys. Letters 74B (1978) 341;
V. Golo and A. Perelomov, Phys. Letters 79B (1978) 112.

6) A.M. Din, P. di Vecchia and W.J. Zakrzewski, Nucl. Phys. B155 (1979) 447.

7) B. Berg and M. Lüscher, Comm. Math. Phys. 69 (1980) 57.

8) A.M. Din and W.J. Zakrzewski, CERN-TH-2722 (1979) (Lett. Nuovo Cim. to appear);
A.M. Din and W.J. Zakrzewski, Nucl. Phys. B168 (1980) 173.

9) A.M. Din and W.J. Zakrzewski, LAPP-TH-17 (1980).

10) A.M. Din and W.J. Zakrzewski, LAPP-TH-21 (1980).

11) V. Glaser and R. Stora, private communication.

12) J.L. Richard and A. Rouet, Marseille preprint 80/P.1191 (1980).

13) A. D'Adda, P. di Vecchia and M. Lüscher, Nucl. Phys. B146 (1978) 63 and
Nucl. Phys. B152 (1978) 125.

14) E. Witten, Nucl. Phys. B149 (1979) 285.

15) I. Affleck, Harvard preprint HUTP-80/A004 (1980).

16) A. Jevicki and H. Levine, Harvard preprint HUTP-80/A017 (1980); Brown preprint
BROWN-HET-418 (1980).

HIGGS REPRESENTATIONS FOR SO(n) GAUGES THEORIES

G. Girardi,
LAPP, Annecy-le-Vieux, France

A. Sciarrino,
Istituto di Fisica Teorica dell'Università
Napoli - I.N.F.N. Sezione di Napoli

and

P. Sorba,
LAPP, Annecy-le-Vieux, France

A B S T R A C T

A characterization of $SO(n)$ representations containing a vector stabilized by subgroups of the type $S[O(n-p) \times O(p)]$ or $SU(k) \times U(1)$ or $SU(k)$ if $n = 2k$ is given. Large use is made of the Gelfand-Zetlin basis, which yields operative methods of some interest in gauge models.

With the attempts at grand unification, particle physics witnesses an intense activity on large gauge symmetry groups. Indeed if we believe that current energy physics is well described by a gauge theory both for weak and electromagnetic interactions $(SU(2) \times U(1))$ and strong interaction ($SU(3)$ colour) we are tempted to embed them into a larger gauge symmetry group, which thereby will unify the coupling constants. The most popular and the simplest candidates in that respect are $SU(5)$ and $SO(10)$[1]. However one may wish to go beyond this grand unification and try to understand the family problem i.e. how it comes that we observe 3 replications of (u, d, e, ν_e). It has also been proposed that symmetry breaking is dynamically realized through superstrong "technicolour" interactions. The unification of all this physics obviously calls for larger and larger groups. For instance an attempt to include the family group in the theory was presented by Gell-Mann, Ramond and Slansky[2] using the $SO(4n+2)$ groups such as $SO(18)$ and $SO(22)$. Likewise embedding of technicolour interaction in a unified theory would be realized with symmetry groups like $SO(14)$.

In this context the aim of this cortribution is to present some operative methods to deal with orthogonal groups. Physically once we choose a symmetry group G, we break the symmetry down to a given (physically realized?) subgroup H : this can be achieved using definite representations of G containing a vector stabilized by H. Necessary and sufficient conditions in the case of $G = SU(n)$ have been given elsewhere[3], here we will concentrate on $G = SO(n)$. In the case where $H = S[O(n-p) \times O(p$ we prove a theorem making extensive use of the Gelfand-Zetlin[4] bases which are very efficient for the study of the chain $SO(n) \supset SO(n-1) \supset \ldots \supset SO(2)$. By this method and also by the use of Schur functions we also obtain characterizations of the $SO(2n)$ representations with a vector of little group $SU(n)$ or $SU(n) \times U(1)$, with explicit calculation of the eigenvalues of the $U(1)$ generator (a relevant information for quantum number attribution).

<u>Basic facts about $O(n)$ and $SO(n)$ groups</u>

Consider $O(n)$ ($SO(n)$) and their covering groups in order to include the spin representations. Then any irreducible representation of $O(n)$ - $n = 2p$ <u>or</u> $2p+1$ - is characterized by p positive ordered numbers - integers (tensor representation) or half-integers (spin representation) - m_j, $1 \leq j \leq p$, which are the components of the highest weight of the representation; we denote the representation $(m_1, \ldots, m_p)$.

Let us now restrict $O(n)$ to $SO(n)$:

i) $n = 2p$: then the representation $(m_1, \ldots, m_p)$ of $O(2p)$ splits into two irreducible $SO(n)$ representations: $(m_1, \ldots, m_p) + (m_1, \ldots, -m_p)$ if $m_p \neq 0$. In the case $m_p = 0$ $(m_1, \ldots, m_{p-1}, 0)$ remains irreducible in $SO(2p)$. It is interesting to note that for $n = 4\nu+2$, the 2 $SO(n)$ representations just obtained are complex conjugate one of each other while for $n = 4\nu$ they are distinct, real or quaternionic real.

ii) $n = 2p+1$: in this case the irreducible representations of $O(2p+1)$ are always irreducible representations of $SO(2p+1)$.

After this reminders, let us present a summary of our results.

Little group of the type $S[O(n-p) \times O(p)]$ in $SO(n)$

One can prove, with the help of the G.Z. basis, a general theorem allowing to select the $SO(n)$ irreducible representations containing a vector invariant under a maximal subgroup of the form $S[O(n-p) \times O(p)]$, $p > 1$, or $SO(n-1)$.

Proposition 1: The irreducible $SO(n)$ representations $(m_1, \ldots, m_k)$ with $n = 2k$ or $n = 2k+1$, which contain a vector stabilized by a subgroup $S[O(n-\ell) \times O(\ell)]$, $\ell > 1$ or $SO(n-1)$ fall in the following classes:

i) $SO(2k) \downarrow S[O[2(k-p)] \times O(2p)]$ with: $k \geq 2p$

$(m_1, \ldots, m_{2p}, 0, \ldots)$

ii) $SO(2k) \downarrow S[O[2(k-p-1) + 1] \times O(2p+1)]$ with: $k \geq 2p+1$

$(m_1, \ldots, m_{2p+1}, 0, \ldots)$

iii) $SO(2k+1) \downarrow S[O[2(k-p)] \times O(2p+1)]$ with: $k \geq 2p$

$(m_1, \ldots, m_{2p+1}, 0, \ldots)$

iv) $SO(2k+1) \downarrow S[O[2(k-p)] \times O(2p+1)]$ with: $2p \geq k$

$(m_1, \ldots, m_{2(k-p)}, 0, \ldots)$

For all these four cases, the m_i's are all odd or all even (0 included).

v) $SO(n) \downarrow SO(n-1)$

$(m_1, 0, \ldots)$ with m_1 even or odd.

Let us note that all the allowed representations correspond to integer values of the m_i, $i = 1, 2, \ldots k$, i.e. that there does not exist any spin representation (m_i half-integer) satisfying such a property.

The proof of this proposition actually consists in the construction of a vector invariant under - and only under - the subgroup $S[O(n-\ell) \times O(\ell)]$ in an irreducible representation R of $SO(n)$. We can distinguish two steps: i) solution of the G.Z. basis vectors in R invariant under $SO(n-\ell)$ if $n \geq 2\ell$. ii) determination of the condition on R allowing to form a linear combination of G.Z. vectors satisfying i) and stabilized under $SO(\ell)$. Because $S[O(n-\ell) \times O(\ell)]$ is a maximal subgroup of $SO(n)$, such a non-zero vector invariant under H in R, if it exists, cannot have a bigger invariant subgroup in $SO(n)$; therefore H will be exactly its little group.

Little group of the type $U(n)$ or $SU(n)$ in $SO(n)$

Another maximal subgroup in $SO(2n)$ which is especially interesting is the group $U(n)$. Let us in particular recall that the $SO(10)$ model of grand unified theory is an extension of the $SU(5)$ model, the latter being sometimes called the minimal model.

It is possible to prove a theorem similar to the previous one in the case of subgroups $S[O(n-p) \times O(p)]$ in $O(n)$, i.e. to characterize the irreducible representations of $SO(2n)$ admitting a vector stabilized by $SU(n) \times U(1)$ or $SU(n)$ in $SO(2n)$

Proposition 2: The irreducible $SO(2n)$ representations $(m_1, m_2, \ldots, m_n)$ which contain a vector the little group of which is $SU(n)$ or $U(n)$ are such that:

i) if $n = 2k+1$:

$SO(4k+2) \downarrow SU(2k+1) \times U(1)$

$$(m_1, m_2, \ldots, m_{2k-1}, m_{2k}, 0) \qquad \text{with: } \begin{array}{l} m_1 = m_2 \\ m_3 = m_4 \\ \vdots \\ m_{2k-1} = m_{2k} \end{array}$$

$SO(4k+2) \downarrow SU(2k+1)$

$$(m_1, m_2, \ldots, m_{2k-1}, m_{2k}, m_{2k+1}) \qquad \text{with: } \begin{array}{l} m_1 = m_2 \\ m_3 = m_4 \\ \vdots \\ m_{2k-1} = m_{2k} \\ m_{2k+1} \neq 0 \end{array}$$

ii) if $n = 2k$

$SO(4k) \downarrow SU(2k) \times U(1)$

$$(m_1, m_2, \ldots, m_{2k-3}, m_{2k-2}, 0, 0) \qquad \text{with: } \begin{array}{l} m_1 = m_2 \\ \vdots \\ m_{2k-3} = m_{2k-2} \end{array}$$

$SO(4k) \downarrow SU(2k)$

$$(m_1, m_2, \ldots, m_{2k-1}, m_{2k}) \qquad \text{with: } \begin{array}{l} m_1 = m_2 \\ \vdots \\ m_{2k-3} = m_{2k-2} \\ m_{2k-1} = m_{2k} \neq 0 \end{array}$$

(Note that in the first and third cases, a certain number of pairs $m_{2k-1} = m_{2k}, \ldots$ can also be zero).

This proposition can be proved either by using the method of King[5)] for reducing a representation of O(2n) with respect to U(n), or once more, with the help of the G.Z. framework. King's techniques are based on the properties of the characters of the O(n) groups, developed by Littlewood[6)] and require the introduction of different tools such as Schur functions.

These 2 propositions enable one to select the suitable representations containing the scalar Higgs field or technicolour condensates to achieve a well defined breaking of an SO(n) gauge symmetry down to maximal subgroups like $S[O(n-p) \times O(p)]$ and $SU(n/2) \times U(1)$. Let us note that the generalization of these theorems to the case of reducible representations $R = \bigoplus_{i=1}^{n} R_i$, with R_i irreducible, is obvious, the resulting stabilizer of a vector $\phi = (\phi,\ldots,\phi_n)$ with $\phi_i \in R_i$ being the intersection: $\mathrm{stab}\,(\phi) = \bigcap_{i=1}^{n} \mathrm{stab}\,(\phi_i)$.

As a final remark, let us note that it is possible, using King's techniques for O(n) groups, to deduce the decomposition of the spinor representations of SO(2n) with respect to SU(n) representations.

For instance, in the case of SO(4n+2) the reduction with respect to SU(2n+1) of the fundamental spinor representations is, in terms of SU(2n+1) Young tableaux:

$$\begin{cases} (1/2,\ldots\ldots,\ 1/2) = 1^0 + 1^2 + 1^4 + \ldots + 1^{2n} \\ (1/2,\ldots\ldots,-1/2) = 1 + 1^3 + 1^5 + \ldots + 1^{2n+1} \end{cases}$$

each one contains an SU(2n+1) singlet 1^0 and 1^{2n+1},

while for SO(4n) containing SU(2n):

$$\begin{cases} (1/2,\ldots\ldots,\ 1/2) = 1^0 + 1^2 + 1^4 + \ldots + 1^{2n} \\ (1/2,\ldots\ldots,-1/2) = 1 + 1^3 + 1^5 + \ldots + 1^{2n-1} \end{cases}$$

the first one contains 2 SU(2n) singlets while the second does not.
(1^k denotes the completely antisymmetric Young tableau with k boxes in one column).

REFERENCES

1) For reviews on the subject see for example:
S.L. Glashow, Cargese Summer Lectures 1979, HUTP preprint 79/A059;
F. Wilczek, Lepton-Photon Conference at Fermilab, 1979;
D.V. Nanopoulos, "Protons are not forever", in High Energy Physics in the Einstein Centenial Year, p. 91 (Ed. by A. Perlmutter, F. Kransz and L. Scott) Plenum Press (1980).

2) M. Gell-Mann, P. Ramond and R. Slansky, Rev. Mod. Phys. 50 (1978) 721.

3) Ph. Combe, A. Sciarrino and P. Sorba, Nucl. Phys. B159 (1979) 452 and Proceedings of the VIIIth International Colloquium on Group Theory, Israel (1979).

4) I.M. Gel'fand and M.L. Zetlin, Dokl. Akad. Nauk SSSR 71 (1950) 825;
I.M. Gel'fand, R.A. Minlos and Z.Ya. Shapiro, "Representations of Rotation and Lorentz Groups", p.353, Pergamon Press, 1963.

5) R.C. King, J. Phys. A: Math. Gen. 8 (1975) 429.

6) D.E. Littlewood, "The Theory of Group Characters", Clarendon Press 1950.

7) G. Girardi, A. Sciarrino and P. Sorba, preprint LAPP-TH-20, May 1980.

MULTIVORTEX CONFIGURATIONS IN THE ABELIAN HIGGS MODEL.

Laurence Jacobs
Instituto de Física
Universidad Nacional Autónoma de México
Apdo. Postal 20-364
México 20, D. F., México

In the past few years, our knowledge of the non-perturbative content of Quantum Field Theory has been greatly enriched. One of the main sources of this knowledge has come from the study of the underlying *classical* field theories. Quantization about solutions of the classical field equations has led to the discovery of a variety of unsuspected phenomena [1].

An outstanding question that must be eventually answered is whether or not the confinement of quarks is a property of the models of current interest- the non-abelian gauge theories. Perhaps the most promising approach to this question comes from the study of the lattice version of these theories. Indeed, encouraging results in this direction are already at hand [2]. Another approach, which is physically very appealing starts from the premise that the mechanism of color confinement is analogous to that of the trapping of magnetic flux in a superconductor. The existence and properties of such vortex configurations in a model theory - the Abelian Higgs model- is the subject of this talk [3]. The results which I will now describe, obtained in colloboration with Claudio Rebbi, show that multivortex configurations exist in this model; moreover, they show that, depending upon the value of a single, dimensionless, coupling constant, λ, two vortices attract or repel each other. For a critical value, $\lambda_c = 1$, however, the interaction energy vanishes, pointing to the existence of a large class of static solutions to the field equations. Since these equations are identical to those of the Ginzburg-Landau theory of superconductivity [4] our results are also relevant in this field.

The model is described by the Lagrangian density

$$L = \frac{1}{2}\,|D_\mu\phi|^2 + c(|\phi^2|-c_0{}^2)^2 - \frac{1}{4}F_{\mu\nu}^2 , \qquad (1)$$

where ϕ is a complex scalar field. In terms of the gauge potential, A_μ,

$$\begin{aligned} D_\mu \phi &= (\partial_\mu - i\,e\,A_\mu)\,\phi , \\ F_{\mu\nu} &= \partial_\mu A_\nu - \partial_\nu A_\mu . \end{aligned} \qquad (2)$$

Being interested in configurations symmetrical under translations in one direction, say $\mu = 3$, the potential energy density becomes

$$E = \frac{c_0 \pi}{e} \, \mathcal{E} \tag{3}$$

with

$$\mathcal{E} = \frac{1}{2\pi} \int dz \, d\bar{z} \, [\, |(\partial - iA)\, \phi|^2 + |(\bar{\partial} - i\bar{A})\, \phi|^2$$

$$+ 2|\, \bar{\partial}A - \partial\bar{A}\, |^2 + \frac{1}{8}\, \lambda \, (|\phi|^2 - 1)^2 \,], \tag{4}$$

where we have rescaled fields and coordinates according to

$$x^i \to (ec_0)^{-1} x^i, \; \phi \to c_0\, \phi \;, \; A_i \to c_0\, A_i, \tag{5}$$

and, since now $A_3 = 0$, have introduced complex quantities

$$A \equiv \frac{1}{2}\, (A_1 - iA_2) \;, \; z \equiv x + iy, (\partial \equiv \partial/\partial z) \;. \tag{6}$$

The parameter remaining in Eq. (4), λ , is related to the original parameters through

$$\lambda^2 = \frac{8c_4}{e^2} \;. \tag{7}$$

Finiteness of $\mathcal{E}$ requires

$$\lim_{|z|\to\infty} |\phi| = 1, \quad \lim_{|z|\to\infty} (\partial - i\, A)\, \phi = 0, \tag{8}$$

therefore we must have, as $|z|\to\infty$,

$$A \to -i\, \partial\, m\, \phi + \theta\, (\frac{1}{z}) \tag{9}$$

and $\quad \phi \to e^{i\, \chi\, (\theta)}$.

From continuity, since (9) applies in some fixed direction, we must have

$$\chi\, (\, \theta + 2\pi\,) = \chi(\theta) + 2\pi\, N, \quad N \in Z. \tag{10}$$

The total magnetic flux through the plane, Φ is then given by Gauss' law:

$$\Phi = -\frac{i}{e}\int dz\, d\bar{z}\, (\partial\bar{A} - \bar{\partial}A)$$

$$= \frac{1}{e} \lim_{|z|\to\infty} \oint (Adz + \bar{A}\, d\bar{z})$$

$$= \frac{1}{e} \oint d\chi = \frac{2\pi}{e}\, N. \tag{11}$$

Notice that N is also given by

$$N = -\frac{i}{2\pi} \oint_{\gamma} d \ \ln \phi \ , \tag{12}$$

where the contour encloses all zeros of ϕ. Thus, finite energy configurations divide the space of fields into classes, labelled by the integer N. Within each class, configurations are homotopically equivalent. Hence the designation of N as a topological number. If the zeros of ϕ occur at the points w_i, and there are n^+ points where $\phi\ (z-w_i^+) = 0$ and n^- points where $\phi(\bar{z}-\bar{w}_i^-) = 0$, $N = n^+ - n^-$ and the configuration describes (n^-) n^+ (anti-)vortices located at (w_i^-) w_i^+.

The field equations for stationarizing E are

$$(\partial - iA)(\bar{\partial} - i\bar{A})\,\phi + (\bar{\partial} - i\bar{A})(\partial - iA)\,\phi - \frac{1}{4}\lambda^2\,\phi\,(\phi\bar{\phi} - 1) = 0\ , \tag{13}$$

$$4\,\partial\,\bar{\partial}A - 4\,\partial^2 A - i\,\bar{\phi}\,\partial\phi + i\phi\,\bar{\partial}\,\phi - 2\,A\,\bar{\phi}\,\phi = 0.$$

The complexity of these equations has prevented an analytic solution for general configurations. However, for configurations of N superimposed vortices (say at r=0) the rotationally symmetric Ansatz (vorticity N),

$$\phi = e^{iN\theta}\,f(r) \qquad A = -\frac{i}{2z}\,N\,a(r), \tag{14}$$

simplifies Eqs. (13) to

$$f'' + \frac{1}{r}\,f' - \frac{N^2}{r^2}(a-1)^2 - \frac{\lambda^2}{2}\,f(f^2-1) = 0$$

$$a'' - \frac{1}{r}\,a' - (a-1)\,f^2 = 0\ . \tag{15}$$

The boundary values a=f=1 are reached asymptotically as

$$f(r) = 1 + \mathcal{O}\,(e^{-\lambda r}), \qquad a(r) = 1 + \mathcal{O}(e^{-r})\ . \tag{16}$$

Thus λ measures the relative rate of decay of the matter field and the gauge field. For the special case $\lambda=1$ discussed in more detail below, a series solution of Eqs. (15) has been given [5]. Alternatively, a constrained varia - tional computation, valid for all λ, which forms the basis for the study of separated vortices has also been performed [6,7]. I shall not have space here to review the details of this computation and will only mention the results. The variational analysis of Ref. [6] done for N = 1, 2 shows that

$$\Delta E\ (\lambda) \equiv E(N = 2,\lambda) - 2E\ (N = 1,\lambda),$$

the interaction energy is negative for $\lambda < 1$ and positive for $\lambda > 1$. Moreover, to the accuracy of the numerical analysis, $\Delta E\ (\lambda = 1) = 0$. Thus, a configuration of two superimposed vortices has a lower energy than that of two asymptotically separated vortices when $\lambda < 1$; the opposite being true for $\lambda > 1$. This suggests that the interaction regime is separated by $\lambda = 1$. To investigate this point further, a calculation for separated vortices is needed. Before describing the results of such a computation, let us mention the existence of bounds on $E(\lambda,N)$. They are [6],

$$E\ (\lambda \leqslant 1,\ N) \geqslant \lambda|N| + \frac{(1-\lambda)}{\pi} < |(\partial - iA)\phi|^2 + |(\bar{\partial} - i\bar{A})\ \phi|^2 > \ \geqslant \lambda|N| \tag{17}$$

and

$$E\ (\lambda \geqslant 1) \geqslant |N| + \frac{\lambda^2 - 1}{8\pi} < (|\phi|^2 - 1)^2 > \ \geqslant |N|. \tag{18}$$

Notice that, except for $\lambda=1$, the bounds cannot be saturated with non-trivial fields. At $\lambda=1$, however, the second term in Eqs. (17,18) vanishes and both bounds bive $E(N,\lambda=1) \geqslant |N|$. It is worthwhile to go through the derivation of the critical bound [8].

Integrating by parts the first term in the expression for E, Eq. (4),

$$\begin{aligned} &\int dz\, d\bar{z}\ [(\partial - iA)\ \phi\ (\bar{\partial} - i\bar{A})\bar{\phi} \\ &= \int dz\, d\bar{z}\ [(\bar{\partial} - i\bar{A})\ \bar{\phi}(\partial - iA)\ \bar{\phi}\ ] \\ &- i \int dz\, d\bar{z}\ (\partial\bar{A} - \bar{\partial}A)\ \bar{\phi}\ \phi\ ; \end{aligned} \tag{19}$$

adding and subtracting $-\frac{i}{2\pi}\int dz\, d\bar{z}\ (\partial\bar{A} - \bar{\partial}A) = N$, the last three terms can be combined into a square, giving

$$E = \frac{1}{\pi}\int dz d\bar{z}\ \{\ |(\bar{\partial} - i\bar{A})\ \phi|^2 + [\tfrac{1}{4}(|\phi|^2 - 1) - i(\partial\bar{A} - \bar{\partial}A)]^2\} + N. \tag{20}$$

Hence E is bounded below by N, the bound being saturated if and only if

$$(\bar{\partial} - i\bar{A})\ \phi = 0 \quad , \tag{21-a}$$

$$\partial\bar{A} - \bar{\partial}A + \frac{i}{4}\ (|\phi|^2 - 1) = 0. \tag{21-b}$$

All solutions of the first-order equations given above clearly solve the field equations, Eqs. (13), but, of course, the converse is not true [9]. To analyze Eqs. (21) further, fix the gauge by setting $A = i\partial\psi$, $\psi \in R$

$(\partial\bar{A} + \bar{\partial}A = \partial_\mu A^\mu = 0)$. Eq. (21-a) then reads

$$(\bar{\partial} - \bar{\partial}\psi)\,\phi = e^{\psi}\,\bar{\partial}\,(e^{-\psi}\,\phi) = 0 \tag{22}$$

which is merely the statement that the function $f = e^{-\psi}\,\phi$ is an analytic function of z. Inserting $\phi(z,\bar{z}) = e^{\psi(z,\bar{z})}\,f(z)$ into (21-b), and redefining $\psi = \chi - \frac{1}{2}\ln(f\bar{f})$ we finally arrive at

$$\partial\,\bar{\partial}\,\chi = \frac{1}{8}\,(e^{2\chi} - 1)\,, \tag{23}$$

with boundary conditions following from those of ϕ: χ should vanish at infinity and diverge logarithmically at the location of the vortices (of course, Eq. (23) should be written more correctly with a sum of δ-sources on the right-hand side).

The analysis of Ref. [6] for two vortices held at a fixed, finite distance shows that vortices attract or repel each other at all distances depending on whether λ is smaller or greater than one. Again, to the accuracy of the approximation, the interaction energy vanishes at $\lambda = 1$. Since at $\lambda=1$ the bound is saturated if Eq. (23) is satisfied, we were led to conjecture that solutions with arbitrary N do indeed exist. This conjecture was supported later by the observation [10] that, if a solution to Eq. (23) exists for some N, it depends only upon 2N parameters (the expected degeneracy for N non-interacting objects in the plane). Finally, in a very recent investigation, the conjecture was proven to be true [11].

In conclusion, the properties of $\lambda = 1$ vortices have been completely understood. For $\lambda \neq 1$, although an analytic solution has not been found, the numerical results of Ref. [6] have led to a good understanding of the properties of vortices in this case as well.

The task of finding explicit solutions of Eq (23) still remains an interesting challenge.

References.

1. There are many reviews on the role played by classical solutions. See, for example, R. Jackiw, Rev. Mod. Phys. 49 (1977).
2. M. Creutz, L. Jacobs, and C. Rebbi; Phys. Rep. C (to appear).
3. The relevance of vortices for particle physics was first suggested by H. B. Nielsen and P. Olesen, Nucl. Phys. B.61 , 45 (1973).
4. V. L. Ginzburg and L. D. Landau, Zh. Eksp. Teor. Fiz. 20, 1064 (1950).
5. H. J. de Vega and F. A. Schaposnik, Phys. Rev. D 14, 1100 (1976).
6. L. Jacobs and C. Rebbi, Phys. Rev. B 19, 4486 (1979).
7. See also C. Rebbi in the proceedings fo the Canadian Mathematical Society Summer Research Institute on Gauge Theories, Springer-Verlag (to appear).

8. This bound was first derived by L. Kramer, Phys. Rev. B 3, 3821 (1971) and has since been rediscovered several times.

9. A fact which has been, however, often overlooked in the literature.

10. E. Weinberg, Columbia University preprint, (1979).

11. C. H. Taubes, Harvard University preprint, (1979).

QUARK CONFINEMENT IN FIELD THEORIES WITH DISCRETE GAUGE SYMMETRY Z(3)

Herbert M. Ruck

Nuclear Science Division, Lawrence Berkeley Laboratory,
University of California, Berkeley, CA. 94720,USA

Field theories with discrete gauge symmetry Z(N) are simple examples of abelian vector field theories with the property to confine fermions. Consider the field theoretical model[1] of two scalar fields $\phi_1(t,x),\phi_2(t,x)$ defined on R^2 with the metric $g_{\mu\nu}=\mathrm{diag}(+1,-1),\mu,\nu=0,1$. The number of scalar fields I consider to be equal to the numbers of "colors". The action defining the model is:

$$S(\phi_1,\phi_2) = \int dt dx(\tfrac{1}{2}|\partial_\mu \phi_1|^2 + \tfrac{1}{2}|\partial_\mu \phi_2|^2 - V(\phi_1,\phi_2)) \qquad (1)$$

with the potential

$$V(\phi_1,\phi_2) = \lambda(\phi_1^2 + \phi_2^2)^2 - \nu(\phi_1^3 - 3\phi_1\phi_2^2) - \mu(\phi_1^2 + \phi_2^2) - \gamma \, , \qquad (2)$$

λ,ν,μ are positive coupling constants and γ is a subtraction constant to make the potential positive definite $0\leq V$.

The action is invariant under a Z(3) gauge transformation of the fields ($\theta=2\pi/3$, n=0,1,2):

$$\phi_1 \to \phi_1\cos\theta n - \phi_2\sin\theta n, \qquad \phi_2 \to \phi_1\sin\theta n + \phi_2\cos\theta n \, . \qquad (3)$$

The potential (2) has three absolute minima-the vacuum states of the model-positioned symmetrically on a circle with radius $\phi_V=[3\nu + (9\nu^2+32\lambda\mu)^{\frac{1}{2}}]/8\lambda$ with an angle of 120° between them: $\Omega_1=(\phi_V,0),\Omega_2=(-\frac{1}{2}\phi_V, +\frac{1}{2}\sqrt{3}\phi_V)$ and $\Omega_3=(-\frac{1}{2}\phi_V,-\frac{1}{2}\sqrt{3}\phi_V)$. The physical vacuum is a coherent superposition of the three vacuum states Ω_1,Ω_2 and Ω_3.

Recently I found a classical solution of the field equations connecting pairwise distinct minima of the potential[2]. The Euler-Lagrange equations

$$\begin{aligned}(\partial_t^2 - \partial_x^2)\phi_1 &= -4\lambda(\phi_1^2 + \phi_2^2)\phi_1 + 3\nu(\phi_1^2 - \phi_2^2) + 2\mu\phi_1 \\ (\partial_t^2 - \partial_x^2)\phi_2 &= -4\lambda(\phi_1^2 + \phi_2^2)\phi_2 - 6\nu\phi_1\phi_2 + 2\mu\phi_2\end{aligned} \qquad (4)$$

admit time independent "soliton" solutions provided that the third coupling constant μ is a function of λ and ν:

$$\mu = 9\nu^2/4\lambda \qquad (5)$$

then $\phi_V=3\nu/2\lambda$ and $\mu=\lambda\phi_V^2$. The soliton trajectory in $\vec{\phi}$ space is a straight line joining the vacuum states.

For instance the soliton moving between the vacua Ω_1 and Ω_2 is:

$$\phi_1(x)=\tfrac{1}{4}\phi_V + (3/4)\phi_V\tanh[(3\lambda/2)^{\frac{1}{2}}\phi_V(x-x_0)], \quad \phi_2(x)=(\phi_V-\phi_1(x))/\sqrt{3}. \qquad (6)$$

The other two solutions are obtained by a rotation according to Eq.(3).

The soliton in x-coordinate can be transformed into an instanton solution in imaginary time $x \to it$, or into a 3-dim. soliton wave in R^4 by $x \to \vec{k}\vec{x}/|\vec{k}|$.

The energy density $E_S(x)$ of a soliton is localized around x_0:

$$E_S(x) = (9/16)\lambda\phi_V^4 \left[\cosh((3\lambda/2)^{\frac{1}{2}}\phi_V(x - x_0))\right]^{-4}, \tag{7}$$

and the total energy of the soliton at rest is

$$\hat{E}_S = \int_{-\infty}^{+\infty} dx E_S(x) = (3\lambda/2)^{\frac{1}{2}}\phi_V^3 . \tag{8}$$

When moving the center of the soliton x_0 along the x-axis with speed v the energy of the soliton increases, its mass being precisely $\hat{E}_S$ (8):

$$\hat{E}_S(v) = \hat{E}_S(v=0) + \tfrac{1}{2}\hat{E}_S(v=0)v^2. \tag{9}$$

The energy density is therefore the matter distribution of the system. The extension of the system is:

$$\langle x^2\rangle = \int_{-\infty}^{+\infty} dx E_S(x)x^2 / \int_{-\infty}^{+\infty} dx E_S(x) = 0.1432\ \lambda^{-1}\phi_V^{-2}. \tag{10}$$

Quantum corrections $\tilde{\phi}(t,x)$ to the classical solutions Eq.(6):

$$\phi_1(t,x) = \phi_1(x) + \tilde{\phi}(t,x);\ \phi_2(t,x) = \phi_2(x) - \tilde{\phi}(t,x)/\sqrt{3} \tag{11}$$

obey the differential equation:

$$(\partial_t^2 - \partial_x^2)\tilde{\phi}(t,x) = -2\lambda\phi_V^2 \{3[\tanh(3\lambda/2)^{\frac{1}{2}}\phi_V(x-x_0)]^2 - 2\}\ \tilde{\phi}(t,x) \tag{12}$$

where $\tilde{\phi}^2$ and $\tilde{\phi}^3$ terms have been neglected on the right hand side of Eq. (12). With $\tilde{\phi}(t,x)=\varphi(x)\exp(iE_Q t)$ Eq.(12) can be solved in closed form[3]. There is a single energy level $(a=(1-3(13-3\sqrt{17})/8)^{\frac{1}{2}}=0.8737...)$:

$$E_Q = (2\lambda)^{\frac{1}{2}} a\phi_V , \tag{13}$$

and the wave function is $(b=(\sqrt{17}-3)/2=0.56...)$

$$\varphi(x) = [\mathrm{sech}((3\lambda/2)^{\frac{1}{2}}\phi_V(x - x_0))]^b \tanh((3\lambda/2)^{\frac{1}{2}}\phi_V(x - x_0)). \tag{14}$$

The field energy is increased mainly by the time derivative:

$$E_Q(x) = (4/6)|\partial_t\tilde{\phi}(t,x)|^2 + ... = (4/3)\lambda a^2\phi_V^2\ \varphi^2(x). \tag{15}$$

The total energy correction to the classical value Eq.(8) equals:

$$\hat{E}_Q = \int_{-\infty}^{+\infty} dx E_Q(x) = 0.7839\ \sqrt{\lambda}\ \phi_V. \tag{16}$$

The real physical problem is the motion of fermions coupled to the scalar fields ϕ_1, ϕ_2 with Z(3) selfinteraction. In order to construct a Z(3) invariant interaction between fermions and the $\vec{\phi}$ field we ought to introduce two Dirac fields ψ_1 and ψ_2 according to the two "colors". Then the Z(3) symmetric interaction is given by the Lagrangian density:

$$L_{int} = gj_1\phi_1 + gj_2\phi_2 \tag{17}$$

with $g > 0$ a coupling constant, and the fermionic currents:

$$j_1 = \bar{\psi}_1\psi_1 - \bar{\psi}_2\psi_2 , \qquad j_2 = - \bar{\psi}_1\psi_2 - \bar{\psi}_2\psi_1 \tag{18}$$

The transformation law for fermions is the same as Eq.(3) written for ψ ($\theta=2\pi/3$, $n=0,1,2$) :

$$\begin{aligned} \psi_1 &\rightarrow \psi_1\cos\theta n - \psi_2\sin\theta n, & \bar{\psi}_1 &\rightarrow \bar{\psi}_1\cos\theta n - \bar{\psi}_2\sin\theta n; \\ \psi_2 &\rightarrow \psi_1\sin\theta n + \psi_2\cos\theta n, & \bar{\psi}_2 &\rightarrow \bar{\psi}_1\sin\theta n + \bar{\psi}_2\cos\theta n . \end{aligned} \tag{19}$$

Because the base space is R^2, the ψ_1,ψ_2 fermion wave functions are two-component spinors and the kinetic part of the Lagrangian is that of the Thirring model[4]. Neglecting possible selfinteraction of the fermions the action is:

$$\begin{aligned} S(\psi_1,\psi_2,\phi_1,\phi_2) = \int dtdx \,[\, & i\bar{\psi}_1\gamma_\mu\overleftrightarrow{\partial}_\mu\psi_1 - m\bar{\psi}_1\psi_1 + i\bar{\psi}_2\gamma_\mu\overleftrightarrow{\partial}_\mu\psi_2 - m\bar{\psi}_2\psi_2 \\ & + g(\bar{\psi}_1\psi_1 - \bar{\psi}_2\psi_2)\phi_1 - g(\bar{\psi}_1\psi_2 + \bar{\psi}_2\psi_1)\phi_2 \,], \end{aligned} \tag{20}$$

where $\mu,\nu=0,1$; $\gamma_0= -\sigma_2$, $\gamma_1= i\sigma_1$, $\{\gamma_\mu,\gamma_\nu\}=2g_{\mu\nu}$; $\bar{\psi}_k=\psi_k^+\sigma_2$ $(k=1,2)$; $\sigma_1,\sigma_2,\sigma_3$ are the Pauli matrices and $\overleftrightarrow{\partial} = \frac{1}{2}\,(\overrightarrow{\partial} - \overleftarrow{\partial})$.

I propose to solve the equations of motion for ψ_1,ψ_2 with ϕ_1,ϕ_2 being the soliton solutions Eq.(6):

$$\begin{aligned} -i\gamma_\mu\partial_\mu\psi_1 - m\psi_1 + g(\, \psi_1\phi_1 - \psi_2\phi_2 \,) &= 0 \\ -i\gamma_\mu\partial_\mu\psi_2 - m\psi_2 + g(-\psi_2\phi_1 - \psi_1\phi_2 \,) &= 0 . \end{aligned} \tag{21}$$

I claim that the system of coupled differential equations (21) has an unique normalizable solution for zero mass fermions ($m=0$) corresponding to a bound state with energy :

$$E_F = \tfrac{1}{2}g\phi_V , \tag{22}$$

determined by the coupling constant g and the value of the vacuum field ϕ_V. The fermion solution is explicitly:

$$\begin{aligned} \psi_1(t,x) &= N \begin{pmatrix} 1 \\ \exp(-i5\pi/6) \end{pmatrix} \exp(-iE_Ft - H(x)) \\ \psi_2(t,x) &= i\sigma_3\, \psi_1(t,x) , \end{aligned} \tag{23}$$

where σ_3 is the third Pauli matrix and $H(x)$ a positive definite function:

$$H(x) = g(2\lambda)^{-\frac{1}{2}} \ln[\cosh((3\lambda/2)^{\frac{1}{2}}\phi_V(x - x_0))]. \tag{24}$$

N is the normalization constant, that I choose to be defined by $\int dx|j_1| = 1$ (from the second of Eqs.(23) follows $\bar{\psi}_1\psi_1=-\bar{\psi}_2\psi_2$) then:

$$N^{-2}= \tfrac{1}{4}\int_0^\infty dx\exp(-2H(x)) = (96\lambda)^{-\frac{1}{2}}\phi_V^{-1}\, B(\tfrac{1}{2},(2\lambda)^{-\frac{1}{2}}g) , \tag{25}$$

where B is the beta-function.

The energy density of the fermions is strongly peaked around the center of the soliton:

$$E_F(x) = 2g\phi_V \, N^2 \exp(-2H(x)), \tag{26}$$

and the total energy

$$\hat{E}_F = g\phi_V \tag{27}$$

is twice the value (22) because there are two fermions occupying the same energy level.

The current densities j_1 and j_2 are proportional to each other:

$$j_1(x) = -2N^2\exp(-2H(x)) := j(x)\cos\pi/3, \quad j_2(x) = j(x)\sin\pi/3. \tag{28}$$

Due to the restricted space in this proceedings I will not discuss the uniqueness of the solution Eq.(23) here, but give a proof elsewhere. For $m \neq 0$ at all energies and $m = 0$ but $E_F < \frac{1}{2}g\phi_V$ or $E_F > \frac{1}{2}g\phi_V$ the solution diverges at infinity $\lim \psi \to \infty$ as $x \to \infty$. This leaves the solution (23) alone as a normalizable solution. ψ_1, ψ_2 given in Eq. (23) describe confined fermions in the sense that (i) the wave function vanishes at infinity and (ii) the wave functions are algebraically related, i.e. the two fermions cannot be separated.

The fact that there are no free (plane wave) solutions of the fermions bound to the soliton can be seen intuitively from the action (20). For plane waves $\psi_1 = u_1\exp(ikx)$ and $\psi_2 = u_2\exp(ikx)$ the currents (18) become constants. Then the part of the action $\int dx(c_1\phi_1 + c_2\phi_2)$ $(j_1=c_1,\ j_2=c_2)$ diverges, and the contribution of this solution to the functional path integration is nil. In contrast in scalar φ^4 theory (opposed to the vector theory presented) a hypotetical coupling to the soliton $\bar{\psi}\psi\tanh(x - x_0)$ escapes such an argument because $\int dxc \tanh(x-x_0)$ vanishes for any x_0.

The confinement is essentially due to the algebraic structure imposed by the Z(3) symmetry reflected in the form of potential energy $V(\phi_1,\phi_2)$ and the coupling between fermion- and the scalar fields, and less to the low dimensionality of the system.

Imagine a particle in R^2 made of two "colored" fermions and the "colored" boson field $\vec{\phi}$ then we make the following concluding remarks:

(i) The average current density of the fermions vanishes:

$$\Sigma\, j_1 = \Sigma\, j_2 = 0 \tag{29}$$

where the sum is taken over all solitons. The average "color" charge is therefore zero.

(ii) The second component ψ_2 is algebraically related to ψ_1 and vanishes if ψ_1 does. This is an example of a "color singlet" state, maintained by the presence of two Dirac fields of opposite "color".

(iii) Both fermions and bosons are massless. The mass of the particle is made of the field energy of both types of fields. The ratio between the soliton energy of the $\vec{\phi}$ field and the interaction energy of the ψ field is (compare Eqs. (8) and (27)):

$$E_S / E_F = (3/2)\, \lambda g^{-1} \phi_V^2 \; . \qquad (30)$$

(iv) The fermions are asymptotically free. This is a trivial consequence of the coupling constants λ, ν, μ having a dimension of mass squared in R^2 [5].

(v) At high temperature the one soliton solution becomes statistically insignificant. Consider a multi-center soliton. If the centers of the individual soliton peaks are further apart than 4: $|x_i - x_j| > 4$, $\forall\; i,j$; the solitons are almost free. This describes the splitting of a single particle into several particles at high temperature. The fermions will be trapped in each produced soliton. At very high temperature where a random field dominates the fermions may be liberated.

This work has been supported by the U.S. Department of Energy.

[1] F. Constantinescu and H.M. Ruck, J. Math. Phys. 19, 2359 (1978), and Ann. Phys. 115, 474 (1978)

[2] H.M. Ruck, Nucl. Phys. B167, 320 (1980)

[3] P.M. Morse and H. Feshbach, "Methods of Theoretical Physics", Mc. Graw-Hill, New York (1953), p. 1651; M.A. Lohe, Phys. Rev. D20, 3120 (1979)

[4] W.E. Thirring, Ann. Phys. 3, 91 (1958)

[5] P. Becher and H. Joos, "1+1 -dimensional Quantum Electrodynamics as an Illustration of the Hypothetical Structure of Quark Field Theory", preprint DESY 77/43, July 1977.

PROLIFERATION OF EFFECTIVE FIELDS BY SPONTANEOUS SYMMETRY BREAKDOWN

Heinrich Saller
Max-Planck-Institut für Physik und Astrophysik
Föhringer Ring 6, 8000 München 40, (Fed. Rep. of Germany)

I. The Problem

experimental situation: huge proliferation of effective fields carrying few basic degrees of freedom (= gauged degrees of freedom)

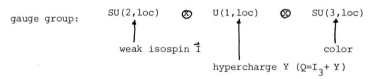

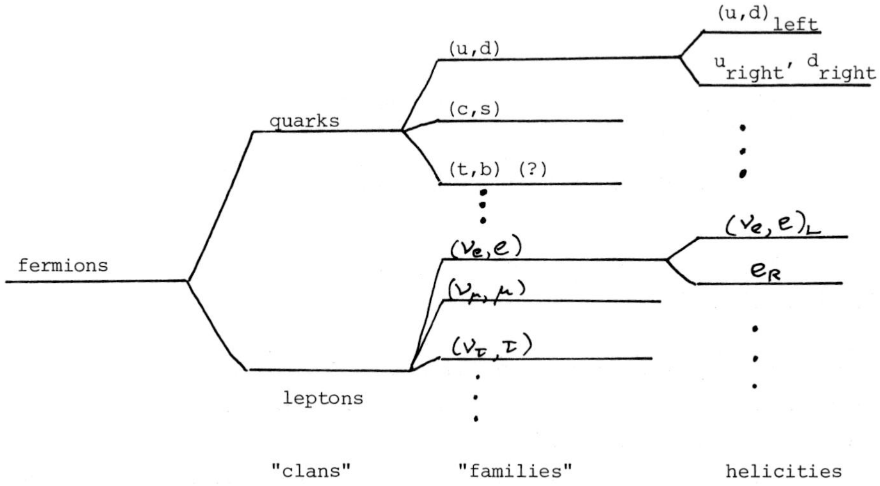

usual theoretical description:

gauge group: SU(2,loc) ⊗ U(1,loc) ⊗ SU(3,loc)

- SU(2,loc): weak isospin $\vec{I}$
- U(1,loc): hypercharge Y ($Q=I_3+Y$)
- SU(3,loc): color

questions:

- Why are there both leptons and quarks ?
- Why do many families exist ?
- Why does there exist the charge relation

$$\sum_{\text{quarks}} Q + \sum_{\text{leptons}} Q = 0 \qquad \text{in each family?}$$

(neutrality of hydrogen)

- Are the neutrinos massless ? If not, why is m_ν small ?
- Is the baryon number conserved ? If not, why is the proton so stable ?
- Why is helicity connected with isospin ?
  (left handed doublets, right handed singlets)
- Why do we have "multiple" nontrivial representations of the basic group
  SU(2) $\otimes$ U(1) $\otimes$ SU(3) even within one family ?
  Minimal nontrivial representation of SU(2) $\otimes$ U(1) $\otimes$ SU(3) given by

$$
\text{(I)} \begin{cases} \begin{pmatrix} \nu \\ e \end{pmatrix}_L & : \quad \text{SU(2) doublet, SU(3) singlet} \\[2ex] (d_b, d_r, d_g)_R & : \quad \text{SU(2) singlet, SU(3) triplet} \end{cases}
$$

or by

$$
\text{(II)} \quad \begin{pmatrix} u_b, u_r, u_g \\ d_b, d_r, d_g \end{pmatrix}_L \quad : \quad \text{SU(2) doublet, SU(3) triplet}
$$

Why are (I) and (II) basic fields ? Are they ?

Those question are tried to answer in the
unification approaches:

- "additive" unifications: large groups e.g. SU(5), SO(10)
  also "horizontal" (gauged) groups leading to SO(4n+2) etc.
  - What is so special about those groups ?
  - Do they answer the above questions or do they only parametrize the situation ?
    (Does SU(5) explain $\sum_{i \in \text{family}} Q_i = 0$?)
- "radical" unification: smaller building blocks (not necessarily particle interpretable) govern a simple basic dynamics and "compose" the effective fields.

II. An Example for Radical Unification: Reducing the Effective Fields in the Glashow Salam Weinberg (GSW) Model

II.1 The GSW Model

Symmetry group in GSW model: $U(1) \otimes SU(2,loc) \otimes U(1,loc)$

- $U(1)$ ↑ fermion number N_F
- $SU(2,loc)$ ↑ weak isospin $\vec{I}$
- $U(1,loc)$ ↑ hypercharge Y

effective fields:

	field	isospin I	hypercharge Y	fermion number N_F
$\mathbb{L}_\alpha(x)$:	left handed leptons $\binom{\nu}{e}_L$	1/2	- 1/2	1
$R(x)$:	right handed lepton e_R	O	- 1	1
$\vec{A}_\mu(x)$:	SU(2) gauge field	1, gauge	O	O
$e_\mu(x)$:	hypercharge gauge field	O	O, gauge	O
$\varphi_\alpha(x)$:	Higgs field $\langle\varphi\rangle = \lambda\binom{0}{1}$	1/2	1/2	O

Lagrangian:

$$
\begin{aligned}
\mathcal{L}^{eff} = & \ \mathbb{L}^* \sigma^\mu \left[\tfrac{i}{2} \overleftrightarrow{\partial}_\mu - \tfrac{\vec{\tau}}{2} \vec{A}_\mu + \tfrac{1}{2} e_\mu \right] \mathbb{L} \\
& + R^* \sigma^\mu \left[\tfrac{i}{2} \overleftrightarrow{\partial}_\mu \qquad + e_\mu \right] R \\
& - \frac{1}{4g^2} \vec{A}^2_{\mu\nu} - \frac{1}{4g'^2} e^2_{\mu\nu} \\
& + \left| \left(i\partial_\mu - \tfrac{\vec{\tau}}{2} \vec{A}_\mu - \tfrac{1}{2} e_\mu \right) \varphi \right|^2 + \frac{h}{2} \left[\varphi^* \varphi - \lambda^2 \right]^2 \\
& + \frac{m_e}{\lambda} \left[(R^* \mathbb{L}) \varphi^* + \varphi (\mathbb{L}^* R) \right]
\end{aligned}
$$

II.2 Puzzling together the Effective Fields

building blocks: needed for spin, for isospin, for hypercharge

	field	I	Y	N_F
$\Psi_\alpha(x)$:	left handed fermion doublet	1/2	O	1/2
$S_\alpha(x)$:	unitary SU(2) dressing operators	1/2	- 1/2	O
$U(x)$:	unitary U(1) dressing operator	O	- 1/2	1/2
$\vec{A}_\mu(x)$:	SU(2) gauge field	1, gauge	O	O

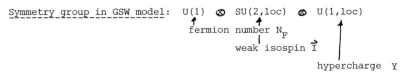

With the building blocks one can construct:

	field	I	Y	N_F
$\tilde{\Psi}_\alpha(x)$:	conjugated right handed fermion doublet	1/2	0	- 1/2
$U_\mu(x) = U i\overleftrightarrow{\partial}_\mu U^*$:	internal derivative of u(x)	0	0,gauge	0
$\vec{B}_\mu(x) = s^* \vec{\tau} i\overleftrightarrow{\partial}_\mu s$:	internal derivative of s(x)	1,gauge	0	0

representation of effective fields (up to constants):

$\mathbb{L}_\alpha(x) \cong U(x)\Psi_\alpha(x)$ (hypercharge dressed)

$\mathcal{R}(x) \cong U^3(x)\, s^{*\alpha}(x)\, \tilde{\Psi}_\alpha(x)$ (SU(2) frozen and hypercharge dressed)

$\varphi_\alpha(x) \cong \tilde{s}_\alpha(x)$ (non radial part of Higgs field is dressing operator)

$\vec{\mathcal{A}}_\mu(x)$ = combination of $\vec{A}_\mu(x), \vec{B}_\mu(x), U_\mu(x)$ } details literature

$\mathcal{C}_\mu(x)$ = combination of $\vec{A}_\mu(x), \vec{B}_\mu(x), U_\mu(x)$

deflated Lagrangian: inserting the representations of effective fields

$$\mathcal{L}^{eff} = \mathcal{L}^{eff}(\Psi, \vec{A}_\mu, U, s, U_\mu, \vec{B}_\mu)$$

II. Origin of the dressing operators

example Higgs model:

$$\mathcal{L}(\phi) = (\partial_\mu\phi)(\partial^\mu\phi^*) + \frac{h}{2}(\phi^*\phi - \lambda^2)^2$$

U(1) symmetry: $\phi(x) \mapsto \exp[-i\alpha]\,\phi(x), \quad \phi^*(x) \mapsto \exp[+i\alpha]\,\phi^*(x)$

U(1) symmetry broken: $\langle\phi(x)\rangle = \lambda$

decomposition
$$\phi(x) = \rho(x)\,U(x) = \lambda\, U(x) + \dots \quad , \quad U^* U(x) = 1$$

radial field: $\rho(x) = \sqrt{\phi^*\phi(x)} \mapsto \rho(x)$

azimuthal field: $U(x) = \sqrt{\phi(x)/\phi^*(x)} \mapsto \exp[-i\alpha]\,U(x)$

$U(x) = \exp[-i\,\theta(x)]$, $\theta(x)$ Goldstone field

U(x) unitary U(1) dressing operator

$\rho(x) = U^*(x)\,\phi(x)$ is built by building blocks U(x), $\phi(x)$ but not "bound" state.

Spontaneous symmetry breakdown leads (via Goldstone fields) to dressing operators connected with the broken group

dressing operators for GWS model via fermion condensation:

Cooper pair condensation of basic left handed doublet $\Psi(x)$ carrying $U_{N_F}(1) \otimes SU_I(2,loc)$ properties

$$\langle \psi\vec{\tau}\psi(x)\rangle = M^3 \begin{pmatrix}1\\ i\\ 0\end{pmatrix} \neq 0 \,, \quad \psi\vec{\tau}\psi(x) = M^3 u^2(x)\, s^*\vec{\tau}\tilde{s}(x) + \dots$$

breaks: $I_1, I_2, N_F + I_3$ } leads to dressing operators { $s(x)$, $u(x)$
leaves intact: $N_F - I_3$

hypercharge Y is given by the I_3 properties of the dressing operators

$$Q = I_3\,(\psi) + I_3\,(s,u) \quad , \quad Y = I_3\,(s,u)$$

II.4 From basic to effective Lagrangian

basic Lagrangian:

$$\mathcal{L}^{\text{basic}} = \psi^*\sigma^\mu\left[\tfrac{i}{2}\overleftrightarrow{\partial}_\mu - \tfrac{\vec{\tau}}{2}\vec{A}_\mu\right]\psi - \frac{G}{M^2}(\psi^*\sigma_\mu\psi)(\psi^*\sigma^\mu\psi) - \frac{1}{4g_0^2}\vec{A}_{\mu\nu}^{\,2}$$

(nonrenormalizability can be avoided (literature))

rearranging $\mathcal{L}^{\text{basic}}$ according to the condensation:

$$\psi\vec{\tau}\psi = M^3 u^2 s^*\vec{\tau}\tilde{s}(x) + \dots$$

$$\left.\begin{matrix}\psi^*\sigma_\mu\psi(x)\\ \psi^*\sigma_\mu\vec{\tau}\psi(x)\end{matrix}\right\} = \text{terms with } U_\mu(x),\ \vec{B}_\mu(x) \qquad + \dots \qquad \text{(literature)}$$

leading to

$$\mathcal{L}^{\text{basic}} = \mathcal{L}^{\text{basic}}(\psi, \vec{A}_\mu, u, s, U_\mu, \vec{B}_\mu)$$

A basic dynamics of a selfinteracting basic left handed SU(2) fermion doublet $\psi(x)$ with a SU(2) gauge field $\vec{A}_\mu(x)$ can lead via a $U(1) \otimes SU(2)$ breakdown to an effective dynamics as given by the GSW-model. To this end a rearrangement of the basic theory with respect to the broken symmetries has to be performed. This rearrangement leads to dressings of the basic field $\psi(x)$. The rearranged theory is linearized using effective fields $\mathbb{L}(x)$, $R(x)$ which incorporate different but well defined dressings.

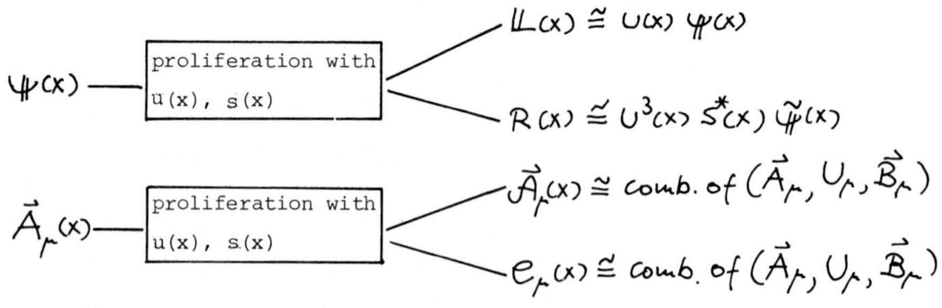

Literature: H.P. Dürr, H. Saller: Is Hypercharge in Weinbergs Model Redundant ?
Phys.Lett. 84B, 336 (1979).

H.P. Dürr, H. Saller: Unification of Isospin and Hypercharge in One Basic SU(2)
Nuov. Cim. 53A, 469 (1979).

EXPLICIT SOLUTIONS FOR THE SCALAR COUPLED YANG-MILLS SYSTEM IN MINKOWSKI SPACE

Luc Vinet
Centre de recherche de mathématiques appliquées
Université de Montréal,
Montréal, Québec, Canada H3C 3J7

The purpose of my contribution is to present exact solutions of the SU(2) Yang-Mills equations in the presence of an isotriplet of massless scalar fields in Minkowski space. These equations may be written in the following fashion:

$$*D*DA = [\Phi, D\Phi]$$

$$*D*D\Phi = \lambda|\Phi|^2\Phi$$

where A is the SU(2) gauge potential 1-form, D denotes the exterior covariant derivative and the scalar field Φ is taken in the adjoint representation of SU(2).

We have obtained solutions to these equations by demanding *a priori* that the fields be invariant under some space-time transformations group. The above equations are conformally invariant. We have used this fact to map the problem onto compactified Minkowski space, isomorphic to the U(2) group manifold and topologically equivalent to $S^3 \times S^1$. This we have done in order to have a well defined and convenient action of the conformal group (locally isomorphic to) SU(2,2).

Let u be a point in U(2). We have considered the following specific conformal actions on U(2): (i) left translations under $SU(2)_L$, $L_w(u) = wu$, $w \in SU(2)$, (ii) right translations under $SU(2)_R$, $R_w(u) = uw$, (iii) left (or right) translations under U(1), $L_\phi u = R_\phi u = e^{i\phi}u$. We looked for invariant fields under these or combinations of these group actions. The problem of characterizing invariant gauge fields had been dealt with by Harnad, Shnider and Vinet [1]. (See also J.Harnad's contribution to this colloquium.) Thus the methods were at hand to determine various symmetric Ansätze. Those were examined all happened to be special cases of the most general $SU(2)_L$-invariant configuration.

Upon insertion of these $SU(2)_L$-invariant fields, the Yang-Mills equations reduce to a system of 15 coupled non-linear second order differential equations for a set of 15 functions depending only on the S^1-angle. This was still a too difficult system to solve. We simplified further by imposing more symmetry on our fields. The natural thing was to require that the gauge field also be invariant under $SU(2)_R$ right translations; the number of unknown functions is then reduced to 4. Unfortunately and contrary to what happens in the absence of scalars, the field equations do not possess solutions with such a high symmetry. Relaxing the $SU(2)_R$-symmetry, in addition to the $SU(2)_L$-symmetry, we impose invariance under the action of U(1) on S^1.

This reduces the Yang-Mills equations to an algebraic system in 15 unknowns. In collaboration with J.Harnad, we had already obtained all solutions to the pure Yang-Mills equations with that symmetry [2]. From these configurations one can obtain solutions to the scalar-coupled system. More interestingly, we have been able to find $SU(2)_L \otimes U(1)_R \otimes U(1)$ invariant solutions that constitute interaction modes for some of the previously known solutions to the pure Yang-Mills equations. For some ranges of the coupling constant λ the solutions are real and their energy is always finite. We refer to ref. [3] for a fuller discussion of the properties of these solutions.

References

[1] J.Harnad, S.Shnider and L.Vinet, "Group Actions on Principal Bundles and Invariance Conditions for Gauge Fields", to appear in J.Math.Phys.

[2] J.Harnad and L.Vinet, Phys.Lett. *76B*, 589 (1978).

[3] L.Vinet, "Some Invariant Solutions to the Yang-Mills Equations in the Presence of Scalar Fields in Minkowski Space", in *Geometrical and Topological Methods in Gauge Theories* edited by J.Harnad and S.Shnider, Lecture Notes in Physics 129, Springer Verlag, New York (1980).

DYNAMICAL SUPERSYMMETRY SU(6/4) IN NUCLEI † *

Itzhak Bars**

J. W. Gibbs Laboratory of Physics, Yale University, New Haven, Ct. 06520, USA.

Abstract

A supersymmetric scheme for classifying the levels of certain odd and even nuclei in a single multiplet of SU(6/4) is presented. Sum rules relating the ground state energies of odd and even nuclei are obtained and shown to be experimentally confirmed by the available data. The scheme makes predictions that can be checked by future experiments. Thus, it appears that broken supersymmetry is relevant to Nature, at least in the nonrelativistic nuclear domain. Comments are included on possible relations to symmetries in particle physics.

† Invited talk presented at (1) Vth Johns Hopkins Workshop on Current Problems in High Energy Particle Theory, Bad Honef, West Germany, June 1980. (2) IXth International Colloquium on Group Theoretical Methods in Physics, Cocoyoc, Mexico, June 1980.

* Research Supported in Part by the U.S. Department of Energy under Contract Numbers EY-76-C-02-3075 and DE-AC02-76ER03074.

** A.P. Sloan Foundation Fellow

I will describe some recent work done in colloboration with F. Iachello and A.B. Balantekin on supersymmetry in Nuclear physics [1] . My talk will consist of the following three parts:

(i) Brief description of U(6) and Spin(6) dynamical symmetries in Nuclei [2,3]

(ii) A quick overview of representations of supergroups [4].

(iii) Supersymmetric scheme of nuclear levels based on SU(6/4), and supporting evidence.

U(6) and Spin(6) in nuclei

Arima and Iachello have proposed[2] a model of low lying levels based on the observation that protons and neutrons tend to pair in large nuclei mainly in $S = 0$ and $L = 0$, $L = 2$ states. Thus they give a description of even-even nuclei in terms of 6 dynamical bosons: one s boson ($L = 0$) and five d - bosons ($L = 2$), which form a multiplet of the group U(6). In the first approximation no distinction is made between proton pairs and neutron pairs, but imporoved results are obtained by treating them separately. Let b_i represent the annihilation operator for these 6 pairs $b_o = s$ ($L = 0$) and $b_i = d_i$, $i = 1,2,..5$ ($L = 2$). Then one may assume a Hamiltonian which contains at most 2-body interactions in the form

$$H = H_o + \alpha_{ij} \, b^{+}_{i} b_{j} + \beta_{ij}{}^{kl} \, b^{+}_{i} \, b^{+}_{j} \, b_{k} \, b_{l} \tag{1}$$

Here H_o is unknown but is taken invariant under U(6) and the coefficients are chosen so as to conserve angular momentum according to the L=0 and L=2 assignments given above. All the levels of a given nucleus will be degenerate if α and β are consistent with U(6) symmetry. Thus, the splitting of the levels within a nucleus is achieved by choosing appropriate values of α_{ij} and β_{ijkl} that break U(6) all the way down to the angular momentum subgroup SO(3). There are only three possible chains of breaking that respect the $L = 0$ and $L = 2$ assignments of the pairs.

These are:

$$U(6) \begin{cases} \nearrow SU(5) \rightarrow SO(5) \rightarrow SO(3) \\ \longrightarrow SU(3) \rightarrow SO(3) \\ \searrow SO(6) \rightarrow SO(5) \rightarrow SO(3) \end{cases} \tag{2}$$

where SU(3) is realized in the 6 - dimensional representation.

Iachello finds examples of nuclei for which the parameters α and β are so special as to realize only one of these chains. For such nuclei the Hamiltonian is re-written in terms of the linear and quadratic casimir operators that appear in the chain. Then H is immediately diagonalizable by appropriately choosing the labels of the states. As an example consider the SO(6) chain. The states can be labeled as

$$|U(6)\text{ labels}\ ;\ SO(6)\text{ labels}\ ;\ SO(5)\text{ labels}\ ;\ So(3)\text{ labels}\ > \tag{3}$$

where a representation of U(6) has 5 labels (N_1 , N_2 , N_3 , N_4 , N_5) , a representation of SO(6) has 3 label { σ_1 , σ_2 , σ_3 } , SO(5) has 2 labels { τ_1 , τ_2 } and finally SO(3) has the usual angular momentum labels (J, M). The only U(6) representations that can be constructed with the bosonic creation operators are the completly symmetric representations (N,0,0,0,0). For these the energy eigenvalue takes the form

$$E = \varepsilon_0 + \varepsilon_1 N + \varepsilon_2 N^2 - \frac{A}{4} \{\sigma_1(\sigma_1+4) + \sigma_2(\sigma_1+2) + \sigma_3^{\,2}\}$$
$$+ \frac{B}{6}[\tau_1(\tau_1+3) + \tau_2(\tau_2+1)] + C\,J(J+1) \tag{4}$$

In this representation the allowed values of σ_i , τ_i and J must clearly depend on N as given in ref. [2], so that all energy splittings are described only by the three parameters A, B, C. A typical level pattern is shown in Fig. 1

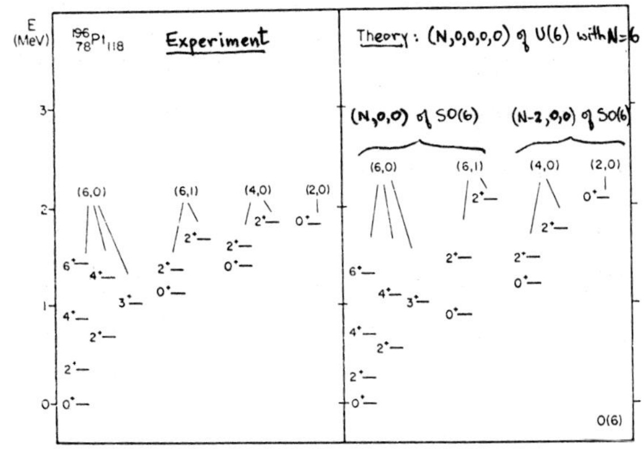

An example of a spectrum with O(6) symmetry: $^{196}_{78}Pt_{118}$.

Fig. 1

It is possible to find many examples of nuclei that fit this pattern. In many cases there is enough data on excited states to test the existence of the second representation of SO(6) (N-2,0,0). Quantitatively the level splittings are described within (10 - 20))% by this model. The scheme was further successfully tested by measuring the E2 radiative transitions and comparing to calculations in which the transition operator is taken as a generator of the group. Many examples of nuclei that fit the other symmetry breaking chains (SU(5) and SU(3)) were also given in ref. [2]. The regions of validity of these breaking schemes are given in Fig. 2. where the closed proton and neutron shells 28, 50, 82, 126 are indicated. It is seen that the SU(5) chains occur in the neighborhood of closed shells, SU(3) chains are found in the center of regions bound by closed shells and finally the SO(6) chains are located near the corners of such regions. A further interesting information is the fact that

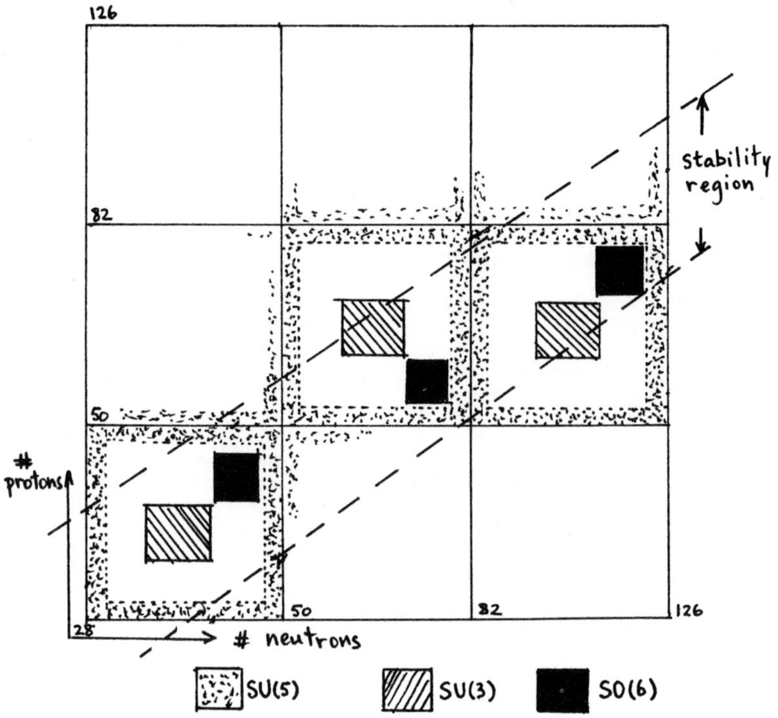

FIG. 2

the shapes of nuclei are corrolated with the type of breaking [5]. SU(5) chains correspond to spherical nuclei, S U(3) chains to ellipsoids with cylindrical symmetry and SO(6) chains to general ellipsoids.

The above results apply to nuclei with an even number of protons and an even number of neutrons. Iachello extended [3] the SO(6) scheme to Spin6 ($\cong$ SU(4)) which includes spinor representations of SO(6). He was then able to find examples of odd nuclei, with ground states in J = 3/2, whose energy levels were described by the same formula (eq. (4) without the ε_1 and ε_2 terms) which applies to even SO(6)-type nuclei. The parameters A, B, C turned out to be identical for both even and odd nuclei provided they were located in the same black regions indicated in Fig. (2). This could be taken as a signal that these even and odd nuclei may belong to the same multiplet of a supergroup.

Representations of Supergroups

A systematic construction of certain representations of supergroups were recently obtained in collobration with B. Balantekin[4] . I will give here just enough information to explain the supersymmetric model of nuclei. More information on representations, characters, dimensions and casimir invariants of the supergroups SU(N/M) and Osp(N/2M) can be found in ref.(4).

The supergroup SU(N/M) has two fundamental representations

$$\xi_A = \begin{pmatrix} \phi_a \\ \psi_\alpha \end{pmatrix} \qquad \tilde{\xi}_A = \begin{pmatrix} \psi_a \\ \phi_\alpha \end{pmatrix} \tag{5}$$

which are called class I and class II respectively. In ξ_A we have N bosons ϕ_a and M fermions ψ_α while in $\tilde{\xi}_A$ we have N fermions ψ_a and M bosons ϕ_α. Both of these basis vectors transform with the same group element

$$U = \begin{pmatrix} \overset{N}{A} & \overset{M}{B} \\ C & D \end{pmatrix} \tag{6}$$

where A and D are bosonic parameters while B and C are fermionic parameters. I will discuss here only the class I representations that can be construct by taking direct products of ξ_A type graded vectors. The supertrace is defined as

$$\mathrm{Str}\, U = \mathrm{Tr}\, A - \mathrm{Tr}\, D \tag{7}$$

Higher dimensional class I representations are associated with super Young tableaux

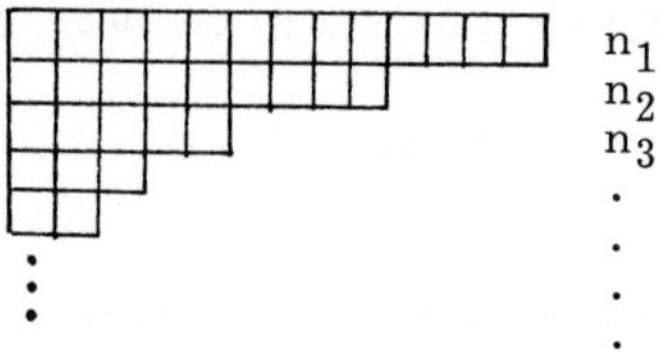

where $(n_1, n_2, \ldots)$ stand for the number of boxes in the first row, second row, etc.. To see the difference between an ordinary Young tableau and a super Young tableau consider the simplest example of 2 boxes which corresponds to the super-symmetrization of 2 fundamental representations $\xi_A^{(1)}$ and $\xi_B^{(2)}$.

$$\square\square \longleftrightarrow \xi_{AB} = \xi_A^{(1)}\,\xi_B^{(2)} + \xi_A^{(2)}\,\xi_B^{(1)} \tag{8}$$

Note that the indices A, B appear in the same order while the wavefunctions are interchanged in the two terms. The various bosonic and fermionic components may be extracted by specializing to the indices $A = a$ or $A = \alpha$ and <u>then</u> putting the wavefunctions in the same order.

$$\text{Bose}\begin{cases} \xi_{ab} = \phi_a^{(1)}\,\phi_b^{(2)} + \phi_b^{(1)}\,\phi_a^{(2)} \rightarrow \tfrac{1}{2}N(N+1) \\ \\ \xi_{\alpha\beta} = \psi_\alpha^{(1)}\,\psi_\beta^{(2)} - \psi_\beta^{(1)}\,\psi_\alpha^{(2)} \rightarrow \tfrac{1}{2}M(M-1) \end{cases} \tag{9}$$

$$\text{Fermi}\left\{ \xi_{a\alpha} = \phi_a^{(1)}\,\psi_\alpha^{(2)} + \psi_\alpha^{(1)}\,\phi_a^{(2)} = \xi_{\alpha a} \rightarrow MN \right.$$

Note that ξ_{ab} is symmetric $\xi_{\alpha\beta}$ is antisymmetric as a result of the bose and fermi properties of ϕ and ψ . The number of components have been indicated, from which we learn the total number of bosons $= \frac{1}{2}N(N+1) + \frac{1}{2}M(M-1)$ and fermions $= MN$.

The representation of the group on this basis is obtained by transforming both $\xi_A^{(1)}$ and $\xi_B^{(2)}$ by the group element U_{AB} . That is

$$\xi'_{AB} = \sum_{A',B'} [U_{AA'}\,\xi_{A'}^{(1)}\,U_{BB'}\,\xi_{B'}^{(2)} + U_{AA'}\,\xi_{A'}^{(2)}\,U_{BB'}\,\xi_{B'}^{(1)}]. \tag{10}$$

The order of the various factors is again important since both ξ_A and $U_{AA'}$ contain anticommuting numbers. By pulling all $U_{AA'}\,U_{BB'}$ to the left and taking into account appropriate minus signs one can write

$$\xi'_{AB} = \sum_{A',B'} U_{AB,\,A'B'}\,\xi_{A'B'} \tag{11}$$

This identifies the representation of the group $U_{AB,A'B'}$.

The character of the representation is defined by

$$K_{(2,0,0,0,..)} = \sum_{A,B} (-1)^{g(A)+g(B)} \; U_{AB,AB} \tag{12}$$

where a minus sign is inserted each time A or B corresponds to a fermionic index $A = \alpha$ etc. Doing the algebra one finds

$$K\,(2,0,0,0,..) = \tfrac{1}{2}\;[\;(\mathrm{Str}\;U)^2 + \mathrm{Str}(U^2)\;] \tag{13}$$

If we had done the corresponding algebra for an ordinary SU(N) group we would have found a similar expression except for traces replacing supertraces. This turns out to be a general observation. In fact the class I characters of SU(N/M) can be obtained from those of SU(N+M) if we first write SU(N+M) characters in terms of traces of the fundamental representation and then replace traces by supertraces. Using this rule we first can write the character for the super Young tableau with a single row of n boxes 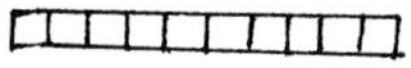

$$K_n\,(U) = \int_0^{2\pi} d\phi \; \frac{e^{-in\phi}}{\mathrm{Sdet}(1 - e^{i\phi}U)} \tag{14}$$

the superdeterminant of a matrix M is defined in terms of the supertrace as

$$\mathrm{S\,det}\;(M) = \exp[\mathrm{Str}\;\ln M] \tag{15}$$

An alternative form is

$$\begin{aligned} K_n(U) &= \frac{(\mathrm{Str}\;U)^n}{n!} + \frac{(\mathrm{Str}\;U)^{n-2}}{(n-2)!}\,\mathrm{Str}\left(\frac{U^2}{2}\right) + \dots + \mathrm{Str}\left(\frac{U^n}{n}\right) \\ &= \sum_{n = k_1+2k_2+..+nk_n} \frac{\left[\mathrm{Str}\left(\frac{U}{1}\right)\right]^{k_1}}{k_1!}\;\frac{\left[\mathrm{Str}\;\frac{U^2}{2}\right]^{k_2}}{k_2!} \;\cdots\; \frac{\left[\mathrm{Str}\;\frac{U^n}{n}\right]^{k_n}}{k_n!} \end{aligned} \tag{16}$$

In terms of Kn we can write an arbitrary character in the form of a determinant

$$K_{(n_1,\,n_2,n_3\,\dots)} = \begin{vmatrix} K_{n_1} & K_{n_2-1} & K_{n_3-2} & \cdots \\ K_{n_1+1} & K_{n_2} & K_{n_3-1} & \cdots \\ K_{n_1+2} & K_{n_2+1} & K_{n_3} & \\ \vdots & \vdots & & \ddots \end{vmatrix} \tag{17}$$

where the rank of the determinant is equal to the number of nontrivial rows in the super Young tableau.

The dimension of a class I representation is obtained by evaluating the character of the special group element

$$J = \left(\begin{array}{c|c} 1 & 0 \\ \hline 0 & -1 \end{array} \right) \tag{18}$$

with (+1) for bosonic indices and (-1) for fermionic indices.

$$D(n_1, n_2 \ldots) = K(n_1, n_2 \ldots)(J) \tag{19}$$

For example the dimension of the fundamental representation will be given by Str J = N + M as expected. explicit forms for the dimensions are founds in ref. (4).

The approach was also generalized to OSp(N/2M) supergroups for which character and dimension formulas were given[4] . Finally Casimir invariants are obtained from the character by differentiating with respect to the bosonic and fermionic "rotation angles" and then setting the "angles" equal to zero[4] .

The decomposition[4] of a representation of SU(N/M) in terms of representations of the subgroups SU(N) x SU(M) is most important for the physical application I will describe shortly. The representation of interest corresponds to a single row super Young tableau with n boxes. Its decomposition is given by

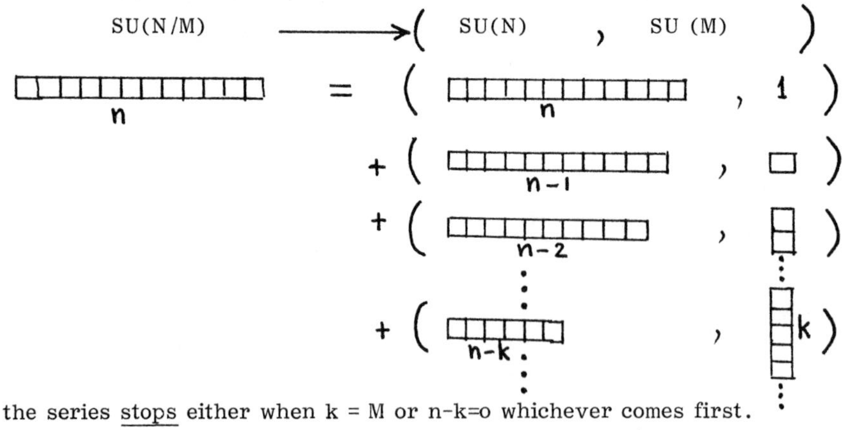

(20)

where the series stops either when k = M or n-k=o whichever comes first.

Supersymmetry Scheme for Nuclei

I will now describe the model of ref. [1[which concentrates on even and odd nuclei that fit the Spin6 scheme described above. Such nuclei would occur when either a proton or neutron shell with spin 3/2 is the next shell to be filled. For example consider $^{190}_{76}Os_{114}$ with 76 protons and 114 neutrons.

If we add 6 more protons we reach close shell 82. The first 4 protons fill a $J = 3/2$ shell and the next 2 fill a $J = \frac{1}{2}$ shell. In the nucleus $^{190}_{76}\mathrm{Os}_{114}$ the total number of protons plus neutrons needed to reach closed shells for both is 6 + 12 = 18 nucleons. We interpret these as holes and furthermore we assume that they form 9 pairs. Thus $^{190}_{76}\mathrm{Os}_{114}$ will be described by 9 dynamical U(6)-bosons of type b_i, $i = o,1,2,\ldots,5$ as described in the first part of the talk. We begin to fill the $J = 3/2$ shell by adding 1 proton. This corresponds to $^{191}_{77}\mathrm{Ir}_{114}$. This nucleus may now be described by 8 dynamical U(6)-bosons (8 pairs of holes) plus one proton in the $J = 3/2$ shell. The $J = 3/2$ proton will be assigned to the fundamental representation of the SU(4) group. Next we add one more proton to the $J = 3/2$ shell reaching $^{194}_{78}\mathrm{Pt}_{114}$ which will be described by 7 U(6)-bosons and 2 U(4) - fermions. We continue the process until the $J = 3/2$ shell is filled. Altogether we have 5 nuclei: ($^{190}_{76}\mathrm{Os}_{114}$, $^{191}_{76}\mathrm{Ir}_{114}$, $^{192}_{78}\mathrm{Pt}_{114}$, $^{193}_{79}\mathrm{Au}_{114}$, $^{194}_{80}\mathrm{Hg}_{114}$). From the description given above it is clear that these must belong the following representations of SU(6) x SU(4) with n = 9

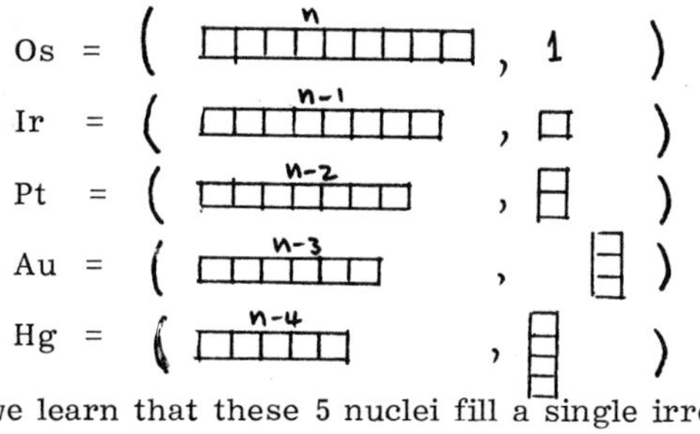

(21)

Comparing to eq. (20) we learn that these 5 nuclei fill a single irreducible representation of the super group SU(6/4) and correspond to the super Young tableau with n = 9. Note that Ir and Au are odd nuclei and have only fermionic states while Os, Pt, Hg are even nuclei and have only bosonic states. They all belong the same supersymmetric representation.

Similarly we can construct new multiplets for isotopes of these nuclei. By adding two more neutrons to the above nuclei (removing one pair of holes) we obtain a new supermultiplet with n=8 consisting of the nuclei ($^{192}_{76}\mathrm{Os}_{116}$, $^{193}_{77}\mathrm{Ir}_{116}$, $^{194}_{78}\mathrm{Pt}_{116}$, $^{195}_{79}\mathrm{Au}_{116}$, $^{196}_{80}\mathrm{Hg}_{116}$)

To make contact with Iachello's previous scheme[3] of Spin6 ($\cong$ SU(4)) we now decompose SU(6/4) according to the following chain of subgroups

$$SU(6/4) \rightarrow SU(6) \times SU(4) \rightarrow SO(6) \times SU(4) \rightarrow \text{Spin } 6 \qquad (22)$$

where the final Spin6 is the diagonal SU(4) subgroup that occurs in the direct product of the bosonic SO(6) ($\cong$ SU(4)) and the fermionic SU(4). Thus the representations in eq. (20) and (21) must be further decomposed according to this group chain. This is described in more detail in refs. [4] and [1]. The final decomposition of Spin 6 to Spin(3)

$$\text{Spin } 6\ (\cong SU(4)) \rightarrow \text{Spin } 5\ (\cong Sp(4)) \rightarrow \text{Spin } 3\ (\cong SU(2)) \qquad (23)$$

follows the lines of ref. [3].

At this point we can completely identify any state of the supermultiplet as follows

$$|\,\text{state}\,\rangle = \left|\begin{matrix}SU(6/4)\\ \text{labels}\end{matrix};\ \begin{matrix}SU(6)\\ \text{labels}\end{matrix};\ \begin{matrix}SU(4)\\ \text{labels}\end{matrix};\ \begin{matrix}SO(6)\\ \text{labels}\end{matrix};\ \begin{matrix}\text{Spin } 6\\ \text{labels}\end{matrix};\ \begin{matrix}\text{Spin}5\\ \text{labels}\end{matrix};\ J,\ M\right\rangle \qquad (24)$$

Once the SU(6/4) representation is chosen it puts restrictions on the remaining labels. In particular we are interested only in the representations of SU(6/4) described above. The Hamiltonian is assummed to be a function of the linear and quadratic casimir operators of these groups. Then the energy eigenvalue of any state takes the form

$$\begin{aligned}
E &= E_o\,(n,m) + \Delta E \\
E_o\,(n,m) &= E_o + (E_1 n + E_2 n^2) + (E_3(n-m) + E_4(n-m)^2) \\
&\qquad + (E_5\, m + E_5\, m^2) \\
\Delta E &= E_7\,[\Sigma_1(\Sigma_1+4) + \Sigma_2(\Sigma_2+2) + \Sigma_3^{\,2}\,] \qquad (25)\\
&\quad - \frac{A}{4}\,[\sigma_1(\sigma_{1+4}) + \sigma_2(\sigma_2+2) + \sigma_3^{\,2}] \\
&\quad + \frac{B}{6}\,[\tau_1(\tau_1 + 3) + \tau_2(\tau_2+1)] \\
&\quad + C\ J\,(J + 1)
\end{aligned}$$

Here ΔE determines the energy splitting <u>within</u> each nucleus and it should be compared to equ.(4), while $E_o\,(n,m)$ determines the energy splitting among <u>different nuclei</u>. The label n determines the SU(6/4) representation as discussed above, the label m is equal to the number of SU(4) fermions (m=o for Os, m=1 for Ir etc.), the other labels are determined in refs. [4] and [1].

We note that there are many more states than parameters (hundreds of states can be measured for this multiplet) so that the scheme makes many predictions. First note that by virtue of being in the same multiplet odd and even nuclei must be described by the same parameters. This was already experimentally observed as discussed in ref.[3]. Second the ΔE part of the spectrum was shown to work in ref. [3]. The new predictions of the supersymmetry scheme come from the form of E_0 (n,m) in which the casimir operators of SU(6) are automatically determined to be a function of n and m. This leads to Gell-Mann-Okubo type sum rules among the ground states of the nuclei in the supermultiplets. These were obtained in ref. [1] after elminating the unknown constants

$$\begin{aligned}
&(1) \quad {}^{190}Os - 2\,{}^{191}Ir + {}^{192}Pt = {}^{192}Os - 2\,{}^{193}Ir + {}^{194}Pt \\
&(2) \quad {}^{190}Os - 2\,{}^{191}Ir + 2\,{}^{193}Au - {}^{194}Hg = A \\
&(3) \quad {}^{192}Os - 2\,{}^{193}Ir + 2\,{}^{195}Au - {}^{196}Hg = A \\
&(4) \quad ({}^{192}Os - {}^{190}Os) - 3({}^{193}Ir - {}^{191}Ir) + 3({}^{194}Pt - {}^{192}Pt) - ({}^{195}Au - {}^{193}Au) = \tfrac{A}{2} \\
&(5) \quad {}^{193}Au - 3\,{}^{192}Pt + 3\,{}^{191}Ir - {}^{190}Os = \tfrac{A}{2}(n-3) + \tfrac{5}{3}B + 15\,C
\end{aligned} \qquad (26)$$

where $\frac{A}{4} \approx 80$ KeV, $\frac{B}{6} \approx 60$ KeV, $C \approx 10$ KeV were determined by examining excited states within a nucleus as in ref. [3].

There is now data to check the first relation while the rest are predictions. Here is the first physical test of supersymmetry. The numbers are taken from data in ref. [6]

$$\begin{aligned}
{}^{190}Os &= -38674 \text{ KeV} & {}^{192}Os &= -35850 \text{ KeV} \\
{}^{191}Ir &= -36672 \text{ KeV} & {}^{193}Ir &= -34499 \text{ KeV} \\
{}^{192}Pt &= -36256 \text{ KeV} & {}^{194}Pt &= -34733 \text{ KeV}.
\end{aligned} \qquad (27)$$

The left hand side of relation (1) in eq. (16) amounts to 1586 KeV, while the right hand side gives1585 KeV. The agreement is remarkable and much better than expected!!!

More data will become available soon to test the remaining relations on the spectrum. More tests are possible although they require an enlargement of the scheme to include interactions with external fields. For example if we assume that the transition operator for electromagnetic transitions (E_2) is a generator of the Spin 6 group then various transitions rates are proportional to Clebsh - Gordon coefficients. This leads to selection rules which already appear to be satisfied. Moreover the transition rate to the ground state is proportional to the label $(\sigma_1)^2$. We recall that σ_1 is completely determined by the supermultiplet and for various nuclei we have $\sigma_1 = n-m$ for m=o,1,2,3,4. Thus the ratios of these transition rates for the various nuclei will test the supersymmetric assignment of the levels. Further test are available from proton and deuteron scattering experiments such as

$$\begin{aligned} &(a) \quad p + A \rightarrow {}^3He + B \\ &(b) \quad d + A \rightarrow {}^3He + B \end{aligned} \tag{28}$$

where the nuclei A and B are members of the two multiplets mentioned above. The (a) and (b) types of reactions are indicated below

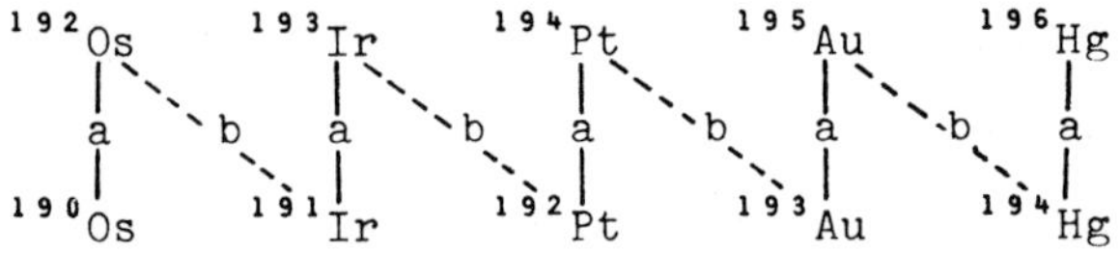

(29)

They all will be related to each other by Clebsh-Gordon coefficients[7] of the supergroup SU(6/4) according to the matrix elements

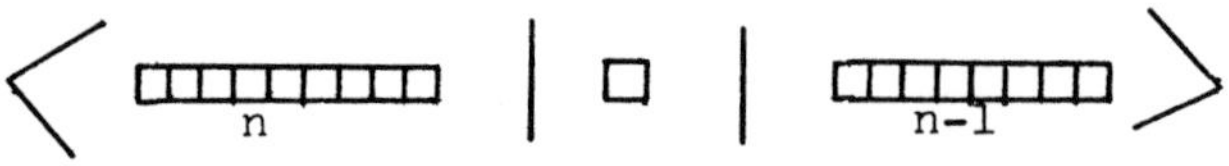

(30)

where the transition operator is assumed to belong to the fundamental representation of SU(6/4). These experiments are expected to be done in the near future.

Conclusions and Speculations

We have shown that broken supersymmetry occurs in Nature. This is the first signal that supersymmetry is physically relevant, at least in the nonrelativistic nuclear domain.

This result raises many interesting questions. First, how can we understand from a more fundamental point of view the occurence of such dynamical symmetries and supersymmetries? Second, is the supersymmetry in nuclei related to the supersymmetry postulated in particle physics? It is, of course, difficult to answer the second question without answering the first, but there are two possibilities: either the supersymmetry in nuclei is accidental and follows from the properties of nuclear forces without any underlying supersymmetry, or there is indeed a connection to some (presently unknown) supersymetric theory of particle physics and individual nucleons "know" about it although it was not discovered until now. We note that what we have found is broken supersymmetry. In particle physics also it is believed that supersymmetry is not exact, if it is there at all. Even in supergravity theories, the local supersymmetry must be broken spontaneously.

Leaving aside the question of how the dynamical symmetry arises, the fact is that it is there provided we look in certain regions of the nuclear chart. Let us compare this situation to particle physics where we have observed at low energies certain broken symmetries such as $SU(2)_W \times U(1)$ and others. The present formulation of particle physics in terms of gauge theories gives a fundamental significance to the underlying symmetries as well as to the gauge bosons, quarks, leptons etc. But, is it possible that the symmetries in particle physics are dynamical symmetries, valid only in a certain energy regime, and could be explained as dynamical symmetries which follow from some underlying more fundamental theory? We will not know until we try some such schemes. A proposal[8] has already been made along such ideas and it seems that such a possibility could be realized.

References

[1] A.B. Balantekin, I. Bars, F. Iachello, Yale preprint 1980, in preparation.

[2) A. Arima and F. Iachello, Am. Phys. (N.Y.) 99, 253 (1976)
A. Arima and F. Iachello, Am. Phys. (N.Y.) 111, 201 (1978)
A. Arima and F. Iachello, Phys. Rev. Lett. 40, 385 (1978).

[3] F. Iachello, Phys. Rev. Lett. 44, 772 (1980)

[4] A.B. Balantekin, I. Bars, Yale preprint YTP80-06, May 1980.

[5] F. Iachello, Yale preprint 1980, in preparation.

[6] A.H. Wapstra and N.B. Gove, Nuclear Data Tables 9, 265 (1971)

[7] S. Kuyucak (Yale) has calculated some of these coefficients.

[8] I. Bars and M. Gunaydin, Yale Preprint YTP 79-05, to be published in Phys. Rev. and YTP - 80 - 14, May 1980

GROUP THEORY OF THE IBA AND ITS APPLICATIONS

Octavio Castaños[+], Pedro Federman[++] and Alejandro Frank[+]

[+]Centro de Estudios Nucleares, UNAM. Apdo. Postal 70-543

[++]Instituto de Física, UNAM. Apdo. Postal 20-364

México 20, D. F., México.

A new approach to the problem of describing in a unified way the collective states observed in medium and heavy even-even nuclei was proposed recently by Arima and Iachello {1}. In this approach, collective states are constructed as states of a system of N bosons which can occupy two levels, a level with L=0 (s-boson) and another with L=2 (d-boson). This model is known as the Interacting Boson Approximation (IBA).

The total number of boson (N) is fixed in each nucleus by the number of nucleon pairs outside closed shells, so in this picture the valence nucleons are paired together with resultant angular momentum either L=0 or L=2. For example, for ${}^{154}_{62}Sm_{92}$ there are 12 protons and 10 neutrons outside the closed shells at nucleon numbers 50 and 82, giving rise to N=11 for this nucleus.

Introducing creation (η_{lm}) and annihilation (ξ_{lm}) boson operators with ℓ=0,2, satisfying the commutation relations

$$\left[\xi^{l'm'}, \eta_{lm}\right] = \delta_{l}^{l'}\,\delta_{m'}^{m} \quad ,$$

$$\left[\xi_{l'm'}, \xi_{lm}\right] = \left[\eta_{l'm'}, \eta_{lm}\right] = 0 \quad , \tag{1}$$

the most general boson Hamiltonian, involving one and two body interactions, can be written in the second quantized form:

$$H = \epsilon_s \bar{\eta}\bar{\xi} + \epsilon_d \sum_m \eta_m \xi^m$$

$$+ \frac{1}{2}\sum_{l_1 l_2}\sum_{l_1' l_2'}\sum_{L} \langle 2l_1\, 2l_2\; L \,|\, V_{12} \,|\, 2l_1'\, 2l_2'\; L\rangle \left[\left[\eta_{l_1} \times \eta_{l_2}\right]^L \times \left[\xi_{l_1'} \times \xi_{l_2'}\right]^L\right]^0_0 \quad , \tag{2}$$

where we have introduced the notation $\eta_{00} \equiv \bar{\eta}$, $\xi_{00} \equiv \bar{\xi}$, $\eta_{2m} \equiv \eta_m$ and $\xi_{2m} \equiv \xi_m$, and $[\eta_{l_1} \times \xi_{l_2}]^L$ denotes angular momentum coupling of ℓ_1 and ℓ_2 to a total L.

The five components η_m and the single $\bar{\eta}$ generate a linear vector space which can be used as a basis for the representations of the U(6) group, and because of relations (1) only the totally symmetric representations are different from zero. The wave functions are thus classified by the partition $[N]$ of U(6).

For $\epsilon \equiv \epsilon_d - \epsilon_s = 0$ and without interactions between the bosons, all states are degenerate. A finite value of ϵ and of the seven independent interactions:

$$
\begin{aligned}
a_L &= \langle d^2 L | V_{12} | d^2 L \rangle \ , \quad L = 0,2,4 \\
b &= \langle d s 2 | V_{12} | d s 2 \rangle \ , \\
c &= \langle s^2 0 | V_{12} | s^2 0 \rangle \ , \\
d &= \langle d^2 0 | V_{12} | s^2 0 \rangle \ , \\
e &= \langle d s 2 | V_{12} | d^2 2 \rangle \ ,
\end{aligned} \tag{3}
$$

breaks the degeneracy and gives rise to a definite spectrum.
For ϵ much bigger than the interactions between s and d bosons, the Hamiltonian is invariant under the separate transformations among the five components of the L=2 state. Thus the states are in this case characterized by the number of d bosons (n_d), and an unbroken U(5) symmetry emerges from the decomposition $U(6) \supset U(1) \times U(5)$. If the interactions between s and d bosons are no longer small compared to ϵ , the above decomposition no longer diagonalizes the IBA Hamiltonian. It turns out however, that the general s-d Hamiltonian can be expressed in terms of the Casimir operators of the three chains of groups {2}

$$
\begin{array}{cccc}
U(6) \supset & U(5) \supset & O(5) \supset & O(3) \ , \\
\hat{N} & \hat{n}_d & \hat{\Lambda}^2 & \hat{L}^2 \\[1em]
U(6) \supset & SU(3) \supset & O(3) & \ , \\
\hat{N} & \hat{G} & \hat{L}^2 & \\[1em]
U(6) \supset & O(6) \supset & O(5) \supset & O(3) \ , \\
\hat{N} & \hat{\Sigma}^2 & \hat{\Lambda}^2 & \hat{L}^2
\end{array} \tag{4}
$$

where underneath each group we have written the corresponding Casimir operator. When the Hamiltonian (2) can be expressed in terms of the operators associated with one of any of the above chains, it is said to exhibit a dynamical symmetry {1}. The form of the eigenvalues, and other spectroscopic quantities in each of these limits is then straightforward to derive {3}. It is remarkable that the three limiting situations, U(5), SU(3) and O(6) dynamical symmetries, have been verified experimentally to a very good approximation in particular regions of the nuclear table.
The question arises, however, whether the most general Hamiltonian for the model can successfully describe the properties of nuclei away from those exhibiting dynamical symmetries.
To investigate this point we have taken advantage of the recently derived analytical expressions for the wave functions of the harmonic oscillator in five dimensions {4}. These solutions, useful in the context of generalized Bohr-Mottelson Hamiltonians {5}, are classified by the irreducible representations (IR) of the chain of groups

$U(5)\supset O(5)\supset O(3)\supset O(2)$, and are associated to the α_{2m} surface coordinates of Bohr-Mottelson. The use of Dragt's theorem {6} allows one to write the wave functions alternatively in terms of the η_{2m} creation operators, related to α_{2m} through the usual expression

$$\eta_{2m} = \frac{1}{\sqrt{2}}(\alpha_{2m} - i\pi_{2m}) \tag{5}$$

Where π_{2m} are the momenta associated to the α_{2m} coordinates. We denote these wave functions by $|\nu\Lambda s LM\rangle$, where

$$\begin{aligned}
\hat{n}_d\,|\nu\Lambda s LM\rangle &= \nu\,|\nu\Lambda s LM\rangle \quad ,\\
\hat{\Lambda}\,|\nu\Lambda s LM\rangle &= \Lambda(\Lambda+3)\,|\nu\Lambda s LM\rangle \quad ,\\
\hat{L}\,|\nu\Lambda s LM\rangle &= L(L+3)\,|\nu\Lambda s LM\rangle \quad ,\\
\hat{L}_3\,|\nu\Lambda s LM\rangle &= M\,|\nu\Lambda s LM\rangle \quad .
\end{aligned} \tag{6}$$

There may be $d(\Lambda,L)$ independent representations {6} L of O(3) contained in a given representation Λ of O(5) and we introduce the index

$$s = 1,\ 2\ ,\ \ldots,\ d(\Lambda,L)\ ,$$

to distinguish them.

To include this chain as a subgroup of U(6) we write the Casimir operator N of the latter group as

$$\hat{N} = \hat{n}_d + \bar{\eta}\xi \quad . \tag{7}$$

Thus if we define

$$|N\nu\Lambda s LM\rangle = |\nu\Lambda s LM\rangle\,|n\rangle \quad , \tag{8}$$

where $|n\rangle$ are the wavefunctions of a one-dimensional harmonic oscillator associated to s bosons, and $N = n+\nu$, we arrive to the solutions of the IBA classified according to the group chain (4.1).

To find the general solutions of (2) we use the complete basis (4.1) to diagonalize the operators $\hat{Q}^2$ and $\hat{P}^2$, associated to $\hat{G}$ and $\hat{\mathcal{L}}^2$ (the Casimir operators of SU(3) and O(6) respectively) through

$$\begin{aligned}
\hat{Q}^2 &= \hat{G} - \tfrac{1}{2}\hat{L}^2 \quad ,\\
\hat{P}^2 &= \tfrac{1}{4}\hat{N}(\hat{N}+4) - \tfrac{1}{4}\hat{\mathcal{L}}^2 \quad .
\end{aligned} \tag{9}$$

This problem was solved in ref {2}, where we have shown, in addition, that except for terms contributing to binding energies, the Hamiltonian (2) can be written in the form:

$$\hat{H} = k_1 \hat{n}_d + k_2 \hat{n}_d \hat{N} + k_3 \hat{n}_d^2 + k_4 \hat{L}^2 + k_5 \hat{\Lambda}^2 + k_6 \hat{P}^2 + k_7 \hat{Q}^2 , \qquad (10)$$

where the ki's are parameters related to the two-body boson-boson interaction matrix elements {8}.

We have investigated the possibility of introducing effective IBA Hamiltonians, which have the attractive feature of simultaneously describing groups of isotopes or isotones.

Proceeding in analogy to the Shell Model case {9}, the parameters ki of the effective IBA Hamiltonian can be determined from a least square fit to experimental energy levels {1}. We have performed calculations in the Z=50-82 Shell with rather impressive fits. We show our results in the table for the Xe isotopes from A=116 to A=132. All experimental levels up to $J=10^+$ were included in the analysis except when unidentified intermediate levels make this impossible. The least square search was performed succesively for one, two, etc. parameters, and the convergent solution with the least r.m.s. was chosen. It turns out that for the Xe isotopes this happens for three parameters: k_1, k_4, and k_6, implying an admixture of U(5) and O(6) symmetries. The best fit is obtained for k_1 = 519 KeV, k_4 = 9.8 KeV, k_6= 38.2 KeV, and the agreement is excellent, since all 70 levels are reproduced with a r.m.s. of less than 120 k.e.v.

In other regions of the shell the results are equally impressive, except for some levels that appear to be associated to other degrees of freedom not included in the IBA space.

TABLE.- Experimental and Calculates Energy Levels for the Xe Isotopes.

A	J	E_{exp} (Mev)	E_{th} (Mev)	A	J	E_{exp} (Mev)	E_{th} (Mev)
116	2	.39	.36	122	5	.77	1.91
	4	.92	.84		6	1.46	1.58
	6	1.53	1.41		6	2.06	2.03
	8	2.21	2.10		8	2.22	2.33
	10	2.96	2.88		10	3.03	3.15
118	0	.83	.83	124	2	.35	.44
	2	.34	.39		2	.85	.86
	2	.93	.74		3	1.25	1.35
	2	1.23	1.31		4	.88	1.00
	3	1.37	1.18		4	1.44	1.43
	4	.81	.88		5	1.84	1.99
	4	1.44	1.26		6	1.55	1.65
	5	1.52	1.76		8	2.33	2.40
	6	1.40	1.48		10	3.23	3.24
	6	2.00	1.88	126	2	.39	.47
	8	2.07	2.17		2	.88	.90
	10	2.81	2.97		3	1.32	1.41
120	0	.91	.87		4	.94	1.03
	2	.32	.40		4	1.49	1.49
	2	.88	.78		5	1.90	2.05
	2	1.27	1.36		6	1.64	1.71
	3	1.27	1.23		6	2.21	2.17
	4	.80	.92		8	2.44	2.47
	4	1.40	1.32		10	3.32	3.33
	4	1.71	1.73	128	2	.44	.48
	5	1.82	1.84		2	.97	.94
	6	1.40	1.54		4	1.03	1.07
	6	1.99	1.96	130	2	.54	.51
	8	2.05	2.25		2	1.12	.97
	10	2.87	3.06		4	1.21	1.12
122	2	.33	.42		6	1.94	1.81
	2	.84	.82	132	2	.67	.53
	3	1.21	1.29		2	1.30	1.01
	4	.83	.96		3	1.80	1.57
	4	1.40	1.37		4	1.44	1.15

References.

1) A. Arima and F. Iachello, Phys. Rev. Lett. 35 (1975) 1069.

2) O. Castaños, E. Chacón, A. Frank and M. Moshinsky, J. Math. Phys. 20 (1979) 35.

3) A. Arima and F. Iachello, Ann. Phys. 99 (1976) 253, 111 (1978) 201, 123 (1979) 468.

4) E. Chacón and M. Moshinsky, J. Math. Phys. 18 , 870 (1977).
E. Chacón, M. Moshinsky and R.T. Sharp J. Math. Phys. 17 , 668 (1976).

5) P. O. Hess and W. Greiner, Sussp. Proc. (August 1977).

6) A. J. Dragt, J. Math. Phys. 6 , 533 (1965).

7) O. Castaños, A. Frank and M. Moshinsky, J. Math. Phys. 19 , 1781 (1978).

8) O. Castaños, P. Federman and A. Frank, Interacting Bose-Fermi Systems in Nuclear Physics Ed. F. Iachello, (Plenum Press) , to be published.

9) P. J. Brussaard and P. W. M. Glaudemans "Shell-Model Applications in Nuclear Spectroscopy" (North-Holland, 1977).

10) M. Sakai and Y. Gono, Quasi-Ground, quasi-beta, and quasi-gamma bands, INS-J-160 University of Tokyo 1979. Table of Isotopes, Ed. by C.M. Lederes and V. S. Shirley, John Wiley & Sons 1978.

BASES FOR IRREPS OF U(6) AND THREE BODY PROBLEMS

Elpidio Chacón
Instituto de Física
Octavio Castaños and Alejandro Frank
Centro de Estudios Nucleares
Universidad Nacional Autónoma de México
México 20, D. F.
México

Introduction.

In the study of systems consisting of 3 identical bodies one frequently uses a set of basis states associated to a symmetric irrep $[N\dot{0}]$ of U(6). In terms of the two relative Jacobi vectors of the system: $\underset{\sim}{x}_1$, $\underset{\sim}{x}_2$ and their canonically conjugate momenta: $\underset{\sim}{p}_1$, $\underset{\sim}{p}_2$, boson creation operators $\underset{\sim}{\eta}_s$ are defined as

$$\underset{\sim}{\eta}_s = \frac{1}{\sqrt{2}} (\underset{\sim}{x}_s - i\underset{\sim}{p}_s) , \quad s = 1,2 \tag{1}$$

and $\underset{\sim}{\xi}_s = \underset{\sim}{\eta}_s^{\dagger}$ are the corresponding annihilation operators which give zero when applied on the vacuum state $|0\rangle$. The basis states mentioned above are homogeneous polynomials of degree N in $\underset{\sim}{\eta}_s$: $\underset{\sim}{P}_N(\underset{\sim}{\eta}_1, \underset{\sim}{\eta}_2)|0\rangle$, and it is known that they represent harmonic oscillator states with N quanta of energy [1]. (We adopt a system of units in which mass of particle = frecuency of oscillation = $\hbar$ = 1).

The purpose of this note is to present explicit realizations for these states with different classifications according to several chains of subgroups of U(6). Connections between the various bases will be mentioned also. The motivation for this work was the fact that some particular aspect of the 3-body problem could be more easily studied in one of the bases considered below. We want to remark that our problem is very similar to the one related to the construction of symmetric irreps of U(6) used in the Interacting Boson-Model [2]. In both cases we have the physical rotation group O(3) as a subgroup of U(6), the main difference being that in the IBM case the 6 boson operators are assembled in two irreducible tensors of rank 0 and 2, while in the present problem we have the two vectors $\underset{\sim}{\eta}_1$, $\underset{\sim}{\eta}_2$. There is one additional peculiarity of the present analysis, namely, that the states must have a definite behaviour under permutations of the 3 bodies. It is known [1] that this part of the problem is best examined in terms of boson operators $\underset{\sim}{\alpha}$, $\underset{\sim}{\beta}$ related to $\underset{\sim}{\eta}_1$, $\underset{\sim}{\eta}_2$ through

$$\underset{\sim}{\alpha} = \frac{1}{\sqrt{2}} (-i\underset{\sim}{\eta}_1+\underset{\sim}{\eta}_2) , \quad \underset{\sim}{\beta} = \frac{1}{2}(i\underset{\sim}{\eta}_1+\underset{\sim}{\eta}_2) . \tag{2}$$

We give below expressions for our states in the $(\underset{\sim}{\eta}_1,\underset{\sim}{\eta}_2)$ or the $(\underset{\sim}{\alpha},\underset{\sim}{\beta})$ notation, according to which is more convenient. Some of the results we are going to mention have been known for sometime already, we quote them with due credit for the sake of completeness.

THE BASIS STATES

i) The 2 oscillator basis [3]

$$(\eta_1,\eta_2 \mid n_1\ell_1 n_2\ell_2\ LM> =$$

$$A_{n_1\ell_1} A_{n_2\ell_2} (\eta_1\cdot\eta_1)^{n_1} (\eta_2\cdot\eta_2)^{n_2} [\mathcal{Y}_{\ell_1}(\eta_1) \times \mathcal{Y}_{\ell_2}(\eta_2)]^L_M \,|0> . \tag{3}$$

This is just the direct product of oscillator states in the vector η_1, and η_2 coupled to angular momentum quantum numbers L, M. The number of quanta of the states is $N=2n_1+\ell_1+2n_2+\ell_2$, and they form an orthonormal basis.

ii) The O(6) Canonical basis [4]

$$(\eta_1,\eta_2|N\lambda\ell_1\ell_2 LM> = A^N_\lambda(\eta_1\cdot\eta_1+\eta_2\cdot\eta_2)^{(N-\lambda)/2}\, P_{\lambda\ell_1\ell_2 LM}|0>,$$

$$P_{\lambda\ell_1\ell_2 LM} \equiv \sum_s b_s^{\lambda\ell_1\ell_2} (\eta_1\cdot\eta_1)^{(\lambda-\ell_1-\ell_2)/2-s} (\eta_2\cdot\eta_2)^s [\mathcal{Y}_{\ell_1}(\eta_1)\times \mathcal{Y}_{\ell_2}(\eta_2)]^L_M \tag{4}$$

This is again an orthonormal basis, with λ characterizing the irrep of the subgroup O(6) of U(6). According to a theorem of Dragt [5], if we replace in $P(\eta_1,\eta_2)$ $\eta_s\to x_s$ the resulting polynomial $P_{\lambda\ell_1\ell_2 LM}(x_1,x_2)$ is a 6-dimensional solid spherical harmonic with the quantum numbers indicated. Attaching to it a well defined hyper-radial function [2],

$$R_{N\lambda}(\rho)\ ,\quad \rho^2 = x_1\cdot x_1 + x_2\cdot x_2 \tag{5}$$

we obtain the basis states of Eq. (4) in terms of coordinates.

iii) The Kramer-Moshinsky basis [6]

$$(\eta_1,\eta_2|k_1k_2fqLM> = (\eta_2\cdot\xi_1)^{(k_1-k_2)/2-f}\, P^{k_1k_2}_{qLM}|0>,$$

$$P^{k_1k_2}_{qLM} \equiv [\mathcal{Y}_{L-k_2+2q+\varepsilon}(\eta_1)\times \mathcal{Y}_{k_2-2q}(\eta')]^L_M\, (\eta_1\cdot\eta_1)^{(k_1-L-\varepsilon)/2-q} (\eta'\cdot\eta')^q \tag{6}$$

In this formula $\eta' = i\,\eta_1\times\eta_2$. This is a nonorthogonal basis, the integer number q not being a true quantum number. The number of quanta of the state (6) in $N=k_1+k_2$.

iv) The Monomial Basis [7]

This basis diagonalizes the number operator:

$$\hat{N} = \eta_1\cdot\xi_1 + \eta_2\cdot\xi_2 = \alpha\cdot\alpha^\dagger + \beta\cdot\beta^\dagger \quad ,$$

the square and third component of the total angular momentum: $\hat{L}^2$, L_3, and a fourth operator

$$\hat{M} = -i(\eta_1 \cdot \xi_2 - \eta_2 \cdot \xi_1) = \alpha \cdot \alpha^\dagger - \beta \cdot \beta^\dagger \quad . \tag{7}$$

It is more convenient in this case to give a realization of the states in terms of α, β. The expression for the states with highest weight in the subgroup O(3) is a monomial formed by products of powers of the 6 "elementary factors"

$$\alpha_{+1} \ , \ \beta_{+1} \ , \ (\alpha \times \beta)_{+1}, \ (\alpha \cdot \alpha) \ , \ (\alpha \cdot \beta) \ , \ (\beta \cdot \beta) \tag{8}$$

where the index +1 means the spherical component +1 of a vector. Owing to the identity

$$4\,[(\alpha \times \beta)_{+1}]^2 = 2\alpha_{+1}\beta_{+1}(\alpha\cdot\beta) - (\alpha_{+1})^2(\beta\cdot\beta) - (\beta_{+1})^2(\alpha\cdot\alpha) \tag{9}$$

the exponent of $(\alpha\times\beta)_{+1}$ in the state is restricted to 0 or 1. We are thus led to

$$(\alpha,\beta|N\nu kqL,\ M=L\rangle =$$

$$(\alpha_{+1})^{L+2q-(k-\nu)/2}\,(\beta_{+1})^{(k-\nu)/2-2q-\varepsilon}\,(\alpha\times\beta)^{\varepsilon}_{+1}\,(\alpha\cdot\alpha)^{(k-L-\varepsilon)/2-q}(\beta\cdot\beta)^{q}\ \times$$

$$\times\ (\alpha\cdot\beta)^{(N-k)/2}\,|0\rangle \tag{10}$$

where here, as in the following, ε can take the values 0 or 1, and all exponents must be non-negative integers.

An equivalent expression for the states, valid for any M, can be seen to be

$$(\alpha,\beta|N\nu kqLM\rangle =$$

$$[\mathcal{Y}_{L+2q-(k-\nu)/2+\varepsilon}(\alpha)\times\mathcal{Y}_{(k-\nu)/2-2q}(\beta)]^L_M\,(\alpha\cdot\alpha)^{(k-L-\varepsilon)/2-q}(\beta\cdot\beta)^q(\alpha\cdot\beta)^{(N-k)/2}|0\rangle . \tag{11}$$

This again is a non-orthogonal basis, the labels k, q not being good quantum numbers.

v) The Dragt Basis [5]

These states are classified by the chain of groups

$$\begin{array}{ccccccc} & & & SO(2)\,\nu & & & \\ & & & \times & & & \\ U(6) & \supset & O(6) & \supset\ SU(3) & \supset & SO(3) & \supset SO(2) \\ N & & \lambda & (\lambda,(\lambda-\nu)/2) & & L & M \end{array} \tag{12}$$

where the labels beside each group specify the irrep of that group. A procedure for the construction of the basis states with $\lambda = N$ consists in the insertion of the group O(6) in the monomial basis (11) by the use of "traceless bosons" [8]. In our case the traceless bosons are defined as

$$a = \alpha - 2(\alpha\cdot\beta)(2\hat{N}+6)^{-1}\beta^\dagger \ , \ b = \beta - 2(\alpha\cdot\beta)(2\hat{N}+6)^{-1}\alpha^\dagger \tag{13}$$

and have the property $a\cdot b|0\rangle = 0$. Under the replacement $\alpha \rightarrow a$, $\beta \rightarrow b$, the state of Eq. (11) with $N = k = \lambda$ becomes a state of an irrep $(\lambda 00)$ of O(6). The scalar part $(a.a)^{-q+(\lambda-L-\varepsilon)/2}\ (b.b)^q|0>$ can be easily written in terms of scalar products of $\underset{\sim}{\alpha}$, $\underset{\sim}{\beta}$, and applying on it the other factor as a differential operator, we are led to

$$(\underset{\sim}{\alpha},\underset{\sim}{\beta}|N=\lambda,\lambda\nu qLM> = \sum_{\sigma\tau} b_{\sigma\tau}\ (\underset{\sim}{\alpha}\cdot\underset{\sim}{\alpha})^{(\lambda-L-\varepsilon)/2-q-\sigma}\ (\underset{\sim}{\beta}\cdot\underset{\sim}{\beta})^{q-\tau}(\underset{\sim}{\alpha}\cdot\underset{\sim}{\beta})^{\sigma+\tau}\ \times$$

$$\times\ [\mathcal{Y}_{L-(\lambda-\nu)/2+2q+\sigma-\tau+\varepsilon}(\underset{\sim}{\alpha})\times\mathcal{Y}_{(\lambda-\nu)/2-2q-\sigma+\tau}(\underset{\sim}{\beta})]^L_M\ |0> \tag{14}$$

with

$$b_{\sigma\tau} = \frac{(-2)^{\sigma+\tau}(\lambda+1-\sigma-\tau)!}{(q-\tau)!((\lambda-L-\varepsilon)/2-q-\sigma)!}\ \sum_r \frac{((L-\lambda-1-\varepsilon)/2)_r}{r!(\tau-r)!(L-(\lambda-\nu)/2+2q-\tau+r)!(\sigma-r)!((\lambda-\nu)/2-2q-\varepsilon-\sigma+r)!}\ . \tag{15}$$

The Dragt basis is a non-orthogonal basis as it carries the same label q which appears in the states (11).

The Dragt states with $N>\lambda$ can be obtained from (14) by multiplication with $(\underset{\sim}{\alpha}\cdot\underset{\sim}{\beta})^{(N-\lambda)/2}$ on the left.

The state (14) can be written in coordinates simply by doing the replacement [5] $\underset{\sim}{\alpha}\rightarrow(-i\underset{\sim}{x}_1+\underset{\sim}{x}_2)/\sqrt{2}$, $\underset{\sim}{\beta}\rightarrow(i\underset{\sim}{x}_1+\underset{\sim}{x}_2)/\sqrt{2}$ thus giving the solid harmonic $\mathcal{Y}_{\lambda\nu qLM}(\underset{\sim}{x}_1,\underset{\sim}{x}_2)$.

If furthermore we introduce new coordinates $\rho_1,\rho_2,\ \vartheta_1,\vartheta_2,\vartheta_3,\ \alpha$ defined through [9]

$$x_{is} = \sum_{k=1}^{2} \rho_k\ D^1_{ki}\ (\vartheta_1,\vartheta_2,\vartheta_3)\ \mathcal{D}^1_{ks}(\alpha)\ , \tag{16}$$

we obtain for the O(6) solid harmonic, when $\varepsilon=0$

$$\rho^{-\lambda}\mathcal{Y}_{\lambda\nu qLM}(\underset{\sim}{x}_1,\underset{\sim}{x}_2) =$$

$$e^{i\nu\alpha}\sum_K{}'(-)^{(\lambda+K)/2}\ [\frac{(L+K)!(L-K)!}{(2L)!}]^{1/2}\ D^{L^*}_{MK}(\vartheta_i)\ \sum_m G^{\lambda\nu qL}_{mK}(\gamma)\ f^{\lambda\nu qL}_m\ (\cos 2\gamma) \tag{17}$$

where $\rho_1=\rho\cos\gamma$, $\rho_2=\rho\sin\gamma$, f is a polynomial in $\cos 2\gamma$ and G is a hypergeometric function in $\sin(\pi/4-\gamma)$, $\cos(\pi/4-\gamma)$. An analogous expression exists for $\varepsilon=1$.

These functions are useful in problems connected with the microscopic approach to collective motions in nuclei.

CONNECTION BETWEEN DIFFERENT BASES.

Expansion coefficients for the states of some of the bases discussed above in terms of another one are now available. This includes the transformation bracket between the orthonormal bases (i) and (ii) [4], the expansion coefficients for (iv) in terms of (i), (v) in terms of (iv), and (iii) in terms of (i).

Details concerning some of the topics discussed above will be presented elsewhere.

References

[1] P. Kramer, M. Moshinsky, article in Group Theory and its Applications, Vol. I,E. Loebl editor, Academic Press, N. Y. 1968.

[2] O. Castaños, E. Chacón, A. Frank, M. Moshinsky, J. Math. Phys. 20, 35 (1979).

[3] T. A. Brody, M. Moshinsky, Tables of Transformation Brackets, Gordon and Breach, N. Y. 1969.

[4] E. Chacón, D. Levi, M. Moshinsky, Rev. Mexicana de Física. 22, 291 (1973).

[5] A. J. Dragt, J. Math. Phys. 6, 533 (1965).

[6] P. Kramer, M. Moshinsky, Nuclear Phys. 82, 241 (1966).

[7] V. C. Aguilera-Navarro, E. Chacón, submitted for publication.

[8] E. Chacón, M. Moshinsky, J. Math. Phys. 18, 870 (1977).

[9] W. Zickendraht, J. Math. Phys. 12, 1663 (1971).
A. YA. Dzublik, V. I. Ovcharenko, A. I. Steshenko, G.F. Filippov, Yad. Fiz. 15, 869 (1972) [Sov. J. Nucl. Phys. 15, 487 (1972)].

INTRINSIC SPHAPE OF NUCLEI

E. Chacõn, M. Moshinsky and S. Rubinstein
Instituto de Física, UNAM
Apdo. Postal 20-364, México 20,D.F.

ABSTRACT

A transformation of coordinates, introduced a decade ago by Dzublik et al. and by Zickendraht, allow us to go from the 3A-3 Jacobi coordinates x_i^s, $i=1,2,3$; $s=1,2,\ldots, A-1$ of an A body system to the Euler angles ϑ_k, $k=1,2,3$ associated with the standard O(3) group, the 3A-9 coordinates α which are a subset of the (A-1)(A-2)/2 associated with an O(A-1) group, and three extra parameters ρ_k, $k=1,2,3$. The latter give a measure of the deformation along the three principal axes in the frame of reference fixed in the body.

In the present note we discuss the deformation of the three body system by considering the expectation values of both the invariants of the inertia tensor and of the ρ_k^2. We compare the results in a basis of harmonic oscillator states for the three body system that reaches up to 22 quanta.

We consider a system of A nucleons whose internal motion is described in terms of the A-1 Jacobi relative vectors $\underline{x}'^s$, $s=1,2,\ldots$, A-1 and their canonically conjugate momenta $\underline{p}'^s$. Each nucleon has a mass M, and we can define dimensionless vectors

$$\underline{x}^s = \sqrt{\frac{M\omega}{\hbar}}\,\underline{x}'^s \quad , \qquad \underline{p}^s = \sqrt{\frac{1}{M\hbar\omega}}\,\underline{p}'^s \tag{1}$$

where $\hbar\omega$ is the energy quantum of an oscillator field to be introduced later.

With respect to an inertial frame with origin at the center of mass of the nucleus the components of the tensor of inertia are

$$I_{ij} = \frac{\hbar}{\omega}\left[\delta_{ij}\sum_{s=1}^{A-1}\underline{x}^s\cdot\underline{x}^s - \sum_{s=1}^{A-1} x_i^s x_j^s\right] , \quad i,j=1,2,3 . \tag{2}$$

We can just as well consider the tensor $\underline{q}$ with components

$$q_{ij} = \sum_{s=1}^{A-1} x_i^s x_j^s \tag{3}$$

as this gives the same information about the mass distribution that the tensor of inertia gives.

The principal values of the tensor $\underline{q}$ are the roots of the secular equation

$$\zeta^3 - Tr\,\underline{q}\,\zeta^2 + \tfrac{1}{2}\left[(Tr\,\underline{q})^2 - Tr\,\underline{q}^2\right]\zeta - Det\,\underline{q} = 0 \qquad (4)$$

where

$$Tr\,\underline{q} = \sum_s \underline{x}^s \cdot \underline{x}^s \qquad (5a)$$

$$Tr\,\underline{q}^2 = \sum_{st} (\underline{x}^s \cdot \underline{x}^t)^2 \qquad (5b)$$

$$Det\,\underline{q} = \frac{1}{6} \sum_{stu} \left[(\underline{x}^s \times \underline{x}^t) \cdot \underline{x}^u\right]^2 \qquad (5c)$$

We take the roots of the secular equation as giving a measure of the shape of the nucleus. We have calculated these roots by replacing the coefficients in the cubic by the mean value of each operator with respect to the ground state wave function of the nucleus, and then solving the equation. The results obtained for the trinucleon will be reported below.

A variant of this approach is discussed next. For this we introduce a set of coordinates especially suited to the analysis of collective properties, namely[1)]

$$x_i^s = \sum_{k=1}^{3} \rho_k\, D^1_{ki}(\vartheta)\, \mathcal{D}^1_{A-4+k,s}(\alpha) \qquad (6)$$

where $D^1(\vartheta)$ is a rotation matrix in 3 dimensions depending on 3 Euler angles ϑ_1, ϑ_2, ϑ_3 and $\mathcal{D}^1(\alpha)$ is a rotation matrix in A-1 dimensions, depending in only 3A-9 angles α_i because only its last 3 rows enter in Eq. (6).

In terms of the new variables we find that

$$q_{ij} = \sum_{k=1}^{3} \rho_k^2\, D^1_{ki}(\vartheta)\, D^1_{kj}(\vartheta) \quad , \quad \text{i.e.}$$

$$\underline{q} = \tilde{D}^1(\vartheta) \begin{bmatrix} \rho_1^2 & 0 & 0 \\ 0 & \rho_2^2 & 0 \\ 0 & 0 & \rho_3^2 \end{bmatrix} D^1(\vartheta) \qquad (7)$$

and thus ρ_1^2, ρ_2^2, ρ_3^2 are the principal values of the tensor $\underline{q}$ in its intrinsic frame. The expectation value of ρ_k^2 with respect to the ground state wave function of the nucleus is a measure of the shape of the nucleus. We have calculated these mean values for the trinucleon and we shall report below the results obtained. We proceed now to give the details of the computation for the case A=3.

The intrinsic hamiltonian of the system is

$$H=\frac{1}{2}\hbar\omega\sum_{s=1}^{2}\left[(\underline{P}^s)^2+(\underline{x}^s)^2\right]+\sum_{s>t=1}^{3}\left[V\left(\sqrt{\tfrac{\hbar}{M\omega}}\,|\underline{R}^s-\underline{R}^t|\right)-\frac{1}{6}\hbar\omega(\underline{R}^s-\underline{R}^t)^2\right] \quad (8)$$

where $\underline{R}^s$, $\underline{R}^t$ are the ordinary position vectors of particles s and t. The matrix of the operator (8) is constructed with respect to a basis of antisymmetric functions formed by sums of products of an orbital part times a spin-isospin part with conjugate permutational symmetries. The orbital functions are chosen as eigenfunctions of the harmonic oscillator hamiltonian given by the first sum in (8). There is a known procedure[2] to adapt these functions to a definite permutational symmetry, we denote the symmetrized functions as $|n_1\, l_1\, n_2\, l_2\, LM, \{f\}r\rangle$.

The two-body interaction was taken as the phenomenological central-plus-exchange soft core potential of Eikemeier-Hackenbroich[3]. As this is a preliminary calculation we took into account only the largest component of the ground state wave-function, namely, the symmetric S state. The lowest eigenvalue of the energy matrix truncated to a certain maximal number $\mathcal{N}$ of quanta gave us the approximate binding energy of the system, and the corresponding eigenvector gave the approximate wave-function of the ground state. We give the binding energies obtained for the optimal value of $\hbar\omega$=10.5 MeV and the dimension of the bases:

$\mathcal{N}$	dim	E(MeV)
18	53	-3.786
20	67	-3.871
22	83	-3.988

For A=3, Det $\underline{q}$ =0, thus the secular equation of the mass distribution tensor is

$$\zeta^2-\mathrm{Tr}\,\underline{q}\,\zeta+\frac{1}{2}\left[(\mathrm{Tr}\,\underline{q})^2-\mathrm{Tr}\,\underline{q}^2\right]=0 \quad . \qquad (10)$$

The mean values of the coefficients in this equation are

N	$\langle \mathrm{Tr}\, \underline{q} \rangle$	$\frac{1}{2}\langle (\mathrm{Tr}\, \underline{q})^2 - \mathrm{Tr}\, \underline{q}^2 \rangle$
18	3.28124	2.18172
20	3.19421	2.03469
22	3.20953	2.10982

(11)

leading to these values for the roots of the quadratic equation (10):

N	ζ_1	ζ_2
18	2.35470	0.92654
20	2.31548	0.87873
22	2.28701	0.92252

These are the numerical values giving the length of the semiaxes of the "mass ellipsoid".

Let us consider now the alternative analysis involving the collective coordinates $\rho_1, \rho_2, \vartheta_1, \vartheta_2, \vartheta_3$ plus an angle α. Instead of ρ_1, ρ_2 we better work with ρ, γ defined through

$$\rho_1 = \rho \cos\gamma \quad , \quad \rho_2 = \rho \sin\gamma \; ; \qquad 0 \leqslant \gamma \leqslant \frac{\pi}{4} \quad . \tag{13}$$

For symmetric S states (L = 0) the eigenfunctions of the oscillator hamiltonian in these coordinates are known to be

$$R_{n\lambda}(\rho) \cos\frac{\mu}{2}\alpha \; d^{\lambda/4}_{\frac{\mu}{2}\,\frac{\mu}{2}}(4\gamma) \; . \tag{14}$$

The transformation coefficient between this basis and the one considered previously, i.e. $|n_1 \ell_1 n_2 \ell_1 ; \{3\} 111\rangle$, was obtained explicitly. This allowed to have the ground state wave-function expressed as a linear combination of the functions (14), and thus the expectation values in the ground state of ρ_1^2, ρ_2^2 or equivalently, $\rho_1^2 + \rho_2^2 = \rho^2$, $\rho_1^2 - \rho_2^2 = \rho^2 \cos 2\gamma$ are given in terms of sums of matrix elements of ρ^2 and $\cos 2\gamma$ with respect to the functions (14), which are easily determined.

Working along these lines we have obtained

N	$\langle \rho_1^2 \rangle$	$\langle \rho_2^2 \rangle$
18	2.82185	0.45940
20	2.74364	0.45057
22	2.76030	0.44923

(15)

We note that both methods give the same order of magnitude for the major and minor axis deformation parameters. There are though differences which may be significant, particularly if we go to nuclei with larger A.

REFERENCES

1. A. Ya. Dzublik, V.I. Ovcharenko, A.I. Steshenko and G.F. Filipov; Yad. Fiz. 15, 869 (1972) [Sov. J. Nuc. Phys. 15, 487 (1972)]. W. Zickendraht, J. Math. Phys., 12,1663 (1971).
2. M. Moshinsky, "The harmonic oscillator in modern physics: From atoms to quarks"; (Gordon E. Breach Science Publishers, New York, 1967).
3. H. Eikemeier and H.H. Hackenbroich, Zeit. f. Physik 195, 412 (1966).

GROUP THEORY AND NUCLEAR HIGH SPIN PHENOMENA

J. P. Draayer[+] and C. S. Han[*]

Department of Physics and Astronomy, Louisiana State University
Baton Rouge, Louisiana 70803

Abstract

A microscopic interpretation of the coherent/critical phenomenon known as backbending, in which certain deformed nuclei, looked upon as rotating spheriods, show a marked increase in their effective moment of inertia at some critical value of the total angular momentum, is presented. As shell-model calculations in a basis constructed from a direct product of single-particle orbitals leads to matrix dimensionalities that are enormous, truncation is required. The complementary roles group theory and methods of statistical spectroscopy play in the selection of a physically significant coupling scheme are illustrated. A weak coupling model of the normal and abnormal parity orbitals organized into SU(3) and R(5) multiplets, respectively, shows pair alignment to be the primary mechanism responsible for backbending but band mixing can be competitive and lead to anomalous E2 behavior. Results for ^{126}Ba are given.

1. Introduction

The phenomenon known as backbending,[1] in which certain deformed nuclei, looked upon as rotating spheriods, show a marked increase in their moment of inertia at some critical value of the total angular momentum, presents a unique challenge to the nuclear theorist. Straightforward shell-model calculations are impossible for a basis built from a direct product of single-particle orbitals leads to matrix dimensionalities that are enormous, too large for modern computers and even if that were not the case, too unwieldy for numerical success to imply an understanding of the physics. Alternatives to the shell model must be provided or justification found for truncating the basis to manageable size.

The simplest suggested explanation is band crossing.[2] Backbending nuclei are assumed to have more than one stable intrinsic shape with the lowest member of the rotational band built from the least deformed intrinsic state lying lowest. Because the moment of inertia $\mathcal{I}$ is proportional to the square of the deformation and $E_I \propto I(I+1)/2\mathcal{I}$, for some critical value of the total spin I a second band will cross the first. The yrast band, which is comprised of states of minimum energy for each spin, thus displays backbending.

[+]Supported in part by the U.S. National Science Foundation.

[*]On leave from National Chiao-Tung University, Hsin Chu, Taiwan, Republic of China.

Perhaps the most intuitive of the alternatives are the various quasiparticle-plus-rotor models.[3] For these, parameters of the core (moment of inertia) and the valence nucleons (pairing gap) are introduced phenomenologically. A dynamics is built in by allowing the parameters to be smoothly varying functions of the total angular momentum and by postulating the existence of a Coriolis force acting between the core and valence quasiparticles. The primary mechanism producing backbending in this case is rotational alignment. For low angular momenta a coupling of the core to paired quasiparticles is energetically favored but for larger angular momenta, because of the negative contribution to the energy from the Coriolis interaction, it is the alignment of the spin of unpaired quasiparticles with the core that leads to favored couplings.

In contrast with the predominately phenomenological models, Hartree-Fock-Bogoliubov mean field theories can start with a realistic nucleon-nucleon interaction and provide therby a look at the microscopic mechanisms responsible for backbending.[4] Nonetheless, cranking is used to insure the intrinsic state has the desired average value of the total angular momentum and this raises doubts about the reliability of the theory through the crucial backbending region where the effective moment of inertia is a double or triple valued function of the total angular momentum. While this difficulty may be overcome by projection after variation with cranking, the implementation of a sound Hartree-Fock-Bogoliubov theory remains a major technical challenge.

Recently we published an article that demonstrates the feasibility of providing a shell-model treatment of high-spin phenomena in strongly deformed nuclei.[5] Group theory and methods of statistical spectroscopy play complementary roles in justifying basis truncation. These features are emphasized in Section 2 where the logic used in selecting a model space is reviewed. In Section 3 we look at the structure of the interaction and comment briefly on the methods and machinery available for carrying out detailed calculations. In Section 4 results for ${}^{126}_{56}Ba_{70}$ are presented which leads one to the conclusion that by-and-large backbending is a pair alignment phenomenon. However, band crossing does occur and may provide a theoretical explanation of the observed pre-backbending reduction in B(E2) values for Ba. The conclusion serves as a critique and a prospectus.

2. Model Space

For heavy nuclei shell closure is anomalous for the magnitude of the single-particle spin-orbit splitting places the highest spin member of an oscillator shell down among levels of the next lower shell. Thus for ${}^{126}_{56}Ba_{70}$ one has a closed Z=50=N core $[(s_{1/2})^4, (p_{3/2}p_{1/2})^{12}, (d_{5/2}s_{1/2}d_{3/2})^{24}, (f_{7/2})^{16}, (f_{5/2}p_{3/2}p_{1/2}g_{9/2})^{44}]$ plus 6 protons (π) and 20 neutrons (ν) distributed among the $(g_{7/2}d_{5/2}h_{11/2}d_{3/2}s_{1/2})$-orbitals. The $h_{11/2}$ negative (unique, abnormal) parity state is an intruder among the other positive (normal) parity levels. A two-body interaction can at most scatter a pair of particles between these two distinct sets of orbitals; the scattering of a single particle from the $h_{11/2}$ to a normal parity orbital or vice versa is forbidden by parity considerations. Our model is based on a weak coupling of normal and abnormal parity orbitals, $[(g_{7/2}d_{5/2}d_{3/2}s_{1/2})^{n_N\Gamma_N}\times(h_{11/2})^{n_A\Gamma_A}]^{\Gamma}$, where Γ refers to angular momentum ($J, \bar{L}+\bar{S} = \bar{J}$) and isospin (T). A savings occurs when a restriction on the occupancies of the normal (n_N) and abnormal (n_A) orbitals can be justified.

Deformation breaks spherical symmetry. The 2j+1 degenerate levels of a single j-orbital are split into j+1/2 doublets, pairs related by time conjugation. A Nilsson level scheme, Figure 1, shows this splitting of single-particle orbitals as a function of deformation. For a deformation of 0.2 (ε=<q_o>, $q_o \propto 2n_3-n_1-n_2$ where n_α is the number of oscillator quanta in the α'th direction) the dominant configuration for ^{126}Ba is $(g_{7/2}d_{5/2}d_{3/2}s_{1/2})^{n_N=18,T_N=3}$ $(h_{11/2})^{n_A=8,T_A=4}$. The leading π and ν scattering configurations which couple to this one are $(g_{7/2}d_{5/2}d_{3/2}s_{1/2})^{16,T_N=4}$ $(h_{11/2})^{10,T_A=3}$ and $(g_{7/2}d_{5/2}d_{3/2}s_{1/2})^{16,T_N=2}$ $(h_{11/2})^{10,T_A=5}$, respectively. However, the dimensionality of a basis built on even the single dominant configuration is huge, for example $(g_{7/2}d_{5/2}d_{3/2}s_{1/2})^{n_N=18,T_N=4,J_N=2}$ alone has a dimension of 5,372,930. Further truncation is required!

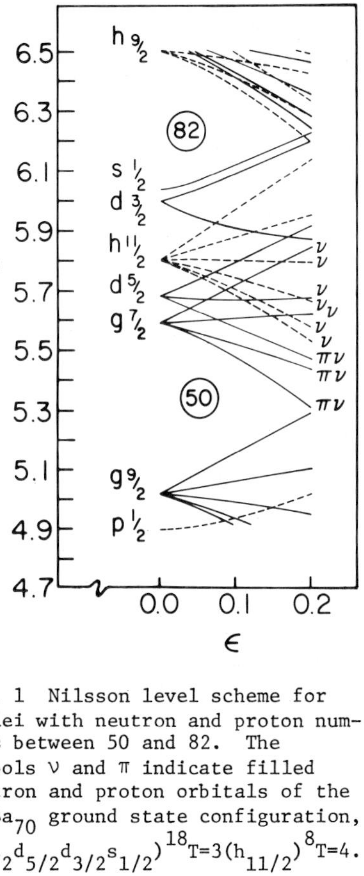

Fig. 1 Nilsson level scheme for nuclei with neutron and proton numbers between 50 and 82. The symbols ν and π indicate filled neutron and proton orbitals of the $^{126}_{56}Ba_{70}$ ground state configuration, $(g_{7/2}d_{5/2}d_{3/2}s_{1/2})^{18}T{=}3(h_{11/2})^{8}T{=}4$.

Without the highest-spin member of a shell the underlying oscillator structure is lost. But this seemingly unhappy situation converts to a potentially happy one upon recognizing that the remaining orbitals can be mapped onto pseudo orbitals which together form a basis for a pseudo oscillator shell of one less quanta.[6] The mapping for $N=4 \rightarrow \tilde{N}=3$ is shown schematically in Figure 2. We use a tilde (~) to denote pseudo quantities. The mapping carries physical significance if the interaction, rewritten in terms of pseudo states, preserves even approximately some symmetry. In the mapping one has the option of fixing relative phases of the single-particle orbitals. It is important to select that choice which optimizes the goodness of a symmetry.

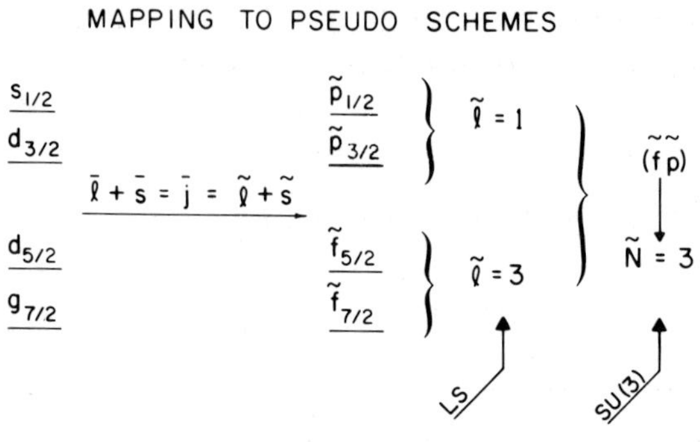

Fig. 2 The mapping of valence gds-shell orbitals onto a pseudo fp-shell

Recent developments in statistical spectroscopy indicate that questions regarding the probable goodness of symmetries can be answered a priori using simple easy-to-apply tests.[7] The tests rely on the elementary notion that two operators are similar if their eigenvalues are strongly correlated. This can be generalized and made quantitative by introducing the notion of operator norms and inner products. Let $\langle\langle\rangle\rangle$ denote a trace and $\langle\rangle = d^{-1}\langle\langle\rangle\rangle$ the corresponding average. Then for an interaction H

$$\varepsilon_H = \langle H\rangle = d^{-1}\langle\langle H\rangle\rangle \rightarrow \text{centroid of H}$$

$$\sigma_H^2 = \langle(H-\varepsilon)^2\rangle \qquad \rightarrow \text{variance of H} \tag{1}$$

The centroid fixes the location of a spectrum and the variance measures its size. The variance is the square of a proper norm, $\sigma_H = ||H||$. The inner product of centered and normalized operators,

$$\cos\theta_{H-K} = (h\cdot k) = \langle(H-\varepsilon_H)(K-\varepsilon_K)\rangle/||H||\ \ ||K|| = \zeta_{H-K} \tag{2}$$

completes the geometry. The inner product of two operators defines an angle between them, $-\pi/2 \leq \theta \leq \pi/2$, and measures their correlation, $-1\leq\zeta\leq 1$. Given a set $\{K_\alpha\}$, H can be expanded as

$$H = \sum_\alpha C_\alpha K_\alpha + H' \tag{3}$$

The inner products $(k_\alpha\cdot h)$ and $(k_\alpha\cdot k_\beta)$ suffice to determine the constants C_α. The ratio $||H-H'||/||H||$ provides a completeness measure. If the K_α are symmetry preserving and the ratio of the norm of $\sum_\alpha C_\alpha K_\alpha$ to the norm of H is large then the symmetry must be a good one. If the contrary is true the symmetry may or may not be good.

Returning to the mapping of real onto pseudo orbitals we assert that the optimum choice for the phases is the one which maximizes the correlation between some realistic H and the pseudo quadrupole-quadrupole operator which itself is simply related ($\tilde{Q}\cdot\tilde{Q} = 4\tilde{C}_2 - 3\tilde{L}^2$) to the second order Casimir invariants of SU(3) and its R(3) subgroup. To illustrate the sensitivity of this prescription, Table 1 gives

Table 1

PARTICLE NUMBER	|CORRELATION COEFFICIENTS| H & $\tilde{Q}\cdot\tilde{Q}$				
m	(ds)	upper (fp) → $(\tilde{d}\tilde{s})$			
		(+−+)	(+−−)	(+++)	(++−)
2	0.41	0.38	0.09	0.03	0.02
4	0.52	0.53	0.19	0.13	0.11
6	0.55	0.57	0.22	0.15	0.14
8	0.56	0.58	0.23	0.16	0.15
10	0.56	0.59	0.23	0.17	0.15
12	0.56	0.59	0.24	0.17	0.15
14	0.55	0.59	0.24	0.17	0.15
16	0.53	0.59	0.24	0.17	0.15
18	0.49	0.58	0.23	0.17	0.15
20	0.44	0.55	0.22	0.16	0.15
22	0.30	0.45	0.18	0.13	0.12

Table 1. Comparison of scalar correlation coefficients between H and $\tilde{Q}\cdot\tilde{Q}$ for ds-shell and four distinct upper fp-shell $(f_{5/2},p_{3/2},p_{1/2})\to(\pm\tilde{d}_{5/2},\pm\tilde{d}_{3/2},\pm\tilde{s}_{1/2})$ mappings. For the ds-shell the interaction is that of Kuo[9] with ^{17}O single particle energies while for the upper fp-shell it is the ASDI of Koops and Glaudemans[8] including their single particle energies.

results for the four distinct choices one can make in mapping the upper fp-shell into a pseudo ds-shell $(f_{5/2},p_{3/2},p_{1/2})\to(\pm\tilde{d}_{5/2},\pm\tilde{d}_{3/2},\pm\tilde{s}_{1/2})$. H is the ASDI (augmented-surface-delta-interaction) of Koops and Glaudemans.[8] On the left are similar results for the ds-shell using the interaction of Kuo.[9] In the latter case the similarity of H and Q·Q accounts for the success of the Elliott Model.[10] Our choice in the $N=4\to\tilde{N}=3$ mapping was determined in this manner, using an SDI (surface-delta-interaction) for H.

Once the mapping is set and the symmetry determined to be a reasonable one the basis can be restricted to the energetically favored representations that are compatible with particle statistics. For SU(3) these can be determined most simply from a Nilsson level scheme by counting oscillator quanta. In so doing its necessary to recall 1) states of maximum intrinsic deformation lie lowest, 2) leading representations have highest weight states of maximum deformation, 3) all states of a given irreducible representation can be obtained by angular momentum projection from the highest weight state. For $(\tilde{f}\tilde{p})^{n_N=18,T_N=3}$, $(\lambda,\mu) = (24,0)$ is the leading SU(3) symmetry while for $(\tilde{f}\tilde{p})^{16,T_N=4}$ and $(\tilde{f}\tilde{p})^{16,T_N=2}$ they are (20,2) and (22,4), respectively. Our calculations have been restricted accordingly.

In a single j-shell all reasonable interactions correlate strongly with pairing. This means that interactions can be written to a good approximation in terms of invariants and generators of quasispin symmetries.[11] For identical particles the underlying group structure is R(3) while for (π,ν) systems it is R(5). Seniority is therefore a reasonably good quantum number and we choose to restrict the abnormal parity part of our basis to v=0 and 2 states with reduced isospin t=1. A complete labeling of the weak-coupled basis states, with the relevant group structure noted below, is

$$|[1^{n_N}](\tilde{f}\tilde{p})\tilde{\alpha}(\tilde{\lambda}\tilde{\mu})\tilde{K}\tilde{L}\tilde{\beta}(\tilde{S}T_N)J_N;\ (h_{11/2})^{n_A}(vt)\beta T_A J_A : T\ \ J> \tag{4}$$

$$U_{40}\quad U_{10}\quad SU_3\quad R_3\quad SU_2^S \times SU_2^T\quad U_{12}\quad R_5\quad SU_2^T\quad SU_2^J$$

In Section 4 we will report on results of various model calculations. Model I results are with n_N=18, n_A=8 and (λ,μ)=(24,0) while Model II(π) and Model II(ν) results include π and ν scattering to n_N=16, n_A=10 configurations with (λ,μ) = (20,2) and (22,4) K=0,2 only, respectively. The Model I, II(π), II(ν) dimensionalities are (6,27,27), (16,94,91), (24,150,146), (30,192,188), (34,220,216), (36,234, 232), (36,236,238), (36,228,239), (35,213,238) for J = 0,2,4,6,8,10,12,14,16, respectively.

3. Tensor Operators

The Hamiltonian has the general form

$$H = C_{\ell^2}\sum_{\alpha}^{n_N}\tilde{\ell}_\alpha^2 + C_{\ell\cdot s}\sum_{\alpha}^{n_N}\tilde{\ell}_\alpha\cdot\tilde{s}_\alpha + \varepsilon_A n_A$$

$$+ G_N H_N + G_A H_A + G_M H_M + G_P H_P \tag{5}$$

The first three terms are one-body potentials which generate, respectively, the separation of the $\tilde{\ell}$=1,3 centroids, the j=$\tilde{\ell}\pm$1/2 splitting, and the placement of the abnormal parity level relative to the normal parity ones. In the restricted bases of our models $\tilde{S}_N$=0 so contributions from the $\tilde{\ell}\cdot\tilde{s}$ term vanish. We set $\varepsilon_{1/2,3/2}$ = -2.02 MeV and $\varepsilon_{5/2,7/2}$ = -4.27 MeV. The single-particle energy of the abnormal parity level is only important when the model includes the possibility of pair scattering. For Model II studies we used $\varepsilon_{11/2}$ = -2.10 MeV. These values are consistent with Goodman's study of single-particle systematics in the region.[12]

The last four terms of (5) are two-body interactions. The form for these was chosen to be a surface delta interaction (SDI) which is known to work well in highly truncated spaces.[13] The labels N and A refer to the normal and abnormal spaces while H_M and H_P are multiple-multiple and pair scattering terms which couple the two. The strength factors were taken to be independent parameters. For Model II(π) results, for example, G_N=0.30, G_A=0.58, G_M=0.12, G_P=0.10 were used with the overall normalization of the SDI interaction set so that G_A=1.0 produces a 6 MeV binding of the $(h_{11/2})^2$ (v,J) = (0,0) configuration.

The procedure used for evaluating matrix elements of operators is straightforward. First the operator is expanded in terms of standard tensors,

$$\theta = \sum_{\substack{x_N^o\Gamma_N^o \\ \Gamma^o \\ x_A^o\Gamma_A^o}} C^{\Gamma^o}(x_N^o\Gamma_N^o, x_A^o\Gamma_A^o)\ [T(x_N^o\Gamma_N^o) \times T(x_A^o\Gamma_A^o)]^{\Gamma^o} \tag{6}$$

We abbreviate the notation by using Γ for T and J and x for additional labels. Basis states, see (4), are designated $|x\Gamma\rangle$, accordingly, or $|(x_N\Gamma_N, x_A\Gamma_A)\Gamma\rangle$ when it is important to emphasize the direct-product structure explicitly. Each tensor is by definition a density operator,

$$T(x^o\Gamma^o) = [|x'\Gamma'\rangle\langle x\Gamma|]^{\Gamma^o} \tag{7}$$

The C's are therefore matrix elements of θ in its defining space. The tensor decomposition of the $\tilde{\ell}^2$ and $\tilde{\ell}\cdot\tilde{s}$ parts of H is illustrated in Reference 6 while Reference 14 gives detailed results for two-body ds-shell interactions. Results for the normal parity part of the E2 operator are given in Reference 15.

For the many-particle reduced or double-barred matrix elements of any one component of θ we have

$$\langle (x_N'\Gamma_N', x_A'\Gamma_A')\Gamma' || [T(x_N^o\Gamma_N^o) \times T(x_A^o\Gamma_A^o)]^{\Gamma^o} || (x_N\Gamma_N, x_A\Gamma_A)\Gamma\rangle$$

$$= \begin{Bmatrix} \Gamma_N & \Gamma_N^o & \Gamma_N' \\ \Gamma_A & \Gamma_A^o & \Gamma_A' \\ \Gamma & \Gamma^o & \Gamma' \end{Bmatrix} \langle x_N'\Gamma_N' || T(x_N^o\Gamma_N^o) || x_N\Gamma_N\rangle \langle x_A'\Gamma_A' || T(x_A^o\Gamma_A^o) || x_A\Gamma_A\rangle \tag{8}$$

The quantity in brackets is a product of T and J normalized SU(2) 9j coefficients. Evaluating the matrix element factors on the right-hand-side of this equation is the major technical challenge. Each counts the frequency with which the structure defined by $T(x\Gamma)$ is found in the corresponding many-particle space. This counting is accomplished by coefficient of fractional parentage technology. Extensive program libraries are available for the N-space coefficients (for the U(N/4)xU(4) parts see Reference 16, for the SU(3)→R(3) parts see Reference 17) while Hemmenger and Hecht[18] give analytic expressions for the A-space parts. To bench mark the complexity of the calculations, we note that to generate a 250x250 energy matrix requires about ten minutes of IBM 3033 CPU time. However, the overhead seems justified by the simplicity gained in an understanding of the microscopic (fermion) mechanism responsible for the dynamics of backbending.

4. Results for ^{126}Ba

Energy spectra from a diagonalization of H in the separate N and A spaces are shown in Figure 3. The effective moment of inertia of the (24,0) representation is

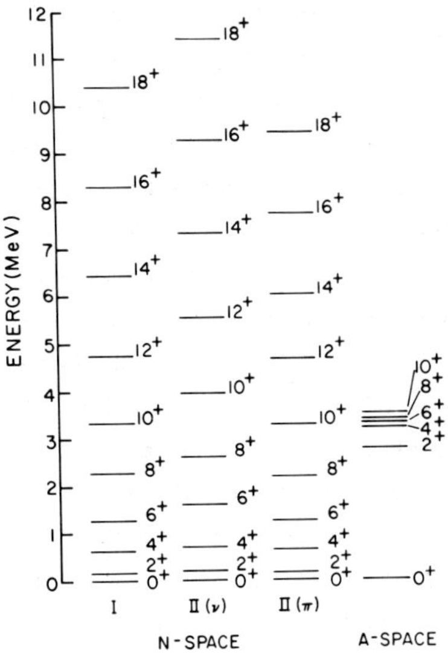

Figure 3 Energy spectra from diagonalizations of a surface delta interaction in normal and abnormal parity parts of model spaces for ^{126}Ba. Model I refers to the $(\tilde{\lambda}\tilde{\mu})$ = (24,0) subspace of the $(\tilde{fp})^{18}$T=3 configuration. Model II(ν) and II(π) include in addition (22,4) and (20,2) neutron and proton scattering representation, respectively. Band crossing accounts for the compression of the II(π) spectrum between spins 8-16.

almost identical to that of the K=0 band of the (22,4) ν-scattering representation but less than the moment of inertia of the K=0 band of the (20,2) π-scattering representation. Thus, while for low yrast states the (24,0) configuration dominates, for higher spins, because of its larger moment of inertia and correspondingly compressed rotational spectra, members of the π-scattering band are energetically favored. With $\varepsilon_{11/2}$ = -2.1 MeV this first occurs for spin 8 and accounts for the non-rotational character of the II(π) spectra between spins 8 and 16. The II(ν) results show no such irregularity. This band-crossing phenomena shows up even more clearly in Figure 4 where ΔE versus I, which for a rotor is a straight line with slope proportional to $1/2\mathcal{I}$, is plotted. Now pseudo SU(3) is not an exact symmetry of the interaction so the effect of the crossing as seen in the spectra appears smooth. The physical E2 operator, on the other hand, is predominantly a generator of the symmetry so transition rates are strongly moderated. We shall return to this point after a discussion of the A-space spectroscopy and pair alignment for this $(\tilde{\lambda}\tilde{\mu})$-mixing may account for the seemingly early onset of a reduction in B(E2) strengths[19] in ^{126}Ba.

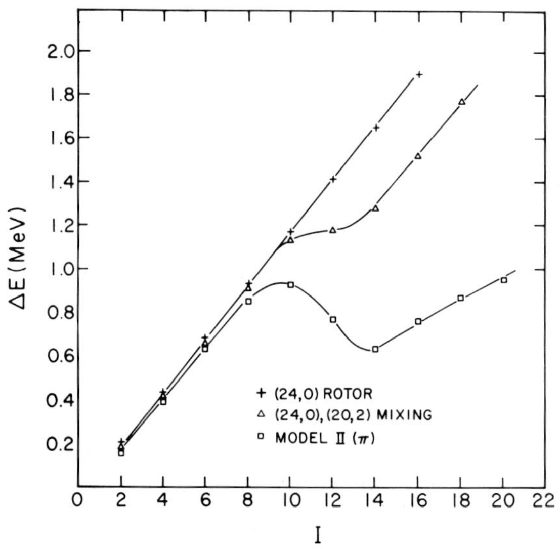

Figure 4 ΔE versus I plots. For a rigid rotor a straight line with slope proportional to $1/2\mathcal{I}$ is expected. The dip in the MIXING results is due to band crossing; the larger dip in the Model II(π) curve is due to pair alignment.

A weak coupling of the N and A spaces leads to a pair alignment phenomena. For an excitation energy less than about 3 MeV in ^{126}Ba the yrast states are built by coupling the paired $J_A=0$ configuration of the A-space to the N-space core. For larger energy values it is a coupling of unpaired ($J_A \neq 0$) configurations to the core that are favored. As 3 MeV corresponds to $J_N^\pi = 10^+$, for I = 12 it is $J_N = 2$ coupled to $J_A = 10$ that dominates; that is, the aligned coupling of an A-space pair coupled to maximum angular momentum to the lowest allowed spin of the core. The interaction term H_M mixes in $(J_N=4)\times(J_A=8)$ et cetera couplings as well. Results for the probability of finding $J_A \neq 0$ pairs coupled to the core are shown in Figure 5. Note that the transition is much sharper for Model I and II(ν) than for Model II(π). This is due to $(\tilde{\lambda}\tilde{\mu})$-mixing for the coupling $[(20,2)J_N=8]\times[J_A = 2]$ makes a significant contribution to the I = 10 state.

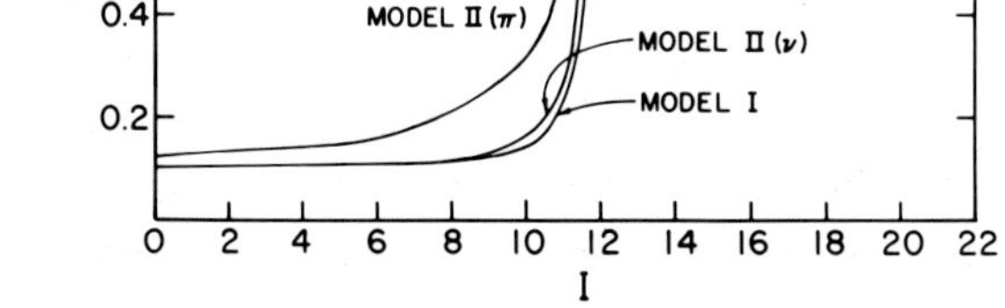

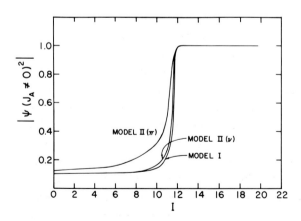

Figure 5 Probability of finding a pair in the abnormal parity space not coupled to angular momentum zero. Band mixing accounts for the greater probabilities at lower spin in the Model II(π) results.

Figure 6 shows on an $\mathcal{I}$ versus ω^2 plot results for the yrast states of ^{126}Ba. The name backbending is derived from displays of this type. The $\mathcal{I}$ versus ω^2 curve of a rigid rotor is a horizontal line. Accordingly, the band-mixing curve (not shown) is simply a smooth curve connecting two such straight lines. The back-sloping character of the theoretical curves is produced by pair alignment. Additional comparisons can be made with the aid of Figure 7. Band mixing is again seen to smooth the

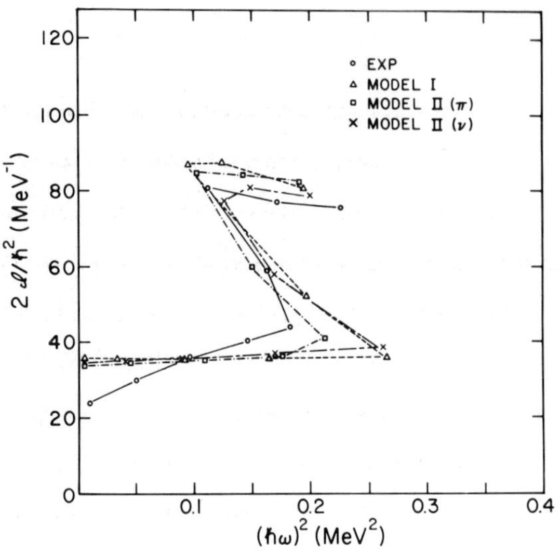

Figure 6 Backbending in ^{126}Ba. The curves were calculated using $2\mathcal{I}/\hbar^2 = (4I-2)/\Delta E$ and $(\hbar\omega)^2 = \Delta E^2/\{[I(I+1)]^{1/2} - [(I-2)(I-1)]^{1/2}\}^2$ with $\Delta E = E_{I+2}-E_I$. Experimental values are from Reference 20.

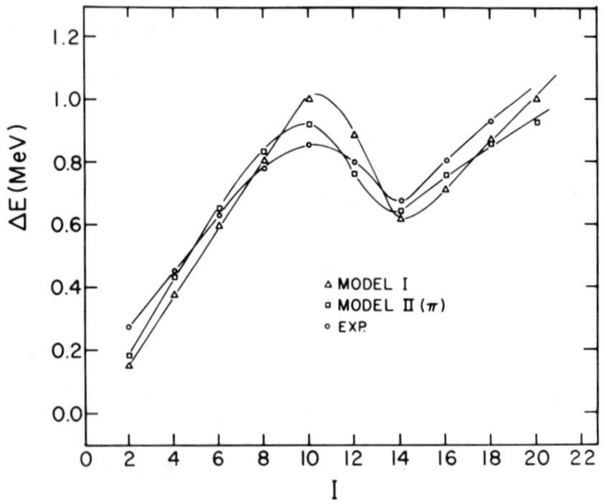

Figure 7 ΔE versus I plots. A departure from linearity signals non-rotational behavior. The abruptness of the pair alignment transition in Model I results is dampened by pre-alignment band mixing in Model II(π) calculations.

transition to the aligned configuration but as Figure 4 shows (Model II(π) results are the same in both) it alone does not explain the experimental data.[20] Differences in the moment of inertia (slope α $1/2\mathcal{I}$) above and below the backbending region are in best agreement with the Model II(π) predictions. A plot of $\hat{I}\cdot\bar{J}_A$, the projection of the angular momentum of the A-space particles onto a unit vector in the direction of the total spin, shows the alignment, either with or without band mixing, to be a very sharp phenomena.

Energy spectra for ^{126}Ba are shown in Figure 8. Experimental numbers are given on the left and calculated Model II(π) results on the right. For Model I studies the parameters G_N of H was adjusted to reproduce the I = 0-10 part of the yrast band, G_A was picked to yield the observed onset of alignment at I = 10, while G_M was selected

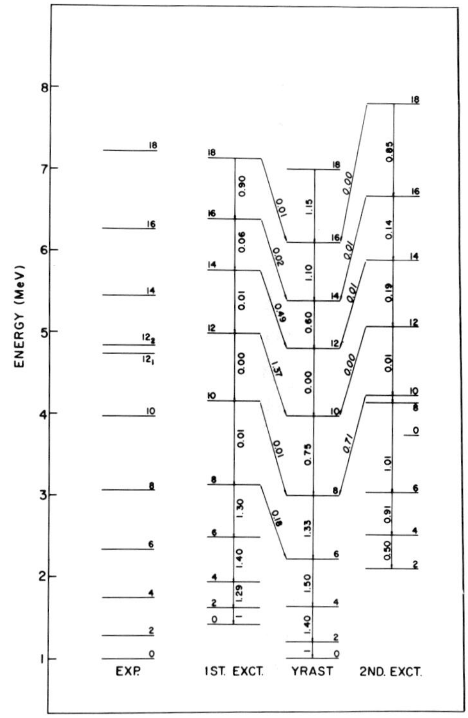

Figure 8 Energy spectra for ^{126}Ba. Experimental results are from Reference 20. Calculated numbers are for Model II(π). B(E2) strengths are for proton and neutron effective charges $e_\pi=(1+0.5)e$ and $e_\nu = 0.5e$. Numbers quoted are relative to the $2^+\to0^+$ transition for which the calculated strength is 0.21×10^{-48} cm^4.

to produce a reasonable fit to the experimental backbending curve. Model II studies include in addition G_P. As the multiple-multiple and pair scattering terms derive from the same N-A space two-body interaction the adjusted G_P value turned out to be nearly the same as G_M. B(E2) strengths for both intraband and interband transitions are also shown. Contributions from particles in the $h_{11/2}$ orbital are small so the yrast strengths for I = 0-8 are nearly identical to those for the N-space alone. Though the $4\to2$ and $6\to4$ strengths are consistent with rigid rotor values, the $8\to6$ and $10\to8$ ratios are significantly less. This pre-alignment reduction in B(E2) strengths is due to band mixing, a feature that is not a part of the Model I picture and hence not previously reported (Reference 5). The yrast I=12 state is an aligned configuration so it does not couple to the yrast I=10 state whose partner is clearly the first excited I=12 state. In general, throughout the backbending region, the assignment of states to bands is at best difficult and probably not a useful exercise as the features that emerge are sensitive to the parameters of H.

5. Conclusion

A weak coupling model based on normal parity configurations organized into multiplets of pseudo SU(3) and abnormal parity configurations organized into multiplets of R(5) has been shown to offer a satisfactory framework for studying backbending and related phenomena in heavy deformed nuclei. Pair alignment emerges as the primary mechanism of backbending but band crossing is competitive and can lead to B(E2) strength anomalies.

The harshest criticism that can be laid against the ^{126}Ba study is the severity of the basis truncation. This could be thwarted by a much expanded effort. But group theoretical arguments and statistical spectroscopy measures together with experience gained in ds-shell studies serve to bolster our confidence in the reliability of the results. We do not claim high accuracy nor expect detailed agreement between theory and experiment; we do believe the essential physics has been bared.

Recent experimental work[21] on lighter nuclei in the Ge region offer a unique challenge for further testing of the model. Forking to non-yrast states in the backbending region has been observed. In this case it is a coupling of a pseudo ds-shell with $g_{9/2}$ configurations that enters so an expanded calculation can be more easily accommodated. We anticipate making such studies.

Finally it is natural to ask about odd-A systems. For normal parity states the odd-A character will be that of the core. To study pair alignment in abnormal parity states one can anticipate the need for including v = 1,3 configurations. Such investigations lie in the future.

The authors are grateful for the interest and input of K. T. Hecht, and assists from Ms. Hortensia V. Delgado.

References

1. A. Bohr and B. R. Mottelson, in Proceedings of the International Conference on Nuclear Structure, edited by T. Marumori (Japan, 1977) J. Phys. Soc. Japan 44, Suppl. 157 (1978) .

2. A. Faessler, W. Greiner and R. K. Sheline, Nucl. Phys. 62 (1965) 241.

3. F. S. Stephens and R. S. Simon, Nucl. Phys. A183 (1972) 257.

4. Alan L. Goodman, in Advances in Nuclear Physics, Vol. 11, edited by J. W. Nagele and Erich Vogt (Plenum, New York, 1978).

5. R. D. Ratna Raju, K. T. Hecht, B. D. Chang and J. P. Draayer, Phys. Rev. 20C (1979) 2397.

6. R. D. Ratna Raju, J. P. Draayer and K. T. Hecht, Nucl. Phys. A202 (1973) 433.

7. T. R. Halemane, K. Kar and J. P. Draayer, Nucl. Phys. A311 (1978) 301.

8. J. E. Koops and P. W. M. Glaudemans, Z. Physik A280 (1977) 181.

9. T. T. S. Kuo, Nucl. Phys. A103 (1967) 71.

10. J. P. Elliott, Proc. Roy. Soc. A245 (1958) 128, 562; J. P. Elliott and M. Harvey, Proc. Roy. Soc. A272 (1963) 557.

11. A. K. Kerman, Ann. Phys. 12 (1961) 300; A. K. Kerman, R. D. Lawson, M. H. Macfarlane, Phys. Rev. 124 (1961) 162.

12. Alan L. Goodman, Nucl. Phys. A331 (1979) 401.

13. J. P. Schiffer, in Two-Body Forces in Nuclei, edited by S. M. Austin and G. M. Crawley (Plenum, New York, 1972).

14. J. P. Draayer, Nucl. Phys. A216 (1973) 457.

15. D. Braunschweig and K. T. Hecht, Phys. Lett. 77B (1978) 33.

16. D. Braunschweig, Comp. Phys. Commun. 14 (1978) 109; 15 (1978) 259.

17. Y. Akiyama and J. P. Draayer, Comp. Phys. Commun. 5 (1973) 405.

18. K. T. Hecht, Nucl. Phys. A102 (1967) 11; R. P. Hemenger and K. T. Hecht, Nucl. Phys. A145 (1970) 468.

19. G. Seiler-Clark, D. Husar, R. Novotny, H. Gräf and D. Pelte, Phys. Lett. 80B (1979) 345.

20. C. Flaum, D. Cline, A. W. Sunyar, O. C. Kistner, Y. K. Lee and J. S. Kim, Nucl. Phys. A264 (1976) 291; C. Flaum and D. Cline, Phys. Rev. 14C (1976) 1224.

21. A. P. deLima, J. H. Hamilton, A. V. Ramayya, B. van Nooijen, R. M. Ronningen, H. Kawakami, R. B. Piercey, E. deLima, R.L. Robinson, H. J. Kim, W. K. Tuttle, L. K. Peker, F. A. Rickey and R. Popli, Phys. Lett. 83B (1979) 43.

SU_3 SYMMETRY IN RESONATING GROUP CALCULATIONS. TOWARD A MICROSCOPIC THEORY OF $^{12}C + ^{12}C$ MOLECULAR RESONANCES .

K. T. Hecht
Max-Planck-Institut für Kernphysik, Heidelberg, Germany

W. Zahn
Institut für Theoretische Physik der Universität Erlangen-Nürnberg, Germany

Although group theory has played an important role in nuclear spectroscopy, detailed applications of group theoretical techniques have nevertheless been quite limited since the needed machinery is difficult to work out for most of the groups of interest in nuclear problems. However, for SU_3, the symmetry group of the 3-dimensional harmonic oscillator, the technology needed for a detailed application of the Wigner-Racah calculus has been developed [1] to a stage that it can be exploited in challenging, practical problems in nuclear physics. In the investigation of nuclear reaction problems by the resonating group method (RGM), e.g., the wave functions are naturally described in a cluster model basis in which it is possible to exploit the SU_3 symmetry of both the internal wave functions of the separate fragments and the wave function which describes the relative motion of the fragments. The norm, overlap, and interaction kernels which are needed in RGM calculations can be evaluated by straightforward techniques only through the evaluation of extremely complicated, multi-dimensional integrals. In a study reported at an earlier conference on the applications of group theoretical techniques to nuclear problems[2] it was shown how the problem of evaluating such kernels can be reduced to purely algebraic techniques involving the algebra of SU_3 recoupling transformations. The method [3] involves the calculation of an integral transform, the Bargmann-Segal (BS) transform, of the kernel and an expansion of this transform in terms of appropriate SU_3-coupled Bargmann space functions. The coefficients in these expansions give the numerical values of the desired multi-dimensional integrals. Further developments have been made in this technique [4] so that it can now be applied to cluster systems made up of two heavy fragments other than closed shell nuclei.

It is our aim to show how such techniques can be used to throw some light on an old and essentially unsolved problem in nuclear reaction theory, the problem of the so-called molecular resonances in the scattering of ^{12}C on ^{12}C and similar heavy ions. Although a fully microscopic multi-channel resonating group calculation for this challenging problem is perhaps not yet within reach, it may be possible to gain some understanding of the observed fine structure of the $^{12}C + ^{12}C$ resonances in terms of predicted reduced width amplitudes for the various observed exit channels. These can now be calculated in terms of a fully microscopic approach for which the proper antisymmetrization of the $^{12}C + ^{12}C$ system is most important. Although a large number of theoretical studies of the $^{12}C + ^{12}C$ resonances seem to confirm the usefulness of the quasi-molecular picture [5], an interpretation of the gross structure of these resonances seems to be possible without invoking molecular pockets in the potentials describing the relative motion of the two ^{12}C ions [6]. Evidence for the nuclear molecular picture must thus be sought in the fine structure of these resonances. The large number of closely spaced, narrow fine structure components are the distinctive feature of the $^{12}C + ^{12}C$ resonances. (A much simpler fine structure is observed in the $^{16}O + ^{16}O$ resonances, e. g.) This may be related to the nonspherical shape of the two ^{12}C nuclei, or, in the language of SU_3 symmetry, to the much richer SU_3 structure of the $^{12}C + ^{12}C$ cluster model wave function.

Such a cluster model wave function for the $^{12}C + ^{12}C$ system can be described by

$$\psi_{CLUSTER} = \mathcal{A}\left(\left(\varphi(^{12}C)^{(04)}_{I_1} \times \varphi(^{12}C)^{(04)}_{I_2}\right)_{I_C} \times \phi(\underset{\sim}{R}_{^{12}C-^{12}C})_L\right)_{JM_J} \tag{1}$$

where $\mathcal{A}$ is the total antisymmetrization operator. The internal wave functions, φ, of the two ^{12}C nuclei, are assumed to have good Wigner supermultiplet and hence good SU_3 symmetry, given by the Elliott quantum numbers $(\lambda\mu) = (04)$, with I_1 (or I_2) = 0, 2, 4. That is, in our model wave function, each ^{12}C nucleus can change its state of orientation but will not change its intrinsic shape. We permit rotational but no vibrational or other intrinsic excitations of the ^{12}C nuclei. In the usual description of such a reaction channel the internal angular momenta I_1 and I_2 of the two ^{12}C nuclei are coupled to channel spin, I_c, which in turn is coupled with the orbital angular momentum of the ^{12}C-^{12}C relative motion function, $\phi(\underset{\sim}{R})$, to total angular momentum J. (The round brackets in eq.(1) denote ordinary angular momentum coupling). It will be useful to expand the relative motion function in terms of harmonic oscillator functions, $\chi(\underset{\sim}{R})^{(Q0)}_{LM}$, carrying Q=2N+L oscillator quanta, and to give the cluster function in SU_3-coupled form

$$\psi = \mathcal{A}\left[\left[\varphi(^{12}C)^{(04)} \times \varphi(^{12}C)^{(04)}\right]^{(\lambda_c\mu_c)} \times \chi(\underset{\sim}{R})^{(Q0)}\right]^{(\lambda\mu)}_{KJM_J} \tag{2}$$

where the square brackets now denote SU_3 coupling. The SU_3 coupling $(04)\times(04)\rightarrow(\lambda_c\mu_c)$ leads to the possibilities $(\lambda_c\mu_c)$ = (08), (16), (24), (32), (40); but, for this system of two identical ^{12}C nuclei, the symmetrically coupled internal functions with $(\lambda_c\mu_c)$ = (08), (24), (40) can couple only with relative motion functions of even Q, hence even L and positive parity, while the antisymmetrically coupled internal functions with $(\lambda_c\mu_c)$ = (16), (32) can couple only with relative motion functions of odd Q and negative parity.

Although matrix elements in the cluster function basis have been calculated as functions of Q, so that Q can be arbitrarily high, and relative motion functions of general shape can be accomodated, the dominant components of the quasi-bound sub-Coulomb resonant states of the $^{12}C + {}^{12}C$ system can be expected to have large overlaps with shell model states of $2\hbar\omega$ oscillator excitation. The projection of the $^{12}C + {}^{12}C$ cluster model functions on states (2) with $Q = 14 = [Q(\text{minimum Pauli-allowed}) + 2]$ correspond to specific shell model states of $2\hbar\omega$ excitation. If the resonances observed in the $^{12}C + {}^{12}C$ excitation functions are of nuclear molecular origin, the quasi-bound states associated with these resonances can thus be expected to be built from fully antisymmetrized, properly normalized basis vectors of the type (2), with Q = 14.

The SU_3-coupled form of the cluster function is extremely useful for identifying the Pauli-allowed state vectors [7]. If internal and relative motion functions, φ and χ are built from oscillator functions of the same length parameter, norm and overlap kernels are simple SU_3-scalars, and the eigenvectors of the antisymmetrizer, $\mathcal{A}$, must be linear combinations of the functions (2) with the same $(\lambda\mu)$. Table 1 lists the possible $(\lambda_c\mu_c)$, $(\lambda\mu)$ combinations for states with Q=14. Most of the states with $2\lambda+\mu \geq 30$ can be identified at once as completely Pauli-forbidden, since the A=24 shell model space of $2\hbar\omega$ oscillator excitation has no states of this high SU_3 symmetry. Only a single core excitation with $(\lambda\mu) = (14,2)$, $|s^4p^{12}[sd^6(82)pf^2(60)](14,2)\rangle$, has SU_3 symmetry with $2\lambda+\mu = 30$. One linear combination of states with $(\lambda_c\mu_c) = (24), (40)$; $(\lambda\mu) = (14,2)$ is therefore Pauli-forbidden. The 2x2 $(\lambda\mu) = (14,2)$ submatrix of the full norm matrix must thus have one eigenvalue of zero, corresponding to a Pauli-forbidden state which is annihilated by $\mathcal{A}$. Symmetrically coupled states of even L, $((I_1I_2)I_c L)J$, number 14, 40, and 57 for J = 0, 2, and 4. Of these only 8, 22, and 29 linear combinations for J = 0, 2, and 4, respectively, are Pauli-allowed in the Q=14 space. These Pauli-allowed vectors, (to be denoted by $|(\lambda\mu)i\rangle$), can be expected to form a good zeroth order basis for quasi-bound states of $^{12}C + {}^{12}C$ resonances; that is, if the molecular picture is a sound one for the description of the observed excitation functions.

For the closely spaced, narrow resonances of the $^{12}C + {}^{12}C$ reactions we assume that the T matrix can be approximated in terms of a few partial width amplitudes $\Gamma^{1/2}_{\nu c}$, by

$$T_{c'c} = \exp(i\delta_c + i\delta_{c'}) \sum_{\nu} \frac{\Gamma^{1/2}_{\nu c'}\Gamma^{1/2}_{\nu c}}{E - \varepsilon_\nu} \tag{3}$$

where the entrance channel, c, consists of $^{12}C + {}^{12}C$ with $I_1 = I_2 = 0$. The important exit channels include: (1) $^{12}C + {}^{12}C$ with $I_1, I_2 \neq 0, 0$; (2) $\alpha + {}^{20}Ne$, where the most important ^{20}Ne excitations can be described in the SU_3 approximation by $(\lambda\mu)_{Ne} = (80)$ for states $0^+, 2^+, 4^+, 6^+, 8^+$ of the ^{20}Ne ground state rotational band, and by $(\lambda\mu)_{Ne} = (82)$ for the $2^-, 3^-, 4^-, \ldots$ band with bandhead at 4.97 MeV; (3) $^{16}O + {}^{8}Be$ with $(\lambda\mu)_{Be} = (40)$; (4) $p + {}^{23}Na$ and $n + {}^{23}Mg$ with ^{23}Na or ^{23}Mg excitations in $K = 3/2^+$ an $1/2^+$ rotational bands which, in the SU_3 approximation, can be described by states with $(\lambda\mu)_{A=23} = (83)$.

Since the observed resonances of the same J^{π} fall within an energy interval of $\sim$ 2 MeV, and since individual fine structure components may correspond to quasi-bound states of very similar structure, the penetrability factors associated with the partial width amplitudes ($\Gamma^{1/2}_{\nu c'}\, \Gamma^{1/2}_{\nu c}$) can be expected to be approximately independent of ν. The relative magnitudes of the ($\Gamma^{1/2}_{\nu c'}\, \Gamma^{1/2}_{\nu c}$) for different ν can thus be expected to be proportional to the relative magnitudes of the corresponding reduced widths or the corresponding spectroscopic amplitudes $A_{\nu c} A_{\nu c'}$.

The calculation of these spectroscopic amplitudes involves two steps: 1) the solution of the quasi-bound state eigenvalue problem, $|\nu\rangle = \sum |(\lambda\mu)i\rangle\langle(\lambda\mu)i|\nu\rangle$, which gives a particular quasi-bound state ν in terms of the molecular basis states $|(\lambda\mu)i\rangle$, where $|(\lambda\mu)i\rangle$ are the Pauli-allowed linear combinations of the states (2); 2) the calculation of the overlaps of states $|(\lambda\mu)i\rangle$ with the appropriate cluster functions for the various exit channels c'.

For the $^{12}C + {}^{12}C$ resonances simple interactions of the quadrupole-quadrupole (Q·Q) type can perhaps be expected to play an important role in the splitting of specific gross structure resonance peaks into different fine structure components. For an interaction of the Q·Q type, it may be sufficient to use a so-called orthogonality condition approximation in which interaction matrix elements are calculated in a basis orthogonal to the Pauli-forbidden states. Matrix elements of the operator, $\mathcal{A}$, between cluster states of type (2) may thus suffice for both steps 1) and 2). Matrix elements of the form

$$\langle [[\varphi(^{12}C)^{(04)} \times \varphi(^{12}C)^{(04)}]^{(\lambda_c\mu_c)} \times \chi(\underset{\sim}{R})^{(Q0)}]^{(\lambda\mu)}_{KJM} \,|\, \mathcal{A} \,|\, [[\varphi(f)^{(\lambda_f\mu_f)} \times \varphi(A-f)^{(\lambda_{A-f}\mu_{A-f})}]^{(\lambda'_c\mu'_c)} \times \chi(\underset{\sim}{R}')^{(Q'0)}]^{(\lambda\mu)}_{KJM} \rangle \qquad (4)$$

can be evaluated by the combination of BS integral transform and SU_3 recoupling techniques. With f = A-f = 12, a $^{12}C + {}^{12}C$ fragment decomposition of the A = 24 system, the above matrix element is needed to find the Pauli-allowed eigenvectors of the operator $\mathcal{A}$. With f $\neq$ 12; e.g., f=4, A-f = 20, the above becomes an overlap matrix element of the type needed in the calculation of a spectroscopic amplitude from a $^{12}C + {}^{12}C$ molecular basis state to a state in an $\alpha + {}^{20}Ne$ exit channel.

<u>The Bargmann-Segal Transform</u>. Norm and Overlap matrix elements of the type (4) can be evaluated most easily in terms of the BS transform [8] of the operator $\mathcal{A}$.

$$\mathcal{H}(\bar{\underset{\sim}{k}}, \underset{\sim}{k}^*) = \int d\underset{\sim}{r}_1 \cdots \int d\underset{\sim}{r}_A \prod_{i=1}^{A} A(\bar{\underset{\sim}{k}}_i, \underset{\sim}{r}_i)\, \mathcal{A} \prod_{i=1}^{A} A(\underset{\sim}{k}^*_i, \underset{\sim}{r}_i) \,. \qquad (5)$$

Here, $A(\underset{\sim}{k}_i, \underset{\sim}{r}_i)$ is the kernel function for the BS transform in single-particle coordinates

$$A(\underset{\sim}{k},\underset{\sim}{r}) = \prod_{\alpha=x,y,z} A(k_\alpha, r_\alpha)\ ;\ A(k_x,x) = \frac{1}{\pi^{1/4}} \exp\left(-\tfrac{1}{2}k_x^2 - \tfrac{1}{2}x^2 + \sqrt{2}k_x x\right) = \sum_{n=0}^{\infty} \chi_n^*(x)\frac{k_x^n}{\sqrt{n!}} \qquad (6)$$

where $\chi_n(x)$ is a normalized, 1-dimensional harmonic oscillator function, and $k_x^n/\sqrt{n!}$ is its BS transform. The BS transform of a 3-dimensional oscillator function $\chi(\underset{\sim}{R})^{(Q0)}_{LM}$, is to be denoted by $P(\underset{\sim}{K})^{(Q0)}_{LM}$, and retains the SU_3 irreducible tensor character (Q0). The evaluation of the above BS transform involves the following steps: 1). the trivial evaluation of the integral in single-particle coordinates; 2). transformation to a set of internal and relative motion coordinates appropriate to the cluster decomposition of the A-particle system, e.g., A=24 → 12+12 for the $\bar{\underset{\sim}{K}}$ Bargmann space variables of the bra and A=24 → f + A-f for the $\underset{\sim}{K}^*$ Bargmann space variables of the ket of eq.(4); and 3). the expansion of the resultant K-space functions in appropriate SU_3-coupled form. Fig. 1 shows the internal and relative motion degrees of freedom for a $^{12}C + {}^{12}C$ cluster system. Oscillator excitations associated with most of the internal degrees of freedom are restricted to 0s states, e.g. all those associated with the α internal degrees of freedom in Fig. 1.

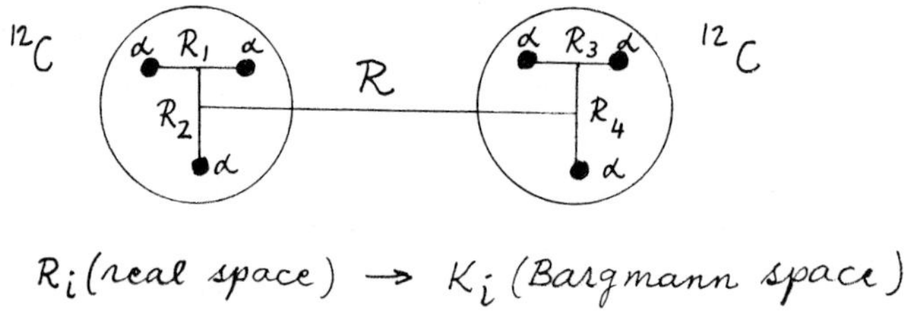

FIG. 1

A few additional internal degrees of freedom are fixed in their lowest Pauli-allowed excitations, those denoted $\underset{\sim}{R}_1$, $\underset{\sim}{R}_2$, $\underset{\sim}{R}_3$, and $\underset{\sim}{R}_4$ in Fig. 1, with Bargmann space transforms $\underset{\sim}{K}_1$, $\underset{\sim}{K}_2$, $\underset{\sim}{K}_3$, and $\underset{\sim}{K}_4$. Bargmann space functions for the internal degrees of freedom of the $^{12}C + {}^{12}C$ system can be expressed as functions of the two pseudo-vectors $\underset{\sim}{K}_{12} = [\underset{\sim}{K}_1 \times \underset{\sim}{K}_2]$ and $\underset{\sim}{K}_{34} = [\underset{\sim}{K}_3 \times \underset{\sim}{K}_4]$ to build K-space internal functions of the form

$$\left[P(\underset{\sim}{K}_{12})^{(04)} \times P(\underset{\sim}{K}_{34})^{(04)}\right]^{(\lambda_c\mu_c)}_{---} \equiv \mathcal{P}(\underset{\sim}{K}_{12},\underset{\sim}{K}_{34})^{(\lambda_c\mu_c)}. \qquad (7)$$

For a $^{12}C + {}^{12}C$ cluster decomposition in both bra and ket the BS transform of the operator, $\mathcal{A}$, can be expanded in the form

$$H(\bar{K},K^*)^{(\lambda_c\mu_c)(\lambda'_c\mu'_c)} = \sum_{Q(\lambda\mu)} I_{(\lambda_c\mu_c)Q,(\lambda'_c\mu'_c)Q;(\lambda\mu)} \left[\dim(\lambda\mu)\right]^{1/2} \times$$

$$\left[\left[P(\bar{K}_{12},\bar{K}_{34})^{(\lambda_c\mu_c)} \times P(\bar{K})^{(Q0)}\right]^{(\lambda\mu)} \times \left[P(K^*_{12},K^*_{34})^{(\mu'_c\lambda'_c)} \times P(K^*)^{(0Q)}\right]^{(\mu\lambda)}\right]^{(00)}_{000} \tag{8}$$

with

$$I_{(\lambda_c\mu_c)Q,(\lambda'_c\mu'_c)Q;(\lambda\mu)} = \langle\left[\left[\varphi(^{12}C)^{(04)} \times \varphi(^{12}C)^{(04)}\right]^{(\lambda_c\mu_c)} \times \chi(R)^{(Q0)}\right]^{(\lambda\mu)}_{KJM}$$

$$|A|\left[\left[\varphi(^{12}C)^{(04)} \times \varphi(^{12}C)^{(04)}\right]^{(\lambda'_c\mu'_c)} \times \chi(R)^{(Q0)}\right]^{(\lambda\mu)}_{KJM}\rangle .$$

The needed norm or overlap matrix elements thus fall out automatically as the coefficients of the appropriate SU_3-coupled K-space expansion. The details of this method have been demonstrated in refs. 3) where the BS transforms were calculated directly for the full A-particle system of arbitrary space symmetry characterized by 4-columned Young tableaux. In cluster systems built from two heavy fragments this direct method becomes somewhat cumbersome. The BS transforms are therefore first calculated for n-particle subsystems of space symmetries characterized by single-columned Young tableaux, (totally antisymmetric n-particle symmetries). The BS transforms for the full A-particle system are then built from these by suitable combinations of these single column functions [4)]. The BS transform for the ^{12}C + ^{12}C norm matrix is built from BS transforms of totally antisymmetric 6-particle subsystems with 3-particle internal functions, each of 3-particle SU_3 symmetry (01). These can couple to resultant 6-particle SU_3 symmetry $(\lambda_c\mu_c)$ = (02) or (10). The BS transform $H^{(02)(10)}$ vanishes because of the identity of the two fragments. The single-column BS transforms with $(\lambda_c\mu_c) = (\lambda'_c\mu'_c)$ are calculated by the techniques of ref. 4). Results are

$$H^{(02)(02)} = e^{\frac{1}{4}(\bar{K}\cdot K^*)}\Big\{\tfrac{1}{2}\left[(1-e')^3-(e'-e^2)(\tfrac{1}{6})^2(\bar{K}\cdot K^*)^2\right] \times$$
$$\left[(\bar{K}_{12}\cdot K^*_{12})(\bar{K}_{34}\cdot K^*_{34})+(\bar{K}_{12}\cdot K^*_{34})(\bar{K}_{34}\cdot K^*_{12})\right]$$
$$-(\tfrac{1}{6})^2(e'-e^2)\Big[(\bar{K}\cdot\bar{K}_{12})(\bar{K}\cdot\bar{K}_{34})(K^*\cdot K^*_{12})(K^*\cdot K^*_{34})-\tfrac{1}{2}(\bar{K}\cdot K^*)\times$$
$$((\bar{K}_{12}\cdot K^*_{12})(\bar{K}\cdot\bar{K}_{34})(K^*\cdot K^*_{34})+(\bar{K}_{34}\cdot K^*_{34})(\bar{K}\cdot\bar{K}_{12})(K^*\cdot K^*_{12})+ \tag{9}$$
$$(\bar{K}_{12}\cdot K^*_{34})(\bar{K}\cdot\bar{K}_{34})(K^*\cdot K^*_{12})+(\bar{K}_{34}\cdot K^*_{12})(\bar{K}\cdot\bar{K}_{12})(K^*\cdot K^*_{34}))\Big]\Big\} ,$$

$$H^{(10)(10)} = e^{\frac{1}{4}(\bar{K}\cdot K^*)}\Big\{(1-e')^2(1+e')(\bar{K}_i\cdot K^*_i)-2e'(1-e')\tfrac{1}{6}\left[(\bar{K}\cdot K^*)(\bar{K}_i\cdot K^*_i)\right.$$
$$\left.-(\bar{K}\cdot K^*_i)(\bar{K}_i\cdot K^*)\right]-e'(1+e')(\tfrac{1}{6})^2(\bar{K}\cdot K^*)(\bar{K}\cdot K^*_i)(\bar{K}_i\cdot K^*)\Big\} \tag{10}$$

with $\underset{\sim}{K}_i = \frac{1}{\sqrt{2}}\left[\underset{\sim}{K}_{12} \times \underset{\sim}{K}_{34}\right]$,

where e^p is shorthand notation for $e^p \equiv \exp\left\{-\frac{p}{6}(\bar{\underset{\sim}{K}}\cdot\underset{\sim}{K}^*)\right\}$. The BS transform for the full 24-particle system is built from the symmetrized combination of these four single-column functions.

$$\mathcal{H}(\bar{\underset{\sim}{K}},\underset{\sim}{K}^*) = \frac{1}{2}\left(\frac{1}{4!}\right)^2 \sum_a \binom{4}{a} \left[\mathcal{H}^{(02)(02)}\right]^{4-a} \left[\mathcal{H}^{(10)(10)}\right]^a . \tag{11}$$

For a system of two identical ^{12}C fragments the sum splits into two separate parts. Terms with a=even lead to 24-particle internal SU_3 symmetries $(\lambda_c\mu_c)$ = (08), (24), (40) only and are thus the only terms needed for states of positive parity, while terms with a=odd lead only to 24-particle $(\lambda_c\mu_c)$ = (16), (32) and are thus needed only for states of negative parity. For the positive parity states of primary interest

$$\mathcal{H}(\bar{\underset{\sim}{K}},\underset{\sim}{K}^*)_+ = \frac{1}{2}\left(\frac{1}{4!}\right)^2 \Big\{\left[\mathcal{H}^{(02)(02)}\right]^4 + 6\left[\mathcal{H}^{(02)(02)}\right]^2\left[\mathcal{H}^{(10)(10)}\right]^2 + \left[\mathcal{H}^{(10)(10)}\right]^4\Big\}. \tag{12}$$

It is now a straightforward exercise in SU_3 recoupling transformations to combine the four factors of each term in eq.(12) and bring them to the appropriate SU_3-coupled form needed for eq.(8). SU_3 coupling rules show at once that the last term of eq.(12) can contribute only to norm matrix elements with $(\lambda_c\mu_c) = (\lambda_c'\mu_c')$ = (40); while the middle term can make contributions to matrix elements with $(\lambda_c\mu_c)$, (and $\lambda_c'\mu_c'$)), = (24) and (40). The first term can contribute to all three $(\lambda_c\mu_c)$ values of (08), (24), and (40). For this term, e.g., the appropriate SU_3 expansion gives [4)]

$$\frac{1}{2}\left(\frac{1}{4!}\right)^2\left[\mathcal{H}^{(02)(02)}\right]^4 = \sum_{\substack{(\lambda_c\mu_c)(\lambda_c'\mu_c')=\\(08)(24)(40)}} \sum_{Q(\lambda\mu)} (-)^{Q+\lambda+\mu}\left[\frac{\dim(Q0)\dim(\lambda_c\mu_c)\dim(\lambda_c'\mu_c')}{\dim(\lambda\mu)}\right]^{1/2} \times$$

$$\sum_{\ell jmn}\sum_{rst}\sum_p \frac{(-)^p}{2^{5+m-j}}\binom{4}{\ell}\binom{\ell}{j}\binom{4-j}{n}\binom{j}{m}\binom{n}{r}\binom{s}{s}\binom{s}{t}\binom{12-2\ell}{p-\ell}\times$$

$$\left(\frac{1}{6}\right)^{2\ell}\left[1-\frac{p}{6}\right]^{Q-2\ell}\frac{Q!}{(Q-2\ell)!}\sum_{(\bar\lambda\bar\mu)}\frac{(-)^{\bar\lambda+\bar\mu}}{[\dim(\bar\lambda\bar\mu)\dim(2j-m,0)]^{1/2}}\times$$

$$\begin{bmatrix}(4-j+m-r+s-t,0) & (j-m+r-s+t,0) & (40)\\ (4-j+r-s+t,0) & (j-r+s-t,0) & (40)\\ (\bar\lambda\bar\mu) & (2j-m,0) & (\mu_c\lambda_c)\end{bmatrix}\begin{bmatrix}(4-j+m-r+t,0) & (4-j+r-t,0) & (\bar\lambda\bar\mu)\\ (j-m+r-t,0) & (j-r+t,0) & (2j-m,0)\\ (40) & (40) & (\mu_c'\lambda_c')\end{bmatrix}\times$$

$$\begin{bmatrix} (4-j+m-n-r,0) & (n+s-t,0) & (4-j+m-r+s-t,0) \\ (n+t,0) & (4-j-n+r-s,0) & (4-j+r-s+t,0) \\ (4-j+m-r+t,0) & (4-j+r-t,0) & (\bar{\lambda}\bar{\mu}) \end{bmatrix} \times \tag{13}$$

$$U((\mu_c'\lambda_c')(0Q)(\bar{\lambda}\bar{\mu})(Q-2j+m,0);(\mu\lambda)(0,2j-m)) \times$$

$$U((\lambda_c\mu_c)(2j-m,0)(\lambda\mu)(Q-2j+m,0);(\bar{\mu}\bar{\lambda})(Q0)) \times [\dim(\lambda\mu)]^{1/2} \times$$

$$\left[\left[\mathcal{P}(\bar{K}_{12},\bar{K}_{34})^{(\lambda_c\mu_c)} \times \mathcal{P}(\bar{K})^{(Q0)}\right]^{(\lambda\mu)} \times \left[\mathcal{P}(K^*_{12},K^*_{34})^{(\mu_c'\lambda_c')} \times \mathcal{P}(K^*)^{(0Q)}\right]^{(\mu\lambda)}\right]^{(00)}_{000}$$

The three 9-$(\lambda\mu)$ coefficients in this expansion are all equivalent to simple SU_2 9-j coefficients. The two U-coefficients are bona fide SU_3 Racah coefficients [1] however, all SU_3 couplings in these coefficients are free of outer multiplicities. Similar expansions of the last two terms of eq. (12) lead to the desired matrix elements (4) via eq.(8).

In the restricted ^{12}C + ^{12}C cluster model basis, the eigenvalue problem for the operator, $\mathcal{A}$, is factored into matrices of dimension $d \leq 3$ in a $|[(\lambda_c\mu_c)(Q0)](\lambda\mu)KJM\rangle$ basis, since the operator $\mathcal{A}$ is diagonal in $(\lambda\mu)$, and its matrix elements are independent of KJM. In the $((I_1I_2)I_cL)JM$ basis of eq.(1), in contrast, the corresponding eigenvalue problem would have led to a matrix of dimension 57 for J = 4, for example, with 29 Pauli-allowed eigenvectors, (with associated eigenvalues different from zero), and with 28 completely Pauli-forbidden eigenvectors, (with zero eigenvalues). In the $|[(\lambda_c\mu_c)(Q0)](\lambda\mu)\rangle$ basis, with Q = 14, only the states with $(\lambda\mu)$ = (10,4) lead to a 3x3 eigenvalue problem, see table 1. The 2x2 matrices with $(\lambda\mu)$ = (14,2), (11,5) and (12,3) have one zero eigenvalue. For each of these only one linear combination of $(\lambda_c\mu_c)$ values leads to a Pauli-allowed eigenvector in the ^{12}C + ^{12}C cluster basis. For $(\lambda\mu)$ = (14,2) and (11,5) only a single shellmodel state exists of this high symmetry, so that only a single Pauli-allowed eigenvector exists for these $(\lambda\mu)$ even in an expanded cluster model basis, including α+ ^{20}Ne and other channels. The eigenvectors of the operator $\mathcal{A}$ are to be denoted by $|(\lambda\mu)i\rangle$, with

$$\mathcal{A}\,|(\lambda\mu)i\rangle = \Lambda_i\,|(\lambda\mu)i\rangle \tag{14}$$

$$|(\lambda\mu)i\rangle = \sum_{(\lambda_c\mu_c)} |[(\lambda_c\mu_c)(Q0)](\lambda\mu)KJM\rangle\langle[(\lambda_c\mu_c)(Q0)](\lambda\mu)|(\lambda\mu)i\rangle .$$

The eigenvectors and eigenvalues for the $^{12}C + {}^{12}C$ states are shown in table 2. (Note that the transformation coefficients are independent of KJM_J). The fully antisymmetrized, normalized state vectors are given by

$$\Lambda_i^{-1/2} \mathcal{A} \,|(\lambda\mu) i K J M\rangle . \qquad (15)$$

The Pauli-allowed eigenvectors $|(\lambda\mu)i\rangle$ can be used as a set of $^{12}C + {}^{12}C$ molecular basis vectors for the quasi-bound state eigenvalue problem, $|\nu JM\rangle = \sum_{(\lambda\mu)i} |(\lambda\mu)i\rangle\langle(\lambda\mu)i|\nu JM\rangle$ for the sub-Coulomb resonances. Simple interactions of Q·Q type can be expected to play an important role in the splitting of the quasibound states into different fine structure components. In addition, the relative motion energy will have an $\underset{\sim}{L}^2$ component which will contribute to this splitting. The follwing simplified Hamiltonian, when diagonalized in an orthgonality condition approximation, may thus give some insight to the structure of the $^{12}C + {}^{12}C$ resonances:

$$H = H_{i_1} + H_{i_2} + H_{rel.} + H_{interaction} . \qquad (16)$$

The H_i for the ^{12}C internal degrees of freedom are restricted to have $J^\pi = 0^+, 2^+, 4^+$ excitations, with internal energies of 0, 4.44, and 14.08 MeV (taken from experiment). The Hamiltonian for the relative motion is taken to have an L(L+1)-dependent term with a rotational constant of the observed ~100 keV. The interaction Hamiltonian is assumed to be dominated by Q·Q terms with Elliott-type eigenvalues

$$(Q_{i_1} \cdot Q_{i_2}) = \frac{1}{2}\left\{C_{(\lambda_c\mu_c)} - 2C_{(04)} - \frac{3}{4}\left[I_c(I_c+1) - I_1(I_1+1) - I_2(I_2+1)\right]\right\} \qquad (17a)$$

$$(Q_{rel} \cdot (Q_{i_1} + Q_{i_2})) = \frac{1}{2}\left\{C_{(\lambda\mu)} - C_{(\lambda_c\mu_c)} - C_{(Q0)} - \frac{3}{4}\left[J(J+1) - L(L+1) - I_c(I_c+1)\right]\right\} \qquad (17b)$$

where the SU_3 Casimir invariants are given by $C_{(\lambda\mu)} = \lambda^2 + \lambda\mu + \mu^2 + 3(\lambda+\mu)$. Even without an $H_{int.}$ of this type, the Pauli principle plays an important role in determining the position and the spacing of individual fine structure components. This is illustrated in Fig. 2 by the 2^+ levels of our molecular basis. Fig. 2(a) shows the 2^+ levels with $H_{int.}$ and the Pauli principle "turned off", (i.e. with antisymmetrization effects neglected). Fig. 2(b) shows the 2^+ eigenvalues in the same molecular basis, restricted by the Pauli principle, with states now properly antisymmetrized. The Pauli principle not only reduces the density of levels but is responsible for large effective energy shifts. Fig.2(c) shows the 2^+ levels with an additional effective interaction, (see Fig. caption), chosen to give several closely spaced 2^+ fine structure components in the energy domain of the observed 2^+ sub-Coulomb resonances. The 0^+ and 4^+ levels for this interaction are also shown. Clearly, a more detailed calculation is needed in which such quasi-bound states are properly coupled to states in the continuum of the entrance and exit channels. A fully microscopic multi-channel resonating group calculation for the ^{12}C + ^{12}C problem is still not quite within reach; but calculations of the type exhibited in Fig. 2 clearly show the importance of antisymmetrization effects.

It may also be possible to gain some understanding of the observed resonance fine structure in terms of predicted reduced widths or spectroscopic amplitudes, A_{ic} for the entrance channel, and $A_{ic'}$ for the many observed exit channels. Moreover, the amplitudes $A_{ic'}$ for the observed exit channels c' may give an indication of the importance of a particular $|i\rangle$ in a specific fine structure component of a resonance. The amplitude $A_{ic'}$ gives the projection of the normalized, antisymmetrized state $|i\rangle$, given by eq.(15), onto a channel state c' characterized by outgoing fragments of mass numbers f and A-f, with SU_3 symmetries $(\lambda_f \mu_f)$ and $(\lambda_{A-f} \mu_{A-f})$, and angular momenta I_f, I_{A-f} coupled to channel spin I_c; see eq.(1). If the internal functions for fragments f and A-f are SU_4-scalars, such an amplitude is given by

$$A_{ic'} = \sum_{(\lambda_c'\mu_c')K_c'} \langle (\lambda_f\mu_f)K_f I_f ; (\lambda_{A-f}\mu_{A-f})K_{A-f} I_{A-f} \| (\lambda_c'\mu_c')K_c' I_c \rangle \times \langle (\lambda_c'\mu_c')K_c' I_c ; (14-\Delta,0)L \| (\lambda\mu)KJ \rangle \; (f+A-f(\lambda_c'\mu_c')|(\lambda\mu)i) \tag{18}$$

where the double-barred coefficients are $SU_3 \supset R_3$ Wigner coefficients [1] and the factor $(f+A-f(\lambda_c'\mu_c')|(\lambda\mu)i)$ is defined by

$$(f+A-f(\lambda_c'\mu_c')|(\lambda\mu)i) \equiv \sum_{(\lambda_c\mu_c)} \Lambda_i^{-1/2} \langle [(\lambda_c\mu_c)(14,0)](\lambda\mu)|(\lambda\mu)i\rangle \times \langle [[\varphi(^{12}C)^{(04)} \times \varphi(^{12}C)^{(04)}]^{(\lambda_c\mu_c)} \times \chi(\underset{\sim}{R})^{(14,0)}]^{(\lambda\mu)}_{KJM} | \mathcal{A} | [[\varphi(f)^{(\lambda_f\mu_f)} \times \varphi(A-f)^{(\lambda_{A-f}\mu_{A-f})}]^{(\lambda_c'\mu_c')} \times \chi(\underset{\sim}{R}')^{(14-\Delta,0)}]^{(\lambda\mu)}_{KJM} \rangle . \tag{19}$$

The i-dependent factors can be read from table 2. The overlap matrix elements of $\mathcal{A}$ are calculated for the most important exit channels by the techniques sketched in connection with eq. (5)-(8). Details can be found in ref. 4). The constant Δ is determined by the total number of internal oscillator excitations in the fragments f and A-f. Outgoing channels in which the fragments carry more complicated SU_4 symmetry and intrinsic spins and isospins require additional isospin and spin and angular momentum dependent factors, (see ref.4)).

With $(\lambda_f\mu_f) = (\lambda_{A-f}\mu_{A-f}) = (04)$, and $I_f = I_{A-f} = 0$, eq.(18) gives the amplitudes A_{ic} for the entrance channel. These amplitudes are shown in table 3 for the states $|(\lambda\mu)iKJM_J\rangle$ with J = 0, 2, 4, the observed J-values of the sub-Coulomb resonances. We note the following properties of these amplitudes: (1) Only states with $K = 0$ (μ= even) and $K = 1$ (μ= odd) have significant amplitudes. Amplitudes for states with $K \geq 2$ are so small that such states can not be expected to play a role in the ^{12}C + ^{12}C resonances. (2) The few most significant amplitudes are those for states with λ= even, μ= even. (3) The large amplitudes for the $K = 0$ band of states in a given $(\lambda\mu)$ are only very mild functions of J.

The factors $(f + A\text{-}f(\lambda_c\mu_c) \| (\lambda\mu)i)$ are shown in table 4 for the most important observed exit channels. They are the primary factors in determining the relative importance of the different exit channels. The combination of the numbers of tables 3 and 4 show that the state (10,4)1 can be expected to make the dominant contribution to the product of partial width amplitudes $\Gamma^{1/2}_{\nu c}\,\Gamma^{1/2}_{\nu c'}$ for most of the observed channels c'. However, states (14,2)1 and (11,5)1 can make significant contributions to the $\alpha + {}^{20}Ne$ channels, and the states (93)1 and (82)1 may be important for the ^{23}Na and ^{23}Mg + nucleon channels.

The amplitudes, $A_{ic'}$, may serve as a diagonostic tool in a quantitative attempt to gain an understanding of the microscopic structure of the $^{12}C + {}^{12}C$ resonances.

This work has been supported by the Deutsche Forschungsgemeinschaft.
One of us (K.T.H.) gratefully acknowledges the support of the Alexander von Humboldt Foundation through its U.S. Senior Scientist Program and extends his thanks to H.A. Weidenmüller for the hospitality of the Max-Planck-Institut für Kernphysik, Heidelberg.

References

1) J.P. Draayer and Y. Akiyama, J. Math. Pyhs. 14(1973)1904, and Comp. Phys. Commun. 5(1973)405.

2) K.T. Hecht and W. Zahn, in Lecture Notes in Physics, Vol. 94, Springer Verlag (1979)408.

3) K.T. Hecht and W. Zahn, Nucl. Phys. A313(1979)77, and A318(1979)1, and Nukleonika, in press.

4) K.T. Hecht, E.J. Reske, T.H. Seligman, and W. Zahn, to be publ.

5) B. Imanishi, Nucl. Phys. A125(1969)33, and H.J. Fink, W. Scheid, and W. Greiner, J. Phys. G1(1975)685, and J.Y. Park, W. Scheid, and W. Greiner, Phys. Rev. C10(1974)967, C16(1977)2276, and Y. Kondo, Y. Abe, and T. Matsuse, Phys. Rev. C19(1979)1356, 1365.

6) R.L. Phillips, K.A. Erb, D.A. Bromley, and J. Weneser, Phys. Rev. Lett. 42(1979)556, and L.E. Cannell, R.W. Zurmühle, and D.P. Balamuth, Phys. Rev. Lett. 43(1979)837, and W.A. Friedman, K.W. McVoy, and M.C. Nemes, Phys. Lett. 87B(1979)179.

7) H. Horiuchi, Progr. Theor. Phys. 51(1974)745, and Suppl. 62(1977)90.

8) T.H. Seligman and W. Zahn, J. Phys. G2(1976)79.

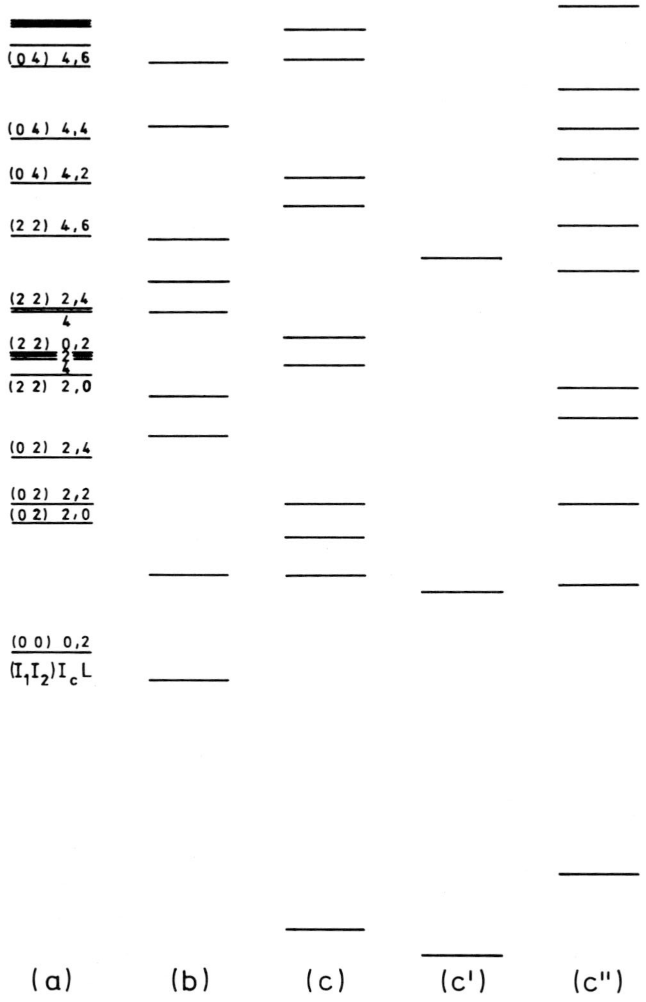

Fig. 2

(a) Energies with interaction and Pauli-principle "turned off"
$E = E_{i_1} + E_{i_2} + 0.1\ L(L+1)$, with $E_{I_i} = 0\,\delta_{I_i 0^+} + 4.44\,\delta_{I_i 2^+} + 14.08\,\delta_{I_i 4}$
$(J^{\pi} = 2^+)$

(b) Energies of the same Hamiltonian, (a), in a fully antisymmetrized basis $(J^{\pi} = 2^+)$.

(c) Energies for the Hamiltonian $H = E_{i_1} + E_{i_2} +$
$+\alpha \left[C_{(\lambda\mu)} - \frac{3}{4}J(J+1)\right] + \beta\,\frac{3}{4}\left[I_1(I_1+1) + I_2(I_2+1)\right]$
$+\gamma\,I_c(I_c+1) + \delta\,L(L+1) + \nu\,C_{(\lambda_c\mu_c)}$
with $\alpha = -.04$, $\beta = -.04$, $\gamma = 0.$, $\delta = .15$, $\nu = .10$
(all numbers are in MeV) $(J^{\pi} = 2^+)$

(c') Same as (c) but $J^{\pi} = 0^+$

(c'') Same as (c) but $J^{\pi} = 4^+$.

Table 1

Possible SU_3 Quantum Numbers for the $^{12}C + ^{12}C$ Cluster Functions with Q = 14 ($2\hbar\omega$ Oscillator Excitations).

$(\lambda_c\mu_c)$	$(\lambda\mu)$	
(08)	(14,8) (13,7) (12,6)	(11,5) (10,4) (93) (82) (71) (60)
(24)	(16,4) (15,3) (14,5) (13,4) (12,6)	(11,5) (14,2) (12,3) (10,4) (93) (82) (13,1) (11,2) (12,0) (10,1)
(40)	(18,0) (16,1)	(14,2) (12,3) (10,4)

↑ Pauli-Forbidden

Table 2

$^{12}C + ^{12}C$ Norm Eigen-values and vectors for Q = 14

$(\lambda\mu)i$	Λ_i	$\langle[(\lambda_c\mu_c)\times(14,0)](\lambda\mu)\mid(\lambda\mu)i\rangle$		
		$(\lambda_c\mu_c)=(08)$	$(\lambda_c\mu_c)=(24)$	$(\lambda_c\mu_c)=(40)$
(14,2)1	.009763	-	.95806	.28656
(11,5)1	.100713	.98029	.19758	-
(12,3)1	.030810	-	.92236	.38633
(13,1)1	.027623	-	1	-
(10,4)1	.316781	.92061	.37059	.12308
(10,4)2	.079319	-.38997	.85606	.33925
(10,4)3	.003156	.02036	-.36033	.93260
(11,2)1	.051464	-	1	-
(12,0)1	.045003	-	1	-
(93) 1	.318246	.97842	.20661	-
(93) 2	.070937	-.20661	.97842	-
(10,1)1	.052000	-	1	-
(82) 1	.304046	.99549	.09492	-
(82) 2	.059327	-.09492	.99549	-
(71) 1	.277100	1	-	-
(60) 1	.252095	1	-	-

Pauli-forbidden states with $\Lambda_i = 0$ are not included.
Eigenvectors for Pauli-forbidden states can be constructed from the above, using their orthogonality with Pauli-allowed eigenvectors.

Table 3

$A_{L=J}((\lambda\mu)iKJ \rightarrow {}^{12}C(0^+) + {}^{12}C(0^+))$

$(\lambda\mu)i$	J = 0	J = 2		J = 4		
	K = 0	K=0(or1)[a]	K = 2	K=0(or1)	K=2(or3)[a]	K = 4
(14,2)1	.02731	.02555	-.00048	.02150	-.00173	*
(11,5)1	*	.03952	*	.06421	-.00494	*
(12,3)1	*	-.01381	*	-.02399	.00113	*
(13,1)1	*	-.01838	*	-.03046	*	*
(10,4)1	.18533	.16686	-.01134	.12512	-.03950	.00119
(10,4)2	.06257	.06246	.00076	.06182	.00229	-.00025
(10,4)3	.01482	.01468	.00009	.01437	.00035	.00000
(11,2)1	*	*	.00393	*	.01468	*
(12,0)1	-.05201	-.04674	*	-.03492	*	*
(93)1	*	-.06505	*	-.10062	.00892	*
(93)2	*	-.00941	*	-.01756	-.00089	*
(10,1)1	*	.02360	*	.03960	*	*
(82)1	-.10945	-.09340	.00951	-.05794	.03267	*
(82)2	-.04527	-.04354	-.00040	-.03958	-.00138	*
(71)1	*	.07302	*	.11396	*	*
(60)1	.11834	.10648	*	.08079	*	*

a) K = 0, 2, ... for states with μ = even
K = 1, 3, ... for states with μ = odd

Table 4

The factors $(f + A - f(\lambda'_c\mu'_c) \mid (\lambda\mu)i)$

$(\lambda\mu)i$	$f + A - f(\lambda'_c\mu'_c)$			
	$\alpha+{}^{20}Ne(80)$	${}^{16}O+{}^{8}Be(40)$	$n+{}^{23}Mg(83)$ $p+{}^{23}Na(83)$	$\alpha+{}^{20}Ne(82)$
(14,2)1	.6209	.1225	*	-.3380
(11,5)1	*	*	*	.3393
(12,3)1	.3235	-.1806	0	-.4125
(13,1)1	*	*	*	.1909
(10,4)1	.3714	.3834	.4054	.0688
(10,4)2	.0736	.1377	-.0313	-.2254
(10,4)3	.0731	.1047	.0268	-.0744
(11,2)1	*	*	0	.0414
(12,0)1	*	*	*	*
(93)1	*	*	.2445	.0341
(93)2	*	*	-.0104	-.0123
(10,1)1	*	*	0	*
(82)1	*	*	.1708	*
(82)2	*	*	-.0034	*
(71)1	*	*	.0836	*
(60)1	*	*	*	*

COLLECTIVE MOTION, COMPOSITE PARTICLE STRUCTURE, AND SYMPLECTIC GROUPS IN NUCLEI

P. Kramer and Z. Papadopolos
Institut für Theoretische Physik
Universität Tübingen, German Federal Republic

We propose to analyze the group theory of nuclear collective motion through representations of the symplectic group $Sp(6n,\mathbb{R})$ of linear canonical transformations of the n-nucleon system. We choose a state description by square-integrable analytic functions $f(x)$ of 3n complex variables $x=\{x_{is}\}$, $i=1,2,3$, $s=12...n$ in Bargmann space /BA 68/. The oscillator creation and annihilation operators X_{is} and D_{is} act on these functions as $(X_{is}f)(x) = x_{is}f(x)$, $(D_{is}f)(x) = \frac{\partial}{\partial x_{is}} f(x)$.

The generators of $Sp(6n,\mathbb{R})$ become

$$A^+_{is,jt} = \frac{1}{2} X_{is} \circ X_{jt}, \quad A_{is,jt} = \frac{1}{2} D_{is} \circ D_{jt}, \quad H_{is,jt} = \frac{1}{4} (X_{is} \circ D_{jt} + D_{jt} \circ X_{is})$$

1. Definition. The collective and the intrinsic symplectic group $\mathcal{S}p(6,\mathbb{R})$ and $Sp(2n,\mathbb{R})$ are the groups generated by

$$\mathcal{S}p(6,\mathbb{R}): \quad A^+_{ij} = \sum_s^n A^+_{is,js}, \quad A_{ij} = \sum_s^n A_{is,js}, \quad H_{ij} = \sum_s^n H_{is,js}$$

$$Sp(2n,\mathbb{R}): \quad A^+_{st} = \sum_j^3 A_{js,jt}, \quad A_{st} = \sum_j^3 A_{js,jt}, \quad H_{st} = \sum_j^3 H_{js,jt}$$

In addition we consider the collective subgroups $\mathcal{U}(3)$ generated by H_{ij} and $\mathcal{O}(3,\mathbb{R})$ generated by $H_{ij}-H_{ji}$, and the intrinsic subgroups $U(n)$ generated by H_{st} and $O(n,\mathbb{R})$ generated by $H_{st}-H_{ts}$.

2. Proposition: The collective and intrinsic subgroups commute with one another according to Fig. 1.

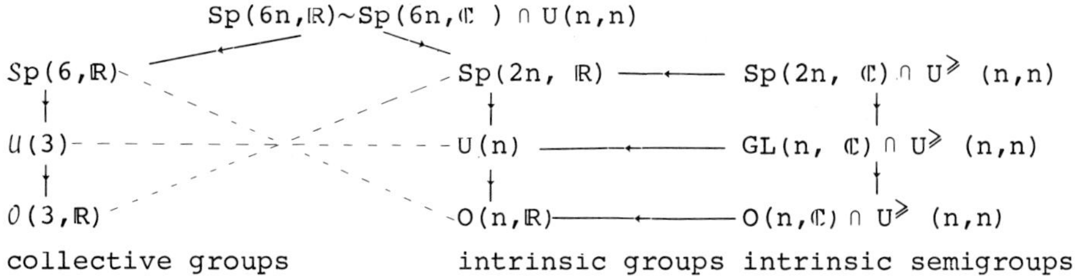

Fig. 1. Subgroup and subsemigroup relations (full lines with arrows) and commutativity (dashed lines) for collective and intrinsic (semi-) groups.

The collective group $Sp(6,\mathbb{R})$ and its subgroups have been considered by various authors with respect to the group theory of collective motion, compare the review by Rowe /RO 78 /.
For a microscopic theory of collective motion we shall now employ the intrinsic groups. Vanagas /VA 77 / proposed to characterize the collective hamiltonian H_{Coll} as the part of the full hamiltonian that is invariant under the intrinsic group $O(n,\mathbb{R})$ or $O(n-1,\mathbb{R})$ after removal of the c.m. motion. In various papers it was shown that a hamiltonian with two-body interactions of Gaussian type is contained in the operator algebra of the symplectic group, provided the corresponding representation is extended to the semigroup $Sp(2n,\mathbb{C}) \cap U^{\geqslant}(n,n)$ /BR 75, BR 80 /. The semigroup $U^{\geqslant}(n,n)$ consists of all length-increasing transformations with respect to the hermitian form underlying $U(n,n)$, it contains the group $Sp(2n,\mathbb{R}) \sim Sp(2n,\mathbb{C}) \cap U(n,n)$. Subsemigroups are indicated in Fig. 1.

We now propose $U(n)$-collectivity by demanding that H_{Coll} be invariant under the intrinsic group $U(n)$. Then, since $U(n)$ contains the generator $H_o = \sum_j \sum_s H_{is,is}$ which counts the total number of quanta, the states belong to an irreducible respresentation $[N]$ of the group $U(1)$ generated by H_o, this fixes the irreducible representation of the group $U(3n)$ generated by all the $H_{is,jt}$ to be $[N\ 0^{3n-1}]$, and complementarity /MO 70 / applied to the intrinsic and collective groups $U(n)$ and $\mathcal{U}(3)$ whose irreducible representations must be of the form $[h_1 h_2 h_3\ 0^{n-3}]$ and $[h_1 h_2 h_3]$ respectively, compare /KR 68 /.

In the next step we combine unitary collectivity with the composite particle model. The analysis of the composite system given in /KR 80 / shows that the interaction operator H of composite particles, when starting from microscopic Gaussian interactions, is again in the operator algebra of a semigroup $Sp(j-1,\mathbb{C}) \cap U(j-1,j-1)$ where j is now the number of composite particles. The reduction from j to j-1 results from the removal of c.m. motion. As an example we choose the oscillator parameter b=1.6 fm and employ a Gaussian two-body interaction with two terms given by /AR 75 / of the form

$$V_i(r) = \left[\alpha_e^{(i)} P_e + \alpha_o^{(i)} P_o\right] \exp\left[-a_i^{-2} r^2\right]$$

where P_e and P_o project on even or odd orbital states. The parameters are given in table 1.

i	a_i [fm]	$\mu_i = b^2[2(a_i)^2 + 4b^2]^{-1}$	$\alpha_e^{(i)}$ [MeV]	$\alpha_o^{(i)}$ [MeV]
1 attractive	1	0,209150	-387,5	-82,45
2 repulsive	2/3	0,230032	753,75	751,05

Table 1 The parameters of the two-body potential taken from AR 75 for an oscillator parameter $b=1,6$ fm corresponding to $\hbar\omega = 16,21$ MeV.

The interaction operator H in Bargmann space becomes an integral operator in the complex relative coordinates $\tilde{s}$, $\bar{s}$. For the system $^4He + {}^4He$ and the corresponding description of states in ^8Be with orbital partition $f=[44]$ one obtains for the potential part K of H

$$K \equiv ((\alpha\gamma)^8[44][1^8]|K|(\alpha\gamma)^8[44][1^8]) = 2\sum_{i=1}^{2} (a_i)^3[(a_i)^2 + 2b^2]^{-3/2}$$

$$\times\Big\{3\alpha_e^{(i)}\Big[[\exp(\tilde{s}.\bar{s}/4) - \exp(-\tilde{s}.\bar{s}/4)]^4 + [\exp(\tilde{s}.\bar{s}/4) - \exp(-\tilde{s}.\bar{s}/4)]^3$$

$$\times\Big\{\exp(\tilde{s}.\bar{s}/4)\exp(-\mu_i(\tilde{s}+\bar{s}).(\tilde{s}+\bar{s})/2) - \exp(-\tilde{s}.\bar{s}/4)$$

$$\times[-\exp(-\mu_i(\tilde{s}-\bar{s}).(\tilde{s}-\bar{s})/2) - 2 + 2\exp(-\mu_i\tilde{s}.\tilde{s}/2) + 2\exp(-\mu_i\bar{s}.\bar{s}/2)]\Big\}$$

$$+ [\exp(\tilde{s}.\bar{s}/4) - \exp(-\tilde{s}.\bar{s}/4)]^2\Big\{\exp(\tilde{s}.\bar{s}/2) + \exp(-\tilde{s}.\bar{s}/2)$$

$$\times[3 - 2\exp(-\mu_i\bar{s}.\bar{s}/2) - 2\exp(-\mu_i\tilde{s}.\tilde{s}/2)$$

$$+ 2\exp(-\mu_i(\tilde{s}-\bar{s}).(\tilde{s}-\bar{s})/2)] - 2\exp(-\mu_i\tilde{s}.\tilde{s}/2)$$

$$- 2\exp(-\mu_i\bar{s}.\bar{s}/2) + 2\exp(-\mu_i(\tilde{s}+\bar{s}).(\tilde{s}+\bar{s})/2)\Big\}\Big]$$

$$+ 5\alpha_o^{(i)}[\exp(\tilde{s}.\bar{s}/4) - \exp(-\tilde{s}.\bar{s}/4)]^3[\exp(\tilde{s}.\bar{s}/4)$$

$$\times\exp(-\mu_i(\tilde{s}+\bar{s}).(\tilde{s}+\bar{s})/2) - \exp(-\tilde{s}.\bar{s}/4)\exp(-\mu_i(\tilde{s}-\bar{s}).(\tilde{s}-\bar{s})/2)]\Big\}$$

The collective hamiltonian H_{Coll} is now obtained by restricting H to its part invariant under the intrinsic group U(1). Since the representation of this group is [N] where N is the total oscillator excitation, the collective states are characterized by the complementary representation [NOO] of the collective group $U(3)$. Explicitly one finds

$$(NL|E_{Coll}|NL) = (NL|T|NL) + [\eta_N]^{-1}(NL|K_{Coll}|NL)$$

where the kinetic energy part is

$$(NL|T|NL) = \frac{1}{2}\,[7(\tfrac{3}{2}\hbar\omega) + N\,\hbar\omega]$$

and the potential energy part is

$$(NL|K_{Coll}|NL)_{N\neq 0} = 2(1+(-)^N)\sum_{i=1}^{2}(a_i)^3[(a_i)^2 + 2\,b^2]^{-3/2}$$

$$\times\Bigg\{6\alpha_e^{(i)}[1-6(1/2)^N] + \sum_{k=0}^{[N(-1)]|2}(\mu_i/2)^{2k}\,J_k(N)$$

$$\times\Big[3\alpha_e^{(i)}\Big\{(1-\mu_i)^{N-2k}+[(1+2\mu_i)/2]^{N-2k}-(-\mu_i)^{N-2k}-[(-1+2\mu_i)/2]^{N-2k}\Big\}$$

$$+5\alpha_o^{(i)}\Big\{(1-\mu_i)^{N-2k}-[(1+2\mu_i)/2]^{N-2k}+3(-\mu_i)^{N-2k}-3[(-1+2\mu_i)/2]^{N-2k}\Big\}\Big]\Bigg\}$$

$$(00|K_{Coll}|00) = 0$$

$$J_k(N) = [k!]^{-2}\,\frac{(N+L+1)!!\,(N-L)!!}{(N+L+1-2k)!!\,(N-L-2k)!!}\qquad k = 0,1,2\ldots\begin{cases}(N-1)/2 & N\text{ odd}\\ N/2 & N\text{ even}\end{cases}$$

$$\eta_N = \sum_{\alpha=0}^{4}(-)^{\alpha}\binom{4}{\alpha}(1-\tfrac{1}{2}\alpha)^N$$

For N=4 and N=6 one finds

$$(4L|E_{Coll}|4L) = [-17{,}03 + 0{,}439\;L(L+1)]\,\mathrm{MeV}$$

$$(6L|E_{Coll}|6L) = [8{,}77 + 0{,}178\;L(L+1) - 0{,}00012\;L^2(L+1)^2]\,\mathrm{MeV}$$

The levels are displayed in Fig. 2.

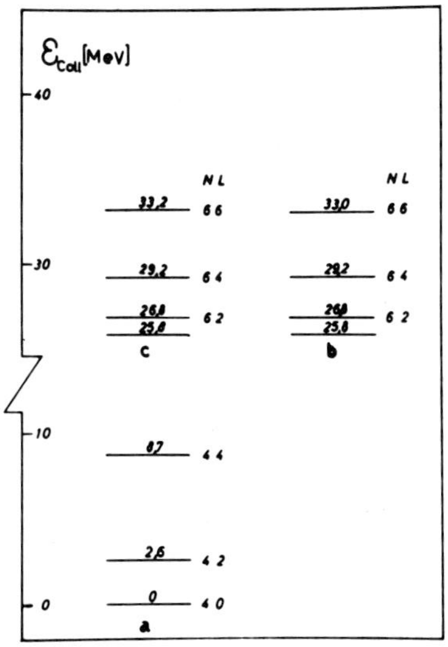

Fig. 2. Collective spectrum in ^{8}Be relativ to the lowest state: (a) for N=4; (b) for N=6; (c) for N=6 without the term proportional to $L^2(L+1)^2$

References

AR 75 F. Arickx, P.Van Leuven and M. Bouten, Z. Phys. A 273 (1975) 2o5

BA 68 V. Bargmann in: Analytic Methods in Mathematical Physics, New York 1968

BR 75 M. Brunet and P. Kramer in: Lecture notes in Physics vol. 50, Berlin 1976

BR 80 M. Brunet and P. Kramer Rep. Math. Phys. 15(1979) 287

KR 68 P. Kramer and M. Moshinsky in: Group Theory and its Applications vol. I, New York 1968

KR 75 P. Kramer, M. Moshinsky and T.H. Seligman in: Group Theory and its Applications vol. III, New York 1975

KR 80 P. Kramer, G. John and D. Schenzle, Group Theory and the Interaction of Composite Nucleon Systems, Braunschweig 1980

MO 70 M. Moshinsky and C. Quesne, J. Math. Phys. 11 (1970) 1631

RO 78 D.J. Rowe in: Lecture Notes in Physics vol. 79, Berlin 1978

VA 77 V.V. Vanagas, Lecture Notes, Univ. of Toronto 1977

GROUP THEORY AND CONFRONTATION OF NUCLEAR COLLECTIVE MODELS

Marcos Moshinsky*

Instituto de Física, UNAM

Apdo. Postal 20-364, México 20, D.F.

1. INTRODUCTION

The purpose of this paper is to establish connections between some of the macroscopic collective models that are being considered at the present time and then indicate how essential features of these models can be derived from microscopic considerations.

The pioneering work in this field was due to Bohr and Mottelson[1] when in the early fifties they introduced collective degrees of freedom in the nuclear Hamiltonian through the liquid drop model.

This Hamiltonian allowed them immediately to discuss vibrational states in nuclei and, with certain assumptions, also rotational ones as well as vibration-rotation interactions[2,3].

A systematic approach to the transitional region between vibrational and rotational limits was achieved by Gneuss and Greiner[4] and by collaborators of the latter in Frankfurt[5]. Their basic assumption was to consider that the Hamiltonian was not limited to quadratic expressions in the collective coordinates and momenta but that higher order terms need to be included as, in fact, would be required if the liquid drop Hamiltonian is not limited to small vibrations. If these higher order terms are static (i.e. momentum independent) then a phenomenological potential[4,5] depending on β^2, $\beta^3 \cos 3\gamma$ can be proposed whose parameters can be fixed from the energy levels and transition probabilities in a given nucleus. The equipotential lines for this potential give information on the deformation of nuclei and show graphically, along the periodic table, the transition between the vibrational limit near closed shells to the rotational one in the middle of the shells. In the present note we shall refer to work along these lines as the Frankfurt model.

More recently another approach has been followed for the discussion of collective phenomena in nuclei: The interacting boson approximation (IBA) model.[6,7]

---

*Member of the Instituto Nacional de Investigaciones Nucleares and El Colegio Nacional.

In it rather than going to higher order terms in the Hamiltonian —as done in the previous paragraph,— one extends the Hilbert space of the collective variables from the d-coordinates α_m, m = 2,1,0,-1,-2 associated with quadrupole vibrations[1] to d-s coordinates and the corresponding six d-s bosons of the IBA model. The group U(6) associated with the latter can accomodate chains of subgroups that reflect the vibrational and rotational limits as well as the transitional region.[7]

Both the Frankfurt and the IBA model seem to describe appropriately the collective behaviour of medium and heavy nuclei throughout the periodic table, despite what seems to be very different theoretical premises. Recently though the author[8] has shown that they must lead to essentially equivalent matrix representations of their Hamiltonians and this will be one of the main points reviewed in the present paper.

Once we establish the relation between some of the current macroscopic collective models of the nucleus the question arises of their connection with a more microscopic approach. In the last decade Zickendraht[9] and Dzublik et al.[10] introduced a simple transformation of the Jacobi coordinates of an A-body system which in a natural way separates the collective coordinates (i.e. the three Euler angles Θ_k and three parameters ρ_k, k = 1,2,3 associated with the deformation along the three principal axis[9,10] from the rest. This suggests the possibility of projecting from an A particle system with two body interactions a microscopic collective Hamiltonian depending only in the six variables ρ_k, Θ_k, k = 1,2,3 and their canonically conjugate momenta.[11,12,13]

An interesting question is whether —at least when we consider only harmonic oscillator interactions— there is a canonical transformation connecting the Hamiltonians of the microscopic collective and IBA model where the latter also depends on six coordinates and their conjugate momenta associated with s+d bosons. We take up this question in the concluding section of this paper to outline a procedure by means of which we could relate the different collective models of the nucleus mentioned in this paper.

2. BOHR-MOTTELSON OSCILLATOR MODEL

As the work of Bohr-Mottelson[1,2] provides the basic framework for the discussion of collective motions in nuclei we begin by reviewing the part of it required in the following analysis.

A liquid drop of irrotational fluid is specified by the form of its surface which, in the case of quadrupole vibrations, is given by the equation[1-3]

$$R = R_o \left[1 + \sum_{m=-2}^{2} \alpha_m Y^*_{2m} (\theta,\phi)\right] . \tag{2.1}$$

For small vibrations the Hamiltonian associated with the collective coordinates α_m, m = 2,1,0,-1,-2 will be

$$H_o = \frac{1}{2} \sum_m (\pi_m \pi^m + \alpha_m \alpha^m) \equiv \hat{n} + (5/2) , \tag{2.2a}$$

$$\pi_m = -i\partial/\partial\alpha^m , \quad \alpha^m = (-1)^m \alpha_{-m} , \tag{2.2b}$$

where we have taken units[3] in which $\hbar = B_2 = C_2 = 1$ and $\hat{n}$ is the operator associated with the number of quanta. We are interested in eigenstates of these Hamiltonians that correspond to definite eigenvalues of the angular momentum where the components of the latter are given by[3]

$$L_\tau = i(10)^{1/2} [\alpha \times \pi]^1_\tau = i(10)^{1/2} \sum_{mm'} \langle 22mm'|1\tau\rangle \alpha_m \pi_{m'} ; \quad \tau = 1,0,-1 \tag{2.3a}$$

Furthermore they can also be characterized by the eigenvalues of the Casimir operator Λ^2 of an 0(5) group given by[14]

$$\Lambda^2 = \frac{1}{2} \sum_{m,m'} \Lambda_{mm'} \Lambda^{m'm} , \tag{2.3b}$$

$$\Lambda_{mm'} = i(\alpha_m \pi_{m'} - \alpha_{m'} \pi_m) . \tag{2.4}$$

In group theoretical language we can say that the states of Bohr-Mottelson five dimensional oscillator Hamiltonian (2.2) are characterized by the irreducible representations (irreps) of the following chain of groups

$$\begin{array}{cccc} U(5) \supset & 0(5) \supset & 0(3) \supset & 0(2) \\ \nu & \Lambda(\Lambda+3) & L(L+1) & M \end{array} \tag{2.5}$$

where under each one of them we have indicated the eigenvalues of the corresponding

Casimir operators. To write these states explicitly we pass from the α_m in a frame of reference fixed in space to the β,γ and Euler angles Θ_k fixed in the body through the relations[1-3]

$$\alpha_m = (1/\sqrt{2})\ \beta \sin\gamma \left[D^{2*}_{m2}(\Theta_k) + D^{2*}_{m-2}(\Theta_k)\right] + \beta\cos\gamma\ D^{2*}_{m0}(\Theta_k)\ . \qquad (2.6)$$

The eigenkets then take the form[1-3]

$$|\nu\Lambda t L M\rangle = F^{\Lambda}_{(\nu-\Lambda)/2}(\beta) \sum_K \Phi_K^{\Lambda t L}(\gamma)\ D^{L*}_{MK}(\Theta_k) \qquad (2.7)$$

where the extra index t distinguishes[14,15] between repeated representations L of O(3) in a given representation Λ of O(5). The expression (2.7) already appears in the original papers of Bohr-Mottelson[1-3] and only $\Phi_K^{\Lambda t L}(\gamma)$ had to wait until recently[14] for its explicit determination through group theoretical methods.[15]

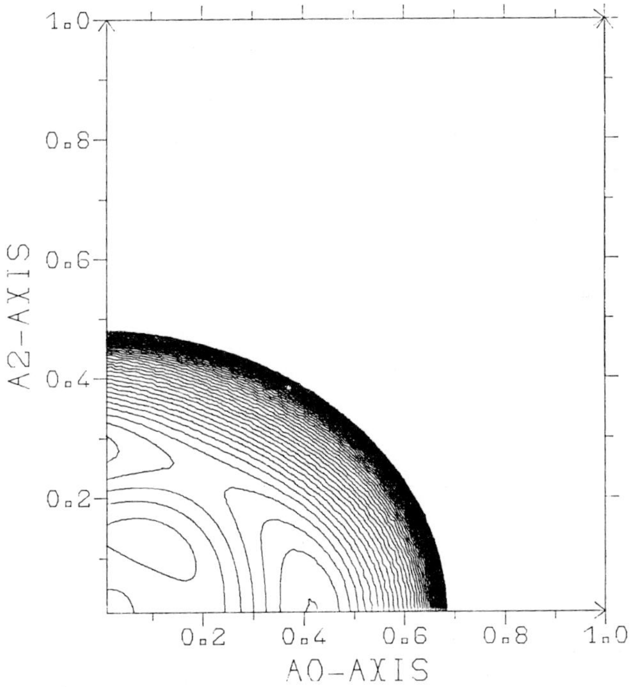

Figure 1 Potential energy surfaces for ^{164}Dy. The parameters U_{rs} in the potential energy $V(\alpha_m) = \sum_{rs} \beta^r(\cos 3\gamma)^s U_{rs}$ with $0 < r \leq 10$, $0 \leq s \leq 1$, were determined from best fits to the energy levels and transition probabilities. The equipotential lines in the plane whose axes are $A0 = \beta\cos\gamma$, $A2 = \beta\sin\gamma$, are then drawn. Two minima appear at $\beta = 0.41$, $\gamma = 0°$ and $\beta = 0.15$, $\gamma = 60°$ corresponding to prolate and oblate deformations where the former is the lowest one and thus associated with the ground state. The unpublished calculations were done by A Frank and P Hess with the type of techniques outlined in section V and they are quoted and analyzed in a thesis by Castaños

3. THE FRANKFURT MODEL

In the Frankfurt model our Hamiltonian H_F is still of the form $H(\alpha_m, \pi^m)$ but is not restricted only to quadratic expressions in these variables as was the case for H_o. If we consider a static Frankfurt model (i.e. independent of the momentum except for the kinetic energy) it takes the form[4,5,15]

$$H_F = H_o + V(\beta^2, \beta^3 \cos 3\gamma) , \tag{3.1}$$

where

$$\beta^2/\sqrt{5} = [\alpha \times \alpha]^0_0 , \quad -(2/35)^{1/2}\beta^3 \cos 3\gamma = \left[[\alpha \times \alpha]^2 \times \alpha\right]^0_0 \tag{3.2a,b}$$

as the potential must be invariant under rotations[4,15].

To determine the eigenvalues and eigenstates of the H_F of (3.1), we can first calculate its matrix elements with respect to the complete set of states of (2.7). The part dealing with the Euler angles Θ_k is trivial. The interesting new part is the one associated with powers of cos 3γ or better still with Legendre polynomials[15] $P_s(\cos 3\gamma) \equiv \Phi^{3s,1,0}_0(\gamma)$. In this case the basic matrix element takes the form

$$\int_o^\pi \sum_{KK'K''} \begin{pmatrix} LL'L'' \\ KK'K'' \end{pmatrix} \Phi^{\Lambda tL}_K(\gamma)\, \Phi^{\Lambda' t' L'}_{K'}(\gamma)\, \Phi^{\Lambda'' t'' L''}_{K''}(\gamma) \sin 3\gamma d\gamma \equiv (\Lambda tL;\ \Lambda' t' L';\ \Lambda'' t'' L''), \tag{3.3}$$

where the round bracket in the integral is a standard 3j symbol. It is possible to see that (3.3) is an isoscalar factor or reduced Wigner coefficient of the $O(5) \supset O(3)$ chain of groups and programs are available for its determination.

Thus all matrix elements required in the static Frankfurt model can be evaluated. In case there are also terms involving π_m one can reduce them to the static ones through the relation[16,8].

$$\pi_m = i(H_o\alpha_m - \alpha_m H_o) = \exp(i\pi\hat{n}/2)\alpha_m \exp(-i\pi\hat{n}/2) . \tag{3.4}$$

We illustrate in Fig. 1 a calculation of potential energy surfaces for ^{164}Dy done by Hess and Frank using the techniques outlined in this section[17]. A Hamiltonian H_F of the type (3.1) was considered whose parameters were determined so as to get the experimental energy levels and transition probabilities[5]. The equipotential lines where then drawn where A0 and A2 axis are respectively β cos γ and β sin γ. Clearly the potential has a minimum at β = 0.41, γ = 0 indicating a pronounced axially symmetric prolate deformation.

We now consider another approach to collective phenomena, the interacting boson approximation model[7].

4. THE IBA MODEL

As mentioned in the introduction, the possibility of encompassing in a single theoretical framework vibrational, rotational and transitional nuclei can be achieved not only by generalizing the Hamiltonian to higher order terms as in (3.1), but also by extending the Hilbert space from the d coordinates α_m to d+s ones i.e. adding a scalar coordinate which we can designate by $\bar{\alpha}$. Thus the operators of the coordinates and momenta are

$$\alpha_{lm} = (\bar{\alpha}, \alpha_m), \; l = 0,2 \; ; \quad \pi_{lm} = -i\partial/\partial\alpha^{lm} , \tag{4.1}$$

and the corresponding creation and annihilation operators are given by

$$\eta_{lm} = (1/\sqrt{2})(\alpha_{lm} - i\pi_{lm}) \; , \; \xi^{lm} = (1/\sqrt{2})(\alpha^{lm} + i\pi^{lm}) \; . \tag{4.2}$$

The Hamiltonian for the oscillator potential in this extended Hilbert space is given by the number operator for the d+s bosons i.e.

$$\hat{N} \equiv \sum_{l,m} \eta_{lm} \xi^{lm} = \hat{n} + \bar{\eta}\bar{\xi} , \tag{4.3}$$

and its symmetry group instead of U(5) of the Bohr-Mottelson oscillator model is given by U(6) whose generators are

$$C_{lm}^{l'm'} = \eta_{lm} \xi^{l'm'} \; . \tag{4.4}$$

This group admits several chains of subgroups which are associated with the different limiting situations for collective motions in nuclei. For example if we take the subgroups[18)]

$$\begin{array}{ccccccccc} U(6) & \supset & U(5) & \supset & O(5) & \supset & O(3) & \supset & O(2) \\ N & & \nu & & \Lambda(\Lambda+3) & & L(L+1) & & M \end{array} , \tag{4.5}$$

(where underneath each group we have given the eigenvalue of its Casimir operator) we include the chain (2.5) of the Bohr-Mottelson oscillator Hamiltonian. Thus the classification of the eigenstates by (4.5) corresponds to the vibrational limit and from (4.3) we immediately see that they are given by[18)]

$$|N\nu\Lambda tLM\rangle = |N - \nu\rangle |\nu\Lambda tLM\rangle , \tag{4.6}$$

where the last ket is (2.7) and $|N - \nu>$ is the wave function of $N - \nu$ quanta of a one dimensional oscillator in the variable $\bar{\alpha}$.

The pioneering work of Elliott[19] shows though that U(6) admits also the chain of groups

$$U(6) \supset U(3) \supset O(3) \supset O(2) \tag{4.7}$$

and eigenstates belonging to it are associated with the rotational limit[19,7,18]. The corresponding eigenkets can be denoted by

$$|[2e_1 2e_2 2e_3]\Omega LM> \quad . \tag{4.8}$$

where the partition of 2N in even integers[7,18] $[2e_1 2e_2 2e_3]$ is associated with an irreducible representation (irrep) of U(3) and the extra index Ω distinguishes between repeated irreps L of O(3) in the irrep of U(3). In reference 18 we showed how to get the eigenkets (4.8) as linear combinations of those of (4.6) with the same N.

Transitional nuclei can be described by the chain of groups[7] $U(6) \supset O(6) \supset O(5) \supset O(3)$ or, more generally, by introducing in the model two body interactions i.e. considering Hamiltonians[7,18] of the form

$$H_I = \varepsilon_d \hat{n} + \varepsilon_s (\hat{N} - \hat{n}) + \frac{1}{2} \sum_{\ell_1 \ell_2} \sum_{\ell'_1 \ell'_2} \sum_{L} \{<\ell_1 \ell_2 L|V|\ell'_1 \ell'_2 L>$$

$$(-1)^{\ell'_1 + \ell'_2} (2L + 1)^{1/2} \left[[\eta_{\ell_1} \times \eta_{\ell_2}]^L \times [\xi_{\ell'_1} \times \xi_{\ell'_2}]^L \right]^0_0 \quad . \tag{4.9}$$

It can be shown[18] that there are seven types of independent two body interactions and a typical one (of relevance in determining the Casimir operator of U(3) and thus the eigenstates (4.8)) has the form[18]

$$E \equiv \left[[\underline{\eta} \times \underline{\eta}]^2 \times \underline{\xi} \right]^0_0 \bar{\xi} + \bar{\eta} \left[\underline{\eta} \times [\underline{\xi} \times \underline{\xi}]^2 \right]^0_0 \tag{4.10}$$

where the operators underlined and overlined are respectively d and s bosons.

The most general IBA Hamiltonian would contain the parameters ε_d, ε_s and the coefficients of the seven independent two body interactions. As shown recently by Castaños, Frank and Federman[20], in many nuclei the more relevant of the latter are $\hat{n}\hat{N}$ and the quadrupole-quadrupole interaction[18] Q^2 related to the Casimir operator of U(3). Thus as the number of s-d bosons is fixed in the IBA model we can in these nuclei restrict ourselves to the Hamiltonian

$$H_I = \varepsilon \hat{n} + a\hat{n}\hat{N} + bQ^2 \tag{4.11}$$

In Fig. 2 we show an analysis of the level distribution for the Samarium isotopes[20] where the parameters ε, a, b where fixed so as to give the best fit for all levels of the set. The number N of s,d bosons is the one that changes with the isotope and clearly shows how we pass from vibrational to rotational level structure as N increases i.e. as we move more towards the middle of the shell.

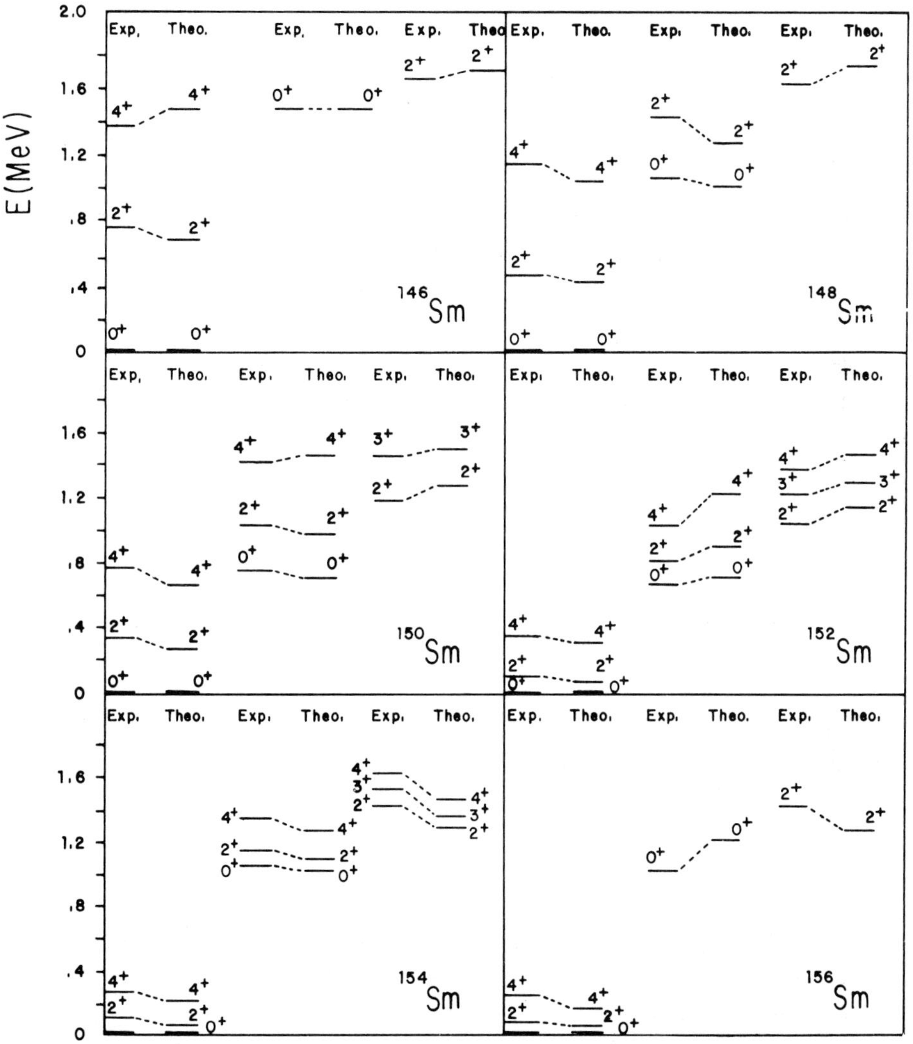

Figure 2 Energy levels for the Samarium isotopes in the interacting boson model. The general Hamiltonian contains two one-body and seven two body interactions. The fit for the energy levels shows though that only the interactions appearing in (4.11) have coefficients significantly different from zero. These coefficients were determined looking for the best fit for the levels in *all* the even-even Samarium isotopes from $A = 146$ to 156. The number N changes from isotope to isotope as it reflects the number of pairs of protons and neutrons above respectively 50 and 82, which are the nearest closed shells. The levels were arranged in different columns to display a band picture and avoid cluttering. A clear change from a vibrational to a rotational regime is observed when we increase the number of neutrons. The calculations were done by Castaños, Frank and Federman (Ref. 20).

5. RELATIONS BETWEEN MACROSCOPIC COLLECTIVE MODELS

In the previous sections we showed how two conceptually very different collective models of the nucleus can describe in a unified manner vibrational, rotational and transitional nuclei.

In this section we wish to indicate[8)] that the matrix elements of Hamiltonians of the type (4.9) with respect to the states (4.6) can be reduced to those of Hamiltonians of the type (3.1) with momentum dependent terms (which can be expressed in terms α_m as in (3.4)) with respect to states (2.7).

For reasons of space we discuss only the interaction of (4.10) though, as shown in reference 8, the arguments can be extended to the other six interactions. To begin with as N is fixed in the IBA model and the state (4.6) contains only the ket $|N - \nu\rangle$ in the $\bar{\alpha}$ variable we get

$$\bar{\xi}|N - \nu\rangle = (N - \nu)^{1/2}|N - \nu - 1\rangle \ . \qquad (5.1)$$

Thus in (4.9) we can replace $\bar{\xi}$ (and by a similar[8)] argument $\bar{\eta}$), by $(N - \hat{n})^{1/2}$. Furthermore as $\alpha_m = (1/\sqrt{2})(\eta_m + \xi_m)$ we have that for matrix elements in which the ket is associated with ν and the bra with $\nu' = \nu + 1$ we can replace[8)]

$$\left[[\eta \times \eta]^2 \times \xi\right]^0_0 \rightarrow (2\sqrt{2/3})\left[[\alpha \times \alpha]^2 \times \alpha\right]^0_0 \ . \qquad (5.2)$$

The selection rule $\nu' = \nu + 1$ can be achieved with help of appropriate functions[8)] of $\hat{n}$ and using (3.2b) we can make the replacement of E by an operator acting on the states (2.7) that has the form

$$-(1/6\sqrt{35})\{\left[\hat{n}\beta^3 \cos 3\gamma - (\beta^3 \cos 3\gamma)(\hat{n} - 3)\right]$$
$$-i \exp(i\pi\hat{n}/2)\left[\hat{n}\beta^3 \cos 3\gamma - (\beta^3 \cos 3\gamma)(\hat{n} - 3)\right]\exp(-i\pi\hat{n}/2)\}(N - \hat{n})^{1/2} + \text{h.c.} \ , \qquad (5.3)$$

where h.c. stands for the hermitian conjugate of the expression preceeding it.

The operator (5.3) is of course restricted to act on states (2.7) in which $0 \leq \nu \leq N$ i.e. the number of d bosons is smaller than the fixed number N of s+d bosons. We could eliminate this restriction if the operator (5.3) is multiplied to the right and left by a step operator $\Theta(N - \hat{n})$ where $\Theta(x) = 1$ for $x \geq 0$, $\Theta(x) = 0$ for $x < 0$. Thus for any Hamiltonian in the IBA model we can find an equivalent one in the Frankfurt model which will lead to the same physical predictions.

Once this relationship is established it becomes interesting to see whether we can cannect the IBA model with a microscopic description as this will also imply a connection with the extended Bohr-Mottelson picture used in the Frankfurt model.

6. TRANSFORMATIONS LEADING TO COLLECTIVE COORDINATES

As indicated in section 1, in the last decade a coordinate transformation[9,10] was introduced that brings out collective features in A body systems. To derive it let us designate x'_{is} the dimensionless coordinates (i.e. in units in which we take $\hbar$, the mass M of the particles and some frequency as 1) of the system, where i = 1,2,3 indicate component and s = 1,.... A particle indices. To eliminate the center of mass we introduce the Jacobi coordinates.

$$x_{is} = [s(s+1)]^{-1/2}\left[\sum_{t=1}^{s} x'_{it} - s x'_{is+1}\right], \quad s = 1,\dots.A-1 \; . \tag{6.1}$$

We then express the latter in the form[9,10]

$$x_{is} = \sum_{k=1}^{3} \rho_k D^1_{ki}(\Theta)\; D^1_{A-4+k,s}(\alpha) \quad , \tag{6.2}$$

where in (6.2) $D^1_{ki}(\Theta_1\Theta_2\Theta_3)$, $D^1_{ts}(\alpha)$ are the defining representations (which is the reason for the upper index 1) of the three dimensional rotation group $O(3)$ in ordinary space and the O(A-1) rotations group associated with the index s = 1,...A-1 of the Jacobi coordinates. The Θ_1, Θ_2, Θ_3 are the standard Euler angles and at first sight it seems that we have (A-1)(A-2)/2 angles α. As in (6.2) we are dealing with the last three rows of the representation[13] it is possible to define the angles so that only 3A-9 α's appear in them. This number plus the three Θ's and three ρ's appearing in (6.2) gives 3A-3, the same number of parameters as of coordinates x_{is}; i = 1,2,3; s = 1,2,...A-1.

We proceed now to indicate the physical meaning of the ρ's. It is easily seen that the inertia tensor —in the present units and when the origin is as the center of mass— can be written as

$$I_{ij} = \rho^2\delta_{ij} - q_{ij} \; , \tag{6.3a}$$

$$\rho^2 = \sum_{s=1}^{A-1}\sum_{i=1}^{3}(x_{is})^2 = \rho_1^2 + \rho_2^2 + \rho_3^2 \; ; \quad q_{ij} = \sum_{s=1}^{A-1} x_{is}x_{js} \; . \tag{6.3b,c}$$

Using the coordinate transformation (6.2) and the orthogonal property of the matrix $\| D^1_{ts}(\alpha) \|$ we see that

$$q_{ij} = \sum_{k=1}^{3} \rho_k^2 D^1_{ki}(\Theta)\; D^1_{kj}(\Theta) \; , \tag{6.4}$$

or in matrix form

$$\|q\| = \|\tilde{D}^1(\Theta)\| \begin{pmatrix} \rho_1^2 & 0 & 0 \\ 0 & \rho_2^2 & 0 \\ 0 & 0 & \rho_3^2 \end{pmatrix} \|D^1(\Theta)\| \quad , \tag{6.5}$$

where $\sim$ indicated transposed. It is then clear from (6.3a) that in the frame of reference fixed in the body —where the Θ_k's are replaced by 0— the principal moments of inertia become $\rho_2^2 + \rho_3^2$, $\rho_1^2 + \rho_3^2$, $\rho_2^2 + \rho_1^2$. Furthermore from (6.3c)(6.5) the expectation values of ρ_1^2, ρ_2^2, ρ_3^2 with respect to nuclear many body states provide a reasonable measure of the intrinsic deformation of these states[21).

Clearly then the six variables ρ_i, Θ_i, $i = 1,2,3$, are related with the collective degrees of freedom of a many body system and it is of interest to obtain some relation between them and those appearing in macroscopic collective models of the nuclei.

7. THE OSCILLATOR HAMILTONIAN AND ITS SYMMETRY GROUPS

We note that the Bohr-Mottelson and IBA models start respectively from five (as in (2.1)) or six (as in (4.3)) dimensional oscillators. It seems therefore of interest to initiate our discussion of microscopic collective models from an A body Hamiltonian with two body oscillator interactions. In Jacobi coordinates and for the units used in the previous section the Hamiltonian —from which the center of mass part was removed— becomes

$$H_0 = \frac{1}{2}\sum_{s=1}^{A-1}\sum_{i=1}^{3}(p_{is}^2 + x_{is}^2) = \sum_{s=1}^{A-1}\sum_{i=1}^{3} b_{is}^{+} b_{is} + [3(A-1)/2] , \qquad (7.1)$$

where the creation and annihilation operators are given by

$$b_{is}^{\dagger} = (1/\sqrt{2})(x_{is} - ip_{is}), \; b_{is} = (1/\sqrt{2})(x_{is} + ip_{is}) . \qquad (7.2)$$

The symmetry group of the Hamiltonian (7.1) is the unitary group of 3A-3 dimensions U(3A-3) whose generators are[22)]

$$b_{is}^{\dagger} b_{jt} \quad i,j = 1,2,3 \; ; \quad s,t = 1,2,\ldots A-1 . \qquad (7.3)$$

As in the previous discussions it is interesting to consider chains of subgroups of U(3A-3) and one that brings out the collective features is[13,22)]

$$\begin{array}{llll} U(3A-3) \supset \mathcal{U}(3) & \times & U(A-1) & [h_1 h_2 h_3] \\ \qquad\qquad\quad \cup & & \cup & \\ \qquad\qquad\quad \mathcal{O}(3) & L & O(A-1) & (\omega_1 \omega_2 \omega_3) \\ \qquad\qquad\quad \cup & & \cup & \\ \qquad\qquad\quad \mathcal{O}(2) & M & D^{[A-1,1]}(S_A) & \{f\} \; , \; (r) \; , \end{array} \qquad (7.3a)$$

where to the right of each subgroup we have indicated the numbers that characterize its irrep.

The generators of the subgroups are respectively

$$\mathcal{C}_{ij} = \sum_s b_{is}^{\dagger} b_{js} \qquad \text{for} \quad \mathcal{U}(3) , \qquad (7.4a)$$

$$C_{st} = \sum_i b_{is}^{\dagger} b_{it} \qquad \text{for} \quad U(A-1), \qquad (7.4b)$$

$$L_{ij} = \mathcal{C}_{ij} - \mathcal{C}_{ji} \qquad \text{for} \quad \mathcal{O}(3) , \qquad (7.4c)$$

$$\Lambda_{st} = C_{st} - C_{ts} \qquad \text{for} \quad O(A-1) \ , \tag{7.4d}$$

and $D^{[A-1,1]}(S_A)$ designates the representation $[A-1,1]$ of the symmetric group of A particles. The irreps of $U(3)$ —and thus also[22] those of U(A-1)— are given by a partition in three numbers $[h_1h_2h_3]$ of the total number of quanta N. Clearly then the irrep of O(A-1) also has only three numbers $(\omega_1\omega_2\omega_3)$, while L,M indicates the total angular momentum and its projection and {f} is the partition of A characterizing the symmetry properties of the state under permutation and (r) is the Yamanouchi symbol.[22]

By an analysis similar to the one that led to the ket (2.7) the eigenstate associated with the chain of groups (7.3) can be written as

$$\left| [h_1h_2h_3]\Omega LM \ ; \ \bar{\Omega}(\omega_1\omega_2\omega_3) \ \phi\{f\}(r) > \right.$$

$$= \sum_{KR} F\left(\begin{matrix} [h_1h_2h_3]\Omega LK \\ \bar{\Omega}(\omega_1\omega_2\omega_3)R \end{matrix} \ ; \ \rho_1\rho_2\rho_3 \right) D^{L*}_{MK}(\Theta) \, D^{(\omega_1\omega_2\omega_3)^*}_{\phi\{f\}(r),R}(\alpha) \tag{7.5}$$

where $\Omega,\bar{\Omega},\phi$ distinguish between repeated irreps of the group on the right appearing in a given irrep of the group on the left. In (7.5) the $D(\Theta)$ is the standard irrep of $O(3)$ characterized by L and depending on the Euler angles Θ_i while $D(\alpha)$ is the irrep of O(A-1) characterized by $(\omega_1\omega_2\omega_3)$ and depending —as indicated in the previous section— on 3A-9 variables α. The column index R of $D(\alpha)$ can be characterized by the irreps of an arbitrary chain of subgroups of O(A-1) and it is convenient[13] to take $O(A-2) \supset O(A-3) \supset O(A-4) \ldots$. One has to sum over this index R, as well as over K appearing in $D(\Theta)$. The functions F depend then only on the collective coordinates ρ_1,ρ_2,ρ_3 appearing in (6.2).

The functions $D(\alpha)$ are, in general, not known explicitly, but this will not be required in the microscopic collective models to be discussed in the next section. The main theoretical problem is the determination of the $F(\rho_k)$ in (7.5) on which work is being done at the present time.[13,23]

8. MICROSCOPIC COLLECTIVE MODELS

The states (7.5) form a complete set for the A body system in configuration space and they could be combined with the spin isospin kets, characterized by the conjugate partition $\{\tilde{f}\}$ and Yamanouchi symbol $(\tilde{r})$ to give a complete set of states satisfying the Pauli principle. Obviously the use of such states is unwidely and has all the difficulties of the complete A body problem. We are interested here though only in the collective features of these states and in the microscopic models mentioned in the introduction, particularly in the one of Vanagas[13], a fundamental restriction is proposed: to limit the states to a single representation $(\omega_1\omega_2\omega_3)$ of the O(A-1) group.

Among the physical reasons behind this restriction we have the following: a) the Slater determinants that represent ground states by filling tightly with particles the oscillator potential levels, can usually be characterized by a given irrep of O(A-1). For example for O^{16} and Ca^{40} we have respectively that $(\omega_1\omega_2\omega_3) = (444)$ and (20,20,20). b) From the ground states one can construct other collective ones by applying C_{ij} of (7.4a) or

$$B^{\dagger}_{lm} = \sum_{i,j} \langle 11ij|lm\rangle \sum_{s=1}^{A-1} b^{\dagger}_{is} b^{\dagger}_{js} \quad ; \; \ell = 0,2 \tag{8.1}$$

where $\langle|\rangle$ is a Clebsch-Gordan coefficient with Cartesian indices i,j = 1,2,3. The operators C_{ij}, $B^{\dagger}_{lm}$ have summations over the index s = 1,...A-1 and thus they are scalars of O(A-1) i.e. they commute with the Λ_{st} of (7.4d). Therefore if we apply them to a state of the form (7.5) they cannot modify the irrep $(\omega_1\omega_2\omega_3)$ of O(A-1) appearing in it.

c) The q_{ij} of (6.3c) involve also summation over s and are scalars of O(A-1). As from (6.5) the collective coordinates ρ_k, Θ_k, k = 1,2,3 can be expressed in terms of the q_{ij}, it is clear that any functions of the former or of their derivatives when applied to (7.5) cannot modify the $(\omega_1\omega_2\omega_3)$ appearing there.

If we take a realistic Hamiltonian H for A particles with two body interactions, we can get in several ways a part of it that acts on a single representation $(\omega_1\omega_2\omega_3)$ of O(A-1). One of them, favored by the Filippov group[11], is to project on a given representation $(\omega_1\omega_2\omega_3)$ of O(A-1) with the help of functions that satisfy the Pauli principle of the form[13]

$$\sum_{r} D^{(\omega_1\omega_2\omega_3)}_{\phi\{f\}(r),R}(\alpha)\, \chi(\{\tilde{f}\}\,(\tilde{r})\, ST) \;, \tag{8.2}$$

where χ is the spin isospin function associated with the conjugate partition $\{\tilde{f}\}$ and we sum over the Yamanouchi symbols (r). In this case though $(\omega_1\omega_2\omega_3)$ <u>are not</u> numbers associated with integrals of motion.

Another, favored by the Vanagas group[13)], is to average the Hamiltonian H over the angles α's of O(A-1) getting in this way a microscopic collective Hamiltonian in which $(\omega_1\omega_2\omega_3)$ are necessarily associated with integrals of motion.

Extensive work is being done at the present time[13,23)] on getting solutions for the Hamiltonians involved in these approaches though lack of space prevents us from discussing them further.

What is important for our objective of confronting microscopic and macroscopic collective models is to note that both can be expressed in terms of six collective coordinates and their corresponding momenta i.e. the ρ_k, Θ_k, k = 1,2,3 for the former and the $\bar{\alpha}$, β, γ, Θ_k for the latter. Thus it is feasible —at least in the case of harmonic oscillator interactions— to relate them through a canonical transformation, as has been attempted recently[23)].

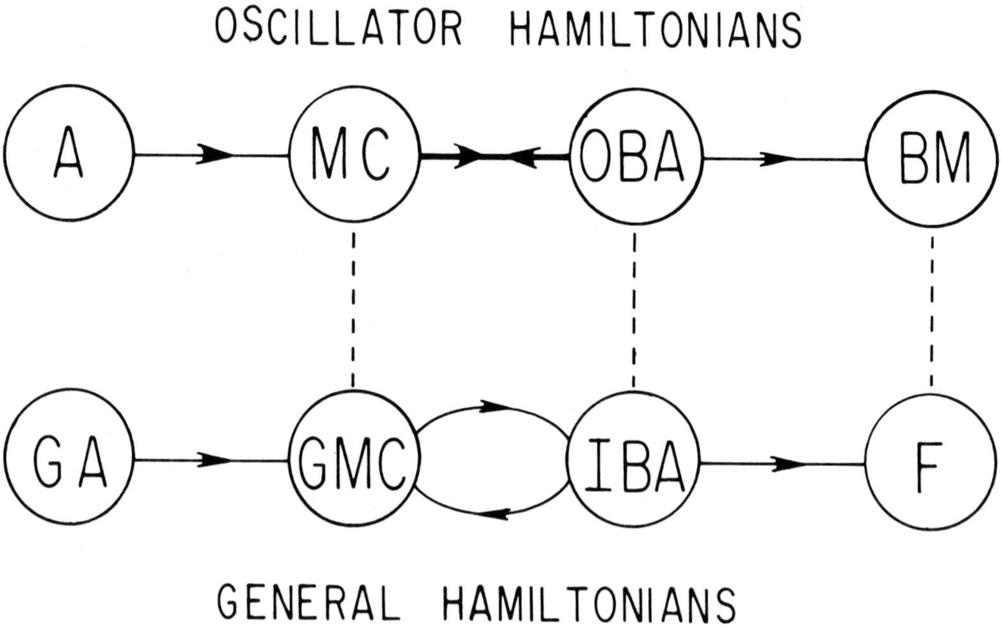

Figure 3. Flow diagram relating the different nuclear collective models. The acronymics in the circles are explained in the concluding section.

9. CONCLUSION

As a last point we summarize the relations between the nuclear models discussed in this paper by a kind of flow diagram given in Fig. 3. We consider first the different oscillator Hamiltonians and their interconnections and then extend the discussion to general Hamiltonians.

We start with the A particle oscillator Hamiltonian (7.1) and denote it by the circle containing A in Fig. 3. We then project from it the microscopic collective (MC) Hamiltonian by the procedure outlined in the previous section. This Hamiltonian is indicated by MC in a circle and, as we obtain it from projection from the A particle oscillator Hamiltonian, we connect A and MC by a line going from the first to the second.

We then consider an oscillator boson approximation (OBA) Hamiltonian (4.3) containing s and d bosons and the discussion in reference 23 indicates that MC and OBA (where the latter also appears in a circle in the figure) are related by a canonical transformation. This fact is shown in the figure by the heavy line connecting MC and OBA with arrows in both directions, as the canonical transformation allows us to go from the MC to the OBA or viceversa. Finally, by projecting out the d-boson part of the OBA we get a Hamiltonian (2.2) which is identical to the oscillator Hamiltonian of Bohr-Mottelson (BM) as indicated by the last circle of the upper line of Fig. 3.

We turn now to the general A particle (GA) Hamiltonian. By the procedure discussed in the previous section we can project out the general microscopic collective (GMC) Hamiltonian. This GMC Hamiltonian can be expressed as a matrix with respect to the eigenstates of the oscillator microscopic collective (MC) Hamiltonian, which is indicated in Fig. 3 by the dotted connection between the two models. These eigenstates can be obtained from (7.5) when we project out the part depending on the α's.

Turning now to the macroscopic s-d interacting boson approximation (IBA) analyzed in section 4, we can express its most general Hamiltonian as a matrix with respect to the eigenstates (4.6) or (4.8) of the oscillator s-d boson approximation (OBA). Thus through the canonical transformation relating MC and OBA we can pass from GMC to a kind of IBA or viceversa as indicated by the two lines with arrows in the diagram. Finally, as discussed in section 5, there is a procedure by which we can relate IBA Hamiltonians with those of a generalized Bohr Mottelson type which, as in section 3, we designate as Frankfurt (F) Hamiltonians. These Hamiltonians can be expressed as matrices with respect to the eigenstates (2.7) of the oscillator Bohr Mottelson (BM) Hamiltonian, which is indicated by the dotted line connecting BM and F.

Thus we have outlined a procedure by which starting from the Hamiltonian of a microscopic system of A nucleons, we can arrive finally to a macroscopic collective model which is a suitable generalization of the one introduced by Bohr and -- Mottelson.

REFERENCES

1. A. Bohr and B. Mottelson, Dan. Mat. Fys. Medd. 27, No. 16, 1 (1953).
2. A. Bohr and B. Mottelson, "Nuclear Structure" Vol. 1 (Benjamin, New York 1969).
3. J.M. Eisenberg and W. Greiner, "Nuclear Models", (North-Holland, Amsterdam, 1970).
4. G. Gneuss and W. Greiner, Nucl. Phys. A171, 449, (1971).
5. R. Sedlmayr et al., Nucl. Phys. A232, 465, (1974);
P.O. Hess and W. Greiner, "Nuclear Collective Models and their Applications", (SUSSP Proc. Scotland, August 1977).
6. R.V. Dzholos, D. Janssen and F. Dönau, Nucl. Phys. A224, 93 (1974);
Yad. Fiz. 22, 965 (1975).
7. A. Arima and F. Iachello, Ann. of Phys. 99, 253 (1976), 111, 201 (1977);
A. Arima, F. Iachello and O. Schotten, 115, 325 (1978).
8. M. Moshinsky, Nucl. Phys. A338, 156 (1980).
9. A.Ya. Dzublik, Preprint ITF-71-122R, Inst. Theoretical Physics, Kiev, USSR;
A.Ya. Dzublik, V.I. Ovcharenko, A.I. Steshenko and G.F. Filippov, Yad. Fiz. 15, 869 (1972), Sov. J. Nucl. Phys. 15, 487 (1972).
10. W. Zickendraht, J. Math. Phys. 12, 1663 (1971).
11. G.F. Filippov, Fiz. Elem. Chastits, At. Yadra 4, 992 (1973), Sov. J. Part. Nucl. 4, 405 (1974).
12. V. Vanagas, R. Kalinauskas, Yad. Fiz. 18, 768 (1973), Sov. J. Nucl. Phys. 18, 395 (1974);
V. Vanagas, Yad. Fiz. 23, 950 (1976), Sov. Phys. 23, 500 (1976).
13. V. Vanagas, "The Microscopic Nuclear Theory", Lecture Notes, Dept. of Physics, University of Toronto, 1977.
14. E. Chacón, M. Moshinsky and R.T. Sharp, J. Math. Phys. 17, 668 (1976).
15. E. Chacón and M. Moshinsky, J. Math. Phys. 18, 870 (1977).
16. P. Hess, J. of Phys. G4, 59 (1978).
17. M. Moshinsky, KINAM (México), 2, 79 (1980).
18. O. Castaños, E. Chacón, A. Frank and M. Moshinsky, J. Math. Phys. 20, 35 (1979).
19. J.P. Elliott, Proc. Roy. Soc. A245, 562 (1958).
20. O. Castaños, A. Frank and P. Federman, Phys. Lett. 88B, 203 (1979).
21. Yu.F. Smirnov and G.F. Filippov, Yad. Fiz. 27, 73 (1978), Sov. J. Nucl. Phys. 27, 39 (1978).
22. P. Kramer and M. Moshinsky, Nucl. Phys. 82, 241 (1966).
23. E. Chacón, M. Moshinsky and V. Vanagas, J. Math. Phys. (submitted for publication).

MATRIX ELEMENTS OF GENERATORS OF U(6) IN A $U(6) \supset U(3) \supset U(2)$ BASIS

Christiane Quesne*

Physique Théorique et Mathématique, CP 229, Université Libre de Bruxelles, Bd. du Triomphe, B 1050 Brussels, Belgium.

In nuclear physics recent developments, much attention has been paid to the basis states of a symmetrical irreducible representation (IR) [N] of U(6) classified by the subgroups $SU(3) \supset SO(3) \supset SO(2)$. For instance they appear in the interacting boson model (IBM) proposed by Arima and Iachello to give a unified description of collective states in nuclei [1] . In the IBM, creation operators of s and d bosons, $\eta_{00} = \bar{\eta}$ and $\eta_{2m} = \eta_m$ ($m = \pm 2, \pm 1, 0$), are introduced as well as their corresponding annihilation operators $\xi^{00} = \bar{\xi}$ and $\xi^{2m} = \xi^m$ ($m = \pm 2, \pm 1, 0$). The operators

$$\mathcal{C}_{lm}^{l'm'} = \eta_{lm} \, \xi^{l'm'} , \qquad l, l' = 0, 2 , \tag{1}$$

are the generators of a U(6) group. N boson states belong to the IR [N] of this group. Three chains of subgroups of U(6) are considered :

$$U(6) \supset U(5) \supset SO(5) \supset SO(3) \supset SO(2) , \tag{2a}$$

$$U(6) \supset SU(3) \supset SO(3) \supset SO(2) , \tag{2b}$$

and

$$U(6) \supset SO(6) \supset SO(5) \supset SO(3) \supset SO(2) . \tag{2c}$$

The chain (2b), which will be of interest here, is connected with the rotational limit of the collective model [2] . As the most general interaction in the IBM can be expressed in terms of the generators of U(6), the knowledge of their matrix elements (M E) with respect to the basis corresponding to the chain (2b) is required.

In a recent work [3] , Castaños, Chacón, Frank, and Moshinsky show that the basis states of the IR [N] classified according to the chain (2b) can be obtained from those classified according to the chain (2a) by diagonalizing the matrix of the operator

$$Q^2 = \sum_{m=-2}^{+2} (-1)^m \, Q_m \, Q_{-m} , \tag{3}$$

where Q_m ($m = \pm 2, \pm 1, 0$) are the components of the quadrupole operator. One can then proceed in the basis corresponding to the chain (2a) , which has been extensively studied, and calculate the M E of all relevant operators in that basis.

* Maître de Recherches F.N.R.S.

In this communication, we propose an alternative approach [4] where the M E of the generators of U(6) are determined in the canonical basis for U(3), corresponding to the following chain of groups

$$U(6) \supset U(3) \supset U(2) \supset U(1). \qquad (4)$$

The M E in the basis corresponding to the chain (2b) can then be deduced with the help of the well known transformation brackets between the two basis [5] .

The boson creation operators classified according to the chain (4) can be written as ζ_μ , $\mu = 1, \ldots, 6$, where [6]

$$\zeta_1 = \eta_2 , \qquad \zeta_2 = \eta_1 , \qquad \zeta_3 = \frac{1}{\sqrt{3}}(\eta_0 + \sqrt{2}\,\bar{\eta}) ,$$
$$\zeta_4 = \frac{1}{\sqrt{3}}(\sqrt{2}\,\eta_0 - \bar{\eta}), \qquad \zeta_5 = \eta_{-1} , \qquad \zeta_6 = \eta_{-2} . \qquad (5)$$

We denote by ζ_μ^+ the corresponding annihilation operators. In terms of them, the generators of U(6) become

$$\mathcal{C}_\mu^{\nu} = \zeta_\mu \zeta_\nu^+ , \qquad \mu, \nu = 1, \ldots, 6 . \qquad (6)$$

Those of U(3) classified according to the chain (4) are denoted by $\mathbb{C}_i^{j}$, i,j = 1, 2, 3.

The basis states of the IR [N] of U(6) classified according to the chain (4) are usual Gel'fand states of the U(3) group

$$\left| \begin{matrix} h_{13} & & h_{23} & & h_{33} \\ & h_{12} & & h_{22} & \\ & & h_{11} & & \end{matrix} \right\rangle , \qquad (7)$$

with the only restrictions that $h_{i3} = h_i$ (i = 1, 2, 3) are even and $h_1+h_2+h_3 = 2N$. The IR $[h_1h_2h_3]$ of U(3) can also be characterized by the three numbers x, y, z, defined by

$$x = \frac{1}{2}(h_1 - h_2) , \qquad y = \frac{1}{2}(h_2 - h_3) , \qquad z = \frac{1}{2} h_3 . \qquad (8)$$

To calculate the M E of the 36 generators of U(6) with respect to the basis (7), it is useful to organize those generators into irreducible tensors (IT) with respect to U(3)⊃ U(2)⊃ U(1) : there are three IT with classification [000] , [10-1] , and [20-2] under U(3) respectively. The [000] IT is just the first-order Casimir operator of U(6), $\sum_{\mu=1}^{6} \mathcal{C}_\mu^{\mu}$. The [10-1] IT is made of the 8 generators of SU(3), obtained from those of U(3) by discarding the trace $\sum_{i=1}^{3} \mathbb{C}_i^{i} = 2 \sum_{\mu=1}^{6} \mathcal{C}_\mu^{\mu}$. Finally, the [20-2] IT is made of the 27 remaining independent linear combinations of the generators. We only need to calculate the M E of the latter, as the M E of the first two are well known. We shall denote its components by $T\begin{pmatrix} 2\,0\,-2 \\ r\ s \\ t \end{pmatrix}$, where the values of r, s, and t are restricted by the usual inequalities satisfied by Gel'fand patterns.

We now turn to that computation. It can be simplified by using the Wigner-Eckart theorem for the SU(3) canonical chain. As the [20-2] IT is a [42] IT with respect to SU(3), the Wigner coefficients needed are those where one of the SU(3) IR is [42]. Using Biedenharn's notations for the Wigner coefficients, the Wigner-Eckart theorem writes

$$\left\langle \begin{matrix} h_1' \ h_2' \ h_3' \\ h_{12}' \ h_{22}' \\ h_{11}' \end{matrix} \right| T\begin{pmatrix} 2 \quad 0 \quad -2 \\ r \quad s \\ t \end{pmatrix} \left| \begin{matrix} h_1 \ h_2 \ h_3 \\ h_{12} \ h_{22} \\ h_{11} \end{matrix} \right\rangle = \sum_{(\gamma)} \langle h_1' h_2' h_3' \| T[20\text{-}2;(\gamma)] \| h_1 h_2 h_3 \rangle$$

$$\times \left\langle \begin{matrix} h_1'+2 \quad h_2'+2 \quad h_3'+2 \\ h_{12}'+2 \quad h_{22}'+2 \\ h_{11}'+2 \end{matrix} \right| \left\langle \begin{matrix} \gamma_{11}+2 \\ \gamma_{12}+2 \quad \gamma_{22}+2 \\ 4 \quad 2 \quad 0 \\ r+2 \quad s+2 \\ t+2 \end{matrix} \right\rangle \left| \begin{matrix} h_1 \quad h_2 \quad h_3 \\ h_{12} \quad h_{22} \\ h_{11} \end{matrix} \right\rangle , \tag{9}$$

where the multiplicity label (γ) is an operator pattern $\begin{pmatrix} \gamma_{11} \\ \gamma_{12} \quad \gamma_{22} \end{pmatrix}$. The SU(3) Wigner coefficient can be factorized into a reduced Wigner coefficient and an ordinary SU(2) Wigner coefficient. Analytical expressions for the reduced Wigner coefficients have been given by Castilho Alcaras, Biedenharn, Hecht, and Neely [7]. Therefore the only quantities left for determination in Eq. (9) are the reduced M E of $T\begin{pmatrix} 2\,0\,-2 \\ r\ s \\ t \end{pmatrix}$.

For that purpose, we use the explicit expression of the highest weight state (hws) of the IR $[h_1 h_2 h_3]$ of U(3) contained in the IR [N] of U(6), that was determined by Pérez some time ago [8] :

$$\left| \begin{matrix} h_1 \quad h_2 \quad h_3 \\ h_1 \quad h_2 \\ h_1 \end{matrix} \right\rangle = \left| \begin{matrix} 2x+2y+2z \quad 2y+2z \quad 2z \\ 2x+2y+2z \quad 2y+2z \\ 2x+2y+2z \end{matrix} \right\rangle = \mathcal{N}(x,y,z)\, \zeta_1^{\,x}\, Y^y Z^z |0\rangle, \tag{10}$$

where

$$Y = \zeta_1 \zeta_4 - \frac{1}{2} \zeta_2^{\,2} , \tag{11}$$

$$Z = \zeta_1 \zeta_4 \zeta_6 - \frac{1}{2} \zeta_1 \zeta_5^{\,2} - \frac{1}{2} \zeta_2^{\,2} \zeta_6 + \frac{1}{\sqrt{2}} \zeta_2 \zeta_3 \zeta_5 - \frac{1}{2} \zeta_3^{\,2} \zeta_4 , \tag{12}$$

and $\mathcal{N}(x, y, z)$ is a normalization factor. By taking hws in both bra and ket in the M E of the left hand side of Eq. (9), one can explicitly calculate that M E. For $[h_1' \, h_2' \, h_3'] = [h_1+2 \; h_2 \; h_3-2]$, $[h_1+2 \; h_2-2 \; h_3]$, $[h_1 \; h_2+2 \; h_3-2]$, $[h_1 \; h_2-2 \; h_3+2]$, $[h_1-2 \; h_2+2 \; h_3]$, and $[h_1-2 \; h_2 \; h_3+2]$, for which the multiplicity is one in the right hand side, one can then determine the corresponding reduced M E in this way. For the remaining case corresponding to $[h_1' \, h_2' \, h_3'] = [h_1 \; h_2 \; h_3]$, the multiplicity is equal to 3 and there are 3 independent reduced M E to be calculated. One then considers the M E between hws of the 3 components $T\begin{pmatrix} 2\;0\;-2 \\ 2\;-2 \\ 0 \end{pmatrix}$, $T\begin{pmatrix} 2\;0\;-2 \\ 1\;-1 \\ 0 \end{pmatrix}$, and $T\begin{pmatrix} 2\;0\;-2 \\ 0\;\;0 \\ 0 \end{pmatrix}$ of the IT, and obtains a system of 3 linear equations for the reduced M E, from which they can be easily calculated. This completes the determination of the M E of the generators of U(6) in the basis corresponding to the chain (4).

An application of the formalism developed in this work to the study of the deformation of nuclei in the rotational model will be published elsewhere [9] .

The author would like to thank E. Chacón , A. Partensky, and especially M. Moshinsky for interesting discussions.

References

1. A. Arima and F. Iachello, Phys. Rev. Lett. 35 (1975) 1069.
2. A. Arima and F. Iachello, Ann. Phys. (N.Y.) 111 (1978) 201.
3. O. Castaños, E. Chacón , A. Frank, and M. Moshinsky, J. Math. Phys. 20 (1979) 35.
4. C. Quesne, to be published.
5. M. Moshinsky, Rev. Mod. Phys. 34 (1962) 813 ;
   M. Moshinsky and V. Syamala Devi, J. Math. Phys. 10 (1969) 455.
6. M. Moshinsky, "Group Theory and the Many Body Problem" (Gordon and Breach, New York, 1967).
7. J.A. Castilho Alcaras, L.C. Biedenharn, K.T. Hecht, and G. Neely, Ann. Phys. (N.Y.) 60 (1970) 85.
8. R. Pérez, private communication from E. Chacón.
9. A. Partensky and C. Quesne , to be published.

HARTREE-FOCK ONE-BODY DYNAMICS
AND U(n) CO-ADJOINT ORBITS

George Rosensteel†
Department of Physics and Quantum Theory Group
Tulane University
New Orleans, Louisiana, U.S.A. 70118

Introduction

Hartree-Fock theory is basic to our microscopic understanding of many-fermion systems [1,2]. Its fundamental approximation restricts the states of A-particle systems to be Slater determinants, i.e. simple vectors $\Phi = \phi_1 \wedge \phi_2 \wedge \cdots \wedge \phi_A$ in the exterior product of A-copies of the single-particle space. The ground state is approximated by the Slater determinant which minimizes the expectation of the energy.

Although the atomic many-electron ground state may be approximated for most purposes quite adequately by a Slater determinant, the same cannot be said for the atomic nucleus. Thus, the restriction to determinantal wavefunctions must be eliminated if a satisfactory microscopic theory of nuclei is to be achieved. However, if the simplifying restriction to determinants is dropped, then we are faced with an intractable many-body problem.

Fortunately, this pessimistic conclusion is unwarranted. To understand how the restriction to determinants can be eliminated entirely, and yet a tractable generalization of HF achieved, it is necessary to appreciate the critical role played by the unitary group U(n) in HF.

The group U(n) of unitary transformations in the n-dimensional single-particle space is known to be important for the many-body problem in general and HF in particular [3,4]. For example, Matsen and Nelin have observed that a Slater determinant is a highest weight vector of U(n) [5,6]. Here, however, we need to exploit two features of U(n).

First note that the group U(n) acts transitively on the Slater determinants, i.e. if Φ is any fixed normalized determinant, then every other normalized determinant is of the form $g\Phi = g\phi_1 \wedge g\phi_2 \wedge \cdots \wedge g\phi_A$. Hence, the group U(n) characterizes the states available in the HF approximation, viz. the HF states constitute a single orbit of U(n) in the exterior product space.

Secondly, recall that the Lie algebra u(n) consists of the hermitian one-body

†Work supported by NSF grant PHY-7906534.

operators, $X = \Sigma X_{\alpha\beta} a_\alpha^\dagger a_\beta$. These Lie algebra elements have the geometrical interpretation of tangent vectors to the determinantal orbit. This viewpoint is rendered by identifying the tangent to the curve of determinants $\gamma_X(t) = \exp(-i tX)g\Phi$ at $g\Phi$ with X itself. A HF solution is a critical point of the energy function $H(g\Phi) = \langle g\Phi|Hg\Phi\rangle$ restricted to the orbit of normalized determinants. Hence, the derivative of H in the direction of every tangent X must vanish,

$$0 = dH(X) = \frac{i dH(\gamma_X(t))}{dt}\Big|_{t=0} = \langle g\Phi|[H,X]g\Phi\rangle \ . \tag{1}$$

<u>Non-Determinantal Orbits</u>

As we have seen, the unitary group provides all the requisite structure for HF. Hence, there is no necessity to consider only the orbit of determinants. Every U(n) orbit in the exterior product space yields a natural generalization of HF.

Suppose Φ is any antisymmetrized wavefunction and consider the orbit of states $g\Phi$ as g ranges over U(n). A generalized HF solution is a critical point of the energy function restricted to this orbit. Thus, its derivative in the direction of every tangent vector $X \in u(n)$ to the orbit is zero. This yields the generalized HF equations via Eq. (1),

$$0 = (\nu_\rho - \nu_\sigma)\,[T(g)_{\rho\sigma} + \sum_\delta V(g)_{\delta\rho\delta\sigma}\,\nu_\delta] + \tfrac{1}{2}\,(\Omega\,(g)_{\rho\sigma} - \Omega\,(g)^*_{\sigma\rho}) \ , \tag{2}$$

where the equations are written with respect to a basis with the density diagonal at Φ, $\langle\Phi|a_\alpha^\dagger a_\beta \Phi\rangle = \nu_\alpha \delta_{\alpha\beta}$; the one-body and two-body parts of H are transformed by g according to

$$\begin{aligned} T(g)_{\alpha\beta} &= \sum_{\alpha'\beta'} g^{-1}_{\alpha\alpha'}\, T_{\alpha'\beta'}\, g_{\beta'\beta} \\ V(g)_{\alpha\beta\gamma\delta} &= \sum_{\alpha'\beta'\gamma'\delta'} g^{-1}_{\alpha\alpha'}\, g^{-1}_{\beta\beta'}\, V_{\alpha'\beta'\gamma'\delta'}\, g_{\gamma'\gamma}\, g_{\delta'\delta} \end{aligned} \tag{3}$$

and

$$\Omega(g)_{\rho\sigma} = \sum_{\alpha\beta\delta} V(g)_{\alpha\beta\sigma\delta}\,[\langle\Phi|a_\alpha^\dagger a_\beta^\dagger a_\delta a_\rho \Phi\rangle - (\delta_{\alpha\rho}\delta_{\beta\delta} - \delta_{\alpha\delta}\delta_{\beta\rho})\,\nu_\rho\,\nu_\delta] . \tag{4}$$

A critical point $g\Phi$ is defined by a g in U(n) satisfying Eq. (2) for every ρ, σ. Note that the term in brackets in Eq. (2) is just the usual HF Hamiltonian, and Ω vanishes if Φ is a determinant.

If an exact eigenstate of the Hamiltonian is in the orbit containing our chosen point Φ, then it is a solution to Eq. (2). This is because an eigenstate is a critical point of the energy whose derivative vanishes in <u>every</u> direction. In addition, Eq. (2) has solutions where the derivative of the energy function vanishes only in the directions of tangents to unitary orbits. Indeed, there is a continuum of the latter solutions, since each orbit has solutions and there is a continuum of orbits.

Symplectic Structure

Time-dependent Hartree-Fock (TDHF) is nothing but classical Hamiltonian dynamics on the orbit of determinants. The relevant phase space structure on the determinantal orbit is given by the non-degenerate symplectic form ω,

$$\omega_{g\Phi}(X,Y) = -i \langle g\Phi|[X,Y]g\Phi\rangle , \tag{5}$$

where $X,\ Y \in u(n)$ are tangent vectors at $g\Phi$. The HF Hamiltonian H_{HF} is the vectorfield on the orbit satisfying the usual classical mechanics relation identifying the form dH with the vectorfield H_{HF} [7],

$$\omega_{g\Phi}(H_{HF}\ (g\Phi),\ X) = dH(X), \tag{6}$$

for every tangent $X \in u(n)$ at $g\Phi$. A TDHF solution is an integral curve of H_{HF}.

One would like to generalize this classical dynamics to every unitary orbit. However, this is not possible because the form is degenerate on the non-determinantal orbits and, hence, Eq. (6) has no unique solution for the HF Hamiltonian. Although the HF Hamiltonian on the general orbits may be constructed from the exact Hamiltonian by projection relative to the non-degenerate Hilbert inner product, the resulting vectorfield does not define Hamiltonian dynamics. Nevertheless, there is an alternative generalization of HF which eliminates the restriction to determinants and yet retains the symplectic structure. This alternative recognizes that the orbit of determinants may be identified with a co-adjoint orbit of $U(n)$.

Co-Adjoint Orbits

The $U(n)$ group action is interrelated with the symplectic geometry since $U(n)$ acts as a group of canonical transformations:

$$\omega_{g\Phi}(g\ X\ g^{-1},\ g\ Y\ g^{-1}) = \omega_{\Phi}\ (X,Y), \tag{7}$$

where the tangent vectors $X,\ Y \in u(n)$ at Φ are transformed by the adjoint action into tangents at $g\Phi$, $g\ X\ g^{-1}$, $g\ Y\ g^{-1} \in u(n)$. Thus, the determinantal orbit is a symplectic manifold on which $U(n)$ acts as a transitive group of canonical transformations. By the Kostant-Souriau classification theorem, the Slater determinants must be in one-to-one correspondence with (a covering space of) a co-adjoint orbit of $U(n)$ [8-12]. This equivalent statement of Hartree-Fock in terms of $U(n)$ co-adjoint orbits turns out to be just the well-known density matrix formulation [13].

A density matrix $\rho_{\alpha\beta} = \langle\Phi|a^{\dagger}_{\beta}a_{\alpha}\ \Phi\rangle$ is an element of the dual space to $u(n)$, i.e. a real-valued linear function of $u(n)$ given by

$$\rho(X) \equiv tr(\rho X) = \langle\Phi|\sum_{\alpha\beta} X_{\alpha\beta}\ a^{\dagger}_{\alpha}\ a_{\beta}\ \Phi\rangle . \tag{8}$$

The co-adjoint action

$$Ad^{*}_{g}\ (\rho) \equiv g\ \rho\ g^{-1} \tag{9}$$

is compatible with the group action on the wavefunctions,

$$Ad^*_g\ (\rho)\ (X) = tr(g\ \rho\ g^{-1}\ X) = \langle g\Phi | \sum_{\alpha\beta} X_{\alpha\beta}\ a^\dagger_\alpha\ a_\beta\ g\Phi \rangle. \tag{10}$$

It is well-known that if ρ is a density corresponding to a determinant, then the co-adjoint orbit containing ρ,

$$\mathcal{O}_\rho = \{\ Ad^*_g\ (\rho)\ |\ g \in U(n)\ \}, \tag{11}$$

is in one-to-one correspondence with the orbit of determinants. On the other hand, if ρ is non-determinantal, then the moment map, $g\Phi \to g\ \rho\ g^{-1}$, from an orbit of non-determinantal wavefunctions onto its associated co-adjoint orbit is many-to-one.

Lie algebra elements may be regarded as tangent vectors to co-adjoint orbits in a similar way as before. But, in contrast to the wavefunction orbits, <u>every</u> co-adjoint orbit is a symplectic manifold equipped with the natural form,

$$\omega_{g\ \rho\ g^{-1}}\ (X,Y) = -i\ tr(g\ \rho\ g^{-1}\ [X,Y]\ ), \tag{12}$$

where $X,\ Y \in u(n)$ are tangent vectors at $g\ \rho\ g^{-1}$. Hence, the co-adjoint orbits provide a natural generalization of Hartree-Fock to non-determinantal densities in which the non-degenerate phase space structure of ordinary HF is retained.

Because of the symplectic structure, the generalized HF Hamiltonian may be constructed immediately from Eq. (6) once the energy function H is defined. A TDHF solution is an integral curve of the HF Hamiltonian vectorfield; a HF stationary state is a point where the HF vectorfield vanishes.

Energy Functions

The energy function on the determinantal densities is identified with the energy function on the Slater determinants. This identification is unambiguous since the moment map onto the determinantal co-adjoint orbit, $g\Phi \to g\ \rho\ g^{-1}$, is one-to-one. However, for all other orbits the moment map is many-to-one. The ambiguity in the correspondence can be expressed best in terms of the differing isotropy subgroups.

Suppose Φ is a wavefunction associated with a diagonal density ρ = diag $(\nu_1,\nu_2,\ldots,\nu_n)$. (Note that every co-adjoint orbit contains a diagonal density.) For the determinantal orbit, ρ = diag $(1,1,\ldots,1,0,\ldots,0)$, and the $U(n)$ elements that fix Φ are the same as those that fix ρ ; the common isotropy subgroup is $U(A) \times U(n-A)$. On the other hand, in the generic case when all occupancies ν_ρ are distinct, the co-adjoint isotropy subgroup is $H_\rho = U(1)\times\ldots\times U(1)$ (n-copies), while the wavefunction is fixed by only the pure phase factors $U(1)$. The moment map is then many-to-one:

$$U(n)/U(1) = SU(n) \to U(n)/U(1)\times\ldots\times U(1) = U(n)/H_\rho\ . \tag{13}$$

Therefore, if the energy function on the co-adjoint orbits is to be inherited from the microscopic interaction, then the ambiguity characterized by H_ρ must be resolved.

One possible choice for the energy is given by

$$H(g\,\rho\,g^{-1}) = \min_{h\varepsilon H_\rho} \langle gh\Phi|Hgh\Phi\rangle. \tag{14}$$

The principle advantage to this energy is that the exact ground state is a critical point for it.

Another energy function is produced by averaging the energy,

$$H(g\,\rho\,g^{-1}) = \int_{H_\rho} d\mu(h)\ \langle gh\Phi|Hgh\Phi\rangle\ , \tag{15}$$

where μ is the invariant measure on H_ρ. From Eq. (6), this energy yields a generalized HF Hamiltonian

$$H_{GHF}(g\,\rho\,g^{-1})_{\rho\sigma} = T(g)_{\rho\sigma} + \sum_\delta V(g)_{\delta\rho\delta\sigma}\ R^\delta_{\rho\sigma}\ , \tag{16}$$

where

$$R^\delta_{\rho\sigma} = (\nu_\rho - \nu_\sigma)^{-1} \langle\Phi|(a^\dagger_\rho a_\rho - a^\dagger_\sigma a_\sigma)\ a^\dagger_\delta a_\delta\ \Phi\rangle\ . \tag{17}$$

One important advantage to this energy is its technical simplicity.

Yet a third choice is defined by weighting the integrand in Eq. (15) with the temperature-dependent Boltzmann factor. This energy is especially appropriate for the description of high energy reactions. Further details will be published elsewhere [14-16].

Acknowledgment: Iam happy to acknowledge collaboration with D.J. Rowe and A. Ryman on this work.

References

1. D.R. Hartree, Proc. Camb. Phil. Soc. 24 (1927-28), 89.
2. V. Fock, Z. Phys. 61 (1930), 126.
3. M. Moshinsky, "Group Theory and the Many-Body Problem", Gordon & Breach, N.Y., 1968
4. F.A. Matsen, Adv. Quantum Chem. 2 (1978), 223.
5. F.A. Matsen and C.J. Nelin, Int. J. Quantum Chem. 15 (1979), 751.
6. C.J. Nelin and F.A. Matsen, in Lecture Notes in Physics, v. 94 (1978) Springer-Verlag, N.Y.
7. R. Abraham, "Foundations of Mechanics", Benjamin, N.Y. 1967, §16.
8. B. Kostant, Lecture Notes in Math., v. 170, Springer-Verlag, N.Y. 1970.
9. J.-M. Souriau, "Structure des Systemes Dynamiques", Dunod, Paris, 1970.
10. D.J. Simms and N.M.J. Woodhouse, Lecture Notes in Physics, v.53, Springer-Verlag N.Y. 1976, §8.
11. V. Guillemin and S. Sternberg, "Geometric Asymplotics", Math. Surveys, No. 14, AMS, Providence, R.I. 1977, §4.
12. J. Sniatycki, "Geometric Quantization and Quantum Mechanics", No. 30 Applied Math. Sci., Springer-Verlag, N.Y. 1980.
13. S. Belyaev, Nucl. Phys. 64 (1965) 17.
14. D.J. Rowe, A. Ryman and G. Rosensteel, "Many-Body quantum mechanics as a symplectic dynamical system, preprint
15. G. Rosensteel and D.J. Rowe, "Non-determinantal Hartree-Fock", preprint.
16. G. Rosensteel, "Hartree-Fock-Bogoliubov Theory without quasiparticle vacua", preprint.

GROUP-THEORETICAL ASPECTS OF THE MANY-BODY PROBLEM IN NUCLEAR THEORY

V. Vanagas
Institute of Physics
Vilnius, Lithuanian SSR, USSR

The microscopic theory of collective motions in nuclei [1,2], based on the clear distinction between the collective and internal degrees of freedom, gives the method for the following expansion of the arbitrary microscopic nuclear Hamiltonian H in series of the 0_{n-1}-irreducible terms $H(\kappa)$

$$H = H(0) + \sum_{\kappa \neq 0} H(\kappa) \tag{1}$$

where n is the number of nucleons, 0_{n-1} is the orthogonal group acting in the (n-1)-dimensional space and (κ) denotes the irreducible representation of 0_{n-1}. This expansion gives the link between the physical problems, connected with the structure of nucleus and the mathematical methods, based on group theory. In particular it has been proven that the 0_{n-1} - scalar terms $H(0) \equiv H_{coll}$ is the microscopic nuclear Hamiltonian. Thus the orthogonal group 0_{n-1} provides the tool to obtain the dynamic equation for the collective degrees of freedom and in this sense the group-theoretical methods play the some role in the nuclear theory as, for example, the variational principle in the Hartree-Fock theory.

The collective Hamiltonian H_{coll} acts within the 0_{n-1}-irreducible space (ω), thus the Schrödinger equation for H_{coll} has the following form:

$$H_{coll}^{(\omega)} \psi_{coll} = E_{coll}(J\ M_T, \omega\lambda T)\psi_{coll}, \tag{2}$$

where J is the total angular momentum, T and M_T are the isospin quantum number and its projection. In (2) λ denotes the S_n-irreducible representation and the symmetric group S_n is imbedded in 0_{n-1} in the chain $0_{n-1} \supset S_n$. It has been proven [1] that in the very particular case when H consists only of the central interaction and H_{coll} acts within the 0_{n-1}-scalar space, i.e. when $(\omega)=(0)$ the equation (2) is isomorphic to the Schrodinger equation

$$H_{R-V}\ \psi_{R-V} = E_{R-V}\ \psi_{R-V} \tag{3}$$

for the Hamiltonian of the phenomenological rotation-vibration model (with the "breathing" degree of freedom), acting in the Hilbert space spanned by the irreducible spaces of the unitary group U_6. Thus the theory of the collective motions in nuclei, based on the Schrödinger equation (2), gives the microscopic foundation of interacting boson model approach [1,2].

The eigenstates ψ_{coll} of H_{coll} give the general description of the collective features of nuclei. Most of them are new ones to be compared with those following from the traditional phenomenological approach. In the microscopic sense the collective excitations are equally important in even and odd mass nuclei and the spectrum of H_{coll} is similar to the rotational-vibrational picture only in the special case of some even-even nuclei. In particular, the equation (2) can be used in the case of odd-odd nuclei in states with the spin-isospin symmetry $\tilde{\lambda} \equiv [k_4+k_2+2, k_4+k_2+1, k_4+1, k_4]$ $(4k_4+2k_2+4=n)$, isospin quantum number $T=k_2+1$, and spin quantum number $S=1$. In such states the vectorial and tensorial terms in H are not supressed and H_{coll}, when it acts on states with $(w)=(0)$ (see (14.2) in [1]) gives the generalization of the phenomenological equation (3).

In the zeroth order approximation the spectrum and eigenstates of H_{coll} can be studied by using the following decomposition of H_{coll} into the U_{n-1}-irreducible terms:

$$H_{coll} = H_{0coll} + \sum_{\beta \neq 0} H_{\beta coll}, \tag{4}$$

where β denotes the U_{n-1}-irreducible term of H_{coll}. The group O_{n-1} is imbeded in the unitary group U_{n-1} by means of the chain $U_{n-1} \supset O_{n-1}$. The first term H_{0coll} of this decomposition gives the U_{n-1}-scalar part of H. In order to take into account some pairing-type features the "anticollective" term $H_{0anticoll}$ must also be included. The Hamiltonian $H_{0coll} + H_{0anticoll}$ acts in the finite dimension space, thus the Schrödinger equation

$$(H_{0coll} + H_{0anticoll})\ \psi_0 = E_0(J\ M_T,\ f\ \lambda\ T)\ \psi_0 \tag{5}$$

is exactly solvable for arbitrary n. In (5) f denotes the integral of motion, given by the irreducible representation $[f_1 f_2 f_3]$ of both groups U_3 and U_{n-1}. The solutions of the equation (5) give the far-reaching generalization of Elliott's model.

For different types of potentials the nucleon-nucleon interaction the spectrum of the equation (5) has been studied in the case of nucleus ^{20}Ne and a remarkable imitation of the experimental data has been obtained, including the level density description and the back-bending type effect.

[1] V. Vanagas. "The Microscopic Nuclear Theory within the Framework of the Restricted Dynamics", Lecture Notes (University of Toronto, 1977).

[2] V. Vanagas. "The Microscopic Theory of the Collective Motion in Nuclei", Lecture Notes (Latin American School of Physics, México, D. F., June 1980)

THE CONTINUING QUEST FOR EXACT SOLUTIONS OF EINSTEIN'S FIELD EQUATIONS--NEW GOALS FOR THE EIGHTIES

Frederick J. Ernst

Department of Physics
Illinois Institute of Technology
Chicago, IL 60616/USA

Abstract: The striking achievements of the last few years in connection with the search for solutions of Einstein's field equations describing the exterior fields of spinning masses give one hope that the next decade will see the solution of additional problems of physical interest.

It has been a long-time goal of people working upon Einstein's general theory of relativity to find some exact solution describing gravitational radiation from a bounded source. Whether or not the decade of the eighties will see the discovery of the first example of this type remains unclear, but surely this goal will be on many people's minds.

The early history of exact solutions provides countless examples of lucky shots in the dark, together with even more numerous fruitless calculations checking whether or not various metrical Ansätze lead to new solutions of the field equations.

During the sixties a new approach to the task of finding exact solutions became very popular; namely, the utilization of null tetrad or spinorial formalisms in conjunction with the assumption of the algebraically special nature of the Weyl conform tensor. The discovery of the Kerr metric, during the course of studying the algebraically special problem, was of course the most exciting result of that era.

In the seventies people became so bold as to tackle algebraically general fields, searching for such solutions with a pair of commuting Killing vectors, especially stationary axially symmetric fields associated with spinning masses. In these investigations the Kerr metric provided a foot in the door, suggesting those mathematical objects which might be usefully employed in the quest for algebraically general solutions.[1] It was during the course of these investigations that the notion of groups of transformations which map one solution into another arose.[2] As will be indicated in the lecture of my colleague, I. Hauser, it is now possible to spell out quite rigorously in what sense one is dealing here with a group. It is quite amazing, but nevertheless anticipated in a conjecture of Geroch[3], that essentially all stationary axially symmetric vacuum and electrovac fields can be constructed from Minkowski space by means of Kinnersley-Chitre transformations.[4] While Hauser's presentation will deal exclusively with the vacuum problem, it is not hard to see how electrovac fields may also be encompassed.

It would be a shame if the talent which enabled the solution of the stationary axially symmetric vacuum and electrovac problems soon dispersed because of a mistaken impression that all has been done which can be done. Even within the context of the stationary axially symmetric vacuum and electrovac problems, there remains much room for clarification of the relationships among various solution-generating techniques.[5] Beyond that I hope my colleagues will not shrink from tackling stationary axially symmetric interior problems, which should be of great interest to astrophysicists.

As at an earlier point in history, where the Kerr solution provided a foot in the door, so now I believe the Wahlquist perfect fluid solution[6] may provide a similar advantage. Like the Kerr solution, it is Petrov type D. Among type D solutions there is also the 7-parameter solution of Plebanski and Demianski.[7] Here too we have a slight glimpse into that which is essentially terra incognita, an example of a solution with nonvanishing cosmological constant. I certainly feel that one should not be content as long as the Wahlquist and Plebanski-Demianski solutions remain outside the bounds of our solution-generating capabilities.

It is not clear that groups will play the same role in the search for perfect fluid solutions or solutions with cosmological constant λ as they did in the vacuum and electrovac studies. That remains to be seen. But it should be recalled that the rigorous identification of true group structure came at the end of that earlier investigation, not at the beginning.

I should also like to mention that in the simplest case of a pressureless perfect fluid (i.e., dust) the general solution is already known, due to some work of Winicour. Here I am, of course, referring to solutions wherein the dust pursues circular orbits about the symmetry axis, with no circulation in the meridian plane.[8]

It is interesting that in the case of such dust solutions, as in the case of vacuum and electrovac solutions, the so-called Weyl canonical coordinates exist. When $\lambda \neq 0$ or the pressure $p \neq 0$, such coordinates are not available. We must learn to cope with that new element of the game.

It is fairly clear that the system of equations with which we shall have to deal either in the $p \neq 0$ or the $\lambda \neq 0$ cases are far more complex than those we have learned to handle in the seventies. The Bianchi identities will surely have to be exploited effectively, and we shall have to use our knowledge of the few existing solutions and the techniques developed during the seventies to crack these more difficult problems.

In the meanwhile there are many things which can be done which would be useful. For example, one can study the joining of Winicour's dust solutions or Wahlquist's perfect fluid solution to exterior fields. This might give us some foretaste of what we shall be able to dc after more general interior solutions are available.

In addition, the knowledge which is currently accruing should be studied carefully to see if it suggests any new insight into the gravitational radiation problem. Recall that even the static C-metric of Levi-Civita has been given a radiation interpretation in the work of Kinnersley and Walker.[9] One must avoid the temptation of dismissing stationary axially symmetric solutions as irrelevant to the gravitational radiation problem.

I hope that in 1990 we shall be able to look back upon as fruitful a decade in the quest for exact solutions as that upon which we now look back, that we shall know then as much about generating interior solutions as we know now about generating vacuum and electrovac solutions. More than that, I hope that during this coming decade someone discovers at least one example of a solution representing gravitational radiation from a bounded source, for I am sure that would provide the encouragement one needs to mount a sustained assault upon that as yet unpenetrated fortress of radiative solutions.

[In his lecture Dr. Ernst described in some detail Winicour's treatment of dust metrics, because he felt that of all things bearing upon the interior problem this was probably the least well known by the relativists in the audience.]

References:

1. F. J. Ernst, Phys. Rev. 167, 1175 and 168, 1415 (1968).
2. R. Geroch, J. Math. Phys. 12, 918 (1971);
W. Kinnersley, J. Math. Phys. 14, 651 (1973).
3. R. Geroch, J. Math. Phys. 13, 394 (1972).
4. W. Kinnersley and D. M. Chitre, J. Math. Phys. 18, 1538 (1977) and 19, 2037 (1978).
5. C. Cosgrove, J. Math. Phys. (in press).
6. H. D. Wahlquist, Phys. Rev. 172, 1291 (1968).
7. J. F. Plebanski and M. Demianski, Ann. of Phys. (N.Y.) 98, 98 (1976).
8. J. Winicour, J. Math. Phys. 16, 1806 (1975).
9. W. Kinnersley and M. Walker, Phys. Rev. D2, 1359 (1970).

ALL NON-TWISTING N's WITH COSMOLOGICAL CONSTANT

A. García Díaz and J.F. Plebañski*
Centro de Investigación y de Estudios Avanzados del I.P.N.
Departamento de Física
México 14, D.F., México, Apartado Postal 14-740

Introduction

The purpose of this work is to present the most general non-twisting N-type solutions with cosmological constant for the case when the distinguished direction e^3 is aligned along the quadruple Debever-Penrose (DP) vector. The congruence of the directions e^3 is then geodetic and shear free.

Methods and Results

The formalism used is the null tetrad approach. Working with the signature (+++-), the metric form is

$$g = 2e^1e^2+2e^3e^4, \quad e^2 = \overline{(e^1)}, \quad e^3 \text{ and } e^4 \text{ are real.} \tag{1}$$

Thus the first Cartan structure equations are

$$de^a = e^b \wedge \Gamma^a{}_b \quad . \tag{2}$$

where $\Gamma^a{}_b$ are the connections 1-forms.

With the direction e^3 oriented along the quadruple DP vector, the Einstein equations $(G_{\mu\nu} = \lambda g_{\mu\nu})$ for the N-type fields can be written in terms of forms as

$$\begin{aligned} d\Gamma_{42}+\Gamma_{42}\wedge(\Gamma_{12}+\Gamma_{34}) &= \frac{\lambda}{3}\, e^3\wedge e^1 \\ d(\Gamma_{12}+\Gamma_{34})+2\Gamma_{42}\wedge\Gamma_{31} &= -\frac{\lambda}{3}\,(e^1\wedge e^2+e^3\wedge e^4) \\ d\Gamma_{31}+(\Gamma_{12}+\Gamma_{34})\wedge\Gamma_{31} &= \frac{\lambda}{3}\, e^4\wedge e^2 + \frac{1}{2}\, C^{(1)} e^3\wedge e^1 \end{aligned} \tag{3}$$

where $C^{(1)}$ is the only non vanishing component of the conformal curvature.

Exploiting the gauges of freedom which maintain invariant the direction e^3, we succeeded to integrate the equations (2) and (3) determining an "optimal" system of coordinates $\{\xi,\bar{\xi},r,t\}$.

In the chart $\{\xi,\bar{\xi},r,t\}$, the class of all non-twisting N-waves is given by

$$\begin{aligned} e^1 &= r d\xi+(\psi_{\bar{\xi}} - rf)dt, \quad e^2 = \overline{(e^1)} \\ e^3 &= \psi dt \\ e^4 &= dr+\left[-\psi_{\xi\bar{\xi}} + \frac{r}{2}\,(f_{\xi}+ \bar{f}_{\bar{\xi}}) + \frac{\lambda}{6}\, r^2\psi\right] dt \end{aligned} \tag{4}$$

where $f = f(\xi,t)$ is an arbitrary complex function depending on ξ and t only.

For future purpose we present also the tangent tetrad associated to (4):

$$r\partial_1 = \partial_\xi \qquad r\partial_2 = \partial_{\bar\xi} \qquad \partial_4 = \partial_r$$

$$\psi\partial_3 = \partial_t - \frac{1}{r}\left[(\psi_{\bar\xi} - rf)\partial_\xi + (\psi_\xi - r\bar f)\partial_{\bar\xi}\right]$$
$$- \left[-\psi_{\xi\bar\xi} + \frac{r}{2}(f_\xi + \bar f_{\bar\xi}) + \frac{\lambda}{6}\psi r^2\right]\partial_r \tag{5}$$

Using appropriately the coordinate freedom, without loose of generality, one can bring the real function ψ to the form

$$\psi = 1 + \varepsilon\xi\bar\xi \tag{6}$$

where the discrete parameter ε takes the values 1, 0, -1.

Optionally, for $\varepsilon = -1$, we represent ψ as

$$\psi = \frac{1}{\sqrt{a}}\,\mathrm{sh}\,x, \quad x := \sqrt{a}\,(\xi + \bar\xi) \quad a = \mathrm{const} \tag{7}$$

(here a is assumed to be positive; for a negative $\psi \to \frac{1}{\sqrt{a}}\,\mathrm{sen}\,x$).

The connections occur to be

$$\Gamma_{42} = -\frac{1}{r}e^1 + \frac{1}{r\psi}\psi_{\bar\xi}e^3$$

$$\Gamma_{12}+\Gamma_{34} = \left[\frac{-\lambda}{3}r + \psi^{-1}f_\xi\right]e^3$$

$$\Gamma_{31} = \frac{1}{r\psi}\psi_{\xi\xi}e^1 + \frac{\lambda}{6}re^2 + \frac{1}{r\psi}\left[-\psi_{\xi\xi\bar\xi} + \frac{r}{2}f_{\xi\xi} + \frac{\lambda}{6}r^2\psi_\xi\right]e^3 \tag{8}$$

The conformal curvatures, corresponding to the ψ's given by (6) and (7), are equal to

$$C^{(1)} = -\frac{f_{\xi\xi\xi}}{r(1+\varepsilon\xi\bar\xi)} \neq 0 \tag{9}$$

and

$$C^{(1)} = -\frac{1}{r}\frac{\sqrt{a}}{\mathrm{sh}x}(\partial_\xi\partial_\xi - 4a)f_\xi \neq 0 \tag{10}$$

respectively.

Note that the tetrad, formulae (4), is linear in the structural function $f(\xi,t)$, note also that the variable t defines the set of characteristic of the metric, t = const.

In the sequel we shall denote this class of solutions by $NT(\lambda,z,\varepsilon)$ where λ stands for cosmological constant, z represents complex expansion $-\Gamma_{421}$ and the parameter ε takes the values 1,0,-1 depending upon whether the "source" lines of the gravitational waves are respectively time-like, null or space-like. The complex expansion

z in the studied problem is real and equal to $1/r$, therefore it represents the divergence of the congruence e^3.

All vacuum non-twisting solutions of the type N, denoted as $NT(0,z,\varepsilon)$, are obtained from the $NT(\lambda,z,\varepsilon)$ by simply equating λ to zero in the expressions (4-5) and (8).

Applying the same methods we have obtained the most general N-solution with cosmological constant but with vanishing complex expansion. This solution, which we call $K(\lambda)$-wave, is the generalization of the Kundt solution /1/. The $K(\lambda)$-wave can be given in terms of the tangent tetrad, $\partial_a = e^{\mu}_{a}\partial_{\mu}$, as follows:

$$\partial_1 = \text{ch } x\ \partial_\xi \qquad \partial_2 = \text{ch } x\ \partial_{\bar\xi} \qquad \partial_4 = \text{ch } x\ \partial_r$$

$$\frac{1}{\sqrt{\mu}}\text{ th } x\ \partial_3 = \partial_t + r\text{ ch } x\ (\partial_\xi + \partial_{\bar\xi}) -$$

$$\text{ch } x\ \{ (\partial_\xi + \partial_\xi - 2\sqrt{\mu}\text{th } x)(f+\bar f) - \sqrt{\mu}r^2\text{ th } x\}\partial_r \tag{11}$$

where $\mu := \lambda/6$, $x := \sqrt{\mu}\ (\xi + \bar\xi)$,

and characterized by the conformal curvature

$$C^{(1)} = -2\sqrt{\lambda/6}\ \frac{\text{ch}^3 x}{\text{sh } x}\ (\partial_\xi\partial_\xi - \frac{2}{3}\lambda) f_\xi \tag{12}$$

(here λ is assumed to be positive, for $\lambda<0$, the hyperbolic functions must be replaced by the trigonometric ones).

From these expressions, by letting λ go to zero and at the same time changing $f_\xi \to f$, we readily obtain the Kundt-waves, (K)

$$\partial_1 = \partial_\xi,\ \partial_2 = \partial_{\bar\xi},\ (\xi+\bar\xi)\partial_3 = \partial_t + r(\partial_\xi+\partial_{\bar\xi}) - (f+\bar f)\partial_r\ ,$$

$$\partial_4 = \partial_r$$

$$C^{(1)} = -2\ \frac{f_{\xi\xi}}{\xi+\bar\xi} \tag{13}$$

Executing the coordinate transformation

$$\xi = \xi' + \frac{1}{2}\epsilon^{-1},\ \bar\xi = \bar\xi' + \frac{1}{2}\epsilon^{-1},\ r = r' \quad t = \epsilon t'$$

and replacing the structural function f by $\epsilon^{-1} f'(\xi', t)$ in the last formulas we let ϵ go to zero obtaining the well known Robinson solution (R) /2/, see also /3/

$$\partial_1 = \partial_\xi \qquad \partial_2 = \partial_{\bar\xi} \qquad \partial_3 = \partial_t - (f+\bar f)\partial_r \qquad \partial_4 = \partial_r$$

$$C^{(1)} = -2f_{\xi\xi} \tag{14}$$

In order to show that the $K(\lambda)$-waves are limiting contractions of the $NT(\lambda,z,\varepsilon)$

solutions, we start from (5) with ψ taken in the form (6). Executing the σ and γ gauges

$$\sigma = \ln \mathrm{ch}\, x \ , \qquad \gamma = -\nu \sqrt{a}\ \mathrm{sh}\, x\ \mathrm{ch}\, x\ r$$

where ν is an arbitrary constant (dropping primes) we arrive at

$$r\partial_1 = \partial_\xi \qquad r\partial_2 = \partial_{\bar\xi} \qquad \partial_4 = \frac{1}{\mathrm{ch}\, x}\,\partial_r$$

$$\frac{1}{\sqrt{a}}\,\mathrm{th}\, x\,\partial_3 = \partial_t - \frac{1}{r}\,(\mathrm{ch}\, x - rf - \nu\,\mathrm{sh}^2 x\, r)\,\partial_\xi - \frac{1}{r}\,(\mathrm{ch}\, x - r\bar f - \nu\,\mathrm{sh}^2 x\, r)\,\partial_{\bar\xi} - \Big[-\sqrt{a}\,\mathrm{sh}\, x + \frac{r}{2}\,(f_\xi + \bar f_{\bar\xi}) + \frac{\lambda}{6}\,\frac{r^2}{\sqrt{a}}\,\mathrm{sh}\, x \Big]\,\partial_r$$

$$C^{(1)} = -\frac{\mathrm{ch}^2 x}{r}\,\frac{\sqrt{a}}{\mathrm{sh}\, x}\,(\partial_\xi\partial_\xi - 4a)\, f_\xi$$

$$x := \sqrt{a}\,(\xi+\bar\xi) \tag{15}$$

We change now the coordinates and other quantities which appear above according to

$$t = \epsilon^{-1}\, t', \quad \xi = \epsilon\xi', \quad \bar\xi = \epsilon\bar\xi', \quad r = \frac{1}{\mathrm{ch}\, x}\,\Big(\frac{r'}{\mathrm{ch}\, x} + \epsilon^{-1}\Big)$$
$$\nu = \epsilon, \quad a = \frac{\lambda}{6}\,\epsilon^{-2}, \quad f = \epsilon(2\epsilon^2 f' + 1) \tag{16}$$

where the function $f'(\xi', t')$ is assumed to be independent of the parameter ϵ. Making ϵ tend to zero we obtain precisely the $K(\lambda)$-solutions given by (11-12).

Let us now establish all the limiting contractions of the $NT(0,z,\varepsilon)$ solutions. The Kundt metrics happens to be the contraction of the $NT(0, z,-1)$-waves. Setting $\varepsilon=-1$ and $\lambda = 0$ in (5) and transforming the coordinates and the structural function according to

$$t = \epsilon t', \quad \xi = \epsilon\xi' - 1, \quad \bar\xi = \epsilon\bar\xi' - 1, \quad r = r' + \epsilon^{-1}$$
$$f = 2\epsilon^3 \int^{\xi'} f'(\xi', t')\, d\xi' - \epsilon^2\xi' + \epsilon \tag{17}$$

by letting ϵ go to zero, we arrive at a slightly modified version of the Kundt-waves:

$$\partial_1 = \partial_\xi, \quad \partial_2 = \partial_{\bar\xi}, \quad \partial_4 = \partial_r$$
$$(\xi+\bar\xi)\,\partial_3 = \partial_t + r(\partial_\xi + \partial_{\bar\xi}) + (r - f - \bar f)\,\partial_r$$
$$C^{(1)} = -2\,\frac{f_{\xi\xi}}{\xi+\bar\xi} \tag{18}$$

(To recover the expressions (13) one must accomplish in (18) a σ-gauge, $\sigma = t$, together with

$$t' = e^t, \ r' = e^{-t} r, \ \xi' = \xi, \ \bar\xi' = \bar\xi, \ f' = e^{-t} f) \ .$$

Finally, to establish that $NT(0,z,\varepsilon) \to R$ we start from (5) with $\lambda = 0$. Performing there the coordinate transformation

$$t = t', \quad \xi = \epsilon\xi', \quad \bar{\xi} = \epsilon\bar{\xi}', \quad r = r' + \epsilon^{-1} \tag{19}$$

accompanied by

$$f = \epsilon^2 \left(\varepsilon\xi' + 2\int^{\xi'} f'(\xi', t')d\xi'\right) \tag{20}$$

and taking ϵ to zero we arrive at (14) i.e. to the Robinson solution.

Conclusion

We would like to emphasize that all non-twisting N-type solutions satisfy the following scheme of contractions

$$\begin{array}{ccccc}
NT(\lambda,z,1) & \to & NT(0,z,1) & & \\
 & & & \searrow & \\
NT(\lambda,z,0) & \to & NT(0,z,0) & \to & R \\
 & & & \nearrow & \\
NT(\lambda,z,-1) & \to & NT(0,z,-1) & & \uparrow \\
\searrow & & & \searrow & \\
K(\lambda) & & \to & & K
\end{array}$$

therefore, if one understands a given set of solutions as containing all its possible contractions, then the $NT(\lambda,z,\varepsilon)$ completely exhausts all N-type solutions with e^3 being surface orthogonal.

* On leave of absence from University of Warsaw, Warsaw, Poland.

References

/1/ W. Kundt, Z. Physik 163, 77 (1961).

/2/ I. Robinson, Lecture at King's College, London (1956). Report to the Rayaumont Conference (1959).

/3/ H.W. Brinkmann, Proc. Nat. Acad. Sci. Wash, 9, 1(1923). J. Helý, C.R. Acad. Sci. Paris 249, 1867 (1959). I. Robinson and A. Trautman, Phys. Rev. Lett. 4, 431 (1960).

CAN QUANTUM EFFECTS PREVENT SPACETIME COLLAPSE?

MARK J. GOTAY

Department of Mathematics
University of Calgary
Calgary, Alberta, Canada

JAMES A. ISENBERG

Department of Mathematics
University of California
Berkeley, California, USA

Introduction

One of the outstanding problems in general relativistic cosmology is whether or not quantum effects can modify the character of -- or perhaps prevent entirely -- the classical final singularity. Although this question was extensively studied during the late 1960's and early 1970's [1], no consensus has emerged. However, much of the work done to date tends to support Misner's 1969 assertion [2] that "quantum effects do not change the nature of the singularity." More recently, Wheeler [3] has proposed a "rule of unanimity" which, if valid, implies that quantum theory can provide no escape from gravitationally induced spacetime collapse.

There are three main obstacles to settling this issue: (*i*) The absence of a complete consistent quantum treatment of the gravitational field and its interactions. (*ii*) Ambiguities inherent in the canonical quantization procedure. (*iii*) The lack of precise general criteria for determining whether or not the quantized system in fact collapses.

These problems, although formidable, can to some extent be circumvented. In particular, it is possible to include the quantum effects of both gravity and matter in the analysis -- while keeping the latter mathematically tractable -- by "freezing out" all but a finite number of degrees of freedom; one is then left with a typical minisuperspace problem [2,4]. Furthermore, many of the difficulties associated with the application of canonical quantization to Hamiltonian cosmology [1,4-8] can be overcome by using instead the geometric quantization procedure of Kostant & Souriau [9]. Indeed, geometric quantization -- essentially a rigorous global generalization of canonical quantization -- has proven to be an effective computational tool in quantum cosmology [5-7]. Finally, in simple cases the last problem (cf. [1,6,8]) can be avoided by directly calculating the asymptotic temporal behavior of certain relevant matrix elements [6,7].

The simplest dynamically nontrivial homogeneous cosmologies are Robertson-Walker universes containing a Klein-Gordon scalar field ("$RW\phi$" models). Recently [6,7] we have geometrically quantized one of the classically collapsing $RW\phi$ universes and have rigorously shown that the quantized model collapses as well. However, there are several important issues that remain to be elucidated, such as the effects of different choices of time and polarization upon the quantum dynamics of these models. To this end, we geometrically quantize here an $RW\phi$ model in a different choice of time and with a topologically different type of polarization than have been used previously. The quantum dynamics so obtained is essentially equivalent to that found in [6,7], as well as that resulting from canonical quantization [8].

There are two primary inferences to be drawn from this work. First, at least in some (highly symmetric) cosmological models, quantum effects do *not* prevent spacetime collapse. Second, whereas different choices of time and polarization may lead to *quantitative* changes in the quantum dynamics of these simple models, such differing choices do not seem to significantly affect the quantized models' *qualitative* behavior.

Classical $RW\phi$ Models [1,6,8]

The homogeneous and isotropic $RW\phi$ universes are described by the metric

$$ds^2 = -N^2(t)dt \otimes dt + R^2(t)\, g_{ij} dx^i \otimes dx^j \;,$$

where $N(t)$ is the "lapse," $R(t)$ is the "radius" and g is the standard metric on S^3. The minisuperphase space for these collapsing models is a 4-manifold with global coordinates R, π_R, ϕ, π_ϕ satisfying $R > 0$ and $\pi_\phi \neq 0$ (the values $R = 0$ and $\pi_\phi = 0$ corresponding to singular states). The Hamiltonian is

$$-NK := -N\left(\frac{1}{24R}\pi_R{}^2 - \frac{1}{2R^3}\pi_\phi{}^2 + 6R\right) \;,$$

and is constrained to vanish: $K = 0$. For convenience, we have chosen the scalar field $\phi(t)$ to be massless.

The vanishing Hamiltonian indicates that the system is in parametrized form and therefore admits a reduction via "choice of time" [1,6,8]. Here, we choose the "extrinsic" time $t = \pi_R$, since (i) π_R-time smoothly covers the entire classical evolution of the model [$\pi_R \in (-\infty, 0)$ is the expansion phase, while $\pi_R \in (0,\infty)$ is the contraction phase], and (ii) with extrinsic time, we may directly quantize the radius R and monitor the asymptotic temporal behavior of its expectation value as a test for collapse.

The unconstrained phase space resulting from this reduction is the disjoint union $\mathbb{R}^2_+ \oplus \mathbb{R}^2_-$ with the standard symplectic structure $d\pi_\phi \wedge d\phi$, where $\mathbb{R}^2_\pm := \{(\phi, \pi_\phi) \mid \pi_\phi \gtrless 0\}$. The choice $t = \pi_R$ has the attractive simplifying feature that the reduced Hamiltonian π_t is just the radius $R(t)$:

$$(1) \qquad R(t) = \alpha\left(\sqrt{\beta^2 t^4 + \gamma\pi_\phi{}^2} - \beta t^2\right)^{\frac{1}{2}} \;,$$

where $\alpha = (24)^{-\frac{1}{2}}$, $\beta = 1/12$ and $\gamma = 48$. Since $\pi_\phi \neq 0$, $R(t)$ is positive definite.

Geometric Quantization of the Extrinsic-Time Model [5,6,9]

The simplest choice of polarization F is the "horizontal" one spanned by the vector field $\partial/\partial\phi$. Since $\mathbb{R}^2_+ \oplus \mathbb{R}^2_-$ is topologically trivial, the various geometric quantization structures (the prequantization line bundle L, the metaplectic frame bundle, and the bundle $\sqrt{\wedge^1 F}$ of half-forms relative to F) are unique and trivial.

Let λ denote a trivializing section of L which is normalized to unity, and ν the appropriate half-form. The quantum Hilbert space H_F defined by the polarization F is the completion of the space of smooth, compactly supported (modulo F) sections of $L \otimes \sqrt{\wedge^1 F}$ of the form

$$\text{(2)} \qquad \psi = \begin{cases} f_+(\pi_\phi) exp[i\,\pi_\phi \phi]\ \lambda \otimes \nu\ , & \text{on } \mathbb{R}^2_+ \\ f_-(\pi_\phi) exp[i\,\pi_\phi \phi]\ \lambda \otimes \nu\ , & \text{on } \mathbb{R}^2_- \end{cases}$$

(where $f_\pm$ are arbitrary, and $\hbar \equiv 1$) with respect to the inner product

$$\text{(3)} \qquad \langle\psi'|\psi\rangle_F = \int_{-\infty}^{0} f'_-(\pi_\phi)\overline{f_-(\pi_\phi)}d\pi_\phi + \int_0^{\infty} f'_+(\pi_\phi)\overline{f_+(\pi_\phi)}d\pi_\phi \quad .$$

Setting $f_* := \chi_- f_- + \chi_+ f_+$, where $\chi_\pm$ are the characteristic functions of $\mathbb{R}^2_\pm$, we can rewrite (2) and (3) as

$$\text{(4)} \qquad \psi = f_*(\pi_\phi) exp[i\,\pi_\phi \phi]\ \lambda \otimes \nu$$

and

$$\text{(5)} \qquad \langle\psi'|\psi\rangle_F = \int_{-\infty}^{+\infty} f'_*(\pi_\phi)\overline{f_*(\pi_\phi)}d\pi_\phi$$

respectively. The association $f_*(\pi_\phi)exp[i\,\pi_\phi\phi]\ \lambda \otimes \nu \to f_*(\pi_\phi)$ defines a unitary isomorphism of H_F with $L^2(\mathbb{R})$.

Quantum Dynamics

Since the polarization F diagonalizes $R(t)$, the Hamiltonian/radius operator $\mathcal{R}(t)$ acts on H_F by multiplication:

$$\text{(6)} \qquad \mathcal{R}(t)[\psi] = R(t)\psi \quad .$$

This positive self-adjoint operator commutes for different times, so that we can solve the Schrödinger equation by expanding in an evolving complete set of states $\{\psi_E\}$ which are simultaneous eigenfunctions of $\mathcal{R}(t)$ at all times. Thus, if

$$\text{(7)} \qquad \mathcal{R}(t_0)[\psi_E] = E\psi_E$$

at some reference time t_0, then there will exist numbers $E(t)$ [with $E(t_0) = E$] such that

$$\text{(8)} \qquad \mathcal{R}(t)[\psi_E] = E(t)\psi_E \quad .$$

Consequently (cf. [8]), the states ψ_E evolve according to

$$\text{(9)} \qquad \psi_E(t) = exp\left[-i\int_{t_0}^{t} E(s)ds\right]\psi_E(t_0) \quad .$$

From (6) and (4) we find that (7) has the distributional solutions [10]

$$\psi_E \sim exp[i\,\pi_\phi \phi]\delta\big(R(t_0) - E\big)\ \lambda \otimes \nu \quad ,$$

from which it follows that $E > 0$. Employing (1) and manipulating the δ-function in the above expression, we obtain

$$\text{(10)} \qquad \psi_E = \left(A_+ e^{i\eta\phi}\delta(\pi_\phi - \eta) + A_- e^{-i\eta\phi}\delta(\pi_\phi + \eta)\right)\ \lambda \otimes \nu \quad ,$$

where

$$\text{(11)} \qquad \eta = \gamma^{-\frac{1}{2}}\left((\alpha^{-2}E^2 + \beta t_0{}^2)^2 - \beta^2 t_0{}^4\right)^{\frac{1}{2}}$$

and $A_\pm$ are normalization constants.

Substituting (10) into (8) and making use of (1) and (11), we calculate

$$\text{(12)} \qquad E(t) = \alpha\left(\sqrt{(\alpha^{-2}E^2 + \beta t_0{}^2)^2 - \beta^2(t_0{}^4 - t^4)} - \beta t^2\right)^{\frac{1}{2}} \quad .$$

Since $E > 0$, this implies that $\mathcal{R}(t)$ has a purely continuous spectrum of $(0,\infty)$ for all t. Finally, substituting (12) into (9), we obtain the evolution (cf. [8])

$$\psi_E(t) = \left[exp\left\{-i\left(\tfrac{1}{2}[tE(t) - t_0E] + \tfrac{1}{4}E(t_0{}^2 + 144E^2)^{\frac{1}{2}} \times \right.\right.\right.$$

(13)
$$\left.\left.\left.\times \ln\left\{\frac{[tE(t) - E(t_0{}^2 + 144E^2)^{\frac{1}{2}}][t_0 + (t_0{}^2 + 144E^2)^{\frac{1}{2}}]}{[tE(t) + E(t_0{}^2 + 144E^2)^{\frac{1}{2}}][t_0 - (t_0{}^2 + 144E^2)^{\frac{1}{2}}]}\right\}\right)\right\}\right]\psi_E(t_0)$$

Gravitational Collapse

We claim that the quantized extrinsic-time $RW\phi$ model, like its classical counterpart, collapses to a singularity in the sense that

(14) $$\lim_{t\to\infty} \langle\psi(t)|\mathcal{R}(t)|\psi(t)\rangle_F = 0$$

for any evolving state $\psi(t)$ in the domain of $\mathcal{R}(t)$ [11].

Since $\mathcal{R}(t)$ commutes for different times we have, upon expanding $\psi(t)$ in terms of the eigenstates $\psi_E(t)$ and applying (13),

$$\langle\psi(t)|\mathcal{R}(t)|\psi(t)\rangle_F = \langle\psi(t_0)|\mathcal{R}(t)|\psi(t_0)\rangle_F$$

for some initial time t_0. From (6), (1) and (5), then,

(15) $$\langle\psi(t_0)|\mathcal{R}(t)|\psi(t_0)\rangle_F = \alpha\int_{-\infty}^{\infty} f_*(\pi_\phi)\overline{f_*(\pi_\phi)}\left(\sqrt{\beta^2t^4 + \gamma\pi_\phi{}^2} - \beta t^2\right)^{\frac{1}{2}}d\pi_\phi$$

for $\psi(t_0)$ of the general form (4). Now, consider the one-parameter family of functions $\{I_t(\pi_\phi)\}$, where I_t is the integrand of (15) at time t. The family $\{I_t(\pi_\phi)\}$ converges pointwise to zero, and $I_t(\pi_\phi) < I_s(\pi_\phi)$ for each π_ϕ if $t > s$. Since in addition $I_t \in L^1(\mathbb{R})$ for each t, the dominated convergence theorem implies that

$$\lim_{t\to\infty}\int_{-\infty}^{\infty} I_t(\pi_\phi)d\pi_\phi = \int_{-\infty}^{\infty}[\lim_{t\to\infty} I_t(\pi_\phi)]d\pi_\phi$$

vanishes. Thus (14) follows.

Since the Hamiltonian $\mathcal{R}(t)$ is self-adjoint, the model *must* evolve to the $t \to \infty$ limit. Consequently, (14) implies that *all physically well-defined states of the extrinsic-time RWϕ model collapse* [12]. Note furthermore that, for large t, the dominant terms in (15) tend to zero as t^{-1}. Since classically $R(t) \to 0$ also as t^{-1}, it follows that the quantum collapse rate matches the classical collapse rate.

Comparison with Canonical Quantization [9]

Blyth & Isham [8], using canonical techniques, have also quantized this extrinsic-time $RW\phi$ model. Since, roughly speaking, canonical quantization is just geometric quantization in the vertical polarization, we may use the Blattner-Kostant-Sternberg ("BKS") transform [9] to compare our respective results.

The "vertical" polarization V is spanned by the vector field $\partial/\partial\pi_\phi$; wave functions relative to V have the general form

(16) $$\sigma = g_*(\phi)\ \lambda \otimes \mu \quad ,$$

where μ is the appropriate half-form. Since $\left(\mathbb{R}^2_+ \oplus \mathbb{R}^2_-\right)/V \approx \mathbb{R} \oplus \mathbb{R}$, the Hilbert space H_V defined by the vertical polarization is unitarily isomorphic to the direct sum $L^2(\mathbb{R}) \oplus L^2(\mathbb{R})$ with the inner product

$$(17) \qquad \langle\sigma'|\sigma\rangle_V = \int_{-\infty}^{\infty} g_+'(\phi)\overline{g_+(\phi)}d\phi + \int_{-\infty}^{\infty} g_-'(\phi)\overline{g_-(\phi)}d\phi \quad .$$

The BKS kernel $K\colon H_V \times H_F \to \mathbb{C}$ is given by

$$(18) \qquad K(\sigma,\psi) = (2\pi)^{-\frac{1}{2}}e^{+i\pi/4}\iint_{-\infty}^{+\infty} g_*(\phi)\overline{f_*(\pi_\phi)}e^{-i\pi_\phi\phi}d\pi_\phi d\phi$$

for $\psi \in H_F$ and $\sigma \in H_V$ of the form (4) and (16) respectively. The kernel K defines a linear operator $U\colon H_F \to H_V$ via $K(\sigma,\psi) = \langle\sigma|U\psi\rangle_V$. Comparing this with (18) and (17), we compute

$$(19) \qquad U\psi = \left\{(2\pi)^{-\frac{1}{2}}e^{-\frac{i\pi}{4}}\sum_{\pm}\left[\chi_\pm\int_{-\infty}^{\infty}\chi_\pm f_\pm(\pi_\phi)e^{i\pi_\phi\phi}d\pi_\phi\right]\right\}\lambda\otimes\mu \quad .$$

Thus, U is "essentially" the Fourier transform.

Applying U to the energy eigendistributions (10), (19) gives

$$(20) \qquad U\psi_E = (2\pi)^{-\frac{1}{2}}e^{-i\pi/4}(\chi_+A_+e^{i\eta\phi} + \chi_-A_-e^{-i\eta\phi})\ \lambda\otimes\mu \quad .$$

This is just Blyth & Isham's result [8], with one major difference [13]: B&I find that the spectrum of the Hamiltonian is not $(0,\infty)$ for all times, but rather $(-\infty,0) \cup (0,\infty)$. Within the canonical framework, the presence of these (unphysical) negative eigenvalues is due to B&I's use of a Klein-Gordon type equation to generate the quantum evolution rather than a true Schrödinger equation. From the standpoint of geometric quantization, on the other hand, this anomaly apparently can be traced to the incompleteness [14] of the vertical polarization [15] -- indeed, such spurious negative eigenvalues do not appear when one quantizes using a complete polarization (such as the horizontal polarization employed in this paper).

Although the BKS transform $U\colon H_F \to H_V$ is an isometry, i.e., $\langle U\psi'|U\psi\rangle_V = \langle\psi'|\psi\rangle_F$, a straightforward calculation shows that U is not invertible. Thus, the BKS transform does *not* unitarily intertwine the quantizations in the horizontal and vertical polarizations. Of course, this is not really surprising since the two polarizations are of topologically different types.

Discussion

In [6,7] we have geometrically quantized a classically collapsing $RW\phi$ model using the "matter-time" $t = \phi$ and a "radial" polarization. The results of that quantization compare favorably with our findings here as well as with Blyth & Isham's work. Indeed, in the quantizations of the positive curvature $RW\phi$ universes (with massless scalar field) studied so far, one finds not only that the quantized models unquestionably collapse, but also that the quantum collapse rate is always exactly the same as the classical collapse rate [16].

Changes in the choices of time and polarization may, as illustrated in the last section, result in modifications to the finer details of the quantum dynamics of these models (e.g., alterations in the spectra of observables). However, such variations do not seem to influence the system's overall characteristics. Hence, it may be that these diverse quantizations will yield at least *qualitatively* equivalent -- if not strictly equivalent -- results. In any case, our experience with the *RWφ* models supports the contention that the final singularity cannot be avoided -- quantum mechanically or otherwise -- and hence [2] must "be treated as an essential element of cosmological theory."

The *RWφ* models considered here furnish a handy laboratory for studying quantum gravity, and the Kostant/Souriau theory allows one to quantize these models in a rather exact fashion. Even so, it is clear that much remains to be done before any agreement can be reached regarding the issue of quantum collapse. One should therefore, as MacCallum [1] points out, properly regard the present work (and, indeed, quantum cosmology in general) as "suggesting possibilities rather than providing definite answers."

Acknowledgments

We thank J. Śniatycki for many enlightening conversations. This work was supported in part by the National Science and Engineering Research Council of Canada.

References & Notes

1. See the review article by M. MacCallum in *Quantum Gravity*, C. Isham, R. Penrose & D. Sciama (eds.), (Clarendon Press, Oxford, 1975).
2. C.W. Misner, Phys. Rev. 186, 1319, 1328 (1969).
3. J.A. Wheeler, Gen. Rel. Grav., 8, 713 (1977).
4. C.W. Misner, in *Magic Without Magic*, J. Klauder (ed.), (Freeman, San Francisco, 1972).
5. M.J. Gotay & J.A. Isenberg, in Lecture Notes in Physics #94, 293 (Springer, Berlin, 1979).
6. M.J. Gotay & J.A. Isenberg, *Geometric Quantization and Gravitational Collapse*, in press (Phys. Rev. D21, 1980).
7. J.A. Isenberg & M.J. Gotay, *Quantum Cosmology and Geometric Quantization*, in press (Gen. Rel. Grav., 1980).
8. W.F. Blyth & C.J. Isham, Phys. Rev. D11, 768 (1975).
9. J. Śniatycki, *Geometric Quantization and Quantum Mechanics*, Applied Mathematics Series 30, (Springer, Berlin, 1980).
10. Note that we have imposed no boundary conditions.
11. Since the domain of the operator $\mathcal{R}(t)$ is time-independent, the expectation value (15) and the limit (14) are both well-defined.
12. Technically, we must temper our enthusiasm somewhat here, for the following reason. Since $\mathcal{R}(t)$ is an unbounded operator (with domain $\mathcal{D}$, say -- cf. [11]), we have really only shown that those states $\psi(t) \in \mathcal{D}$ undergo quantum collapse. However, as $\mathcal{D}$ is dense in $L^2(\mathbb{R})$, this is probably "good enough". It is not clear what can be said regarding those states $\psi(t) \notin \mathcal{D}$, since such states have no well-defined "radius".
13. A minor discrepancy concerns the presence of the characteristic functions in (20). They appear here because the classical phase space is disconnected, but not in B&I's treatment where the phase space is assumed connected.
14. A polarization is *complete* if the canonical vectorfields spanning it are complete (cf. [9]).
15. J. Sniatycki & B. Kostant (private communications).
16. This remark can be shown to hold true also for the canonically quantized models of Blyth & Isham.

COMPLEX PLANE REPRESENTATION OF THE GEROCH GROUP AND A PROOF OF A GEROCH CONJECTURE*

I. Hauser
Illinois Institute of Technology
Chicago, Illinois 60616

1. Introduction

One of the striking ideas in general relativity over the past decade has been the conjecture[1] that the stationary axisymmetric vacuum spacetimes (or, at the very least, the asymptotically flat ones) can be generated from Minkowski space by applying the Geroch[1] group K of transformations. The work of Kinnersley and Chitre[2] and of Hoenselaers, Kinnersley, and Xanthopoulos[3] had made this conjecture of Geroch highly plausible, and Xanthopoulos[4] has come very close to a proof of the conjecture for the asymptotically flat case.

F. Ernst and the author[5] have recently proven the following version of the Geroch conjecture.

> Let V_0^ω denote the set of all stationary axisymmetric vacuum spacetimes whose manifolds each includes at least one point of the axis at which one of the two Killing vectors characterizing the spacetime is timelike (while, of course, the exterior product of the Killing vectors vanishes at the point). There exists a subgroup $\mathcal{K}$ of K such that any given member of V_0^ω can be generated from Minkowski space by at least one member of $\mathcal{K}$; conversely, any given member of $\mathcal{K}$ maps Minkowski space into a member of V_0^ω.

V_0^ω includes all asymptotically flat stationary axisymmetric vacuum spacetimes, but it excludes, for example, all Weyl static vacuums for which the Weyl harmonic function has a logarithmic singularity at all points of the axis. Our principal objective is to prove the above statement after sketching (without proof) some new developments concerning that complex representation of K which was introduced by Ernst and Hauser[5-7] and which is used in the proof. We start with the basic concepts of the $\mathcal{E}$-potential of Ernst,[8] the H-potential of K-C (Kinnersley and Chitre),[2] and the group K of Geroch.[1]

2. $\mathcal{E}$-Potentials and H-Potentials

We shall be discussing a class V^ω of solutions of the Einstein vacuum field equations. This class includes those gravitational fields $g_{\alpha\beta}$ $(\alpha,\beta = 1,2,3,4)$ whose sources are bounded axisymmetric masses which are spinning at a constant rate. We may define V^ω as that class of vacuum spacetimes for which there exist coordinates $x^1 = z$, $x^2 = \rho$, x^3, x^4 such that the line element has the form

$$g_{\alpha\beta}\delta x^\alpha \delta x^\beta = g_{11}(\delta z^2 + \delta\rho^2) + g_{ab}\delta x^a \delta x^b,$$

$$g_{11} > 0, \ (a = 3,4), \ (b = 3,4), \ \rho = [(g_{34})^2 - g_{33}g_{44}]^{\frac{1}{2}},$$

and g_{11} and g_{ab} depend at most on z and ρ. We often loosely use the term "stationary axisymmetric vacuum spacetime", abbreviated SAV, for any member of V^ω, but it should be kept in mind that not all members of V^ω fit the connotations of that term.

It is a remarkable fact which was first discovered by F. Ernst[8] that the metric components g_{11} and g_{ab} can all be simply derived from a single complex scalar field $\mathcal{E}$, the $\mathcal{E}$-potential as we call it, which is a solution of the field equation

$$\nabla^2\mathcal{E} = f^{-1}(\nabla\mathcal{E})^2, \qquad f := \mathrm{Re}\,\mathcal{E}. \tag{1}$$

In the above equation, $\mathcal{E}$ is to be formally treated as an axisymmetric field in Euclidean 3-space E^3, where z, ρ, ϕ are cylindrical coordinates, and ∇ and ∇^2 are the gradient and Laplacian operators for E^3. Let us trace the origins of $\mathcal{E}$ by considering that 2-form W_b which is an eigenvector of the duality operator belonging to the eigenvalue i and which is related to the exterior derivative of $g_{ba}dx^a$ by

$$d(g_{ba}dx^a) = -\tfrac{1}{2}(W_b + W_b^*), \quad W_b^* := \text{c.c. of } W_b.$$

The condition that the Ricci tensor components R_{ab} vanish turns out to be equivalent to the statement that the components $W_{\alpha\beta b}$ of W_b satisfy the condition that $dx^\alpha W_{\alpha ab}$ is a closed 1-form,[7] and the definition of W_b implies that the real part of this 1-form equals $-dg_{ab}$. So, there exist scalar fields H_{ab} such that

$$dH_{ab} = dx^\alpha W_{\alpha ab}, \quad \mathrm{Re}H_{ab} = -g_{ab}.$$

The condition that W_b is a duality eigenvector is then equivalent to the pair of equations[7]

$$2r_{\pm}\frac{\partial H}{\partial r_{\pm}} = (H + H^\dagger)\Omega\frac{\partial H}{\partial r_{\pm}}, \tag{2}$$

where $$r_{\pm} := z \pm i\rho, \qquad \Omega := \begin{pmatrix} 0 & i \\ -i & 0 \end{pmatrix}. \tag{3}$$

H, or the H-potential as we call it, is the 2×2 matrix whose elements are H_{ab}, and $H^\dagger$ is the h.c. of H.

Now, $\mathcal{E}$ is defined by

$$\mathcal{E} := H_{44}, \tag{4}$$

whereupon the integrability condition for Eq. (2) implies that $\mathcal{E}$ is a solution of Eq. (1).

Conversely, as Ernst[8] discovered, if $\mathcal{E}$ is any non-trivial ($\mathcal{E} \neq$ imaginary constant) axisymmetric solution of Eq. (1) over some region (connected open set) in E^3, then there exists a unique member of V^{ω} whose $\mathcal{E}$-potential is $\mathcal{E}$, and the other H-potential components and the metric $g_{\alpha\beta}$ are easily computed[8,5] from $\mathcal{E}$. The $\mathcal{E}$-potential thus completely characterizes any given SAV in V^{ω}. Therefore, the study of V^{ω} is reduced to the study of all solutions of Eq. (1).

3. The Geroch Group K

In the early history of the subject, it was found that certain transformations, e.g.,

$$\mathcal{E} \rightarrow \mathcal{E}' = \mathcal{E}^{-1} \quad \text{and} \quad \mathcal{E} \rightarrow \mathcal{E}' = \mathcal{E} + i\beta \quad (\beta = \text{real const.}), \tag{5}$$

yield new solutions of Eq. (1). It was R. Geroch[1] who first grasped the full implications of these facts. We shall now attempt to explain his ideas in terms which are somewhat different than those originally used[1] by him.

In effect, Geroch noted that the group $SL(2,R)_1$ of rational linear transformations ($\alpha, \beta, \gamma, \delta$ = real constants)

$$\mathcal{E} \rightarrow \mathcal{E}' = \frac{\alpha\mathcal{E} + i\beta}{-i\gamma\mathcal{E} + \delta}, \quad \alpha\delta - \beta\gamma = 1,$$

which are generated by repeated applications of the transformations in Eq. (5) is a representation of $SU(1,1) = SL(2,R)$. Some of these transformations, e.g. those due to Ehlers,[9] yield new SAVS, but the resulting metrics are (with trivial exceptions) not asymptotically flat if the original one is asymptotically flat, and the result has only one additional essential parameter. (The other two parameters arise from gauge transformations.)

However, Geroch[1] noted that there is also the group $SL(2,R)_2$ of transformations of the ignorable coordinates (S is a 2×2 constant unimodular matrix)

$$dx \rightarrow dx' = dx\, S^{-1}, \quad dx := (dx^3\ dx^4),$$

which induces the transformation

$$H \rightarrow H' = SHS^T$$

of the H-potential. The group K may now be formally defined as the infinite direct product

$$K = SL(2,R)_1 \times SL(2,R)_2 \times SL(2,R)_1 \times \quad \ldots \quad .$$

Technically, in terms of concepts which were later introduced by K-C,[2] the operand of each element of K is not the metric, is not $\mathcal{E}$, and is not H; it is an infinite family of 2×2 matrix potentials which are all defined (up to a gauge transformation) in terms of H and which include H as a member. Geroch found that K is infinite dimensional, and it is this fact which led him to introduce the conjecture mentioned in Sec. 1.

The ideas of Geroch did not bear fruit until the tremendous breakthrough[2] of K-C in which they constructed a viable representation of the Lie algebra of K and introduced techniques which were actually used to transform known members of V^{ω} into <u>interesting</u> hitherto unknown members of V^{ω}.

4. Complex Plane Representation of K

On the basis of the work of K-C, Ernst and Hauser[6-7] introduced a complex plane representation of the K-C transformations which seems to have some merits and which we shall now describe. In this formulation, there is a representation K_L of K which is associated with each smooth positively oriented contour L about the origin of the complex plane C such that L is symmetric with respect to the real axis. K_L is the set of all holomorphic 2×2 matrix functions u each of which is defined over a region (connected open set) of C which covers L such that[7]

$$\begin{aligned} &u(t)^{\dagger} \Omega \, u(t) = \Omega, \qquad \det u(t) = 1, \\ &u(t)^{\dagger} := \text{h.c. of } u(t^*), \end{aligned} \tag{6}$$

and such that

$$\begin{pmatrix} t & 0 \\ 0 & 1 \end{pmatrix} u(t) \begin{pmatrix} t^{-1} & 0 \\ 0 & 1 \end{pmatrix} \tag{7}$$

is holomorphic at $t = \infty$.

The domain transformed by use of K_L is a set $\mathfrak{F}$ of 2×2 matrix functions $F(z,\rho,t)$ of z, ρ, and the complex variable t. The members of $\mathfrak{F}$ will be called F-<u>potentials</u>. They were originally introduced by K-C[2] as generating functions for part of their hierarchy of potentials. In terms of our conventions,[7] they are defined (for given H) as any solution of the equation

$$dF = \Gamma \Omega F, \qquad \Gamma := -\frac{1}{2}\left(\frac{dr_+}{r_+ - \tau} \frac{\partial H}{\partial r_+} + \frac{dr_-}{r_- - \tau} \frac{\partial H}{\partial r_-} \right) , \tag{8}$$

$$\tau := (2t)^{-1}, \quad r_{\pm} := z \pm i\rho,$$

which is holomorphic (for given z,ρ) in a neighborhood of $t = 0$ and which satisfies[7] (we suppress z,ρ in $F(t)$)

$$F(0) = \Omega, \quad \dot{F}(0) = H, \quad \dot{F}(t) := \partial F(t)/\partial t,$$

$$\det F(t) = -[(1-2tz)^2 + (2t\rho)^2]^{-1/2}, \tag{9}$$

$$F(t)^{\dagger}[\Omega - t\Omega(H + H^{\dagger})\Omega]\, F(t) = \Omega .$$

Equations (9) are consistent with Eq. (8), and the complete integrability condition for Eq. (2) is equivalent to the complete integrability condition for Eq. (8).[7]

5. SAVS in V_0^{ω}

Our actual fruitful results concern a subgroup $\mathcal{K}_L$ of K_L and a subset V_0^{ω} of V^{ω}.

$\mathcal{K}_L$ is the set of all members of K_L for which the expression (7) is holomorphic in the region L_- which is outside L and includes $t = \infty$. (L_+ denotes the region inside L, and $C = L + L_+ + L_-$.)

V_0^{ω} was defined in Sec. 1. Equivalently, V_0^{ω} is the set of all SAVS in V^{ω} for which $\mathcal{E}$ expressed as a function of (z,ρ) has a domain which is a region in the space (with the relative topology induced by R^2)

$$R^2_+ := \{(z,\rho) : -\infty < z < \infty, \ \rho \geq 0\}$$

such that $\mathcal{E}$ is C^3 and such that the domain of $\mathcal{E}$ covers at least one point $(z_0,0)$ of the axis at which $f := \mathrm{Re}\,\mathcal{E} > 0$. From now on, we shall find it convenient to restrict $\mathcal{E}$ to a domain U such that $f > 0$ at all points of U. The theory of elliptic equations such as Eq. (1) then enables us to prove[5] that $\mathcal{E}$ is holomorphic at all points of U (including those on the axis), by which we mean that $\mathcal{E}$ has a holomorphic extension to a region in $C \times C$.

Also, for V_0^{ω}, we have proven the following theorem concerning F-potentials.[5]

Consider any given member of V_0^{ω} with a given $\mathcal{E}$ whose domain U covers a point $(z_0,0)$ of the axis, and choose the arbitrary constant in z so that $z_0 = 0$. Let (z_1,ρ_1) be any given point in U. Then there exists at least one solution F of Eqs. (8) and (9) with a domain Δ such that the following statements (a) to (e) hold:

(a) Δ is a simply connected region in $U \times C$, and $[\underset{\sim}{x} := (z,\rho)]$

$$U_\Delta := \{\underset{\sim}{x} \in U : \exists t \in C, (\underset{\sim}{x},t) \in \Delta\}$$

is a simply connected region in R^2_+ such that U_Δ covers (0,0) and (z_1,ρ_1).

(b) For each $\underset{\sim}{x} \in U_\Delta$,

$$\Delta_{\underset{\sim}{x}} := \{t \in C : (\underset{\sim}{x},t) \in \Delta\}$$

equals C minus the points $(r_\pm := z \pm i\rho)$

$$t(\underset{\sim}{x}) := (2r_-)^{-1}, \quad t(\underset{\sim}{x})^* := (2r_+)^{-1},$$

and minus the union $K_{\underset{\sim}{x}}$ of two simple smooth line segments which join $t(\underset{\sim}{x})$ and $t(\underset{\sim}{x})^*$ to a common point on the negative real axis and which (exclusive of their end points) lie in the upper and lower half-planes, respectively. When $\rho = 0$ and $z \leq 0$, $K_{\underset{\sim}{x}}$ degenerates to the single point $(2z)^{-1}$. Also, $K_{\underset{\sim}{x}}$ is symmetric with respect to the real axis and varies smoothly with $\underset{\sim}{x}$.

(c) F is holomorphic at all points of Δ, i.e. F has a holomorphic extension to a region in $C \times C \times C$. For fixed $\underset{\sim}{x}$ such that $\rho > 0$, $t(\underset{\sim}{x})$ and $t(\underset{\sim}{x})^*$ are branch points of $F(\underset{\sim}{x},t)$ (as a function of t) with indices -1/2.

(d) For any contour L such as previously defined, there is an open interval I_L of the axis such that $(0,0) \in I_L \subset U_\Delta$, and for all $t \in L + L_+$ and $(z,0) \in I_L$:

$$F(z,0,t) = \begin{pmatrix} 0 & i \\ -i(1-2tz)^{-1} & t\mathcal{E}(z,0)(1-2tz)^{-1} \end{pmatrix} .$$

(e) For given $\mathcal{E}$ with domain U, given choice of the axis point $(z_0,0)$ in U, and given Δ, the F which satisfies (a) to (d) above is unique.

We shall let $\mathcal{D}_0$ denote the set of all F-potentials of members of V_0^ω for which the conditions (a) to (d) of the above theorem hold regardless of the choices of $\mathcal{E}$, z_0, and Δ.

6. Complex Plane Representation of K-C Transformations

Consider any given u in K_L and any given F-potential $F^{(0)}$ for any given SAV in V^ω. Ernst and Hauser introduced a function $F(\underset{\sim}{x},t)$ which was defined by the homogeneous Hilbert-Riemann problem[7]

$$F(\underset{\sim}{x},t)u(t)\,F^{(0)}(\underset{\sim}{x},t) = X_-(\underset{\sim}{x},t), \quad \underset{\sim}{x} \in \Delta_L, \tag{10}$$

where Δ_L is the set of all $\underset{\sim}{x}$ for which $F^{(0)}(\underset{\sim}{x},t)$ is holomorphic in $L + L_+$ and for which (F,X_-) exists subject to the conditions that (for fixed $\underset{\sim}{x}$) $F(\underset{\sim}{x},t)$ is holomorphic in $L + L_+$, and $X_-(\underset{\sim}{x},t)$ is holomorphic in $L + L_-$ including at $t = \infty$, and $F(\underset{\sim}{x},0) = \Omega$. A recent formulation[5] which is equivalent to the above HHP (homogeneous Hilbert problem) is the Fredholm equation $(s,t \in L)$

$$F(t) - \frac{1}{2\pi i}\int_L ds\, F(s)K(s,t) = F^{(0)}(t), \tag{11}$$

$$K(s,t) := \frac{t}{s(s-t)}\left[F^{(0)}(s)F^{(0)}(t)^{-1} - u(s)F^{(0)}(s)F^{(0)}(t)^{-1}u(t)^{-1}\right].$$

With the aid of the above Eqs. (10) and (11), we have been able to prove[5] the following theorems for any given $u \in \mathcal{K}_L$ and $F^{(0)} \in \mathcal{G}_0$.

(1) A unique solution (F,X_-) of Eq. (10) exists such that the domain of F is

$$\Delta = \left\{(\underset{\sim}{x},t) : \underset{\sim}{x} \in \Delta_L,\ t \in (L + L_+) + (L_- \cap \Delta_{\underset{\sim}{x}}^{(0)}\right\},$$

where

$$\Delta_{\underset{\sim}{x}}^{(0)} := \left\{t \in C : (\underset{\sim}{x},t) \in \Delta^{(0)}\right\}, \quad \Delta^{(0)} := \mathrm{dom}\, F^{(0)}.$$

Δ is an open subset of $R_+^2 \times C$, and Δ_L is an open subset of R_+^2 such that $(0,0) \in \Delta_L$. Moreover, F is holomorphic at all points of Δ and is the F-potential of a member of V_0^{ω}; F is an extension of at least one member of $\mathcal{G}_0$.

(2) The elements $u^a_b(t)$ (row 2 and column b) of $u(t)$ are related to the axis values $\mathcal{E}^{(0)}(z,0)$ and $\mathcal{E}(z,0)$ of the initial and final $\mathcal{E}$-potentials in a neighborhood of $z = 0$ by

$$\begin{aligned}&\mathcal{E}^{(0)}(z,0)\,\mathcal{E}(z,0)\,k\,u^4_3(k) + k^{-1}\,u^3_4(k)\\ &+ i\,\mathcal{E}(z,0)\,u^4_4(k) - i\,\mathcal{E}^{(0)}(z,0)\,u^3_3(k) = 0,\end{aligned} \tag{12}$$

where

$$k := (2z)^{-1}.$$

7. Application to Geroch Conjecture

We now consider any given SAV in V_0^{ω}, say $S_g = S_{given}$, with an $\mathcal{E}$-potential $\mathcal{E}_g$. We use Eq. (12) to define a $u(t)$ by first letting $\mathcal{E}^{(0)} = 1$, which is the $\mathcal{E}$-potential for Minkowski space, and then replacing $\mathcal{E}$ by $\mathcal{E}_g$ in Eq. (12). This supplies u(k) for real k in a neighborhood of $k = \infty$. We then replace k by t and conduct a maximal holomorphic continuation of u(t).

We next substitute this u(t) and the $F^{(0)}(t)$ for Minkowski space into the HHP of Eq. (10). The solution supplies us, according to Theorem (1) in Sec. 6, with an F-potential F for a member S of V_0^{ω} such that $F \in \mathfrak{F}_0$ after a suitable restriction. From Eq. (12) and from the way u(t) was chosen, the $\mathcal{E}$-potential $\mathcal{E}$ for S is related to the $\mathcal{E}$-potential $\mathcal{E}_g$ for S_g by

$$\mathcal{E}(z,0) = \mathcal{E}_g(z,0) \tag{13}$$

in at least one open interval containing z = 0. However, from Eq. (1) and the analyticity of solutions of Eq. (1), we then obtain[5] $\mathcal{E}(z,\rho) = \mathcal{E}_g(z,\rho)$ over some neighborhood of (0,0) in R_+^2; hence $S = S_g$, <u>and the Geroch conjecture is proven</u>.

References

* Work supported in part by National Science Foundation Grant PHY79-08627.

1 R. Geroch, J. Math. Phys. <u>12</u>, 918 (1971); <u>13</u>, 394 (1972).

2 W. Kinnersley and D. M. Chitre, J. Math. Phys. <u>18</u>, 1538 (1977); <u>19</u>, 1926 (1978); <u>19</u>, 2037 (1978).

3 C. Hoenselaers, W. Kinnersley, and B. Xanthopoulos, J. Math. Phys. <u>20</u>, 2530 (1979).

4 B. Xanthopoulos, preprint.

5 I. Hauser and F. J. Ernst, submitted for publication.

6 I. Hauser and F. J. Ernst, Phys. Rev. D<u>20</u>, 362 (1979).

7 I. Hauser and F. J. Ernst, J. Math. Phys. <u>21</u>, 1126 (1980).

8 F. J. Ernst, Phys. Rev. <u>167</u>, 1175 (1968); J. Math. Phys. <u>15</u>, 1409 (1974).

9 J. Ehlers, <u>Les Théories Relativistes de la Gravitation</u> (CNRS, Paris, 1959).

THE GEROCH GROUP AND SOLITON SOLUTIONS OF THE STATIONARY AXIALLY SYMMETRIC EINSTEIN'S EQUATIONS

William M. Kinnersley

Physics Department
Montana State University
Bozeman, Montana 59717 USA

1. Introduction

In contrast with the gauge theories discussed elsewhere in these proceedings, General Relativity is an old theory. One might suppose that in three-quarters of a century, most of the important features of a theory would be thoroughly understood. On the contrary, Einstein's theory is still greeting us with surprises. Part of the difficulty has been a lack of realistic exact solutions. However we are glad to report that that situation has seen a drastic change in the last few years. The solution-generating methods we will discuss now enable us to write down large numbers of new solutions, all obeying the physical boundary condition of asymptotic flatness. Recall that all of the interesting results of Black Hole physics have emerged from the study of only one exact solution, the Kerr metric.

2. The Geroch Group In A Nutshell

We concentrate on the gravitational fields which are stationary and axially symmetric. We use coordinates t, ϕ, ρ, z which would reduce to the usual cylindrical coordinates if the space were flat. $\rho = 0$ is the symmetry axis. The metric we begin with is the one introduced by Lewis,

$$ds^2 = f(dt - \omega\, d\phi)^2 - f^{-1}[e^{2\gamma}(d\rho^2 + dz^2) + \rho^2 d\phi^2] \tag{2.1}$$

where f, ω, γ are functions of ρ and z.

The vacuum Einstein equations fall naturally into two parts. Some of the equations determine γ in terms of f and ω (and so we ignore γ completely in the subsequent discussion). The others are quite simple:

$$\nabla\cdot(\rho^{-2}f^2\nabla\omega) = 0 \tag{2.2}$$

$$\nabla\cdot(f^{-1}\nabla f + \rho^2 f^2\omega\nabla\omega) = 0 \tag{2.3}$$

There is an SL(2,R) invariance group G which arises from linear transformations of the two ignorable coordinates t,ϕ. For example, one of the infinitesimal transformations of G is

$$\begin{aligned} t &\to t \\ \phi &\to \phi + \varepsilon t \\ f &\to f + \varepsilon(2f\omega) + O(\varepsilon^2) \\ \omega &\to \omega - \varepsilon(\omega^2 + \rho^2 f^{-2}) + O(\varepsilon^2) \end{aligned} \tag{2.4}$$

The Lewis parameterization f,ω is not well adapted to G. The symmetry of Eqs. (2.2), (2.3) is "hidden", and the action on f,ω is nonlinear.

Equations (2.2), (2.3) are both divergence equations, and this is the essential ingredient that leads to the Geroch group. Just as $\nabla\cdot\underset{\sim}{B} = 0$ is solved identically by $\underset{\sim}{B} = \nabla \times \underset{\sim}{A}$, Eq. (2.2) may be solved by introduction of a "twist potential" ψ. Eq. (2.2) implies

$$\rho^{-2} f^2 \nabla\omega = \rho^{-1}\tilde{\nabla}\psi \tag{2.5}$$

where $\tilde{\nabla}$ is a convenient substitute for the curl operator,

$$\tilde{\nabla} = \underset{\sim}{e}_\phi \times \nabla \tag{2.6}$$

The twist potential ψ is not a completely arbitrary function, because Eq. (2.5) must be integrable for ω. Eq. (2.5) can be rewritten as

$$\rho^{-1}\tilde{\nabla}\omega = -f^2\nabla\psi \tag{2.7}$$

which shows that

$$\nabla\cdot(f^2\nabla\psi) = 0 \tag{2.8}$$

Eq. (2.8) involves only f and ψ. Likewise ω may be eliminated from Eq. (2.3), yielding

$$\nabla\cdot(f^{-1}\nabla f + f^{-2}\psi\nabla\psi) = 0 \tag{2.9}$$

Eqs. (2.8), (2.9) are a pair of equations for f,ψ alone. They may be solved directly, completely replacing Eqs. (2.2), (2.3), and ω treated as an auxiliary function which is defined by Eq. (2.7).

Furthermore, Eqs. (2.8), (2.9) also have a hidden SL(2,R) symmetry group H, called the NUT group. In particular, one of the infinitesimal transformations of H is

$$f \to f + \varepsilon(2f\psi)$$
$$\psi \to \psi + \varepsilon(\psi^2 - f^2) \qquad (2.10)$$

or in terms of the complex combination $\mathcal{E} = f + i\psi$, (the "Ernst potential")

$$\mathcal{E} \to \mathcal{E} - i\varepsilon\mathcal{E}^2$$

Unlike G, which is merely a coordinate gauge group, the group H produces a physical change on the solution, leading to a physically different spacetime. However the new solutions produced in this way by the action of H are not asymptotically flat.

The full Geroch group arises when we try to complete the picture, and ask what effect the action of G has on ψ. Since Eqs. (2.2), (2.3) transform into one another under G, clearly we must treat them on the same footing, and solve Eq. (2.3) by introducing a new potential ψ_2:

$$f^{-1}\nabla f + \rho^{-2}f^2\omega\nabla\omega = \rho^{-1}\tilde{\nabla}\psi_2 \qquad (2.11)$$

Then we can write the action of G as

$$\psi \to \psi + \varepsilon\,\psi_2 \qquad (2.12)$$

But now what is the action of H on ψ_2?

$$\psi_2 \to \psi_2 + \varepsilon(\ ?\ ) \qquad (2.13)$$

Writing this step requires the introduction of yet another new potential. The process will continue in this way indefinitely, and we are led to an infinite hierarchy of potentials. The Geroch group K acts transitively on the entire hierarchy. K consists of G and H, plus all possible commutators and multiple commutators among them.

We now terminate this preliminary discussion and turn to a properly systematic approach.

3. Covariant Reformulation

We introduce a metric which is better adapted to the symmetry,

$$ds^2 = f_{AB}dx^A dx^B - e^{2\Gamma}\delta_{MN}dx^M dx^N \qquad A,B = 1,2 \quad M,N = 3,4 \qquad (3.1)$$

where f_{AB} and Γ are functions of x^3, x^4 alone. Indexed quantities such as f_{AB} are tensors under G. We raise and lower indices using the symplectic metric

$$V^A = \varepsilon^{AX} V_X$$

and make the following definitions:

$$\mathrm{Det}(f_{AB}) = -\rho^2$$

$$\nabla = (\partial_3, \partial_4) \qquad (3.2)$$

$$\tilde{\nabla} = (\partial_4, -\partial_3)$$

We treat the more general case in which both gravity and electromagnetism are present, since it is easy to do so, and the results obtained fit in quite naturally. We describe the electromagnetic field by means of a complex vector potential

$$\Phi_A = A_A + i\, B_A \qquad (3.3)$$

where A_A is the usual vector potential and B_A is its magnetic analog. One does not usually use both, since A_A alone is sufficient. However this redundant description enables us to write Maxwell's equations in a simpler form. Maxwell's equations

$$F^{AM}{}_{;M} = 0 \qquad (3.4)$$

would be second order when written out in terms of A_A alone. By introducing B_A as well, we double the number of variables and reduce Eq. (3.4) to a set of coupled first order differential equations, which are

$$\nabla \Phi_A = -i\rho^{-1} f_A{}^X \tilde{\nabla} \Phi_X \qquad (3.5)$$

The gravitational field is similarly described by means of a complex <u>tensor</u> potential

$$H_{AB} = (f_{AB} - \bar{\Phi}_A \Phi_B + \varepsilon_{AB} K) + i\, \Omega_{AB} \qquad (3.6)$$

where

$$\nabla K = \bar{\Phi}_X \nabla \Phi^X \qquad (3.7)$$

$$\nabla \Omega_{AB} = -\rho^{-1} f_A{}^X (\tilde{\nabla} f_{XB} - 2A_B \tilde{\nabla} A_X - 2B_B \tilde{\nabla} B_X) \qquad (3.8)$$

(Both Eqs. (3.7), (3.8) are integrable by virtue of Maxwell's equations.) H_{AB} is a tensor generalization of the Ernst potential, and in fact $H_{11} = \mathcal{E}$.

The field equations are

$$R_{AB} = -8\pi\, T_{AB} \tag{3.9}$$

where T_{AB} is the electromagnetic stress-energy tensor. When written out, Eqs. (3.9) are usually extremely complicated. However, in the present case they become remarkably simple when expressed in terms of H_{AB}:

$$\nabla H_{AB} = -i\rho^{-1} f_A{}^X \tilde{\nabla} H_{XB} \tag{3.10}$$

Note that Eqs. (3.5) and (3.10) have exactly the same form, except for the presence of an extra index. (Since we are dealing with the full nonlinear theory, this symmetry between electromagnetism and gravity goes deeper than the usual platitude about spin 1 and spin 2.)

4. The Infinite Hierarchies

Suppose we have an infinite sequence of fields $\Phi_A^{(n)}$, $H_{AB}^{(n)}$, all obeying the same field equations:

$$\begin{aligned} \nabla\Phi_A^{(n)} &= -i\rho^{-1} f_A{}^X \tilde{\nabla}\Phi_X^{(n)} \\ \nabla H_{AB}^{(n)} &= -i\rho^{-1} f_A{}^X \tilde{\nabla} H_{XB}^{(n)} \end{aligned} \qquad n = 1,2,\ldots \tag{4.1}$$

From the fields we define four families of potentials:

$$\begin{aligned} \nabla K^{(m,n)} &= \bar{\Phi}_X^{(m)}\ \nabla\Phi^{X(n)} \\ \nabla L_B^{(m,n)} &= \bar{\Phi}_X^{(m)}\ \nabla H^{X(n)}{}_B \\ \nabla M_A^{(m,n)} &= \bar{H}_{XA}^{(m)}\ \nabla\Phi^{X(n)} \\ \nabla N_{AB}^{(m,n)} &= \bar{H}_{XA}^{(m)}\ \nabla H^{X(n)}{}_B \end{aligned} \qquad m,n = 1,2,\ldots \tag{4.2}$$

Conversely, from the potentials we construct solutions of the field equations:

$$\begin{aligned} \Phi_A^{(n+1)} &= i\,[M_A^{(1,n)} + 2\Phi_A K^{(1,n)} + H_{AX}\Phi^{X(n)}] \\ H_{AB}^{(n+1)} &= i\,[N_{AB}^{(1,n)} + 2\Phi_A L_B^{(1,n)} + H_{AX} H^{X(n)}{}_B] \end{aligned} \tag{4.3}$$

The fields and potentials are thus defined recursively, starting with

$$\begin{aligned} \Phi_A^{(1)} &= \Phi_A \\ H_{AB}^{(1)} &= H_{AB} \end{aligned} \tag{4.4}$$

In fact by means of an artifice it is convenient to combine fields and potentials together into an "extended hierarchy". Define

$$H_{AB}^{(0)} = i\varepsilon_{AB} \tag{4.5}$$

Then from Eq. (4.2), we may consistently define

$$N_{AB}^{(0,n)} = -iH_{AB}^{(n)}$$
$$M_{A}^{(0,n)} = -i\Phi_{A}^{(n)} \tag{4.6}$$

The fields thus appear as an extra row added on to the hierarchy of potentials. (In fact they do transform this way, as part of the same representation of K.)

5. Generating Functions

Given any stationary axially symmetric spacetime which is a solution of the Einstein-Maxwell equations, there is an infinite set of potential functions which must be explicitly calculated before the Geroch group can be applied. For such practical purposes it is more convenient to work with the generating functions

$$F_{AB}(t) = \sum_{n=0}^{\infty} t^n H_{AB}^{(n)}$$

$$G_{AB}(s,t) = \sum_{m,n=0}^{\infty} s^m t^n N_{AB}^{(m,n)} \tag{5.1}$$

$$G_{AB}(0,t) = -iF_{AB}(t)$$

As an example we give the results of the calculation for flat space.

$$F_{AB}(t) = \begin{pmatrix} \frac{t}{S} & \frac{i}{S} \\ \frac{-i(1 - 2tz + S)}{S} & \frac{1 - 2tz - S}{2tS} \end{pmatrix} \tag{5.2}$$

$$G_{11}(s,t) = \frac{-it}{2S(t)}\left(1 + \frac{s + t - 4stz}{sS(t) + tS(s)}\right) \tag{5.3}$$

where

$$1 - 2tz - 2it\rho = S(t)e^{i\varphi(t)} \tag{5.4}$$

($S(t)$, $\varphi(t)$ are geometrical factors which arise repeatedly in applications. They are essentially the polar coordinates based on the point $z = \frac{1}{2t}$, $\rho = 0$.) The flat-space potentials may of course be obtained by expanding the above generating functions in powers of s and t.

6. The Geroch Group

We are now ready for a complete description of the Geroch group K. It has the following properties:

1) K preserves the Maxwell and Einstein-Maxwell equations.
2) K maps old (stationary and axially symmetric) solutions into new ones.
3) K acts on the extended hierarchy of potentials.
4) K contains an infinite number of continuous parameters:
   a) $\gamma_{AB}^{(k)}$, $k = 0,\pm 1,\pm 2,\ldots$ real and symmetric. All the γ's preserve vacuum. $\gamma_{12}^{(k)}$ also preserve static fields, whereas $\gamma_{11}^{(k)}$, $\gamma_{22}^{(k)}$ produce rotation.
   b) $c_A^{(k)}$, $k = 0,\pm 1,\pm 2,\ldots$ complex. The $c_A^{(k)}$'s generate electromagnetic fields out of vacuum.
   c) $\sigma^{(k)}$, $k = 0,\pm 1,\pm 2,\ldots$ real. The $\sigma^{(k)}$'s are generalizations of the electromagnetic "duality" rotation which turns E into B. They vanish identically for vacuum.

Next we write out the action of each generator on the potentials.

$$\begin{aligned}
\gamma_{AB}^{(k)}:\quad & K^{(m,n)} \to K^{(m,n)} + \gamma^{XY}\Big(\sum_s L_X^{(m,s)} M_Y^{(k-s,n)}\Big) \\
& L_B^{(m,n)} \to L_B^{(m,n)} - \gamma^X{}_B L_X^{(m,n+k)} + \gamma^{XY}\Big(\sum_s L_X^{(m,s)} N_{YB}^{(k-s,n)}\Big) \\
& M_A^{(m,n)} \to M_A^{(m,n)} - \gamma^X{}_A M_X^{(m+k,n)} + \gamma^{XY}\Big(\sum_s N_{AX}^{(m,s)} M_Y^{(k-s,n)}\Big) \\
& N_{AB}^{(m,n)} \to N_{AB}^{(m,n)} - \gamma^X{}_A N_{XB}^{(m+k,n)} - \gamma^X{}_B N_{AX}^{(m,n+k)} + \gamma^{XY}\Big(\sum_s N_{AX}^{(m,s)} N_{YB}^{(k-s,n)}\Big)
\end{aligned} \tag{6.1}$$

$$\begin{aligned}
c_A^{(k)}:\quad K^{(m,n)} \to K^{(m,n)} & + \bar{c}^X\Big[M_X^{(m+k-1,n)} + 2i\Big(\sum_s K^{(m,s)} M_X^{(k-s,n)}\Big)\Big] \\
& + c^X\Big[L_X^{(m,n+k-1)} - 2i\Big(\sum_s L_X^{(m,s)} K^{(k-s,n)}\Big)\Big]
\end{aligned}$$

$$
\begin{aligned}
L_B^{(m,n)} &\to L_B^{(m,n)} + \bar{c}^X[N_{XB}^{(m+k-1,n)} + 2i(\sum_s K^{(m,s)}N_{XB}^{(k-s,n)})] \\
&\quad - 2i\bar{c}_B K^{(m,n+k)} + c^X[-2i(\sum_s L_X^{(m,s)}L_B^{(k-s,n)})] \\
M_A^{(m,n)} &\to M_A^{(m,n)} + \bar{c}^X[2i(\sum_s M_A^{(m,s)}M_X^{(k-s,n)})] + 2i\ c_A K^{(m+k,n)} \\
&\quad + c^X[N_{AX}^{(m,n+k-1)} - 2i(\sum_s N_{AX}^{(m,s)}K^{(k-s,n)})] \\
N_{AB}^{(m,n)} &\to N_{AB}^{(m,n)} + \bar{c}^X[2i(\sum_s M_A^{(m,s)}N_{XB}^{(k-s,n)})] - 2i\bar{c}_B M_A^{(m,n+k)} \\
&\quad + c^X[-2i(\sum_s N_{AX}^{(m,s)}L_B^{(k-s,n)})] + 2ic_A L_B^{(m+k,n)}
\end{aligned}
\tag{6.2}
$$

$$
\begin{aligned}
\sigma^{(k)}:\quad K^{(m,n)} &\to K^{(m,n)} + i\sigma K^{(m+k,n)} - i\sigma K^{(m,n+k)} - 2\sigma(\sum_s K^{(m,s)}K^{(k-s+1,n)}) \\
L_B^{(m,n)} &\to L_B^{(m,n)} + i\sigma L_B^{(m+k,n)} - 2\sigma(\sum_s K^{(m,s)}L_B^{(k-s+1,n)}) \\
M_A^{(m,n)} &\to M_A^{(m,n)} - i\sigma M_A^{(m,n+k)} - 2\sigma(\sum_s M_A^{(m,s)}K^{(k-s+1,n)}) \\
N_{AB}^{(m,n)} &\to N_{AB}^{(m,n)} - 2\sigma(\sum_s M_A^{(m,s)}L_B^{(k-s+1,n)})
\end{aligned}
\tag{6.3}
$$

In each case, these are the <u>infinitesimal</u> actions only, and terms of $\mathcal{O}(\gamma^2)$, etc. have been omitted.

Note that the structure throughout is the same. There are linear terms where the indices m, n have been raised by k, and there are quadratic terms involving a finite sum. Since the action of the generators is nonlinear, we have a <u>nonlinear realization of an infinite parameter Lie Algebra</u>.

Exponentiation of the generators must be done to produce a finite transformation of the group. Successive terms in the power series in γ will contain cubic, quartic, etc. products of the potentials. For a great many transformations of K this exponentiation has been carried out. However for simplicity we restrict ourselves in the present discussion to the generators alone.

The commutators of K are

$$
\begin{aligned}
[\gamma_{AB}^{(k)}, \gamma_{CD}^{(\ell)}] &= -\varepsilon_{A(C}\gamma_{D)B}^{(k+\ell)} - \varepsilon_{B(C}\gamma_{D)A}^{(k+\ell)} \\
[c_A^{(k)}, \gamma_{BC}^{(\ell)}] &= -\varepsilon_{A(B}c_{C)}^{(k+\ell)} \\
[c_A^{(k)}, \bar{c}_B^{(\ell)}] &= 2i\gamma_{AB}^{(k+\ell-1)} - \varepsilon_{AB}\sigma^{(k+\ell-1)}
\end{aligned}
\tag{6.4}
$$

$$[\sigma^{(k)}, c_A^{(\ell)}] = 3ic_A$$

$$[c_A^{(k)}, c_B^{(\ell)}] = [\sigma^{(k)}, \gamma_{AB}^{(\ell)}] = [\sigma^{(k)}, \sigma^{(\ell)}] = 0$$

The root diagram of K may be drawn in three dimensions as an infinite hexagonal cylinder extending indefinitely to the left and to the right (see Fig. 1).

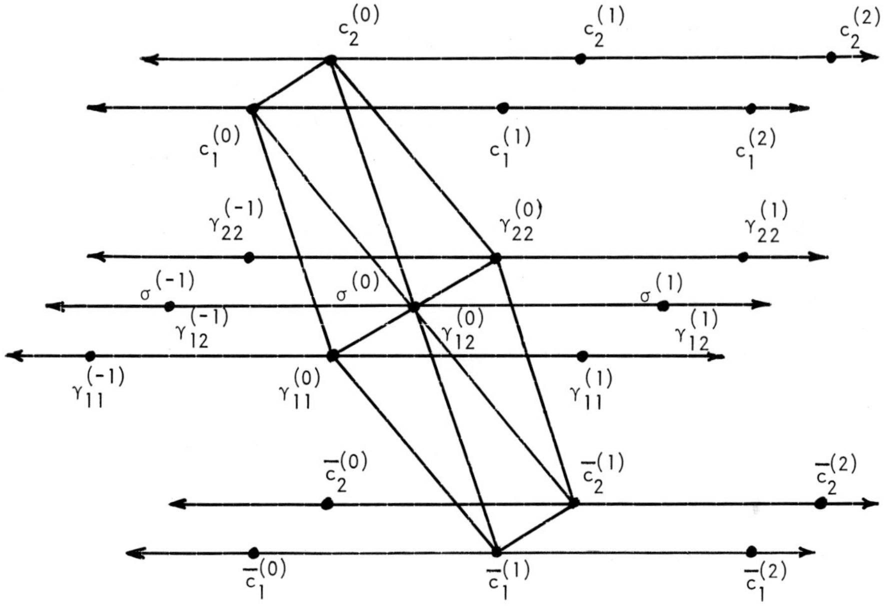

Fig. 1

Various slices through the origin as shown correspond to subsets of eight generators forming closed subgroups isomorphic to SU(2,1).

The subgroups are

$$G^{(m,n)} = \{\gamma_{11}^{(-m)}, \gamma_{12}^{(0)}, \gamma_{22}^{(m)}, c_1^{(n)}, c_2^{(n+m)}, \bar{c}_1^{(1-m-n)}, \bar{c}_2^{(1-n)}, \sigma^{(0)}\} \tag{6.5}$$

Corresponding to each $G^{(m,n)}$ there is a particular finite subset of potentials on which the action of $G^{(m,n)}$ is closed. (Examples were (f, ω) for $G = G^{(0,0)}$ and (f, ψ) for $H = G^{(1,0)}$ discussed in Section 2.)

Furthermore there exists a family of involutive automorphisms which map the subgroups into one another:

$$\begin{aligned} I^{(r,s)}:\ & K \to K \\ & G^{(m,n)} \to G^{(r-m,s-n)} \end{aligned}$$

$$\begin{aligned}
\gamma_{11}^{(k)} &\to i\gamma_{22}^{(k+r)} & c_1^{(k)} &\to \bar{c}_2^{(k+1-s)} \\
\gamma_{12}^{(k)} &\to -\gamma_{12}^{(k)} & c_2^{(k)} &\to i\bar{c}_1^{(k+1-r-s)} \\
\gamma_{22}^{(k)} &\to -i\gamma_{11}^{(k-r)} & \bar{c}_1^{(k)} &\to -ic_2^{(k+s+r-1)} \\
\sigma^{(k)} &\to -\sigma^{(k)} & \bar{c}_2^{(k)} &\to c_1^{(k+s-1)}
\end{aligned} \tag{6.6}$$

One must always bear in mind when applying $I^{(r,s)}$ that they do not respect the reality of either the group parameters, the potentials or the metric itself. This is due to the explicit factors of i contained in them, as well as the dissimilar transformations of $c_A^{(k)}$ and $\bar{c}_A^{(k)}$. In such cases one must usually return to a real metric by finding an analytic continuation, or by some other special device.

The particular automorphism $I^{(1,0)}$ is called the Neugebauer-Kramer map:

$$\begin{aligned}
I:\quad & G \to H \\
& f \to \rho f^{-1} \\
& \omega \to i\psi
\end{aligned} \tag{6.7}$$

The reality is violated, since both ω and ψ are normally expected to be real.

7. What the Generators Really Do

To understand what the generators of K actually do to a solution it is most convenient to study their action on flat space ($H_{11} = 1$). We find for $k \geq 1$,

$$\begin{aligned}
\gamma_{11}^{(k)}:\quad & H_{11} \to 1 + i\gamma(2r)^{k+1}P_{k+1}(\cos\theta) \\
\gamma_{12}^{(k)}:\quad & H_{11} \to 1 - 2\gamma(2r)^{k}P_{k}(\cos\theta) \\
\gamma_{22}^{(k)}:\quad & H_{11} \to 1 - i\gamma(2r)^{k-1}P_{k-1}(\cos\theta)
\end{aligned} \tag{7.1}$$

where r, θ are the usual spherical coordinates.

The physical interpretation is that the infinitesimal transformations create linearized multipole gravitational fields. The real term created by γ_{12} is a mass multipole, and the imaginary one created by γ_{11}, γ_{22} is a multipole of angular momentum. Note that these are all "inner" multipoles, as would be produced by sources at infinity, and the gravitational fields are not asymptotically flat.

Likewise, for infinitesimal $c_A^{(k)}$, the transformations generate infinitesimal multipole electromagnetic fields, both real and imaginary (electric and magnetic).

$$c_1^{(k)}: \quad \Phi_1 \to -\, ic(2r)^k P_k(\cos\theta)$$
$$c_2^{(k)}: \quad \Phi_1 \to c(2r)^{k-1} P_{k-1}(\cos\theta) \tag{7.2}$$

From this point on in the discussion, we will drop consideration of electromagnetism, and concentrate on the vacuum case and the vacuum subgroup of K generated by $\gamma_{AB}^{(k)}$.

From the preceeding discussion we see that $\gamma_{11}^{(k)}$ and $\gamma_{22}^{(k+2)}$ produce identically the same action on flat space. This motivates us to consider in place of $\gamma_{AB}^{(k)}$ another set of generators which are the linear combinations

$$\begin{aligned} \alpha^{(k)} &= \gamma_{11}^{(k)} - \gamma_{22}^{(k+2)} \\ \beta^{(k)} &= \gamma_{11}^{(k)} + \gamma_{22}^{(k+2)} \\ \gamma^{(k)} &= \gamma_{12}^{(k)} \end{aligned} \tag{7.3}$$

In this basis the commutators are

$$\begin{aligned} [\alpha^{(k)}, \beta^{(\ell)}] &= 2\gamma^{(k+\ell)} & \qquad [\alpha^{(k)}, \alpha^{(\ell)}] &= 0 \\ [\beta^{(k)}, \gamma^{(\ell)}] &= 2\beta^{(k+\ell)} & \qquad [\beta^{(k)}, \beta^{(\ell)}] &= 0 \\ [\gamma^{(k)}, \alpha^{(\ell)}] &= -\,2\alpha^{(k+\ell)} & \qquad [\gamma^{(k)}, \gamma^{(\ell)}] &= 0 \end{aligned} \tag{7.4}$$

Thus $\{\alpha^{(k)}\}$, $\{\beta^{(k)}\}$, $\{\gamma^{(k)}\}$ generate three Abelian subgroups of K. The commutation relations among different subgroups resemble those of SO(2,1).

Applied to flat space,

$$\begin{aligned} \alpha^{(k)}&: \quad H_{11} \to 1 + 2i\varepsilon(2r)^{k+1} P_{k+1}(\cos\theta) \\ \beta^{(k)}&: \quad H_{11} \to 1 \\ \gamma^{(k)}&: \quad H_{11} \to 1 - 2\varepsilon(2r)^k P_k(\cos\theta) \end{aligned} \tag{7.5}$$

The subgroups $\{\alpha^{(k)}\}$, $\{\gamma^{(k)}\}$ produce real and imaginary multipoles as before. The $\{\beta^{(k)}\}$ leave flat space invariant. In other words, if we regard each solution of Einstein's equations to be a point in a "superspace" on which K acts, then $\{\beta^{(k)}\}$ is the little group of the flat space solution. Every other solution will have its own little group, obtained from $\{\beta^{(k)}\}$ by conjugation. The action of K on the superspace is thus multiply transitive. From a practical point of view, this means that many different elements of K exist which will generate a given solution, and one of our jobs is to identify the simplest transformation which involves the least calculation.

On solutions other than flat space of course $\{\beta^{(k)}\}$ does produce an action. From the commutation relations we see that $\{\beta^{(k)}\}$ turns real and imaginary multipoles into each other. For example, $\{\beta^{(k)}\}$ may be used to turn the Schwarzschild metric into Kerr.

8. The HKX Transformation

To produce asymptotically flat solutions we need a way of generating outer multipole fields. The simplest approach is to use a linear combination of $\gamma_{22}^{(k)}$. The action of $\gamma_{22}^{(k)}$ was

$$\gamma_{22}^{(k)}: \quad H_{11} \to 1 - i\varepsilon(2r)^{k-1}P_{k-1}(\cos\theta) \tag{8.1}$$

Introducing an arbitrary parameter u and forming the infinite sum,

$$\sum_{k=0}^{\infty} u^k \gamma_{22}^{(k)}: \quad H_{11} \to 1 - i\varepsilon\sum_{k=1}^{\infty} (2ur)^{k-1}P_{k-1}(\cos\theta) \tag{8.2}$$

we recognize the generating function of the P_k. The result is

$$H_{11} \to 1 - \frac{i\varepsilon}{2R} \tag{8.3}$$

where R is the radial distance from a point on the axis, $\rho = 0$, $z = \frac{1}{2u}$. (See Fig. 2)

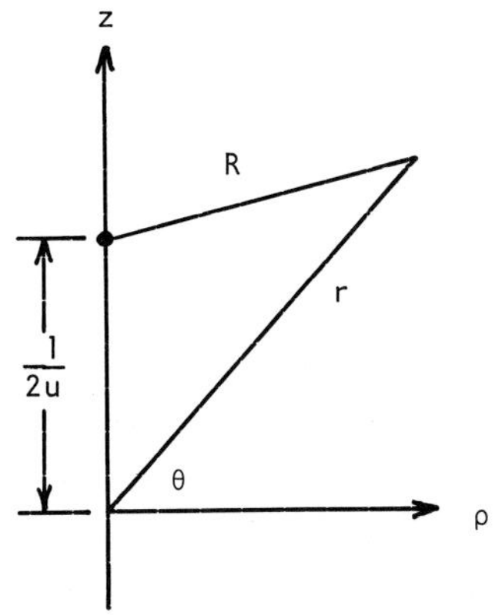

Fig. 2

The infinitesimal HKX transformation creates an angular momentum monopole of strength $\frac{\varepsilon}{4}$, displaced away from the origin a distance $\frac{1}{2u}$.

Infinite sums of generators can sometimes cause convergence difficulties, and indeed the sum in Eq. (8.2) diverges for $r > \frac{1}{2u}$. However Eq. (8.3) is valid everywhere and amounts to an analytic continuation. In this latter form the HKX transformation is legitimate even in the limit $u \to \infty$.

The HKX transformation can be generalized to a set of transformations which generate the entire class of outer multipole fields. The transformations of rank ℓ are defined as

$$\begin{aligned} D_{11}^{(\ell)}(u) &= \sum_{k=0}^{\infty} \binom{k+\ell}{\ell} u^{k+\ell+1} \gamma_{11}^{(k-1)} \\ D_{12}^{(\ell)}(u) &= \sum_{k=0}^{\infty} \binom{k+\ell}{\ell} u^{k+\ell+1} \gamma_{12}^{(k)} \\ D_{22}^{(\ell)}(u) &= \sum_{k=0}^{\infty} \binom{k+\ell}{\ell} u^{k+\ell+1} \gamma_{22}^{(k+1)} \end{aligned} \tag{8.4}$$

Applied to flat space they create mass and angular momentum multipoles located at $z = \frac{1}{2u}$

$$\begin{aligned} D_{11}^{(\ell)}: \quad H_{11} &\to 1 + i\varepsilon\left(\frac{1}{2R}\right)^{\ell+1} P_\ell(\cos\theta) \\ D_{12}^{(\ell)}: \quad H_{11} &\to 1 - 2\varepsilon\left(\frac{1}{2R}\right)^{\ell+1} P_\ell(\cos\theta) \\ D_{22}^{(\ell)}: \quad H_{11} &\to 1 - i\varepsilon\left(\frac{1}{2R}\right)^{\ell+1} P_\ell(\cos\theta) \end{aligned} \tag{8.5}$$

In actual practice the mass multipoles do not need to be generated in this way, since the static (nonrotating) Weyl solutions are all known. We can start with a Weyl solution having the desired mass moments, and add angular momentum multipoles using $D_{22}^{(\ell)}(u)$ alone.

We will briefly consider the exponentiation of the HKX transformation to finite values of the parameter ε. This can be written down very easily using the generating function $G_{11}(s,t)$ and its derivatives. For rank zero the finite transformation is

$$G_{11}(s,t) \to G_{11}(s,t) + \frac{\varepsilon G_{11}(s,u)\, G_{11}(u,t)}{1 - \varepsilon G_{11}(u,u)} \tag{8.6}$$

Applied to flat space the finite HKX transformation creates a solution known as "extreme Kerr-NUT", centered once again at $z = \frac{1}{2u}$.

If HKX is applied repeatedly to a given initial Weyl solution (Fig. 3), the result is a nonlinear superposition of n extreme Kerr-NUT particles with the existing Weyl sources.

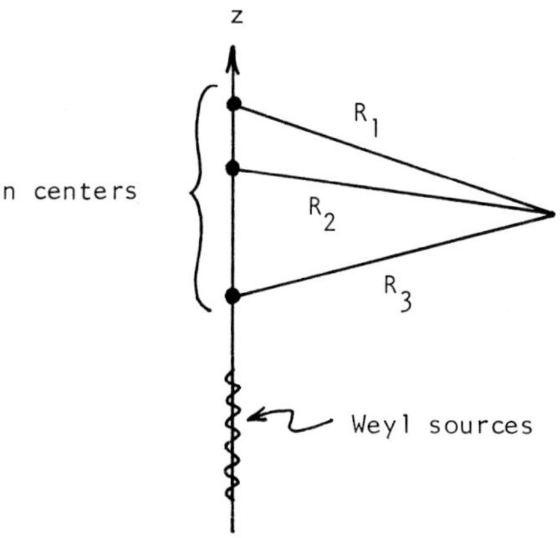

Fig. 3

A simple way of producing nonextreme Kerr-NUT is to start with a Schwarzschild rod of mass -m and apply two HKX transformations centered at both ends of the rod. Starting with n such Schwarzschild rods and applying 2n HKX transformations at all the ends produces Neugebauer's "multisoliton" solution with n Kerr-NUT particles.

9. Symmetries Which Change the Coordinates

More recently, additional symmetry transformations of the stationary Einstein's equations have been found, by permitting the coordinates ρ, z to be transformed at the same time. These transformations lie outside the Geroch group, but do bear a very simple relationship to it.

Cosgrove's transformation $\tilde{Q}$ is

$$\tilde{Q}: \quad N_{AB}^{(m,n)} \to N_{AB}^{(m,n)} + \varepsilon[(m+1)N_{AB}^{(m+1,n)} + nN_{AB}^{(m,n+1)}]$$

$$\begin{aligned} \rho &\to \rho + \varepsilon(4\rho z) \\ z &\to z + \varepsilon(2z^2 - 2\rho^2) \\ \nabla &\to \nabla + \varepsilon(4\rho\nabla - 4z\tilde{\nabla}) \end{aligned} \qquad (9.1)$$

One can show the commutators of $\tilde{Q}$ with the vacuum Geroch group are

$$[\tilde{Q}, \gamma_{AB}^{(k)}] = k\gamma_{AB}^{(k+1)} \qquad (9.2)$$

In particular

$$[\tilde{Q}, \gamma_{AB}^{(0)}] = 0$$

which shows that $\tilde{Q}$ is a scalar under G.

Using the Neugebauer-Kramer duality map $G \to H$, Eq. (6.7), we construct

$$Q = (I)(\tilde{Q})(I) \tag{9.3}$$

which must be a scalar under H. In terms of solution generation, Q has more desirable properties than $\tilde{Q}$.

Q maps:

flat space → flat space
asymptotically flat spaces → asymptotically flat spaces
static spaces → static spaces
Tomimatsu-Sato → Tomimatsu-Sato.

On the other hand, $\tilde{Q}$ (by itself) does not preserve asymptotic flatness.

Although we have presented Q and $\tilde{Q}$ on pretty much the same footing as the elements of the Geroch group, they are really Backlund transformations in disguise, as will be shown in the next section.

10. Backlund Transformations

Several authors have recently shown that the stationary Einstein's equations possess certain solutions which can be interpreted as solitons. The transformations which generate these solutions can be derived using many of the same techniques which work for other soliton equations.

Define a "pseudopotential"

$$q(c,t) = \frac{F_{11}(t) + cF_{12}(t)}{\bar{F}_{11}(t) + c\bar{F}_{12}(t)} \tag{10.1}$$

where c is an arbitrary real parameter. One can show that q satisfies a total Riccati equation,

$$\nabla q = \underset{\sim}{A} + \underset{\sim}{B}q + \underset{\sim}{C}q^2 \tag{10.2}$$

where $\underset{\sim}{A}$, $\underset{\sim}{B}$, $\underset{\sim}{C}$ are simple vector functions depending on c, t, and the Ernst potential.

Once Eq. (10.2) is solved for q, we write down the Backlund transformation

$$T_1: \quad f^{-1}\nabla\mathcal{E} \to q[(f^{-1}\nabla\mathcal{E})\cos\varphi + (f^{-1}\tilde{\nabla}\mathcal{E})\sin\varphi] \tag{10.3}$$

$$z \to S^{-1}[z \cos \varphi + \rho \sin \varphi]$$

$$\rho \to S^{-1}[\rho \cos \varphi - z \sin \varphi]$$

where as before

$$1 - 2tz - 2it\rho = S(t)e^{i\varphi(t)}$$

We also define, by duality,

$$T_2 = (I)(T_1)(I) \qquad (10.4)$$

Neugebauer has shown that T_1, T_2 obey a Commutation Theorem and a Composition Theorem, analogous to those found in the study of the sine-Gordon equation. The Commutation Theorem is illustrated in Fig. 4.

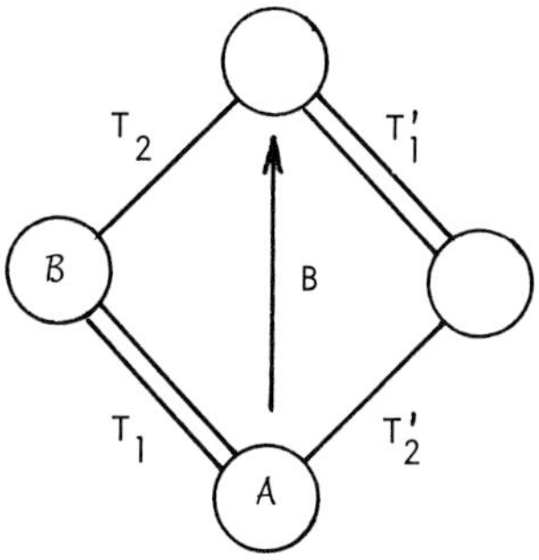

Fig. 4

Start with any solution A. Choose any T_1. Then transformations T_2, T_1', T_2' uniquely exist such that $T_2'T_1'T_2T_1$ is the identity transformation, returning to the same solution A.

The Composition Theorem (Fig. 5) enables one to start from solution A and the "one-soliton" solutions $\mathcal{B}$, $\mathcal{B}'$, $\mathcal{B}''$, ... generated by T_1, T_1', T_1'', ... and construct multi-soliton solutions $\mathcal{C}$, $\mathcal{D}$, ... algebraically (i.e. without solving Eq. (10.2) for the new pseudopotential q at each stage).

11. How Do Backlund and Geroch Get Along Together?

Different soliton transformations for the Einstein equations have been discovered independently by various authors in the last year or so. However, they are very simply related to each other, and to the transformations in the Geroch group. These relations have been worked out by Cosgrove.

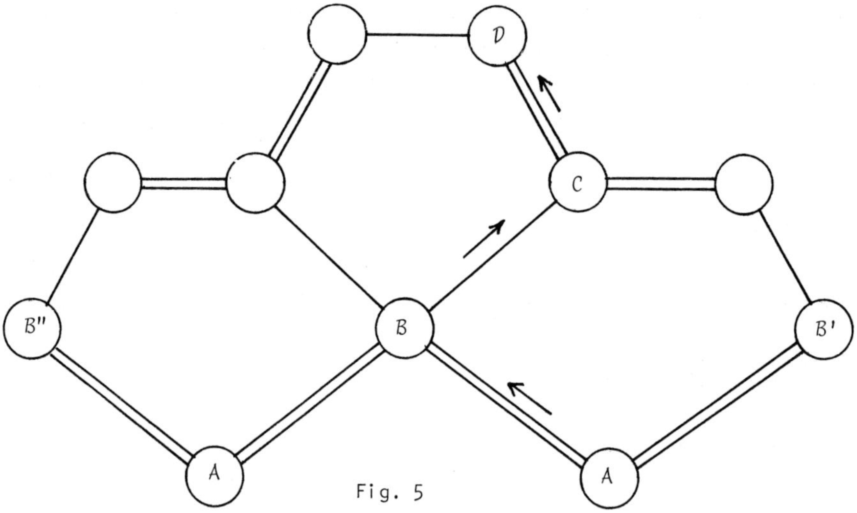

Fig. 5

T_1 and T_2 are essentially the same as Q and $\tilde{Q}$ (exponentiated to finite parameter) accompanied by additional rotations from the subgroups G and H. Harrison's B is two such transformations in succession, namely, the product T_2T_1 which appears in Fig. 4. Belinsky-Zakharov's Z adds to this an application of the duality mapping. (See Table 1)

Neugebauer's T_1	=	$H\, e^{sQ}$
Neugebauer's T_2 / Maison's μ	=	$G\, e^{s\tilde{Q}}$
Harrison's B	=	$H\, e^{-sQ}\, e^{s\tilde{Q}}\, G$
Belinsky-Zakharov's Z	=	(I)(B)

Table 1

All of the soliton transformations listed above have two nontrivial parameters. Among them, only T_1 and B preserve asymptotic flatness. B, Z, and the HKX transformation have the useful properties of being self-dual and commuting.

When we come to ask how many of these tramsformations actually lie outside the Geroch group, it turns out that there is only one independent outsider, which may be taken to be Q. Any known symmetry of the stationary axially symmetric Einstein's equations can be decomposed as

$$T = K\, e^{sQ} \tag{11.1}$$

where the parameter s is chosen to reproduce the desired coordinate transformation, and K belongs to the Geroch group, which fixes the coordinates.

For example, Harrison's transformation B fixes the coordinates, and so is actually itself a member of the Geroch group. One can show this from the relation between the generators,

$$Q - \tilde{Q} = \gamma_{12}^{(1)} \tag{11.2}$$

By the Baker-Hausdorff Theorem and the commutators of K, it then follows that

$$e^{-sQ} e^{s\tilde{Q}} = \sum_{n=1}^{\infty} b_n s^n \gamma_{12}^{(n)} \tag{11.3}$$

and so B is a product of elements of K.

Eq. (11.3) is a particular Harrison transformation that maps Weyl solutions to Weyl solutions. One can show that the HKX transformation can be written as a product

$$HKX = (B)_{General} (B)_{Weyl\text{-}to\text{-}Weyl} \tag{11.4}$$

Thus all of these approaches will ultimately lead to the same set of exact solutions.

This work was supported by National Science Foundation grant PHY79-19565.

References

1. R. Geroch, J. Math. Phys. 13, 394 (1972)
2. W. Kinnersley and D.M. Chitre, J. Math. Phys. 18, 1538 (1977)
3. W. Kinnersley and D.M. Chitre, J. Math. Phys. 19, 2037 (1978)
4. C. Hoenselaers, W. Kinnersley, and B.C. Xanthopoulos, J. Math. Phys. 20, 2530 (1979)
5. C.M. Cosgrove, J. Math. Phys. to appear
6. G. Neugebauer, J. Phys. A12, L67 (1979)
7. D. Maison, J. Math. Phys. 20, 871 (1979)
8. B.K. Harrison, Phys. Rev. Lett. 41, 1197 (1978)
9. V.A. Belinsky and V.E. Zakharov, Sov. Phys. JETP 48, 985 (1978)

GEOMETRY OF PROJECTIVE ASYMPTOTIC TWISTOR SPACE

Gabriel G. Lugo
Department of Mathematics
University of Kansas*
Lawrence, Ka 66045

ABSTRACT

We show that asymptotic twistor space PJ^+ is an Einstein Kähler manifold of positive curvature. We relate the curvature of PJ^+ to the CR-curvature of its boundary and we show that the function defining the boundary satisfies the complex Monge-Ampere equations.

Introduction

During the past few years a great deal of work has been done in the applications of complex manifolds to the theory of general relativity. In particular, much attention has been paid to half flat space-times; that is, four-dimensional complex manifolds with Ricci flat, self-dual curvatures [1,-4]. It has been shown by Newman and his coworkers [2] that given an asymptotically flat space-time (M,g) there exists a naturally associated half-flat manifold called an $\tilde{\mathcal{H}}$-space.

It is also possible to construct from (M,g) an asymptotic projective twistor space PJ. The space PJ is a three-dimensional complex manifold endowed with a hermitian inner product. The set of vectors which are null with respect to this inner product defines in PJ a real five dimensional hypersurface PN. In this paper we show that the region PJ^+ lying inside PN is an Einstein Kähler manifold of positive scalar curvature. We also relate the Bochner tensor of PJ^+ to the fourth order Chern Moser tensor of PN. Finally, we show that solutions to Newman's good cut equation, roughly speaking, satisfy the complex Monge Ampere equation.

$\mathcal{H}$-space

Let (M,g) be an asymptotically flat space-time with future null infinity I^+. As usual we assume that I^+ has topology $S^2 \times R$ and that the space is analytic in a neighborhood of I^+. Denote by CI^+ the complexification of I^+ and let ζ and $\tilde{\zeta}$ be the stereographic coordinates on the complexified two sphere. We define a quantity P by

$$P = \frac{1}{2}(1 + \zeta\tilde{\zeta}). \qquad (1)$$

*Part of this work was done at the University of California at Berkeley.

By a good cut one means a cross-section $u = X(\zeta,\tilde{\zeta})$ of CI^+ satisfying the differential equation [2]

$$\tilde{\partial}^2 X = \tilde{\sigma}^o(X,\zeta,\tilde{\zeta}). \tag{2}$$

Here the operators $\tilde{\partial}$ and its dual ∂ acting on a function η of spin weight s [2] are defined by

$$\begin{aligned} \partial\eta &= 2P^{1-s}\frac{\partial}{\partial\zeta}(P^s\eta) \\ \tilde{\partial}\eta &= 2P^{1+s}\frac{\partial}{\partial\tilde{\zeta}}(P^{-s}\eta) \end{aligned} \tag{3}$$

and $\tilde{\sigma}^o$ is the asymptotic shear of a Bondi family at null hypersurfaces [2].

The non-linear differential equation (2) has a four parameter family of solutions for $\tilde{\sigma}^o$ sufficiently close to zero. The manifold at solutions is called the $\tilde{\mathcal{H}}$-space associated with (M,g).

Asymptotic Twistor Space

Consider now a particular surface $u = X(\zeta,\tilde{\zeta})$ satisfying the good cut equation (2). For fixed value of ζ the equation becomes an ordinary second order differential equation. The solution is therefore a curve $X(\tilde{\zeta}) = X(\omega^o,\omega^1,\zeta,\tilde{\zeta})$ lying on $X(\zeta,\tilde{\zeta})$ where we think of $\omega^A = (\omega^o,\omega^1)$ as the two constants of integration. These so-called twistor curves turn out in fact to be null geodesics in CI^+.

The space of twistor curves is a three dimensional complex manifold called asymptotic projective twistor space PJ. We may think of the quantities $t^\alpha = (\omega^o,\omega^1,\zeta)$ as being the local coordinates of a point C in PJ representing a particular twistor curve in CI^+. The space PJ is C^2-fibered over CP_1. The quantities ω^A are the fiber coordinates and ζ represents the base coordinate.

It has been shown [3] that PJ arises also as a deformation of the complex structure of a tubular neighborhood of a projective line in CP_3. Furthermore, there exists in PJ a four dimensional half-flat manifold of compact holomorphic curves isomorphic to $\tilde{\mathcal{H}}(M)$. We should remark that in the case when the original space time is Minkowski space $M_{1,3}$ then the corresponding asymptotic twistor space is isomorphic to CP_3 and the $\tilde{\mathcal{H}}$-space is flat.

In the space of asymptotic twistors it is possible to define a scalar product [4]. The scalar product is in some sense a deformation of the scalar product induced on CP_3 by a flat hermitian form in C^4 of signature (++ --). At present, however, we do not know of a canonical way to obtain the inner product directly from the deformation data. In the coordinates t^α, the inner product has the form [5]

$$K(t^\alpha,t^{\bar{\alpha}}) = 2iP(X(\omega^o,\omega^1,\zeta,\bar{\zeta}) - \bar{X}(\omega^o,\omega^1,\zeta,\bar{\zeta}).$$

The space PJ may thus be divided into three regions PJ^+, PN and PJ^- according to whether $K > 0$, $K = 0$, or $K < 0$. The region PJ^+ has been suggested as representing a nonlinear graviton state of positive helicity. The region PN is a five dimensional real hypersurface and it is the common boundary between the other two regions.

The existence of the inner product gives rise to a non-degenerate metric on PJ^+ defined by taking

$$ds^2 = 2g_{\alpha\bar{\beta}}dt^\alpha dt^{\bar{\beta}} = 4\frac{\partial^2 \log K}{\partial t^\alpha \partial t^{\bar{\beta}}} dt^\alpha dt^{\bar{\beta}} . \tag{4}$$

The corresponding Kähler two form Φ is given by

$$\Phi = -2ig_{\alpha\bar{\beta}}\, dt^\alpha \wedge dt^{\bar{\beta}} . \tag{5}$$

We now have the following

<u>Theorem 1</u>. The space PJ^+ with the metric (4) is an Einstein Kähler manifold of positive scalar curvature.

The proof of the theorem involves a long manipulation with the good cut equation and the details may be found somewhere else [6].

In analogy with Riemannian geometry the curvature tensor of a Kähler manifold of dimension n admits a decomposition of the form

$$\begin{aligned} R_{\alpha\bar{\beta}\gamma\bar{\delta}} = C_{\alpha\bar{\beta}\gamma\bar{\delta}} &+ (n+2)^{-1}(R_{\alpha\bar{\beta}}g_{\gamma\bar{\delta}} + R_{\gamma\bar{\beta}}g_{\alpha\bar{\delta}} + g_{\alpha\bar{\beta}}R_{\gamma\bar{\delta}} + g_{\gamma\bar{\beta}}R_{\alpha\bar{\delta}}) \\ &- R(n+1)^{-1}(n+2)^{-1}(g_{\alpha\bar{\beta}}g_{\gamma\bar{\delta}} + g_{\gamma\bar{\beta}}g_{\alpha\bar{\delta}}). \end{aligned} \tag{6}$$

where R is the scalar curvature, $R_{\alpha\bar{\beta}}$ is the Ricci tensor and $C_{\alpha\bar{\beta}\gamma\bar{\delta}}$ is a fourth order tensor called the <u>Bochner tensor</u>. The theorem states that the non-trivial part of the curvature is contained in the Bochner tensor. If the original space-time is $M_{1,3}$ then the Bochner tensor of the corresponding twistor space vanishes. For a general space-time, the Bochner tensor of PJ is expressible in terms of the space-time Weyl tensor components Ψ_4, Ψ_3 and Im Ψ_2. In other words, the gravitational radiation data is coded into the Bochner tensor of the asymptotic twistor space.

<u>CR Structure</u>

A real submanifold $\hat{N}$ of a complex manifold inherits from the ambient manifold an intrinsic structure called a Cauchy-Riemann or <u>CR structure</u>. That is, there exists a subbundle $B(\hat{N})$ of the holomorphic tangent bundle $T(\hat{N})$ having a complex vector space structure on each fiber. When the real submanifold is a hypersurface as in the case of PN the induced CR structure is called a pseudoconformal structure.

In the later case the subbundle $B(\hat{N})$ defines a codimension one distribution. The annihilator $E(\hat{N})$ of $B(\hat{N})$ is called the real transversal. The theory of pseudo-conformal structures has been used by Chern-Moser [7] to study the invariants of strongly pseudoconvex hypersurfaces (i.e. hypersurfaces with positive definite Levi form) under biholomorphic mappings. The known invariants include a curvature tensor and certain real curves called chains.

Biholomorphic invariants have also been independently studied by Fefferman [8] by constructing a defining function for the hypersurface which is an approximate solution to the complex Monge-Ampere equations. The invariants found by Fefferman have been related to those of Chern-Moser by work of Burns Shnider [9] and Webster [10].

Although the hypersurfaces which arise in twistor theory are not favorable ones in the sense that they are not strongly pseudoconvex, it is still possible to extend some of the known results to our situation.

Given the projective twistor space PJ^+ it is possible to construct from it in a natural way a real hypersurface.

Let $\{t^\alpha\}$ be the coordinates in an open set u_1 of PJ^+. The forms $\{\theta^\alpha,\theta^{\bar{\alpha}}\}$ where $\theta^\alpha = dt^{\bar{\alpha}}$ define a basis for the cotangent space over u. It terms of this coframe the Kähler metric (4) and the corresponding Kahler form Φ can be written as

$$ds^2 = 2g_{\alpha\bar{\beta}}\theta^\alpha\theta^\beta \tag{7}$$

and

$$\Phi = -4i\partial\bar{\partial}\log K(t^\alpha,t^{\bar{\alpha}}). \tag{8}$$

Let λ be a complex coordinate in C and consider the real valued function on $u_1 \times C$ defined by

$$r = \lambda\lambda K(t^\alpha,t^{\bar{\alpha}}) - 1 \tag{9}$$

The surface $\hat{N}$ defined by the equation $r = 0$ is a trivial circle bundle over the open set u_1. We now define a one-form θ by the quantity $i\partial\gamma$ restricted to $\hat{N}$. The form θ is the annihilator of the subbundle $B(\hat{N})$ spanned by $\{\theta^\alpha,\theta^{\bar{\alpha}}\}$.

It is not difficult to show that if we take another open set u_2 with coordinates $\{t^{\alpha'}\}$ then over the intersection $u_1 \cap u_2$ the CR structures of the two surfaces built over u_1 and u_2 are equivalent. Thus $\hat{N}$ is a CR manifold with a preferred choice of one form θ which depends on the defining function for the hypersurface. The Levi form of $\hat{N}$ is simply the lift of the Kähler form to the bundle.

Of particular importance in this paper is the fourth order Chern-Moser invariant $S_{\alpha\bar{\beta}\gamma\bar{\delta}}$. This quantity is a tensor over the hypersurface which depends on the first four derivatives of the defining function. All the other invariants depend on higher order derivatives. By an extension of the results of Webster [10a] to indefinite cases we then have

Theorem 2. The Bochner tensor of PJ^+ is equal to the fourth order Chern-Moser tensor of $\hat{N}$.

It is the subject of a later paper to demonstrate that the information contained in the Chern-Moser invariants of $\hat{N}$ is essentially the same as that of the invariants on the hypersurface PN. These results are perhaps not surprising since all these tensors are ultimately built out of the first four derivatives of K.

The conclusion of Theorem 2 is that the radiation data of the space-time is also coded into the CR structure of the hypersurface of null twistors. We expect to exploit this fact in the future to see what we can learn about the asymptotics of $\mathcal{H}$-spaces purely from the study of twistors. The final results that we will mention in this paper is the following

Theorem 3. The function K defining the null twistors satisfies the complex Monge-Ampere equations.

Sketch of Proof. Define a new function R on $u \times C$ where u is an open set in PN by the equation

$$R = (z^0 \bar{z}^0)^p K \tag{10}$$

Here $z^0 \in C$ and p is a positive constant. One can then define a Kähler metric on a circle bundle over PN using R as the Kähler potential. By a long computation again using the good cut equation we find that the Ricci tensor of this metric vanishes identically. By the results of [10b] these are the necessary and sufficient conditions to guarantee the assertion of the theorem.

Conclusion

The results in this paper, together with a new intepretation (Tod and Sparling, [11]) of the good cut equation as the Dolbeault version of the deformation of twistor spaces, seems to indicate that the theory of the Bergman Kernel and bounded domains may have some relevance in this context. We also would like to explore the connection, if any, between the complex Monge-Ampere equation as it enters in Theorem 3 and the key equations used by Plebanski [1] to describe half flat spaces.

References

[1] C. P. Boyer, J. D. Finley and J. F. Plebanski, "Complex general relativity, $\mathcal{H}$ and $\mathcal{HH}$-spaces." Einstein Memorial Volume. Plenum, N. Y. 1979.

[2] R. O. Hansen, E. T. Newman, R. Penrose and K. P. Tod, "The metric and curvature properties of $\mathcal{H}$-space," Proc. Roy. Soc. Lond. A363(1978) 445-468.

[3] R. Penrose, "Non-linear gravitons and curved twistor theory." G.R.G. 7, 1 (1976) 35-52.

[4] R. S. Ward, Ph.D. Thesis. Oxford, 1977.

[5] M. Ko, E. T. Newman and R. Penrose, "The Kähler structure of asymptotic twistor space." J.M.P. 18, 1(1977) 58-64.

[6] G. Lugo, "Structure of asymptotic twistor space," submitted to J.M.P. (1980).

[7] S. S. Chern and J. K. Moser, "Real hypersurfaces in complex manifolds," Acta. Math. 133 (1974) 219-270.

[8] C. L. Fefferman, "Mange-Ampere equations, the Bergman kernel and geometry of pseudoconvex domains." Annals. Math. 103 (1976) 395-416.

[9] D. Burns and S. Shnider, "Real hypersurfaces in complex manifolds." Proc. Symp. Pure. Math. 30(1977) 141-167.

[10] a) S. Webster, "On the pseudoconformal geometry of Kähler manifolds," Math. Zeitschrift 157 (1977) 265-270.

b) ________ , "Kähler metrics associated with real hypersurfaces." Comm. Math. Helv. 52 (1977) 235-250.

[11] P. Tod. Twistor Newsletter 9. Oxford preprint (1979).

A TRULY RELATIVISTIC APPROACH OF THE CONCEPT OF TIME.

J.P. Provost.
Physique Théorique, Université de Nice, Parc Valrose,
06034 Nice Cedex, France.

The standard presentations of space-time and (special) relativity proceed along the following theoretical two steps: first one defines some particular frames of reference, called inertial frames, where space appears to be homogeneous and isotropic and time appears to be homogeneous. Second one introduces relativity by telling (or deriving) what is the correct transformation of coordinates when two inertial frames have a definite (constant) relative velocity. Although we are fully aware that most text book authors discuss at length, how to measure space and time or how to characterise in practice inertial frames, we consider that their theoretical background is rather poor, even from a physicist's point of view. Our main criticism is the following: this approach does not consider the concept of motion as a key concept, theoretically introduced from the beginning in the definition of inertial frames. In fact they define motion in a very old way (as the ratio of space and time intervals in the way Galilei did), whereas one knows that the general structure underlying relativity theories is the group structure; we think that the concept of time is the less obvious one and that it would be better to derive it from the other two concepts of space and motion (and of course from some physical assumptions). As we show, not only such a theoretical approach is possible, but the way time is obtained immediatly gives information on which group (i.e. which relativity, Galilei or Einstein relativity) is involved. Therefore it is a very instructive presentation and trivialisation of relativity in teaching [1].

For the sake of simplicity, let us sketch our approach for a one dimensional space (a generalization to 3-dimensional space is possible). We define inertial frames by the following three properties

i) space is described by a group of translations $x \in R$;

ii) inertial frames are related to each other by a one parameter group; the corresponding additive parameter [2] φ will be called the "rapidity" (we do not say speed or velocity because we have not defined time); the composition of rapidities reads: $\varphi_1, \varphi_2 \to \varphi_1 + \varphi_2$ (inertia therefore has been defined from a group hypothesis without calling for forces);

iii) relative rapidities of reference frames change in sign when their respective x axes are all changed into -x (vectorial character of the 3-dimensional rapidity).

One main idea [1, 3] of this introduction is to consider that:

iv) any translation x in some frame R may be realized by (the motion of) any inertial frame with rapidity φ. We shall speak of (x, φ) as a "dynamical" translation (whereas x is a "geometrical" translation).

In a frame R' whose rapidity with respect to R is $\in$, the dynamical translation is described by (x', φ') with:

$$\varphi' = \varphi - \varepsilon \; ; \; x' = x - \varepsilon \; f(x, \varphi) = x - \varepsilon \, x \; f(\varphi) \qquad (1)$$

$(f(x, \varphi) = x\, f(\varphi)$ because φ is a dimensionaless group parameter). Moreover, due to iii), f is an odd function;

v) the concept of time is introduced through the following demand. If in any reference frame three dynamical translations satisfy

$$x_3 = x_1 + x_2 \qquad (2)$$

the translations on both sides of the equation have been realized in equal times. This introduction is an authentically relativistic one. A reflexion on the concept of simultaneity shows that it is a very natural one. From the relation between x and x', one deduces that the time equality reads:

$$x_3 \; f(\varphi_3) = x_1 \; f(\varphi_1) + x_2 \; f(\varphi_2) \qquad (3)$$

((2) and (3) may be thought as conservation relations). Let us note that (3) should not depend on the frame R and should be also true for primed quantities. From (1) one obtains:

$$x' \; f(\varphi') = x \; f(\varphi) - \varepsilon \; x \; (f^2(\varphi) + f'(\varphi)) \qquad (4)$$

and from (3) and (4) one deduces a new relation:

$$x_3 \; (f^2(\varphi_3) + f'(\varphi_3)) = (1) + (2) \qquad (5)$$

If $x_1, x_2, \varphi_1, \varphi_2$ are fixed, equations (2), (3) and (5) determine φ_3 and must be compatible. It is easy to realize that this is possible only if:

$$f^2(\varphi) + f'(\varphi) = \lambda \; f(\varphi) + \mu \qquad (6)$$

According to the oddness property of f, λ is equal to zero and without loss of gener ality we can consider the following cases:

1) $\mu = 0 \; ; \; f(\varphi) = \varphi^{-1}$ or $x = \varphi \; t$ (Galilei)

2) $\mu = 1 \; ; \; f(\varphi) = (\mathrm{th}\,\varphi)^{-1}$ or $x = (\mathrm{th}\,\varphi) t$ (Lorentz)

(The concept of speed $v = \mathrm{th}\,\varphi$ is therefore different from the concept of rapidity φ in that case).

3) $\mu = -1 \; ; \; f(\varphi) = (\mathrm{tg}\,\varphi)^{-1}$

Coming back to equations (1) and (4) one obtains the relativity group structure:

indeed, due to (6) these equations may be written

$$x' = x - \varepsilon t \quad ; \quad t' = t - \varepsilon \mu x$$

These relations clearly are the infinitesimal version of the group laws for Galilei ($\mu = 0$), Lorentz ($\mu = 1$), rotation in space time ($\mu = -1$), this last case being eliminated by the following demand;

vi) there exist space time intervals whose time coordinate keep a constant sign (causality condition).

References.

[1] For more details see "Temps et mouvement, les fantômes de la géométrie" J.P. Provost, Publication of the I.R.E.M. of Nice, (1979).

[2] J.M. Lévy-Leblond, J.P. Provost, Am. Jour. Phys. 47, 1045 (1979).

[3] A similar approach (apart from group ideas) due to Lange (1885) may be found in "Relativity and Cosmology" (Robertson and Noonan), W.B. Saunders Company, Philadelphia-London-Toronto (1968).

BASES IN Sp(2n) REPRESENTATION SPACE

Adam M. Bincer
Physics Department
University of Wisconsin-Madison
Madison, Wisconsin, 53706 USA

1. Introduction: Because the natural reduction chain for the symplectic group is not multiplicity-free, the standard techniques for constructing bases in representation space yield basis states that are either labeled rationally but are not orthogonal or are orthogonal but labeled nonrationally. I have developed an algorithm for constructing bases such that basis states are both mutually orthogonal and labeled rationally. The essential new ingredient involves the concept of quasimaximal states defined as semimaximal states subject to a set of subsidiary conditions. The quasimaximal states are mutually orthogonal and orthogonal to the remaining semimaximal states needed to complete the basis. These remaining semimaximal states can be made mutually orthogonal by a procedure similar to, but not suffering from the arbitrariness inherent in, the Schmidt orthogonalization procedure.

2. Generators, Tensor Operators and Weights: I denote the generators of Sp(2n) by

$$G^a_b = -\varepsilon^a\varepsilon^b G^{\bar b}_{\bar a}, \quad \bar a \equiv -a, \quad \varepsilon^a = a/|a| \ , \tag{1}$$

with the indices ranging from $\bar n$ to n, zero excluded. The Sp(2n-2t), $0\leq t<n$, subgroup is obtained by omitting from the range of the indices the values from $\bar t$ to t. In a unitary representation the G's have the hermiticity property

$$G^{a\dagger}_b = G^b_a \ . \tag{2}$$

Their commutation relations are

$$[G^a_b, G^c_d] = \delta^c_b G^a_d - \delta^a_d G^c_b + \varepsilon^a\varepsilon^b (\delta^{\bar b}_d G^c_{\bar a} - \delta^c_{\bar a} G^{\bar b}_d) \ . \tag{3}$$

I define an Sp(2n-2t) tensor operator T, with $(2n-2t)^2$ components T^a_b, $t+1 \leq |a|, |b| \leq n$, by the transformation properties

$$[G^a_b, T^c_d] = \delta^c_b T^a_d - \delta^a_d T^c_b + \varepsilon^a \varepsilon^b (\delta^{\bar b}_d T^c_{\bar a} - \delta^c_{\bar a} T^{\bar b}_d). \tag{4}$$

It is obvious that G, whose components are the generators, is a tensor operator. Also, if T and W are two tensor operators then so is TW, where

$$(TW)^a_b \equiv \sum_c T^a_c W^c_b \ . \tag{5}$$

with all indices in the same range.

Let $|\underset{\sim}{w}>$ denote a simultaneous eigenstate of the n generators G^a_a, $1 \leq a \leq n$, from the Cartan subalgebra:

$$G^a_a|\underset{\sim}{w}> = w_a|\underset{\sim}{w}> \ , \qquad w_{\bar a} = -w_a \ , \tag{6}$$

where $\underset{\sim}{w} \equiv (w_n, w_{n-1},\ldots,w_1)$ is called the weight. The weight $\underset{\sim}{w}$ will be called higher than $\underset{\sim}{w}'$ if in the expression

$$\underset{\sim}{w} - \underset{\sim}{w}' = (w_n - w_n', w_{n-1} - w_{n-1}', \ldots, w_1 - w_1') \tag{7}$$

the first nonzero entry starting from the left is positive (if all entries vanish then $\underset{\sim}{w} = \underset{\sim}{w}'$).

It follows from Eq. (4) that

$$T^a_b|\underset{\sim}{w}> \sim |\underset{\sim}{w}'> \;, \quad w_c' = w_c + \delta^a_c - \delta^c_b + \delta^{\bar{b}}_c - \delta^c_{\bar{a}} \tag{8}$$

and therefore

$$\underset{\sim}{w}' \gtreqless \underset{\sim}{w} \quad \text{if} \quad a \gtreqless b \;, \tag{9}$$

which leads to the classification of T^a_b as a raising, or weight, or lowering operator according as $a > b$, or $a = b$, or $a < b$.

3. Semimaximal States: Since all unitary irreducible representations (unireps) of Sp(2n) are finite-dimensional they necessarily contain a basis state of highest weight, the so-called maximal state $|\underset{\sim}{h}>$ (known to be unique):

$$G^a_b|\underset{\sim}{h}> = \delta^a_b h_a|\underset{\sim}{h}> \;, \quad h_{\bar{a}} = -h_a, \; \bar{n} \leq b \leq a \leq n, \tag{10}$$

where $\underset{\sim}{h} \equiv (h_n, h_{n-1},\ldots,h_1)$ is the highest weight and specifies this unirep. Since a unirep of Sp(2n) is necessarily a representation (possibly reducible) of the Sp(2n-2)xU(1) subgroup, the representation space of Sp(2n) specified by $\underset{\sim}{h}$ must also contain so-called semimaximal states $|s.m.>$, which are maximal states of Sp(2n-2)x U(1) specified by $\underset{\sim}{r}$ and w_1:

$$G^a_b|s.m.> = \delta^a_b r_a|s.m.> \;, \quad r_{\bar{a}} = r_a, \quad b \leq a, \tag{11}$$

$$G^1_1|s.m.> = w_1|s.m.> \;. \tag{12}$$

Here the indices a, b range as is appropriate to Sp(2n-2), i.e., from $\bar{n}$ to n with -1, 0, 1 excluded, the U(1) is generated by G^1_1, and the weight of $|s.m.>$ is given by $(r_n, r_{n-1},\ldots,r_2, w_1)$.

Using the known branching law [1] for the splitting of a unirep of Sp(2n) under restriction to Sp(2n-2) x U(1) I have

$$|s.m.> = \left| \begin{matrix} h_n, & h_{n-1},\ldots, & h_2, & h_1 \\ q_n, & q_{n-1},\ldots, & q_2, & q_1 \\ r_n, & r_{n-1},\ldots, & r_2 & \end{matrix} \right\rangle \tag{13}$$

where all the labels are nonnegative integers satisfying the betweenness conditions

$$h_n \geq q_n \geq h_{n-1} \geq \cdots \geq q_2 \geq h_1 \geq q_1 \geq 0 \;, \tag{14}$$

$$q_n \geq r_n \geq q_{n-1} \geq \cdots \geq r_2 \geq q_1 \;. \tag{15}$$

Whereas the labels $\underset{\sim}{h}$ and $\underset{\sim}{r}$ have an obvious group-theoretical significance, the n components of $\underset{\sim}{q}$ have no such interpretation except for the one relation among them:

$$w_1 = 2q_1 - h_1 + \sum_{a=2}^{n} (2q_a - r_a - h_a) . \tag{16}$$

Clearly these $|s.m.\rangle$ provide the first step in obtaining a basis with a rational labeling scheme, similar to the Gel'fand-Zetlin scheme for the U(n) and O(n) groups [2]. However, whereas two $|s.m.\rangle$ with different $\underset{\sim}{r}$ or w_1 are clearly orthogonal (being eigenstates of hermitian operators to different eigenvalues) this is not necessarily the case if $\underset{\sim}{r}$ and w_1 are the same, but the $\underset{\sim}{q}$ differ.

4. Quasimaximal States: As a first step in obtaining an orthogonal scheme I define quasimaximal states $|\underset{\sim}{q}\rangle$ by [a, b in Sp(2n-2)]:

$$G^a_b|\underset{\sim}{q}\rangle = \delta^a_b\, q_a|\underset{\sim}{q}\rangle \quad , \quad q_{\bar{a}} = -q_a \; , \quad b \leq a, \tag{17}$$

$$G^a_1|\underset{\sim}{q}\rangle = 0, \qquad 2 \leq a \leq n, \tag{18}$$

$$G^1_1|\underset{\sim}{q}\rangle = w_1|\underset{\sim}{q}\rangle \; , \qquad w_1 = 2q_1 - h_1 + \sum_{a=2}^{n} (q_a - h_a) . \tag{19}$$

Except for the additional n-1 conditions, Eq. (18), these defining equations are the same as for semimaximal states with $r_a = q_a$. I will show that for these $|\underset{\sim}{q}\rangle$ Eq. (14) holds and therefore quasimaximals are in fact semimaximals with $r_a = q_a$. Most importantly, however, it follows from Eq. (17) that

$$\langle \underset{\sim}{q}|\underset{\sim}{q}'\rangle = 0 \quad \text{for} \quad \underset{\sim}{q} \neq \underset{\sim}{q}' \quad . \tag{20}$$

To obtain the $|\underset{\sim}{q}\rangle$ I define shift operators P_k and P^k, $1 \leq k \leq n$, that respectively lower and raise the value of the label q_k by one unit:

$$P_k|\underset{\sim}{q}\rangle = |\underset{\sim}{q} - \delta^k\rangle \; , \qquad P^k|\underset{\sim}{q}\rangle = |\underset{\sim}{q} + \delta^k\rangle \tag{21}$$

$$|\underset{\sim}{q} \pm \delta^k\rangle \equiv |q_n, \ldots, q_{k+1}, q_k \pm 1, q_{k-1}, \ldots, q_1\rangle \tag{22}$$

It follows that an arbitrary quasimaximal state can be formed by applying to $|\underset{\sim}{h}\rangle$ the various lowering operators P_k an appropriate number of times. Since the states on both sides of Eq. (21) are quasimaximals an equivalent definition of P_k is given by

$$[G^a_b, P_k]|\underset{\sim}{q}\rangle = \delta^a_b(\delta^a_{\bar{k}} - \delta^a_k)\, P_k|\underset{\sim}{q}\rangle \; , \qquad b \leq a \text{ in Sp(2n-2)}, \tag{23}$$

$$[G^a_1, P_k]|\underset{\sim}{q}\rangle = -\delta^a_1(1 + \delta^k_1)\, P_k|\underset{\sim}{q}\rangle \; , \quad 1 \leq a \leq n, \tag{24}$$

and similarly for P^k.

For k=1 a solution is obviously given by

$$P_1 \sim G^{\bar{1}}_1 \quad . \tag{25}$$

For $2 \leq k \leq n$ I have found solutions for P_k and P^k in the form

$$P_k \sim T(k)_1^{\bar{k}} \quad , \quad P^k \sim [G\,U(k)]_{\bar{k}}^1 \quad , \tag{26}$$

where the T(k) are Sp(2n) tensor operators and the U(k) are Sp(2n-2) tensor operators given by (where I is the unit operator)

$$T(k) = \sum_{\rho=\bar{n}}^{n} \prod_{\sigma=\bar{n}}^{\rho-1} (G - \lambda_\sigma I) \prod_{\tau=\rho+1}^{n} (\kappa_{\bar{k}} - \lambda_\tau + 1), \tag{27}$$

$$U(k) = \prod_{j=\bar{n}}^{\bar{k}-1} (G - \kappa_j I) \quad , \tag{28}$$

$$\lambda_\tau = h_\tau + \tau + n, \tag{29}$$

$$\kappa_j = q_j + j + n - 2\theta_j \tag{30}$$

with $\theta_j=1$ for $j>0$ and $\theta_j=0$ for $j<0$. Also greek letters denote indices ranging over Sp(2n), latin letters over Sp(2n-2). I have also determined an algorithm for obtaining P^1 but am unable to exhibit P^1 for general n explicitly.

These shift operators have the crucial commutativity property

$$P^k P^{k'} |\underset{\sim}{q}\rangle = P^{k'} P^k |\underset{\sim}{q}\rangle \quad , \quad P_k P_{k'} |\underset{\sim}{q}\rangle = P_{k'} P_k |\underset{\sim}{q}\rangle \quad ,$$
$$P^k P_{k'} |\underset{\sim}{q}\rangle = P_{k'} P^k |\underset{\sim}{q}\rangle \tag{31}$$

for $1 \leq k \neq k' \leq n$, which enables one to show that the $\underset{\sim}{q}$ must obey Eq. (14) and that the $|\underset{\sim}{q}\rangle$ are unambiguously labeled.

5. Orthogonalization: Consider now a semimaximal state, Eq. (13), which will be abbreviated by $|\underset{\sim}{r}\rangle$. With the lowering operator S_k defined by

$$S_k |\underset{\sim}{r}\rangle = |\underset{\sim}{r} - \delta^k\rangle \, , \qquad 2 \leq k \leq n \quad , \tag{32}$$

I find the solution

$$S_k \sim [GW(k)]_k^1 \tag{33}$$

where the W(k) are Sp(2n-2) tensor operators given by

$$W(k) = \prod_{j=\bar{n}}^{k-1} (G - \mu_j I) \quad , \qquad \mu_j = r_j + j + n - 2\theta_j \tag{34}$$

To avoid some of the complexity I demonstrate the orthogonalization procedure on the example of Sp(4). The quasimaximals are given by $|q_2, q_1\rangle$ and semimaximals are obtained by repeated application of S_2. Among the semimaximals so obtained those with different weights are orthogonal. However, the following states, at most q_1+1 in number,

$$(S_2)^p \, |r_2 + p, q_1 - p\rangle \quad , \qquad 0 \leq p \leq q_1, \tag{35}$$

all have the same weight and are, in general, not orthogonal. However, because of the way the quasimaximals were defined the p=0 state in Eq. (35) is,in fact, orthogonal to the others. Thus I may form the sequence

$$\left| r_2, q_1 \right\rangle \quad , \tag{36}$$

$$\left| \begin{matrix} h_2, h_1 \\ r_2+1, \; q_1-1 \\ r_2 \end{matrix} \right\rangle \sim S_2 \left| r_2+1, q_1-1 \right\rangle \quad , \tag{37}$$

$$\left| \begin{matrix} h_2, h_1 \\ r_2+2, \; q_1-2 \\ r_2 \end{matrix} \right\rangle \sim S_2^2 \left| r_2+2, q_1-2 \right\rangle + \alpha \left| \begin{matrix} h_2, h_1 \\ r_2+1, \; q_1-1 \\ r_2 \end{matrix} \right\rangle \quad , \tag{38}$$

$$\left| \begin{matrix} h_2, h_1 \\ r_2+3, \; q_1-3 \\ r_2 \end{matrix} \right\rangle \sim S_2^3 \left| r_2+3, q_1-3 \right\rangle + \beta \left| \begin{matrix} h_2, h_1 \\ r_2+1, \; q_1-1 \\ r_2 \end{matrix} \right\rangle + \gamma \left| \begin{matrix} h_2, h_1 \\ r_2+2, \; q_1-2 \\ r_2 \end{matrix} \right\rangle \quad , \tag{39}$$

etc. Here (36) is orthogonal to all the others, α in (38) is determined by demanding orthogonality with (37), β and γ in (39) are determined by demanding orthogonality with (37) and (38), and so forth.

This, of course, is nothing but the Schmidt procedure except that the above-proposed sequence seems logically well-founded in contrast to the arbitrariness usually present in this procedure.

[1] G. C. Hegerfeldt, J. Math. Phys. 8, 1195 (1967); J. Mickelsson, Rep. Math. Phys. 3, 193 (1972).

[2] I. M. Gel'fand and M. L. Zetlin, Dokl. Akad. Nauk SSSR 71, 825 (1950); 1017 (1950).

SEMIUNITARY PROJECTIVE REPRESENTATIONS OF NON-CONNECTED LIE GROUPS: A criterion for the choice of the unitary subgroup.

José F. Cariñena, Dpto. de Física Teórica, Universidad de Zaragoza (Spain).
Mariano Santander, Dpto. de Física Teórica, Universidad de Valladolid (Spain).

It is well known that in the usual formulation of Quantum Mechanics in Hilbert spaces, the symmetries of a quantum system are realized as unitary or antiunitary projective transformations and a symmetry group G will be realized in the space of states by means of a semiunitary projective representation (hereafter SUPR). The set of symmetries realized in a unitary way is a subgroup G_U (of index one or two) which is called the unitary subgroup. In particular, when G is a connected Lie Group, $G_U = G$. The criterion for the choice of the unitary subgroup, we are going to give is based on the following mathematical point: every SUPR of (G,G_U) subduces a UPR of G_U and therefore a UPR of G_0 (connected component of G), but not every UPR of G_0 arises in such subduction process; moreover, which UPR's arise depends on the choice for G_U. So, a natural criterion for the choice of the unitary subgroup is the following one:

Criterion The choice for G_U must be such that every "relevant" UPR of G_0 arises in the above subduction process.

Physical assumptions are in the word "relevant". In general, the UPR's of G_0 are easy to find and to give a physical interpretation to some of them. This point will be clarified with a simple example: Galilei group.

To show the mathematical comments previous the above criterion, we must study the factor systems $Z^2_*(G,T)$, where T is the torus group and the action of G on T is given by $\lambda^g=\lambda$ if $g\in G_U$, $\lambda^g=\lambda^*$ if $g\in G-G_U$, the asterisk standing for complex conjugation. We will consider here the case where the Lie group G is a semidirect product group $G = H \circledS V$, that is $G=\{(a,\alpha) \mid a\in H,\ \alpha\in V\}$ and the composition law in G is given by $(a,\alpha).(b,\beta) = (ab^{\alpha},\alpha\beta)$. In this case a straightforward generalization of Mackey's theorem[1] leads to the following theorem

Theorem 1 Let ω' be a factor system of G with respect to the sybgroup G_U and let H_U and V_U denote the intersections $H_U = H \cap G_U$, $V_U = V \cap G_U$ respectively: There exist an equivalent factor system $\omega \varepsilon Z^2_*(G,T)$ which decomposes as a product

$$\omega[(a,\alpha),(b,\beta)] = \xi(a,b^{\alpha})\,[\eta(\alpha,\beta)]^{ab^{\alpha}}\,[\Lambda(b,\alpha)]^{a} \tag{1}$$

with $\xi \varepsilon Z^2_*(H,T)$, $\eta \varepsilon Z^2_*(V,T)$, while $\Lambda : H \times V \to T$ is a Borel function such that

$$\xi(a^{\alpha}, b^{\alpha}) = [\xi(a,b)]^{\alpha} \cdot \frac{\Lambda(ab,\alpha)}{\Lambda(a,\alpha)\,[\Lambda(b,\alpha)]^{ab^{\alpha}}} \tag{2.a}$$

$$\Lambda(a,\alpha\beta) = \Lambda(a^{\beta},\alpha)\,[\Lambda(a,\beta)]^{\alpha}\,\frac{[\eta(\alpha,\beta)]^{a^{\alpha\beta}}}{\eta(\alpha,\beta)} \tag{2.b}$$

The proof of this theorem follows the pattern of that of Parthasarathy[2] and it will not be given here. Conversely, if actions of H and V on T are given, let G_U be the subgroup of index one or two generated by the kernels of ineffectiveness of each action:

Theorem 2 if $\xi \varepsilon Z^2_*(H,T)$, $\eta \varepsilon Z^2_*(V,T)$ and Λ is a Borel function satisfying the above relations (2), then ω defined by (1) is a cocycle $\omega \varepsilon Z^2_*(G,T)$.

The proof of this theorem is a cumbersome but straightforward calculation, so that it is omitted.

The two theorems are very useful in the case of a nonconnected Lie group which is a semidirect product group such that $G = G_o \odot \pi_o(G)$, G_o being the connected component of G. The normal subgroup G_o is always contained in G_U and therefore in this case the decomposition (1) reduces to ($V = \pi_o(G)$)

$$\omega[(a,\alpha),(b,\beta)] = \xi(a,b^{\alpha}) \quad \eta(\alpha,\beta) \quad \Lambda(b,\alpha) \tag{3}$$

where $\Lambda : G \times V \to T$ is a Borel function satisfying

$$\xi(a^{\alpha}, b^{\alpha}) = [\xi(a,b)]^{\alpha} \, \frac{\Lambda(ab,\alpha)}{\Lambda(a,\alpha)\Lambda(b,\alpha)} \tag{4.a}$$

$$\Lambda(a,\alpha\beta) = \Lambda(a^{\beta},\alpha)\,[\Lambda(a,\beta)]^{\alpha} \tag{4.b}$$

Once ξ and η have been chosen, how many solutions Λ do exist for these equations? It would be interesting to know a necessary condition for the existence of (at least) one such Λ. The first equation provides such a condition: every $\alpha \varepsilon V$ defines an application $\tau_\alpha : Z^2(G_\circ,T) \to Z^2(G_\circ,T)$, $(\tau_\alpha \xi)(a,b) = \xi(a^{\alpha^{-1}}, b^{\alpha^{-1}})$; this mapping is an endomorphism such that $\tau_\alpha [B^2(G_\circ,T)] \subset B^2(G_\circ,T)$, so that there is an induced homomorphis, $\bar{\tau}_\alpha : H^2(G_\circ,T) \to H^2(G_\circ,T)$. With this notation the relation (2.4,a) may be rewritten as

$$\left[(\tau_{\alpha^{-1}} \xi).(\xi^\alpha)^{-1}\right](a,b) = \frac{\Lambda(ab,\alpha)}{\Lambda(a,\alpha)\Lambda(b,\alpha)}$$

Therefore, in order that a solution may exist, the factor system $(\tau_{\alpha^{-1}}\xi)(\xi^\alpha)^{-1}$ must be a trivial factor system. This fact is independent of the choice for η . The preceding results can be summarized in

Theorem 3 With the above conditions and notations, the existence of a class $\bar{\omega}$ such that its restriction to $G_\circ$ and $V=\pi_\circ(G)$ are $\bar{\xi}$ and $\bar{\eta}$ respectively, does not depend on $\bar{\eta}$. A necessary condition for the existence of such $\bar{\omega}$ is that

$$\bar{\tau}_{\alpha^{-1}} \bar{\xi} = \bar{\xi}^\alpha \qquad \forall \alpha \varepsilon V.$$

Finally we are going to see how this criterion works in the case of the complete Galitie group. We will use througout the notation of Lévy-Leblond's paper[3]. There are four candidates for the unitary subgroup

i) $G_U = G$ ii) $G_U = G_\circ \cup I_S G_\circ$ iii) $G_U = G_\circ \cup I_T G_\circ$ iv) $G_U = G_\circ \cup I_{ST} G_\circ$

The second cohomology group of $G_\circ$ is[4] $H^2(G_\circ,T) = \mathbb{R} \otimes C_2$. The element $[M,1]$ will be the class of the factor system

$$\xi_{M,1}(g',g) = \zeta_1(R',R) \exp iM\left\{\frac{1}{2} bv'^2 + v'R'a\right\}$$

where $[\zeta_1(R',R)]^2 = +1$. The parameter M is to be identified with the mass of the elementary system described by the corresponding projective representation.

We can use the above mentioned method of finding $H^2_*(G,T)$: once a factor system $\xi \varepsilon Z^2(G_\circ,T)$ has been chosen, is there any Borel function $\Lambda: G \to T$ satisfying the relation (4-a)? The factor systems $\tau_{I_S}\xi$, $\tau_{I_T}\xi$ and $\tau_{I_{ST}}\xi$ are given respectively by $(\tau_{I_S}\xi)(g',g) = \xi(g',g)$, $(\tau_{I_T}\xi)(g',g) = [\xi(g',g)]^*$, $(\tau_{I_{ST}}\xi)(g',g) = [\xi(g',g)]^*$

Therefore, a direct use of Theorem 3 shows that if the class $[M,1]$ of the factor system $\xi_{M,1}$ with $M\neq 0$ is considered, then in order that a solution Λ to exist $I_S \in G_U$, $I_T \notin G_U$ and $I_{ST} \notin G_U$. In other words, G_U must be the subgroup generated by $G_\circ$ and I_S: with a different choice for G_U the restriction to $G_\circ$ of any SUPR of (G,G_U) would correspond to a "massless representation" of $G_\circ$; these representations have been shown to be unphysical[3]. The representations of $G_\circ$ corresponding to massive systems would not arise.

In the case of Poincaré group the former criterion gives us no information because $H^2(P_\circ,T)$ reduces to C_2. What about other kinematic groups[5]? The criterion is useful in the case of "absolute-time groups": in these groups $H^2(G_\circ,T) = \mathbb{R} \otimes C_2$. A more detailed study will be given elsewhere.

REFERENCES

1.- Mackey, G.W., Acta Math. 99, 265 (1958)

2.- Parthasarathy, K.R., Multipliers on locally compact groups, Lect. Not. in Math. 93, Springer (1969)

3.- Lévy-Leblond, J.M., "Galilei group and Galilean INvariance" in Group Theory and Its Applications, Vol II, ed. by Loebl, Academic (1971)

4.- Bargmann, V., Ann.Math. 59,1 (1954)

5.- Bacry, H and Lévy-Leblond, J.M., J. Math. Phys. 9, 1605 (1968)

$SU(mn) \supset SU(m) \times SU(n)$ ISOSCALAR FACTORS AND $S(f_1+f_2) \supset S(f_1) \times S(f_2)$ ISOSCALAR FACTORS

Jin-Quan Chen
Department of Physics, Nanjing University
Nanjing, People's Republic of China†
and
School of Physics and Astronomy, University of Minnesota
Minneapolis, Minnesota 55455

1. Introduction

In a series of papers[1-2] we have proposed a new approach to group representation theory. Three kinds of complete sets of commuting operators, denoted by CSCO-I, II and III, were introduced, whose eigenvalues can uniquely label the irreps, irreducible bases and irreducible matrix elements of a group G, respectively. They are CSCO in the class space, irreducible space and group space of G respectively. The problems of calculating (I) characters and isoscalar factors (ISF) (II) irreducible bases and the Clebsch-Gordan coefficients (CGC) (III) irreducible matrix elements etc. are all simplified to that of solving the eigen function of the CSCO-I, II and III respectively. Therefore we call it the Eigen-function Method. A fundamental theroem is:

A necessary and sufficient condition for $\psi^{(\nu)}$ to belong to the irrep (ν) of a group G is that $\psi^{(\nu)}$ is an eigen-function of the CSCO-I of G.

For finite groups, CSCO-I consists of a few class operators. The CSCO-I of the permutation group S(f) is $C(f) = C_{(2)}(f)$ for $f = 2\text{-}5, 7$, and $C(f) = (C_{(2)}(f), C_{(3)}(f))$ for $f = 6$, $8 \leq f \leq 14$, where $C_{(i)}(f)$ is the i-cycle class operator of S(f). The CSCO-II of S_f consists of f-1 2-cycle class operators

$$(C_{(2)}(f),\ C_{(2)}(f-1)\ \ldots\ C_{(2)}(2))$$

whose simultaneous eigen-functions give the Yamanouchi bases.

Since the Eigen function Method proves to be very successful for calculating the CGC and the outer-product reduction coefficients of the permutation group[3], this method is now used to calculate the $S(f_1+f_2) \supset S(f_1) \times S(f_2)$ ISF. We will also prove that the $SU(mn) \supset SU(m) \times SU(n)$ ISF is equal to the $S(f_1+f_2) \supset S(f_1) \times S(f_2)$ ISF, therefore the former is independent of m and n, and can be calculated for all m and n at a single stroke instead of one m and one n at a time.

2. $S(f) \supset S(f_1) \times S(f_2)$ ISF

Suppose the coordinates q of a particle are divided into two parts, χ and ξ,

---

† permanent address

and $q=(\chi,\xi)$. For example, χ may represent the orbital coordinates and ξ may represent the spin-isospin coordinates. Thus for a system with f particles we have three realizations of the permutation group S(f) i.e. $S^{\chi}(f)$, $S^{\xi}(f)$ and $S^{q}(f)$ which permute the indices of χ_i's, ξ_i's and q_i's respectively.

Let $S(f_1)$ and $S(f_2)$ be the permutation groups for particles $1,2\ldots f_1$ and f_1+1, $f_1+2\ldots f_1+f_2$, respectively, with $f = f_1+f_2$. The irreducible bases classified according to the irreps of the group chain $S(f) \supset S(f_1) \times S(f_2)$ in the χ, ξ and q space are denoted by

$$\left|\begin{matrix}[\sigma]\\ \theta[\sigma']m_1'[\sigma'']m_1''\end{matrix}\right\rangle , \quad \left|\begin{matrix}[\mu]\\ \phi[\mu']m_2'[\mu'']m_2''\end{matrix}\right\rangle , \quad \left|\begin{matrix}[\nu]\\ \tau[\nu']m'[\nu'']m''\end{matrix}\right\rangle \tag{1}$$
$$\theta=1,2,\ldots\{\sigma'\sigma''\sigma\}, \qquad \phi=1,2,\ldots\{\mu'\mu''\mu\}, \qquad \tau=1,2,\ldots\{\nu'\nu''\nu\} ,$$

where θ, ϕ and τ are multiplicity labels. For example $\left|\begin{matrix}[\sigma]\\ \theta[\sigma']m_1'[\sigma'']m''_1\end{matrix}\right\rangle$ belongs to the irrep $[\sigma]$ of $S^{\chi}(f)$ and at the same time is the Yamanouchi basis $[\sigma']m_1'$ and $[\sigma'']m_1''$ of the group $S^{\chi}(f_1)$ and $S^{\chi}(f_2)$ respectively.

The $S(f) \supset S(f_1) \times S(f_2)$ ISF are defined as the coefficients in the following expansion:

$$\left|\begin{matrix}[\nu]\beta\\ \tau[\nu']m'[\nu'']m''\end{matrix}\right\rangle = \sum_{\substack{\sigma\,\sigma'\theta\,\beta'\\ \mu'\mu''\phi\,\beta''}} C^{[\nu]\beta,\ \tau[\nu']\beta'\ [\nu'']\beta''}_{[\sigma]\theta\sigma'\sigma'',\ [\mu]\phi\ \mu'\mu''} \left|(\sigma'\sigma'')\theta(\mu'\mu'')\phi\beta'\beta''\right\rangle , \tag{2}$$

$$\left|(\sigma'\sigma'')\theta(\mu'\mu'')\phi\beta'\beta''\right\rangle = \left[\left|\begin{matrix}[\sigma]\\ \theta[\sigma'][\sigma'']\end{matrix}\right\rangle \left|\begin{matrix}[\mu]\\ \phi[\mu'][\mu'']\end{matrix}\right\rangle\right]^{[\nu']\beta'\ [\nu'']\beta''}_{m'\qquad m''} \tag{3}$$

where the square bracket indicates that the bases are to be combined into the irreducible basis $[\nu']m'$ and $[\nu'']m''$ of $S^{q}(f_1)$ and $S^{q}(f_2)$ by means of the CGC $C^{[\nu']\beta',m'}_{\sigma'm_1',\mu'm_2'}$ and $C^{[\nu'']\beta'',m''}_{\sigma''m''_1,\mu''m_2''}$ of $S(f_1)$ and $S(f_2)$ respectively. According to the fundamental theorem in section 1, Eq. (2) must be an eigenfunction of the CSCO-I of $S^{q}(f)$:

$$C(f)\left|\begin{matrix}[\nu]\\ \tau[\nu']m'[\nu'']m''\end{matrix}\right\rangle = \lambda^{(\nu)}\left|\begin{matrix}[\nu]\\ \tau[\nu']m'[\nu'']m''\end{matrix}\right\rangle . \tag{4}$$

From Eqs. (2) and (4), it follows that

$$\sum_{\substack{\sigma'\sigma''\theta\beta'\\ \mu'\mu''\phi\beta'}} \left(\langle(\bar\sigma'\bar\sigma'')\bar\theta(\bar\mu'\bar\mu'')\bar\phi\ \bar\beta'\bar\beta''|\ C(f)\,|(\sigma'\sigma'')\theta(\mu'\mu'')\phi\ \beta'\beta''\rangle - \lambda^{(\nu)}\delta_{\bar K K}\right)$$
$$\times\ C^{[\nu]\beta,\ \tau[\nu']\beta'[\nu'']\beta''}_{[\sigma]\theta\ \sigma'\sigma'',[\mu]\phi\ \mu'\mu''} = 0 \tag{5}$$

$$\delta_{\bar K K} \equiv \delta_{\bar\beta'\beta'}\delta_{\bar\beta''\beta''}\delta_{\bar\sigma'\sigma'}\delta_{\bar\mu'\mu'}\ \delta_{\bar\sigma''\sigma''}\ \sigma_{\bar\mu''\mu''}\delta_{\bar\theta\ \theta}\ \delta_{\bar\phi\ \phi} .$$

The $S^{\chi}(f) \supset S^{\chi}(f_1) \times S^{\chi}(f_2)$ and $S^{\xi}(f) \supset S^{\xi}(f_1) \times S^{\xi}(f_2)$ basis in the right hand side of Eq. (3) can be transformed into the Yamanouchi basis of $S^{\chi}(f)$ and $S^{\xi}(f)$, respectively, by means of the transformation coefficients of the permutation group[4-5]. The matrix element of any permutation p of $S^{q}(f)$ between states of Eq. (3) can be expressed as

$$\langle(\bar\sigma'\bar\sigma'')\bar\theta(\bar\mu'\bar\mu'')\bar\phi\ \bar\beta'\bar\beta''|P|(\sigma'\sigma'')\theta(\mu'\mu'')\phi\beta'\beta''\rangle =$$

$$\sum\ D^{[\sigma]}_{\bar m_1 m_1}(P)\ D^{[\mu]}_{\bar m_2 m_2}(P)\ C^{[\nu']\bar\beta',m'}_{\bar\sigma'\bar m'_1,\bar\mu'\bar m_2'}\ C^{[\nu'']\bar\beta'',m''}_{\bar\sigma''\bar m_1'',\ \bar\mu''\bar m_2''}\ C^{[\nu']\beta',m'}_{\sigma m_1',\ \mu'm_2'}\ C^{[\nu'']\beta'',m''}_{\sigma''m_1'',\ \mu''m_2''}$$

$$\times \begin{pmatrix} [\sigma] \\ \overline{m}_1 \end{pmatrix} \Big| [\sigma], \begin{matrix} \overline{\theta}[\overline{\sigma}'][\overline{\sigma}''] \\ \overline{m}_1' \;\; \overline{m}_1'' \end{matrix} \Big) \Big(\begin{matrix} [\mu] \\ \overline{m}_2 \end{matrix} \Big| [\mu], \begin{matrix} \overline{\phi}[\overline{\mu}'][\overline{\mu}''] \\ \overline{m}'_2 \;\; \overline{m}_2'' \end{matrix} \Big) \Big(\begin{matrix} [\sigma] \\ m_1 \end{matrix} \Big| [\sigma], \begin{matrix} \theta[\sigma'][\sigma''] \\ m_1' \;\; m_1'' \end{matrix} \Big)$$

$$\times \Big(\begin{matrix} [\mu] \\ m_2 \end{matrix} \Big| [\mu] \begin{matrix} \phi[\mu'][\mu''] \\ m'_2 \;\; m_2'' \end{matrix} \Big) \tag{6}$$

where the sum runs over $\overline{m}_1'$ $\overline{m}_2'$ $\overline{m}_1''$ $\overline{m}_2''$ m_1' m_2' m_1'' m_2'' $\overline{m}_1$ m_1 $\overline{m}_2$ and m_2; $D^{[\;]}_{\overline{m}_1 m_1}(P)$ etc are the irreducible matrix elements of S(f) and $\Big(\begin{matrix} [\sigma] \\ m_1 \end{matrix} \Big| [\sigma], \begin{matrix} \theta[\sigma'][\sigma''] \\ m_1' \;\; m_1'' \end{matrix} \Big)$ etc. are the transformation coefficients of the permutation group.

With the help of Eq. (6), one can calculate the matrix elements of the CSCO-I of the permutation group $S^q(f)$. Therefore from the CGC of $S(f_1)$ and $S(f_2)$ and the transformation coefficients of the permutation group, which can also be evaluated by the Eigenfunction Method[5], one obtains the matrix elements occurring in Eq. (5). Solving the eigenequation (5), one gets the $S(f) \supset S(f_1) \times S(f_2)$ ISF.

A computer program has already been set up[3] for calculating the CGC of the permutation group by the Eigenfunction Method. It is straightforward to transplant this program to the case of eigenequation (5) for the $S(f) \supset S(f_1) \times S(f_2)$ ISF.

3. SU(mn) $\supset$ SU(m) x SU(n) ISF

Let

$$\Big| \begin{matrix} [\nu'] \\ \beta'[\sigma']W_1'[\mu']W_2' \end{matrix} \Big\rangle, \quad \Big| \begin{matrix} [\nu''] \\ \beta''[\sigma'']W_1''[\mu'']W_2'' \end{matrix} \Big\rangle, \quad \Big| \begin{matrix} [\nu] \\ \beta[\sigma]W_1[\mu]W_2 \end{matrix} \Big\rangle, \tag{7}$$

be the SU(mn) $\supset$ SU(m) x SU(n) irreducible bases in the q-space for particles $(1,2...f_1)$ $(f_1+1,...f)$ and $(1,2...f)$ respectively, and W_1' (W_2') etc. be the component indices of the irreps of SU(m) (SU(n)). The SU(mn) $\supset$ SU(m) x SU(n) ISF are defined as the coefficients in the following expansion:

$$\Big| \begin{matrix} [\nu]\tau \\ \beta[\sigma]W_1[\mu]W_2 \end{matrix} \Big\rangle = \sum_{\substack{\beta'\sigma'\mu'\theta \\ \beta''\sigma''\mu''\phi}} C^{[\nu]\tau,\beta[\sigma]\theta \quad [\mu]\phi}_{[\nu']\beta'\sigma'\mu', \;\; [\nu'']\beta''\sigma''\mu''}$$

$$\times \Bigg[\Big| \begin{matrix} [\nu'] \\ \beta'[\sigma'][\mu'] \end{matrix} \Big\rangle \Big| \begin{matrix} [\nu''] \\ \beta''[\sigma''][\mu''] \end{matrix} \Big\rangle \Bigg]^{[\sigma]\theta \;\; [\mu]\phi}_{W_1 \;\; W_2}, \tag{8}$$

where $\tau = 1,2,...\{\nu'\nu''\nu\}$ is the multiplicity label, and the square bracket indicates that the bases are to be combined into the irreducible bases $[\sigma_1]W_1$ and $[\sigma_2]W_2$ of SU(m) and SU(n) in terms of the CGC of SU(m) and SU(n), respectively.

Attaching the Young tableaux $Y^{[\nu']}_{m'}(\omega_1)$ and $Y^{[\nu'']}_{m''}(\omega_2)$ with $(\omega_1) = (1,2,...f_1)$ and $(\omega_2) = (f_1+1,...f)$ to the two irreducible bases in the right-hand side of the Eq. (8), it reads

$$\Big| \tau[\nu']m'[\nu'']m'', \begin{matrix} [\nu] \\ \beta[\sigma]W_1[\mu]W_2 \end{matrix} \Big\rangle \tag{9a}$$

$$= \sum_{\substack{\beta'\sigma'\mu'\theta \\ \beta''\sigma''\mu''\phi}} C^{[\nu]\tau \;\beta[\sigma]\theta[\mu]\phi}_{[\nu']\beta'\sigma'\mu', \;\; [\nu'']\beta''\sigma''\mu''} \Bigg[\Big| m'\omega_1, \begin{matrix} [\nu'] \\ \beta'[\sigma'][\mu'] \end{matrix} \Big\rangle \Big| m''\omega_2, \begin{matrix} [\nu''] \\ \beta''[\sigma'][\mu'] \end{matrix} \Big\rangle \Bigg]^{[\sigma]_\theta [\mu]_\phi}_{W_1 \;\; W_2}$$

The left-hand side of Eq. (9a) is still the SU(mn) $\supset$ SU(m) x SU(n) basis. It belongs to the irrep [ν] of SU(mn), therefore it must also belong to irrep [ν] of the permutation group $S^q(f)$[1a]. In other words it is also a $S^q(f) \supset S^q(f_1) \times S^q(f_2)$ basis.

The CGC of the permutation group are known[1a] as the coupling coefficients which couple the irreducible bases of SU(m) and SU(n) into those of SU(mn) and the CGC of the unitary group are the coupling coeficients which couple the irredicible basis of $S(f_1)$ and $S(f_2)$ into those of S(f). Using these facts the last factor in Eq. (9a) can be put into the form

$$\left[\left|\begin{matrix}[\nu']\\ m'\omega_1^\circ,\ \beta'[\sigma'][\mu']\end{matrix}\right\rangle \left|\begin{matrix}[\nu'']\\ m''\omega_2^\circ,\ \beta''[\sigma''][\mu'']\end{matrix}\right\rangle\right]^{[\sigma]\theta\ [\mu]\phi}_{W_1\quad W_2} \tag{9b}$$

$$= \sum_{m_1' m_2' m_1'' m_2''} C^{[\nu']\beta',m'}_{\sigma' m_1',\ \mu' m_2'}\ C^{[\nu'']\beta'',m''}_{\sigma'' m_1'',\ \mu'' m_2''} \left|\begin{matrix}[\sigma]\\ \theta[\sigma']m_1'[\sigma'']m_1'',W_1\end{matrix}\right\rangle \left|\begin{matrix}[\mu]\\ \phi[\mu']m_2'[\mu'']m_2'',W_2\end{matrix}\right\rangle$$

Comparing Eq. (9) with Eq. (2) one gets an important relation

$$C^{[\nu]\tau,\beta[\sigma]\theta[\mu]\phi}_{[\nu']\beta'\sigma'\mu',\ [\nu'']\beta''\sigma''\mu''} = C^{[\nu]\beta,\tau[\nu']\beta'[\nu'']\beta''}_{[\sigma]\theta\sigma'\sigma'',\ [\mu]\phi\mu'\mu''} \tag{10}$$

namely the SU(mn) ⊃ SU(m) x SU(n) ISF (or the f_2 - particle CFP (coefficients of fractional parentage)) are equal to the $S(f_1+f_2) \supset S(f_1) \times S(f_2)$ ISF.

Furthermore, since the value of $S(f_1+f_2) \supset S(f_1) \times S(f_2)$ ISF is independent of m and n, one arrives at the conclusion that the value of SU(mn) ⊃ SU(m) x SU(n) ISF is independent of m and n. The reason we failed to realize this obvious fact for so long is because we usually use concrete quantum numbers for a given m and n rather than the partitions to represent the irreps of SU(m) and SU(n). For example, in the case of SU(3), we use (λμ) (corresponding to partition [λ+μ,μ]) or the dimension of the irrep; for SU(2) we use the quantum number S or T. As a test of the above conclusion, in Table 1 we list the SU(6) ⊃ SU(3) x SU(2) ISF for $[\nu]=[21^3]$, (λμ)=(12), S=1/2 calculated by Chang et al.[6] and the SU(4) ⊃ SU(2) x SU(2) ISF for $[\nu]=[21^3]$, S=T=1/2 given by Jahn[7]. They are exactly the same.

Table 1. SU(mn) ⊃ SU(m) x SU(n) ISF $C^{[21^3],\ [32]\ [32]}_{[211][\sigma'][\mu'],\ [1][1][1]}$

		[22][31] $^{13}\Gamma$ (02)1	[31][22] $^{31}\Gamma$ (21)0	[31][31] $^{33}\Gamma$ (21)1	[σ'][μ'] $^{2T'+1\ 2S'+1}\Gamma$ (λ'μ')S'
[ν][σ][μ] $[\nu]^{2T+1\ 2S+1}\Gamma$ [ν](λμ)S	$[21^3][32][32]$ $[21^3]\ ^{22}\Gamma$ $[21^3](12)1/2$	$-\sqrt{\frac{1}{5}}$	$-\sqrt{\frac{1}{5}}$	$-\sqrt{\frac{3}{5}}$	

Therefore, every SU(mn) ⊃ SU(m) x SU(n) ISF with a particular m and n gives an infinite number of SU(m'n') ⊃ SU(m') x SU(n') ISF with m'=m, m+1,... and n'=n, n+1,....

Another point worth mentioning is that not every SU(mn) ⊃ SU(m) x SU(n) ISF can be deduced from the SU((m-1)n) ⊃ SU(m-1) x SU(n) ISF or the SU(m(n-1)) ⊃ SU(m) x SU(n-1) ISF. The reason is that the Young diagrams [σ'], [σ''] and [σ] of SU(m-1) can have at most m-1 rows, and the SU(mn) ⊃ SU(m) x SU(n) ISF with the Young diagrams [σ'], [σ''] and [σ] of m rows can not be deduced from the SU((m-1)n) ⊃ SU(m-1) x SU(n) ISF.

The relations between the SU(m+n) ⊃ SU(m) x SU(n) ISF and the outer-product reduction coefficients are very similar to those between the SU(mn) ⊃ SU(m) x SU(n) ISF and the CGC of the permutation group, which will be the subject of our next paper.

References

1. Jin-Quan Chen, Fang Wang and Mei-Juan Gao, Acta. Phys. Sinica 26, 307 (1977); 26, 427 (1977); 27, 31 (1978); 27, 203 (1978); 27, 237 (1978).
2. Jin-Quan Chen, Fan Wang and Mei-Juan Gao, Journal of Nanjing University, No. 2, 1977. ibid, No. 2 1978.
3. Jin-Quan Chen and Mei-Juan Gao, "Reduction Coefficients of Permutation Groups and Their Applications" (to be published by Beijing Academy Pub. Co.)
4. I. G. Kaplan, Zh. Eksp. Teor. Fiz. 41, 560 (1961).
5. Jin-Quan Chen, "A New Approach to Group Representation Theory" (to be published by Shanghai Science Pub. Co.).
6. Z. Y. Zhang and G. L. Li, Acta Phys. Sinica 26, 467 (1977).
7. H. A. Jahn, Proc. Roy. Soc. A 209, 502 (1951).

STRUCTURE OF ENVELOPING ALGEBRAS OF LOW-RANK GROUPS

M. Couture and R. T. Sharp
Physics Department, McGill University
Montreal, Quebec, Canada

1. Introduction

The structure of the enveloping algebras of Lie groups has been the subject of many investigations. It has long been known[1,9] that for a group G of rank ℓ there are just ℓ functionally independent invariant polynomials in the generators, or Casimir invariants. Over 15 years ago, Kostant[6] proved several important theorems on the structure of enveloping algebras; I shall refer to some of them during the talk. More recently[7], a complete description (degree and explicit algebraic forms) of a basis for vector operators in the enveloping algebras of A_n, B_n, C_n, D_n and G_2, in any given irreducible representation, was given by Okubo. The problem of labelling states in terms of a complete set of commuting operators, in the case where one uses a non canonical chain of groups (the missing label problem), has motivated the search for subgroup scalars in the enveloping algebra of a group; solutions[4,5] have been given for various group-subgroup combinations.

The object of our work is the reduction of enveloping algebras considered as group modules, in other words, to enumerate and find a basis for all tensors in the enveloping algebras of simple compact groups of rank $\leqslant 3$, that is: SU(2), SU(3), SO(5), G_2, SU(4), Sp(6) and SO(7).

2. Our approach to the reduction problem

In order to give a complete description of a basis for tensors in the enveloping algebra U of a group G, one must answer the following questions concerning any of its elements

(1) By which representation does it transform?
(2) What is its degree (its components being polynomials in the generators)?
(3) What is its multiplicity?
(4) How do we construct it?

Now, there follows, from the Poincaré-Birkhoff-Witt theorem and from the fact that the order in a product of generators does not affect its transformation properties under G, a one to one correspondence between the basis for tensors in the enveloping algebra of a group and the basis for tensors whose components are polynomials in the components of a tensor A that transforms by the adjoint representation of G (they will be referred to as polynomial tensors).

Answering the above four questions for polynomial tensors is therefore answering them for tensors in U since, once one knows the algebraic form of a polynomial tensor, the corresponding tensor in U is obtained through symmetrization (with respect to order).

It turns out that the answer to all these questions is given in terms of a generating function (GF). Before discussing methods of constructing these functions, let us look at the GF for SO(5).

3. Structure of the enveloping algebra of SO(5)

The GF giving a basis for tensors in the enveloping algebra of SO(5) (which is identical to the GF for polynomial tensors based on a tensor that transforms by the (2,0) representation (Cartan labels) is[3]

$$G(U;\Lambda_1,\Lambda_2) = \frac{1+U^4\Lambda_1^2\Lambda_2}{(1-U^2)(1-U^4)(1-U\Lambda_1^2)(1-U^2\Lambda_2)(1-U^2\Lambda_2^2)(1-U^3\Lambda_1^2)}; \qquad (3.1)$$

U carries the degree, and Λ_1,Λ_2 carry the SO(5) representation labels (Cartan labels) of the tensors as their exponents, A term $c^u_{\lambda_1\lambda_2}U^u\Lambda_1^{\lambda_1}\Lambda_2^{\lambda_2}$ in the expansion of (3.1) informs us that the number of linearly independent irreducible SO(5) tensors in the enveloping algebra of SO(5), which transform by the (λ_1,λ_2) representation and whose components are symmetric homogeneous polynomials of degree u in the generators is $c^u_{\lambda_1\lambda_2}$. For example, collecting all terms (omitting products with Casimirs) containing $\Lambda_1^2\Lambda_2^2$ in the expansion of (3.1) we get

$$\Lambda_1^2\Lambda_2^2(U^3+2U^5+U^6+U^7), \qquad (3.2)$$

which indicates the presence of five linearly independent tensors of degree 3,5 (multiplicity 2), 6 and 7; this agrees with a theorem due to Kostant which states,that the number p_λ of independent λ tensors (which transform by the (λ) representation) in the enveloping algebra is equal to the number of states of zero weight in the representation (λ); the representation (2,2) has 5 states of zero weight. Kostant also shows that the highest degree of a λ tensor (modulo multiplying it by Casimir operators) is the sum of the coefficients of the simple roots in the highest weight of (λ); in the case of SO(5) and in terms of Cartan labels, this highest degree is $3/2\lambda_1+2\lambda_2$, which predicts for the (2,2) tensor a highest degree of 7.

The GF(3.1) not only enumerate all tensors but also suggests an integrity basis, i.e., a finite set of elementary tensors in terms of which all may be obtained as stretched tensor products. Denoting by (u,λ_1,λ_2) the tensors enumerated in (3.1), where u is the degree and λ_1,λ_2 the representation labels, the integrity basis consists of the quadratic and quartic Casimir invariants (2,0,0) and (4,0,0), two decuplets of degree 1 and 3, (1,2,0) and (3,2,0), a quintet (2,0,1) and a 14-plet (2,0,2) each of degree 2, and a 35-plet (4,2,1) of degree 4. The structure of (3.1) tell us that any element of the basis may be obtained by the following stretched products of powers (representation labels and degrees are additive) of the elementary tensor operators

$$(2,0,0)^a.(4,0,0)^b.(1,2,0)^c.(2,0,1)^d.(2,0,2)^e.(3,2,0)^f.(4,2,1)^g \qquad (3.3)$$

where a, b, c, d, e and f may take values from 0 to ∞ and g may be only 0 or 1. For example the two (2,2) tensors of degree 5 are obtained by the following stretched products (2,0,2).(3,2,0) and $(1,2,0).(2,0,1)^2$.

From the general product (3.3) it is easy to show that the formula giving the degree m_{ij} (they correspond to Kostant's generalized exponents) for all independent tensors in the enveloping algebra of SO(5) (modulo multiplying it by Casimir operators) is

$$m^{(1)}_{ij} = \tfrac{1}{2}\lambda_1 + 2\lambda_2 + 2i-2j \qquad i = 0,1,\ldots,\tfrac{1}{2}\lambda_1$$
$$\qquad j = 0,1,\ldots,\tfrac{1}{2}\lambda_2 \quad (\tfrac{1}{2}(\lambda_2-1) \text{ for } \lambda_2 \text{ odd}),$$
$$m^{(2)}_{ij} = \tfrac{1}{2}\lambda_1 + 2\lambda_2 + 2i-2j+1 \qquad i = 0,1,\ldots,\tfrac{1}{2}\lambda_1-1$$
$$\qquad j = 0,1,\ldots,\tfrac{1}{2}\lambda_2-1 \quad (\tfrac{1}{2}(\lambda_2-1) \text{ for } \lambda_2 \text{ odd}).$$

There are no tensors with odd λ_1.

4. Methods of constructing generating functions for polynomial tensors

A possible approach to this problem is the one proposed by Gaskell, Peccia and Sharp[4] which starts with a weight GF. Unfortunately the tedium of the method increases rapidly with the number of generators. Two alternative approaches are considered.

The first one makes use of a larger group. The polynomial tensors of degree u in the components of a tensor A (A transforms by the adjoint representation of G) are precisely the multiplets (representations of G) contained in the representation (u 0....0) of SU(r), r being the order G, i.e., the number of its generators. Hence the GF for polynomial tensors based on A is that for the branching rules SU(r) G, restricted to one-rowed representations of SU(r). A technique, first proposed by Patera[8] and Sharp in the evaluation of GF's for general plethysms, simplifies the calculations: it consists, whenever possible of inserting a group G' in the chain $SU(r)\supset G$, i.e., to consider the chain $SU(r)\supset G'\supset G$; one finds the GF for $SU(r)\supset G'$ and for $G'\supset G$ and substitutes the latter in the former. For example, in the case of SO(5) one may use the chain $SU(10)\supset SU(5)\supset SO(5)$; the embedding is such that (10,...0) of SU(10) contains (0100) of SU(5) which contains (20) of SO(5). An alternative chain is

$SU(10) \supset SU(4) \supset SO(5)$ with the embedding $(10\ldots0) \supset (200) \supset (20)$.

The other approach, which we believe to be novel is to work through a subgroup H of G. The tensor A is a reducible tensor of H; denoting by $\Gamma_H^1, \Gamma_H^2, \ldots, \Gamma_H^n$ the n irreducible subgroup tensors into which A reduces, it may be relatively easy to construct the GF for H tensors based on the n tensors Γ_H^i. Under certain circumstances it may be possible to convert that subgroup GF into the corresponding GF for G tensors based on A. A necessary tool in doing this conversion is the group-subgroup characteristic function; the role it plays in the conversion parallels that if Weyl's characteristic function in converting a weight GF into the corresponding GF for group tensors. Details may be found in our paper[3] on the reduction of enveloping algebras or in the thesis[2] of one of us (M.C.).

5. Realizations and representations of tensors in enveloping algebras

Can one find a representation in which all (λ_1,λ_2) tensors (modulo multiplying them by Casimirs) enumerated in a GF for tensors in the enveloping algebra of a group, for any fixed value of λ_1 and λ_2, exist and have components whose matrix elements are linearly independent? Kostant has proven that such a representation exists and that actually, there are an infinite number of them for any fixed value of λ_1 and λ_2. However, this isn't true in all representations: he showed that the multiplicity of a λ tensor (modulo multiplying it by Casimirs) in a representation (ν) is equal to the multiplicity of (ν) in the Clebsch Gordan series of $(\lambda)x(\nu)$, which implies that the multiplicity of a λ tensor isn't the same in all representations). Taking for example the SO(5) group, certain tensors enumerated in the GF(3.1), in the representation $(\lambda,0)$ or $(0,\lambda)$, no longer exist (their matrix elements are zero) or are no longer linearly independent. In order to find if a tensor should be omitted from the GF, one constructs the algebraic form (homogeneous symmetric polynomials in the generators) of its highest component, and then substitutes for the generators a certain realization of them (differential operators) proper to the $(\lambda,0)$ or $(0,\lambda)$ representation. When one does this for each elementary tensor, relations appear among certain of them. Here one must make a distinction between linear independence in a certain realization and linear independence in terms of matrix elements: two tensors may be linearly independent in a certain realization (the algebraic forms in tèrms of differential operators of their components, are linearly independent) although their matrix elements aren't; this may happen when they differ by some group scalar which isn't in the enveloping algebra. Consequently, for certain groups (such as SO(3)), one is led to two types of GF's: one enumerating all linearly independent tensors in a certain realization and a different one enumerating all tensors (modulo multiplication by Casimirs) whose matrix elements are linearly independent. No such distinction is required for SO(5) and the GF (3.1) reduces to the following GF.

$$\frac{1 + U^2\Lambda_2}{(1-U^2)(1-U\Lambda_1^2)(1-U^2\Lambda_2^2)} \qquad \text{for } (\lambda,0)$$

and

$$\frac{1}{(1-U^2)(1-U\Lambda_1^2)(1-U^2\Lambda_2^2} \qquad \text{for } (0,\lambda)\,.$$

REFERENCES

1. Borel A, and Chevalley C, 1955 Amer. Math. Soc. Mem. 14 1.
2. Couture M, 1980 Ph.D. Thesis McGill University, to be submitted.
3. Couture M, and Sharp R T, 1980 J. Phys. A: Math. Gen. 13 1925.
4. Gaskell R, Peccia A, and Sharp R T, 1978 J. Math. Phys. 19 727.
5. Judd B R, Miller W, Patera J, and Winternitz P, 1974, J. Math. Phys. 15, 1787; Quesne C, J. Math. Phys. 17, 1452, 18, 1210; Sharp R T, 1975, J. Math. Phys. 16 2050.
6. Kostant B, 1963 Amer. J. Math. 85, 327.
7. Okubo S, 1977, J. Math. Phys. 18, 2382.
8. Patera J, and Sharp R T, J. Phys. A: Math. Gen 13, 397.
9. Samuelson H, 1941, Ann. Math. 42, 1091.

COMPUTER PROGRAMS FOR THE REDUCTION OF SYMMETRISED nth POWERS OF SPACE GROUP IRREDUCIBLE REPRESENTATIONS USING GARD'S SUBGROUP METHOD.

B.L. Davies
School of Mathematics and Computer Science, University College of North Wales, BANGOR LL57 2UW, Wales, U.K.

and

A.P. Cracknell
Carnegie Laboratory of Physics, University of Dundee, DUNDEE DD1 4HN, Scotland, U.K.

1. Introduction.

At several previous Colloquia in this series we have reported on the work which we have in progress to reduce the Kronecker products of all the (induced) irreducible representations of the 230 space groups into their irreducible component representations using a computer. This work has now been completed and the results have recently been published (1 - 3). Our present work is concerned with the reduction of symmetrised powers of (induced) irreducible representations of space groups.

2. Summary of Theory.

The Kronecker nth power of an induced representation of a group $\underset{\sim}{G}$ has been shown by Gard (4) to be capable of reduction in terms of the symmetric group ($\underset{\sim}{S}_n$) of degree n so that, within each symmetry class, the reduction is expressed as a sum of induced representations. The symmetry classes Ω^ν, which are in one-to-one correspondence with the unitary irreducible representations $[\nu]$ of $\underset{\sim}{S}_n$, are carrier spaces for the corresponding symmetrised nth powers. By a well known result due to Mackey (5) a partial decomposition of the Kronecker nth power of an induced representation, carried by the space Ω, may be obtained through double coset decompositions. Thus Ω may be decomposed into a direct sum of subspaces $\Omega_{(\alpha)}$ where (α) denotes an ordered n-tuple of double coset representatives. An action of the elements of $\underset{\sim}{S}_n$ on each subspace $\Omega_{(\alpha)}$ is defined by permuting the order of the double coset representatives in the n-tuple (α). The n! subspaces obtained, which may not all be distinct, are all present in the direct sum decomposition of Ω and carry equivalent representations. The set of distinct subspaces obtained is said to form an orbit under $\underset{\sim}{S}_n$ and all the subspaces $\Omega_{(\alpha)}$ are partitioned into disjoint orbits. The direct sum of the subspaces in an orbit, denoted by $T(\alpha)$, is invariant under both $\underset{\sim}{G}$ and $\underset{\sim}{S}_n$ and so $T(\alpha)$ may be separately decomposed into its symmetrised parts $T^\nu(\alpha)$ where $[\nu]$ runs through the unitary irreducible representations of $\underset{\sim}{S}_n$. Furthermore, the representation carried by $T^\nu(\alpha)$ is induced from a representation $\Gamma_\nu(\alpha)$ of a subgroup $\underset{\sim}{M}(\alpha)$ of $\underset{\sim}{G}$. The character χ_ν of $\Gamma_\nu(\alpha)$ is rather complicated and is given by equations (6.14) and 6.15) of Gard (4) as amended by Backhouse and Gard (6). Thus the symmetrised components of the Kronecker nth power of an induced representation are obtained as the direct sum (over disjoint orbits) of induced representations.

A unitary irreducible representation $\Delta^{\underset{\sim}{k}_i}_p = (\Gamma^{\underset{\sim}{k}_i}_p \uparrow \underset{\sim}{G})$ of a space group $\underset{\sim}{G}$ is induced from a small (or allowed) unitary representation $\Gamma^{\underset{\sim}{k}_i}_p$ of a subgroup of $\underset{\sim}{G}$ called the little group $\underset{\sim}{G}^{\underset{\sim}{k}_i}$, where $\underset{\sim}{k}_i$ is a vector in the representation domain Φ of the first Brillouin zone and p is a label used to distinguish among the different small representations labelled by $\underset{\sim}{k}_i$. The Kronecker nth power of $\Delta^{\underset{\sim}{k}_i}_p$ is a direct sum of symmetrised nth powers $(\Delta^{\underset{\sim}{k}_i}_p)^{\nu}$, where $[\nu]$ runs through the unitary irreducible representations of $\underset{\sim}{S}_n$. Each $(\Delta^{\underset{\sim}{k}_i}_p)^{\nu}$ may be decomposed into a direct sum of irreducible representations, so that

$$(\Delta^{\underset{\sim}{k}_i}_p)^{\nu} = \sum_{\gamma,r} C^{\underset{\sim}{k}_i,\underset{\sim}{k}^{\gamma}_{\ell}}_{p,r}(\nu)\ \Delta^{\underset{\sim}{k}^{\gamma}_{\ell}}_r \tag{1}$$

where $\underset{\sim}{k}^{\gamma}_{\ell}\varepsilon\Phi$, r indexes the different small representations labelled by $\underset{\sim}{k}^{\gamma}_{\ell}$ and γ runs through the orbits. The superscript γ is omitted in the following for convenience.

Let (α) be a representative n-tuple labelling an orbit then

$$(\alpha) = (d_{\alpha_{n-1}},\ldots,d_{\alpha_1},1) \tag{2}$$

where $d_{\alpha_i} = \{T_i|\underset{\sim}{x}_i\}$, $(i = 1,2,\ldots,n-1)$ and $1 = \{E|\underset{\sim}{O}\}$. The group $\underset{\sim}{M}(\alpha)$ is a subgroup of $\underset{\sim}{G}^{\underset{\sim}{q}}$ where

$$\underset{\sim}{q} = \underset{\sim}{k}_i + T_1\underset{\sim}{k}_i + \ldots + T_{n-1}\underset{\sim}{k}_i . \tag{3}$$

The vector $\underset{\sim}{q}$ may or may not lie in Φ; however, there exists an element $\{R|\underset{\sim}{v}\}\varepsilon\underset{\sim}{G}$ and $\underset{\sim}{k}_{\ell}\ \varepsilon\Phi$ such that

$$R\underset{\sim}{q} \cong \underset{\sim}{k}_{\ell} \tag{4}$$

where $\cong$ denotes equivalence. The vectors $\underset{\sim}{k}_{\ell}$ in equation (4) are precisely those that appear on the right hand side of equation (1). Equations (3) and (4) lead to the definition of a symmetrised wave vector selection rule (SWVSR) given by

$$R_o\underset{\sim}{k}_i + R_1\underset{\sim}{k}_i + \ldots + R_{n-1}\underset{\sim}{k}_i \cong \underset{\sim}{k}_{\ell} \tag{5}$$

where $R_o = R$ and $R_i = RT_i$ $(i = 1,2,\ldots,n-1)$. Thus there is a one-to-one correspondence between the orbits and the SWVSRs.

Equation (5) implies that the determination of the SWVSRs need only be carried out for the symmorphic space groups. The SWVSRs for a non-symmorphic space group $\underset{\sim}{G}$ are identical to those for the symmorphic space group that is obtained by replacing $\underset{\sim}{v}_i$ by $\underset{\sim}{O}$ in each of the elements $\{R_i|\underset{\sim}{v}_i\}$ of $\underset{\sim}{G}$.

Having found the vectors $\underset{\sim}{k}_{\ell}\ \varepsilon\Phi$ for each orbit, the coefficients $C^{\underset{\sim}{k}_i,\underset{\sim}{k}_{\ell}}_{p,r}(\nu)$ in equation (1) are found as follows

$$C^{\underset{\sim}{k}_i,\, \underset{\sim}{k}_\ell}_{p,r}(\nu) = \frac{|\underset{\sim}{T}|}{|\underset{\sim}{G}^{\underset{\sim}{k}_\ell}|} \sum_{\{S|\underset{\sim}{w}\}\in \underset{\sim}{G}^{\underset{\sim}{k}_\ell}/\underset{\sim}{T}} \Theta_\nu^{\underset{\sim}{k}_\ell}(\{S|\underset{\sim}{w}\})\chi_r^{\underset{\sim}{k}_\ell *}(\{S|\underset{\sim}{w}\}) \tag{6}$$

where $\underset{\sim}{T}$ denotes the invariant abelian subgroup of translations and the summation is over the left coset representatives $\{S|\underset{\sim}{w}\}$ of $\underset{\sim}{T}$ in $\underset{\sim}{G}^{\underset{\sim}{k}_\ell}$. The asterisk denotes complex conjugation. The character $\Theta_\nu^{\underset{\sim}{k}_\ell}(\{S|\underset{\sim}{w}\})$ is given by

$$\Theta_\nu^{\underset{\sim}{k}_\ell}(\{S|\underset{\sim}{w}\}) = \Theta_\nu^{\underset{\sim}{q}}(\{R|\underset{\sim}{v}\}^{-1}\{S|\underset{\sim}{w}\}\{R|\underset{\sim}{v}\}) \tag{7}$$

for the $\{S|\underset{\sim}{w}\}$ in equation (6) where R is given by equation (4). The character of $\underset{\sim}{G}^{\underset{\sim}{q}}$ in the right hand side of equation (7) is induced from the character χ_ν of the representation $\Gamma_\nu(\alpha)$ of $\underset{\sim}{M}(\alpha)$.

3. Computer Programs and Results.

Two computer programs were written in ALGOL-60 for a DEC-system-10 computer. The first program, called the SWVSR program, determined for each special $\underset{\sim}{k}$-vector $\underset{\sim}{k}_i \varepsilon \Phi$ of a given symmorphic space group, the SWVSRs given by equation (5) which label each orbit. The second program, called the SKP program, was run for all special $\underset{\sim}{k}$-vectors $\underset{\sim}{k}_i \varepsilon \Phi$ for all space groups and determined the reduction of the SKPs (symmetrised Kronecker powers) $(\Delta_p^{\underset{\sim}{k}_i})^\nu$ given by equation (1) using the formula for the coefficients given by equation (6).

The programs are quite general so that the value of n is read in as part of the input data. Complete reductions for n = 2,3 and 4 have been obtained for all 230 space groups. Gard's method is a generalisation of the work of Mackey (7) and Bradley and Davies (8) for n = 2. At the Seventh Group Theory Colloquium in Austin we reported on the adaptation of our Kronecker products program to yield the reductions of the totally symmetrised and totally antisymmetrised squares (9). The complete reductions obtained for all space groups checked exactly with those obtained by the SKP program for n = 2. We have also checked our results, wherever possible, against tables of symmetrised powers of the irreducible representations of individual space groups in the literature.

The reductions for n = 2,3 have direct application to the Landau theory of second order phase transitions and these results will constitute the fourth volume of the Kronecker products tables series (10). It is planned to deposit the reductions for n = 4 in the British Library under the Supplementary Publications Scheme (11).

References

1. A.P. Cracknell, B.L. Davies, S.C. Miller and W.F. Love, 1979, Kronecker product tables, volume 1, General introduction and tables of irreducible representations of space groups, Plenum Press, New York.

2. B.L. Davies and A.P. Cracknell, 1979, Kronecker product tables, volume 2, Wave vector selection rules and reductions of Kronecker products for irreducible representations of orthorhombic and cubic space groups, Plenum Press, New York.

3. A.P. Cracknell and B.L. Davies, 1979, Kronecker product tables, volume 3, Wave vector selection rules and reductions of Kronecker products for irreducible representations of triclinic, monoclinic, tetragonal, trigonal, and hexagonal space groups, Plenum Press, New York.

4. P. Gard, 1973, J. Phys. A: Math., Nucl. Gen., 6, 1807-1828.

5. G.W. Mackey, 1951, Am. J. Math., 73, 576-92.

6. N.B. Backhouse and P. Gard, 1974, J. Phys. A: Math., Nucl. Gen., 7, 1239-1250.

7. G.W. Mackey, 1953, Am. J. Math., 75, 387-405.

8. C.J. Bradley and B.L. Davies, 1970, J. Math. Phys., 11, 1536-1552.

9. B.L. Davies and A.P. Cracknell, 1979, Lecture notes in physics, 94, Group theoretical methods in physics, Seventh International Colloquium and Integrative Conference on Group Theory and Mathematical Physics, Austin, 1978, 67-69 (Springer, Berlin).

10. B.L. Davies and A.P. Cracknell, Kronecker product tables, volume 4, Symmetrized powers of irreducible representations of space groups, Plenum, New York.

11. B.L. Davies and A.P. Cracknell, British Library, Lending Division, Boston Spa, Wetherby, West Yorks, LS23 7BQ, U.K., Supplementary Publications Scheme, No. SUP 90047.

SOME REMARKS ON THE UNIQUENESS AND REALITY OF CLEBSCH-GORDAN COEFFICIENTS FOR COREPRESENTATIONS

Rainer Dirl
Institut für Theoretische Physik, TU Wien
A-1040 Wien, Karlsplatz 13; Austria

It is reported on, that CG-coefficients for ordinary representations are unique up to arbitrary unitary transformations concerning their multiplicity, whilst such coefficients for corepresentations are unique up to three different groups of transformations (real orthogonal, real symplectic and unitary) according to their respective irreducible corepresentations. Moreover restrictive conditions are specified, which guarantee, that CG-coefficients for ordinary and in particular for corepresentations can be transformed into some real ones.

Preliminary remarks: Let $G = H \cup \{s\}H$ be a finite group, which contain H as subgroup of index two. A unitary matrix representation $B(g);g \in G$ of G over $\mathbb{C}$ is called a unitary corepresentation, if

$$B(g)\,B(g')^{g} = B(gg') \qquad \text{for all } g,g' \in G \tag{1}$$

holds, where the superscript g implies complex conjugation of $B(g')$, if $g \in \{s\}H$, respectively leaves $B(g')$ unchanged, if $g \in H$. One distinguishes three different types of counirreps , at which the following convention is adopted: We denote counirreps of G by $B^{\alpha}(g);g \in G$ and unirreps of H by $D^{\alpha}(h);h \in H$. Furthermore "complex conjugation by s" being inherent to the definition of counirreps of G, devides the set A_H of equivalence classes of H into three disjoint subsets $A_{K(s)}$; K = I,II,III.

$$Z^{\alpha\dagger} D^{\bar{\alpha}}(h)\, Z^{\alpha} = D^{\alpha}(s^{-1}hs)^{*} = D^{\alpha(1)}(h) \qquad \text{for all } h \in H \tag{2}$$

$$\begin{aligned} &\underline{\text{type I(s)}}: && Z^{\alpha} Z^{\alpha *} = +\,D^{\alpha}(s^2) && (\alpha = \bar{\alpha} \in A_{I(s)}) \\ &\underline{\text{type II(s)}}: && Z^{\alpha} Z^{\alpha *} = -\,D^{\alpha}(s^2) && (\alpha = \bar{\alpha} \in A_{II(s)}) \\ &\underline{\text{type III(s)}}: && Z^{\alpha} Z^{\alpha \dagger} = 1_{\alpha} && (\alpha \neq \bar{\alpha} \in A_{III(s)}) \end{aligned} \tag{3}$$

For the sake of clearness we use in (3) a somewhat extended notation for the equivalence classes, whose meaning should be $\alpha(0) = \alpha$ and $\alpha(1) = \bar{\alpha}$. Moreover it is assumed, that counirreps of G are always given in "standard form".

On the uniqueness of CG-coefficients for G: Due to a general method of calculating CG-coefficients for ordinary and in particular for corepresentations [1], one considers the columns of unitary CG-matrices as orthonormalized vectors of appropriated defined Euclidean spaces, which have to transform with respect to the Kronecker products $B^{\alpha\alpha'}(g) = B^{\alpha}(g) \otimes B^{\alpha'}(g);g \in G$ according to

$$B^{\alpha\alpha'}(g)(\vec{W}_{A''}^{\alpha\alpha';\alpha''w})^g = \sum_{B''} B^{\alpha''}_{B''A''}(g)\, \vec{W}_{B''}^{\alpha\alpha';\alpha''w} \quad ; \quad \alpha'' \in A_{K''(s)}, \quad K'' = I,II,III \qquad w = 1,2, \ldots M_{\alpha\alpha';\alpha''} \tag{4}$$

and whose components are nothing else than the corresponding CG-coefficients for G. In particular A" is a double index, if the corresponding counirrep is of type II(s) or type III(s). Collecting on the other hand "convenient" CG-coefficients for H (whose precise definition is given in Ref.[1]), which are denoted hereafter by

$$\begin{pmatrix} \alpha(k) & \alpha'(k') \\ j & j' \end{pmatrix}\Bigg| \begin{matrix} \alpha'' & v \\ j'' & \end{matrix}\Bigg) = \{\vec{C}_{j''}^{\alpha(k),\alpha'(k');\alpha''v}\}_{jj'} \quad ; \quad k,k' = 0,1 \qquad v = 1,2, \ldots m(\alpha(k),\alpha'(k');\alpha'') \qquad j(j',j'') = 1,2, \ldots n_\alpha(n_{\alpha'},n_{\alpha''}) \tag{5}$$

to suitable defined vectors $\vec{P}_{j''}^{\alpha\alpha';\alpha''Mv}$ which have to transform according to

$$B^{\alpha\alpha'}(h)\, \vec{P}_{j''}^{\alpha\alpha';\alpha''Mv} = \sum_{k''} D^{\alpha''}_{k''j''}(h)\, \vec{P}_{k''}^{\alpha\alpha';\alpha''Mv} \tag{6}$$

Schurs Lemma with respect to H gives rise to identities linking type K(s)-CG-coefficients for G (K = I,II,III) with convenient ones for H. In particular M is in principle the double index (k,k');k = 0,1 , but whose actual meaning depends on the considered Kronecker products of counirreps. M is superfluous, if $\alpha,\alpha' \in A_{I(s)}$, a single index, if $\alpha \in A_{I(s)}$ and $\alpha' \in A_{II(s)} \cup A_{III(s)}$ (and conversly) and indeed a double index, if $\alpha,\alpha' \in A_{II(s)} \cup A_{III(s)}$.

<u>type I(s)</u>: $$\vec{W}_{j''}^{\alpha\alpha';\alpha''w} = \sum_{Mv} (B_o)_{Mv;w}\, \vec{P}_{j''}^{\alpha\alpha';\alpha''Mv} \tag{7}$$

<u>type II(s)</u>: $$\vec{W}_{a'',j''}^{\alpha\alpha';\alpha''w} = \sum_{Mv} (B_o)_{Mv;a''w}\, \vec{P}_{j''}^{\alpha\alpha';\alpha''Mv} \quad ; \quad a'' = 1,2 \tag{8}$$

<u>type III(s)</u>: $$\vec{W}_{1,j''}^{\alpha\alpha';\alpha''w} = \sum_{Mv} (B_o)_{Mv;w}\, \vec{P}_{j''}^{\alpha\alpha';\alpha''Mv}$$
$$\vec{W}_{2,j''}^{\alpha\alpha';\alpha''w} = \sum_{Mv} (F\, B_o^*)_{Mv;w}\, \vec{P}_{j''}^{\alpha\alpha';\bar{\alpha}''Mv} \tag{9}$$

Thereby the matrices B_o are special solutions of the following equations

<u>type I(s)</u>: $F\,B^* = B\,;\qquad F\,F^* = +\,1_{M\ldots}$

<u>type II(s)</u>: $F\,B^* = B\,G^t\,;\qquad F\,F^* = -\,1_{2M\ldots}\quad (G^2 = -\,1_{2M\ldots})$

<u>type III(s)</u>: $F\,B^* = C$ (10)

where the respective matrices F are uniquely determined through the corresponding "convenient" CG-coefficients for H and G is a special real skew-symmetric matrix.

In order to be able to make some statements concerning the uniqueness of type K(s)-CG-coefficients for G, it is reasonable to consider at first this problem for the "convenient" CG-coefficients for H, which are the non-zero components of the vectors $\vec{P}_{j''}^{\alpha\alpha';\alpha''Mv}$. Obviously CG-coefficients for H are unique up to arbitrary unitary transformations, which concern the "multiplicity index" v, i.e.

$$\vec{\tilde{C}}_{j}^{\alpha\alpha';\alpha''v} = \sum_{w} \beta_{wv}\, \vec{C}_{j}^{\alpha\alpha';\alpha''w} \quad ; \quad \beta \in U(n) \quad \text{with } n = m_{\alpha\alpha';\alpha''} \tag{11}$$

Moreover, if assuming that the "convenient" CG-coefficients for H are fixed, it follows immediately from (10)

$$\begin{aligned} &\underline{\text{type I(s)}}: && B = B_o\, R\ ; && R \in O(n, \mathbb{R}) \\ &\underline{\text{type II(s)}}: && B = B_o\, S\ ; && S \in Sp(2n, \mathbb{R}) \\ &\underline{\text{type III(s)}}: && C = F\, B^*\ ; && B \in U(n) \end{aligned} \tag{12}$$

at which $n = M_{\alpha\alpha';\alpha''}$, that type I(s)-CG-coefficients for G are unique up to arbitrary $M_{\alpha\alpha';\alpha''}$-dimensional real orthogonal transformations, type II(s)-CG-coefficients are unique up to arbitrary $2M_{\alpha\alpha';\alpha''}$-dimensional real symplectic transformations and type III(s)-CG-coefficients are unique up to arbitrary $M_{\alpha\alpha';\alpha''}$-dimensional unitary transformations, where additional the standard form of the counirreps is retained.

On the reality of CG-coefficients for G: For obvious reasons, it is worthwhile to consider this problem at first for CG-coefficients for H. For this purpose we define an auxiliary direct product group $H^{*} = H \times \{e,c\}$, at which in particular the "antiunitary" group element c should generate "ordinary complex conjugation". Quite analogous to "complex conjugation by s", we have

$$U^{\alpha\dagger}\ \overset{\star}{D}{}^{\alpha}(h)\ U^{\alpha} = D^{\alpha}(h)^{*} \qquad \text{for all } h \in H \tag{13}$$

$$\begin{aligned} &\underline{\text{type I}(*)}: && U^{\alpha}\, U^{\alpha *} = +\, 1_{\alpha} \\ &\underline{\text{type II}(*)}: && U^{\alpha}\, U^{\alpha *} = -\, 1_{\alpha} \\ &\underline{\text{type III}(*)}: && U^{\alpha}\, U^{\alpha \dagger} = 1_{\alpha} \end{aligned} \tag{14}$$

which devides A_H into three disjoint subsets $A_{K(*)}$; K = I,II,III. However one must be aware that these sets are in general quite different to $A_{K(s)}$. Utilizing the operation "ordinary complex conjugation", we obtain the following symmetry relations

$$\overset{\star}{\vec{C}}{}_{j}^{\alpha\alpha';\alpha''v} = \sum_{w} f_{wv}\ (U^{\alpha} \otimes U^{\alpha'}) \sum (U^{\alpha''}_{jk}\ \vec{C}_{k}^{\alpha\alpha';\alpha''w})^{*} \tag{15},$$

if $\alpha,\alpha',\alpha'' \in A_{I(*)} \cup A_{II(*)}$, respectively the following "generating relations"

$$\overset{\star\star}{\vec{C}}{}_{j}^{\overset{\star}{\alpha\alpha'};\alpha''v} = (U^{\alpha} \otimes U^{\alpha'}) \sum_{k} (U^{\alpha''}_{jk}\ \vec{C}_{k}^{\alpha\alpha';\alpha''v})^{*} \tag{16},$$

if at least one of the quantities α,α',α'' belongs to $A_{III(*)}$. Thereby the $m_{\alpha\alpha';\alpha''}$-dimensional unitary matrix f is uniquely determined through the corresponding CG-coefficients for H and has to satisfy additionally

$$f\ f^{*} = (-1)^{\gamma(\alpha)+\gamma(\alpha')+\gamma(\alpha'')}\ 1_{m\ldots} \tag{17},$$

at which $\gamma(\alpha) = 0$ for $\alpha \in A_{I(*)}$ and $\gamma(\alpha) = 1$ for $\alpha \in A_{II(*)}$. If carrying out a unitary similarity transformation (11), f transforms into $f' = \beta^{t}\, f\, \beta$, which presents a congruence transformation. Due to well known theorems of matrix calculus, f can be transformed by means of appropriated defined unitary β's either into the unit matrix 1_m, or into a special real skew-symmteric matrix $J_{m'}$; $m' = m_{\alpha\alpha';\alpha''}/2$, depending on the sign of the phase factor which appears in (17). Moreover, if passing over from $D^{\alpha}(h)$ by means of unitary similarity transformations Q^{α} to some new unirreps $F^{\alpha}(h)$, i.e.

$$D^{\alpha}(h) = Q^{\alpha\dagger} F^{\alpha}(h) Q^{\alpha} \quad ; \quad V^{\alpha} = \overset{*}{Q^{\alpha}} U^{\alpha} Q^{\alpha t} \tag{18}$$

CG-coefficients for H with respect to the new unirreps are reasonably defined by

$$\vec{A}_j^{\alpha\alpha';\alpha''v} = (Q^{\alpha} \otimes Q^{\alpha'}) \sum_k (Q_{jk}^{\alpha''})^* \vec{C}_k^{\alpha\alpha';\alpha''v} \tag{19}$$

Moreover, it is readily verified that the new CG-coefficients for H are satisfying relations which are identical with (15) and (16). Now we are in the position to give an answer to the question, whether CG-coefficients for H can be transformed by means of appropriated defined similarity transformations Q^{α} into some real ones. In virtue of the fact that for $\alpha \in A_{I(*)} \cup A_{II(*)}$ the n_{α}-dimensional (symmetric or skew-symmetric, depending on $\alpha \in A_{I(*)}$ or $\alpha \in A_{II(*)}$) matrices V^{α} always can be transformed either into the unit matrix 1_n; $n = n_{\alpha}$, or a special real skew-symmetric matrix $J_{n'}$; $n' = n_{\alpha}/2$, it can be verified by means of correspondigly transformed Eq.(15), that CG-coefficients for H can always be transformed into some real ones, if $\alpha,\alpha',\alpha'' \in A_{I(*)} \cup A_{II(*)}$. However each other case has to be decided on its own merits.

In view of the question of reality of CG-coefficients for G, it is very useful to consider also the operation "complex conjugation by s", which gives rise either to further symmetry relations for CG-coefficients for H, namely

$$\vec{C}_j^{\alpha\alpha';\alpha''v} = \sum_w g_{wv} (Z^{\alpha} \otimes Z^{\alpha'}) \sum_k (Z_{jk}^{\alpha''} \vec{C}_k^{\alpha\alpha';\alpha''w})^* \tag{20}$$,

if $\alpha,\alpha',\alpha'' \in A_{I(s)} \cup A_{II(s)}$, or to further "generating relations" for "convenient" CG-coefficients for H, if at least one of the quantities α,α',α'' belongs to $A_{III(s)}$. Obviously these coefficients are especially suited for the definition of corresponding CG-coefficients for G. Apart from this, similar arguments hold for the $m_{\alpha\alpha';\alpha''}$-dimensional unitary matrices g, which are uniquely determined through the corresponding CG-coefficients for H and have to satisfy additionally

$$g\, g^* = (-1)^{\gamma(\alpha)+\gamma(\alpha')+\gamma(\alpha'')}\, 1_{m\ldots} \tag{21}$$,

where however $\gamma(\alpha) = 0$ for $\alpha \in A_{I(s)}$ and $\gamma(\alpha) = 1$ for $\alpha \in A_{II(s)}$ means which should not be confused with the analogous Eq.(17). Consequently g is either symmetric or skew-symmetric and can therefore be transformed either into the unit matrix or the special real skew-symmetric matrix $J_{n'}$. Hence for the sake of convenience, CG-coefficients for H are called "canonical with respect to complex conjugation by s", if they are either satisfying (20) with $g = 1_n$, or $g = J_{n'}$ (which can always be achieved by means of appropriated defined similarity transformations (11)), or are defined by the above mentioned "generating relations", which however have not been written down for the sake of shortness. Assuming that the "convenient" CG-coefficients for H are additionally "canonical with respect to complex conjugation by s", very simple solutions of (10) can be derived, which define "convenient" CG-coefficients for G, if inserting these special matrices into (7), (8) and (9) respectively.

Concerning the problem, whether CG-coefficients for G can also be transformed into some real ones, it can be verified on hand of the previously mentioned special solutions of (10), that type I(s)-CG-coefficients for G never can be made real (apart from I ⊗ I:I-CG-coefficients, if further restrictive conditions are satisfied), whereas type II(s)- and type III(s)-CG-coefficients for G can be transformed into some real ones, if the corresponding "convenient, canonical" CG-coefficients for H can be made real. This reality can be achieved in any case, if $\alpha(k),\alpha'(k'),\alpha''(k'') \in A_{I(*)} \cup A_{II(*)}$; $k,k',k'' = 0,1$ holds, where in particular the difference of the sets $A_{I(s)} \cup A_{II(s)}$ and $A_{I(*)} \cup A_{II(*)}$ has to be noted. Besides this, each other case has to be decided on its own merits. Finally for a detailed discussion the reader is referred to Ref.[2].

[1] R. Dirl, J.Math.Phys.21(1980)961,968,975,983,989,997

[2] R. Dirl: "On the uniqueness and reality of Clebsch-Gordan coefficients for ordinary and corepresentations" (submitted for publication)

ON SOME SPECIAL RELATIONS INVOLVING 3-jm SYMBOLS

Jean Pierre Gazeau
Centre de Recherche de Mathématiques Appliquées
Université de Montréal, Montréal, Québec, Canada

Maurice Kibler*
Institut de Physique Nucléaire (et IN2P3)
Université Lyon-1, 69622 Villeurbanne Cedex, France

The Bander-Itzykson-Talman polynomials arise out from the study of the irreducible representations of the group SO_4 in a $SO_3 \otimes SO_3 \supset SO_3$ basis. The connection between the latter polynomials and the Gegenbauer polynomials, on one hand, and the hyperspherical harmonics, on the other hand, enables to generate particular relations involving 3-jm, 6-j, and 9-j symbols for the chain $SU_2 \supset U_1$. We illustrate this way of producing (known and unknown) relations with some examples.

1. Introduction

In the recent years, many special relations involving 3-jm symbols for the chain $SU_2 \supset U_1$ have been casually discovered in connection with investigations of the hydrogen and helium atoms. The relevant (dynamical) symmetry group for such systems turns out to be O_4 so that it should be possible to derive these and other relations by directly starting from the representation theory of SO_4. Along this line, we further note that the local isomorphism $SO_4 \sim SU_2 \otimes SU_2$ should be of interest for obtaining relations between the basic ingredients of the Wigner-Racah algebra of $SU_2 \supset U_1$ from the representation theory of SO_4.

The representation theory of the group SO_4 is well-known [1-4]. The irreducible representations of SO_4 can be built in a basis adapted to the chain $SO_3 \otimes SO_3 \supset SO_3$ from the ones of the group SO_3. Specialization of the matrix elements of the irreducible representations of SO_4 in a $SO_3 \otimes SO_3 \supset SO_3$ basis yields polynomials $H_{j,k}$, we refer to as the Bander-Itzykson-Talman (BIT) polynomials, defined by [2-3] :

$$H_{j,k}(\Phi) = \sum_m (-1)^{j-m+k} \begin{pmatrix} j & k & j \\ -m & 0 & m \end{pmatrix} e^{-im2\Phi} \qquad (1)$$

As a matter of fact, the BIT polynomials $H_{j,k}$ are connected to the commoner Gegenbauer polynomials C^{k+1}_{2j-k} via [3] :

* to whom correspondance should be adressed

$$H_{j,k}(\Phi) = k\,!\left[\frac{(2j-k)\,!}{(2j+k+1)\,!}\right]^{1/2} (2i\sin\Phi)^{k}\; C^{k+1}_{2j-k}(\cos\Phi)\,. \qquad (2)$$

Therefore, we may redefine the $(SO_4 \supset SO_3 \supset SO_2)$ hyperspherical harmonic $Y_{n\ell m}$ through :

$$Y_{n\ell m}(\Phi,\theta,\varphi) = \left(\frac{2n}{\pi}\right)^{1/2} H_{\frac{n-1}{2},\ell}(\Phi)\; Y_{\ell m}(\theta,\varphi)\,, \qquad (3)$$

where $Y_{\ell m}$ stands for a $(SO_3 \supset SO_2)$ ordinary spherical harmonic. The properties of the Gegenbauer polynomials and of the hyperspherical harmonics are well-known. Consequently, we may expect obtaining particular relations involving 3-jm symbols by playing with Eqs. (1) - (3).

It is the aim of this note to show how to derive some prototype relations. A more complete study will be the subject of a forthcoming paper.

2. Trivial relations

By combining Eqs. (1) and (2) in the case $\Phi = \pi/2$, we get

$$\sum_m \begin{pmatrix} j & k & j \\ -m & 0 & m \end{pmatrix} = 2^k\,\frac{(j+k/2)\,!}{(j-k/2)\,!}\left[\frac{(2j-k)\,!}{(2j+k+1)\,!}\right]^{1/2} \quad \text{for } 2j-k \text{ even ;}$$

the right-hand side of the latter relation has to be replaced by 0 when $2j-k$ is odd. Other sum rules may be obtained, from the same stategy, for other particular values of Φ. The case $\Phi = 0$ leads to what is known as the barycenter theorem in atomic and molecular spectroscopy.

As a second example , from the Taylor development of Eqs. (1) and (2), we have

$$\sum_m (-1)^{j-m}\, m^k \begin{pmatrix} j & k & j \\ -m & 0 & m \end{pmatrix} = \frac{(k\,!)^2}{(2k+1)\,!}\left[\frac{(2j+k+1)\,!}{(2j-k)\,!}\right]^{1/2} .$$

(A similar relation also holds when replacing m^k by m^q with $q < k$.) Specialization to the case $k = 2j$ produces the combinational relation

$$\sum_{m=-j}^{j} (-1)^{j-m}\, m^{2j}/(j-m)\,!\,(j+m)\,! = 1\,.$$

3. Less trivial relations

The combination of Eqs. (1) and (2) used in conjunction with the orthogonality property of the Gegenbauer polynomials gives

$$\sum_{mm'} \frac{\sin (m'-m)\pi}{(m'-m)[1-(m'-m)^2]} \begin{pmatrix} j & k & j \\ -m & 0 & m \end{pmatrix} \begin{pmatrix} j' & k & j' \\ -m' & 0 & m' \end{pmatrix} = \delta(j'j)\,\delta(jkj)\,\pi/[j]$$

or alternatively

$$\sum_{mm'} \begin{pmatrix} j & k & j \\ -m & 0 & m \end{pmatrix} \begin{pmatrix} j' & k & j' \\ -m' & 0 & m' \end{pmatrix} / \Gamma(2+m-m')\ \Gamma(2-m+m') = \delta(j'j)\,\delta(jkj)/[j] .$$

It should be stressed that the latter relation is in agreement with the Clebsch-Gordan identity derived by Best [5] although the phase factor in the Best identity (apparently) differs from our's.

Indeed, the two preceding relations are nothing but a consequence of the known (cf. Ref. [3]) orthogonality property on $[0,\pi]$ with the weight $\sin^2\Phi$ of the polynomials $H_{j,k}$ and $H_{j',k}$.

4. Less and less trivial relations

Fourier analysis of the product of two BIT polynomials allows one to write

$$H_{j,\ell}(\Phi)\ H_{j',\ell'}(\Phi) = \sum_M (j\ell, j'\ell' ; M)\, e^{iM2\Phi}$$

with

$$(j\ell, j'\ell' ; M) = \frac{1}{2\pi} \int_0^{2\pi} e^{-iM2\Phi}\ H_{j,\ell}(\Phi)\ H_{j',\ell'}(\Phi)\, d\Phi .$$

The expansion coefficients (, ;) can be calculated in two different ways. The use of Eqs. (1) and (2) leads to

$$(j\ell, j'\ell' ; M) = (-1)^{j+j'-M} \sum_m \begin{pmatrix} j & \ell & j \\ -m & 0 & m \end{pmatrix} \begin{pmatrix} j' & \ell' & j' \\ -M+m & 0 & M-m \end{pmatrix} \tag{4}$$

and

$$(j\ell, j'\ell' ; M) = (2i)^{\ell+\ell'}\, \ell\,!\ \ell'\,! \left[\frac{(2j-\ell)\,!\ (2j'-\ell')\,!}{(2j+\ell+1)\,!\ (2j'+\ell'+1)\,!} \right]^{1/2}$$

$$\frac{1}{2\pi} \int_0^{2\pi} C^{\ell+1}_{2j-\ell}(\cos\Phi)\ C^{\ell'+1}_{2j'-\ell'}(\cos\Phi)\ (\sin\Phi)^{\ell+\ell'}\ e^{-iM2\Phi}\, d\Phi, \tag{5}$$

respectively. (In fact, the integral is amenable to an integral over $[0,\pi]$ by standard manipulation.) Equations (4) and (5) may be interpreted as giving the value of an integral of two Gegenbauer polynomials, with an unusual measure, in terms of 3-jm symbols. On the other hand, the integral under consideration can be easily calculated in some particular cases and Eqs. (4) and (5) then provide special sum rules for the product of two 3-jm symbols.

We continue with the case where $\ell' = \ell$. By making use of angular momentum recoupling techniques, Eq. (4) with $\ell' = \ell$ can be transformed as

$$(j\ell,\ j'\ell\ ;\ M) = \sum_L (-1)^{j+j'+L} \begin{Bmatrix} j & j & \ell \\ j' & j' & L \end{Bmatrix} \quad \text{for} \quad |M| \le |j-j'| ,$$

which indicates that the value of $(j\ell,\ j'\ell\ ;\ M)$ is independent of M once $|M| \le |j-j'|$. Therefore,

$$(j\ell,\ j'\ell\ ;\ M) = (j\ell,\ j'\ell\ ;\ 0) \quad \text{for} \quad |M| \le |j-j'| .$$

The integral in Eq. (5) with $\ell' = \ell$ and $M = 0$ can be easily calculated. As a final result, we get

$$(-1)^{j+j'-M} \sum_m \begin{pmatrix} j & \ell & j \\ -m & 0 & m \end{pmatrix} \begin{pmatrix} j' & \ell & j' \\ -M+m & 0 & M-m \end{pmatrix}$$

$$= \frac{(-1)^\ell}{[\ell]} \left[\frac{(2j-\ell)!\,(2j'+\ell+1)!}{(2j'-\ell)!\,(2j+\ell+1)!} \right]^{1/2} \quad \text{for} \quad |M| \le j-j' ,$$

a relation recently obtained by Dunlap and Judd by means of combinational analysis [6].

5. Miscellaneous comments

(i) The present work lies on an investigation of the BIT polynomials. Such polynomials arise from the SO_4 irreducible representation matrix element [3]:

$$D^{jj'}(A,I)_{LM,\ JM} = \sum_k (-1)^{j+j'+k+M} [LJ]^{1/2} [k] \begin{pmatrix} L & J & k \\ M & -M & 0 \end{pmatrix} \begin{Bmatrix} L & j & j' \\ j & J & k \end{Bmatrix} H_{j,k}(\Phi) .$$

We note that

$$D^{jj'}(A,I)_{LM,\ LM} = \sum_k (-1)^{2L} [kL] \begin{Bmatrix} j & j' & k \\ j & j' & L \end{Bmatrix} G^{jj'}_{kM}(\Phi) ,$$

where

$$G^{jj'}_{kM}(\Phi) = \sum_{mm'} \begin{pmatrix} j & j' & k \\ m & -m' & M+2m' \end{pmatrix}^2 e^{-im2\Phi} .$$

The pending polynomials $G^{jj'}_{kM}$ might be of interest too. In that direction, the defining relation of the polynomials $G^{jj'}_{kM}$ may be inverted to give

$$\sum_{L} (-1)^{2L} \begin{Bmatrix} j & j' & L \\ j & j' & k \end{Bmatrix} D^{jj'}(A, I)_{LM,\, LM} = G^{jj'}_{kM}(\Phi) .$$

Specialization to the case $\Phi = 0$ yields

$$\sum_{L} (-1)^{2L} \begin{Bmatrix} j & j' & L \\ j & j' & k \end{Bmatrix} = \sum_{mq} \begin{pmatrix} j & k & j' \\ -m & q & M-m \end{pmatrix}^2 \quad \text{for } |M| \leq |j - j'| ,$$

to be compared with Eq. (4) of Ref. [6].

(ii) It is hoped that this study and related works shall be of interest for generating combinational identities.

(iii) The BIT polynomials have been recently (implicitly) used by Daumens and Winternitz in expansions of scattering amplitudes for particles with spin [7].

Acknowledgments

One of the authors (M.K.) is grateful to R. Gilmore for pointing him out, some ten years ago, the work by M.E. Best (cited as Red. [5]).

References

[1] L.C. Biedenharn, J. Math. Phys. 2, 433 (1961).

[2] M. Bander and C. Itzykson, Rev. Mod. Phys. 38, 330, 346 (1966).

[3] J.D. Talman, Special functions (Benjamin, New York, 1968).

[4] R.T. Sharp, J. Math. and Phys. 47, 359 (1968).

[5] M.E. Best, Evaluation of certain spherical integrals using the irreducible representations of O(4), McGill University preprint (1970).

[6] B.I. Dunlap and B.R. Judd, J. Math. Phys. 16, 318 (1975).

[7] M. Daumens and P. Winternitz, Phys. Rev. D 21, 1919 (1980).

TWO PROBLEMS RELATED TO THE STONE-VON NEUMANN-MACKEY UNIQUENESS THEOREM

J.C.H. GODDARD

DEPARTAMENTO DE MATEMATICAS
UNIVERSIDAD AUTONOMA METROPOLITANA - IZTAPALAPA
MEXICO 13, D.F.
MEXICO.

The Stone-von Neumann-Mackey uniqueness theorem and generalizations of it have played an interesting rôle in the developement of group representations. In fact it is a key result for several theorems of harmonic analysis such as Pontrjagin's duality theorem and Plancherel's theorem c.f. | 1 | , | 2 | . Furthermore in abstract formulations of representations of the canonical commutation relations (C.C.R.) and anti-commutation relations (C.A.R.) a guiding light has been played by Weyl's formulation c.f. | 3 | , | 4 |.

We shall discuss two problems related to this theorem for which we have only partial answers. These problems could be of use in the structure of self dual groups and of certain nilpotent locally compact groups.

To begin with we shall recall a special case of the Stone-von Neumann-Mackey theorem c.f. | 5 | .

<u>Theorem</u> Let G be a locally compact abelian group with dual group $\hat{G}$. Let U,V be two unitary representations of G, $\hat{G}$ respectively, acting in the same Hilbert space H . Suppose that (U,V) form an irreducible pair such that

$$V(\alpha)U(x) = \alpha(x)U(x)V(\alpha)$$

for all $x \in G$, $\alpha \in \hat{G}$.

Then there exists a linear isometry S of H onto $L^2(G)$ such that

$$U(x) = S^{-1}\lambda(x)S \quad , \quad V(\alpha) = S^{-1}M_\alpha S$$

where λ is the left regular representation of G and M the multiplication representation of $\hat{G}$.

We shall recall now Weyl's formulation c.f. | 6 |, in which we recast the above in terms of a projective representation. For this we define, with the same notation as above, a projective representation W on $G \times \hat{G}$ by:

$$W(x,\alpha) = U(x)V(\alpha)$$

We observe that the multiplier σ corresponding to W is then:

$$\sigma((x,\alpha),(y,\beta)) = \alpha(y)$$

This multiplier has the following property of being totally skew. We may define this as:

<u>Definition</u> A multiplier ψ on an abelian group K is totally skew if whenever $\psi(x,y) = \psi(y,x)$ for all $y \in K$ then $x = e$ (the identity element of K).

Let us examine a simple example to illustrate this concept. If $K = \mathbb{R}^2$ (two copies of the real numbers) then for each $a \in \mathbb{R}$ we can define:

$$\delta_a((x,y),(u,v)) = e^{2\Pi ia(xv-yu)}$$

This realizes the multiplier group $H^2(\mathbb{R}^2)$ of $\mathbb{R}^2$ as $\mathbb{R}$. We can see that δ_a is totally skew if and only if $a \neq 0$.

With this concept we can state a generalization of the Stone-von Neumann - Mackey theorem obtained in | 7 |.

<u>Theorem</u> Let K be a locally compact abelian group with a totally skew, type I multiplier ψ. Then K has an essentially unique irreducible ψ - representation.

We remark that in this context, ψ type I means that all the ψ - representations are type I. Baggett and Klepner have also shown that for a totally skew multiplier ψ, ψ is type I if and only if the map h_ψ of K to $\hat{K}$ defined by:

$$h_\psi(x)(y) = \psi(x,y)\ \overline{\psi(y,x)}$$

is a bicontinuous isomorphism.

This leads to our first problem:

Problem If K is a locally compact abelian group with a totally skew type I multiplier then is $K \simeq G \times \hat{G}$, for some locally compact abelian group G?

We know this is true for certain classes of abelian groups such as compactly generated ones (which include finite, compact and vector groups).

We mention that the above conditions always imply the existence of a short exact sequence:

$$G \to K \to \hat{G}$$

Having considered locally compact abelian groups with a totally skew multiplier, it is natural to ask "how many" of these multipliers can there be? The above example tells us that for $K = \mathbb{R}^2$, all but the identity multiplier have this property. We shall change the concept of 'how many' by using a device of Edwards and Lewis c.f. | 8 | , | 9 | . We define an action of the automorphism group $H^2(K)$ by essentially defining:

$$(\theta\cdot\psi)(x,y) = \psi(\theta^{-1}(x),\ \theta^{-1}(y))$$

We notice that $H^2(\mathbb{R}^2)$ has precisely two orbits under this action. This leads to the following:

Problem Do all the totally skew type I multipliers of a locally compact abelian group belong to the same orbit? what can we say in general about the orbits?

We again remark that this is indeed true for compactly generated groups.

We can look at this problem from a different point of view c.f. | 9 | . Namely, we can define a twisted group algebra $L^1(K,\psi)$ (for each multiplier ψ on K) which will be a Banach* - algebra whose nondegenerate representations are in 1 - 1 correspondence with the ψ - representations of K . $L^1(K,\psi)$ is, as a Banach space the same as $L^1(K)$. However we redefine multiplication and involution as:

$$f\, x_\psi\, g(x) = \int f(y)\, g(y^{-1}x)\, \psi(y,y^{-1}x)\,dy$$

$$f^\sim(x) = \overline{f(x^{-1})}\; \overline{\psi(x, x^{-1})}\,\Delta(x^{-1})$$

We then find that for two multipliers ψ_1 ψ^1 on K, $L'(K,\psi)$, $L'(K,\psi')$ are isometric * - isomorphic if andonly if ψ ,ψ' belong to the same orbit.

The C* - enveloping algebras associated with these twisted group algebras have been used often in problems related to the C.C.R. and C.A.R. c.f. |10| , |11|.

If one takes the discrete group $K = \bigoplus_1^\infty (Z_2 \times \hat{Z}_2)$, where Z_2 is the cyclic group of order 2, and the multiplier $\psi = \prod_1^\infty \psi_n$ where $\psi_n((x,\alpha), (y,\beta)) = \alpha(y)$, then we have a bijection between representations of the C.A.R., ψ-representations of K and nondegenerate representations of $L'(K, \psi)$ or its enveloping C*-algebra $C^*(K,\psi)$.

We finally mention that it is often more natural to encounter representations of C.A.R. by looking at ψ-representations of K and especially at projective positive definite functions on K (a natural analogue of positive definitive functions to the projective case). In fact the generating functions of Segal are examples of these projective positive definite functions c.f. |12|.

References

|1| Sankaran, S., Math. Zeitschr. 98 (1967) 387-390

|2| Nakamura, M., and H. Umegaki. Proc. Japan Acad. 37 (1961) 239-242

|3| Slawny, J., Commun. Math. Phys. 24(1972) 151-170

|4| Guichardet, A., Algébres d'Observables Associées aux Relations

de Commutations, (Colin, Paris, 1968)

|5| Mackey, G.W., Duke Math. J. 16(1949) 313-325

|6| Weyl, H., Theory of groups and Quantum Mechanics, (Dover, 1931)

|7| Baggett, L., and A. Kleppner., Jour. of Funct. Anal. 14(1973) 299-324.

|8| Edwards, C.M., and J.T. Lewis., Commun. Math. Phys. 13(1969) 131-141.

|9| Edwards, C.M., Quart. J. Math Oxford 22(1971) 197-220.

|10| J. Manuceau, M. Sirugue, D. Testard, and A. Verbeure., Commun. Math. Phys. 32 (1973) 231-243.

|11| A.L. Carey, J.M. Gaffney and C.A.Hurst., Jour-Math. Phys. 18(1977)629-640.

|12| Emch, G.G., Algebraic Methods in Statistical Mechanics and Quantum Field Theory (Wiley-Interscience, New York 1972)

NON-COMPLEX REPRESENTATIONS AND THEIR RELATION TO ANTIUNITARY SYMMETRY

Peter Kasperkovitz and Gerhard Kahl
Institut für theoretische Physik, TU Wien, A-1040 Karlsplatz 13, Austria

1.Consider the following problem (lattice dynamics) : Given a real symmetric matrix H , dim H $< \infty$, and a finite group of (real) orthogonal matrices O(x) , $x \varepsilon G$, commuting with H ; how can this symmetry group be used to simplify the eigenvalue problem of H and to make predictions on the minimal degeneracies ?

The usual answer [1] is given in two steps : In the first step an orthonormalized set of complex column vectors transforming according to complex irreducible matrix representations C^λ is constructed . The action of H on this symmetry adapted basis is given by a direct sum of Hermitean matrices H^λ ,

$$\dim H^\lambda = m^\lambda_{\mathbb{C}} \, , \tag{1}$$

collected into chains of $d^\lambda_{\mathbb{C}}$ identical copies where

$$d^\lambda_{\mathbb{C}} = \dim C^\lambda(x) \, . \tag{2}$$

Using the notation

$$\sigma(A) = \text{set of eigenvalues of } A = A^\dagger \, ,$$
$$\deg(e,A) = \text{degeneracy of the eigenvalue } e \text{ of } A = 0 \text{ if } e \text{ is not an eigenvalue} \tag{3}$$

one obtains after the first step

$$\sigma(H) = \cup_\lambda \sigma(H^\lambda) \, , \; \deg(e,H) = \textstyle\sum_\lambda d^\lambda_{\mathbb{C}} . \deg(e,H^\lambda) \, . \tag{4}$$

In the second step the reality of the original matrices H and O(x) is taken into account . If θ is the operation of complex conjugation of the elements of a column vector then

$$\theta H \theta = H \, , \; \theta O(x) \theta = O(x) \, , \; \theta^2 = E \text{ (1-matrix)} \, . \tag{5}$$

To discuss the consequences of (5) the matrix representations C^λ are partitioned into three classes called 'kinds' .

1st kind : C^λ is equivalent to a real representation .
2nd kind : C^λ is equivalent to $C^{\lambda *}$ but inequivalent to a real representation .
3rd kind : C^λ is inequivalent to $C^{\lambda *}$. (6)

For representations of the 1st kind there are no new results on $\deg(e,H^\lambda)$ but

$$C^\lambda \text{ is of } 2^{nd} \text{ kind} \Rightarrow \deg(e,H^\lambda) \text{ is even} \, ,$$
$$C^{\lambda +} = (C^{\lambda -})^* \text{ is of } 3^{rd} \text{ kind} \Rightarrow \deg(e,H^{\lambda +}) = \deg(e,H^{\lambda -}) \, . \tag{7}$$

Accordingly collecting representations of the 3rd kind into pairs $\lambda \pm$

$$\deg(e,H) = \sum_{\lambda}^{(1)} d_{\mathbb{C}}^{\lambda}.\deg(e,H^{\lambda}) + \sum_{\lambda+}^{(3)} 2d_{\mathbb{C}}^{\lambda}.\deg(e,H^{\lambda+}) + \sum_{\lambda}^{(2)} 2d_{\mathbb{C}}^{\lambda}.(1/2)\deg(e,H^{\lambda}) \quad (8)$$

We now give a direct derivation of (8) dispensing with the antiunitary operator θ . Our approach is based on matrix representations of G irreducible over the reals. They can be determined from first principles or from the complex representations C^{λ}. Because of the structure of the matrices $R^{\lambda}(x)$ the rows are labelled by double indices

$$r = 0,\dots,r^{\lambda}-1 \ ; \ j = 0,\dots,d^{\lambda}-1 \ . \quad (9)$$

Like the complex representations the representations R^{λ} divide into three classes called 'types' ; they are related to the kinds according to the following scheme :

$\mathbb{R}$-type : R^{λ} is a real form of a representation C^{λ} of the 1st kind ; $r^{\lambda}=1$, $d^{\lambda}=d_{\mathbb{C}}^{\lambda}$. (10)

$\mathbb{C}$-type : $R^{\lambda}_{jr,JR}(x) = R^{\mathbb{C}}_{rR}\{C^{\lambda+}_{jJ}(x)\}$, $R^{\mathbb{C}}\{a+ib\} = \begin{pmatrix} a & -b \\ b & a \end{pmatrix}$, $r^{\lambda}=2$, $d^{\lambda}=d_{\mathbb{C}}^{\lambda+}$; C^{λ} is of the 3rd kind . (11)

$\mathbb{Q}$-type : $R^{\lambda}_{jr,JR}(x) = R^{\mathbb{Q}}_{rR}\{Q^{\lambda}_{jJ}(x)\}$, $R^{\mathbb{Q}}\{a+ib+jc+kd\} = \begin{pmatrix} a & -b & -c & -d \\ b & a & -d & c \\ c & d & a & -b \\ d & -c & b & a \end{pmatrix}$, $r^{\lambda}=4$, $d^{\lambda}=(1/2)d_{\mathbb{C}}^{\lambda}$; the quaternionic irreducible representation Q^{λ} is related to a complex irreducible representation C^{λ} by $C^{\lambda}_{js,JS}(x) = C^{\mathbb{Q}}_{sS}\{Q^{\lambda}_{jJ}(x)\}$, $C^{\mathbb{Q}}\{a+ib+jc+kd\} = \begin{pmatrix} a+ib & c+id \\ -c+id & a-ib \end{pmatrix}$; every representation C^{λ} of the 2nd kind can be brought into this form by a unitary transformation [2] . (12)

In (10) to (12) the same index λ can be used for the equivalence classes of the real irreducible representations as for the complex ones if the symbols $\lambda\pm$ characterizing conjugate pairs of representations of the 3rd kind are replaced by one single symbol λ .

Knowing the representation R^{λ} one forms the matrices

$$E^{\lambda}_{jJ} = |G|^{-1} \sum_x r^{\lambda}d^{\lambda} R^{\lambda}_{j0,J0}(x)\, O(x) \ , \quad (13)$$

$$F^{\lambda}_{r} = |G|^{-1} \sum_x r^{\lambda}d^{\lambda} \sum_j R^{\lambda}_{jr,J0}(x)\, O(x) \ . \quad (14)$$

The matrices E^{λ}_{jj} are projection matrices and it is possible to find for each E^{λ}_{00} a set of real column vectors $w^{\lambda m}_{0r}$ satisfying

$$E^{\lambda}_{00}w^{\lambda m}_{0r} = w^{\lambda m}_{0r} \ , \ F^{\lambda}_{r}w^{\lambda m}_{00} = w^{\lambda m}_{0r} \ , \ \langle w^{\lambda m}_{0r}, w^{\Lambda M}_{0R}\rangle = \delta_{\lambda\Lambda}\delta_{mM}\delta_{rR} \ ;$$
$$r = 0,\dots,r^{\lambda}-1 \ ; \ m = 0,\dots,m^{\lambda}_{\mathbb{R}}-1 \ ; \ r^{\lambda}m^{\lambda}_{\mathbb{R}} = \mathrm{trace}\, E^{\lambda}_{00} \ . \quad (15)$$

By means of the shift operators E^{λ}_{j0} , $j \neq 0$, these subbases may be extended to a basis of the Hilbert space of real column vectors which is again called symmetry adapted since it transforms according to the representation R^λ .

$$w_{jr}^{\lambda m} = E_{j0}^\lambda w_{0r}^{\lambda m} \; ; \; O(x) w_{JR}^{\lambda m} = \textstyle\sum_{jr} w_{jr}^{\lambda m} R_{jr,JR}^\lambda(x) \; . \qquad (16)$$

The action of H on the bases $\{w_{jr}^{\lambda m}\}$ is given by the direct sum $\sum_\lambda \oplus d^\lambda H^\lambda$, the real submatrices H^λ showing a similar structure as the matrices $R^\lambda(x)$.

$$\underline{\mathbb{R}\text{-type}} : H_{j0,J0}^\lambda = H_{jJ}^{\lambda(\mathbb{R})} \; , \; H^{\lambda(\mathbb{R})} = H^{\lambda(\mathbb{R})\dagger} \text{ real} \; , \; \dim H^{\lambda(\mathbb{R})} = m_{\mathbb{R}}^\lambda = m_{\mathbb{C}}^\lambda \qquad (17)$$

$$\underline{\mathbb{C}\text{-type}} : H_{jr,JR}^\lambda = R_{rR}^{\mathbb{C}}\{H_{jJ}^{\lambda(\mathbb{C})}\} \; , \; H^{\lambda(\mathbb{C})} = H^{\lambda(\mathbb{C})\dagger} \text{ complex} \; , \; \dim H^{\lambda(\mathbb{C})} = m_{\mathbb{R}}^\lambda = m_{\mathbb{C}}^\lambda \qquad (18)$$

$$\underline{\mathbb{Q}\text{-type}} : H_{jr,JR}^\lambda = \bar{R}_{rR}^{\mathbb{Q}}\{H_{jJ}^{\lambda(\mathbb{Q})}\} \; , \; \bar{R}^{\mathbb{Q}}\{a+ib+jc+kd\} = \begin{pmatrix} a & b & c & d \\ -b & a & -d & c \\ -c & d & a & -b \\ -d & -c & b & a \end{pmatrix} ,$$

$$H^{\lambda(\mathbb{Q})} = H^{\lambda(\mathbb{Q})\dagger} \text{ quaternionic} \; , \; \dim H^{\lambda(\mathbb{Q})} = m_{\mathbb{R}}^\lambda = (1/2) m_{\mathbb{C}}^\lambda \; . \qquad (19)$$

For every real matrix H^λ of $\mathbb{F}$-type there exists a norm-preserving matrix with elements from $\mathbb{F}$ transforming H^λ into a direct sum of r^λ copies of $H^{\lambda(\mathbb{F})}$. Therefore

$$\sigma(H) = \cup_\lambda \, \sigma(H^{\lambda(\mathbb{F})}) \; , \; \deg(e,H) = \textstyle\sum_\lambda r^\lambda d^\lambda . \deg(e, H^{\lambda(\mathbb{F})}) \; . \qquad (20)$$

(20) is equivalent to (4) and (8) . The matrices to be diagonalized have the same dimension ($\mathbb{R}$- and $\mathbb{C}$-type) or half the dimension ($\mathbb{Q}$-type) as the matrices appearing in the usual approach and are characterized by the same number of real parameters ($\mathbb{C}$-type) or by half the number ($\mathbb{R}$- and $\mathbb{Q}$-type) . The minimal degeneracies $r^\lambda d^\lambda$ are seen to be given by the dimensions of the real irreducible matrix representations of the symmetry group .

2.Next consider the following problem (level splitting due to crystal field and spin-orbit interaction , 1 electron) : Given a Hermitean Matrix H with $\dim H < \infty$, a symmetry group of unitary matrices U(x) , $x \in G$, and an antiunitary operator θ satisfying

$$\theta H \theta = H \; , \; \theta U(x) \theta = U(x) \; , \; \theta^2 = -E \; ; \qquad (21)$$

how can the eigenvalue problem be simplified and what can be predicted on the minimal degeneracies ?

We skip the usual solution [1] showing immediately the alternative approach . Relations (21) ensure that H and U(x) can be unitarily transformed into matrices H' , U'(x) which are composed of 2x2 submatrices of the form $C^{\mathbb{Q}}$ /see (10)/ . Therefore these complex matrices can be replaced by quaternionic matrices of half the dimension say $\bar{H}$ and $\bar{U}(x)$, without altering the algebraic structure of the real algebra generated by these matrices . Then the quaternionic matrices

$$\bar{E}_{jJ}^\lambda = |G|^{-1} \textstyle\sum_x r^\lambda d^\lambda \, R_{j0,J0}^\lambda(x) \, \bar{U}(x) \; , \qquad (22)$$

$$\bar{F}_r^\lambda = |G|^{-1} \textstyle\sum_x r^\lambda d^\lambda \sum_j R_{jr,j0}^\lambda(x) \, \bar{U}(x) \; , \qquad (23)$$

differing from (13) and (14) only by the substitution $O(x) \to \bar{U}(x)$ are built and used to define a symmetry adapted bysis of the space of quaternionic column vectors. To this end one starts again constructing subbases $\{w_0^{\lambda m}\}$ which are reproduced by the projection matrices $\bar{E}_{00}^{\lambda}$:

$$\bar{E}_{00}^{\lambda} w_0^{\lambda m} = w_0^{\lambda m} \; ; \; \langle w_0^{\lambda m}, w_0^{\Lambda M}\rangle = \delta_{\lambda\Lambda}\delta_{mM} \; ; \; m = 0,\ldots,m_{\mathbb{Q}}^{\lambda}-1 \; , \; m_{\mathbb{Q}}^{\lambda} = \text{trace } \bar{E}_{00}^{\lambda} \; . \tag{24}$$

These vectors can be chosen to satisfy r^{λ} eigenvalue equations of the form

$$\bar{F}_r^{\lambda} w_0^{\lambda m} = w_0^{\lambda m} f_r \; ; \; f_0 = 1 \; , \; f_1 = i \; , \; f_2 = j \; , \; f_3 = k \tag{25}$$

and are extended via

$$w_j^{\lambda m} = \bar{E}_{j0}^{\lambda} w_0^{\lambda m} \tag{26}$$

to a basis transforming under the matrices $\bar{U}(x)$ according to quaternionic irreducible matrix representations Q^{λ} .

$$\bar{U}(x) w_J^{\lambda m} = \textstyle\sum_j w_j^{\lambda m} Q_{jJ}^{\lambda}(x) \; , \; Q_{jJ}^{\lambda}(x) = \sum_r f_r R_{jr,J0}^{\lambda}(x) \; ;$$

$$\mathbb{R}\text{-type} : Q^{\lambda} \text{ real} \; ; \; \mathbb{C}\text{-type} : Q^{\lambda} \text{ complex} \; ; \; \mathbb{Q}\text{-type} : Q^{\lambda} \text{ quaternionic} \; . \tag{27}$$

The action of $\bar{H}$ on this basis is given by the direct sum $\sum_{\lambda}\oplus\, d^{\lambda}\bar{H}^{\lambda}$ with

$$\begin{aligned}
&\mathbb{R}\text{-type} : \bar{H}^{\lambda} = \bar{H}^{\lambda\dagger} \text{ quaternionic} \; , \; \dim \bar{H}^{\lambda} = m_{\mathbb{Q}}^{\lambda} = (1/2)m_{\mathbb{C}}^{\lambda} \; ; \\
&\mathbb{C}\text{-type} : \bar{H}^{\lambda} = \bar{H}^{\lambda\dagger} \text{ complex} \; , \; \dim \bar{H}^{\lambda} = m_{\mathbb{Q}}^{\lambda} = m_{\mathbb{C}}^{\lambda +} \; ; \\
&\mathbb{Q}\text{-type} : \bar{H}^{\lambda} = \bar{H}^{\lambda\dagger} \text{ real} \; , \; \dim \bar{H}^{\lambda} = m_{\mathbb{Q}}^{\lambda} = m_{\mathbb{C}}^{\lambda} \; .
\end{aligned} \tag{28}$$

Therefore taking into account that $\bar{H}$ and H satisfy the same minimal polynomial but that H has twice the dimension of $\bar{H}$ one finds

$$\sigma(H) = \sigma(\bar{H}) = \textstyle\bigcup_{\lambda} \sigma(\bar{H}^{\lambda}) \; ; \; \deg(e,H) = 2.\deg(e,\bar{H}) = \sum_{\lambda} 2d^{\lambda}.\deg(e,\bar{H}^{\lambda}) \; . \tag{29}$$

For the matrices to be diagonalized the remarks made for the first example apply again with the roles of the $\mathbb{R}$- and the $\mathbb{Q}$-types interchanged . The minimal degeneracies are twice the dimensions of the quaternionic irreducible matrix representations of the symmetry group .

3.The above results are not accidential but based on operator isomorphisms studied by other authors for different purposes [3] . That the partitioning of the representations into three classes is intimately related to the three fields $\mathbb{R}$, $\mathbb{C}$, $\mathbb{Q}$ becomes more obvious if the representation theory over these fields is studied for a fixed compact group [4] .

4.Conclusion : Non-complex representations can be useful and are by no means more difficult to handle than the familiar complex ones .

[1] L.Jansen and M.Boon,Theory of Finite Groups . Applications in Physics . North-Holland , Amsterdam 1967 . See especially Ch.II , Sec.5.9. , and

Ch.III , Sec.5.1.

[2] G.Frobenius und I.Schur , Sitzgsber. Preuss. Akad. B II (1906) 186 .
D.Finkelstein , J.M.Jauch , and D.Speiser , J. Math. Phys. 4 (1963) 136 .

[3] E.C.G.Stueckelberg , Helv. Phys. Acta 33 (1960) 727 .
E.C.G.Stueckelberg and M.Guenin , Helv. Phys. Acta 34 (1961) 621 .
C.Chevalley , Theory of Lie Groups . Princeton Univ. Press , Princeton 1946 , pp. 16 - 24 .
D.Finkelstein , J.M.Jauch , S.Schiminovich , and D.Speiser , J. Math. Phys. 3 (1962) 207 .

[4] P.Kasperkovitz (preprint)

REPRESENTATION THEORY OF COMPACT GROUPS

William H. Klink and Tuong Ton-That
The University of Iowa
Iowa City, Iowa 52242 USA

I. Introduction

The theory of compact Lie groups has had a long and distinguished history, so it may seem strange to devote much time to topics such as concrete realizations of orthogonal bases or matrix elements of representations of compact Lie groups, topics that are in some instances already well known. There are several reasons for looking at these topics. First, there remain a number of problems that are only partially solved, such as the multiplicity problem occurring in the decomposition of tensor products of irreducible representations, or the explicit computation of the Racah coefficients. Secondly, our point of departure in this paper is the Borel Weil theory of holomorphic induction, which we use to realize all irreducible representations of U(n), Sp(2n) and SO(n) as polynomials over the complexification of these groups. Since this theory is not as well known as some of the more traditional ways of realizing the irreducible representations of the compact groups, it is worthwhile spelling out the ways in which one can use this theory to compute even well-known quantities of interest, in order to see if simplifications or new insights occur. Finally, the Borel Weil theory has certain similarities with the Mackey induced representation theory, and by applying some of the ideas of the Mackey theory, such as double cosets or the subgroup theorem, it is possible to gain a new insight into properties of the representations of the compact groups.

The goal of this paper is to present a brief account of our work on polynomial representations of the compact groups using the Borel Weil holomorphic induction theory. In Section II the basic polynomial spaces are defined and double coset maps introduced to relate the representations of groups to representations of their subgroups. In Section III attention is focused on $GL(n,\mathbb{C})$, and explicit orthogonal polynomial bases and matrix elements are given in terms of tensor products of the so-called fundamental representations. Finally, in Section IV certain results of Weyl are used to discuss the multiplicity problem arising in the direct sum decomposition of an r-fold tensor product space.

II. Irreducible Polynomial Representation of U(n), SO(n), and Sp(n)

The unitary irreducible representations of U(n) can be realized as polynomial representations of the complexification of U(n), namely $GL(n,\mathbb{C})$. Polynomials over $GL(n,\mathbb{C})$ form irreducible representation spaces defined by

$$V^{(m)} = \{f: g \in GL(n,\mathbb{C}) \to \mathbb{C}, f(bg) = \pi^{(m)}(b)f(g), \text{ f polynomial in } GL(n,\mathbb{C})\} \quad . \tag{II.1}$$

Here $\pi^{(m)}(b)$ is a character of $b \in B$, the Borel subgroup of $GL(n,\mathbb{C})$ consisting of lower triangular matrices, and $(m) = (m_1 \ldots m_n)$ is a dominant weight satisfying $m_1 \geq m_2 \geq m_n \geq 0$. All of the finite dimensional irreducible representations of $GL(n,\mathbb{C})$ are given by right translation,

$$(R^{(m)}_{g_o} f)(g) = f(gg_o) \quad , \quad f \in V^{(m)} \quad , \quad g_o \in GL(n,\mathbb{C}) \quad . \tag{II.2}$$

If g_o is restricted to the unitary subgroup $U(n)$ of $GL(n,\mathbb{C})$, the representation (II.2) remains irreducible. Further, $R^{(m)}_{g_o}$, $g_o \in U(n)$, becomes a unitary representation with respect to the following "differentiation" inner product:

$$(f,f') = f(D)\,\overline{f'(\overline{g})}\big|_{g=0} \qquad f,f' \in V^{(m)} \tag{II.3}$$

where $f(D)$ is the differential operator obtained from the polynomial $f(g)$ by replacing the entries g_{ij} with the corresponding partial derivatives $\partial/\partial g_{ij}$ [1].

A double coset map [2] can be used to obtain irreducible polynomial representations of $Sp(2n)$ and $SO(n)$. To keep the computations as short as possible we will work out the polynomial representations only for $Sp(2n)$ and $SO(2n)$. As with the unitary groups we deal with the complexification of the various groups, namely

$$Sp(2n,\mathbb{C}) = \{h \in GL(2n,\mathbb{C}) \ , \ h\,\sigma_-\,h^T = \sigma_-\}$$

$$SO(2n,\mathbb{C}) = \{h \in GL(2n,\mathbb{C}) \ , \ h\,\sigma_+\,h^T = \sigma_+\} \tag{II.4}$$

where

$$\sigma_\pm = \begin{pmatrix} 0 & \pm I_n \\ I_n & 0 \end{pmatrix} \quad , \quad \text{and T means transpose} \quad .$$

Then general elements of $Sp(2n,\mathbb{C})$ and $SO(2n,\mathbb{C})$ can be written as

$$s_1\,\hat{g}\,s_2 \quad \in \quad Sp(2n,\mathbb{C})$$

$$a_1\,\hat{g}\,a_2 \quad \in \quad SO(2n,\mathbb{C}) \tag{II.5}$$

where

$$\hat{g} = \begin{pmatrix} g & 0 \\ 0 & g^{-1T} \end{pmatrix} \ , \quad g \in GL(n,\mathbb{C}) \ , \quad s_1 = \begin{pmatrix} I_n & 0 \\ s & I_n \end{pmatrix} \ , \quad s_2 = \begin{pmatrix} I_n & s' \\ 0 & I_n \end{pmatrix} \ ,$$

$$a_1 = \begin{pmatrix} I_n & 0 \\ a & I_n \end{pmatrix} \ , \quad \text{and} \quad a_2 = \begin{pmatrix} I_n & a' \\ 0 & I_n \end{pmatrix} \ .$$

Here s, s' are symmetric $n \times n$ matrices, while a, a' are antisymmetric matrices, and I_n is the n dimensional identity matrix.

Given this parametrization of the Sp and SO groups we define maps from $V^{(m)}$ to representation spaces of $Sp(2n,\mathbb{C})$ and $SO(2n,\mathbb{C})$ via the identity double cosets of $B \backslash G / Sp(2n,\mathbb{C})$ and $B \backslash G / SO(2n,\mathbb{C})$:

$$(\Phi_e f)(h) = f(h) \quad , \quad h \in Sp(2n,\mathbb{C}) \quad \text{or} \quad SO(2n,\mathbb{C}) \quad . \tag{II.6}$$

These maps generate irreducible representations since the elements $\Phi_e f$ transform to the left properly:

$$(\Phi_e f)(\hat{b}\, h) = f(\hat{b}\, h) = \pi^{(m)}(\hat{b})(\Phi_e f)(h) \quad , \tag{II.7}$$

where the Borel subgroup $\hat{B}$ of $Sp(2n,\mathbb{C})$ or $SO(2n,\mathbb{C})$ is given by $\hat{B} = H \cap B$, $H = Sp(2n,\mathbb{C})$ or $SO(2n,\mathbb{C})$ and B is the Borel subgroup of $GL(2n,\mathbb{C})$. Thus, the irreducible representations of $Sp(2n,\mathbb{C})$ or $SO(2n,\mathbb{C})$ are concretely realized on polynomial spaces and as with $GL(n,\mathbb{C})$, if $h_o \in Sp(2n,\mathbb{C})$ or $SO(2n,\mathbb{C})$ is restricted to the compact subgroup, the representation

$$(R^{(m)}_{h_o} \Phi_e f)(h) = (\Phi_e f)(hh_o) \quad , \quad f \in V^{(m)} \quad ,$$

remains irreducible and becomes unitary with respect to the inner product (II.3).

III. Orthogonal Polynomial Bases and Matrix Elements of the Irreducible Representations of $GL(n,\mathbb{C})$

We are interested in obtaining the matrix elements of the irreducible unitary representations of U(n), and of computing the Clebsch-Gordan coefficients that occur in the decomposition of tensor products of irreducible representations. The strategy we follow is to write all irreducible representations as tensor products of fundamental representations and then build the matrix elements of the irreducible representations out of the matrix elements of the fundamental representations. Thus, in this section we first turn our attention to an analysis of the fundamental representations of $GL(n,\mathbb{C})$. The fundamental representations of $GL(n,\mathbb{C})$ are those representations of the form $(1 \ldots 10 \ldots 0)$ with s ones and n - s zeros $(n \geq s \geq 1)$. The orthogonal basis elements of the fundamental representations are of the form [3]

$$h^{(m)}_{[k]}(g) = \Delta^{1\ldots s}_{k_1\ldots k_s}(g) \tag{III.1}$$

where $\Delta^{\cdots}_{\cdots}$ is the minor of $g \in GL(n,\mathbb{C})$ formed from rows $1\ldots s$ and columns $k_1 \ldots k_s$ with $1 \leq k_1 < \ldots < k_s \leq n$. The relation of $k_1 \ldots k_s$ to a Gelfand pattern [k] is given in Reference [3]. With respect to the inner product $(\cdot,\cdot)$ of (II.3) it is easy to verify that

$$(\Delta^{1\ldots s}_{k_1\ldots k_s}, \Delta^{1\ldots s}_{k'_1\ldots k'_s}) = \delta_{k_1 k'_1} \cdots \delta_{k_s k'_s}\, s! \tag{III.2}$$

so that the basis elements (III.1) are indeed orthogonal (but not normalized to one). The matrix elements of the fundamental representations are obtained by right translating the basis elements:

$$(R_{g_o} h^{(m)}_{[k]})(g) = \Delta^{1\ldots s}_{k_1\ldots k_s}(gg_o) = \sum_{j_1\ldots j_s} \Delta^{1\ldots s}_{j_1\ldots j_s}(g)\, \Delta^{j_1\ldots j_s}_{k_1\ldots k_s}(g_o) \quad . \tag{III.3}$$

Thus, the matrix elements are given by

$$D^{(m)}_{[j][k]}(g) = \Delta^{j_1\ldots j_s}_{k_1\ldots k_s}(g) \quad , \quad (m) \text{ a fundamental representation} \quad . \tag{III.4}$$

To obtain matrix elements for the nonfundamental irreducible representations of $GL(n,\mathbb{C})$, we make use of the fact that any dominant weight can be viewed as the highest weight in an r-fold tensor product of fundamental representations. In this way we are able to explictly exhibit orthogonal polynomial bases $h^{(m)}_{[k]}(g)$. Thus, consider the irreducible representation (m) as arising from the r-fold tensor product of fundamental representations, in which the fundamental representation (10 ... 0) occurs $m_1 - m_2$ times, the fundamental representation (110 ... 0) occurs $m_2 - m_3$ times etc. For example, for the nonfundamental representation (311) of SU(3), (100) occurs two times while (110) does not occur at all, and (111) occurs once.

Let F be an element of the r-fold tensor product space $V^{(m^1)} \otimes \ldots \otimes V^{(m^r)}$ where the fundamental representations $(m^1) \ldots (m^r)$ are determined by the representation (m). Let the identity double coset map be given by $(\Phi_e F)(g) = F(g\ldots g)$; under Φ_e the element F is sent into the irreducible space $V^{(m)}$, since

$$(\Phi_e F)(bg) = F(bg\ldots bg) = \pi^{(m)}(b)(\Phi_e F)(g) \quad .$$

So the identity double coset map generates an irreducible representation, the highest weight representation obtained by adding all the fundamental representations together. However, in general, orthogonal basis elements $h^{(m^1)}_{[k^1]} \otimes \ldots \otimes h^{(m^r)}_{[k^r]}$ do not remain orthogonal under Φ_e, so that $\Phi_e h^{(m^1)}_{[k^1]} \otimes \ldots \otimes h^{(m^r)}_{[k^r]}$ is not orthogonal in $V^{(m)}$. Reference [4] shows that an orthogonal basis is formed under pattern addition; that is

$$\begin{aligned} h^{(m)}_{[k]}(g) &= \sum_{[k^1]+\ldots+[k^r]=[k]} (\Phi_e h^{(m^1)}_{[k^1]} \otimes \ldots \otimes h^{(m^r)}_{[k^r]})(g) \\ &= \sum_{[k^1]+\ldots+[k^r]} h^{(m^1)}_{[k^1]}(g) \ldots h^{(m^r)}_{[k^r]}(g) \end{aligned} \tag{III.5}$$

does form an orthogonal basis in $V^{(m)}$.

As shown in Reference [4], a matrix element for a nonfundamental representation can be written as

$$D^{(m)}_{[k'][k]}(g) = \sum_{\substack{[k^1]\ldots[k^r] \\ [k^{1'}]\ldots[k^{r'}]}} \langle (m)[k'] | (m^1)[k^{1'}] \ldots (m^r)[k^{r'}] \rangle$$

$$\times\, D^{(m^1)}_{[k^{1'}][k^1]}(g) \ldots D^{(m^r)}_{[k^{r'}][k^r]}(g) \langle (m^1)[k^1] \ldots (m^r)[k^r] | (m)[k] \rangle \quad . \qquad \text{(III.6)}$$

Thus, to compute the matrix element $D^{(m)}_{[k'][k]}(g)$, it is necessary to know the matrix elements of the fundamental representations $D^{(m^i)}_{[k^{i'}][k^i]}(g)$ and the highest weight Wigner coefficients relating $(m) = \Sigma_{i=1}^{r} (m^i)$ to the fundamental representations. But the matrix elements of the fundamental representations are given in Eq. (III.4). It remains to compute the highest weight Wigner coefficients.

Reference [4] shows that up to a normalization factor (which can also be explicitly computed),

$$\langle (m)[k] | (m^1)[k^1] \ldots (m^r)[k^r] \rangle = (h^{(m)}_{[k]}, \Phi_e h^{(m^1)}_{[k^1]} \otimes \ldots \otimes h^{(m^r)}_{[k^r]}) \quad . \qquad \text{(III.7)}$$

Since the fundamental bases are known from Eq. (III.1) and the nonfundamental bases are given by pattern addition, Eq. (III.5), it follows that the Clebsch-Gordan coefficients can always be computed using the inner product, (II.3). Hence, the matrix elements $D^{(m)}_{[k][k']}(g)$ can be computed for any representation (m), using Eq. (III.6).

IV. The Multiplicity Problem for $GL(n,\mathbb{C})$

We now turn to a problem of central importance in the representation theory of compact groups, the decomposition of an r-fold tensor product into a direct sum of irreducible representations. Let V^r denote the r-fold tensor product space, $V^r = V^{(m^1)} \otimes \ldots \otimes V^{(m^r)}$, where now the irreducible representations (m^i) are not necessarily fundamental representations. We want to construct maps $\Lambda^{(m)}_{\eta}$ from the irreducible space $V^{(m)}$ to V^r; η is a multiplicity label that distinguishes between equivalent representations in the direct sum decomposition of V^r. The idea is to first construct maps from $V^{(m)}$ to a space denoted P_f by Weyl, and then map elements from P_f to V^r. Let (1) denote the fundamental representation (10 ... 0), with n - 1 zeros. Then $P_f \equiv V^{(1)} \otimes \ldots \otimes V^{(1)}$, an f-fold tensor product space, where

$$f = \sum_{i=1}^{n} m_i^1 + \ldots + \sum_{i=1}^{n} m_i^r \quad .$$

Note that a necessary (but not sufficient) condition that the representation (m) be contained in V^r is that $\Sigma_{i=1}^{n} m_i = f$.

It is straightforward to construct a map from P_f to a space containing V^r; such a map is a double coset map of the form

$$(\Phi\, F)(g_1 \ldots g_r) = F(p_1 g_1, p_2 g_1 \ldots p_f g_r) \quad , \quad F \in P_f \qquad \text{(IV.1)}$$

where the p_i are elements of the permutation group S_n chosen so as to make Φ F transform to the left as (m^1), (m^2) ... (m^r) with respect to the diagonal subgroup of $GL(n,\mathbb{C})$. A subspace of elements Φ F, $F \in P_f$, will be those elements that transform to the left as

$$(\Phi\ F)(b_1 g_1 \ldots b_r g_r) = \pi^{(m^1)}(b_1) \ldots \pi^{(m^r)}(b_r)(\Phi\ F)(g_1 \ldots g_r) \quad ; \qquad \text{(IV.2)}$$

these elements will generate the tensor product space of interest, namely $V^{(m^1)} \otimes \ldots \otimes V^{(m^r)}$.

So what must be done is to construct maps $\Omega_\alpha^{(m)}$ from $V^{(m)}$ to P_f. This will be done with the help of several theorems of Weyl [5]:

Theorem 1.--Every dominant weight (m) of $GL(n,\mathbb{C})$ satisfying $\Sigma_{i=1}^n m_i = f$ is in 1 - 1 correspondence with an irreducible representation of S_f, the symmetric group on f numbers.

Theorem 2.--The number of times the irreducible representation (m) is contained in P_f equals the degree of the associated representation of S_f.

The map $\Omega_\alpha^{(m)}$ can be written as

$$\Omega_\alpha^{(m)} h_{[k]}^{(m)} = \Sigma\ C_{[k][k^1]\ldots[k^f]}^{(m)(1)\ldots(1)\ \alpha}\ h_{[k^1]}^{(1)} \otimes \ldots \otimes h_{[k^f]}^{(1)} \in P_f \qquad \text{(IV.3)}$$

and with the help of the Weyl theorems, the coefficients $C^{\cdots}_{\cdots}$ may be computed. From Theorem 2 it follows that the multiplicity label α comes from the irreducible representation of S_f associated with (m). If $\Omega_\alpha^{(m)}$ is composed with Φ, the Clebsch-Gordan coefficients become (up to a normalization factor)

$$\langle (m^1)[k^1] \ldots (m^r)[k^r] | (m)\alpha[k] \rangle = (h_{[k^1]}^{(m^1)} \otimes \ldots \otimes h_{[k^r]}^{(m^r)}, \Phi\Omega_\alpha^{(m)} h_{[k]}^{(m)}) \qquad \text{(IV.4)}$$

and it is seen that the multiplicity labels needed to distinguish equivalent representations (m) occurring in V^r are obtained from the representations of the symmetric group that label equivalent representations in P_f.

References

[1] Proofs and further references can be found in W. H. Klink and T. Ton-That, Holomorphic Induction and the Tensor Product Decomposition of Irreducible Representations of Compact Groups. I. SU(n) Groups, Ann. Inst. Henri Poincaré 31 (1979) 77-97.

[2] W. H. Klink and T. Ton-That, Orthogonal Polynomial Bases for Holomorphically Induced Representations of the General Linear Groups, Ann. Inst. Henri Poincaré 31 (1979) 99-113.

[3] W. H. Klink and T. Ton-That, Construction Explicite Non Itérative des Bases de GL(n,¢)-modules, C. R. Acad. Sc. Paris, Série B 289 (1979) 115-118.

[4] W. H. Klink and T. Ton-That, Matrix Elements of the General Linear Groups, to be submitted for publication.

[5] H. Weyl, The Classical Groups: Their Invariants and Representations, Princeton University Press, Princeton, NJ, 1939, ch. IV.

SHIFT OPERATORS FOR DISCRETE UNITARY REPRESENTATIONS OF O(p,q)

Timothy C. Murphy
Department of Physics, University of Wisconsin-Madison
Madison, WI 53706/USA

Introduction

The orthogonal groups arise in connection with many physical problems, ranging from the study of angular momentum to the construction of grand unified theories. A powerful approach to the study of the representations of these groups is the use of shift operators, an idea introduced by Pang and Hecht[1] and developed by Wong[2] and Bincer.[3] In this paper I extend Bincer's technique to those non-compact orthogonal groups possessing discrete representations.

1. Definitions and Notation

The construction of representations using shift operators relies on the use of Gel'fand-Tsetlin[4] subgroup chains and their corresponding Gel'fand-Tsetlin patterns. This technique is discussed at length by Bincer,[3] and I shall restrict myself to a brief survey concentrating on the points where the non-compact algebras differ from the compact ones.

I shall consider the algebra O(p,q) where p is even. The algebra is generated by operators G^a_b satisfying the commutation relations

$$[G^a_b, G^c_d] = \delta^c_b G^a_d - \delta^a_d G^c_b + \delta^{\bar{b}}_d G^{\bar{c}}_{\bar{a}} - \delta^{\bar{c}}_{\bar{a}} G^{\bar{b}}_d \quad \text{with } \bar{a} = -a \tag{1.1}$$

The range of the indices is given by

$$\bar{v} \le a \le v \qquad \text{for } p + q = 2v + 1 \tag{1.2}$$

$$\bar{v} \le a \le v \, , \quad a \ne 0 \qquad \text{for } p + q = 2v \, . \tag{1.3}$$

For the non-compact algebras the indices are divided into two blocks:

$$\text{Block 1} = \{a | 1 \le |a| \le \tfrac{p}{2}\} \tag{1.4}$$

$$\text{Block 2} = \{0\} \cup \{a | \tfrac{p}{2} + 1 \le |a| \le v\} \, . \tag{1.5}$$

In a unitary representation I require

$$(G^a_b)^\dagger = \begin{cases} G^b_a & \text{if a and b are in the same block} \\ -G^b_a & \text{if a and b are in different blocks.} \end{cases} \tag{1.6}$$

The ν operators G_a^a, $a = 1...\nu$, may be chosen to be simultaneously diagonal and their eigenstates form the basis for the representation, as described by Bincer. The eigenvalues of these generators are called the weight of the representation.

In constructing a Gel'fand-Tsetlin basis one must proceed down a "canonical" subgroup chain. For the non-compact algebra I use the chain

$$0(p,q) \supset 0(p,q-1) \supset 0(p,q-2) \supset ... \supset 0(p,1) \supset 0(p) \supset ... \supset 0(2) \qquad (1.7)$$

which has the property that each subgroup in the chain has a discrete representation. The generators of the subgroups are defined by Eq. (1.1) - (1.3) as before, except that the range of the indices is reduced. For a non-compact subgroup of rank ℓ the allowed range is

$$\ell \text{ odd} \quad |a| \in \{0\} \cup \{1... \tfrac{p}{2}\} \cup \{\tfrac{p}{2} +1+\nu- \tfrac{\ell-1}{2} ,...\nu\} \qquad (1.8)$$

$$\ell \text{ even} \quad |a| \in \qquad \{1... \tfrac{p}{2}\} \cup \{\tfrac{p}{2} +1+\nu- \tfrac{\ell}{2} ,...\nu\} . \qquad (1.9)$$

For a compact subgroup of rank ℓ the allowed range is

$$\ell \text{ odd} \quad |a| \in \{0\} \cup \{1... \tfrac{\ell-1}{2}\} \qquad (1.10)$$

$$\ell \text{ even} \quad |a| \in \qquad \{1... \tfrac{\ell}{2}\} . \qquad (1.11)$$

A state in the completed Gel'fand-Tsetlin pattern is specified by a set of labels m_i^ℓ, where m_i^ℓ is the eigenvalue of the operator G_i^i acting on the highest weight state of the rank ℓ subgroup.

2. Shift Operators

All the states in a discrete representation may be generated from the highest weight state by the application of appropriate shift operators. The action of such an operator is given by

$$^\ell S_a \left| \begin{array}{c} m_i^{p+q} \\ \vdots \\ m_i^\ell \\ \vdots \\ m_i^{\ell-1} \end{array} \right\rangle = \left(\begin{array}{c} m_i^{\ell-1} \\ \vdots \\ m_i^{\ell-1} + \delta_{i\bar{a}} - \delta_{ia} \end{array} \right) \left| \begin{array}{c} m_i^{p+q} \\ \vdots \\ m_i^\ell \\ \vdots \\ m_i^{\ell-1} + \delta_{i\bar{a}} - \delta_{ia} \end{array} \right\rangle \qquad (2.1)$$

The operator $^\ell S_a$ changes the $|a|$ component of the weight of the rank $\ell-1$ subgroup by one unit up or down, depending on the sign of a. The commutator of two such operators is zero, indicating that each component of the weight may be shifted independently. The highest weight state has the property

$$ {}^{\ell}S_{\bar a} \left| m_i^{\ell} \right\rangle_{max} = 0 \quad \text{for all} \quad a > 0 \ . \tag{2.2} $$

The functional form of the shift operators is the same as in the compact case:

$$ {}^{\ell}S_{a} = \{V(\bar v) \prod_{j=\bar v}^{a-1}{}' (G - c_j^{\ell-1} \mathbb{1})\}_a \tag{2.3} $$

$$ \begin{aligned} V(\bar v)_a &= G_a^{o} && \text{for } \ell \text{ odd} \\ &= \frac{1}{\sqrt{2}} (G_a^{\delta} - G_a^{\bar\delta}) && \text{for } \ell \text{ even .} \end{aligned} \tag{2.4} $$

The prime in Eq. (2.3) indicates the range of j is restricted to that of the rank ℓ-1 subgroup, and in Eq. (2.4) I have defined $\delta = \frac{p}{2} + 1 + v - \frac{\ell}{2}$. The constants $c_j^{\ell-1}$ are given by

$$ c_j^{\ell-1} = m_j^{\ell-1} + \sum_{b=\bar v}^{j} (1-\delta_{\bar b}^{j}) \ . \tag{2.5} $$

The normalization coefficient in Eq. (2.1) may be evaluated in terms of the m_i^{ℓ} and the value of the quadratic casimir operator. The resulting expressions are quite lengthy and not at all transparent, so I shall not reproduce them here. The requirement that these coefficients be real places restrictions on the values of the m_i^{ℓ} as follows:

$$ \ell > p \qquad m_i^{\ell} \geq m_i^{\ell-1} \geq m_{i-1}^{\ell} \tag{2.6} $$

$$ m_i^{\ell-1} \geq v - \frac{\ell-1}{2} + 1 - i \qquad \text{for } i > \frac{p}{2},\ \ell \text{ odd} \tag{2.7} $$

$$ m_i^{\ell-1} \leq 1 - i \qquad \text{for } i \leq \frac{p}{2},\ \ell \text{ odd} \tag{2.8} $$

$$ m_i^{\ell-1} \geq v - \frac{\ell}{2} + 2 - i + |m_\delta^{\ell}| \qquad \text{for } i > \frac{p}{2},\ \ell \text{ even} \tag{2.9} $$

$$ m_i^{\ell-1} \leq -1 - i - |m_\delta^{\ell}| \qquad \text{for } i \leq \frac{p}{2},\ \ell \text{ even} \tag{2.10} $$

$$ \ell \leq p \qquad m_i^{\ell} \leq m_i^{\ell-1} \leq m_{i+1}^{\ell} \leq 0 \ . \tag{2.11} $$

Notice that the label $m_i^{\ell-1}$ has no lower bound for $\ell > p$, so the representation is (denumerably) infinite dimensional.

For the compact orthogonal algebras a representation is uniquely specified by the values of the casimir operators. For the non-compact algebras, however, one may construct two inequivalent discrete representations with the same casimir values, one based on a highest weight state as above and one constructed similarly but using a lowest weight state. Furthermore, if both p and q are even, one may interchange their roles and construct a representation using the chain

$$O(p,q) \supset O(p-1,q) \supset O(p-2,q) \supset \dots \supset O(2) \ . \tag{2.12}$$

Such a representation would have the same highest (or lowest) weight state as that described previously, but would not be equivalent to it.

Acknowledgement

I would like to thank Dr. A. M. Bincer for patient instruction, assistance and encouragement in this work.

References

1. S. C. Pang and K. T. Hecht, J. Math. Phys. 8, 1233 (1967).
2. M. K. F. Wong, J. Math. Phys. 8, 1899 (1967).
3. A. M. Bincer, J. Math. Phys. 19, 1173 (1978).
4. I. Gel'fand and M. Tsetlin, Dok. Akad. Navk. USSR 71, 1017 (1951).

MAXIMAL ABELIAN SUBALGEBRAS OF sp(2n,$\mathbb{R}$) AND THEIR APPLICATIONS IN PHYSICS

J. Patera and P. Winternitz
Centre de recherches de mathématiques appliquées
Université de Montréal
Montréal, Québec, Canada

and

H. Zassenhaus

Department of Mathematics
Ohio State University
Columbus, Ohio

1. INTRODUCTION

The classification and construction of maximal abelian subgroups of the classical Lie groups and maximal abelian subalgebras (MASAs) of the classical Lie algebras is an important problem both from the point of view of mathematics and of physical applications. The Lie algebra of a compact simple Lie group has, upto conjugation under the group, a single MASA, its Cartan subalgebra. In the noncompact case the situation is quite different. A MASA is a Cartan subalgebra only if it is selfnormalizing in the entire Lie algebra. Each complex simple Lie algebra has, upto conjugacy, a single Cartan subalgebra. Each simple real Lie algebra has a finite number of classes of Cartan subalgebras, classified by Kostant[1] and Sugiura[2]. A considerable amount of work has been devoted to MASAs of $g\ell(n,C)$ and $s\ell(n,C)$ (the early work of Schur and of Kravchuk has been reviewed by Suprunenko and Tyshkevich[3]. The highest dimensional MASAs of the classical complex Lie algebras were discussed by Maltsev[4].

We have studied the problem of classifying all MASAs of all classical Lie algebras over arbitrary fields. A publication in book form is in preparation[5]. Here we shall restrict ourselves to one specific real Lie algebra, namely the symplectic Lie algebra sp(2n,$\mathbb{R}$) and present results without any attempt at providing proofs.

2. GENERAL STRUCTURE OF MASAs

We shall work in the defining representation of sp(2n,$\mathbb{R}$), i.e. $X \in \mathbb{R}^{2n\times 2n}$

$$XK + KX^T = 0, \quad K = -K^T. \tag{1}$$

A set of matrices $\{X\}$ is orthogonally decomposable if there exists a nonsingular matrix G simultaneously transforming K and all $X \in \{X\}$ into the block diagonal form

$$\begin{gathered} GXG^{-1} = \oplus_{i=1}^{s} X_i \quad , \quad GKG^T = \oplus_{i=1}^{s} K_i, \quad 1 \le i \le s \\ X_i, K_i \in \mathbb{R}^{2n_i\times 2n_i}, \quad \Sigma_{i=1}^{s} n_i = n \quad , \quad X_i K_i + K_i X_i^T = 0 \\ K_i = -K_i^T. \end{gathered} \tag{2}$$

If s is maximal, we have an orthogonal *Remak decomposition*; if s=1 is maximal, then the set {X} is *orthogonally indecomposable*.

For each orthogonal Remak decomposition we obtain a MASA of $sp(2n,\mathbb{R})$ by taking a MASA of each $sp(2n_i,\mathbb{R})$ in the decomposition. At most one of these MASAs of $sp(2n_i,\mathbb{R})$ is allowed to be a maximal abelian nilpotent subalgebra (MANS), i.e. a MASA consisting entirely of nilpotent matrices. Each of the MASAs of $sp(2n_i,\mathbb{R})$ is itself orthogonally indecomposable.

The classification of MASAs of $sp(2n,\mathbb{R})$ is thus reduced to the classification of orthogonally indecomposable MASAs of $sp(2n_i,\mathbb{R})$ with $1 \le n_i \le n$.

The following types of orthogonally indecomposable MASAs of $sp(2n,\mathbb{R})$ occur.

1. Maximal abelian nilpotent subalgebras

These consist entirely of nilpotent matrices (all eigenvalues equal to zero). Each MANS of $sp(2n,\mathbb{R})$ is characterized by a Kravchuk signature, i.e. a triplet of integers

$$(\lambda,2\mu,\lambda), \quad \lambda+\mu = n, \quad 1 \le \lambda \le n, \quad 0 \le \mu \le n-1. \tag{3}$$

The matrices X of a given MANS can be simultaneously transformed to the form

$$X = \begin{pmatrix} 0^{\lambda} & B & D & Y \\ 0 & R & S & JD^T \\ 0 & 0^{\mu} & -JR^TJ & -JB^T \\ 0 & 0 & 0 & 0^{\lambda} \end{pmatrix}, \quad \begin{array}{l} B,D \in \mathbb{R}^{\lambda\times\mu}, \\ Y = Y^T \in \mathbb{R}^{\lambda\times\lambda}, \\ R,S \in \mathbb{R}^{\mu\times\mu}, \ \ R \text{ nilpotent} \end{array} \tag{4}$$

satisfying $XK_{\lambda\mu} + K_{\lambda\mu}X^T = 0$, with

$$K_{\lambda\mu} = \begin{pmatrix} & & & I_\lambda \\ & & J_\mu & \\ & -J_\mu & & \\ -I_\lambda & & & \end{pmatrix}, \quad K_{\lambda 0} \equiv K_\lambda = \begin{pmatrix} 0 & I_\lambda \\ -I_\lambda & 0 \end{pmatrix}, \quad J_\mu = \begin{pmatrix} & & 1 \\ & 1 & \\ \cdot^{\cdot^{\cdot}} & & \\ 1 & & \end{pmatrix} \in \mathbb{R}^{\mu\times\mu}. \tag{5}$$

The matrices X are nilpotent and symplectic. The fact that they form a MASA implies that D, R and S depend linearly on the matrix B; $Y=Y^T$ is arbitrary. The matrix B has μ free entries, all other entries depend linearly on these μ independent ones. For a discussion of the possible forms of the entries in (4) we refer to our detailed publication[5]. Let us just mention that the dimension of a MANS of signature $(\lambda,2\mu,\lambda)$ is $d_{n\lambda} = n + \frac{\lambda(\lambda-1)}{2}$ and that the highest possible dimension is $d_{max} = d_{nn} = n(n+1)/2$, the lowest one is $d_{min} = d_{n1} = n$. Up to conjugacy, just one MANS corresponds to the Kravchuk signature $(n,0,n)$, two to $(n-1,2,n-1)$ (for $n\ge2$) and 36 to $(n-2,4,n-2)$ (for $n\ge6$).

2. Indecomposable MASAs that are not MANS

By definition, these MASAs must contain at least one nonnilpotent element. We have shown[5] that they contain precisely one nonnilpotent element X_0, satisfying

$$X_0^2 + 1 = 0$$

(X_0 represents the imaginary unit i). The matrix X_0 can always be written in the same standard form, the matrix K of (1) in one of $[\frac{n}{2}] + 1$ different forms

$$X_0 = \begin{pmatrix} K_2 & & & \\ & K_2 & & \\ & & \ddots & \\ & & & K_2 \end{pmatrix} \qquad K = \begin{pmatrix} \left.\begin{matrix} K_4 & & \\ & \ddots & \\ & & K_4 \end{matrix}\right\} k & \\ & \left.\begin{matrix} K_2 & & \\ & \ddots & \\ & & K_2 \end{matrix}\right\} \ell \end{pmatrix}, \quad n = 2k+\ell. \tag{6}$$

The corresponding MASAs are obtained by taking the centralizer of X_0 in $sp(2n,\mathbb{R})$ and finding all of its indecomposable MASAs. The centralizer of X_0 in the realization with $n = 2k+\ell$ is

$$\text{cent } X_0 = u(\ell+k,k) \tag{7}$$

so that the problem reduces to that of finding all indecomposable MASAs of the pseudo-unitary Lie algebra $u(\ell+k,k)$. This in turn essentially reduces to a classification of MANSs of $su(\ell+k,k)$.

3. Decomposable MASAs

A decomposable MASA that is orthogonally indecomposable consists of matrices that can be written in the form

$$X = \begin{pmatrix} A & 0 \\ 0 & -A^T \end{pmatrix}, \quad K = \begin{pmatrix} 0 & I_n \\ -I_n & 0 \end{pmatrix}, \quad A \in \mathbb{R}^{n\times n}. \tag{8}$$

To obtain all such decomposable MASAs of $sp(2n,\mathbb{R})$ we must find all indecomposable MASAs of $g\ell(n,\mathbb{R})$ which again essentially reduces to a classification of MANS of $s\ell(n,\mathbb{R})$ and of $s\ell(\frac{n}{2},C)$ (for n even).

We see that classification of MASAs of $sp(2n,\mathbb{R})$ is intimately related to a classification of MASAs of other classical Lie algebras and that the essence of the task is a classification of MANSs.

3. EXAMPLE: MASAs OF $sp(4,\mathbb{R})$

Low dimensional cases have been treated explicitly and completely. Thus, $sp(2,\mathbb{R})$ has 3 classes of MASAs, $sp(4,\mathbb{R})$ has 10, $sp(6,\mathbb{R})$ has 29. For $sp(4,\mathbb{R})$ they can be represented as follows.

(i) Orthogonally decomposable MASAs

$$S_1 = \begin{pmatrix} a & & & \\ & -a & & \\ & & b & \\ & & & -b \end{pmatrix}, \quad S_2 = \begin{pmatrix} a & 0 & & \\ 0 & -a & & \\ & & 0 & b \\ & & -b & 0 \end{pmatrix}, \quad S_3 = \begin{pmatrix} 0 & a & & \\ -a & 0 & & \\ & & 0 & b \\ & & -b & 0 \end{pmatrix}$$

$$S_4 = \begin{pmatrix} 0 & a & & \\ -a & 0 & & \\ & & 0 & b \\ & & 0 & 0 \end{pmatrix}, S_5 = \begin{pmatrix} a & 0 & & \\ 0 & -a & & \\ & & 0 & b \\ & & 0 & 0 \end{pmatrix}, \text{with } K = \begin{pmatrix} K_2 & 0 \\ 0 & K_2 \end{pmatrix}.$$

(ii) MANSs with Kravchuk signature $(\lambda, 2\mu, \lambda)$

(202): $$S_6 = \begin{pmatrix} 0 & Y \\ 0 & 0 \end{pmatrix}, \quad Y = \begin{pmatrix} a & b \\ b & c \end{pmatrix}, \quad K = \begin{pmatrix} 0 & I_2 \\ -I_2 & 0 \end{pmatrix}$$

(121): $$S_7 = \begin{pmatrix} 0 & b & 0 & c \\ 0 & 0 & b & 0 \\ 0 & 0 & 0 & -b \\ 0 & 0 & 0 & 0 \end{pmatrix} \quad K = \begin{pmatrix} 0 & J_2 \\ -J_2 & 0 \end{pmatrix}$$

(iii) Indecomposable, not MANSs

$$S_8 = \begin{pmatrix} 0 & b & c & 0 \\ -b & 0 & 0 & c \\ 0 & 0 & 0 & b \\ 0 & 0 & -b & 0 \end{pmatrix} \quad K = \begin{pmatrix} 0 & I_2 \\ -I_2 & 0 \end{pmatrix}$$

(iv) Decomposable orthogonally indecomposable MASAs

$$S_9 = \begin{pmatrix} a & b & & \\ -b & a & & \\ & & -a & b \\ & & -b & -a \end{pmatrix} \quad S_{10} = \begin{pmatrix} a & b & & \\ 0 & a & & \\ & & -a & 0 \\ & & -b & -a \end{pmatrix} \quad K = \begin{pmatrix} 0 & I_2 \\ -I_2 & 0 \end{pmatrix}$$

4. APPLICATIONS

We would like to mention several applications of the MASA classification.

(i) The solution of the Schrödinger equation $H\psi = E\psi$ where H is a quadratic Hamiltonian in phase space: $H = \sum_{i,k=1}^{n} (A_{ik}x_i x_k + B_{ik}p_i p_k + C_{ik}(x_i p_k + p_k x_i))$. The group $Sp(2n,\mathbb{R})$ of canonical transformations was used to simplify H, the simplified H was embedded into a complete set of commuting quadratic operators in phase space (a MASA of $sp(2n,\mathbb{R})$). This was used to separate variables and solve the equation[6].

(ii) The separation of variables in Helmholtz and Hamilton Jacobi Equations in spaces of constant curvature[7] (using MASAs and other abelian subalgebras e.g. of o(p,q) algebras).

(iii) The derivation of superposition principles for coupled sets of Riccati equations[8] (using MASAs of o(p,q) and $s\ell(n,\mathbb{R})$).

Other applications are in progress.

REFERENCES

1. B. Kostant, Proc. Nat. Acad. Sci. USA 41, 967 (1955).
2. M. Sugiura, J. Math. Soc. Japan 11, 374 (1959).
3. D.A. Suprunenko and R.I. Tyshkevich, *Commutative Matrices*, Academic Press, New York (1968).
4. A.I. Mal'tsev, Izv. Akademii Nauk, SSSR, Ser. Mat. 9, 291 (1945) [Am. Math. Soc. Transl. Ser. 1, Vol. 9, 214 (1962)].
5. J. Patera, P. Winternitz and H. Zassenhaus, *Maximal Abelian Subalgebras of the Classical Lie Algebras*, Preprint CRMA-934, Montréal, 1980.
6. M. Moshinsky and P. Winternitz, J. Math. Phys. 21, 1667 (1980).
7. W. Miller Jr, J. Patera and P. Winternitz, J. Math. Phys. 21, XXXX (1980).
8. R.L. Anderson and P. Winternitz, these Proceedings.

POLYNOMIAL SPACE GROUP TENSORS

D. Phaneuf and R. T. Sharp
Physics Department, McGill University,
Montreal, Quebec, Canada

1. Introduction

An (ℓ,m) tensor of a group G is one whose components transform by the representation ℓ of G and are homogeneous polynomials in the components of a tensor which transforms by the representation m. One knows how to find them for point groups[1,2]. The method applies to any finite group and it is our purpose to show that it may be used to solve the same problem for space groups.

The first step in enumerating (ℓ,m) tensors of a finite group G is to calculate the generating function

$$B_{\ell,m}(\lambda) = \sum_{n=o} C^{(n)}_{\ell,m} \lambda^n \tag{1,1}$$

for the pair (ℓ,m) of irreducible representations (IR's). The coefficient $C^{(n)}_{\ell,m}$ in (1,1) is the number of linearly independent (ℓ,m) tensors of degree n. It turns out that $B_{\ell,m}(\lambda)$ is a rational function

$$B_{\ell,m}(\lambda) = \sum_s h^{(s)}_{\ell,m} \lambda^s / \prod_t (1-\lambda^t). \tag{1,2}$$

The sum over s and the product over t are both finite. The coefficients $h^{(s)}_{\ell,m}$ are non-negative integers; the denominator factors are equal in number to the dimension of the IR m and are the same for all generating functions with the same m. The denominator factors correspond to functionally independent scalars, of degrees t. The numerator terms correspond to (ℓ,m) tensors, of degrees s, which are linearly independent even when their coefficients belong to the ring of denominator scalars. The numerator tensors and denominator scalars constitute an integrity basis for (ℓ,m) tensors; their algebraic form may be determined by imposing that they transform properly under the generators of the group.

2. Space group tensors

The (unitary) IR's of a space group G[3,4] are labelled $(\overline{k})_m$. The vector $\overline{k}$ lies in or on the boundary of a sector of reciprocal space comprising 1/g of the Brillouin zone, where g is the order of the point group of G; the integer m takes a finite number of values and distinguishes inequivalent IR's belonging to $\overline{k}$.

Write $\bar{k}$ as a linear combination of the reciprocal lattice vectors $\bar{K}_i$ and suppose for now that the coefficients are rational fractions p_i/q:

$$\bar{k} = q^{-1} \ \Sigma_i p_i \bar{K}_i$$

If $\bar{a}_i$ are the primitive lattice translations, it is apparent that translations differing by $q\bar{a}_i$ have the same effect on a state (tensor) and need not be distinguished. The number of effective translations is finite and space group problems map on those of a finite group. The crystal lattice is infinite; no periodic boundary conditions are imposed. But we restrict our attention to those representations whose κ vector has as components rational fractions with common denominator q (or a divisor of q). The number of IR's is then finite, as befits a finite group. In forming tensor products, of which polynomial tensors are a special case, the $\bar{k}$-vectors are additive, so one never goes outside the space of IR's with fixed denominator q.

There are straightforward algorithms for constructing the matrices of all (unitary) IR's of a space group G[3,5]. Starting with a one-demensional IR of the translation group, one induces IR's of higher groups till one arrives at the group of $\bar{k}$, and, in a final step, at G. When the matrices are known, the methods of Ref. 1 may be used to find the generating function $B_{\ell m}(\lambda)$ for any pair of IR's. Miller and Love[3] have tabulated the matrices for most IR's of 3-dimensional space groups.

Jarić and Birman have given two algorithms[6] for computing Molien functions for space groups (a Molien function is a generating function $B_{\ell m}(\lambda)$ in which ℓ is the scalar IR of the group). They find[7] a number of Molien functions for the nonsymmorphic space group Pm3n based on IR's whose $\bar{k}$'s are at points of special symmetry in the Brillouin zone. Our work was inspired by their papers and by a talk by one of the authors. We do not follow their methods, but our results agree where they overlap. Birman's articles contain discussions of the application of space group methods to physical problems, in particular to Raman and infrared transitions in cyrstals[4] and to crystal phase transitions[8].

3. Conclusions

We do not reproduce here explicit formulas for generating functions $B_{\ell m}(\lambda)$. They are numerous, even for a single space group; typical examples for the two-dimensional space group p4m (square lattice with reflections) and Pm3n will be contained in the M.Sc. thesis of one of us (D.P.) and in a forthcoming publication.

The finite group on which a space group maps can often be given a geometrical interpretation. Thus p4m IR's with denominator q are those of the symmetry group of a 4-dimensional figure consisting of two 2q-sided regular polygons centered at the origin and lying in the 12 and 34 planes. Rotations of the 12 and 34 planes correspond to x and y translations; rotation of the 12 plane into 34 corresponds to rotation of x into y; inversions of the polygons correspond to x- and y- reflections.

The restriction to rational points $\bar{k}$ with denominator q can be removed by letting $q\to\infty$. Dimensionality considerations show that nothing is lost in this process.

Polynomial tensors correspond to completely symmetric plethysms (one-rowed Young diagrams). Generating functions for space group plethysms of all exchange symmetries can be derived by the methods of Ref. 9. Generating functions based on irreducible tensors can be combined to form generating functions based on reducible tensors.

References

1. J. Patera, R. T. Sharp, and P. Winternitz, J. Math. Phys. 19, 2362 (1978).
2. P. E. Desmier and R. T. Sharp, J. Math. Phys. 20, 74 (1979).
3. S. C. Miller and W. F. Love, "Tables of IR's of space groups and co-representations of magnetic space groups", (Pruett Press, Boulder, Col. 1967).
4. J. L. Birman, "Theory of crystal space groups and infra-red and Raman processes in crystals", Handbuch der Physik 25/2b (Springer-Verlag, Berlin 1974).
5. I.V.V. Raghavacharyulu, Can J. Phys. 39, 830 (1961).
6. M. V. Jarić and J. L. Birman, J. Math. Phys. 18, 1456 (1977).
7. M. V. Jarić and J. L. Birman, J. Math. Phys. 18, 1459 (1977).
8. J. L. Birman, "Symmetry change in continuous phase transitions in crystals", Second international colloquium on group theoretical methods in physics. (Nijmegen, 1972) ed. A. Janner.
9. J. Patera and R. T. Sharp, J. Phys. A: Math. Gen. 13 (397) 1980.

A PERMUTATION CONVENTION FOR SUn CONSISTENT WITH DUALITY

John J. Sullivan
Physics Department
University of New Orleans

New Orleans, LA 70122

I Review of Duality

The works of Frobenius and Schur[1] and later Weyl[2] identified the fundamental duality between the representation theory of the symmetric group S_L and tensor representations of the general linear group Gln and its unitary Un and unitary unimodular SUn subgroups. The most frequent exploitation of this duality has been to analyze the decomposition series of various tensor products. In a series of papers[3] we have developed a fairly complete correspondence for outer product coupling in S_L and inner product coupling in Un or for the basis adaption of Un to the subgroup sequence $U(n_1+n_2)/\ Un_1 \times Un_2$. Duality also exists between inner product coupling in S_L and basis adaption to the subgroup sequence $Un_1n_2/Un_1 \times Un_2$, but we have not examined this aspect in any detail.

The basic result of Schur and Weyl is that Lth rank tensors of a basic n dimensional defining vector space may be reduced w.r.t. the action of $S_L \times Un$ into symmetrized tensors labeled by a common irrep λ with indexes m and M describing the action of the two groups.

$$\pi\ \mu \left|\begin{array}{c} m \\ \lambda\ n \\ M \end{array}\right\rangle = \left|\begin{array}{c} m' \\ \lambda\ n \\ M' \end{array}\right\rangle \begin{bmatrix} \lambda & \pi \\ m' & m \end{bmatrix} \begin{bmatrix} \lambda & \mu \\ M' & M \end{bmatrix} \qquad \pi \in S_L\ ,\quad \mu \in Un\ . \tag{1.1}$$

(The convention of summing repeated Roman indices is adopted.) Tensors can be coupled by two dual procedures: using the Wigner coefficients of Un (or identically SUn)

$$\left|\begin{array}{c} {}_i^m \\ {}_i^\lambda \\ \lambda\ n \\ M \end{array}\right\rangle = \left|\begin{array}{c} {}_i^m \\ {}_i\lambda n \\ {}_i^M \end{array}\right\rangle \left\langle\begin{array}{c} {}_i^\lambda \\ {}_i^M \end{array}\right|\left.\begin{array}{c} \lambda \\ M \end{array}\right\rangle_n \tag{1.2}$$

and using projection operators in S_L

$$(\lambda|m,\lambda_j m_j) \left|\begin{array}{c} m_j \\ \lambda_j n_j \\ M_j \end{array}\right\rangle = \left|\begin{array}{c} m \\ \lambda n \\ M \end{array}\right\rangle \left\langle\begin{array}{c} \lambda \\ M \end{array}\right|\left.\begin{array}{c} \lambda_j \\ M_j \end{array}\right\rangle_n \tag{1.3}$$

General subscripts (pre or post) are meant to be taken over a range of values, usually two. Hence (1.2) in extended notation reads

$$\left|\begin{array}{cc} {}_1^m & {}_2^m \\ {}_1^\lambda & {}_2^\lambda \\ \lambda & n \\ M & \end{array}\right\rangle = \sum_{{}_1^M,\ {}_2^M} \left|\begin{array}{c} {}_1^m \\ {}_1^\lambda\ n \\ {}_1^M \end{array}\right\rangle \left|\begin{array}{c} {}_1^m \\ {}_1^\lambda\ n \\ {}_1^m \end{array}\right\rangle \left\langle\begin{array}{cc} {}_1^\lambda & {}_2^\lambda \\ {}_1^M & {}_2^M \end{array}\right|\left.\begin{array}{c} \lambda \\ M \end{array}\right\rangle_n \tag{1.2'}$$

The defining spaces n_j of the subtensors on the l.h.s. of (1.3) must be such that their union spans the defining space n.

For considering tensor factorization we have used the double coset (DC) decomposition $S_{{}_iL} \backslash S_L / S_{L_j}$. The DC representatives (DCR) q are in one to one correspondence with the symbol $q = \begin{bmatrix} L & L_j \\ {}_iL & {}_iL_j \end{bmatrix}$ the ${}_iL_j$ taking all possible positive values such that

${}_i\Sigma_i L_j = L_j$ and ${}_j\Sigma_i L_j = {}_iL$. A particular DCR can be considered to be the permutation

leaving the first ${}_1L_1$ and the last ${}_2L_2$ ordered elements invariant while repeating the cyclic permutation of the intermediate $({}_1L_2+{}_2L_1)$ ordered elements ${}_2L_1$ times and hence a member of the class $(1)^{({}_1L_1+{}_2L_2)}\,({}_1L_2+{}_2L_1)^{{}_2L_1}$. The DCR has matrix element (DCME) in an irrep with mixed basis symmetry adapted to $S_{{}_iL_j} \backslash S_{{}_iL} \backslash S_L / S_{L_j} / q^{-1} S_{{}_iL_j} q$ given by

$$\begin{bmatrix} \lambda & & q \\ r' & {}_i\lambda_i r_i \lambda'_{ji} m'_j & r\lambda_j r_{ji} \lambda_{ji} m_j \end{bmatrix} = \delta({}_i\lambda_{ji}\lambda'_j)\,\delta({}_im_{ji}m'_j) \begin{matrix} \\ r \\ {}_ir \end{matrix} \overset{\begin{matrix} r' & r_j \end{matrix}}{\begin{bmatrix} \lambda & \lambda_j \\ {}_i\lambda & {}_i\lambda_j \end{bmatrix}} \qquad (1.4)$$

The rows and columns of the DCME must couple by the Littlewood-Richardson rules[4] for outer product coupling. Because it is a matrix element of S_L the DCME can always be chosen orthogonal on the indexes $(r\ {}_i\lambda_i r,\ r\lambda_j r_j)$ with $(\lambda, {}_i\lambda_j)$ acting as fixed parameters. The group orthogonality completeness condition in S_L requires the weighted double coset matrix element (WDCME) defined by

$$\begin{matrix} \\ r \\ {}_ir \end{matrix} \overset{\begin{matrix} r' & r_j \end{matrix}}{\begin{Bmatrix} \lambda & \lambda_j \\ {}_i\lambda & {}_i\lambda_j \end{Bmatrix}} \equiv \left(\frac{|\lambda|\,{}_iL!\,L_j!\,|{}_i\lambda_j|}{L!\,|{}_i\lambda|\,|\lambda_j|\,{}_iL_j!} \right)^{\frac{1}{2}} \begin{matrix} \\ r \\ {}_ir \end{matrix} \overset{\begin{matrix} r' & r_j \end{matrix}}{\begin{bmatrix} \lambda & \lambda_j \\ {}_i\lambda & {}_i\lambda_j \end{bmatrix}} \qquad (1.5)$$

(where $|\lambda|$ is the dimension of λ in S_L) be orthogonal on the indexes $(r\lambda r', {}_ir_i\lambda_j r_j)$ the indexes $({}_i\lambda, \lambda_j)$ acting as fixed parameters.

By considering the DC decomposition of a scalar product over a projection operator in S_L

$$\left\langle \begin{matrix} {}_im \\ {}_i\lambda_i n \\ {}_iM \end{matrix} \middle| (\lambda | r'\,{}_i\lambda_i m, r\lambda_j m_j) \middle| \begin{matrix} m_j \\ \lambda_j n_j \\ M_j \end{matrix} \right\rangle = \left\langle \begin{matrix} {}_i\lambda \\ {}_iM \end{matrix} \middle| \begin{matrix} r'\lambda \\ M \end{matrix} \right\rangle_n \left\langle \begin{matrix} r\ \lambda \\ M \end{matrix} \middle| \begin{matrix} \lambda j \\ M_j \end{matrix} \right\rangle_n \qquad (1.6)$$

we have shown the following identities hold depending on the overlap of the defining vector spaces:

Case I:

${}_in = n = n_j$ The DCME is identical to a recoupling coefficient in Un (or SUn).

Case II:

${}_in = n = n_1 + n_2$ The WDCME is identical to an isoscalar factor for Un/Un_j (or SUn/SUn_j).

Case III:

${}_1n + {}_2n = n_1 + n_2$ with ${}_in_j = {}_in \cap n_j$ The DCME is a matrix element for the operator

$q = \begin{bmatrix} n & n_j \\ {}_in & {}_in_j \end{bmatrix}$ that permutes the dimensions (so that $q \in Sn \subset Un$) when expressed in basis symmetry adapted to $U_i n_j \backslash U_i n \backslash U/Un_j/qU_i n_j q^{-1}$. These identifications not only reverify the usual orthogonality relations for these coefficients, but also imply the coefficients may be assummed real and surprising symmetries (from the representation theory of S_L and Un) exist for the magnitudes and phases of these coefficients. The objective of this paper is to specify these symmetries and to examine the requirements they impose on a phase convention to be consistent with duality.

II Symmetries of Duality

Irreps $[\ell^n]$ (to avoid complicated subscripting of powers we will write $[\ell^n] \to [\ell n]$) with rectangular young patterns play an important role in the theory for they are pseudoscalars in Un (scalars in SUn) such that $\mu \in Un$, $\mu|[\ell^n]\rangle = \det(\mu)^{\ell}|[\ell n]\rangle$. Complex conjugation in Un is the isomorphism $\lambda \to [\ell n-\lambda]$. Equivalence in SUn identifies the irreps $\lambda = [\lambda \pm \ell^n]$. Thus for the three cases of Section I we have

Coefficient Equivalence Complex Conjugation

Case I

$$ {}^{r}_{{}_i r}\begin{bmatrix} {}^{r'}\lambda & {}^{r_j}\lambda_j \\ {}_i\lambda & {}_i\lambda_j \end{bmatrix} = {}^{r}_{{}_i r}\begin{bmatrix} {}^{r'}[\lambda\pm\ell n] & {}^{r_j}[\lambda_j\pm\ell_j n] \\ [{}_i\lambda\pm{}_i\ell n] & [{}_i\lambda_j\pm{}_i\ell_j n] \end{bmatrix} = {}^{r*}_{{}_i r*}\begin{bmatrix} {}^{r'*}[\ell n-\lambda] & {}^{r_j*}[\ell_j n-\lambda] \\ [{}_i\ell n-{}_i\lambda] & [{}_i\ell_j n-{}_i\lambda_j] \end{bmatrix} A_n\, {}^{r}_{{}_i r}\begin{pmatrix} {}^{r'}\lambda & {}^{r_j}\lambda_j \\ {}_i\lambda & {}_i\lambda_j \end{pmatrix} \tag{2.1} $$

Case II

$$ {}^{r}_{{}_i r}\begin{Bmatrix} {}^{r'}\lambda & {}^{r_j}\lambda_j \\ {}_i\lambda & {}_i\lambda_j \end{Bmatrix} = {}^{r}_{{}_i r}\begin{Bmatrix} [\lambda\pm\ell n] & [\lambda_j\pm\ell n_j] \\ [{}_i\lambda\pm{}_i\ell n] & [{}_i\lambda_j\pm{}_i\ell n_j] \end{Bmatrix} = {}^{r*}_{{}_i r*}\begin{Bmatrix} {}^{r'*}[\ell n-\lambda] & {}^{r_j*}[\ell n_j-\lambda_j] \\ [{}_i\ell n-{}_i\lambda] & [{}_i\ell n_j-{}_i\lambda_j] \end{Bmatrix} \begin{matrix} A(n\lambda:{}_i\lambda r')\ A(n_j\lambda_j:{}_i\lambda_j r_j) \\ x(n\lambda:n_j\lambda_j r)\ (n_i\lambda:n_{ji}\lambda_{ji} r) \end{matrix} \tag{2.2} $$

Case III

$$ {}^{r}_{{}_i r}\begin{bmatrix} {}^{r'}\lambda & {}^{r_j}\lambda_j \\ {}_i\lambda & {}_i\lambda_j \end{bmatrix} = (-1)^{\ell_1 n_2\ 2n_1}\, {}^{r}_{{}_i r}\begin{bmatrix} {}^{r'}[\lambda\pm\ell n] & {}^{r_j}[\lambda_j\pm\ell n_j] \\ [{}_i\lambda\pm\ell_i n] & [{}_i\lambda_j\pm\ell_i n_j] \end{bmatrix} = (-1)^{\ell_1 n_2\ 2n_1}\, {}^{r*}_{{}_i r*}\begin{bmatrix} [\ell n-\lambda] & [\ell n-\lambda_j] \\ [\ell_i n-{}_i\lambda] & [\ell_i n_j-{}_i\lambda_j] \end{bmatrix} $$

$$ {}^{r}_{{}_i r}\begin{pmatrix} {}^{r'}\lambda & {}^{r_j}\lambda_j \\ {}_i\lambda & {}_i\lambda_j \end{pmatrix} \tag{2.3} $$

The explicit phase in (2.3) enters because det $q = (-1)^{1^{n}2\ 2^{n}1}$. There are two types of phase factors that must be considered because of complex conjugation[5]. The transformation from the complex conjugate of a representation to the complex conjugate representation is accomplished by the 1-λM matrix

$$ \begin{bmatrix} \lambda^* & \mu \\ M^* & M'^* \end{bmatrix} = (\lambda)^*_{MM^*} \begin{bmatrix} \lambda & \mu \\ M & M' \end{bmatrix}^* (\lambda)_{M'M'^*}. \tag{2.4} $$

Schur and Frobenius have shown for $\lambda = \lambda^*$ the 1-λM matrix is necessarily symmetric or skew symmetric depending on whether the scalar irrep occurs in the symmetric or antisymmetric square of the irrep λ. Upon factoring to the subgroup sequence Un/Un_j we obtain the 1-λM factor defined by

$$ (n\lambda;n_j\lambda_j r) \equiv (\lambda)_{r\lambda_j M_j\ r^*\lambda_j^* M_j}\ (\lambda_j)_{M_j M_j^*} \tag{2.5} $$

Our notation assumes there is a unique correspondence in multiplicity label $r \leftrightarrow r^*$ a question examined more thoroughly in a more extensive report on this work.[6]

Thus the 1-λM factor carries one multiplicity label rather than a pair i.e. the 1-λM matrix has elements zero and a ±1 in any row or column. The product of 1-λM factors corresponding to the elements of a DCME has been designated in (2.3) by the array.

$$\begin{matrix} & r' & r_j \\ r & \lambda & \lambda_j \\ {}_ir & {}_i\lambda & {}_i\lambda_j \end{matrix} \equiv (n\lambda:{}_in\ {}_i\lambda r')({}_in_i\lambda:{}_in_{j\,i}\lambda_{j\,i}r)(n_j\lambda_j:{}_in_{j\,i}\lambda_j r_j)(n\lambda:n_j\lambda_j r) \quad (2.6)$$

On transforming the complex conjugate of a Wigner (Clebsch-Gordan) coefficient to the Wigner coefficient involving the complex conjugate irreps one must introduce the Derome-Sharp [7] phase factor A as

$$\left\langle \begin{matrix}\lambda^* & r^*\\ M^* & \end{matrix}\middle| \begin{matrix}\lambda_j^*\\ M_j^*\end{matrix}\right\rangle_n = \left\langle \begin{matrix}\lambda r\\ M\end{matrix}\middle| \begin{matrix}\lambda_j\\ M_j\end{matrix}\right\rangle_n^* (\lambda)^*_{MM^*}\ (\lambda_j)_{M_j M_j^*}\ A(n\lambda:\lambda_j r) \quad (2.7)$$

A remark on the multiplicity label and the meaning of the A array in (2.1) hold as above.

Association in S_L w.r.t. the alternating group A_L, a normal subgroup of index two, introduces a third unitary transformation relating the matrix representative and its associate in S_L by

$$\begin{bmatrix}\lambda & \pi\\ m & m\end{bmatrix} = (-1)^p\ (\lambda)_{m\tilde{m}} \begin{bmatrix}\tilde{\lambda} & \pi\\ \tilde{m} & m\end{bmatrix} (\lambda)_{m'\tilde{m}'} \quad (2.8)$$

where p is the parity of the permutation π. For $\lambda = \tilde{\lambda}$ the association matrix is of two types depending on whether the antisymmetric irrep $[1^L]$ occurs in the symmetric or antisymmetric square of the irrep λ in S_L. In either case the matrix $(\lambda)_{m\tilde{m}}$ is equivalent to the symmetric traceless form $\begin{pmatrix}D & E\\ E & O\end{pmatrix}$ where E is the unit matrix of dimension $\frac{|\lambda|}{2}$. On factoring to the subgroup sequence S_L/S_{L_j} we obtain the association factor

$$(\lambda:\lambda_j r) \equiv (\lambda)_{\lambda_j m,\, \tilde{\lambda}_j \tilde{m}}\ (\lambda_j)_{m\tilde{m}'}$$

Each of these unitary transformations is an involution so one obtains

$$(n\lambda:n_j\lambda_j r)(n\,[\ell n-\lambda]:n_j\,[\ell n_j-\lambda_j]r^*)=\phi(\lambda n)\ \phi(\lambda_j n_j)=(-1)^{L_1 n_2+L_2 n_1}. \quad (2.10)$$

$$A\ (n\lambda:\lambda_j\ )\ A(n\,[\ell n-\lambda]:[\ell_j n-\lambda_j]\ r^*)=\phi(\lambda n)\phi(\lambda_j n)=+1 \quad (2.11)$$

and

$$(\lambda:\lambda_j r)\ (\tilde{\lambda}:\tilde{\lambda}_j\tilde{r})=\theta(\lambda)\theta(\lambda_j) \quad (2.12)$$

where in (2.10) and (2.11) we have used the phase convention of the next section to reduce the r.h.s. 1-λ phases using $\phi(\lambda n)= -1^{L(n-1)}$. The association phase $\theta(\lambda)=\theta(\tilde{\lambda})$ is easily evaluated by taking the young tableau for λ in standard order, interchanging rows and columns and determining the parity of the permutation necessary to bring the tableau to standard order for $\tilde{\lambda}$. By their definitions (2.5) and (2.6) and using the duality identifications and their associates we must require

$$(n\lambda:n_j\lambda_j r) = (-1)^{L_2(\ell n_1-L_1)}\ \begin{matrix} r \\ r^* \end{matrix}\begin{pmatrix}[\ell n] & [\ell n_j]\\ \lambda & \lambda_j\\ [\ell n-\lambda] & [\ell n_j-\lambda_j]\end{pmatrix} A(\ell\tilde{\lambda}:\tilde{\lambda}_j\tilde{r}). \quad (2.13)$$

The round bracket array is defined similar to (2.6) as that product of association phase factors related to a given DCME. Association does not commute with complex conjugation. In fact complex conjugation w.r.t. Un/Un_j gives the chain $[\ell n-\lambda]$, $[\ell n_j-\lambda_j]$ the associate of which is $[n\ell-\tilde{\lambda}]\,[n_j\ell-\tilde{\lambda}_j]$, which is the result of complex conjugating $\tilde{\lambda},\tilde{\lambda}_j$ w.r.t. U_ℓ. This is expressed in (2.13). Moreover using

(2.13) and association, (2.3) can be shown to be the expression associate to (2.1) while (2.2) is shown to be self-associate.

While there still exists sufficient freedom to choose $A(n\lambda;\lambda_j r)=+1$ for most cases, it is clear from (2.13) that in some cases (e.g. $\lambda_j=[\ell n_j-\lambda_j]$ for which (2.13) becomes

$$(n\lambda:n_j\lambda_j r) = (-1)^{L_1 L_2}\,([\ell n]:\lambda\otimes\nu)\,([\ell n_j]:\lambda_j\otimes\nu_j)\,A(\ell\tilde{\lambda}:\tilde{\lambda}_j r),\ \text{where}\ \nu=\nu_1\nu_2=\begin{matrix}[2]\\ [1^2]\end{matrix} \qquad (2.13')$$

the choice A= -1 must enter if the relations (2.1) and (2.3) are required of the phase and reality of the coupling coefficients is assumed.

III Permutation Convention

One must also adopt a phase convention on permuting the order of the component irreps of a DCME or a WDCME that is consistent with their dual role in S_L and Un. The basic permutation is a transpose of row or column in the identity DCME giving

$$\begin{matrix} & r' \\ r & \end{matrix}\begin{bmatrix}\lambda & \lambda_1 & \lambda_2\\ \lambda_2 & 0 & \lambda_2\\ \lambda_1 & \lambda_1 & 0\end{bmatrix} \equiv \delta rr'\Phi(\lambda:\lambda_1\lambda_2 r)=\delta rr'\Phi(\lambda:\lambda_2\lambda_1 r)=(-1)^{L_1L_2}(\lambda:\lambda_1\lambda_2 r)(\lambda:\lambda_2\lambda_1 r')\delta rr'$$
$$\chi\Phi(\tilde{\lambda}:\tilde{\lambda}_1\tilde{\lambda}_2\tilde{r}) \qquad (3.1)$$

Diagonalization in the mulitnlicity index (rr') is always possible but requires for the case $\lambda_1=\lambda_2$ use of symmetrized and antisymmetrized outer products. By successive transformations of the basis one shows.

$$\begin{matrix} & r' & r_1 & r_2\\ r & \lambda & \lambda_1 & \lambda_2\\ {}_1r & {}_1\lambda & {}_1\lambda_1 & {}_1\lambda_2\\ {}_2r & {}_2\lambda & {}_2\lambda_1 & {}_2\lambda_2\end{matrix} = \Phi(\lambda:\lambda_j r)\ \Phi({}_i\lambda:{}_i\lambda_{ji} r)\ \begin{matrix} & r' & r_2 & r_1\\ r & \lambda & \lambda_2 & \lambda_1\\ {}_1r & {}_1\lambda & {}_1\lambda_2 & {}_1\lambda_1\\ {}_2r & {}_2\lambda & {}_2\lambda_2 & {}_2\lambda_1\end{matrix} \qquad (3.2)$$

(and similarly for a transposition of the rows). The permutation phase is easily shown to satisfy all the requirements of the basic identifications of duality. However it also must be invariant under the symmetries of equivalence and complex conjugation of (2.1,2, and 3) and it must satisfy the symmetry under association expressed in (3.1). The symmetries of (2.1) require

$$\Phi(\lambda:\lambda_j r)= \Phi([\lambda\pm\ell n]:[\lambda_j\pm\ell_j n]r)=\Phi([\ell n-\lambda]:[\ell_j n-\lambda_j]r^*), \qquad (3.3)$$

while the symmetries of (2.2) require

$$\Phi(\lambda:\lambda_j r)=(-1)^{\ell n_1 n_2}\Phi([\lambda\pm\ell n]:[\lambda_j\pm\ell n_j]r)=(-1)^{\ell n_1 n_2}$$
$$\times\,\Phi([\ell n-\lambda]:[\ell n_j-\lambda_j]r^*)\ (n\lambda:n_1\lambda_1 n_2\lambda_2 r)\ (n\lambda:n_2\lambda_2 n_1\lambda_1 r) \qquad (3.4)$$

The last equality of (3.4) is the associate relation via (2.13) of the last equality of (3.3).

If m_{ij} designates the number of nodes transferred from row i of a component irrep to the row j of the composite irrep $i\leq j$ in the process of carrying out the i th step in the Littlewood-Richardson rules for an outer product, we propose the phase convention

$$\Phi(\lambda:\lambda_j r) = (-1)^P \ \text{where}\ P\equiv\sum_{i<j}(j-i)\,m_{ij} \qquad (3.5)$$

In a more extensive report on this work we show this permutation phase convention satisfies all the above requirements and is indenendent of the multiplicity index r. In case $\lambda_1=\lambda_2$ and λ belongs uniquely to $\lambda_1\otimes[2]$ or $\lambda_1\otimes[1^2]$ the phase convention coincides with $\Phi(\lambda:\lambda_1 \otimes [2]) = +1$ and $\Phi(\lambda:\lambda_1 \otimes [1^2]) = -1$. If λ is contained both in $\lambda_1 \otimes [2]$ and $\lambda_1 \otimes [1^2]$, the phase convention will give the same sign for both occurrences. It is exactly in this case that the plethysm separation must be used to diagonalize the (transpose) permutation matrix giving the above assignment of phase based on the plethysm.

Duality via (2.13) relates the two unitary transformations required for complex conjugation in Un. It also requires the same permutation phase convention be used in S_L and Un. We have shown that a phase convention consistent with duality and orthogonal matrix irreps of S_L is possible. Relation (2.13) still allows a freedom of absolute sign choice for the association factor and for either the 1-λM factor or the Derome-Sharp factor, but not both.

References:
1. I. Schur and Frobenius, Berlin Berichte 186 (1906); I. Schur, Sitzgsber. Preuss. Akad. Wiss. 58 (1927); 100 (1928).
2. H. Weyl, The Classical Groups, Princeton (1946).
3. J. J. Sullivan, a) J. Math. Phys. 14, 387 (1973)· b) Proc. Int. Symp. Math. Phys. Mexico City Jan. 5-8, 1976 Vol. I pg. 253; c) J. Math. Phys. 16, 756 (1975).
4. D. E. Littlewood, The Theory of Group Characters, 2nd Ed., Oxford (1950)·
5. P. H. Butler, Phil. Trans. Roy. Soc. Lon. 277, 545 (1975).
6. J. J. Sullivan, to be published J. Math. Phys.
7. J. R. Derome and W. T. Sharp, J. Math. Phys. 6, 1584 (1965).

INTEGRAL TRANSFORM REPRESENTATIONS OF SL(2,R)

Kurt Bernardo Wolf
Instituto de Investigaciones en Matemáticas Aplicadas
y en Sistemas,
Universidad Nacional Autónoma de México,
Apartado Postal 20-726, México 20, D.F. MEXICO.

Introduction. Consider the three formal differential operators

$$J_1^{(\mu)} = \frac{1}{4}\left(-\frac{d^2}{dx^2} + \frac{\mu}{x^2} - x^2\right),\quad J_2^{(\mu)} = -\frac{i}{2}\left(x\frac{d}{dx} + \frac{1}{2}\right),\quad J_3^{(\mu)} = \frac{1}{4}\left(-\frac{d^2}{dx^2} + \frac{\mu}{x^2} + x^2\right), \tag{1}$$

which are the Schrödinger Hamiltonians of the centrifugal barrier or well (with constant $\mu > 0$ or $\mu < 0$), with a repulsive or attractive harmonic oscillator potential. The pure barrier or well case is given by $J_1^{(\mu)} + J_3^{(\mu)}$ while $\mu = 0$ includes the repulsive or atractive ordinary oscillator. The operator $J_2^{(\mu)}$ is the generator of dilatations.

It is commonly known that these operators follow under commutation the Lie bracket of the sl(2,R) algebra, and belong to representation(s) of sl(2,R) given by the Casimir operator $J_1^{(\mu)2} + J_2^{(\mu)2} - J_3^{(\mu)2} = q\mathbb{1}$, as [1]

$$q = -\frac{1}{4}\mu + \frac{3}{16} = k\,(1-k),\quad k = \frac{1}{2}\left(1 \pm [\mu + 1/4]^{1/2}\right) \tag{2}$$

It seems to be less commonly appreciated that the $J_k^{(\mu)}$ are rather unique in being the only second-order operators in one dimension to do so, and that they close precisely into sl(2,R). Second-order operators do not appear in ordinary Lie transformation theory, as they are *not* generators of a group action on some homegeneous space. Instead, they are generators of a group of *unitary integral transform* actions on Hilbert spaces of functions on R^+:

$$f(x) \xrightarrow{g = \binom{ab}{cd}} f_g(x) = C^k\binom{ab}{cd} f\,(x) = \int_{R^+} dx'\, C_g^k(x, x')\, f(x'),\quad ad-bc = 1, \tag{3a}$$

with the composition property common to any representation,

$$C^k\begin{pmatrix} a_1 b_1 \\ c_1 d_1\end{pmatrix} \bullet C^k\begin{pmatrix} a_2 b_2 \\ c_2 d_2\end{pmatrix} = C^k\left[\begin{pmatrix} a_1 b_1 \\ c_1 d_1\end{pmatrix}\begin{pmatrix} a_2 b_2 \\ c_2 d_2\end{pmatrix}\right] \tag{3b}$$

with the identity $C^k\,\underset{\sim}{1} = \mathbb{1}$ and continuous in the group parameters. For the inverse of a given transformation $\underset{\sim}{g}$, unitarity implies

$$C_{g^{-1}}^k(x, x') = [C_g^k(x', x)]^*. \tag{3c}$$

In this contribution we present a brief but complete account of the Hilbert spaces and of the integral kernel functions $C_g^k(x, x')$, for all SL(2,R) representation series. Among the points in this manifold of integral transforms there appear some well-known ones. For the D_k^+ series, the Fourier and Hankel transforms $\begin{pmatrix} 0 & 1 \\ -1 & 0 \end{pmatrix}$, the bilateral Laplace $\begin{pmatrix} 0 & i \\ i & 0 \end{pmatrix}$, diffusion $\begin{pmatrix} 1 & -2it \\ 0 & 1 \end{pmatrix}$ $t \geqslant 0$, Bargmann [2] and Barut-Girardello transforms $2^{-1/2}\begin{pmatrix} 1 & -i \\ -i & 1 \end{pmatrix}$. In the principal and exceptional series, there appear transforms related to some of those studied by Titchmarsh.

In prior work, Moshinsky, Quesne [3] and the author [4] among others have studied the *oscillator* (*metaplectic*) representation provided by (1) on $L^2(R)$, with $\mu = 0$ (i.e. $q = 3/16$, $k = 1/2 \pm 1/4$), $D_{1/4}^+ \dotplus D_{3/4}^+$ [1]. An expository account of this subject at graduate level, can be found in the author's recent book [5].

In Section 2 we gather the elements from the oscillator representation and indicate how to use it to obtain all other representations. Section 3 covers the 'discrete' series $D_k^\pm$, $k > 0$, while Section 4 deals with the principal and exceptional continuous series C_k^ε, for $q > 1/4$ and $0 < q < 1/4$ respectively. Further comments on group theoretical methods in integral transform theory are given in Section 5.

2. The Oscillator Representation.

The integral transform representation kernel (3) for the oscillator representation $\mu = 0$ is [3, 4, 5]

$$C_g^0(x, x') = e^{-i\pi/4}(2\pi b)^{-1/2} \exp i([ax'^2 - 2x'x + dx^2]/2b), \tag{4}$$

following (3a) with R for R^+. It obeys (3b) becomming $\delta(x - x')$ as $\underset{\sim}{g} \to \underset{\sim}{1}$, and is unitary from $L^2(R)$ onto $L^2(R)$. The subgroup of lower-triangular transformations ($b = 0$) can be shown to collapse (4) into a Dirac δ with a weight function

$$C_{g(b=0)}^0(x,x') = a^{-1/2} \exp(icx^2/2a)\ \delta(x' - x/a). \tag{5}$$

This is the subgroup generated by the subalgebra of (1) consisting of up-to-first order differential operators, and consists of all multiplier Lie actions.

The group parameters of SL(2,R) may be complexified [4] to the subsemigroup HSL(2,C) ⊃ SL(2,R) of complex 2 x 2 unimodular matrices with $\mathrm{Im}(a/b) \geqslant 0$ and, if $a = 0$, $\mathrm{Im}\, b = 0$. The integral kernels (4) are not unitary from $L^2(R)$ onto itself, but are unitary as transformations from $L^2(R)$ onto Hilbert spaces of analytic functions B_g^0 defined through the inner product over the complex $\mathbf{x}$-plane

$$(f_g, h_g)_g^0 = \int_C \nu_g(x, x^*)\, d\,\mathrm{Re}\,\mathbf{x}\, d\,\mathrm{Im}\, x\ f_g(x)^* h_g(x) = (f,h)_{L^2}, \tag{6a}$$

$$\nu_g\ (x, x^*) = (\pi v/2)^{-1/2}\ \exp\ [ux^2 - 2x^*\, x + u^* x^{*2})/2v], \tag{6b}$$

$$u = a^*\, d - b^*\, c, \qquad v = 2\ \mathrm{Im}\ (b^*\, a) \geqslant 0 \tag{6c}$$

Among the integral transforms in HSL(2,C) we have the well-known Gauss-Weierstrass and Bargmann transforms. In the latter case B_g^o is Bargmann's Hilbert space [2] of analytic functions of growth (2, 1/2).

The oscillator integral transform representation (4) - (6) of HSL(2,C) has been extended to the W ^ HSL(2,C) semigroup of inhomogeneous complex linear transformations and has been applied to a number of problems [5-10]. We shall use the oscillator representation (4) to obtain all other SL(2,R) representation series. For this we work in two variables.

Out of the quantum mechanical position and momentum operators we build the so (3, 2) $\simeq$ sp (4,R) algebra of operators on $L^2(R^2)$:

$$\begin{aligned}
M_{12} &= \tfrac{1}{2}\,(Q_1P_2 - Q_2P_1), \qquad M_{13} = -\tfrac{1}{2}\,(P_1P_2 + Q_1Q_2),\\
M_{14} &= -\tfrac{1}{2}\,(Q_1P_2 + Q_2P_1), \qquad M_{15} = -\tfrac{1}{2}\,(P_1P_2 - Q_1Q_2),\\
M_{23} &= \tfrac{1}{4}\,(P_1^2 - P_2^2 + Q_1^2 - Q_2^2) = \tfrac{1}{2}\,(H_1^{(h)} - H_2^{(h)}),\\
M_{24} &= \tfrac{1}{2}\,(Q_1P_1 - Q_2P_2),\\
M_{25} &= \tfrac{1}{4}\,(P_1^2 - P_2^2 - Q_1^2 + Q_2^2) = \tfrac{1}{2}\,(H_1^{(r)} - H_2^{(r)}),\\
M_{34} &= -\tfrac{1}{4}\,(P_1^2 + P_2^2 - Q_1^2 - Q_2^2) = -\tfrac{1}{2}\,(H_1^{(r)} + H_2^{(r)})\\
M_{35} &= \tfrac{1}{4}\,(\{Q_1, P_1\}_+ + \{Q_2, P_2\}_+)\\
M_{45} &= \tfrac{1}{4}\,(P_1^2 + P_2^2 + Q_1^2 + Q_2^2) = \tfrac{1}{2}\,(H_1^{(h)} + H_2^{(h)}),
\end{aligned} \tag{7}$$

where $H_n^{(h)}$ and $H_n^{(r)}$ are the harmonic and repulsive oscillator Hamiltonians in the nth coordinate. The SO(2) $\otimes$ SO(3) $\subset$ SO(3,2) compact generators are $\{M_{45};\ M_{12}, M_{13}, M_{23}\}$. The following subsets generate subgroups containing SO(2,1):

$$\{M_{12};\ M_{34}, M_{35}, M_{45}\} \quad : \quad \mathrm{SO}(2) \otimes \mathrm{SO}(2,1), \tag{8a}$$

$$\{M_{14};\ M_{23}, M_{25}, M_{35}\} \quad : \quad \mathrm{SO}(1,1) \otimes \mathrm{SO}(2,1). \tag{8b}$$

3. The 'Discrete' Series.

In (8a) the subalgebra so(2,1) generates identical transformations (3) in x_1 and x_2 coordinates, while M_{12} rotates. We parametrize the plane in polar coordinates [11-12]: $x_1 = x \cos \phi$, $x_2 = x \sin \phi$, $x \in (0, \infty)$, $\phi \in S_1$, and introduce the projection operator of $L^2(R^2)$ onto the $L^2(R^+)$ irreducible subspace belon—

ging to the m eigenvalue of M_{12}:

$$f^m(x) = (T^m f)(x) = (2\pi)^{-1/2}\, x^{1/2} \int_{S_1} d\phi\, f(\underset{\sim}{x}(x,\phi))\, e^{-im\phi}, \qquad (9)$$

which effects the identification of (8a) with (1) through $T^m\{M_{12}, -M_{34}, M_{35}, M_{45}\} = \{m; J_1^{(\mu)}, J_2^{(\mu)}, J_3^{(\mu)}\}\, T^m$, and $\mu = m^2 - 1/4 \geqslant -1/4$, $q = (1/4)(1-m^2) \leqslant 1/4$, $k = 1/2 + |m|/2$, $m \in Z$, placing us in the $k = (1/2)(1+|m|)$ representations $\mathcal{D}_k^+$ of $sl(2,R)$.

We may now T^m-project the integral transform action of (4) and find the integral kernel

$$C_g^k(x, x') = e^{-ik\pi}\, b^{-1}\, (x\, x')^{1/2} \quad \exp i([ax'^2 + dx^2]/2b)\, J_{2k-1}(x\, x'/b). \qquad (10)$$

The punctured plane (9) may be made to contain an infinity of Reimann sheets, allowing us to cover $m \in R$. The $\mathcal{D}_k^-$ representations may be obtained through a different $\overline{T}^m$ which precedes (11b) with the exterior automorphism $J_3^{(\mu)} \leftrightarrow -J_3^{(\mu)}$, $J_1^{(\mu)} \leftrightarrow -J_1^{(\mu)}$. The corresponding group automorphism is best described in Euler angles through a change of sign in the parameters. The kernel (12) collapses to a Dirac δ as in (5) for the $b = 0$ subgroup of Lie transformations.

The singularity of $C_g^k(x, x')$ at the integration interval endpoint $x' = 0$ is acceptable for integration with $L^2(R^+)$ functions for $k > 0$. We recall that the interval $1 > k > 0$ corresponds to $q \in (0, 1/4]$ where the 'discrete' series and the exceptional continuous series overlap.

The complex extension of (10) to HSL(2,C) may be implemented through complex radial coordinates in the complex right-half plane C^+. For the m-irreducible space $\mathcal{D}_k^+$, $k = (1/2)(m+1)$, we have the inner product over the complex radial plane

$$(f_g^m, h_g^m)_g^m = \int_{C^+} \nu_g^m(x, x^*)\, d\,\mathrm{Re}x\; d\,\mathrm{Im}x\; f_g^m(x)^*\, h_g^m(x) = (f^m, h^m)_{L^2(R^+)} \qquad (11a)$$

$$\nu_g^m(x, x^*) = (2/\pi v)\,|x|\, \exp\,[(ux^2 + u^* x^{*2})/2v]\; K_m(x\, x^*/v), \qquad (11b)$$

where u and v are given by (6c). The inner product (11) defines, for each g, a Hilbert space B_g^m of analytic functions in C^+ with certain growth conditions. This turns the complexified (10) into a unitary transform between $L^2(R^+)$ and B_g^m. The Barut-Girardello transform corresponds to the particular semigroup element $\underset{\sim}{g} = 2^{-1/2}\begin{pmatrix} 1 & -i \\ -i & 1\end{pmatrix}$. As we move from HSL(2,C) to its SL(2,R) boundary, the B_g^m spaces collapse to $L^2(R^+)$ through a limit where $\nu_g^m(x, x^*)$ becomes a Dirac δ on Re x.

4. The Continuous Series.

We turn now to the subalgebra [13] so(2,1) $\subset$ so(3,2) in (8b), the first two generators being now the difference of generators of canonical transformations along the x_1 and x_2 axes. The transformation kernel in the $x_1 - x_2$ plane will thus be $\begin{pmatrix} a & b \\ c & d\end{pmatrix}$ in x_1, $\begin{pmatrix} a & be^{-i\pi} \\ ce^{i\pi} & d\end{pmatrix}$ in x_2. We parametrize the plane in hyperbolic coordinates given by the triad (y, β, σ) for $x_1^2 - x_2^2 > 0$, $\sigma = +1$, $x_1 = y \operatorname{ch} \beta$, $x_2 = y \operatorname{sh} \beta$, while for $x_1^2 - x_2^2 < 0$, $\sigma = -1$, $x_1 = y \operatorname{sh} \beta$, $x_2 = y \operatorname{ch} \beta$ with the ranges $y \in R$, $\beta \in R$, $\sigma \in \{1,-1\}$.

The elements $f(\underset{\sim}{x})$ of $L^2(R^2)$ will be represented as pairs of functions $f_\sigma(y,\beta) = f(\underset{\sim}{x}(y, \beta\ \sigma))$.

We now Fourier/parity-decompose $L^2(R^2)$ into eigenspaces of the hyperbolic rotation generator M_{14} with eigenvalue $\lambda/2 \in R$, and of the inversion operator I of the plane with eigenvalue p. The projection operator onto this subspace is

$$f_\sigma^{p,\lambda}(y) = (I^{p,\lambda} f_\sigma)(y) = y^{1/2} \int_R d\beta\, [f_\sigma(y,\beta) + p\, f_\sigma(-y,\beta)]\, e^{-i\lambda\beta} , \tag{12}$$

in analogy with (11). As the hyperbolic coordinate system has two parts $\sigma = \pm 1$, we have two-component functions. The formal operators (1) become similarly doubled into diagonal 2×2 matrices, with operator entries $T^{p\lambda}\{I;\ M_{14};\ M_{25}, M_{35}, M_{23}\} = \{p,\ \lambda/2,\ \sigma J_1^{(\mu)},\ J_2^{(\mu)},\ \sigma J_3^{(\mu)}\}\ T^{p\lambda}$, and $\mu = -(1/4 + \lambda^2) \leqslant -1/4, q = (1/4)(1+\lambda^2) \geqslant 1/4$, $k = 1/2 + i\lambda/2$.

As M_{23} is the difference of harmonic oscillators in the x_1 and x_2 directions, the eigenfunction set is reduced into two irreducible subsets, that of integer and that of half-integer eigenvalues, through the eigenvalues of I, $p = 1$ ($\varepsilon = 0$) for the former, and $p = -1$ ($\varepsilon = 1/2$) for the latter. Hence $T^{p\lambda}$ places us in the $k = 1/2 + i\lambda/2$ continuous series representation C_q^ε with $q \geqslant 1/4$. Note that the exceptional interval C_q^0, $0 < q < 1/4$ is not contained in this decomposition.

As before, we may project the integral transform action (4) on each of the (p,λ) - irreducible subspaces. In each of these subspaces the integral transform (3) has an integral kernel arranged as a 2×2 matrix

$$(\underset{\sim}{C}_g^{p\lambda})_{\sigma\sigma'}(x,x') = [2\pi|b|]^{-1} (xx')^{1/2} G^g_{\sigma\sigma'}(x,x')\, H^{p\lambda}_{\sigma\sigma'}(xx'/b), \tag{13a}$$

where the entries decompose as Gaussians times cylinder functions of imaginary index:

$$G^g_{\sigma\sigma'}(x,x') = \exp\ i\,([d\sigma\, x^2 + a\sigma'\, x'^2]/2b), \tag{13b}$$

$$H^{p\lambda}_{11}(z) = pH^{p\lambda}_{-1-1}(z) = pH^{p\lambda}_{11}(-z) = H^{p,-\lambda}_{11}(z) = i\pi[pe^{-\lambda\pi/2}H^{(1)}_{i\lambda}(z) - e^{\lambda\pi/2}H^{(2)}_{i\lambda}(z) =$$

$$= 2\pi\,[i \begin{Bmatrix}\operatorname{sh} \lambda\pi/2\\ \operatorname{ch} \lambda\pi/2\end{Bmatrix}^p J_{i\lambda}(z) + \begin{Bmatrix}\operatorname{ch} \lambda\pi/2\\ \operatorname{sh} \lambda\pi/2\end{Bmatrix}^p N_{i\lambda}(z)], \tag{13c}$$

$$H^{p\lambda}_{1-1}(z) = pH^{p\lambda}_{-11}(z) = pH^{p\lambda}_{1-1}(-z) = pH^{p,-\lambda}_{1-1}(z) = 4\begin{Bmatrix}\operatorname{ch} \lambda\pi/2\\ \operatorname{sign} z \cdot \operatorname{sh} \lambda\pi/2\end{Bmatrix}^p K_{i\lambda}(|z|), \tag{13d}$$

where the upper entry of $\{:\}^p$ is taken for $p = 1$ and the lower one for $p = -1$. The square-integrability properties of the above kernel, in company with L^2-functions are proper at the origin for all $\lambda \in R$. They constitute unitary transformations between $L^2_{\sigma=+1}(R^+) + L^2_{\sigma=-1}(R^+)$ and itself, representing the principal continuous series of SL(2,R). It is moreover possible to analytically continue in the λ variable to values $i\lambda = 2k-1$. The square-integrability of (17) at the origin is satisfied for

$0 < k < 1$, $p = 1$ corresponding to the exceptional continuous representations C_q^0, $q \in (0, 1/4]$.

The 'hyperbolic Fourier transform' element $\mathfrak{f} = \begin{pmatrix} 0 & 1 \\ -1 & 0 \end{pmatrix}$ has a 2×2 kernel matrix which may be diagonalized. The upper and lower components of the $\varepsilon = 0$ case then become uncouppled, in the form (3a), with the kernel.

$$\hat{C}^k_{\mathfrak{f}}(x,x') = (xx')^{1/2}\{\cos \pi k\, J_{2k-1}(xx') - \sin \pi k\, [N_{2k-1}(xx') \mp 2/\pi\, K_{2k-1}(xx')]\} \quad (14)$$

where $2k-1 = i\lambda$, $\lambda \in R$ for the principal and $0 < k < 1$ for the exceptional continuous series. When $k = 1/2$, (18) is an integral transform studied by Titchmarsh. The class (14) for the full range of k can be shown to be the only class of Fourier kernels involving these functions.

5. Closing Comments.

We have given the integral transform kernels of representations of $\overline{SL(2,R)}$ generated by (1). The uses which group theory makes of unitary irreducible representations take also a special meaning for this realization, since the basis functions include harmonic and repulsive oscillator wavefunctions, Whittaker, Bessel and the Mellin expansion functions $(2\pi)^{-1/2}x^{1/2+i\nu}$. As for the oscillator representation [5] these elements allow us to solve problems of self-reciprocal functions [8] under integral transforms, of separation of variables and similarity solutions [6], evolution of coherent states [10] and a unitary description of diffusion problems [9], as well as hyperdifferential and special-function relations.

We have also the intertwining operators [13, 14] between our integral representation and Bargmann's realizations [1] of SL(2,R) through multiplier action on the unit circle, disk or half-plane.

References.

[1] V. Bargmann, Ann, Math. 48, 568 (1947).

[2] V. Bargmann, Comm. Pure Appl. Math. 14, 187 (1961).

[3] M. Moshinsky and C. Quesne, J. Math. Phys. 12, 1772,1780 (1971).

[4] K.B. Wolf, J. Math. Phys. 15, 1295 (1974).

[5] K.B. Wolf, *Integral Transforms in Science and Engineering* (Plenum Publ. Corp., New York, 1979).

[6] K.B. Wolf, J. Math. Phys. 17, 601 (1976).

[7] C.P. Boyer and K.B. Wolf, Rev. Mex. Física 25, 31 (1976).

[8] K.B. Wolf, J. Math. Phys. 18, 1046 (1977).

[9] S. Steinberg and K.B. Wolf, N. Cimento 53A, 149 (1980).

[10] K.B. Wolf, On Time-Dependent Quadratic Quantum Hamiltonians, to appear in SIAM J. Appl. Math.

[11] M. Moshinsky, T.H. Seligman and K.B. Wolf, J. Math. Phys. 13, 1634 (1972).

[12] K.B. Wolf, J. Math. Phys. 15, 2101 (1974).

[13] K.B. Wolf, J. Math. Phys. 21, 680 (1980).

[14] C.P. Boyer and K.B. Wolf, J. Math. Phys. 16, 1493 (1975).

THE THEORY OF G^∞ SUPERMANIFOLDS.

Charles P. Boyer
Instituto de Investigaciones en Matemáticas Aplicadas y en Sistemas.
Universidad Nacional Autónoma de México.

Samuel Gitler
Departamento de Matemáticas
Centro de Investigaciones y Estudios Avanzados.
Instituto Politécnico Nacional de México.

Various approaches for describing super- or graded manifolds have been developing within the last few years. The first approach of F. A. Berezin[1] and his collaborators in the Soviet Union is equivalent to viewing a supermanifold as a topological space locally homeomorphic to the product of m copies of the even part $\Lambda_0(R^N)$ of a Grassmann algebra $\Lambda(R^N)$ and n copies of the odd part $\Lambda_1(R^N)$. This space $E_N^{m,n} \approx \Lambda_0(R^N) \times \Lambda_1(R^N)$ is called super Euclidean space. The topology on $E_N^{m,n}$ is that induced by the augmentation map $\varepsilon: E_N^{m,n} \to R^m$. The topology is thus very coarse.

The second technique is that of B. Kostant[2] who replaced the sheaf of germs of C^∞ functions over a manifold by a sheaf of Z_2- graded commutative algebras, and in this way defined a graded manifold. Kostant's theory is general which allowed him to define graded Lie groups in terms of Hopf algebras and to describe graded homogeneous spaces for the first time. M. Bachelor[3] has established an algebra isomorphism between Kostant's approach and Berezin's approach. Since a sheaf isomorphism is possible only for a restricted class of Kostant's graded manifolds, Berezin's is a less general approach than Kostant's.

A third approach has been formulated by A. Rodgers[4]. In this approach super-manifolds are again modelled after $E_N^{m,n} = \Lambda_0(R^N)\times \Lambda_1(R^N)$, but the usual Euclidean topology is put on $\Lambda_0(R^N)$ and $\Lambda_1(R^N)$ viewed as vector spaces of dimension 2^{N-1}. Therefore, super manifolds can be considered as ordinary manifolds with an added structure, just as complex manifolds are even dimensional real manifolds with an added complex structure. Rodgers' approach is followed in this announcement (All proofs will appear in our forthcoming paper). As with the Berezin theory it is less general than Kostant's theory; however, it has the advantage of allowing a broader, and so, more interesting topological description. Such a description is more conventional from a physicist's point of view. Moreover, a theory of instantons in the setting of supergravity seems to require a broader topology.

Let us now fix some notation. The Grassmann algebra over R^N is denoted by Λ. It has a natural Z grading with homogeneous parts Λ^p, $p=0,\dots,N$. There is also a Z_2 grading with $\Lambda=\Lambda_0+\Lambda_1$ induced by the even and odd p's. Elements of Λ will usually be Z_2 homogeneous in which case the Z_2 degree is denoted by $|\lambda|=\alpha$ for

$\lambda\varepsilon\Lambda_\alpha$. Let $\{e^a\}$ be a basis for R^N, then using the standard multi-index notation $e^\mu = e^{\mu_1}\ldots e^{\mu_k}$, $\{e^\mu\}$ is a basis for Λ. We use $V=V_0+V_1$ to denote a Z_2 graded vector space with dim $V_0=m$ and dim $V_1=n$. If v_i $i=1,\ldots,m+n$ is a basis of V which repects the grading, we denote the degree of v_i by $|i|$ which is 0 for $i=1,\ldots,m$ and 1 for $i=m+1,\ldots,m+n$. Similarly we put $|\mu|=|e^\mu|$.

Now we put the usual Euclidean topology on $\Lambda(R^N)$ inducing the Euclidean topology on $E_N^{m,n} \approx R^{2^{N-1}(m+n)}$. Denote by $C^r(U,\Lambda)$ the C^r maps of $U \subset E_N^{m,n}$ into Λ for $r=0,\ldots,\infty$. Following Rodgers[4] an $f\varepsilon C^\circ(U,\Lambda)$ is called $\underline{G^1}$ if there are $(m+n)$ functions $(G_if)\,\varepsilon\, C^\circ(U,\Lambda)$ and a map $\eta: U \to \Lambda$ such that

$$f(x+h)-f(x) = \sum_{i+1}^{m+n} h^i (G_if)(x) + \eta(h)\,\|h\|$$

where $\eta \to 0$ as $\|h\| \to 0$, $x,x+h\ \varepsilon\ U$.

Define $\underline{G^p}$, $p=2,\ldots,\infty$ by induction.

<u>Remark</u>. For $|i|=1$ (G_if) is unique mod Λ^N.

One easily shows that $f\varepsilon G^p(U)$ implies $f\varepsilon C^p(U,\Lambda)$. Moreover, one can relate the ordinary total derivative to the above G-derivative to obtain.

$$\frac{\partial f}{\partial x^i_\mu} = e^\mu (G_if) \qquad |\mu|=|i|$$

<u>Theorem 1</u>: Let $U \subset E_N^{m,n}$ be open and $f\varepsilon C^\infty(U,\Lambda)$ then $f\varepsilon G^\infty(U)$ if and only if

$$\frac{\partial f}{\partial x^i_\mu} = e^\mu \frac{\partial f}{\partial x^i_0} \qquad |i|=|\mu|=0$$

$$e^a \frac{\partial f}{\partial x^i_\mu} + e^\mu \frac{\partial f}{\partial x^i_a}=0 \qquad |i|=|\mu|=1$$

<u>Remark</u>. These equations are the analogues of the Cauchy-Riemann equations.

Using this theorem one proves the composite mapping and inverse mapping theorems. Indeed,

<u>Theorem 2</u>: The set S of local G^∞ - diffeomorphisms of $E_N^{m,n}$ is a pseudogroup.

<u>Definition 1</u>: A G^∞ - (super) manifold of type (m,n) is a topological space which is locally homeomorphic to $E_N^{m,n}$ and whose transition functions are G^∞ diffeomorphisms, i.e belong to the pseudogroup S.

One easily sees that a G^∞ manifold of type (m,n) is an ordinary C^∞ manifold of dimension $2^{N-1}(m+n)$.

The study of G^∞ manifolds entails a study of a certain group. To this end consider the free Λ-module $B=\Lambda\otimes V$. The Z_2 gradings on Λ and V induce a Z_2 grading on B. Denote by $(\mathrm{End}_\Lambda B)_0$ the degree preserving Λ endomorphisms of B.

Define

$$GL_N(m/n) = \{A\varepsilon(\mathrm{End}_\Lambda B)_0 :\ A \text{ is invertible}\}$$

It is easy to show that $GL_N(m/n)$ is a group. In fact group composition and the inverse map are G^∞ and is thus an example of a G^∞ Lie supergroup[5]. We have

Theorem 3: $GL_N(m/n)$ is homeomorphic to $O(m) \times O(n) \times R^{N'}$ where $N'=(n+m)^2\, 2^{N-1} - \frac{n(n-1)}{2} - \frac{m(m-1)}{2}$.

If follows that $GL_N(m/n)$ has the homotopy type of $O(m) \times O(n)$.

Now let M be a C^∞ manifold of dimension $2^{N-1}(m+n)$. Then the tangent space $T_p(M)$ at $p\varepsilon M$ can be given the structure of $E_N^{m,n} \approx (\Lambda\otimes V)_0$. If this can be done smoothly for all of $T(M)$, we say that M is an almost G^∞ manifold. Now $GL_N(m/n)$ acts on $(\Lambda\otimes V)_0$ but not effectively. The kernel of the homomorphism $\phi: GL_N(m/n) \longrightarrow GL(2^{N-1}(m+n))$ is an invariant subgroup $\mathcal{R}$ of dimension m^2 if N is even and n^2 if N is odd. Then $GL_N(m/n)/\mathcal{R}$ sits naturally in $GL(2^{N-1}(m+n))$ and an almost G^∞ manifold is a reduction of the bundle of linear frames $L(M)$ to the subgroup $GL_N(m/n)/\mathcal{R}$. From theorem 3 we have

Theorem 4: A $2^{N-1}(m+n)$ dimensional manifold M admits the structure of an almost G^∞ manifold if and only if there are vector bundles α and β of dimension m and n respectively such that

$$T(M) \approx 2^{N-1}\alpha + 2^{N-1}\beta$$

As an immediate corollary one has M is orientable if $N \geqslant 2$ and M has a spin structure if $N \geqslant 3$.

Using theorem 1 it is not difficult to show that the pseudogroup S is isomorphic to the pseudogroup $P_{GL_N(m/n)/\mathcal{R}}$ of local diffeomorphisms of $R^{2^{N-1}(m+n)}$ whose Jacobian matrices at every point belong to $GL_N(m/n)/\mathcal{R}$. It follows that

Theorem 5: A G^∞ manifold is an almost G^∞ - manifold. Moreover, the G^∞ structures on M are in bijective correspondence with the integrable $GL_N(m/n)/\mathcal{R}$ structures.

Theorem 4 can be used to give nontrivial examples of almost G^∞ manifolds. For appropriate values of k and ℓ it can be shown that $RP^k \times RP^\ell \times S^1 \times S^1$ and $CP^k \times CP^\ell \times S^3 \times S^3$ have the structure of almost G^∞ manifolds, where RP^k and CP^k are real and complex projective planes, respectively. As an example of a noncompact almost G^∞ manifold consider any vector bundle ξ over a C^∞ manifold M and let $T^{(N-1)}(M)$ denote the $(N-1)^{th}$ iterated tangent bundle over M. Denote by $\xi^{(N-1)}$ the pullback of ξ over $T^{(N-1)}(M)$. Then the total space of the bundle $2^{N-1}\xi^{(N-1)}$ has an almost G^∞ structure.

It seems unlikely that these examples admit a true G^∞ structure, but this is really unknown at this time. One reason for believing this is that G^∞- manifolds admit many foliations. In fact let I^μ be the principal ideal generated by e^μ and set $V^\mu=(I^\mu\otimes V)_0$. If we identify $(\Lambda\otimes V)_0$ with $T_p(M)$, then if M is a G^∞ -manifold it can be shown using the pseudogroup $P_{GL_N(m/n)/\mathcal{R}}$ that V^μ generates an integrable subbundle of $T(M)$ for any sequence μ. Moreover, any sum (not direct) $V^\mu + V^\nu$ also gives rise to a foliation of M. The foliation of lowest codimension (equal to m) is described by the integrable subbundle generated by $\sum_{a=1}^{N} V^a$. This foliation can equally well be described by the submersion $\varepsilon\colon M \longrightarrow \varepsilon(M)$ where ε is the map induced by the augmentation $\varepsilon\colon E_N^{m,n} \longrightarrow R^m$. The manifold $M_0 = \varepsilon(M)$ is the space of leaves of the codimension m foliation above. It is called the core manifold. Rodgers[4] has shown how G^∞ functions can be uniquely constructed from $C^\infty(\varepsilon(U),\Lambda)$ functions on the core.

On any G^∞- manifold there is a naturally defined free Λ-module bundle. Denote by $G^\infty(p)$ the Z_2 graded algebra of germs of G^∞ functions at $p\varepsilon M$.

Definition 2: A superderivation at $p\varepsilon M$ is a Z_2 graded map $D\colon G^\infty(p) \longrightarrow \Lambda$ which satisfies $D(\alpha f+\beta g) = (-1)^{|\alpha||D|}\alpha Df+(-1)^{|\beta||D|}\beta Dg$ and $D(fg)=(Df)g + (-1)^{|f||D|}f(Dg)$ where $\alpha,\beta \varepsilon \Lambda$ and $f,g\varepsilon G^\infty(p)$. The set $ST_p(M)$ of superderivations at p is called the super tangent space at p. Moreover, $ST(M) = \bigcup_p ST_p(M)$ is called the super-tangent bundle.

Theorem 6: $ST(M)$ is a G^∞ manifold of type $(2m+n,m+2n)$ and a bundle of free Λ-modules over M. Moreover, $ST(M)_0$ is isomorphic to $T(M)$.

A G^∞ section X of $ST(M)$ is called a G^∞ super vector field on M. The set of G^∞ super vector fields on M forms a Lie superalgebra[6] under the Z_2 graded bracket $[X,Y] = X\,Y - (-1)^{|X||Y|}Y\,X$. We shall describe the "integral curves" of a G^∞ super vector field X.

Let $I_\alpha \subset \Lambda_\alpha$ be a neighborhood of 0 in Λ_α. I_α is a local group under addition. Let $\phi\colon I_\alpha \times M \to M$ be the corresponding local transformation group of local G^∞ diffeomorphisms on M. Put $\phi_t(p) = \phi(t,p)$. We call ϕ_t a local 1-parameter supergroup on M if ϕ_t is G^∞ in t.

Theorem 7: Let X be a G^∞ super vector field on M. For every $p\varepsilon M$ there are neighborhoods U of p, I_α of $0\varepsilon\Lambda_\alpha$, and a unique local 1-parameter supergroup ϕ_t on U satisfying $G\phi= X \circ \phi$ if X is even and $G\phi= X \circ \phi_0$ if X is odd. Moreover, X_q, $q\,\varepsilon U$, is tangent to the curve $t \longrightarrow \phi(t,p)$.

Remark. Odd curves are straight lines.

Theorem 7 is a basic theorem for analysis on G^∞ manifolds. As usual we can use it to obtain the G^∞ version of Frobenius' theorem. Let $W \subset V_\alpha$ with $W=W_0+W_1$ and $r=\dim W_0$, $s=\dim W_1$, then $\Lambda\otimes W$ is a free Λ-module of type (r,s). Let $D(M)$ be subbundle of $ST(M)$ which is a bundle of free Λ modules of type (r,s). $D(M)$ is involutive if for every point $p\varepsilon M$ and local G^∞ sections X, Y of $D(M)$, the Z_2 graded bracket $[X,Y]_{(p)} \varepsilon D_p(M)$. $D(M)$ is G^∞ integrable if for each point $p\varepsilon M$ there is a G^∞ submanifold $i\colon N \to M$ of type (r,s) with $p\varepsilon i(N)$ such that $i^* D(M)$ is isomorphic as free Λ module bundles to $ST(N)$.

Theorem 8: $D(M)$ is integrable if and only if it is involutive.

Finally we would like to say a few words about the integrability problem for almost G^∞ manifolds which is part of a classical problem of E. Cartān known as the general equivalence problem. The integrability problem entails identifying the necessary and sufficient conditions for the integrability of the given structure. For analytic manifolds it is a consequence of the Cartan-Kähler theorem[7]. It has recently[8] been shown that the integrability problem can be solved for all G-structures in the case of C^∞ manifolds. Our research with almost G^∞ manifolds indicates that a direct proof using theorem 8 is possible, at least when assuming analyticity in the "odd coordinates". Some results concerning formal structures were given in reference 9.

Acknowledgement. The authors would like to thank B. Kostant for discussions which clarified the differences between our approaches.

References

1. F. A. Berezin and G. I. Kac, Math. USSR Sborni 11, 311(1970); F. A. Berezin and D. A. Leites, Sov. Math. Dokl. 16, 1218 (1975).
2. B. Kostant, in Differential Geometric Methods in Mathematical Physics, Lecture Notes in Mathematics 570 (Springer-Verlag, New York, 1977).
3. M. Bachelor, Trans. Amer. Math. Soc. 253, 329 (1979); ibid ,258, (1980); in Group Theoretical Methods in Physics, Lecture Notes in Physics 94 (Springer-Verlag, New York, 1979).
4. A. Rodgers, Imperial College preprint ICTP/78-79/15.
5. V. Rittenberg and M. Scheunert, J. Math. Phys. 19, 709(1978).
6. L. Corwin, Y. Ne'eman, and S. Sternberg, Rev. Mod. Phys. 47, 573(1975); V. G. Kac, Adv. Math. 26, 8 (1977); M. Scheunert, The Theory of Lie Superalgebras, Lecture Notes in Mathematics 716 (Springer-Verlag, New York, 1979).
7. I. M. Singer and S. Sternberg, J. Anal. Math. 15, 1 (1965).
8. H. Goldschmidt and D. C. Spencer, Acta Math. 136, 103(1976); H. Goldschmidt, Bull. Amer. Math. Soc. 84, 531 (1978).
9. C. P. Boyer, J. Pure Appl. Alg. 18, 1 (1980).

CASIMIR OPERATORS AND THE RELATIONS OF THE SIMPLEST SUPERSYMMETRIC SUPERALGEBRA

Ladislav Hlavatý and Jiří Niederle.
Institute of Physics,
Czechoslovak Academy of Sciences,
18040 Prague 8, Czechoslovakia.

It is argued there exist supersymmetric Lie superalgebras not contained in [1] which may be of some interest for supergravity theories. As an example, the Lie superalgebra first discovered by Konopelchenko [2] is discussed. It consists of 10 generators of the Poincaré algebra, J_{AB}, $\bar{J}_{\dot{A}\dot{B}}$, $P_{A\dot{B}}$, with the usual commutation relations and 4 odd generators Q_A, $\bar{Q}_{\dot{A}}$ with the remaining relations of the form

$$[J_{AB}, Q_C] = \varepsilon_{AC} Q_B + \varepsilon_{BC} Q_A \quad , \quad [\bar{J}_{\dot{A}\dot{B}}, Q_C] = 0, \tag{1}$$

$$[J_{AB}, \bar{Q}_{\dot{C}}] = 0 \quad , \quad [\bar{J}_{\dot{A}\dot{B}}, \bar{Q}_{\dot{C}}] = \varepsilon_{\dot{A}\dot{C}} \bar{Q}_{\dot{B}} + \varepsilon_{\dot{B}\dot{C}} \bar{Q}_{\dot{A}}, \tag{2}$$

$$[P_{A\dot{B}}, Q_C] = \varepsilon_{AC} \bar{Q}_{\dot{B}}, \tag{3}$$

$$[P_{A\dot{B}}, \bar{Q}_{\dot{C}}] = 0, \tag{4}$$

$$\{Q_A, Q_B\} = J_{AB}, \tag{5}$$

$$\{\bar{Q}_{\dot{A}}, \bar{Q}_{\dot{B}}\} = 0 \tag{6}$$

$$\{Q_A, \bar{Q}_{\dot{B}}\} = P_{A\dot{B}} \tag{7}$$

where $J_{AB} = J_{BA}$, $\bar{J}_{\dot{A}\dot{B}} = \bar{J}_{\dot{B}\dot{A}}$ and $\varepsilon_{AB} = \varepsilon_{\dot{A}\dot{B}} = -\varepsilon^{AB} = -\varepsilon^{\dot{A}\dot{B}} = \begin{pmatrix} 0 & -1 \\ 1 & 0 \end{pmatrix}$.

Notice that the often used Lie superalgebra in simple supergravity, further denoted as the GL superalgebra since it was first discovered by Gol'fand and Lichtman [3] (and then rediscovered by Wess, Zumino, Volkov, Akulov and others), differs from the Konopelchenko superalgebra (K superalgebra) in relations (3) and (5) that have to be replaced by

$$[P_{A\dot{B}}, Q_C] = 0 , \tag{8}$$

and $$\{Q_A, Q_B\} = 0 \tag{9}$$

respectively.

The Casimir operators of the K superalgebra are given by [4]

$$C_2^{(K)} = P_{A\dot{B}} \; P^{A\dot{B}} - \bar{Q}_{\dot{A}} \, \bar{Q}^{\dot{A}} \,, \tag{10}$$

$$C_4^{(K)} = K_{A\dot{B}} \; K^{A\dot{B}} + \frac{1}{2} [Q_A \, Q^A \, \bar{Q}_{\dot{B}} \, \bar{Q}^{\dot{B}} + \bar{Q}_{\dot{B}} \, \bar{Q}^{\dot{B}} \, Q_A \, Q^A] + [2 + \frac{1}{2} (J_{AB} \, J^{AB} - \bar{J}_{\dot{A}\dot{B}} \, \bar{J}^{\dot{A}\dot{B}})] \, \bar{Q}_{\dot{A}} \, \bar{Q}^{\dot{A}} \,, \tag{11}$$

with

$$K_{A\dot{B}} = J_{AC} \, P^C_{\dot{B}} - \bar{J}_{\dot{B}\dot{C}} \, P_A^{\dot{C}} - [Q_A, \bar{Q}_{\dot{B}}] \tag{12}$$

Operator $C_2^{(K)}$ combines the mass operator $P_{A\dot{B}} \, P^{A\dot{B}}$ with $\bar{Q}_{\dot{A}} \, \bar{Q}^{\dot{A}}$. Thus the *K* superalgebra might have irreducible multiplets consisting of particles with different masses. The price we pay for this possibility is that there is no involution in the *K* superalgebra. These questions and the relevance of the remarkable *K* super—algebra for particle physics will be studied elsewhere.

The Casimir operators of the *GL* superalgebra were derived in [4], too. They are of the form

$$C_2^{(GL)} = P_{A\dot{B}} \; P^{A\dot{B}} \tag{13}$$

and $$C_4^{(GL)} = K_{A\dot{B}} \; K^{A\dot{B}} + \frac{1}{2} [Q_A \, Q^A \, \bar{Q}_{\dot{B}} \, \bar{Q}^{\dot{B}} + \bar{Q}_{\dot{B}} \, \bar{Q}^{\dot{B}} \, Q_A \, Q^A] \tag{14}$$

with $K_{A\dot{B}}$ defined by *(12)*.

Notice that the operator $C_2^{(GL)}$ is well known but, instead of $C_4^{(GL)}$, the sixth order Casimir operator is used (see e.g. [5])

$$C_6^{(GL)} = K_{\mu\nu} \; K^{\mu\nu}, \tag{15}$$

where $$K_{\mu\nu} = P_\mu \, K_\nu - P_\nu \, K_\mu \tag{16}$$

with $$K_\mu = W_\mu - \frac{i}{4} \bar{S} \, \gamma_\mu \, \gamma_5 \, S \tag{17}$$

Here, W_μ is the Pauli-Lubanski vector and S is the Majorana bispinor formed of spinors Q, $\overline{Q}$. Our operator $C_4^{(GL)}$ is closer to the fourth order Casimir operator $W_\mu W^\mu$ of the Poincaré algebra than $C_6^{(GL)}$. In fact

$$C_6^{GL} = C_2^{GL} \cdot C_4^{GL} \qquad (18)$$

Finally let us remark that the K and GL superalgebras as well as their Casimir operators can be obtained from the corresponding quantities of the ortho-symplectic superalgebra $OSp(1,4)$ by contraction (for details see [4]).

[1] Haag R., Lopuszanski J., Sohnius M., Nucl. Phys. B88 (1975), 257.

[2] Konopelchenko B.G., "Letters JETP", (in Russian) 20 (1974), 608; 21 (1975), 612.

[3] Gol'fand Yu A., Lichtman E.P., "Letters JETP" (in Russian) 13 (1971), 452.

[4] Hlavatý L., Niederle J., Letters in Math. Physics (in press).

[5] Ogievetsky V.I., Mezinchescu, Uspekhi Fizicheskikh Nauk (in Russian) 117 (1975), 637.

ON REALIZATIONS OF GRADED LIE ALGEBRAS AND ON PROPERTIES OF d-POLYNOMIALS

Yehiel Ilamed
Soreq Nuclear Research Centre
Yavne, Israel

1. INTRODUCTION

In previous papers [1a-1e] we tried to understand intrinsic properties of polynomials that generalize the commutator and the anticommutator in free associative algebras. In this paper we define d-polynomials that are noncommutative polynomials with noncommutative coefficients inserted between the variables. We obtain realizations of graded Lie algebras using d-polynomials that are a natural extension of the usual symmetric and alternating polynomials. We show that d-polynomials have a universal derivative property and that some d-polynomials may be used as an extension of the brackets of Nambu. As a result we obtain that the following sets can be considered in a unified way as canonical sets of variables: i) n Heisenberg-Dirac canonical pairs, ii) the identity and n Heisenberg-Dirac canonical pairs, iii) the spin matrices of Pauli and iv) the octet matrices of Gell-Mann.

2. DEFINITIONS

2.1. Let $x,y,z,x_1,y_1,z_1,x_2,y_2,z_2,\ldots$ denote associative indeterminates, let I denote the unity and let F denote the field of real or complex numbers.

2.2. A <u>d-monomial</u> of length $2n-1$, $x_1y_1x_2 \ldots x_{n-1}y_{n-1}x_n$, is defined by two sets of variables: the <u>f-variables</u> $x_1,x_2,\ldots,x_n$ and the <u>p-variables</u> $y_1,\ldots,y_{n-1}$ that are inserted between the f-variables. We say that the f-variables are dressing the p-variables and that the p-variables are gluing the f-variables. A sum of d-monomials with cofficients in F is called a <u>d-polynomial</u>.

2.3. The <u>trace</u> of a polynomial q denoted by $/q/$ is defined by: i) $/q/ = 0$ if the polynomial q is a sum of commutators and ii) the polynomials q_1 and q_2 have the same trace if $/q_1 - q_2/ = 0$.

2.4. The polynomial p is <u>orthogonal</u> to the polynomial q if $/pq/ = 0$.

2.5. Let us define $h_n^+=h_n^+(x_1,\ldots,x_n;y_1,\ldots,y_{n-1})$ and $h_n^-=h_n^-(x_1,\ldots,x_n;y_1,\ldots,y_{n-1})$ by

$$h_n^+(x_1,\ldots,x_n;y_1,\ldots,y_{n-1}) = S_{n,x}^+ (x_1y_1x_2y_2 \ldots x_{n-1}y_{n-1}x_n) \quad (1)$$

$$h_n^-(x_1,\ldots,x_n;y_1,\ldots,y_{n-1}) = S_{n,x}^- (x_1y_1x_2y_2 \ldots x_{n-1}y_{n-1}x_n) \quad (2)$$

where $S_{n,x}^+$ means the action of summation over the $n!$ permutations of the factors

$x_1,\dots,x_n$ keeping the factors $y_1,\dots,y_{n-1}$ fixed, and $S^-_{n,x}$ means the same summation as in $S^+_{n,x}$ only that signs are alternating following the signs of the corresponding permutations. For example: $h^-_3(x_1,x_2,x_3;y_1,y_2) = S^-_{3,x}(x_1y_1x_2y_2x_3) = x_1y_1x_2y_2x_3 + x_2y_1x_3y_2x_1 + x_3y_1x_1y_2x_2 - x_1y_1x_3y_2x_2 - x_2y_1x_1y_2x_3 - x_3y_1x_2y_2x_1$. The d-polynomials $h^+_n(x_1,\dots,x_n;y_1,\dots,y_{n-1})$ and $h^-_n(x_1,\dots,x_n;y_1,\dots,y_{n-1})$ are respectively n-th degree symmetric and alternating polynomials in the f-variables $x_1,\dots,x_n$; the p-variables $y_1,\dots,y_{n-1}$ may be considered as parameters specifying the corresponding symmetric or alternating functions.

2.6. The usual symmetric and alternating functions, c_n and s_n, defined by

$$c_n(x_1,\dots,x_n) = S^+_{n,x}(x_1x_2 \dots x_n) \quad \text{and} \quad s_n(x_1,\dots,x_n) = S^-_{n,x}(x_1x_2\dots x_n) \tag{3}$$

can also be defined by

$$c_n = h^+_n(x_1,\dots,x_n;I,\dots,I) \quad \text{and} \quad s_n = h^-_n(x_1,\dots,x_n;I,\dots,I) \tag{3'}$$

2.7. Let $p(x_1,\dots,x_m;y_1,\dots,y_{m-1})$ and $q(x_1,\dots,x_n;z_1,\dots,z_{n-1})$ be homogeneous d-polynomials linear in each variable. We define the compositions $\underset{0}{*}$ and $\underset{1}{*}$ by

$$p\underset{i}{*}q\,(x_1,\dots,x_{m+n-1};y_1,\dots,y_{m-1},z_1,\dots,z_{n-1}) =$$
$$\Sigma_\tau(\mathrm{sg}\tau)^i p(q(x_{\tau 1},\dots,x_{\tau n};z_1,\dots,z_{n-1}),x_{\tau(n+1)},\dots,x_{\tau(n+m-1)},y_1,\dots,y_{m-1}) \tag{4}$$

where $i = 0$ or 1, $\mathrm{sg}\tau$ means the sign of τ and Σ_τ means the action of summation over all the permutations on $\{1,\dots,m+n-1\}$ restricted to $\tau 1<\tau 2<\dots<\tau n$ and $\tau(n+1)<\tau(n+2)<\dots<\tau(m+n-1)$. For $y_1=\dots=y_{m-1}=z_1=\dots=z_{n-1}=I$, one obtains that the compositions $\underset{0}{*}$ and $\underset{1}{*}$ are identical, respectively, with $*$ and $\underline{*}$ defined in [1e].

2.8. Let us define $g^-_n(x_1,\dots,x_n;y_1,\dots,y_{n+1}) = S^-_{n,x}(y_1x_1y_2x_2\dots y_nx_ny_{n+1})$ (5)
where $S^-_{n,x}$ is defined in &2.5. ; here $x_1,\dots,x_n$ are p-variables and $y_1,\dots,y_{n+1}$ are f-variables.

2.9. Let us define $g^{-,c}_{n,n+1}(x_1,\dots,x_n;y_1,\dots,y_{n+1}) = C^-_{n+1,y}g^-_n(x_1,\dots,x_n;y_1,\dots,y_{n+1})$ (6)

where $C^-_{n+1,y}$ means the action of summation, with the sign of the corresponding permutation, over the n+1 cyclic permutations of $y_1,\dots,y_{n+1}$ keeping $x_1,\dots,x_n$ fixed. For example: $C^-_{4,y}g^-_3(x_1,x_2,x_3;y_1,y_2,y_3,y_4) = g^-_3(x_1,x_2,x_3;y_1,y_2,y_3,y_4) - g^-_3(x_1,x_2,x_3;y_2,y_3,y_4,y_1) + g^-_3(x_1,x_2,x_3;y_3,y_4,y_1,y_2) - g^-_3(x_1,x_2,x_3;y_4,y_1,y_2,y_3)$.

2.10. Let L_z and R_z denote respectively the usual left and right multiplication by z when acting on f-variables or on d-polynomials. The action of L_z and R_z on p-variables is defined by

$$L_z(\text{p-variable}) = -(\text{p-variable})z \quad \text{and} \quad R_z(\text{p-variable}) = -z(\text{p-variable}) \tag{7}$$

This means that the action of L_z and R_z on p-variables is with a minus sign and on the opposite side; on the right and on the left side respectively.

3. REALIZATIONS OF GRADED LIE ALGEBRAS

3.1. The following multiplication tables are valid, [1e],

$$c_m \underset{0}{*} c_n = mc_{m+n-1} \tag{8}$$

for $c_1, c_2, \ldots$, and

$$s_{2m+1} \underset{1}{*} s_{2n+1} = (2m+1)s_{2m+2n+1}, \quad s_{2m+1} \underset{1}{*} s_{2n} = s_{2m+2n}$$
$$s_{2m} \underset{1}{*} s_{2n+1} = 2ms_{2m+2n} \quad \text{and} \quad s_{2m} \underset{1}{*} s_{2n} = 0 \tag{9}$$

for $s_1, s_2, \ldots$.

We have shown that substituting $h_n^+(x_1,\ldots,x_n;y,y,\ldots,y)$ for c_n in Eq. (8) and $h_n^-(x_1,\ldots,x_n;y,y,\ldots,y)$ for s_n in Eq. (9), we obtain valid equations . The result is that the vector spaces over F spanned by $h_n^+(x_1,\ldots,x_n;y,y,\ldots,y)$ and $h_n^-(x_1,\ldots,x_n;y,y,\ldots,y)$, $n=1,2,\ldots$, with the multiplication laws $\underset{0}{*}$ and $\underset{1}{*}$ respectively are Lie admissible algebras. The associated Lie algebras are connected with string algebras appearing in physics, [2].

3.2. Using the composition law defined by Eq.(4) we obtain by direct calculations

$$h_3^+(\ ;z_1,z_2)\underset{0}{*}h_2^+(\ ;y) = h_4^+(\ ;y,z_1,z_2) + h_4^+(\ ;z_1,y,z_2) + h_4^+(\ ;z_1,z_2,y) \quad \text{and}$$
$$h_3^-(\ ;z_1,z_2)\underset{1}{*}h_2^-(\ ;y) = h_4^-(\ ;y,z_1,z_2) - h_4^-(\ ;z_1,y,z_2) + h_4^-(\ ;z_1,z_2,y) \tag{10}$$

Substituting I for $y=z_1=z_2$ in Eq.(10) we verify the Eqs. (8) and (9) for one case. In a similar way the second proposition of &3.1. can be verified.

4. PROPERTIES OF d-POLYNOMIALS

4.1. The $g^{-,c}_{n,n+1}$ polynomial defined in &2.9. is orthogonal to its p-arguments [3],

$$/x_i g^{-,c}_{n,n+1}(x_1,\ldots,x_n;y_1,\ldots,y_{n+1})/ = 0 \quad , \quad i=1,\ldots,n \tag{11}.$$

4.2. The action of L_z or R_z on d-polynomials is a derivation. For example, using &2.10. , it is easy to verify

$$L_z(x_1yx_2) = (L_zx_1)yx_2 + x_1(L_zy)x_2 + x_1y(L_zx_2) \tag{12}$$

4.2. A corollary of the derivative property of the d-polynomials is the following proposition. Let $p(x_1,\ldots,x_n;y_1,\ldots,y_{n-1})$ be any given d-polynomial and let $p(a_1,\ldots,a_n;a_{n+1},\ldots,a_{2n-1})$ denote the evaluation of $p(x_1,\ldots,x_n;y_1,\ldots,y_{n-1})$ for $a_1,\ldots,a_{2n-1}$, a (2n-1)-list of elements in any given algebra A over F. Let V be the vector space over F spanned by all the evaluations of $p(x_1,\ldots,x_n;y_1,\ldots,y_{n-1})$ in A . Then V is an ideal in A .

5. ON CANONICAL SETS OF VARIABLES

5.1. Nambu [4] has suggested a generalization of the classical mechanics taking the Liouville theorem as a guiding principle. The problem of quantization of this generalized mechanics "is not an easy task" ,[4], since the derivative property of the Poisson brackets is difficult to realize.

5.2. The polynomial $s_n(x_1,\ldots,x_n)$ defined in &2.6. , called the standard polynomial of degree n ,[5], is also called the bracket of Nambu of degree n ,[6].

A canonical set of variables $a_1,\ldots,a_n$, elements of an algebra A over F, may be defined by the condition (see Refs. [4] and [6])

$$s_n(a_1,\ldots,a_n) = \alpha I \quad , \quad 0 \neq \alpha \text{ in } F \tag{13}$$

It is natural to suggest the following extension of the above definition of a canonical set of variables.

5.3. Let A be an algebra over F. The list $a_1,\ldots,a_n$ of n elements in A is a canonical set of elements if there exist a list $b_1,\ldots,b_{n+1}$ of n+1 elements in A so that

$$g^{-,c}_{n,n+1}(a_1,\ldots,a_n;b_1,\ldots,b_{n+1}) = \alpha I \quad , \quad 0 \neq \alpha \text{ in } F \tag{14}$$

We say that the canonical set of variables $a_1,\ldots,a_n$ is associated to the set $b_1,\ldots,b_{n+1}$. For example the sets $\{p_1,q_1\}$, $\{p_1,q_1,I\}$ and $\{p_1,q_1,p_2,q_2\}$,... (where p_i,q_i , i=1,2,... are mutually commutative Heisenberg canonical pairs) are canonical sets of variables ,[7], associated,respectively, to the sets {I,I},{I,I,I} and {I,I,I,I},... .

5.4. The spin matrices of Pauli , $\sigma_1,\sigma_2,\sigma_3$, and the octet matrices of Gell-Mann $\lambda_1,\ldots,\lambda_8$ are canonical sets of variables associated respectively to b_1,b_2,b_3,b_4 four generic 2 by 2 matrices with entries in F and to $c_1,\ldots,c_9$ nine generic 3 by 3 matrices with entries in F.

5.5. The previous sentence, &5.4. , is a corollary of the following proposition.

Let $h = n^2-1$. The polynomial $g^{-,c}_{h,h+1}(x_1,\ldots,x_h;y_1,\ldots,y_{h+1})$ is a central polynomial (its evaluations are in the center) for $x_1,\ldots,x_h$ generic $n \times n$ matrices of trace zero and $y_1,\ldots,y_{h+1}$ generic $n \times n$ matrices with entries in F.

This theorem is a corollary of the proposition &4.1. ; for the proof of these theorems see Refs. [3] and [8] .

6. NOTE

Complete proofs of the results of this communication will be given in a forthcoming paper.

REFERENCES

1. Y. Ilamed, a) A characterization of the even degree standard polynomial, in Proc.Fifth Int. Coll. G.T.M. in Physics, Montreal 1976, Academic Press, N. Y. 1977, p. 623-626.

   b) Lie elements, the Killing form and trace identities, in Proc. Sixth Int. Coll.G.T.M. in Physics, Tübingen 1977, L.N. in Physics #79, Springer 1978, p. 494-496.

   c) Dual sets in associative algebras and generalized Lie algebras, in Proc. Seventh Int. Coll. G.T.M. in Phys., Austin 1978, Springer L.N. in Phys. # 94, 1979, p. 446-447.

   d) On realizations of infinite dimensional Lie algebras, in Proc. Eight Int. Coll. G.T.M. in Phys., Kiriat Anavim 1979, Israel Phys. Soc. Annals #3, 1980, p. 323-324.

   e) On realizations of infinite dimensional Lie admissible algebras, Hadronic J. 3, (1979) 327-338.

2. J. Schwartz, Dual resonance theory,, Phys. Rep. 8c, (1973) 269-335.

3. Y. Ilamed, On identities for matrix rings and polynomials orthogonal to their arguments, in Ring Theory, Proc.of the 1978 Antwerp Conf. Ed. F. van Oystaeyen M. Dekker Inc. 1979, p. 81-86 .

4. Y. Nambu, Generalized Hamiltonian dynamics, Phys. Rev. D7 (1973) 2405-2412.

5. S. A. Amitsur and J. Levitzki, Minimal identities for algebras, Proc. A.M.S. 1, (1950) 449-463 .

6. J. A. Kalnay, On the new Nambu mechanics, its classical partners and its quantization, in Coll. Int. C.N.R.S. # 237, Geometrie symplectique et Physique Mathematique, 1975, p. 401-410.

7. Y. Ilamed, Heisenberg algebras, n-Lie algebras and alternating polynomials that are central, A.M.S. Notices, 1978, 78T-A 111.

8. S. A. Amitsur, On a central identity for matrix rings, J. London Math. Soc. (2) 14, (1976) 1-6.

ON GAUGE FORMULATIONS OF GRAVITATION THEORIES.

E. A. Ivanov *) and J. Niederle.
Institute of Physics.
Czechoslovak Academy of Sciences.
18040 Prague 8, Czechoslovakia.

1. Introduction.

As was pointed out first by Utiyama [1] in 1956 and then by a number of physicists (see [2]-[7] and the references therein), gravitation theory may be looked upon as a gauge theory. However, diverse answers can be found to the questions:

- What is the corresponding gauge group?
- What are the gauge fields?
- What is the form of the associated Lagrangian?
- What about the metric tensor $g_{\mu\nu}$?

In order to clarify the situation we recall first a few facts about the standard Yang-Mills gauge theories. The minimal Yang-Mills structure of these theories is fully determined by the group of global internal symmetry transformations g. The fields in the Lagragian transform according to representations of g and the Lagrangian is explicitely invariant with respect to the group

$$K_{YM} = g^{loc} \otimes \tilde{P}. \tag{1.1}$$

Here g^{loc} is an ∞-parameter group of gauge (local) transformations uniquely determined by g, and $\tilde{P}$ denotes the physical Poincaré group. Note that K_{YM}, in general, may contain still another internal symmetry group and the conformal or other space-time symmetry groups instead of $\tilde{P}$.

Many physicists have tried to treat gravity in an analogous manner. Namely they put $g^{loc} = \mathcal{D}iff\ R^4$ -the group of general covariant coordinates transformations. This setting has led to several difficulties which will be discussed in Section 2 and which essentialy arise from the fact that there does not exist one special finite-parameter subgroup in $\mathcal{D}iff\ R^4$ playing the rôle of g. As a consequence, there is no unique recipe how to get $g^{loc} = \mathcal{D}iff\ R^4$. (In fact $\mathcal{D}iff\ R^4$ can be obtained from its many subgroups, e.g. from the Poincaré subgroup or form the de Sitter one or even from the Galilei one.)

We, therefore, suggest another approach [6] constructed in the spirit of [4] in which we have the group Diff R^4 acting on x_μ from the very beginning and in which we introduce a tangent space with the group of action $\mathcal{H}$ in every x_μ. The group $\mathcal{H}$ does

---

*) Permanent address: Joint Institute for Nuclear Research in Dubna,
P. O. Box 79, 101000 Moscow, NSSR.

not act on x_μ and plays the rôle of g. The matter fields in the theory transform according to (even non-linear) representations of $\mathcal{H}$. The full invariance group of the Lagrangian is then of the form

$$K = \mathcal{H}^{loc} \otimes \mathcal{D}iff\, \mathbb{R}^4 \qquad (1.2)$$

where $\mathcal{D}iff\, \mathbb{R}^4$ plays the rôle of $\tilde{P}$ in (1.1) and $\mathcal{H}^{loc}$ is obtained from $\mathcal{H}$ exactly as g^{loc} from g. In order to find the explicit form of K we remark three things:

1) The group of global symmetries of the theory (i.e. the stability group of the classical vacuum) consists of generators which are linear combinations of some generators from $\mathcal{H}^{loc}$ and some from $\mathcal{D}iff\, \mathbb{R}^4$. This group is identified with the group of motion of the background. It turns out it is determined by $\mathcal{H}$. To different $\mathcal{H}$ there correspond different theories of gravity. Thus if $\mathcal{H} = P$, (the Poincaré group) we get the Einstein theory of gravity in the Minkowski space as a background. If $\mathcal{H} = SO(3,2)$ ($SO(4,1)$) we obtain the Einstein theory with a negative (positive) cosmological term in the corresponding de Sitter space and if $\mathcal{H} = C$ (i.e. to the conformal group) we get the Cartan-Weyl theory.

2) The structure of $\mathcal{H}$ is very restricted by requiring: i) The Lorentz invariance of the theory and ii) The construction of the theory by means of $g_{\mu\nu}$ (expressed in terms of the vierbein. From the first requirement we see that $\mathcal{H}$ contains the Lorentz group $SO(3,1)$. The generators of the physical Lorentz group $\tilde{L}_{\mu\nu}$ are of the usual form $i(x_\mu \partial_\nu - x_\nu \partial_\mu) + L_{\mu\nu}$. The first term corresponds to the Lorentz generators in $\mathcal{D}iff\, \mathbb{R}^4$ and the second to those in $\mathcal{H}$. The gauge field corresponding to $L_{\mu\nu}$ will be the gravitation connection. From the second requirement it follows that we have to add still other generators to $\mathcal{H}$ in order to get the vierbein as a gauge field. The simplest choice is to add to $L_{\mu\nu}$ the generators P_a which transform as a $SO(3,1)$-four-vector. As a consequence, we obtain $\mathcal{H}$ equal to the Poincaré group $P = J^4 \otimes SO(3,1)$ which leads, as already mentioned, to the usual Einstein theory of gravity. (The other choices lead to $\mathcal{H} = SO(3,2)$, $SO(4,1)$ or C).

3) The group $\mathcal{H}^{loc}$ obtained by the standard way from $\mathcal{H}$ must be spontaneously broken (in the sense of [8]) in order to exclude a redundant gauge field (the gravitation connection) from the theory in an invariant way.

In [6] all cases of $\mathcal{H}$ were studied in detail because of the recent revival of interest in gauge theoretic formulations of gravity in connection with developments in supergravity. The case $\mathcal{H} = P$ was briefly discussed in [4] and the case $\mathcal{H} = SO(3,2)$ in [5]. Here we shall illustrate the approach on the simplest case $\mathcal{H} = P$ and clarify the connection with the other approaches. For the remaining cases of $\mathcal{H}$ we refer to [6].

2. The Einstein theory of gravity.

In this case $\mathcal{H} = P = T^4 \otimes SO(3,1)$ generated by P_a and L_{ab} satisfying the usual commutation relations

$$[L_{ab}, L_{cd}] = i(\eta_{ad}L_{bc} + \eta_{bc}L_{ad} - \eta_{ac}L_{bd} - \eta_{bd}L_{ac}), \tag{2,1}$$

$$[L_{ab}, P_c] = i(\eta_{bc}P_a - \eta_{ad}P_b), \tag{2.2}$$

$$[P_a, P_b] = 0 \tag{2.3}$$

with diag $\eta_{ab} = (1,-1,-1,-1)$. The matter fields $\phi_t(x)$ (world scalars) transform with respect to $\mathcal{H}$-transformations as

$$\delta_L \phi_t(x) = \frac{i}{2} a^{ab}(x)(\bar{L}_{ab})_{tm} \phi_m(x), \tag{2.4}$$

$$\delta_P \phi_t = 0 \tag{2.5}$$

and with respect to $\mathcal{D}iff\ \mathrm{R}^4$ - transformations as

$$\delta_r \phi_t(x) = -\lambda^\mu(x)\partial_\mu \phi_t(x), \tag{2.6}$$

where $a^{ab}(x)$ and $\lambda^\mu(x)$ are arbitrary functions. Notice that the physical Poincaré group $\tilde{P}$ contained in K is obtained by setting

$$\lambda^\mu(x) = \tilde{c}^\mu - \tilde{\lambda}^{\mu\nu}x_\nu, \quad a^{ab}(x) = \tilde{\lambda}^{ab} \tag{2.7}$$

where $\tilde{c}^\mu$ and $\tilde{\lambda}^{\mu\nu}$ are constants. The Poincaré group $\tilde{P}$ does not distinguish the world indices (denoted by Greek letters) and the vierbein indices (denoted by Latin letters).

As usually to L_{ab} and P_a these correspond the gauge fields (the covariant world vectors) $\Omega^{ab}_\mu(x)$ and $\ell^a_\mu(x)$ respectively. Their transformation properties are given by,

$$\delta_L \Omega^{ab}_\mu(x) = -a^{an}(x)\Omega_{\mu n}{}^b(x) - a^{bn}(x)\Omega^a_{\mu n}(x) - \frac{1}{g}\partial_\mu a^{ab}(x), \tag{2.8}$$

$$\delta_L \ell^a_\mu(x) = -a^{an}(x)\,\ell_{\mu n}(x), \tag{2.9}$$

$$\delta_P \Omega^{ab}_\mu(x) = 0, \tag{2.10}$$

$$\delta_P \ell^a_\mu(x) = \Omega^{am}_\mu(x)\,c_m(x) - \frac{1}{g}\partial_\mu c^a(x), \tag{2.11}$$

$$\delta_R \Omega^{ab}_\mu(x) = -\lambda^\rho(x)\partial_\rho \Omega^{ab}_\mu(x) - \partial_\mu\lambda^\rho(x)\,\Omega^{ab}_\rho(x), \tag{2.12}$$

$$\delta_R \ell^a_\mu(x) = -\lambda^\rho(x)\partial_\rho \ell^a_\mu(x) - \partial_\mu\lambda^\rho(x)\,\ell^a_\rho(x) \tag{2.13}$$

In (2.11) $c^a(x)$ are parameters associated with P_a.

Following the standard procedure we introduce the covariant derivation $\mathcal{D}_\rho$ by means of the gauge field (with the values in the algebra of $\mathcal{H}$) $A_\mu(x)$,

$$A_\mu(x) = \ell^a_\mu(x)\, P_a + \frac{1}{2}\, \Omega_\mu^{\ ab}\, L_{ab}, \tag{2.14}$$

in the form

$$\mathcal{D}_\rho\, \phi_a(x) = \partial_\rho\, \phi_a(x) + \frac{i}{2} g\, \Omega_\rho^{\ mn}(x)\, (L_{mn})_{ab}\, \phi_b(x), \tag{2,15}$$

The covariant curl $C^a_{\mu\nu}(x)$ $(R^{ab}_{\mu\nu}(x))$ of gauge field $\ell^a_\mu(x)$ $(\Omega^{ab}_\mu(x))$ is defined as the projection of the quantity

$$A_{\mu\nu}(x) = \partial_\mu A_\nu(x) - \partial_\nu A_\mu(x) + ig\,[A_\mu(x), A_\nu(x)] \tag{2.16}$$

on P_a (L_{ab}), i.e.

$$C^{\ a}_{\mu\nu}(x) = \partial_\mu\, \ell^a_\nu(x) - \partial_\nu\, \ell^a_\mu(x) - g\, \Omega^{ab}_\mu(x)\, \ell_{\nu b}(x) + g\Omega^{ab}_\nu(x)\ell_{\mu b}(x) = \tag{2.17}$$

$$= \mathcal{D}_\mu\, \ell^a_\nu(x) - \mathcal{D}_\nu\, \ell^a_\mu(x)$$

$$R^{ab}_{\mu\nu}(x) = \partial_\nu\, \Omega^{ab}_\nu(x) - \partial_\nu \Omega_\mu^{\ ab}(x) - g\,[\Omega^a_{\mu m}(x)\Omega^{mb}_\nu(x) - \Omega^a_{\nu m}(x)\,\Omega^{mb}_\mu(x)]. \tag{2,18}$$

Their transformation properties follow from those of $\ell^a_\mu(x)$ and $\Omega_\mu^{\ ab}(x)$. Here, we recall only the property

$$\delta_P\, R^{\ ab}_{\mu\nu}(x) = 0 \tag{2,19}$$

$$\delta_P\, C^{\ a}_{\mu\nu}(x) = R^{\ am}_{\mu\nu}(x)\, c_m(x) \tag{2.20}$$

The physicists usually interpret gauge fields $\ell^a_\mu(x)$ as a vierbein and express $\Omega^{ab}_\mu(x)$ in terms of $\ell^a_\mu(x)$ and its inverse by setting the constraint on $C^{\ a}_{\mu\nu}(x)$ namely

$$C^{\ a}_{\mu\nu}(x) = 0$$

(cf. [5]). In this connection let us note two things:

1) In order to have the inverse of $\ell^a_\mu(x)$ we have to assume that $\ell^a_\mu(x)$ contain a constant term $\sim \delta^a_\mu$. However, since $\ell^a_\mu(x)$ by definition is a field, its from is determined by solving the corresponding equation of motion. Thus to assume the particular form of $\ell^a_\mu(x)$ apriori is not justified.

2) Condition (2 21) is explicitely non-invariant. Consequently, taking P-transformations in order (2.21) to be true we must put $R^{\ ab}_{\mu\nu} = 0$. Thus, $\ell^a_\mu(x)$ and $\Omega^{ab}_\mu(x)$ correspond to a choi e of the gauge and, what is more important, $\Omega^{ab}_\mu(x)$ is not transforming with respect to P-transformations according to (2.10) as should have provided $\ell^a_\mu(x)$ is transforming according to (2 11). In orden to rid of this difficulty various authors

have modified the transformation rule (2.11).

We think a correct and more natural treatment consists of the assumption that the group $\mathcal{H} = P$ is spontaneously broken down to the Lorentz group $SO(3,1)$. This leads to the solution of both difficulties mentioned above simultaneously.

A mimimal way of breaking P down to $SO(3,1)$ is via non-linear realizations (see [8]). In our case this leads to introducing four Goldstone fields $y^a(x)$ with the property

$$\delta_P\, y^a(x) = c^a(x). \tag{2.22}$$

Fields $y^a(x)$ serve as coordinates in the factor space $P/SO(3,1) \simeq$ Minkowski space. In other words to any x_μ we associate the internal Minkowki space with the coordinates $y^a(x)$ and with the group of motion P.

By applying the standard technique we may introduce the Cartan forms ω_μ^a and ω_μ^{ab} via

$$e^{-iy^a(x)P_a}\{\partial_\mu + ig\, A_\mu(x)\}\, e^{iy^a(x)P_a} = i\,\omega_\mu^a(x)P_a + \frac{1}{2}\, i\,\omega_\mu^{ab}\, L_{ab} \tag{2.23}$$

so that their explicit forms in terms of the fields are given by:

$$\omega_\mu^{ab}(x) = g\,\Omega_\mu^{ab}(x), \tag{2.24}$$

$$\omega_\mu^a\,(x) = \partial_\mu y^a(x) + g\,\ell_\mu^a(x) - g\,\Omega_\mu^{an}(x)\, y_n(x). \tag{2.25}$$

They have the property,

$$\delta_p\,\omega_\mu^a(x) = 0, \qquad \delta_p\,\omega_\mu^{ab}(x) = 0. \tag{2.26}$$

The form of $\omega_\mu^a(x)$ can be simplified by setting

$$\ell_\mu^a(x) - \Omega_\mu^{an}(x)\, y_n(x) = \tilde{\ell}_\mu^a. \tag{2.27}$$

By using the relation

$$\tilde{C}_{\mu\nu}{}^a(x)P_a + \frac{1}{2}\,\tilde{R}_{\mu\nu}^{ab}(x)L_{ab} = e^{-iy^a(x)P_a}\{C_{\mu\nu}{}^a(x)P_a + \frac{1}{2}\,R_{\mu\nu}^{ab}(x)L_{ab}\}\, e^{iy^a(x)P_a} \tag{2.28}$$

we can introduce the new curls $\tilde{C}_{\mu\nu}{}^a(x)$ and $\tilde{R}_{\mu\nu}^{ab}(x)$ which are invariant with respect to P-transformations. They have the form

$$\tilde{R}_{\mu\nu}{}^{ab}(x) = R_{\mu\nu}{}^{ab}(x), \tag{2.29}$$

$$\begin{aligned}\tilde{C}_{\mu\nu}{}^a(x) &= C_{\mu\nu}{}^a(x) - R_{\mu\nu}{}^{an}(x)\, y_n(x) = \\ &= \mathcal{D}_\mu\,\omega_\nu{}^a(x) - \mathcal{D}_\nu\,\omega_\mu{}^a(x).\end{aligned} \tag{2.30}$$

Now, $\omega_\mu^a\,(x)$ can be identified with the vierbein due to its transformation properties. Since $y^a(x)$ form four independent functions, det $\partial_\mu y^a(x) \neq 0$ and it is clear that there exists the inverse vierbein $\omega^{\mu a}(x)$ satisfying the relations

$$\omega^{\mu a}\ \omega_\mu{}^b = \eta^{ab}, \tag{2.31}$$

$$\omega_\mu{}^a\ \omega^\rho{}_a = \delta_\mu{}^\rho . \tag{2.32}$$

Then the metric can be written as usually, i. e.

$$g_{\mu\nu}(x) = \omega_\mu^a(x)\ \omega_{\nu a}(x), \tag{2.33}$$

$$g^{\mu\nu}(x) = \omega^{\mu a}(x)\ \omega^{\nu a}(x). \tag{2.34}$$

It is easy to check that the condition

$$\tilde{C}_{\mu\nu}{}^a = 0 \tag{2.35}$$

i invariant with respect to all symmetry transformations and hence its solution

$$\Omega_\mu{}^{mn}(x) = \frac{1}{2g}\{\omega^{\rho n}(x) - [\partial_\mu \omega_\rho{}^m(x) - \partial_\rho \omega_\mu{}^m(x)] + \omega_\mu^b(x)\omega^{\rho n}(x)\omega^{\nu m}(x)\partial_\nu \omega_{\rho b}(x) - (m \Leftrightarrow n)\} \tag{2.36}$$

has the right transformation properties (2.8), (2.10) and (2.12)

Geometrically $\tilde{C}_{\mu\nu}{}^a(x)$ and $\tilde{R}_{\mu\nu}{}^{ab}(x)$ are torsion and curvature respec - tively in the fibre bundle space with the base space isomorphic to the Minkowski space $\{x_\mu\}$ and the fibre isomorphic to $P/SO(3,1)$. Condition (2.35) defines a connected subspace in the fibre bundle span with torsion equal to zero. Thus, we see that there is only one real gauge field in the theory ($\omega_\mu^a(x)$ or $\tilde{\ell}_\mu{}^a(x)$).

The simplest invariant with respecto to $K' = P^{loc} \otimes Diff\ R^4$ which yields the second order equation for $\omega_\mu^a(x)$, is the scalar curvature

$$R = \omega^{\mu a}(x)\ \omega^{\rho b}(x)\ R_{\mu\rho ba}(x). \tag{2.37}$$

The corresponding action is of the familiar form

$$S = \frac{1}{16\pi G}\int d^4x \quad R.\ det\ \omega , \qquad G\text{- the Newton constant} \tag{2.38}$$

which by variation leads to the Einstein equation

$$R_{\mu\nu} - \frac{1}{2}\ g_{\mu\nu}\ R=0 \tag{2.39}$$

where

$$R_{\mu\nu} = \omega_\nu^a\ \omega^{\rho b}\ R_{\mu\rho ba} \tag{2.40}$$

3.Conclusion and discussion.

We have shown in the preceding section that the minimal dynamics associated with the group $K = P^{loc} \otimes \mathcal{D}iff\, R^4$ is the dynamics of the usual Einstein theory of gravity.

Finally let us make several remarks concerning the structure of the theory:

1) If $\Omega^{ab}_{\mu}(x)$ is considered as the gauge field independent on the vierbein then variation of S and so on leads to the Einstein-Cartan theory which involves spin and torsion and hence is not equivalent to the Einstein theory of gravity constructed above.

2) It is easy to show that by variation of S with respect to four independent $y^a(x)$ we obtain the equations of motion which are identically satisfied (provided (2.39) holds). Consequently, $y^a(x)$ do not appear neither in the Lagrangian nor in the equations of motion (for details see [6]). Notice that in our case $y^a(x)$ are not making $\omega^a_\mu(x)$ massive but just redefine their transformation properties so that $\Omega^{ab}_\mu(x)$ can be expressed in terms of $\omega^a_\mu(x)$ in an invariant way (cf. (2.28), (2.36)).

3) We may consider $y^a(x)$ also as not being independent, e. g. by setting

$$y^a(x) = \delta^a_\mu\, x^\mu \tag{3.1}$$

in each x_μ (i.e. by chossing the gauge). Condition (3.1) appears to be invariant under the physical Poincaré group $\tilde{P}$ (2.7) contained in K defined by (1.2), which does not distinguish the world and vierbein indices (for details see [6]). It is precisely this Poincaré group which was discussed in [2], [4] etc. By using it(i.e. $\tilde{P}$) we can rid off the antisymmetric part of the vierbein. Then the group which leaves invariant this procedure is a group called in [9] the minimal group of general relativity (for details see [6]).

[1] Utiyama R., Phys. Rev. 101 (1956), 1597.
[2] Kibble T. W. B., J. Math. Phys. 2(1961), 212.
[3] Sciama D. W., in Recent Developments in General Relativity, Pergamon Press and PWN, Oxford and Warsaw, 1962, p. 415.
Mehl F. W. et al., Rev. Mod. Phys. 48 (1976), 393;
Carmeli M.Group Theory and General Relativity, McGraw Hill Co.,New York,1977;
Macrae K.I.,Phys.Rev. D18(1978),3737,3777,3761;
Trautman A., preprint of the Inst. of Theoretical Physics,IFT/10/78,Warsaw,1978;
Niederle J,ICTP preprint,IC/79/1 , Trieste, 1979
[4] Volkov,D.V.,Soroka V.A. JETP Letters 18(1973),312;Theor Mat.Fiz 20(1974),829 (in Russian)
Volkov D.V.,CERN preprint TH 2288,1977
[5] Chamseddine A.H.,West P.C.,Nucl.Phys. B129 (1977), 39
[6] Ivanov E.A., Niederle J.,Czech. J. Phys (to be published)
[7] Stelle K.S., West P.C.,Phys,Rev. D21(1980), 1466
[8] Coleman S.,Wess J.,Zumino B.,Phys. Rev. 177 (1969),2247; Callan C.G., Coleman S, Wess J. Zumino B.,Phys. Rev. 177 (1969), 2247; Isham C.J. Nuovo Cimento 59A (1969), 356.- Salam A., Strathdee J., Phys. Rev. 184(1969),1750; 1760; Volkov D.V.,Sov.J. Particles and Nuclei 4(1973),3.
[9] Ogieretsky V., Polubarinov I., Ann. Phys. (N.Y) 35 (1965),167; JETP 48(1965), 1625 (in Russian).

LAGRANGIAN FORMALISM FOR SUPERFIELDS WITH INTERNAL SYMMETRY

Jue Changkeun,[1,2] C. C. Chiang,[3,4] and E. C. G. Sudarshan[4]
Center for Particle Theory, The University of Texas at Austin
Austin, Texas 78712 USA

One of the problems in the Lagrangian formalism for superfields with internal symmetry is the presence of higher order derivatives in the Lagrangian. For SU(2) internal symmetry, Capper and Leibbrandt [1] have constructed a Lagrangian in which higher order derivatives for spinless fields are explicitly eliminated and ghosts for spin-half fields do not contribute to the S-matrix. Dondi [2] has been able to eliminate the ghosts completely by imposing a constraint condition on the superfields. In fact Dondi has succeeded in incorporating the constraint condition into the Lagrangian by properly redefining the constituent fields to obtain a manifestly ghost-free Lagrangian. Wess has also constructed a similar Lagrangian manifestly free of ghosts by the same method used by Dondi.

In this note we show that in fact for SU(2) the constraint condition of the superfields proposed by Dondi [2] is not needed; it is implied by the superfield equations. For SU(N) with $N \geqq 3$ the situation is quite different; a constraint condition is still needed in order to eliminate the ghosts.

Let us first consider the case of SU(2). The superfield Φ_{12} with SU(2) internal symmetry can be expanded as follows

$$\Phi_{12} = A + \bar{\theta}^A \psi_A + \frac{1}{2}\bar{\theta}^A\bar{\theta}^B B_{[AB]} + \bar{\lambda}^A \chi_A + FU_2 \tag{1}$$

where

$$\chi_A = \frac{1}{3!}\varepsilon_{ABCD}\bar{\theta}^B\bar{\theta}^C\bar{\theta}^D \tag{2a}$$

$$U_2 = \frac{1}{4!}\varepsilon_{ABCD}\bar{\theta}^A\bar{\theta}^B\bar{\theta}^C\bar{\theta}^D \tag{2b}$$

A suitable action might be

$$A_2 = -\frac{1}{2}\int d^4x \left(\frac{d}{d\theta}\right)^4 \left(\frac{d}{d\bar{\theta}}\right)^4 [2\Phi_{12}^+ e^{i\bar{\theta}\not{\partial}\theta}\Phi_{12} + \Phi_{12}(\Box + 2M^2)\Phi_{12}U_{12}^+ + \Phi_{12}^+(\Box + 2M^2)\Phi_{12}^+ U_2] \tag{3}$$

The action (3) can be rewritten as follows

$$A_2 = -\frac{1}{2}\int d^4x d^4\theta d^4\bar{\theta}[\Phi_{12}^+ e^{i\bar{\theta}\not{\partial}\theta}\Phi_{12} + \Phi_{12}(\Box + 2M^2)\Phi_{12}U_2^+ + \Phi_{12}^+(\Box + 2M^2)\Phi_{12}^+ U_2] \tag{4}$$

The action (4) leads to the simple superfield equation

---

[1]On leave of absence from the Department of Physics, Kyungpook National University, Taegu, Korea.
[2]Supported in part by the Ministry of Education of the Korean Government.
[3]On leave of absence from the Institute of Physics, National Taiwan Normal University, Taipei, Taiwan, Republic of China.
[4]Supported in part by DE-AS05-76ER03992.

$$\Phi_2^s + (\Box + 2M^2)\Phi_{12}^+ = 0 \tag{5}$$

where

$$\Phi_2^s = \left(\frac{d}{d\bar\theta}\right)^4 e^{i\bar\theta\partial\!\!\!/\theta}\Phi_{12} \tag{6}$$

The superfield equation (5) gives us the following field equations for the constituent fields

$$F = -(\Box + 2M^2)A^+ \tag{7a}$$

$$i\partial\!\!\!/_{A\dot A}\bar\lambda^A = (\Box + 2M^2)\bar\psi^{\dot A} \tag{7b}$$

$$\partial\!\!\!/_{A\dot A}\partial\!\!\!/_{B\dot B}\tilde B^{[AB]} = (\Box + 2M^2)\bar B^{[\dot A\dot B]} \tag{7c}$$

$$i\Box\partial\!\!\!/_{\dot A A}\psi_A = (\Box + 2M^2)\lambda_{\dot A} \tag{7d}$$

$$\Box^2 A = -(\Box + 2M^2)F^+ \tag{7e}$$

The constituent field $B_{[AB]}$ can be decomposed into an isovector scalar field B_k and an isoscalar vector field $B^{\alpha\beta}$. From eqs. (7a-e) it can be shown that

$$(\Box + M^2)A = 0 \tag{8a}$$

$$F = -M^2A^+ \tag{8b}$$

$$i\partial\!\!\!/_{\dot A A}\psi_A + \lambda_{\dot A} = 0 \tag{8c}$$

$$i\partial\!\!\!/_{A\dot A}\lambda_{\dot A} + M^2\psi_A = 0 \tag{8d}$$

$$(\Box + M^2)B_k = 0 \tag{8e}$$

$$(-i\partial\!\!\!/ + M)_{\alpha\alpha'}\psi_{\alpha'\beta} = 0 \tag{8f}$$

$$(-i\partial\!\!\!/ + M)_{\beta\beta'}\psi_{\alpha\beta'} = 0 \tag{8g}$$

where

$$\psi_{\alpha\beta} = B^{\alpha\beta} + B_{\alpha\beta} + B^{\alpha}{}_{\beta} + B_{\alpha}{}^{\beta} + B_{\alpha\beta} \tag{9}$$

The detailed derivations of eqs. (8a-g) are given elsewhere [4]. Here we note that there are no ghosts for the ordinary fields. Eqs. (8f) and (8g) are the Bargmann-Wigner equations for a vector field.

For SU(3) if we start with an action similar to eq. (4), it can be explicitly shown that there are higher order derivatives for the ordinary fields. To eliminate the ghosts, we may impose a constraint condition on superfields.

Come back to the SU(2) case. Instead of the action (4), we may choose the action

$$A_2' = -\frac{1}{2}\int d^4x d^4\theta d^4\bar\theta[2\Phi_{12}e^{i\bar\theta\partial\!\!\!/\theta}\Phi_{12} + M^2\Phi_{12}^2U_2^+ + M^2\Phi_{12}^{+2}U_2] \tag{10}$$

It is not difficult to see that if we assume the constraint condition suggested by Dondi

$$\Phi_2^s = \Box\Phi_{12}^+ \quad , \tag{11}$$

then eq. (10) leads to eq. (3).

We may now generalize the action (10) and the constraint condition (11) for SU(N) and we have the action

$$A_N = -\frac{M^{2(1-N)}}{2}\int d^4x d^{2N}\theta d^{2N}\bar{\theta}[\Phi^+_{1N} e^{i\bar{\theta}\not{\partial}\theta}\Phi_{1N} + M^N\Phi^2_{1N}U^+_N + M^N\Phi^{+2}_{1N}U_N] \tag{12}$$

and the constraint condition

$$\Phi^S_N = M^{N-2}\Box\Phi^+_{1N} \quad . \tag{13}$$

The action (12) leads to higher order derivative field equations for the ordinary fields. However, one of the solutions for each of the higher order derivative field equations satisfies the Klein-Gordon equation. The constraint condition (13) is nothing but a restriction to those solutions satisfying the Klein-Gordon equation. The equations of motion for the ordinary fields have been explicitly derived for SU(3) and SU(4) elsewhere [5]. It is found that the scalar fields satisfy the Klein-Gordon equation, the spin-half fields satisfy the Dirac equation and the higher spin fields satisfy the Bargmann-Wigner equation.

References

[1] D. M. Capper and G. Leibbrandt, Nucl. Phys. B85 (1975), 503.
[2] P. H. Dondi, J. Phys. A8 (1975) 1298.
[3] J. Wess, Act. Phys. Aust. 41 (1975) 409.
[4] Jue Changkeun, C. C. Chiang, and E.C.G. Sudarshan, University of Texas preprint DOE-394 (1980).
[5] Jue Changkeun, C. C. Chiang, and E.C.G. Sudarshan, University of Texas preprint DOE-395 (1980), DOE-396 (1980).

CAUSAL PROPAGATION IN SECOND ORDER LORENTZ-INVARIANT WAVE EQUATIONS

Vittorio Amar, Ugo Dozzio and Claudio Oleari
Istituto di Fisica dell'Università , Parma
Istituto di Fisica Nucleare, Sezione di Milano, Milano / Italy

Our aim is to construct relativistic linear wave equations for spin greater than one, which are physically meaningful in presence of minimal electromagnetic coupling . Let us initially consider the first order case , i.e. finite dimensional relativistic wave equations of the form

$$\left[\Gamma^{\mu}\left(\partial_{\mu}-ie A_{\mu}(x)\right)+i\chi\right]\varphi(x)=0 \qquad (1)$$

with

$$\Gamma'^{\mu} \equiv t^{\mu}_{\ \nu}\Gamma^{\nu} = T\,\Gamma^{\mu}T^{-1} ,$$

where t is the matrix of a Lorentz transformation and T is the corresponding representation .

It is known that, when the spin is greater than one , such equations exhibit some anomalies, namely :

a) the existence of signals propagating faster than light (Velo and Zwanziger patology, 1969 [1]);

b) the non-positive definiteness of the charge density ρ or of the energy density ω (*) . This anomaly implies problems in defining a Hilbert space of second quantization, suitable for consistent particle interpretation(Whigtman 1976 [2]).

Hence it is worthwhile to point out what is known about causality and positive definiteness of the densities . It has been shown that :

α) a sufficient condition for causal propagation in an equation of the form (1) is that the submatrix of Γ^{0} relative to null eigenvalue be diagonalizable (Amar , Dozzio 1972,1975 [3]).

β) if ρ or ω are positive definite, the submatrix of Γ^{0} relative to non-zero eigenvalues must be diagonalizable (Speer 1969 [4]).

Then it follows that a diagonalizable Γ^{0} is a choice suitable for having causal propagation and positive definiteness of the density .

However a well known theorem (Wild 1947 Gel'fand-Yaglom 1948 [5]) proves that an

(*) In this connection it is well known the following result due to Pauli : for equations of the form (1) ρ cannot be positive definite if the spin is integer and ω cannot be positive definite if the spin is half-odd integer (I.M.Gel'fand, R. A. Minlos and Z.Ya.Shapiro : Representations of the Rotations and Lorentz Group , London 1963, p. 350).

equation of the form (1) with diagonalizable Γ^0 and ρ or ω positive definite must have spin ≤ 1 .

Therefore if we want construct physically meaningful equations for spin > 1 it is necessary to generalize conditions α) and β). In this connection, the following results have been proved :

α') A sufficient condition for causal propagation in an equation of the form (1) is that

$$\Big[\sum_{P(\mu_1\mu_2\mu_3)} (\Gamma^{\mu_1}\Gamma^{\mu_2} - g^{\mu_1\mu_2})\Gamma^{\mu_3}\Big]\Gamma^{\mu_4}\ldots\Gamma^{\mu_n} = 0 \ , \tag{2}$$

where $\sum_{P(\mu_1\mu_2\mu_3)}$ denotes a sum over any permutation of the indices μ_1, μ_2 and μ_3 (Khalil 1977 [6]).

β') Consider an equation of the form (1) with unique mass, unique half-odd spin ℓ and the submatrix Γ^0_ℓ diagonalizable (**);necessary and sufficient condition in order to have $\rho \geq 0$, is that the Gel'fand chain of representations is a connected subchain of the following reducible Lorentz representation $\mathcal{R}$

$$\mathcal{R} = (\ell,\tfrac{1}{2}) \oplus (\ell,-\tfrac{1}{2}) \oplus \sum_{\substack{j\leq \ell-1\\ |k|\leq j}}^{\oplus} (j,k)\, n_j \ ,$$

where n_j is the multiplicity of the (j,k) representation (Amar–Dozzio 1972 [7]). An analogous result can be shown for integer spin .

Unfortunately , a solution satisfying both conditions α') and β') is unknown because the algebraic structure of the relations (2) is very complicated . It is also unknown whether the two conditions α') and β') are mutually compatible . Other sufficient algebraic conditions have been found (Amar,Dozzio and Oleari ,unpublished), which are different from Khalil's ones and also do not imply the diaganalizability of the submatrix of Γ^0 relative to zero eigenvalue ;however these conditions are more cumbersome than conditions (2) .

Therefore it is worth investigating relativistic wave equations of order greater than the first . It is easy to verify that, if the order is greater than two, imaginary masses cannot be avoided . Hence we take into account only second-order equations .

Let us consider the equations

(**) The matrix Γ^0 can always be written as follows

$$\Gamma^0 = \sum_r^{\oplus} \underbrace{(\Gamma^0_r \oplus \Gamma^0_r \oplus \ldots \oplus \Gamma^0_r)}_{(2r+1)\ \text{times}} \ ,$$

where $\sum_r^{\oplus}$ denotes a sum over any representation of the rotation group, which is contained in the Lorentz group representation $\mathcal{R}$.

$$\begin{cases} (\Gamma^{\mu\nu}\partial_\mu\partial_\nu + \chi^2)\,\varphi(x) = 0 \\ \Gamma^{\mu\nu} = \Gamma^{\nu\mu}\,, \quad \Gamma'^{\mu\nu} = t^\mu_\rho t^\nu_\sigma \Gamma^{\rho\sigma} = T\,\Gamma^{\mu\nu}\,T^{-1}\,. \end{cases} \tag{3}$$

Following a procedure like that of Haris-Chandra [8] ,it can be proved that

$$\sum_{P(\alpha_1\beta_1\ldots\alpha_n\beta_n)} \left(\Gamma^{\alpha_1\beta_1}\cdots\Gamma^{\alpha_n\beta_n} + a_1 g^{\alpha_1\beta_1}\Gamma^{\alpha_2\beta_2}\cdots\Gamma^{\alpha_n\beta_n} + \ldots \right. \\ \left. \ldots + a_n g^{\alpha_1\beta_1}\cdots g^{\alpha_n\beta_n}\right) = 0 \tag{4}$$

and consequently :

$$(\partial_\rho\partial_\sigma\Gamma^{\rho\sigma})^n + a_1\Box(\partial_\rho\partial_\sigma\Gamma^{\rho\sigma})^{n-1} + \ldots + a_n\Box^n = 0\,. \tag{5}$$

By relations (4) and (5) it can be easly shown that in the free case no problem exists related to causality .

Now let us consider equations (3) with minimal electromagnetic coupling , i.e.

$$(\Gamma^{\mu\nu}D_\mu D_\nu + \chi^2)\,\varphi(x) = 0\,, \tag{6}$$

where $D_\mu \equiv \partial_\mu - ieA_\mu(x)$.

If Γ^{00} satisfies the minimal equation

$$\Gamma^{00}(\Gamma^{00} - 1) = 0\,, \tag{7}$$

by relations (4) and (5) the following result can be proved : <u>a sufficient condition in order to have causal propagation for equations (6) with condition (7) is that</u>

$$(\Gamma^{\mu\nu} - g^{\mu\nu})\Gamma^{\rho\sigma} + (\Gamma^{\rho\sigma} - g^{\rho\sigma})\Gamma^{\mu\nu} + 4\,(\Gamma^{\mu\rho} - g^{\mu\rho})\Gamma^{\nu\sigma} = 0 \tag{8}$$

(Amar,Dozzio and Oleari 1980 [9]).

The relations (8) are the generating rules of an infinite-dimensional algebra , whose representations are difficult to find directly . However we can note that the subrules

$$(\Gamma^{\mu\nu} - g^{\mu\nu})\,\Gamma^{\rho\sigma} = 0 \tag{9}$$

are compatible with the (8)'s . The (9)'s are the generating rules of a ten-dimensional algebra with non-zero radical and consequently with irreducible and reducible but non completely reducible representations . The only irreducible representation is the trivial one, in which $\Gamma^{\mu\nu} = g^{\mu\nu}$. Reducible representations of (9) can be directly found but they are not a-priori Lorentz covariant ; in order to find covariant representations we assume

$$\Gamma^{\mu\nu} \equiv \frac{1}{2}(\Gamma^\mu\Gamma^\nu + \Gamma^\nu\Gamma^\mu)$$

and by putting this in (9) we obtain

$$(\Gamma^\mu\Gamma^\nu + \Gamma^\nu\Gamma^\mu - 2g^{\mu\nu})(\Gamma^\rho\Gamma^\sigma + \Gamma^\sigma\Gamma^\rho) = 0\,. \tag{10}$$

A covariant representation of (10) has been found in literature [10] and is the following one

$$\Gamma^0 = -(\tilde{\Gamma}^0)^+\,, \quad \Gamma^k = (\tilde{\Gamma}^k)^+\,,$$

where

$$\tilde{\Gamma}^{0} = \left|\begin{array}{c|cc|cc|c}
 & & & 0 & 0 & \\
 & & & \sqrt{2}C & 0 & \\
 & & & -C & 0 & \\
 & & & 0 & C & \\
 & & & 0 & -\sqrt{2}C & \\
 & & & 0 & 0 & \\
\cline{4-6}
 & & & -1 & 0 & \\
 & & & 0 & -1 & \\
\hline
 & -1 & 0 & & & \\
 & 0 & -1 & & & \\
\cline{1-3}
 & 0 & 0 & & & \\
 & -C & 0 & & & \\
 & 0 & -\sqrt{2}C & & & \\
 & \sqrt{2}C & 0 & & & \\
 & 0 & C & & & \\
 & 0 & 0 & & &
\end{array}\right|,$$

$$\tilde{\Gamma}^{1} = \left|\begin{array}{c|cc|cc|c}
 & & & \sqrt{2}C & 0 & \\
 & & & 0 & 0 & \\
 & & & 0 & C & \\
 & & & -C & 0 & \\
 & & & 0 & 0 & \\
 & & & 0 & -\sqrt{2}C & \\
\cline{4-6}
 & & & 0 & -1 & \\
 & & & -1 & 0 & \\
\hline
 & 0 & 1 & & & \\
 & 1 & 0 & & & \\
\cline{1-3}
 & -\sqrt{2}C & 0 & & & \\
 & 0 & -C & & & \\
 & 0 & 0 & & & \\
 & 0 & 0 & & & \\
 & C & 0 & & & \\
 & 0 & \sqrt{2}C & & &
\end{array}\right|,$$

$$\tilde{\Gamma}^{2} = \left|\begin{array}{c|cc|cc|c}
 & & & \sqrt{2}iC & 0 & \\
 & & & 0 & 0 & \\
 & & & 0 & iC & \\
 & & & iC & 0 & \\
 & & & 0 & 0 & \\
 & & & 0 & \sqrt{2}iC & \\
\cline{4-6}
 & & & 0 & -i & \\
 & & & i & 0 & \\
\hline
 & 0 & i & & & \\
 & -i & 0 & & & \\
\cline{1-3}
 & -\sqrt{2}iC & 0 & & & \\
 & 0 & -iC & & & \\
 & 0 & 0 & & & \\
 & 0 & 0 & & & \\
 & -iC & 0 & & & \\
 & 0 & -\sqrt{2}iC & & &
\end{array}\right|,$$

$$
\tilde{\Gamma}^3 = \left|\begin{array}{c|cc|cc|c}
 & & & 0 & 0 & \\
 & & & -\sqrt{2}C & 0 & \\
 & & & -C & 0 & \\
 & & & 0 & -C & \\
 & & & 0 & -\sqrt{2}C & \\
 & & & 0 & 0 & \\
\hline
 & & & -1 & 0 & \\
 & & & 0 & 1 & \\
\hline
 & 1 & 0 & & & \\
 & 0 & -1 & & & \\
\hline
 & 0 & 0 & & & \\
 & C & 0 & & & \\
 & 0 & \sqrt{2}C & & & \\
 & \sqrt{2}C & 0 & & & \\
 & 0 & C & & & \\
 & 0 & 0 & & & \\
\end{array}\right| .
$$

The existence of a non trivial representation of the rules (8) is a first step towards the understanding of the causal propagation for the second order relativistic equations with minimal electromagnetic coupling .

REFERENCES

[1] G.Velo and D.Zwanziger : Phys.Rev. 186 1337 (1969)

[2] A.S.Whigtman : p.423 Studies in Math.Phys. (Princeton,N.J.1976)

[3] V.Amar,U.Dozzio: Lett.Nuovo Cimento 5,355 (1972)
Lett.Nuovo Cimento 12,659 (1975)

[4] C.R.Speer :-Generalized Feynman amplitudes- in Ann.Math.Studies n.62 (Princeton University Press, Princeton N.J. 1969)

[5] E.Wild : Proc.Royal Soc. A191 253 (1947)
I.M.Gel'fand and A.M.Yaglom : J.E.P.T. 18 1096 (1948)

[6] M.A.K.Khalil : Phys.Rev. D115 ,1532 (1977)

[7] V.Amar and U.Dozzio : Nuovo Cimento 9B 53 (1972)
Nuovo Cimento 11A 87 (1972)

[8] Haris-Chandra : Phys.Rev. 71 793 (1947)

[9] V.Amar,U.Dozzio and C.Oleari : Nuovo Cimento , in press

[10] A.Z.Capri : Phys.Rev. 187 1811 (1969)

THE EXCEPTIONAL GROUP E_8 FOR GRAND UNIFICATION*

ITZHAK BARS**
YALE UNIVERSITY, PHYSICS DEPARTMENT

ABSTRACT

A grand unified model based on E_8 is described in which maximum unification within GUTS is achieved by classifying the fermions as well as the Higgs bosons in single and smallest possible representations. This provides a theory of flavor and of families and predicts, depending on the symmetry breaking chain, that the next three SU(5) families or the next two E_6 families should be of V+A type and lie below 1 TeV.

I will describe a grand unified model, proposed in collaboration with M. Gunaydin[1,2] which is based on E_8 - the largest exceptional group. The successful grand unified models based on SU(5), SO(10), E_6 can be described in terms of subgroups of E_8. If one notes that SU(5) is isomorphic to E_4, and SO(10) is isomorphic to E_5[3], it is inevitable to follow the "E-chain" to its end to explore unification ideas. All E_n groups ($n \geq 4$) can be used successfully to describe grand unification (including E_7 provided one uses two 56-dimensional representations). However, none of them, except E_8, is capable of maximum unification and providing a theory for the number of flavors and of families.

Is there a secret physical reason that explains why Nature seems to choose the "E-series"? Perhaps octonions play a role in physics? If so, it is worth noting that E_8 makes maximum use of octonions: It requires two octonions, not one, as described by the magic square and explicit constructions in terms of ternary algebras[4]. This fact gives it many interesting unique properties not shared by other groups.

In E_8 all fundamental fermions are placed in the 248 dimensional adjoint representation and all Higgs bosons belong to the 3875 dimensional representation. These are the smallest possible irreducible representations that can be used with E_8 to classify these fields. Irreducibility and single representations are the keys for a theory of flavors and families, and the use of the smallest possible representations eliminates arbitrariness.

E_8 has many interesting subgroups: SO(16), SU(9), SU(8)xSU(2), E_7xSU(2), E_6xSU(3), F_4xG_2, SU(5)xSU(5), SO(14)xU(1), SO(10)xSU(4), etc. Which of these subgroups is chosen in the symmetry breaking chain depends on the details of the Higgs potential. Some remarks on different discrete symmetry schemes which can influence the symmetry breaking chain can be found in ref. (1). The details of low energy physics depends on the chain chosen, and this can make the difference on e.g. whether the top quark exists, etc. We have produced E_8 models with varying low energy degrees of freedom[1,2]. In some chains which include the SU(5) group of Georgi and Glashow the third family includes a top quark, while in other chains which include the E_6 group of Gursey and collaborators there is no top quark, but instead there is a sixth quark with (-1/3) charge. In our scheme both possibilities predict as many (V+A) families as there are (V-A) families at low energies (three SU(5) families or two E_6 families). The (V+A) families can be arranged to be heavier than the (V-A)'s, but they must lie below 1 TeV. Therefore, independent of details, the prediction of E_8 is that the next family to be discovered (SU(5) or E_6) is of (V+A) type. There is one catch: this prediction was based on the assumption that the right-handed partners of low energy neutrinos are superheavy. If this is not so, then a somewhat different symmetry breaking scheme could allow one more (V-A) family as well as one more (V+A) family.

I will describe, briefly, the E_8→[SO(16)]→SO(10)xSU(4)→SU(5)xSU(3)$_F$ xU(1)xU(1) symmetry breaking chain[1] where the last SU(3)$_F$ is a family group and the SU(5) is the Georgi-Glashow group which includes the SU(2)$_W$xU(1)$_W$ of the Weinberg-Salam-Glashow model as well as the SU(3)$_c$ of color. More details and other symmetry breaking chains are given in refs. (1,2).

First, let us note the decomposition of the 248 and 3875 dimensional representations(5) according to this chain as shown in Table 1, The superscripts on the SU(5)representations are quantum numbers for the U(1) that occurs in SO(10)→SU(5)xU(1) and the superscripts on the SU(3) representation denotes the U(1) quantum numbers for SU(4)→SU(3)xU(1).

$E_8 \rightarrow SO(16) \rightarrow (SO(10)xSU(4)) \rightarrow ([SU(5)xU(1)]x[SU(3)xU(1)])$

248	120	(45,1)	$= (1^0+10^{+4}+\overline{10}^{-4}+24^0,1^0)$
		(1,15)	$= (1^0,1^0+3^{+4}+\bar{3}^{-4}+8^0)$
		(10,6)	$= (5^{+2}+\bar{5}^{-2},3^{-2}+\bar{3}^{+2})$
	128	$(16,\bar{4})$	$= (1^{-5}+\bar{5}^{+3}+10^{-1},1^{+3}+\bar{3}^{-1})$
		$(\overline{16},4)$	$= (1^{+5}+5^{-3}+\overline{10}^{+1},1^{-3}+3^{+1})$
3875	135	(1,1)	$= (1^0,1^0)$
		(1,20')	$= (1^0,6^{-4}+\bar{6}^{+4}+8^0)$
		(54,1)	$= (15^{+4}+\overline{15}^{-4}+24^0,1^0)$
		(10,6)	$= (5^{+2}+\bar{5}^{-2},3^{-2}+\bar{3}^{+2})$
	1820	(10,10)	$= (5^{+2}+\bar{5}^{-2},1^{-6}+3^{-2}+6^{+2})$
		$(10,\overline{10})$	$= (5^{+2}+\bar{5}^{-2},1^{+6}+\bar{3}^{+2}+\bar{6}^{-2})$
		(1,15)	$= (1^0,1^0+3^{+4}+\bar{3}^{-4}+8^0)$
		(210,1)	$= (1^0+5^{-8}+\bar{5}^{+8}+10^{+4}+\overline{10}^{-4}+24^0+40^{-4}+\overline{40}^{+4}+75^0,1^0)$
		(45,15)	$= (1^0+10^{+4}+\overline{10}^{-4}+24^0,1^0+3^{4}+\bar{3}^{-4}+8^0)$
		(120,6)	$= (5^{+2}+\bar{5}^{-2}+10^{-6}+\overline{10}^{+6}+45^{-2}+\overline{45}^{+2},3^{-2}+\bar{3}^{+2})$
	1920	$(16,\bar{4})$	$= (1^{-5}+\bar{5}^{+3}+10^{-1},1^{+3}+\bar{3}^{-1})$
		$(\overline{16},4)$	$= (1^{+5}+5^{-3}+\overline{10}^{+1},1^{-3}+3^{+1})$
		(16,20)	$= (1^{-5}+\bar{5}^{+3}+10^{-1},3^{-5}+\bar{3}^{-1}+6^{-1}+8^{3})$
		$(\overline{16},\overline{20})$	$= (1^{+5}+5^{-3}+10^{+1},\bar{3}^{+5}+3^{+1}+\bar{6}^{+1}+8^{-3})$
		$(144,\bar{4})$	$= (5^{+7}+\bar{5}^{+3}+10^{-1}+24^{-5}+15^{-1}+\overline{40}^{-1}+\overline{45}^{+3},1^{+3}+\bar{3}^{-1})$
		$(\overline{144},4)$	$= (5^{-3}+\bar{5}^{-7}+\overline{10}^{+1}+24^{+5}+\overline{15}^{+1}+40^{+1}+45^{-3},1^{-3}+3^{+1})$

TABLE 1 - DECOMPOSITION OF 248 AND 3875 REPRESENTATIONS

In this table the SO(16) representations can be described as follows: 120 is the adjoint representation corresponding to an antisymmetric tensor with two indices, $128=2^7$ is the spinor, 135 is the symmetric traceless tensor, 1820 is the antisymmetric tensor with 4 indices, 1920 is a higher dimensional spinor representation occuring in 16x128'=1920+128. The 1920 can be represented by a field $S^{i\alpha}$(n=1,...,16 and α=1,...,128) with the condition $\gamma^i_{\alpha\beta}S^{i\beta}=0$ which eliminates 128 components by using the γ-matrices $\gamma^i_{\alpha\beta}$ of SO(16).

Denoting the gauge bosons, fermions and Higgs bosons by A^a_μ, $\psi^a_{L\alpha}$, ϕ_{ab} where a,b=1,...,248, we can write the Lagrangian density as

$$L = \psi^a_L \not{D} \psi^a_L + (D_\mu\phi^\dagger)_{ab}(D^\mu\phi)_{ab} + h\psi^a_L C\psi^b_L\phi_{ab} - V(\phi) \quad .$$

This generates the following gauge and Yukawa vertices for Higgs bosons. The couplings are decomposed according to the SO(16) subgroup.

Yukawa couplings $\phi\psi\psi$

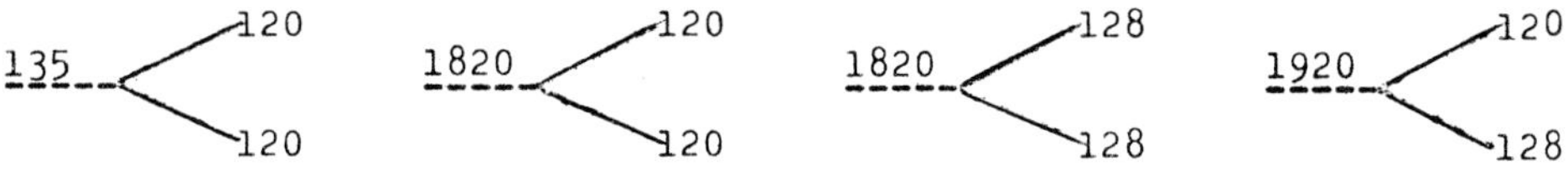

$A_\mu\phi^\dagger\partial_\mu\phi$ gauge couplings

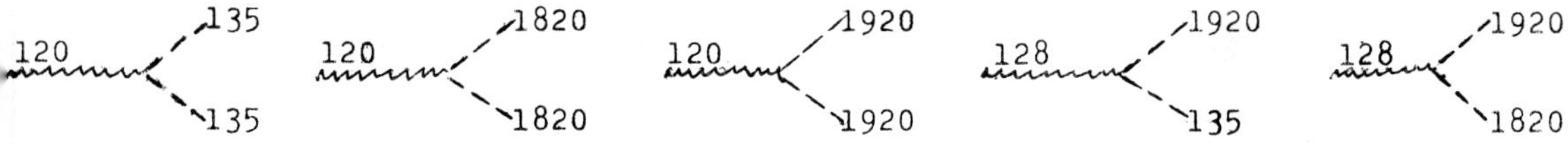

$A_\mu A_\mu\phi^\dagger\phi$ gauge couplings

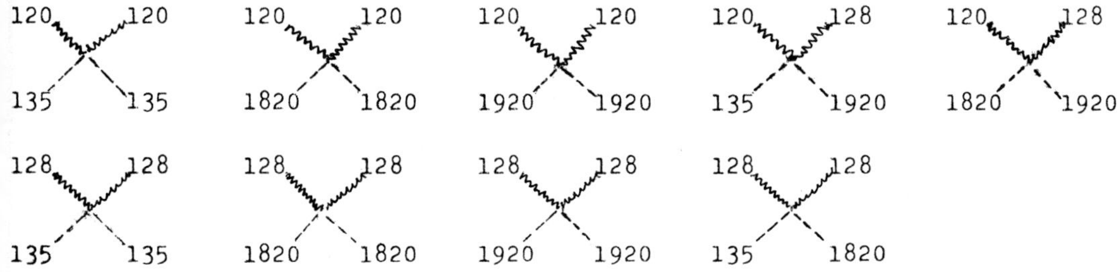

For later reference I note the following Kronecher products for SO(16)[6]

$120\times120 = \{1+135+1820+5304\}_S + [120+7020]_A$

$128\times128 = \{1+1820+6435\}_S + [120+8008]_A$

$120\times128 = 128+1920+13312$

$135\times135 = \{1+135+3740+5304\}_S + [120+8925]_A$

$135\times1820 = 1820+7020+60060+176800$

$135\times1920 = 128+1920+13312+15360+87040+141440$

$1820\times1820 = \{1+135+1820+5304+6435+6435'+60060+89760+716040+771120\}_S + +[120+7020+8008+141372+162162+1336608]_A$

$1820\times1920 = 128+1920+13312+56320+161280+326144+898560+2036736$

$1920\times1920 = \{1+135+2\times1820+5304+6435+6435'+7020+60060+162162+176800+ +700128+716040\}_S + [2\times120+7020+2\times8008+8925+60060+141372+ +162162+595595+850850]_A$

From Table 1 we see that there is no singlet Higgs boson at the level of SO(16). Therefore SO(16) is not one of the stages of symmetry breaking, we must break to one of its subgroups. A physically relevant subgroup is SO(10)xSU(4). From Table 1 we see that only the 135 has a singlet with respect to SO(10)xSU(4). By giving a large vacuum expectation value (V.e.V) to this single component many gauge bosons and fermions become simultaneously heavy, with masses proportional to this V.e.V. The gauge bosons that remain massless belong to the adjoint representation of SO(10)xSU(4), i.e. (45,1)+(1,15) which sit in the 120 with respect to SO(16), as seen in Table 1. The remaining massless fermions transform as spinors of SO(10)xSU(4), i.e. $(16,\bar{4})+(\overline{16},4)$, and they belong to 128 of SO(16) as shown in Table 1. These fermions remain massless because they have no Yukawa coupling to the 135 as indicated in the figure.

The symmetry breaking down to [SU(5)xU(1)]x[SU(3)xU(1)] proceeds via the Higgses which are singlets with respect to this group. There are 3 such singlets as seen in Table 1 and they occur as members of (1,15)+(210,1)+(45,15) which belong to the 1820 of SO(16). We allow only one combination of these 3

Higgses to develop a v.e.v. in such a way that all the extra neutrino components in $(16,\bar{4})+(\bar{16},4)$ become heavy. This was shown explicitly in ref. 1. But we find that simultaneously one (V-A) and one (V+A) family combine with each other and become superheavy. The massless fermions that remain transform according to $(\bar{5}^{+3}+10^{-1},3^{-1})$ and $(5^{-3}+\bar{10}^{+1},3^{+1})$ as seen from Table 1. These are the three (V-A) and the three (V+A) families which survive all the way down to low energies.

The further breaking of the symmetry down to $SU(3)_c \times U(1)$ was outlined in ref. (1) and I will not repeat it here. The major point is that the (V+A) families, which are so far not observed, can be pushed up to mass levels beyond presently explored regions. However, they cannot be made heavier than 0.5-1.0 TeV, because otherwise, through radiative corrections, they would destroy successful low energy relations such as $m_W^2=m_Z^2 \cos^2\theta_W$. Therefore, these (V+A) families of quarks and leptons should lie in the region $\approx$30-1000 GeV.

Details of the symmetry breaking for other subgroup chains are described in ref. (2) where we show that the low energy physics depends on the chain. This includes schemes in which the top quark is present as well as other schemes in which it is absent but it is replaced by a second bottom quark. Experiment will show which of these possibilities is correct.

The virtue of our model is that it reduced by a great deal the arbitrariness in constructing a grand unified theory. Thus, we have provided a theory of flavor and of families. If this scheme is experimentally confirmed then it would be tempting to explore the possibility that octonions play a fundamental role in Nature. This possibility has been already suggested[(7)] and is being revived again in the context of ternary algebras[(8)].

References

* Research supported by DOE contract No. EY-76-C-02-3075
** Alfred P. Sloan Foundation Fellow

(1) I. Bars and M. Günaydin, Yale preprint YTP80-09, to be published in Phys. Rev. Lett. Sept. 9, 1980.

(2) I. Bars and M. Günaydin, in preparation.

(3) This remark is due to F. Gürsey.

(4) J. L. Kantor, Sov. Math. Dokl. 44, 254 (1973).
(5) W. McKay, J. Patera and R. T. Sharp, J.M.P. 17, 1371 (1976); B. G. Wybourne and M. J. Bowick, Aust. J. Phys. 30, 259 (1977).
(6) Many of these Kronecker products are not found elsewhere.
(7) M. Günaydin and F. Gürsey, Phys. Rev. D9, 3387 (1974); F. Gürsey, Proc. Workshop on Nonassociative Algebras, U. Virginia 1978, and Proc. Johns Hopkins Workshops 1975-1978, M. Günaydin, J. Math. Phys. 17, 1975 (1975) and Johns Hopkins Workshop (1978).
(8) For a brief review and latest proposals see I. Bars, Yale Preprint YTP80-21 July 1980, to appear in Proc. XXth Intl. Conf. High En. Phys. Madison, Wisconsin, July 1980, and Proc. IXth Intl. Colloq. on Group Theor. Methods in Phys. Cocoyoc, Mexico, June 1980.

COMPLETELY INTEGRABLE N-BODY PROBLEMS IN THREE-DIMENSIONS AND THEIR RELATIVISTIC GENERALIZATION

A. O. Barut
Department of Physics
The University of Colorado
Boulder, CO. 80309

Dynamical algebras (Lie algebras) define in a global way integrable physical dynamical systems once the Hamiltonian and other integrals have been physically identified. The dynamical variables (coordinates and momenta) are secondary. On this basis, a class of integrable 3-dimensional N-body problems are studied which can approximate many-body problems in atomic, nuclear and particle physics.

1. Introduction

We present here a class of completely integrable N-body problems with N-body forces in three-dimensions. These models provide an alternative departure point to the physics of N-body atomic or nuclear systems. In the standard approach, one begins from an independent particle model (i.e. direct product of N one-body wave functions each in some potential), and then introduces configuration mixing (or interparticle interactions) to approximate the real problem. Here we shall start from an exactly soluble situation with N-body forces, and then introduce a configuration decoupling to approximate the real problem (Fig. 1)

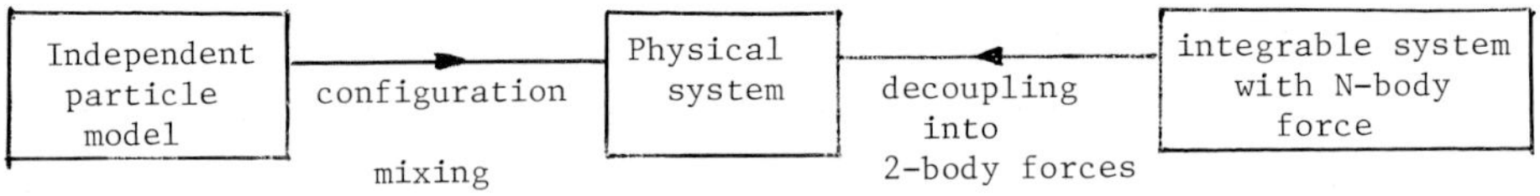

Fig. 1

2. Preliminaries

A Hamiltonian system of n degrees of freedom (n = 3N for N-bodies) is integrable if there are n constants of motion J_m, m = 1...n, which are single-valued, differentiable, analytic functions of the dynamical variables (p.q) in the phase space , and which are furthermore linearly independent and existing globally, and in involution, with respect to Poisson bracket, i.e. $\{J_n, J_m\} = 0$. The motion is then restricted to invariant tori of dimension n in $\Gamma(2n)$. According to an old theorem of Liouville, the existence of n integrals in involution of a Hamiltonion System then assures its explicit solution by quadrature.

Very few integrable systems in 3-dimensions are known explicitly; 2-body Kepler problem, N-dimensional oscillator. In one dimension the situation is better; a larger class of integrable many-body problems are known(1).

If integrable systems are on one extreme of all dynamical systems, the ergodic systems are on the other extreme, where the trajectory winds around densely through the whole of (2n-1)-dimensional energy surface in $\Gamma(2n)$ Now N-body atomic and nuclear systems have relatively stable stationary states. Even if their assumed theoretical Hamiltonian does not look like to be integrable (e.g. the N-body Coulomb Hamiltonian), we may expect that "near" to such Hamiltonians, others should exist which would be integrable. This expectation is based on Kolmogorov-Arnold. Moser theorem[2], according which for slightly perturbed integrable systems, even if there are no integrals other than the energy, enough n-dimensional submanifolds in $\Gamma(2n)$ exist so that the system in most cases acts like an integrable system. In such a situation, most orbits are slightly modified; there are quasi-periodic solutions (although some small stochastic regions appear), hence we have reasonable stability. Thus the above expectation is a converse of this theorem.

The significance of exactly soluble non-trivial N-body problems in the quantum theory of atomic, nuclear and hadronic models, is that they would provide us with a complete set of exact quantum numbers. As is well-known, so far most of our intuition in this area comes from the independent particle models which provide also the quantum numbers, or labeling of states. Accordingly we label N-particle states by the quantum numbers of the non-interacting particles which must then be considerable mixed. That is the observed state is a superposition of states labelled by the quantum numbers of the independent particles. In contrast in the case of an integrable system each state is labelled by the so called "good" quantum numbers: energy, total angular momentum, plus the new global quantum numbers of the system.

3. Dynamical Groups as Dynamical Systems

Every Lie-algebra (rather its representation) determines in principle a class of dynamical systems, once the Hamiltonian and other integrals have been physically identified. This idea grew originally from the study of known physical systems[3], then formulated as the "dynamical group quantization"[4]. Here we show how such systems are obtained abstractly from the Lie groups, and then realized physically.

For simplicity, we consider a semi-simple Lie algebra, L, $\{X_i\}$, $i = 1,\cdots n$, a basis, $H = \{H_i\}$, $i = 1\cdots\ell$, a cartan subalgebra in L. We have then the decomposition

$$L = H + \sum_{\alpha\in\Delta} E_\alpha \ ,$$

where Δ is the set of non-zero roots, and the commutation relations

$$[H_i, E_\alpha] = \alpha(H_i)E\alpha$$

$$[E_\alpha, E_\beta] = \begin{cases} 0 \ , & \text{if } \alpha+\beta \neq 0, \alpha+\beta\notin\Delta \\ H_\alpha, & \text{if } \alpha+\beta = 0 \\ N_{\alpha,\beta}E_{\alpha+\beta}, & \text{if } \alpha+\beta\in\Delta \ . \end{cases}$$

Let the enveloping algebra of L be $\mathcal{E}(L)$. We may now identify physically any element of L, or $\mathcal{E}(L)$, to be the "Hamiltonian" of a physical system. From the commutation relations we can then determine the complete set of commuting operators (CSCO), i.e. the integrals of motion J_m. Furthermore, from the representation $\Pi(L)$ of L (or of $\mathcal{E}(L)$) we know in principle the spectrum of $\mathcal{H}$ or J_m, and can label the states in the representation space by the eigenvalues of the CSCO. In this way, we have automatically infinitely many integrable dynamical systems. It remains then to see, when necessary, if this representation $\Pi(L)$ can be realized in some underlying phase space $\Gamma(p,q)$, or more generally in phase-space plus spin-variables, etc. Finite dimensional representations can correspond to "finite" quantum mechanics"[5].

Actually in quantum physics the underlying phase-space is not always known. In macroscopic physics we may proceed from a set of interacting particles, hence the phase-space is a priori given. In microscopic physics, however, we probe a system by external agents (photons, electrons, neutrons) and try to infer a possible internal dynamics - the quantum numbers come first.

In these cases, and in the general abstract case the approach via dynamical group has the following features:

(i) it provides a coordinate independent treatment

(ii) all integrals of motion, including Hamiltonian, are on the same footing (any integral can be considered as the "Hamiltonian", and one can then define a symmetry relative to this quantity)[6]

(iii) it provides a unified treatment of classical and quantum physics,

(IV) it provides a method of passage to relativistic dynamics, once the Lorentz group within the dynamical group is physically identified.

(V) the passage to field theory can also be done by the limit $n \to \infty$. But here other functional methods are more useful.

4. Different Realizations of the Dynamical Groups

The Lie algebra $L = \{X_i\}$, and hence the enveloping algebra $\mathcal{E}(L)$, has the following realizations in terms of the dynamical variables $\{p_n\}$, $\{q_n\}$ (including spin variables such as, $\sigma_u, \gamma_\mu, \cdots$):

(I) Phase space functions. $X_i = X_i(p,q)$ under Poisson Brackets,

(II) Phase space functions $X_i = X_i(p,q)$ under the Heisenberg algebra relation $[p_i,q_j] = i\delta_{ij}\bar{h}$,

(III) Schrödinger representations: Differential operators either in q-space, or in p-space, $X_i = X_i(q,-i\frac{\partial}{\partial q})$, or $X_i = X_i(p,-i\frac{\partial}{\partial p})$.

(IV) Vector fields over phase space $\Gamma(n)$: First order differential operators of the form $\mathcal{X}_i = \mathcal{X}_i(p,\frac{\partial}{\partial p},q,\frac{\partial}{\partial q})$.

The last one, perhaps less familiar to physicists, is defined by the relation

$$\mathcal{X}_i f(p,q) \equiv \{f,X_i\},$$

where $X_i(p,q)$ is the phase space-function (see (I)) and $\mathcal{X}_i$ the corresponding first order operator acting on phase-space-functions $f(p,q)$. For example, for SL(2,R) with $X_1 = \Gamma_o = \frac{1}{4}(\vec{p}^2+\vec{q}^2)$, $X_2=\Gamma_4 = \frac{1}{4}(\vec{p}^2-\vec{q}^2)$ and $X_3 = T = \frac{1}{2}\vec{q}\cdot\vec{p}$ we obtain

$$\mathcal{X}_1 = \frac{1}{2}(p_n\frac{\partial}{\partial q_n} - q_n\frac{\partial}{\partial p_n})$$

$$\mathcal{X}_2 = \frac{1}{2}(p_n\frac{\partial}{\partial q_n} + q_n\frac{\partial}{\partial p_n})$$

$$\mathcal{X}_3 = \frac{1}{2}(q_n\frac{\partial}{\partial q_n} - p_n\frac{\partial}{\partial p_n}),$$

where we used the following convention for Poisson brackets

$$\{f,g\} = \frac{\partial f}{\partial q_n}\frac{\partial g}{\partial p_n} - \frac{\partial f}{\partial p_n}\frac{\partial g}{\partial q_n}.$$

5. N-Body Integrable Systems

We consider the following two classes of realizations of the dynamical algebra SL(2,R):

$$\begin{aligned}\Gamma o &= \frac{1}{4}(\vec{p}^2+\vec{q}^2) + V(q)\\ \Gamma 4 &= \frac{1}{4}(\vec{p}^2-\vec{q}^2) + V(q)\\ T &= \frac{1}{2}\vec{q}\cdot\vec{p},\end{aligned} \tag{1}$$

and

$$\begin{aligned}\Gamma o &= \frac{1}{2}(rp^2+r) + rV(r)\\ \Gamma 4 &= \frac{1}{2}(rp^2 - r) + rV(r)\\ T &= \vec{r}\cdot\vec{p},\end{aligned} \tag{2}$$

where in both cases

$$p_k \varepsilon R^n, \quad r_k \varepsilon R^n,$$

$$p^2 = \sum_{k=1}^{n} p_k^2 \ , \ r^2 = \sum_{k=1}^{n} r_k{}^2, \ n = 3N.$$

which we interpret as the coordinates and momenta of N particles in 3-dimensions.

The two algebras in (1) and (2) are closed if the potential V satisfies

$$[V,T] = -2V, \tag{3}$$

i.e. if V is homogeneous of degree -2,

or $$r^k \frac{\partial V}{\partial r^k} = -2V. \tag{4}$$

Examples of potentials satisfying (4) are

$$V = \frac{a}{r^2}\,[1 + A_k \frac{r_k}{r} + A_{ik} \frac{r_i r_k}{r^2} + \cdots] \tag{5}$$

where a, A_k A_{ik} ··· are constants parameters of the potential,

or,

$$V = \frac{a}{r^2}\,[1 + [\frac{\vec{S}\cdot\vec{r}}{r} + \Sigma \frac{\vec{S}_i\cdot\vec{r}_j S_n\cdot\vec{r}_n}{r^2} + \ldots] , \tag{5^1}$$

where S_i are the quantum mechanical spin variables, which we can treat as additional dynamical variables and satisfy additional Heisenberg equations of motion of the form $\dot{S} = i[H,S]$.

The spectrum of $\Gamma_o + \Gamma_4 + K = \frac{1}{2}p^2 + 2V + K$, in case (1), and of $(\Gamma_o - \Gamma_4)^{-1}(\Gamma_o + \Gamma_4 + K) = \frac{1}{2}p^2 + V + \frac{K}{r}$, in case (2), where K is a constant, can be obtained from the representation of SL(2,R) in exactly the same way as in the well known 2-body case[(7)].

In order to describe all states of the system, i.e. all energy levels (discrete and continuous) with their multiplicities, the algebras (1) or (2) must be generalized to the full dynamical algebra $\mathcal{G}$. It is here that the N-body problems differs from the 1-body case. The algebra (2) in the case V = 0, can be embedded in the algebra SO(3N+1,2) of dimension $\frac{1}{2}$(3N+3)(3N+2) with the following additional elements

	number of elements	
$L_{ij} = r_i \Pi_j - r_j \Pi_i$	$\frac{1}{2}3N(3N-1)$	
$A_i = \frac{1}{2} r_i \Pi^2 - \Pi_i (\vec{r}_k \cdot \vec{\Pi}_k) - \frac{1}{2} r_i$	3N	
$M_i = \frac{1}{2} r_i \Pi^2 - \Pi_i (\vec{r}_k \cdot \vec{\Pi}_k) + \frac{1}{2} r_i$	3N	(6)
$\Gamma_i = r\Pi_i$	3N	

Here we have used a more general realization with $P_i \to \Pi_i$ such that $[\Pi_i, \Pi_j] = \mu i \varepsilon_{ijk} \frac{r_k}{r^3}$.

We have further the isomorphisms

Dimensions of Cartan Subalgebra

$$SO(3N+1,2) \sim B_{\frac{3N}{2}+1}, \text{ if } 3N \text{ even} \qquad \ell = \frac{3N}{2}+1$$

$$\sim D_{\frac{3N+3}{2}}, \text{ if } 3N \text{ odd} \qquad \ell' = \frac{3N+3}{2}.$$

The number of quantum numbers in addition to the Cartan subalgebra are, in the general case, $\ell(\ell-1) = \frac{3N(3N+2)}{4}$, if 3N even

$\ell'(\ell'-2) = \frac{(3N+3)(3N-1)}{4}$, if 3N odd,

which gives altogether $\frac{(3N+2)^2}{4}$ quantum numbers, if 3N is even and $\frac{(3N+2)^2-1}{4}$ if 3N is odd. In the special case, when all Casimir operators are zero except one, the number of quantum numbers is equal to 3N, the number of degrees of freedom - which is the case for the spinless realization (6). Thus with the representations of SO(3N+1,2) the integrable quantum system is completely specified. Physically, we have in these systems an N-body potential depending on $r = \sqrt{r_1^2+r_2^2+\cdots+r_{3N}^2}$, for example, a collective Coulomb potential of the form $V = a/\sqrt{r_1^2+\cdots+r_{3N}^2}$. In atomic problems, the sum of two-body potentials is on the other hand, given by $V = \sum_{j<i} \frac{e_i e_j}{|r_i-r_j|}$.

If we use hyperspherical coordinates[8] in the 3N-dim. space:

$$r_1 = r\cos\theta_1,\ r_2 = r\sin\theta_1\cos\theta_2,\ r_3 = r\sin\theta_1\sin\theta_2\cos\theta_3$$

$$r_4 = r\sin\theta_1\sin\theta_2\sin\theta_3\cos\theta_4$$

$$- - - - - - - - - - - - - - -$$

$$r_{3N-1} = r\sin\theta_1\cdots\sin\theta_{3N-2}\cos\theta_{3N-1}$$

$$r_{3N} = r\sin\theta_1\cdots\sin\theta_{3N-2}\sin\theta_{3N-1},$$

with $dV = r^{3N-1}dr d\Omega$, and the separation of the Laplace-Beltrami operator $\Delta = \Delta(r) + \frac{1}{r^2}\Delta^*(\Omega)$, into a radial part $\Delta(r)$ and an angular part $\Delta^*(\Omega)$ with eigenvalues $-K(K+3N-2)$, $K = 0,1,2,\cdots$, one can write the realistic Coulomb potential as [9]

$$V = \sum_{j<i} \frac{\ell_i \ell_i}{|r_i-r_j|} = \frac{Z(\Omega)}{r}.$$

Thus, our soluble model treats the angular factor $Z(\Omega)$ as a constant. Details of the comparison of the two cases and further details of the physics of the soluble problem will be reported elsewhere[10].

References:

(1) For integrable many-body problems in one-dimension and related studies see

P. D. Lax, Comm. Pure and Appl Math. 21, 467 (1968)

F. Calogero, J. Math. Phys. 10,2197 (1969); 12,419 (1971); 15,1420 (1974);

M. Toda, Progr. Theor. Phys. Suppl. 45,1974 (1970),

J. Moser, Advances in Math. 16, 197 (1975),

M. A. Olshanetsky, A. M. Perelomov, Invent. Math. 37,93 (1976),

Lett. Math. Phys. 2, 7 (1977);

H. Airault, H. P. McKean, J. Moser, Comm. Pure and Appl. Math. 30, 95 (1977).

B. Kostant, Advances in Math. 34, 195-338 (1980).

(2) V. I. Arnold, Proof of a Theorem of A. N. Kolmogorov on the invariance of quasi-periodic Motions under small perturbations of the Hamiltonian, Russian Math. Surveys 18, 9-36 (1963)

J. Moser, Nachr. Akad. Wiss. Göttingen, Math-Phys. Kl. II a, 1-20 (1962).

(3) A. O. Barut, Phys. Rev. 133,B839 (1964)

A. O. Barut and A. Böhm, Phys. Rev. 139, B1107 (1965)

(4) A. O. Barut, Reports in Math. Phys. 11, 401 (1977)

(5) H. Weyl, The Theory of Groups and Quantum Mechanics, (Dover, N.Y. 1950)

(6) A. O. Barut, in Non-compact Group in Particle Physics, Edit. Y. Chow, (W. A. Benjamin, N.Y. 1966) p. 1-22.

(7) A. O. Barut, in Groups, Systems and Many-Body Physics, (Edit. by P. Kramer and M. Dal Cin, Vieweg 1980) ,Ch. 6, p. 285-317.

(8) cf. for example, A. O. Barut and R. Raczka, Group Representations and Applications, Warsaw 1977.

(9) R. Peterhop, Theory of Ionisation of Atoms by Electron Impact, Color. Ass. Univ. Press, Boulder, 1977.

(10) A. O. Barut and Y. Kitagawara, (to be published)

WARD-TAKAHASHI IDENTITIES
AS A NECESSARY AND SUFFICIENT CONDITION FOR GAUGE INVARIANCE

by

J.C. HOUARD and M. IRAC-ASTAUD

Laboratoire de Physique Théorique et Mathématique
Université Paris VII - 75221 PARIS CEDEX 05 - FRANCE

The original Ward identity is a relation between the self-energy function of the electron and the vertex function in Quantum Electrodynamics, mainly used to prove the equality $Z_1 = Z_2$ for the renormalization constants. When the Ward proof uses perturbation theory [1], a global demonstration together with a generalization was later given by Takahashi [2]. The result is a set of identities between Green functions, which essentially express the conservation of the electric current. Although these proofs, as also the subsequent ones[x], do not refer to the gauge invariance - understood as the gauge invariance of the second kind of the Lagrangian - these identities were rapidly considered as characterizing it. However this assertion does not seem to have been proved, the more as it cannot hold exactly in this form, since, as is known, for the need of quantization the Lagrangian cannot be strictly invariant. It is the purpose of the present note to give, in the framework of the canonical formalism, a precise statement of a necessary and sufficient condition connecting the existence of the Ward-Takahashi identities with the gauge invariance properties of the Lagrangian. Of course, such a demonstration is possible only by considering a wide class of theories, a priori not necessarily invariant, and not by only reasoning upon a single model as Electrodynamics.

1. A basic identity

We assume the validity of the Schwinger action principle

$$[\delta Q(\sigma), \phi(x)] = -i\,\delta\phi(x) \quad , \quad x \in \sigma \tag{1}$$

where $\delta Q(\sigma)$ is the charge on the hypersurface σ associated with the varia-

---

x For a brief review see [3]

tions $\delta\phi$ of the fields ϕ . Furthermore, we require the relations

$$\delta Q(\mp\infty)\,|0, {}^{in}_{out}\rangle = \lambda\,|0, {}^{in}_{out}\rangle \quad , \quad \lambda \in \mathbb{R} \tag{2}$$

In particular, the relation (1) may be demonstrated for local transformations, the condition (2) being automatically satisfied when the variations have compact supports. By introducing the graphical notations

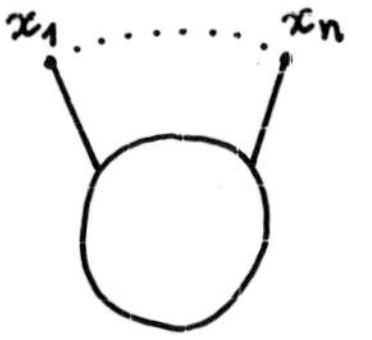

$$= \frac{\langle 0, out|\, T\big(\phi(x_1)\cdots\phi(x_n)\big)\,|0, in\rangle}{\langle 0, out\,|\,0, in\rangle} \tag{3}$$

and

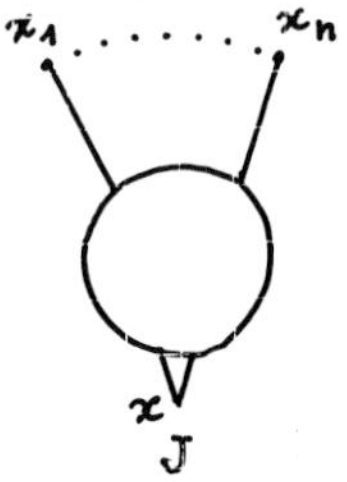

$$= \frac{\langle 0, out|\, T\big(J(x)\,\phi(x_1)\cdots\phi(x_n)\big)\,|0, in\rangle}{\langle 0, out\,|\,0, in\rangle} \tag{4}$$

we obtain the basic identity

δ 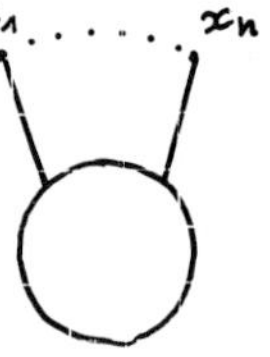 $= -i\int dx$ 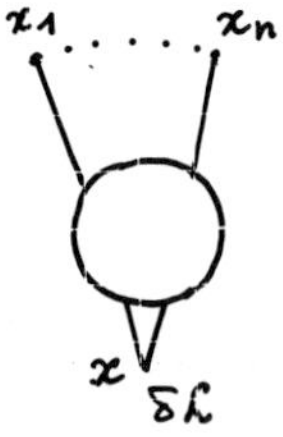 (5)

where the left hand side is the variation of (3) for a variation $\delta\phi$ of ϕ , and where $\delta\mathcal{L}$ is the corresponding variation of the Lagrangian.

In the special case of a boson fields translation $\delta\phi = \delta\alpha$, where $\delta\alpha$ represent arbitrary functions, the relation (5) is equivalent to the equations of motion combined with the canonical commutation relations. In that case, another relation holds, that will be useful in the following,

δ 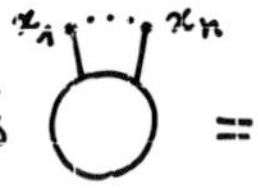 $= -i\int dx$ 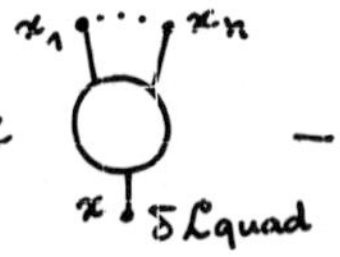 $- \int dx\, \delta\alpha(x)$ 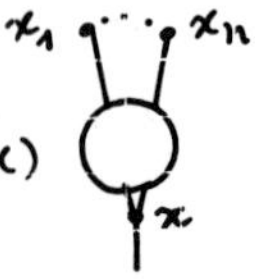 (6)

in which $\mathcal{L}_{quad}$ is the quadratic part of $\mathcal{L}$, the additional line in the last term entering at the point x by the full interaction [3] .

2. Abelian gauge transformations

The preceding results are now applied to the case of the fields $\phi = (A_\mu, \psi)$ where A_μ is a vector field and ψ a Dirac field, subject to the transformations

$$
\begin{cases}
\delta \psi = i e_0 \, \delta \omega \, \alpha_1 \, \psi \\
\delta A_\mu = \delta \omega \, \partial_\mu \alpha_2
\end{cases}
\tag{7}
$$

where α_1 and α_2 are arbitrary real functions. The gauge transformations are obtained for $\alpha_1 = \alpha_2$. By denoting the corresponding variations by δ_1 and δ_2 , the relation (5) reads

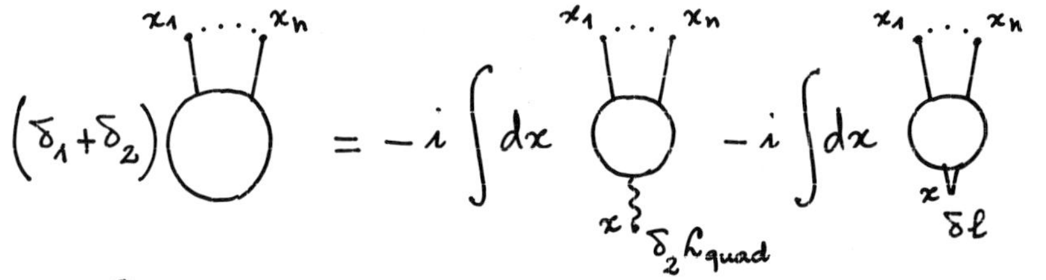

$$
(\delta_1 + \delta_2) \; \raisebox{-1em}{\scriptsize $x_1 \ldots x_n$} \; \bigcirc \;\; = -i \int dx \;\; \bigcirc_{x \; \delta_2 \mathcal{L}_{quad}}^{x_1 \ldots x_n} \; -i \int dx \;\; \bigcirc_{x \; \delta \ell}^{x_1 \ldots x_n}
\tag{8}
$$

where $\mathcal{L}_{quad}$ is here the quadratic part of $\mathcal{L}$ with respect to A_μ only, and $\ell = \mathcal{L} - \mathcal{L}_{quad}$. Since δ_2 represents a translation, the formula (6) can be used to reduce (8) into

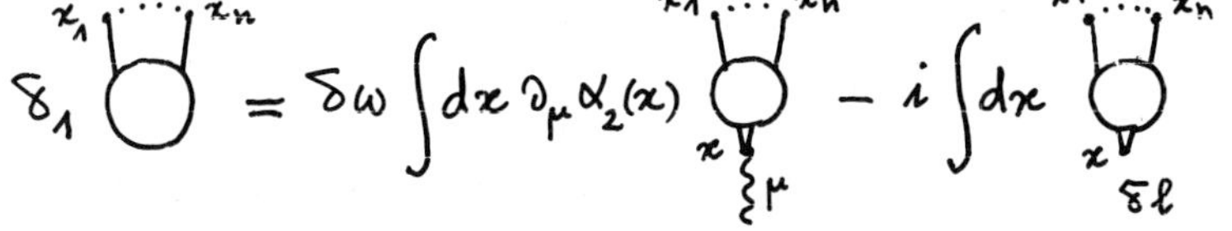

$$
\delta_1 \;\; \raisebox{-1em}{\scriptsize $x_1 \ldots x_n$} \bigcirc \;\; = \delta \omega \int dx \, \partial_\mu \alpha_2(x) \;\; \bigcirc_{x \; \mu}^{x_1 \ldots x_n} \; - i \int dx \;\; \bigcirc_{x \; \delta \ell}^{x_1 \ldots x_n}
\tag{9}
$$

where the weavy line is a A_μ-line. The last term represents the first order variations of the Green functions deduced from the Lagrangian $\mathcal{L}$. From the group property of the transformations, its vanishing implies that the Green functions for $\mathcal{L}$ and its transformed are identical, and then that $\delta \ell$ reduces to a divergence. Taking $\alpha_1 = \alpha_2 = \alpha$, it results the following theorem :

Theorem : The part $\ell = \mathcal{L} - \mathcal{L}_{quad}$ of the Lagrangian is gauge invariant up to a divergence, if and only if the following relations are satisfied

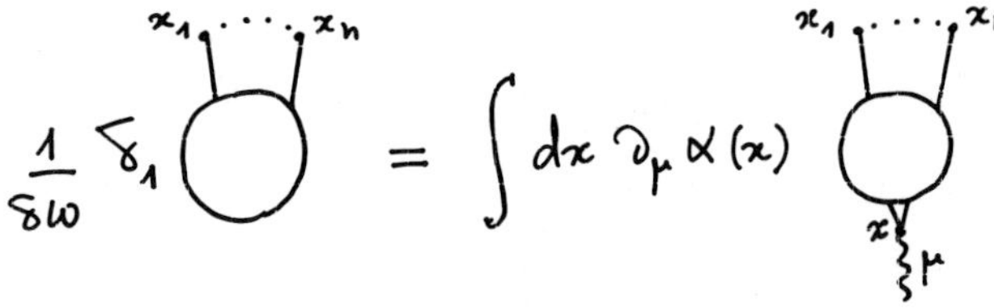

$$\frac{1}{\delta \omega} \delta_1 \bigcirc \overset{x_1 \cdots x_n}{} = \int dx \, \partial_\mu \alpha(x) \bigcirc \overset{x_1 \cdots x_n}{}_{\substack{x \\ \big\{\mu}} \tag{10}$$

These relations are the <u>Ward-Takahashi identities</u>.

Therefore the Ward-Takahashi identities do not restrict in any way the quadratic part $\mathcal{L}_{quad}$ (provided that the theory be canonical), thus allowing arbitrary mass and gauge fixing terms. Let us also note that the identities (10) may be replaced by the relations having the same form but containing only proper diagrams with respect to the additional A_μ-line.

3. Non-Abelian gauge transformations

An analogous result holds for the non-Abelian transformations

$$\begin{cases} \delta \psi = - i \, \delta \omega_1^a \, \mathcal{C}_a \psi \\ \delta A_\mu^a = C_{bc}^a \, \delta \omega_1^b A_\mu^c - \frac{1}{\varepsilon} \partial_\mu \delta \omega_2^a \end{cases} \tag{11}$$

Let us at first observe that $\delta_2 \mathcal{L}_{quad} = \delta_2 \overline{\mathcal{L}}_{quad}$ if only $\mathcal{L}_{quad} - \overline{\mathcal{L}}_{quad}$ is proportional to $(\partial_\mu A_\nu - \partial_\nu A_\mu)(\partial^\mu A^\nu - \partial^\nu A^\mu)$. So, writing now $\mathcal{L} = \overline{\mathcal{L}}_{quad} + \ell$, the formula (8) is modified by an additional term containing $\delta_1 \overline{\mathcal{L}}_{quad}$ (that does not occur in the Abelian case for which $\delta_1 A_\mu = 0$). The same argument as in the previous section then leads to

<u>Theorem</u>   The part $\ell = \mathcal{L} - \overline{\mathcal{L}}_{quad}$ of the Lagrangian is gauge invariant up to a divergence, if and only if the following relations are satisfied

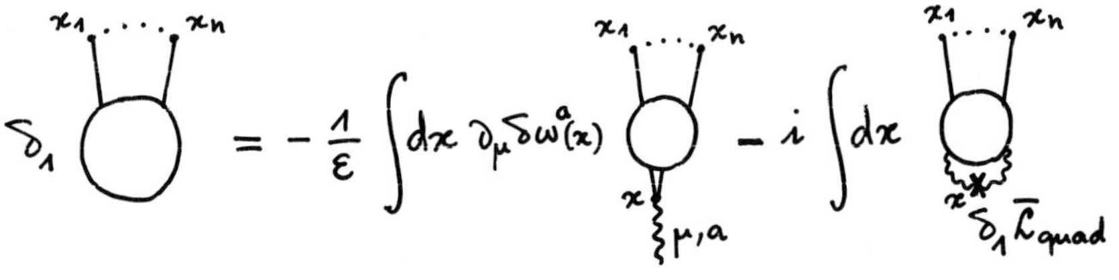

$$\delta_1 \bigcirc \overset{x_1 \cdots x_n}{} = - \frac{1}{\varepsilon} \int dx \, \partial_\mu \delta \omega^a(x) \bigcirc \overset{x_1 \cdots x_n}{}_{\substack{x \\ \big\{\mu, a}} - i \int dx \, \bigcirc \overset{x_1 \cdots x_n}{}_{\substack{x \\ \delta_1 \overline{\mathcal{L}}_{quad}}} \tag{12}$$

Contrarily to the Abelian case, the form of the Ward-Takahashi relations (12) does depend on the choice of $\mathcal{L}_{quad}$. Conversely, giving the last term in

(12), determines $\overline{\mathcal{L}}_{quad}$ except for the mass term. The non-vanishing of $\delta_1 \overline{\mathcal{L}}_{quad}$ is responsible for the non-unitarity of the S-matrix between physical states. As is known, in the case of the usual Yang-Mills Lagrangian, this problem is solved by the introduction of the Faddeev-Popov ghosts [4 - 5]. The identities which then seem to be relevant for obtaining relations between renormalization constants are those obtained by Slavnov [6] rather than (12) modified by adding the terms containing the ghosts. Let us stress that the transformation used by Slavnov seems to go beyond the preceding framework since the parameters $\delta\omega$ defining it are non-local functions of the fields. On the contrary, the Becchi-Rouet-Stora transformation [7] could offer a more suitable way to get the Slavnov identities in the canonical formalism. It is in fact local, satisfies the Schwinger principle, and one can check that a direct application of the preceding method indeed gives the Slavnov identities. However the algebraic aspects of this transformation, essentially based on the properties of anticommuting scalars, need further investigations.

References

[1] J.C. Ward, Phys. Rev., 78, 182 (1950)

[2] Y. Takahashi, Nuovo Cim., 6, 371 (1957)

[3] J.C. Houard and M. Irac-Astaud, Ward-Takahashi identities and gauge invariance to appear in "Field Theory, Quantization and Statistical Physics" (Jouvet memorial book), E. Tirapegui, ed.

[4] R.P. Feynman, Acta Physica Polonica, 24, 697 (1963)

[5] L.D. Faddeev, V.N. Popov, Phys. Letters, 25B, 29 (1967)

[6] A.A. Slavnov, Theor. Math. Phys., English translation, 10, 99 (1972)

[7] C. Becchi, A. Rouet, R. Stora, Comm. Math. Phys., 42 , 127 (1975).

SUPERSYMMETRIC GENERALIZATION OF RIEMANNIAN SYMMETRIC PAIRS

Jerzy Lukierski
Institute for Theoretical Physics
University of Wrocław
50-205 Wrocław, ul.Cybulskiego 36, Poland

1. The nonlinear representations of Lie group G linear on its closed subgroup H (see e.g. [1]) are described in a compact way by Cartan calculus of differential forms [2,3] if the pair $(H, \frac{G}{H})$ is the symmetric Riemannian space (see e.g. [4]) . In such a case there exists in Lie algebra g an involution $\hat{\eta}$ $(\hat{\eta}^2 = 1)$ which describes by its two eigenstates

$$\hat{\eta}\, h = h \qquad \hat{\eta}\, k = -k \tag{1}$$

the Z_2- grading of the algebra $g = h \oplus k$

$$[h,h] \subset h \qquad [h,k] \subset k \qquad [k,k] \subset h \tag{2}$$

If we consider the superalgebra $S = g \oplus f$, its bosonic sector $g = g_1 \oplus g_2$ is the sum of geometric and internal symmetry algebra. Introducing in g_1 and g_2 respectively the symmetric Riemannian pairs (h_1,k_1) , (h_2,k_2) ($g_i = h_i \oplus k_i$, $i = 1,2$) one can consider the algebraic framework for the nonlinear realizations of the superalgebra S which are linear only on $h = h_1 \oplus h_2$. One can proceed in the following two different ways:

a) to consider such a nonlinear realization of S as a composition of three nonlinear realizations described by the symmetric pairs $(g_1 \oplus g_2, f) \times (h_1,k_1) \times (h_2,k_2)$.

This method was applied e.g. by Zumino [5] to the description of $OSp(4;1)$ realizations linear on $S\ell(2;c)$. In such a way one obtains always all the Grassmann parameters (Goldstone spinors) for f as independent fundamental coordinates.

b) one can look also for the possible splits of the fermionic sector $f = (f_+, f_-)$ and look for the super-Riemannian quadruples

$$S = (h_1 \oplus h_2, k_1 \oplus k_2, f_+, f_-) \tag{3}$$

endowed with four-dimensional graded structure.

In such a case only the part of fermionic sector (e.g. f_+) describes the fundamental fermionic coordinates. The nonlinear realizations of such a type we shall call supernonlinear. From the importance of chiral superspace coordinates in supergravity (see e.g. [6,7]) and reducible structure of fundamental SU(5) fermionic multiplets $\underline{5} + \underline{10}$, (where it would be desirable to have $\underline{10}$ composite) such realizations might be very useful for the description of supersymmetric unification of elementary forces.

2. The super-Riemannian quadruple (3) can be endowed with the following two types of fourdimensional gradings:

a) Z_4 - grading

$$\begin{array}{cccc} L_0 & L_1 & L_2 & L_3 \\ h_1 \oplus h_2 & f_+ & k_1 \oplus k_2 & f_- \end{array} \tag{4}$$

where the generalized Lie bracket $\langle \cdot,\cdot \rangle$ satisfies the following relations written modulo 4 :

$$\langle L_i , L_j \rangle \subset L_{i+j} \tag{5a}$$

$$\langle L_i , L_j \rangle = (-1)^{ij+1} \langle L_j , L_i \rangle \tag{5b}$$

$$\langle\langle L_i, \langle L_j , L_k \rangle\rangle (-1)^{ik} + \text{cycl perm} = 0 \tag{5c}$$

The formulae (5a-c) describe Z_2- graded superalgebra with the additional structure expressed by the relation (5a).

If we describe the superalgebra S using real generators f, in general the decomposition (4) leads to complex conjugated supercharges $f_+ = (f_-)^*$. One can perform therefore the following rescaling of the generators

$$L_0' = L_0 \qquad L_1' = \frac{1}{R} L_1 \qquad L_2' = \frac{1}{R^2} L_2 \qquad L_3' = \frac{1}{R} L_3 \tag{6}$$

and in the contraction limit $R \to \infty$ one gets:

$$\{L_1', L_3'\} \subset L_0' \qquad [L_2', L_j'] = 0 \qquad (j \neq 0)$$

$$[L_0', L_0'] \subset L_0' \qquad [L_0', L_i'] \subset L_i' \tag{7}$$

$$\{L_1', L_1'\}, \{L_3', L_3'\} \subset L_2'$$

We see that the sector $L_2' = k_1' \oplus k_2'$ forms Abelian invariant subalgebra.

In general case the generators k_2' form a linear representation of the internal symmetry h_2'. If we restrict h_2' to the subalgebra h_2'' (the stability group of k_2'), we obtain the theory with the internal symmetry reduced to h_2'' and central charges.

b) $Z_2 \times Z_2$ grading

$$\begin{array}{cccc} L_{0,0} & L_{1,0} & L_{0,1} & L_{1,1} \\ h_1 \oplus h_2 & f_+ & f_- & k_1 \oplus k_2 \end{array} \tag{8}$$

where

$$\langle L_{i,j}, L_{k,\ell}\rangle \subset L_{i+k,\,j+\ell} \qquad \text{mod } 2 \tag{9a}$$

$$\langle L_{i,j}, L_{k,\ell}\rangle = (-1)^{(i+j)(k+\ell)+1} \langle L_{k,\ell}, L_{i,j}\rangle \tag{9b}$$

$$\langle\langle L_{i,j} \langle L_{k,\ell}, L_{m,n}\rangle\rangle (-1)^{(i+j)(m+n)} + \text{cycl. perm. of pairs} = 0 \tag{9c}$$

It is easy to check that the additional structure of superalgebra is provided only by the relation (9a).

In the case of $Z_2 \times Z_2$ grading the superalgebra S contains two conjugated supersymmetric subalgebras

$$S_\pm = (\tilde{f}_\pm, \{\tilde{f}_\pm, \tilde{f}_\pm\}) \tag{10}$$

which can be used as generating the stability supergroup in the construction of supersymmetric coset spaces.

3. We would like to add the following comments:

a) Our Z_4 and $Z_2 \times Z_2$ gradings describe "internal structure" of

Z_2- graded superalgebra. Recently Bars and Günaydin [8] considered the algebras and superalgebras with two-sided odd-dimensional gradings. It is interesting to study how these two types of grading are related.

b) We expect the following applications of the presented scheme:

i) consideration of linear multiplets in extended supergravity with $N > 8$ with the composite components.

ii) construction of fundamental fermionic multiplets in grand unification theories starting from supersymmetry.

iii) discussion of spontaneous symmetry breaking and central charges in supersymmetric theories.

c) The examples of $Z_2 \times Z_2$ and Z_4 - graded Riemannian quadruples for physically interesting groups OSp(N;4) (extended graded de-Sitter geometry), SU(2,2;N) (extended graded conformal geometry) and OSp(8;1) (fermionic twistor geometry [9]) has been given by L. Rytel and the author in [10].

References

1. S. Coleman, J. Wess and B. Zumino, Phys.Rev. 177, 2239 (1969).
2. D.V. Volkov, Sov.J. Particles and Nuclei 4, 3(1973).
3. V.I. Ogievetski, Proc. X-th Winter School of Theor.Physics, Vol. I, p. 117 (1974).
4. S. Helgason, "Differential Geometry and Symmetric Spaces", Academic Press, New York 1962.
5. B. Zumino, Nucl.Phys. B127, 189 (1977).
6. V.I. Ogievetski and E. Sokhaczev, Yad.Fiz. 28, 1633 (1978).
7. W. Siegel and S.J. Gates, Nucl.Phys. B147, 77 (1979).
8. I. Bars and M. Günaydin, J.Math.Phys. 20, 1977 (1979).
9. J. Lukierski, Nuovo Cim.Lett. 24, 309 (1979).
10. J. Lukierski and L. Rytel , "Algebraic Framework for Super Nonlinear Realizations" , Leuven Univ. preprint KUL - TF -80/11, April 1980.

COMPOSITE GAUGE FIELDS AND RIEMANNIAN GEOMETRY

Jerzy Lukierski
Institute for Theoretical Physics
University of Wrocław
50-205 Wrocław, ul.Cybulskiego 36, Poland

1. Introduction

The conventional gauge theory is formulated in terms of the elementary potentials $A_\mu(x)$ ($A_\mu \in \mathrm{adj}\, G$) which locally describe the connection one - forms $\omega_1 = A_\mu dx^\mu$ in the principal G-bundle over the coordinate manifold $\mathcal{S}$ $(x \in \mathcal{S})$. If the principal G-bundle is trivial (bundle space isometric to $G \times \mathcal{S}$) the potentials $A_\mu(x)$ can be defined globally and are homotopic to vanishing connection. The topologically interesting examples of gauge fields (e.g. instanton and monopole solutions) are described by the choice of nontrivial principal G-bundle.

Recently several authors arrived at the notion of composite gauge potentials starting from different mathematical ideas. For simplicity we shall assume that $G = U(n)$ or $SU(n)$. It was observed that

i) Any sufficiently smooth $U(n)$ gauge potential over compact $\mathcal{S}$ is homotopically equivalent to a natural connection on the following universal Stiefel bundle([1,2] ; for application to composite gauge theory see [3,4]):

$$S_{N,n}(C) = \frac{U(N)}{U(N-n)} \quad \text{(Stiefel manifold)} \qquad G_{N,n}(C) = \frac{U(N)}{U(N-n)\times U(n)} \quad \text{(Grassmann manifold)} \tag{1}$$

provided that we take N sufficiently large. Parametrizing $S_{N,n}(C)$ globally as follows:

$$S_{N,n}(C) \ : \qquad \bar{Z}^{iJ} Z^{Jk} = \delta^{ik} \qquad \begin{matrix} i=1\ldots n \\ J=1\ldots N \end{matrix} \tag{2}$$

one obtains the natural choice of the composite $U(n)$ gauge potentials :

$$A^{ij} = \frac{i}{2} \bar{Z}^{iJ} \overleftrightarrow{\partial_\mu} Z^{Jj} \qquad (Z^{iJ} \equiv Z^{iJ}(x)) \qquad (3)$$

induced by the Hermitean metric on the total bundle space $S_{N,n}(C)$.

ii) one can consider the field - theoretic realizations of the ternary algebras, describing locally the symmetric coset spaces of the Lie groups and supergroups ([5,6] ; for applications to composite gauge theory see [7-9])

iii) for large class of Riemannian manifolds $\mathcal{M}$ endoved with additional structures (complex, quaternionic, symmetric) one can introduce a "natural" principal bundle with its gauge - invariant properties determined by the metric structure on $\mathcal{M}$. These manifolds are Kählerian ; the "natural" bundles have the curvature two-forms satisfying Bianchi identities, and subsequently define the composite gauge potentials.

It appears in such a framework that the composite gauge theory describes a generalization of conformal-invariant nonlinear σ-models from two to four dimensions [11-14] , with the unconventional Lagrangeans fourlinear in the first σ-field derivatives.

We shall consider here three classes of Riemannian manifolds:

a) complex Kähler manifolds, with holonomy group $U(m)=U(1)\times SU(m)$ (2m real dimensions)

b) quaternionic Kähler manifolds, with holonomy group $Sp(1)\times Sp(m)$ ($Sp(1)= SU(2)$), (4m real dimensions)

c) Riemannian manifolds with the holonomy group $U(n)\times U(m)$ ($n\geqslant 2$, 2nm real dimensions). It can be shown that due to the classification theorem of Riemannian geometries [15] these manifolds are necessarily symmetric, i.e. isometric to the complex Grassmannians $G_{N,n}(C)$ (see (1); $N = n+m$).

We shall present here some new aspects of the approach iii), providing a geometric background for the choice of the fourlinear Lagrangeans for σ - fields. We shall show that the composite gauge theory can be considered as a well defined generalization of the theory of strings moving in particularly chosen Riemannian manifolds. Such an interpretation was pointed out in [16] and provides the relation between the composite gauge theory and the generalized σ-models.

2. From strings to fourlinear σ-models.

We define σ- field $\xi_i(x)$ as a mapping

$$\xi_i(x): \qquad \mathcal{S}\ (x\in \mathcal{S}) \longrightarrow \mathcal{M}\ (\xi_i \in \mathcal{M}) \qquad (4)$$

where $\mathcal{S}$ is the coordinate manifold (Euclidean k-dimensional space R^k, compactified Euclidean space $R^k \cup \infty$ or pseudoEuclidean k-dimensional space) and $\mathcal{M}$ denotes arbitrary Riemannian manifold.

In order to determine the dynamics of σ-field we should choose the Lagrangean. One can introduce two classes of σ-models:

a) Conventional σ-models.

The action is the "energy integral" of the harmonic mapping $\mathcal{S} \to \mathcal{M}$. One can construct the following sequence of Lagrangeans, leading to k-dimensional conventional σ-models:

i) Free point particle in R^N in proper time parametrization (we put $c=1$):

$$S = m \int d\tau \left(\frac{dY_\mu}{d\tau} \right)^2 \qquad \mu = 1 \ldots N \tag{5}$$

ii) One-dimensional σ-model which describes geodesic motions of a point particle on the Riemannian manifold $\mathcal{M}$ with the metric tensor g_{ik}. We replace in (5) differentials by one-forms

$$dY_\mu \longrightarrow \theta^a = E^a_i(\xi)\, d\xi_i \in T^*(\mathcal{M}) \tag{6}$$

where E^a_i denote the "generalized vierbeins" for the Riemannian manifold $\mathcal{M}$, i.e.

$$g^{ik} E^a_i E^b_k = \delta^{ab} \tag{7a}$$

$$g_{ik} = E^a_i E^a_k \qquad \text{or} \qquad ds^2 = \theta^a \cdot \theta^a \tag{7b}$$

One obtains

$$S = m \int d\tau\, g_{ik}(\xi) \frac{d\xi_i}{d\tau} \frac{d\xi_k}{d\tau} \tag{8}$$

iii) k-dimensional σ-model is obtained by the replacement of one-dimensionsl trajectories $\xi_i(\tau)$ in $\mathcal{M}$ by σ-field submanifolds (4) (K-dimensional cycles for compact $\mathcal{S}$):

$$S = m^{2-k} \int d^k x\, g_{ik}(\xi) \frac{\partial \xi_i}{\partial x_\mu} \frac{\partial \xi_k}{\partial x^\mu} \tag{9}$$

We see that in the Lagrangeans of conventional σ-models there is "free" bilinear part describing the flat limit on $\mathcal{M}$. If $k=2$ the models (9) are conformal - invariant and have interesting classical and quantum properties.

b) Generalized fourlinear σ-models.

i) The prototype for fourlinear σ-models, analogous to (5) for conventional ones, is the action describing the motion of free string in R^N, parametrized invariantly by two parameters τ_1, τ_2 [17,18]. We get in place of (5)

$$S = M^2 \int d\tau_1 \, d\tau_2 \left(\frac{\partial(Y_\mu, Y_\nu)}{\partial(\tau_1,\tau_2)} \right) = M^2 \int d^2 s \tag{10}$$

where the Jacobian

$$\frac{dS_{\mu\nu}}{dS} = \frac{\partial(Y_\mu, Y_\nu)}{\partial(\tau_1,\tau_2)} = \frac{\partial Y_\mu}{\partial\tau_1}\frac{\partial Y_\nu}{\partial\tau_2} - \frac{\partial Y_\nu}{\partial\tau_1}\frac{\partial Y_\mu}{\partial\tau_2} \tag{11}$$

describes the two-dimensional "area" derivative (see e.g. [19]). and the orthonormal coordinates (τ_1, τ_2) parametrize the surface swept by the motion of a string in such a way that

$$d^2 s = d\tau_1 \, d\tau_2 = (dS_{\mu\nu} \, dS^{\mu\nu})^{\frac{1}{2}} \tag{12a}$$

with

$$dS_{\mu\nu} = dY_\mu \wedge dY_\nu \tag{12b}$$

denoting the projection of an infinitesimal area on $\frac{1}{2}N(N-1)$ coordinate planes (μ, ν).

It should be mentioned that for the action (10) one can introduce the generalized Hamiltonian formalism ([20]; see also [21])

ii) By the replacement (6) in formula (10) one obtains the one-dimensional generalized σ-models. Because

$$dY_\mu \wedge dY_\nu \rightarrow \theta^a \wedge \theta^b \in \bigwedge_2 T^*(M) \tag{13a}$$

one gets the action

$$S = M^2 \int d\tau_1 d\tau_2 \, g_{ik}(\xi) g_{jl}(\xi) \frac{\partial(\xi_i,\xi_j)}{\partial(\tau_1,\tau_2)} \frac{\partial(\xi_k,\xi_l)}{\partial(\tau_1,\tau_2)} \tag{13b}$$

describing the motion of string $\xi_i(\tau_1, \tau_2)$ on the Riemannian manifold $\mathcal{M}$ with metric g_{ik}.

One can show that the two-time Voltera-Nambu Hamiltonian formalism [20,21] can be generalized to the class of actions (13).

iii) k-dimensional generalized σ-model is obtained by the replacement of two-dimensional string submanifold $\xi_i(\tau_1, \tau_2)$ by k-dimensional $(k > 2)$ σ-field $\xi_i(x_1 \ldots x_k)$. We obtain

$$S = M^{4-k} \int d^k x \, g_{ik}(\xi) g_{jl}(\xi) \frac{\partial(\xi_i,\xi_j)}{\partial(x_\mu, x_\nu)} \cdot \frac{\partial(\xi_k,\xi_l)}{\partial(x^\mu, x^\nu)} = \tag{14}$$

$$= M^{4-k} \int d^k x \; F^{ab}_{\mu\nu} \; F^{ab\,\mu\nu}$$

where

$$F^{ab}_{\mu\nu} = E^a_i(\xi)\, E^b_\gamma(\xi)\, \frac{\partial(\xi_i, \xi_\gamma)}{\partial(x_\mu, x_\nu)} \tag{15a}$$

The action (14) is conformal - invariant in four dimensions - - similarly like gauge theories.

The indices a,b are in tangent bundle T(M) and are rotated by the holonomy group transformations $H \in SO(n)$ (dim $\mathcal{M}$ =n). We shall consider $F^{ab}_{\mu\nu}$ as a candidate for composite field strenght in H - gauge theory. If $H = H_1 \times H_2$ we should replace $a \to (a_1, a_2)$ and

$$F^{a\,b}_{\mu\nu} \longrightarrow F^{(a_1,a_2)(b_1,b_2)}_{\mu\nu} \tag{15b}$$

One can consider also $F^{(a_1,a_2)(b_1,a_2)}$ or $F^{(a_1,a_2)(a_1,b_2)}$ as possible choices for the composite H_1 or H_2 gauge field strenghts.

3. From fourlinear σ -models to composite gauge theory.

The action (14) describes the composite Yang-Mills theory if the candidate for composite field strenght (15a) satisfies Bianchi identities; in such a case one can introduce the composite potentials. If the holonomy group of $\mathcal{M}$ is not reduced we shall obtain the O(N) composite gauge fields. The examples of real Riemannian manifolds defining composite O(N) field strenght (15a) satisfying Bianchi identities are $\frac{O(N+1)}{O(N)}$ (O(N) -model) and $\frac{O(N) \times O(N)}{O(N)}$.

It appears however that for the construction of the composite gauge fields it is convenient to use the Riemannian manifolds with additional complex, quaternionic or symmetric coset space structure and reduced holonomy group $H = H_1 \times H_2$. We shall discuss below these three cases separately.

a) Complex Kähler manifolds ($H = SU(n) \times U(1) = U(n)$)

The complex metric $h_{i\bar{j}} = h^*_{j\bar{i}}$ is expressed by complexified one-forms (see(6)).

$$h_{i\bar{j}} = E^a_i\, \bar{E}^a_\gamma \qquad \text{or} \qquad ds^2 = \bar{\theta}^a_c\, \theta^a_c \tag{16}$$

We postulate also the Kähler property

$$d\,\omega_2 = d(\bar{\theta}_c^{\;a} \wedge \theta_c^{\;a}) = 0 \tag{17}$$

where the fundamental two-form ω_2 is not degenerate.

N^2 projections of two-dimensional infinitesimal area on the complex coordinate planes (i,j) can be described by the $U(n)$ - valued two-form:

$$\omega_2^{ab} = \frac{i}{2}\, \bar{\theta}^a \wedge \theta^b \tag{18}$$

where $U(n)$ is the reduced holonomy group. Using (17) we see that from the Kähler-manifold - valued σ-fields one can construct only the $U(1)$ composite gauge fields. Pulling back the closed two-form $\omega_2 = \bar{\theta}^a \wedge \theta^a$ onto σ-field submanifold in one gets

$$F_{\mu\nu} = \frac{i}{2}\, h_{i\bar{j}}\, \frac{\partial(z_i, \bar{z}_j)}{\partial(x_\mu, x_\nu)} \tag{19}$$

and from (17) it follows that

$$\partial^\mu\, {}^*F_{\mu\nu} = 0 \qquad \left({}^*F_{\mu\nu} = \frac{1}{2}\, \epsilon_{\mu\nu\rho\tau}\, F^{\rho\tau} \right) \tag{20}$$

The relation (20) implies that one can introduce the composite $U(1)$ potentials.

b) Quaternionic Kähler geometry $(H = Sp(n) \times Sp(1))$

We introduce quaternionic - Hermitean metric

$$H_{\bar{i}j} = G_{ij} + e_r\, H_{ij}^{(r)} \tag{21}$$

where $e_r e_s = -\delta_{rs} + \epsilon_{rst}\, e_t$ and

$$G_{ij} = G_{ji} \qquad H_{ij}^{(r)} = -H_{ji}^{(r)} \qquad \text{(real)} \tag{22}$$

One can introduce the quaternion-valued one-forms

$$\theta_H^a = E_i^a\, dq_i \tag{23}$$

which transform under the holonomy group $Sp(n) \times Sp(1)$ as follows:

$$\theta_H^{a'} = A^{ab}\, \theta_H^b\, a \qquad A \in Sp(n) \qquad a \in Sp(1) \tag{24}$$

We define the $Sp(n) \times Sp(1)$ -invariant Riemannian line element

$$ds^2 = d\bar{q}_i\, H_{\bar{i}j}\, dq_j = \bar{\theta}_H^a \cdot \theta_H^a \tag{25}$$

and introduce the following imaginary quaternion - valued two-forms

$$\omega_2^{ab} = \bar{\theta}_H^a \wedge \theta_H^b = \omega_2^{ab;r}\, e_r \tag{26}$$

describing the projections of the area element of quaternionic manifold $\mathcal{M}_H$ on the quaternionic coordinate planes (a,b).

The Kähler property for the quaternionic manifolds has a form

$$d\,\omega_4 = 0 \tag{27a}$$

where

$$\omega_4 = \sum_{r=1}^{3} \omega_2^r \wedge \omega_2^r \qquad (\omega_2 \equiv \omega_2^{aa}) \tag{27b}$$

The relations (27) describe also the $Sp(1) = SU(2)$ vector bundle V over any point of a quaternionic Kähler manifold (see e.g. [22]). The curvature two-forms in V projected on the σ-field submanifold $q_i(x)$

$$F^r_{\mu\nu} = \frac{1}{4}\,\mathrm{Tr}\left\{ H_{ij}\left(\frac{\partial q_j}{\partial x_\mu}\, e_r\, \frac{\partial \bar q_i}{\partial x_\nu} - \frac{\partial q_j}{\partial x_\nu}\, e_r\, \frac{\partial \bar q_i}{\partial x_\mu} \right)\right\} \tag{28}$$

are possible candidates for the description of composite $Sp(1) = SU(2)$ non Abelian gauge theory.

However the Bianchi identities for the field streght (28) are not a part of the definition of quaternionic Kähler manifold, they provide the only explicitely known solutions of the relation (27a).

c) $G_{N,n}(C)$ as an example of symmetric coset space.

The independent coordinates on $G_{N,n}(C)$ are described by $n \cdot m$ complex variables z_i , and the complex Hermitean metric $h_{i\alpha, \bar\gamma\bar\beta}$ can be expressed in terms of the one - forms

$$\theta_C^{ka} = E^{ka}_{i\alpha}\, dz_{i\alpha} \qquad \begin{array}{l} k = 1\ldots n \\ \alpha = 1\ldots m \quad (m=N-n) \end{array} \tag{29}$$

as follows

$$ds^2 = \bar\theta_C^{ka}\,\theta^{ka} = h_{i\alpha,\bar\gamma\bar\beta}\, dz_{i\alpha}\, d\bar z_{\gamma\beta} \tag{30}$$

One can introduce the following two families of two-form describing infinitesimal area elements on $G_{N,n}(C)$

$$\omega_2^{kj} = \bar\theta_C^{ka} \wedge \theta_C^{ja} \qquad (\,U(n) - \text{valued}) \tag{31a}$$

$$\omega_2^{ab} = \bar\theta_C^{ka} \wedge \theta_C^{kb} \qquad (\,U(m) - \text{valued}) \tag{31b}$$

Using Maurer - Cartan equations for the group $U(n+m)$ one can show that the following Bianchi identities are valid

$$d\,\omega_2^{kj} + \omega_1^{kl} \wedge \omega_2^{lj} = 0 \tag{32a}$$

$$d\,\omega_2^{ab} + \omega_1^{ac} \wedge \omega_2^{cb} = 0 \tag{32b}$$

where the one-forms $\omega_1^{kl}\delta^{ab} + \delta^{kl}\omega_1^{ab}$ define the Lewi-Civita connection on $G_{N,n}(C)$ and determine by the pull-back of the map (4) the composite $U(n) \times U(m)$ gauge potentials. The relations (32 a-b) can be considered as the solutions of the following two generalized Kähler conditions:

$$d\omega_4 = d\,(\omega_2^{ij} \wedge \omega_2^{ij}) = 0 \tag{33a}$$

$$d\omega_4 = d\,(\omega_2^{ab} \wedge \omega_2^{ab}) = 0 \tag{33b}$$

for the $U(n)$ and $U(m)$ - invariant elements of $H^4(G_{N,n}(C))$.

4. From composite gauge theories to physical models.

We shall present only some remarks which indicate that the composite gauge fields might become important for physical considerations in the future.

a) Composite gauge potentials occur in extended supergravity [24] ; their existence allowed recently to relate N=8 supergravity with grand unification theories (see e.g. [25])

b) By introducing σ -superfields one can extend the composite gauge theory to fermions [26,8,9] ; also using the formalism of one-forms for super Riemannian geometries with additional structures [27] one can generalize our present discussion to the generalized σ -models with the field values in super-coset spaces [12,28-31] . The fermionic fields in the formalism with σ -superfields have the naive scale dimension $d=\frac{1}{2}$: in two - dimensional theories such a choice is canonical, but in four dimensions we obtain subcanonical fermionic fields. Recently following Heisenberg ideas we argued [26] that such subcanonical fermionic fields can be used for the description of quark degrees of freedom.

c) For the quantization of composite gauge theory one can use Feynman path integral. In order to perform any calculations one should introduce background field or auxiliary composite fields leading to the appearence of Gaussian functional integrals. It has been already shown [9] that the composite gauge theory can be described as a conventional Yang-Mills theory with some gauge-invariant correction terms. It is now an important task to calculate these corrections explicitely in renormalized form; the work in this dirrection is now under progress [32] .

References

1. M.S. Narasimhan and S. Ramanan, Amer.J.Phys. 83, 563 (1961); 85 233 (1963),
2. J. Nowakowski and A. Trautman, Journ.Math.Phys.19,1100(1978).
3. J. Fröhlich, talk at Fielefeld Symposium, December 1978; " A new look at generalized nonlinear σ -models and Yang-Mills", IHES preprint, 1979.
4. M. Dubois-Violette and Y. Georgelin, Phys.Lett.82B, 251(1979).
5. J.L. Kantor, Trudy Sem.Vector. Analiza, 16, 407 (1972).
6. I. Bars and M. Günaydin , Journ.Math.Phys. 20, 1977 (1979).
7. I. Bars, Proc. VIIIth Int.Colloquium on Group- Theor.Methods in Physics, Kiriat Anavin (Israel), March 1979.
8. I. Bars and M Günaydin, "Theory of Ternons", Yale Univ. preprint YTP 79-05, Phys.Rev. D. , in press under the title "Dynamical Theory of subconstituents based on ternary algebras".
9. I. Bars and M. Günaydin, Yale Univ. preprint YTP 80-15, May 1980
10. M.A. Semenov-Tjan-Szamski and L.D. Faddeev, Westnik Leningr. Univ. 13, 81 (1977).
11. J. Lukierski, CERN preprint TH 2678, May 1979, improved version publ. , in Proceeding of Summer Institute, Kaiserslautern, August 1979, ed. Springer under the title "Field-Theoretical Methods in Particle Physics", 1980, p. 361.
12. F. Gürsey and H.C. Tze, "Complex and Quaternionic Analyticity in Chiral and Gauge Theories", Yale Univ.preprint , August 1979; Annals of Physics, in press.
13. J. Lukierski, "Quaternionic and Supersymmetric σ -models", lecture at Symposium on Differential-Geometric Methods in Physics, Aix-en-Provence, September, 1979, in press, Proceedings, publ. Springer , 1980.
14. D. Maison, Max Planck Inst. preprint, November 1979.
15. M. Berger, Bull.Soc.Math.France 83, 279 (1955).
16. J. Lukierski, "Supersymmetric pre-QCD dynamics", talk at Erice Workshop, March 1980, to be publ. in Proceedings (Plenum Press).
17. A. Schild, Phys.Rev. D16, 1722 (1977)
18. T. Eguchi, Phys.Rev. Lett. 44, 126 (1980).
19. T. Rado, "Lenght and Area", AMS publications, vol.XXX,New York, 1948.
20. Y. Nambu, Physics Letters B, in press
21. M. Voltera , Rend.Acad. dei Lincei, 6, 127 (1890).

22. S. Ishihara, Jour.Diff. Geom. 9, 483 (1974)
23. J. Harnod, J.Tafel and S. Shnider, University of Montreal preprint, CRM 922; Journ.Math.Phys., in press
24. Cremmer and B. Julia, Nucl.Phys. B159, 141 (1979)
25. J. Ellis , M.K. Gaillard and B. Zumino, CERN, preprint TH-2842 and LAPPTH-16.
26. J. Lukierski and B. Milewski, Phys. Lett. 93B, 91 (1980).
27. J. Lukierski in "Supergravity", ed. P. van Nieuvenhuizen and D.Z. Freedman, North-Holland 1979, p. 301.
28. F. Gürsey and L. Marchildon, Phys. Rev. D17, 2038 (1978); Journ. Math.Phys. 19, 942 (1978).
29. J. Lukierski, Lett. in Math.Phys. 3, 135 (1979)
30. R. d'Auria and S. Sciuto , Geneve University preprint UGVA-DPT 1980/01-230,
31. J. Lukierski, ICTP preprint IC/80/48; to be published in Proc. of XVII Winter School in Karpacz (Poland) February 1980.
32. J. Lukierski and B. Milewski, to be published.

GAUGE-THEORY GHOSTS AND UNITARITY - PROGRESS IN THE GEOMETRICAL TREATMENT

Yuval Ne'eman*+
Tel Aviv University, Tel Aviv, Israel[x]
and
C.P.T., University of Texas, Austin, Texas (USA)

Abstract

We review the geometric unitarity equations for an Internal gauge theory, for a supergauge, for the Kalb Ramond field, and for a Non-internal gauge.

* Supported in part by the U.S.-Israel Binational Science Foundation
+ Supported in part by the U.S. DOE, Grant EY-76-S-05-3992
x Wolfson Chair Extraordinary of Theoretical Phsyics

Introduction

After the success of Relativistic Quantum Field Theory in 1948, with the completion of the renormalization program for Quantum Electrodynamics (QED), attempts were made to extend the results to both Strong and Weak interactions. Both programs failed in the fifties. Pseudoscalar mesons, which were thought to mediate the basic Strong Interaction, happened to involve couplings of the order of $g^2_{\pi \overline{N}N}/4\pi = 14.5$, defeating the perturbation approach. Weak interactions were either unrenormalizable due to their dimensionality if treated as a four-fermion interaction, or symmetry (and gauge) breaking due to the Intermediate Boson masses. In the Western U.S. and much of the rest of the Physics world, Field theory was discredited beyond the "lucky accident" of QED. Mainly it was said to involve a breakdown of Unitarity off-mass-shell.

Happily for the evolution of Physics, we still live in a somewhat heterogeneous world. Field theory and the Unitarity problem continued to be treated by non-mainstream groups or individuals. Feynman made the first step[1], when treating the Yang Mills gauge (as a pilot-project on the way to Gravity -- a suggestion of Gell-Mann's). Feynman was, of course, beyond being influenced by "the concensus"... B. de Witt continued the project[2]; as a General Relativist, he did not know that Field Theory was "out". The Unitarity issue was indeed resolved in the USSR, Fradkin, Faddeev and Popov[3] perfected the method; and in Holland G. 't Hooft[4] achieved the final breakthrough, by adding Regularization to Unitarity, and doing it even for the case of Spontaneously broken gauge theories, i.e. massive vector mesons.

1. The Conventional Method

The procedure was based upon the introduction of ghost fields $X^a(x)$. These were scalar fields with Fermi statistics. Those fields are set to cancel the unphysical contributions of redundant components of the gauge fields. Fermi statistics provide the necessary minus signs to the closed loops. The Yang Mills invariant Lagrangian

$$L_{INV} = -\tfrac{1}{4} R^a \, {}^*R_a \tag{1.1}$$

does not possess an invertible Fourier transform,needed to construct a propagator. To obviate this difficulty, a gauge-fixing term is added (in the linear treatment)

$$L_{FIX} = - \sigma \, \Sigma \tag{1.2}$$

with σ a Lagrange multiplier ensuring $\Sigma \overset{\Omega}{=} 0$. ($\overset{\Omega}{=}$ implies equality resulting from an equation of motion). This is the gauge fixed by L_{GF}. Alternatively, one may use a quadratic L_{FIX} ($C^a = \partial_\mu A^a_\mu$ for example, in QED)

$$L'_{FIX} = -\tfrac{1}{2} C^a C_a \tag{1.2'}$$

One then adds the ghost Lagrangian

$$L_{GH} = \overline{X}_a \hat{m} X \tag{1.3}$$

where we use the gauge transformation of the Yang Mills field-potential

$$\delta A^a_\mu = (D_\mu \varepsilon)^a \qquad \text{or } \delta\phi = D\varepsilon, \quad \phi = A^a_\mu \lambda_a \, dx^\mu \tag{1.4}$$

ϕ in the Lie-algebra valued horizontal component of the connection ω.

This transformation defines the $\hat{m}$ in (1.3)

$$\delta\Sigma \, (A^a_\mu) = \hat{m}\varepsilon^a \qquad (\text{or } \delta C^a = \hat{m}\varepsilon^a) \tag{1.5}$$

$\overline{X}$ is the "anti-ghost", another Fermi scalar (or X read from future to past). With the spin-statistics theorem broken, $\delta\overline{X} \neq \pm \delta X$. The new Lagranian

$$L = L_{INV} + L_{FIX} + L_{GH} \tag{1.6}$$

obeys Slavnor-Taylor invariance[5]. As a result of laborious efforts, this produces Ward-Takahashi identities.

After 't Hooft had achieved the final aim, the procedure was greatly simplified[6] by Becchi, Rouet and Stora and by Tyutin (BRST). Defining

$$\varepsilon^a = \chi^a \Lambda \qquad s = -\frac{\partial}{\partial\Lambda}\delta \tag{1.7}$$

with Λ a constant Grassmann element, anticommuting with itself, with χ^a and with $\overline{\chi}^a$, we have

$$s\phi = D\chi \tag{1.8}$$

$$s\chi = -\tfrac{1}{2}[\chi,\chi] \tag{1.9}$$

$$s\Sigma = \hat{m}\,\chi \qquad \text{or } sC = \hat{m}\,\chi \tag{1.10}$$

$$s\psi = [\,\chi,\psi] = \zeta \tag{1.11}$$

Equations (1.9) and (1.11) represent the homogeneous group action under (1.7). Adding

$$s\overline{\chi} = \sigma \qquad , \qquad (s\overline{\chi} = C^a) \tag{1.12}$$

$$s\sigma = o \tag{1.13}$$

ensures that

$$sL = o$$

since

$$sL_{FIX} = -\sigma s\Sigma \quad , \quad sL_{GH} = \sigma s\Sigma - \overline{\chi}s^2\Sigma$$

provided that $s^2\Sigma = o$, which is guaranteed by

$$s^2 A^a_\mu = s^2\phi^a = o \tag{1.14}$$

Using the Jacobi identity ("algebraic closure") we also find

$$s^2\,\chi = o \tag{1.15}$$

and (1.12) - (1.13) imply

$$s^2\overline{\chi} = o \tag{1.16}$$

Alternatively, using L'_{FIX} yields

$$s^2\overline{\chi} = \hat{m}\chi \neq o \tag{1.16'}$$

These equations guarantee Unitarity[6].

2. The Geometric Derivation

Using a Principal Bundle P $(P,M,\pi,G,\cdot)$ to represent a Yang-Mills theory, it was shown [7,8,9,10,11] that X is the "vertical" piece of the connection ω, that s is the vertical piece of the exterior derivative and that the BRST equations are the Cartan-Maurer structural equations on the Bundle. M is the base manifold, π the projection, G the structure group and the dot is the right action.

$$\left.\begin{array}{ll}\pi(p\cdot g) = \pi(p) & (2.1)\\ (p\cdot g)\cdot g' = p\cdot(g\,g') & (2.2)\end{array}\right\} \quad \forall\, p \in P;\ \forall g,g' \in G$$

$$(\cdot) : P \times G \to P \tag{2.3}$$

Using (locally) co-ordinates $M(x^{\mu})$ and $G(\alpha^{i})$ we lift them onto Σ, a section now described as

$$\Sigma : \alpha^{i} = o \tag{2.4}$$

We refer the reader to the 3 different proofs in references [8,9,12]. Here we only write

$$\omega^{a} = \Xi^{a}_{i}\, d\alpha^{i} + A^{a}_{\mu}\, dx^{\mu} = \Xi^{a} + \phi^{a} \tag{2.5}$$

and remind the reader that for a flat and trivial section, Ξ is the Cartan left-invariant one form. Under a gauge transformation, Ξ and ϕ become both α and x dependent. Denoting the exterior derivative by $\tilde{d}$, we define

$$\left.\begin{array}{l}\tilde{d}\mathcal{f} = z\mathcal{f} + d\mathcal{f} \quad , \quad z\mathcal{f} = d\alpha^{i}\frac{\partial}{\partial\alpha^{i}}\mathcal{f},\ d\mathcal{f} = dx^{\mu}\partial_{\mu}\mathcal{f}\\ \tilde{d}^{2} = o \to d^{2} = z^{2} = zd + dz = o\end{array}\right\} \tag{2.6}$$

Defining the curvature 2-form

$$R = \tilde{d}\omega + \tfrac{1}{2}\,[\omega,\omega] \tag{2.7}$$

the Cartan-Maurer equations state that R has only horizontal components (i.e. in $dx^{\mu}{}_{\wedge}dx^{\nu}$),

$$z\,\phi = DX \tag{2.8}$$

$$z\,\Xi = -\tfrac{1}{2}\,[\Xi,\Xi] \tag{2.9}$$

$$z\,\psi = [\Xi\,,\psi] = \zeta \tag{2.10}$$

Equations (2.8)-(2.10), together with the anticommuting properties of Ξ as a 1-form solve (1.8)-(1.11),

$$s = z \quad , \quad X = \Xi \tag{2.11}$$

As to the antighost $\overline{X}$, it is clearly non-geometric in (1.16'), but (1.16) puts it as an element of a cohomology class of P. Note that since $sL = o$, and $dL = o$ by saturation of the horizontal submanifold, we have a geometric closure[11] of the quantum gauge-Lagrangian,

$$\tilde{d}L = o \qquad (2.12)$$

The precise role of $\overline{X}$ is unclear. Ojima[13] has recently suggested complexification of the fiber G, so that X and $\overline{X}$ make up the vertical piece of the connection

$$\omega = \phi + X + \overline{X}$$

and with $\tilde{d} = d + s + \overline{s}$

To see that the Fermionic properties of X correspond to the anticommutation of one-forms in the exterior calculus (or to differentials in the measure) we have to display a geometric functional integral. Indeed, using the conventional generating functions Γ, we write the Faddeev-Popov expression

$$e^{i\Gamma} = G \int \mathcal{D}(\phi)\; e^{i\, S_{INV}(\phi)} \quad , \quad S_{INV} = \int F \wedge {}^*F \qquad (2.13)$$

Using [8,9]

$$1 = \int \delta(\Sigma)\, \tilde{d}\,\Sigma = \int \delta\;(\Sigma)\; s\,\Sigma \;, \qquad \delta(\Sigma) = \int d\,\sigma\; e^{i\sigma\Sigma}$$

and the Berezin integral $\int \overline{d}\,\theta\, f(\theta) = -i \frac{\partial}{\partial\theta} f(\theta)\Big|_{\theta = o}$

$$s\Sigma = \int \overline{d}\,\overline{X}\; e^{i\overline{X}s\Sigma}$$

and remembering that the group volume G in front of the integral is the integral over $d\alpha$ or over X (see ref[17] for a more formal demonstration) we have,

$$e^{i\Gamma} = \int \mathcal{D}\,(\phi, X, \overline{X}, \sigma)\; e^{i\,L} \qquad (2.14)$$

a geometric integration.

3. Supergroups

We thus observe that any Yang-Mills theory has an "enlarged" set of gauge-fields, providing a non-linear representation of the non-simple internal supergroup G x s (with s^{-1} defined by integration),

$$(A^a_\mu \ , \ D_\mu X^a)$$

and a linear representation for the matter fields

$$(\psi^n \ , \ \zeta^n)$$

Note that G x s commutes with the Poincaré group.

We now take G itself to be a supergroup[11,12], with H its even subgroup. The indices a,b,.. cover the range of H, while i,j,.. cover G/H. The connection is now locally

$$\begin{aligned} \omega^a &= X^a + dx^\mu A^a_\mu \\ \omega^i &= \eta^i + dx^\mu \xi^i_\mu \end{aligned} \qquad (3.1)$$

Fields and ghosts now appear at cross positions. We have demonstrated [12,13] that η^i is a Goldstone meson multiplet. Replacing G by $G' \supset H$ an even group conjugate to G by analytical continuation or otherwise, the η^i provide for a non-linear realization of G' over its linear subgroup H. Alternatively, in local gauge theories, the η^i become Higgs fields [15,16] as demonstrated in our suggested super-unification of SU(2) x U(1) of the electroweak interactions in SU(2/1). ξ^i_μ is a set of vector ghost fields, like in gravity. The correct method of gauging an internal super-group is not yet clarified, and it is possible that G should be treated globally only, whereas H is local. The BRST equations for H are

$$s_H \, A^a_\mu = D^H_\mu X \quad , \quad s_H \, \xi^i_\mu = D^H_\mu \eta^i \qquad (3.2)$$

and for a matter supermultiplet (ψ,μ)

$$s_H \, (\psi,\mu) = [X, (\psi,\mu)] = (\zeta,\nu)$$

where (ζ,ν) is a supermultiplet with inverted statistics. Such representations exist in the SU(m/n) or the Q(m) supergroups. For another example of an internal supergroup, see Sternberg's presentation in these proceedings of SU(5+k/1) as a super unification of the electroweak SU(2/1) with $SU(3)_{color}$ and seriality (2^k or 2^{k+1} "families" or "generations").

4. The Antisymmetric Tensor Gauge Field

The geometrical treatment recently scored a hit[17] in providing Unitarity for the Kalb-Ramond field[18] or its non-Abelian extension[19]. This theory has as Lagrangian,

$$L = B^a \wedge F_a^{A+H} - \tfrac{1}{2} m^2 \, {}^*H^a \wedge H_a - \tfrac{1}{2} \, {}^*F_A^i \wedge F_i^A \qquad (4.1)$$

$$\{a\} \in G \quad , \quad \{i\} \in G' \subset G \quad , \quad \{r\} \in G/G'$$

$$B = \tfrac{1}{2} B^a_{\mu\nu} Y_a \; dx^\mu \wedge dx^\nu \qquad (4.2)$$

$$H = H^a_\mu Y_a \; dx^\mu \qquad (4.3)$$

$$A = A^i_\mu Y_i \; dx^\mu \quad , \quad A^r = o \qquad (4.4)$$

G' is conserved locally, with A as gauge field.

$$\delta A = D\varepsilon = d\varepsilon + [A,\varepsilon] \quad , \quad \delta H = [\varepsilon,H] \quad , \quad \delta B = [\varepsilon,B] \qquad (4.5)$$

For G, A+H serves as an Abelian gauge connection, in gauging locally with a Lorentz 4-vector ξ_μ (or a one-form $\xi = \xi^a_\mu Y_a \, dx^\mu$). This is a hybrid system, since the B field is needed in addition, transforming like a gauge field,

$$\delta_G A = o \quad , \quad \delta_G H = o \quad , \quad \delta_G B = D^{A+H} \xi = d\,\xi + [A+H \, , \, \xi] \qquad (4.6)$$

The curvatures

$$F^a = dA + \tfrac{1}{2} [A,A] \qquad \text{(over } G'\text{)} \qquad (4.7)$$

$$F^{A+H} = dA + dH + \tfrac{1}{2} [A+H, A+H] \qquad \text{(over G)} \qquad (4.8)$$

transform homogeneously. $\delta_G(B^a \wedge F_a^{A+H})$ vanishes by integration by parts and the Bianchi identity $D^{A+H}F^{A+H} = o$. Varying B we get the Euler-Lagrange equation $F^{A+H} \stackrel{\Omega}{=} o$, which yields

$$D^2_{A+H} (A+H) \stackrel{\Omega}{=} o \qquad (4.9)$$

by explicit calculations. This defines a cohomology class on shell, and the physical field is just a corresponding scalar field. Solving the equation of motion by taking A+H as a Left-invariant one form $(\theta(x) \varepsilon G)$

$$H = -A + \theta^{-1} d\theta = -\theta \, (\theta^{-1} A\theta + \theta^{-1} \; d\,\theta) \, \theta^{-1}$$

produces as a 2nd order theory

$$L = \tfrac{1}{2} m^2 \, \mathrm{Tr} \, (\theta^{-1} D_\mu \theta)^2 - \tfrac{1}{4} (F^A_{\mu\nu})^2 \qquad (4.10)$$

which is a generalized sigma model with gauge. In writing the BRST equations for this theory, difficulties were originally encountered. Thierry Mieg showed[17] that the geometric approach can be applied in a straightforward manner. The ghosts arise in G'

$$\omega = A_\mu \, dx^\mu + X \qquad (4.11)$$

$$\left.\begin{array}{l} s'A = D\chi \quad , \quad s'\chi = \tfrac{1}{2}\,[\chi\,,\chi] \\ \qquad\qquad s'H = [\chi, H] \end{array}\right\} \qquad (4.12)$$

and for G (y^i is the group co-ordinate)

$$\left.\begin{array}{l} \tilde{B} = \tfrac{1}{2}\, B_{\mu\nu} dx^\mu \wedge dx^\nu + \tfrac{1}{2}\, B_{\mu i}\, dx^\mu \wedge dy^i + \tfrac{1}{2}\, B_{ij}\, dy^i \wedge dy^i \\ B_{\mu i}\, dy^i = \xi_\mu \quad , \tfrac{1}{2}\, B_{ij}\, dy^i \wedge dy^j = \eta \end{array}\right\} \qquad (4.13)$$

ξ_μ is a 4-vector (Fermi) ghost, η a scalar (Bose) "non-ghostly" ghost. Setting to zero everything but the fully horizontal components of $\tilde{D}\tilde{B}$, we get the BRST equations,

$$\left.\begin{array}{l} S'B + D^{A+H}\,\xi = o \\ S'\xi + D^{A+H}\,\eta = o \\ S'\eta = o \end{array}\right\} \qquad (4.14)$$

with S' the vertical covariant derivative in G'

$$S' = s' + [\chi,\ldots]$$

$$(s')^2 = o \quad \text{yielding} \quad F = o$$

5. The Soft Group Manifold (SGM) and Unitarity

Gravity, Supergravity and their extensions involve a geometrical construct more general than a Principal Bundle. This is G, the Soft Group Manifold (SGM) [20,21,10,11] its dimensionality is that of the gauged Lie group G (10-for Poincaré, 14-for Supersymmetry) and its tangent at any point is the group manifold G itself. G is a space in which finite motions are specified, up to topological considerations, by the Lie algebra A, with structure constants $f^a{}_{bc}$. The connections ω are the Cartan left-invariant forms, providing a rigid triangulation and a vanishing curvature

$$d\omega + \tfrac{1}{2}[\omega,\omega] = 0 \tag{5.1}$$

a condition which held only (eq. 2.8-2.11) for the $s\phi$, s_X and d_X, but not the $d\phi$ in P. P was rigid in the fiber G direction, but Soft horizontally, so that two covariant derivatives $[D_\mu, D_\nu]$ did not commute, yielding curvature terms

$$F^a_{\mu\nu} = \partial_\mu A^a_\nu - \partial_\nu A^a_\mu + [A_\mu, A_\nu]^a \tag{5.2}$$

The SGM is a triplet (G,G,σ). The group action provides a map

$$G\cdot \mathrm{G} \overset{\Omega}{\to} G, \quad g'' = \Omega(g'g) \; ; \quad g', g'' \in G, \quad g \in \mathrm{G} \tag{5.3}$$

and we have "associativity"

$$g'(g_1 g_2) = (g'g_1)g_2, \quad g' \in G; \qquad g_1, g_2 \in \mathrm{G} \tag{5.4}$$

The connection ρ is an A-valued one-form over G. Besides providing a set of frames, it provides a map from $G_\star$ onto A

$$\rho : G_\star \to A \; ; \; \forall\, \mathfrak{t} \in G \quad \mathfrak{t} \,\lrcorner\, \rho = \rho(\mathfrak{t}) = \Lambda, \quad \Lambda \in \mathcal{A} \tag{5.5}$$

Note that whereas the right group action by G still involves A, the connection yields operations in $\mathcal{A} \neq$ A. We define a basis in $P_\star$ or $G_\star$ (see examples in section 3 of ref.[21])

$$\rho^A(\mathfrak{X}_B) = \delta^A_B, \quad \mathfrak{X} \in G_\star \tag{5.6}$$

The structural equations now provide a right-hand-side to (5.1),

$$d\rho + \tfrac{1}{2}[\rho,\rho] \equiv R \tag{5.7a}$$

or in components

$$R^A = d\rho^A + \tfrac{1}{2} f^A{}_{BC}\, \rho^B \wedge \rho^C \tag{5.7b}$$

Projecting R^A onto the ρ, since they provide a basis in G^*, we can also write

$$d\rho^A + \tfrac{1}{2}(f^A{}_{BC} - R^A{}_{BC})\, \rho^B \wedge \rho^C = 0 \tag{5.7c}$$

The Bianchi identity holds,

$$(DR)^A = 0 \tag{5.8}$$

where the covariant derivatives are defined over the entire G,

$$(D\eta)^A = d\eta^A + f^A{}_{BC}\,\rho^B \wedge \rho^C \tag{5.9a}$$

$$(D\eta)_A = d\eta_A - f^C{}_{BA}\,\rho^B \wedge \eta_C \tag{5.9b}$$

$$(D(D\eta))^A = f^A{}_{BC}\,R^B \wedge \eta^C \tag{5.10}$$

From (5.6) and (5.7c) we get $\tilde{\lambda}$

$$[\tilde{\lambda}_B, \tilde{\lambda}_C] = (f^A{}_{BC} - R^A{}_{BC})\tilde{\lambda}_A \tag{5.11}$$

reducing to $\tilde{\lambda}$ when $R^A = 0$.

We now evaluate $\delta\rho^A$ for a local transformation. Denoting by z^M $(M = 1\ldots n)$ a local co-ordinate patch over G, we have[21]

$$\delta\rho^A = \delta(dz^M \rho^A_M) = dz^M\ \delta\rho^A_M + \delta dz^M\ \rho^A_M \tag{5.12a}$$

$$\delta\rho^A = dz^N \frac{\partial\varepsilon^M}{\partial z^N}\rho_M{}^A + dz^M \varepsilon^N \frac{\partial\rho^A_M}{\partial z^N} \tag{5.12b}$$

where we define an Anholonomized General Co-ordinate Transformation (AGCT)

$$(\delta z)^A = \varepsilon^A(z) \tag{5.13}$$

Expanding the $\tilde{\lambda}_A$ basis over the $(\partial/\partial z^M)$, and the ρ^A over the dz^M,

$$\tilde{\lambda}_A = \tilde{\lambda}_A{}^M \frac{\partial}{\partial z^M}\ , \qquad \rho^A = \rho^A{}_M dz^M$$

$$\rho^A{}_M \tilde{\lambda}_B{}^M = \delta^A{}_B\ , \qquad \rho^A{}_M \tilde{\lambda}_A{}^N = \delta_M{}^N$$

($\rho^A{}_M(z)$ and $\tilde{\lambda}_A{}^M(z)$ act like generalized tetrad fields and their inverses.)

The holonomic variations can be deduced from (5.13) by multiplying both sides by $\tilde{\lambda}_A{}^M(z)$,

$$\delta z^M = \varepsilon^M(z) = \varepsilon^A \tilde{\lambda}_A{}^M$$

$$\varepsilon^A(z) = \varepsilon^M \rho^A{}_M$$

$$\delta dz^M = d\varepsilon^M(z)$$

so that (5.12a) can be rewritten as

$$\delta\rho^A = dz^M \delta\rho^A_M + d\varepsilon^M \rho^A_M \tag{5.13a}$$

Returning to (5.12), we may now write

$$\delta\rho^A = dz^N \left[\frac{\partial\varepsilon^A}{\partial z^N} + \varepsilon^M \left(\frac{\partial\rho^A_N}{\partial z^M} - \frac{\partial\rho^A_M}{\partial z^N} \right) \right]$$

$$= d\,\varepsilon^A + \tilde{\varepsilon} \lrcorner\, d\rho^A\ , \tag{5.14}$$

where $\tilde{\varepsilon} = \varepsilon^M(\partial/\partial z^M)$. Using

$$\delta\rho^A = D\varepsilon^A + \tilde{\varepsilon} \lrcorner R^A = D\varepsilon^A + \varepsilon^B \rho^C R_{BC}{}^A \qquad (5.15)$$

This action on ρ^A is by definition the Lie-derivative with respect to the vector field $\tilde{\varepsilon}$

$$L_{\tilde{\varepsilon}}\,\rho = d\,\varepsilon + \tilde{\varepsilon} \lrcorner d\,\rho = D\,\varepsilon + \tilde{\varepsilon} \lrcorner R \equiv \delta\rho \qquad (5.16)$$

and we have

$$L_{\tilde{\varepsilon}}\,R = D\,(\tilde{\varepsilon} \lrcorner R) - [\varepsilon, R] \qquad (5.17)$$

Local AGCT "gauge" transformations on G are thus represented by the Lie derivatives and A or $\tilde{A}$ will have the structure functions $F^A{}_{BC}(A^D_\mu)$

$$F^A{}_{BC}\,(A_\mu) = f^A{}_{BC} - R^A{}_{BC}$$

as determined by (5.6) and (5.11). The transformations of (5.16) have been used in Gravity, with $\mathcal{J}^A$ restricted to space-time. In Supergravity[20,21], these are the "local supersymmetry transformations", when $\tilde{\varepsilon}$ is a spinor $\tilde{\varepsilon}^\alpha$.

We now have to check the group property ("closure") for these variations, since many of the difficulties in Supergravity resulted from the "lack of closure" of the algebra. We extract from (5.12, 5.16) the variation for the gauge field ρ^A_M, more commonly required by physicists, (though still unprojected over space-time). This is the usual result for a general co-ordinate transformation (GCT), including an added passive transformation $\delta\mathcal{J}^A = 0$ guaranteeing a return to the original value of z^M. This was not needed in (5.12) where we were dealing with a form. Here it yields the first term $\delta^0 \mathcal{J}^A_M$ in (5.18)

$$\delta\rho^A{}_M = \delta^0 \mathcal{J}^A_M + \delta\,\mathcal{J}^A_M = (\partial_M \varepsilon^N)\rho^A{}_N + \varepsilon^N \partial_N \rho^A{}_M \qquad (5.18a)$$

We calculate the commutator

$$\tilde{\delta}_3\, \rho_M{}^A \qquad (5.19)$$

Using

$$\tilde{\delta}\varepsilon^A = \delta(\varepsilon^N \rho_N{}^A) = 0 \qquad (5.20)$$

the closure of the algebra is indeed given by (5.11)

$$\varepsilon^A_3 = \varepsilon^N_3\, \rho^A_N = (f^A{}_{BC} - R^A{}_{BC})\, \varepsilon^B_2\, \varepsilon^C_1 \qquad (5.20a)$$

which confirms the identification of $L_{\tilde{A}}$ with $\tilde{\lambda}_A$ even when "gauged" according to (5.18). Note that "non-closure" in Supergravity is a reading of (5.20) in which the $R^A{}_{BC}$ have not been identified, partly because of incomplete dimensionality.

Algebraic closure, which we achieved in (5.20) for the algebra (5.11) is essential to "geometric" closure, i.e. Poincaré conditions such as $d^2=0$ or $s^2=0$/

Space-time (the parameter space of translations P_a in the Poincaré group) is a submanifold of the SGM. Gauging has been understood in those cases in which A, the algebra of G, is Weakly-Reducible. This implies the existence of a WR decomposition

$$\left.\begin{array}{l} A = F + H \;;\;\; [F,F] \subset F \\ \quad [F,H] \subset H \end{array}\right\} \qquad (5.21)$$

In such a geometry, one can impose the Cartan-Maurer equations for the subgroup F as a constraint. All curvatures are reduced to 2-forms over G/F. The coefficients $R^A_{||}$ and $R^A_{|-}$ of $\mathcal{J}^f \wedge \mathcal{J}^{f'}$ and $\mathcal{J}^f \wedge \mathcal{J}^h (f,f' \in \{F\}\;,\; h \in \{H\})$ can be made to vanish, where $\mathcal{J}^{ab}$ represents an (anholonomic) Lorentz (or F-) connection and $\mathcal{J}^a$ stands for a horizontal connection. The algebra in (5.11) for $A,B,C \in \{F\}$ is identical to F itself.

In $\overline{ISO}$ (1,3), $G/F = M^{1,3}$ (Minkowski space), but in $g\overline{ISO}$ (1,3), $G/F = M^{1,3/4}$ known as Superspace. It has four Grassmann (and Majorana spinor component) dimensions besides space-time. The nilpotence of Grassmann elements allows an expansion in powers of the Grassmann generating element θ^α for any "superfield" so that one can still deal unambiguously with fields over space time as coefficients of the $(\theta^\alpha)^r$. For extended supersymmetry the nilpotence appears only at $r > 4n$, n being the dimension of the internal symmetry O(n) vector. Our treatment is limited at this stage to cases in which the bosonic submanifold of G/F is $M^{1,3}$, or S^4 in pre-Wigner-Inönu contraction situations (with G a simple group such as $\overline{SO}$ (1,4) or OSp (4/N).

In the renormalization of quantum field theories, constraints such as explicit symmetry violations have caused difficulties. These were removed by <u>spontaneous</u> symmetry breakdown, i.e. when the breakdown resulted from the equations of motion (and described the on-mass shell situation only). Rather than apply SGM factorization as a constraint, we have described the conditions allowing for its occurring spontaneously[11]. This requires A to obey in addition to (5.21), a (so-called) "symmetric decomposition" WRS,

$$[H,H] \subset F \qquad (5.22)$$

If in addition A is Simple (WRSS), and if the Lagrangian L_{INV} is F-gauge invariant (but not G-gauge invariant) and of the form (A,B span the entire A)

$$L_{INV} = R^A \wedge h_{AB} R^B \;, \qquad D^{(F)} h_{AB} = 0 \qquad (5.23)$$

where h_{AB} is a regular (F-invariant) 2-tensor under G one has pseudo-closure (in the geometric sense)

$$dL_{INV} = DL_{INV} = D^{(F)}\, L_{INV} \overset{0}{=} 0 \qquad (5.24)$$

where $\overset{0}{=}$ denotes equality modulo the equations of motion. In addition,

$$R^h \overset{0}{=} 0 \;,\quad h \in \{H\} \qquad (5.25)$$

This vanishing of the torsions (curvatures in the H directions) is always true for $R^h_{||}$ i.e. the $R^h{}_{f'f}$ where $h \in H$; f, $f' \in F$, all anholonomic. It is true for $\overline{O}(1,4)$ and OSp(4/N) for all components, including $R^h_{|-}$ and $R^h_{=}$. Moreover, for WRS in which [H,H] spans the entire F ($R^A_{|-}$ is $R^A{}_{fh}$),

$$R^A_{||} \mathrel{\underset{\sim}{\Omega}} 0\,, \qquad R^A_{|-} \mathrel{\underset{\sim}{\Omega}} 0\,, \qquad A \in \{F\} \tag{5.26}$$

i.e. we get factorization, all relevant forms being only over H.

Gravity and Supergravity can be written as in (5.23) in an uncontracted form and arise in their precise form through a Wigner-Inönu contraction. This discussion then fits both cases and one can write on-shell BRST equations as for a Principal Bundle. However, BRST are mainly needed to prove Unitarity and this is an off-mass-shell problem (see for example ref.[22])

Off-mass-shell, the SGM is not a Principal Bundle and we are not allowed to use the results of sections 2-3 whether or not Spontaneous Fibration has occurred.

We shall apply the algebra $\tilde{A}$ (5.22) of Lie derivatives (5.16) as the off-mass-shell local gauge (for the horizontal directions $H \subset G$, this will remain different from $\tilde{A}$ even after fibration). We treat the connections and curvatures as forms over the full dimensionality of G. The indices U,V and R,S respectively denote holonomic variables over G/F and over F. G/F is a submanifold larger or equal to space-time. We expand $\mathcal{J}^A$

$$\mathcal{J}^A = dz^U \tau_U^A + dz^R \psi_R^A = \tau^A + \psi^A \tag{5.27}$$

$$\mathcal{J}^M = \lambda^M_A \mathcal{J}^A = \lambda^M_A (dz^U \tau_U^A + dz^R \psi_R^A)$$

and since

$$\mathcal{J}^M = dz^N \lambda^M_A \mathcal{J}^A_N = dz^N \delta^M_N = dz^U \delta_U^M + dz^R \delta_R^M$$

we have

$$\lambda^U_A \tau^A_V = \delta^U_V\,, \quad \lambda^R_A \psi^A_S = \delta^R_S$$

so that

$$\left.\begin{aligned} \tau^U &= dz^U, \ \tau^R = 0 \\ \psi^U &= 0\,, \ \psi^R = dz^R \end{aligned}\right\} \tag{5.28}$$

The structural equations for τ^A and ψ^A can be read directly from (5.7c)

$$\left.\begin{aligned} z\psi^A &= -\tfrac{1}{2}[\psi,\psi]^A + R^A{}_{BC}\,\psi^B \psi^C \\ z\tau_U^A &= D_U \psi^A - R^A{}_{BC}\,\psi^B \tau^C_U \end{aligned}\right\} \tag{5.29}$$

All indices are anholonomic except for U. Note that just as in an internal gauge, the action of s is of an "alibi" type, i.e. it represents an active vertical group translation.

We have proved[23] that the correspondence we had in an internal gauge theory between structural equations and BRST transformations is preserved: that equations (5.29) reproduce the BRST equations fitting the Lie-derivative (or AGCT) gauge of (5.15), provided we use the identification (1.7).

Note that the curvature term $R^{f}_{BC}\ \tau^{C}_{U} = R^{f}_{BU}$ can be considered as an auxiliary field over G/F.

We note that for any form σ

$$L_{\tilde{\varepsilon}}\ \sigma = \tilde{\varepsilon} \lrcorner\ d\sigma + d\ (\tilde{\varepsilon} \lrcorner\ \sigma)$$

Thus for the Lagrangian

$$L_{\tilde{\varepsilon}} \mathcal{L} = \tilde{\varepsilon} \lrcorner\ d\mathcal{L} + d\ (\tilde{\varepsilon} \lrcorner \mathcal{L}) \qquad (5.30)$$

The second term vanishes upon integration (as a divergence) but the first term vanishes only after the fibration, when $d\mathcal{L}$ vanishes because it becomes a 5-form over $M^{1,3}$. This thus weakens the applicability of (5.29) in the geometric approach, since it requires fibration anyhow. However, this does not interfere with our first aim, which was to check that the BRST equations on the SGM are indeed given again by the structural equations.

Alternatively, one has added to $\mathcal{L}$ a quartic ghost term[24,22], apparently cancelling the vertical piece of $\mathcal{L}_{INV}$ over the SGM. We have not yet studied this method geometrically.

References

1. R.P.Feynman, Acta Phys. Polon. 26, 697 (1963).
2. B.S.De Witt, "Dynamical Theory of Groups and Fields", Gordon & Breach Pub., N.Y./London/Paris, 1965.
3. L.D.Faddeev and V.N.Popov, Phys. Lett. B25, 29 (1967).
4. G.'t Hooft, Nucl. Phys. B33, 436 (1971).
5. A.A.Slavnov, Teor. Mat. Fiz. 10, 153 (1972)
   J.C.Taylor, Nucl. Phys. B33, 436 (1971).
6. C.Becchi, A.Rouet and R.Stora, Com. Math. Phys. 42, 127 (1975)
   I.V.Tyutin, rep. FIAN 39 (1975).
7. J.Thierry-Mieg, These de Doctorat d'Etat (Paris-Sud) 1978.
8. J.Thierry-Mieg, J.Maths. Phys.
9. J.Thierry-Mieg, Nuovo Cim. A.
10. Y.Ne'eman, Proc. XIX Int. Conf. HEP (1978), S.Homma et al eds., Phys. Soc. Japan Pub. Tokyo (1979), p.552.
11. J.Thierry-Mieg and Y.Ne'eman, Ann. of Phys. (N.Y.) 123, 247 (1979).
12. Y.Ne'eman and J.Thierry-Mieg, Proc. Nat. Acad. Sci. USA 77. 720 (1980).
13. I.Ojima, preprint.
14. Y.Ne'eman and J.Thierry-Mieg, Proc. VIII Int. Conf. on App. of Group Theory to Phys. (Kiryat Anavim 1979), Ann. Isr. Phys. Soc. 3, 100 (1980).
15. Y.Ne'eman, Phys. Lett. B81, 190 (1979).
16. Y.Ne'eman and J.Thierry-Mieg, Proc. Salamanca (1979). Int. Conf. Diff. Geom. Methods in Phys. A.Perez-Rendon ed., Springer Verlag L.N. in Math.
17. J.Thierry-Mieg, Harvard rep. HUTMP 79/B86.
18. M.Kalb and P.Ramond, Phys. Rev. D9, 2273 (1974)
    E.Cremmer and J.Scherk, Nucl. Phys. B72, 117 (1974).
19. D.Z.Freedman, rep. CALT 68-624 (1977) unpub.
    P.K.Townsend, CERN th. 2753 (1979).
20. Y.Ne'eman and T.Regge, Phys. Lett. 74B, 54 (1978).
21. Y.Ne'eman and T.Regge, Rivista d. Nuovo Cm. III, 1 n.5 (1978).
22. G.Sterman, P.K.Townsend and P.van Nieuwenhuizen, Phys. Rev. D17, 1501 (1978).
23. Y.Ne'eman, E.Takasugi and J.Thierry-Mieg, Phys. Rev. D (to be pub.).
24. R.E.Kallosh, JETP Lett. 26, 575 (1977).

COLOR IN INTERNAL SUPERSYMMETRY UNIFICATION.

by

Yuval Ne'eman*↑ and Shlomo Sternberg†↑

Tel Aviv University

Tel Aviv

Israel

*also University of Texas, Austin

†also Harvard University, Cambridge, Mass.

Einstein Fellow, Israel National Academy of Sciences and Humanities.

↑Supported in part by the Wolfson Chair Extraordinary in Theoretical Physics at Tel Aviv University, by the United States-Israel Binational Science Foundation, by the USDOE contract DE/AS0278ER04742 and EY-76-S-05-3992 and by the Israel National Academy of Sciences and Humanities.

In [1], a scheme was introduced for determining the charge of the quarks and leptons in successive generations: A family of representations of the superalgebra sl(n/1) was constructed, depending on a complex parameter, b. For each b, the underlying space of the representation is $\Lambda(\mathbb{C}^n)$, the exterior algebra of the standard n dimensional vector space. Under the even part of $s\ell(n/1)$ the space $\Lambda(\mathbb{C}^n)$ decomposes into the direct sum of the i-th exterior powers, $\Lambda^i(\mathbb{C}^n)$, with $0 \leq i \leq n$. Let us write a diagonal matrix, U, in this even part as

$$U = \mathrm{diag}\,(c_1, \ldots c_n/T)$$

where the c_i and T are the diagonal entries of U and the condition that U be supertraceless requires that

(1) $$T = c_1 + \ldots + c_n$$

Then the eigenvalues of the operator corresponding to U on $\Lambda^i(\mathbb{C}^n)$ are

(2) $$c_{j_1} + \ldots + c_{j_i} + (b-i)T \quad , \quad j_1 < \ldots < j_i \quad .$$

The third component of the weak Isospin was introduced in the form

(3) $$I_3 = \mathrm{diag}\,(\tfrac{1}{2}, -\tfrac{1}{2}, 0, \cdots, 0/0)$$

We choose the weak hypercharge U to commute with I , with the electric charge given by the standard formula $Q = I_3 + \frac{1}{2}U$. To reproduce the observed values of electric charge, the eigenvalues of U must be restricted to 2,4/3,1,2/3,1/3,0, -1/3,-2/3,-1,-4/3 and -2. In 1 it was shown how to choose U with these properties and the eigenvalue 2 in Λ^0 and the eigenvalue -2 in Λ^n, under the hypothesis of three colors, i.e. r = 3. We show here that the number of colors is fixed by the observed charges and the hypothesis that U is 2 in Λ^0 and -2 in Λ^n. Indeed, it follows from (1) and the construction in 1 , that the eigenvalues of U on Λ^0 and Λ^n are bT and (b - (n-1)T respectively. If 2 is to occur in Λ^0 and -2 in Λ^n this requires that (n-1)T = 4 and bT = 2 so

(4) $$T = \frac{4}{n-1}$$

and

(5) $$b = \tfrac{1}{2}(n-1)$$

The leptonic eigenvalues of U, 1 and -1 must reccur in the I doublets in ΛV, to accommodate the e^+ and e^- and ν_L, $\overline{\nu}_L$ of $1 = \frac{1}{2} + \frac{1}{2}$, $0 = \frac{1}{2} - \frac{1}{2}$, $0 = -\frac{1}{2} + \frac{1}{2}$, $-1 = -\frac{1}{2} - \frac{1}{2}$. The eigenvalues 4/3 and 2/3 must occur as I_3 singlets, to accomodate the quark charges of 2/3 and 1/3, Similarly, 0 must occur as an I_3 singlet.

Let us now examine the eigenvalues on Λ^1, reflecting by (2) the linear action of (3). An I_3 doublet appears. It must correspond to U = 1 eigenvalues, since the value 1 must occur as an I_3 doublet somewhere, and is not allowed to occur for an isosinglet. Using eq. (2) systematically, one finds that there is no choice of the eigenvalues for Λ^1 isosinglets that would produce U = 1 for an isodoublet anywhere else in ΛV, if we select in Λ^1 any one of the alternative allowed isodoublet assignments (1/3, -1/3 or -1). Thus the I_3 doublet eigenvalues of U in Λ^1 are 1.

The eigenvalue 1/3 is not allowed to occur as a singlet in Λ^1 since there is no corresponding physical system. The eigenvalue 4/3 must occur, reasoning as in the case of U = 1. No eigenvalues smaller than 1/3 can occur as eigenvalues for Λ^1 because this would imply wrong eigenvalues somewhere else in ΛV.

The eigenvalues of U are thus:

(1,1)	occuring once
(2)	occuring say s times, $s \geq 1$
(4/3)	occuring say k times, $k \geq 1$
(2/3)	occuring say l times, $l \geq 0$

yielding for the trace of U on Λ^1 the value

$$\mathrm{tr}(U)_{\Lambda^1} = 2 + 2s + 4/3\, k + 2/3\, l = (n(b-1)+1)T$$

but $2 + s + k + l = n$, and $(n(b-1)+1)T = 2n - 4$ by equations (4) and (5).

As a result:

(6) $2/3\ k + 4/3\ l = 2$

The only solutions of eq.(6) for non-negative integer k and l are

(7) $k = l = 1$

or

(8) $k = 3, \quad l = 0$

If we interpret k as the number of colors, then solution (7) corresponds to only one color degree of freedom (i.e. color is reduced to quark number, at most). This solution can occur for example in $s\ell(4/1)$: Taking $b = 3/2$ and

$$U = \text{diag}\ (1/3,1/3,2/3,0/4/3)\ .$$

The eigenvalues of U are then

$\Lambda 0$	$\Lambda 1$				$\Lambda 2$						$\Lambda 3$				$\Lambda 4$
2	1	1	4/3	2/3	0	1/3	1/3	-1/3	-1/3	0	-1	-1	-2/3	-4/3	-2

Thus $s = 0$ and $r = 1$ and all the right eigenvalues occur.

The case (8) corresponds to 3 colors, i.e. we interpret U as commuting with an su(3) which sits as a single block-diagonal in $s\ell(n/1)$. There are then altogether only three distinct eigenvalues for U acting on V, with multiplicities 2, s and r respectively. Furthermore, s gives rise to 2^s generations.

The charge-symmetric subset of the representations we constructed in [1] thus allows "one" or three colors only. The number three can of course be traced to the denominator of the quark charges i.e. to the original classification su(3). Ofer Gabber has pointed out to us a more fundamental reason why we should expect (5) to hold: the structure of the anticommutation relations for the creation and annihilation operators for the particles, together with the requirement that conjugate particles occur in Λ^i and Λ^{n-i}, requires that there exist a bilinear pairing between Λ^i and Λ^{n-i}, invariant under the even part of $s\ell(n/1)$. In particular, eigenvalues of diag (1,1, ..., 1/n) on these two spaces must be the negatives of one another. It follows from (2) that $b = \frac{1}{2}(n-1)$. We should remark, that under this condition, there exists a bilinear form invariant under the whole superalgebra. Indeed, as discussed in [1], the space Λ^i is really $\Lambda^i \otimes F^{b-i}$ where F^{b-i} denotes the one dimensional space of homogeneous functions of degree b-i. Exterior multiplication on the Λ factor and ordinary multiplication on the F factor gives a bilinear map of $(\Lambda^i \otimes F^{b-i}) \times (\Lambda^{n-i} \otimes F^{b-(n-i)})$ into $\Lambda^n \otimes F^{-1}$ if (5) holds. But, under the action of the even part of $s\ell(n/1)$, this space has a canonical identification with the complex numbers. We thus get an invariant bilinear pairing on the representation space $\Lambda \otimes F$. It is straightforward to check that this is invariant (in the superalgebra sense) under the entire superalgebra $s\ell(n/1)$.

[1] Y. Ne'eman and S. Sternberg, "Internal supersymmetry and unificiation", to appear in the Proc. of the Nat. Acad. of Sci.

CANONICAL (POSSIBLY LAGRANGIAN) REALIZATIONS OF THE POINCARE' GROUP WITH INCREASING MASS-SPIN TRAJECTORIES

Massimo Pauri

Istituto di Fisica della Università - 43100 PARMA (Italy)
Istituto Nazionale di Fisica Nucleare - Sezione di Milano

I summarize here the main results concerning a class of phase-space realizations of the Poincarè group (P) which, as far as I know, have never been described in the literature explicitly. The present approach shares the form of a "traditional" exposition in classical relativistic mechanics so it might appear quite naive from the point of view of modern global techniques. A thorough discussion of the whole matter from the geometrical point of view will be given subsequently with particular attention to the possible definition of a global Lagrangian structure, relevant for a sound quantization procedure. The work is based on a technique for the local construction of the most general canonical realization (C. R.) of a Lie group expounded by the author in a series of papers [1]. This technique generalizes to an arbitrary number of degrees of freedom a method (essentially due to S.Lie) for the construction of local charts of symplectic coordinates over the orbits of the co-adjoint representation in the dual space of the Lie algebra. The basic framework here is the Dirac "instant form" of relativistic dynamics; dynamical variables (D.V.) are then defined at coordinate time $t=0$ for all the equivalent Lorentz observers and their time evolution is derived by the Hamilton equations. $H, \vec{P}, \vec{J}, \vec{K}$ denote the infinitesimal canonical generators of P (P.C.G.) with the usual meaning of symbols.

<u>Statement of the problem</u>: Find (if any) the most general, minimal (minimum number of degrees of freedom), canonical realization of P (P.C.R.) compatible with the existence of a phase-space D.V. $\vec{x}(q,p)$ for the relativistic position such that it

a- is a 3-vector under space rotations, space translations, canonical space reflection and anti-canonical time reflection [1]

$$\{J^i, x^j\} = \varepsilon^{ijk} x^k \quad , \quad \{P^i, x^j\} = -\delta^{ij} \quad , \quad i,j,k = 1,2,3 \; ;$$
$$I_s \vec{x} = -\vec{x} \quad , \quad I_t^* \vec{x} = \vec{x} \quad ;$$

b- satisfies the "world-line condition", i.e. it transforms as the $t=0$ section, for each Lorentz frame, of an invariantly defined world-line (w.l.)

$$\{K^i, x^j\} = \frac{1}{c^2} x^i \{x^j, H\} \quad , \quad i,j = 1,2,3 \; ;$$

c- is "Lagrangian" in the sense that

$$\{x^i, x^j\} = 0 \quad , \quad i,j = 1,2,3 .$$

A more or less similar program has been considered by Jordan and Mukunda [2] in connection with the finite, manifestly covariant, unitary reps. of P.

Recall the status of the basic assumptions b and c with reference to the relevant, already known, relativistic position 3-vectors (they are functions of the P.C.G. only and are consequently definable in all P.C.R.), actually the Møller center-of-mass $\vec{q}$, the Fokker-Pryce center-of-inertia $\vec{X}$ and the Newton-Wigner-Pryce center-of-spin $\vec{Q}$ see Refs. [3,4,5]. If $Mc = (P_\lambda P^\lambda)^{1/2}$ and $\vec{S}$ is the Poincarè canonical spin (constant of the motion)

$$\vec{S} = \vec{J} - \vec{Q}\wedge\vec{P} = \frac{H}{Mc^2}\vec{J} + \frac{\vec{K}\wedge\vec{P}}{M} - \frac{c^2\,\vec{J}\cdot\vec{P}}{Mc^2(Mc^2+H)}\vec{P},$$

we have

$$\vec{X} = -\frac{c^2}{H}\vec{K} + \frac{\vec{S}\wedge\vec{P}}{MH} = \vec{q} + \frac{\vec{S}\wedge\vec{P}}{MH} = \vec{Q} + \frac{\vec{S}\wedge\vec{P}}{M(Mc^2+H)}$$

and

D.V.	$\vec{q}$	$\vec{X}$	$\vec{Q}$
condition b satisfied	no	yes	no
condition c satisfied	no	no	yes

.

Note that the non Lagrangian character of the "covariant" $\vec{X}$ is a spin effect

$$\{X^i, X^j\} = \frac{1}{MH}\varepsilon^{ijk}\left[S^k + \frac{\vec{S}\cdot\vec{P}}{M(Mc^2+H)}P^k\right], \quad i,j,k = 1,2,3.$$

It can be shown [3] that the (non invariant) w.ls. associated to $\vec{q}$ by the whole set of Lorentz observers fill a world-tube with the $\vec{X}$ w.l. as axis and radius S/Mc . This non-locality, typical of the irreducible P.C.R., suggests that these are actually providing only a mean description of something like a hidden structure, just as it happens at the quantum level.

The problem stated above has a definite solution: the main properties of the resulting class of P.C.R., which I would call Minkowski P.C.R., can be summarized as follows:

1) The Minkowski P.C.R. describe a relativistic obiect having 5 degrees of freedom. The magnitude S of the canonical spin does not have a fixed value. The 10-dim. phase-space is invariantly foliated in submanifolds S = const.,i.e. these P.C.R. are necessarily non-irreducible in the sense of Refs.[1]. There exists a local chart in which S itself plays the role of a canonical variable conjugated to an "internal" angle α which is functionally independent of the P.C.G. [1](it is a coordinate over the dual of the SO(2) subalgebra of the Cartan subalgebra).

2) A definite mass-spin relation $M = M(S)$ must necessarily exist. Since the canonical Hamiltonian has the form

$$cP_0 = H = c\sqrt{M^2(S)c^2 + \vec{P}^2},$$

the angle $\alpha(t)$ evolves linearly with time.

3) The explicit expression of the new D.V. satisfying conditions a,b,c is

$$\vec{x} = \vec{X} + \lambda_0(S)\vec{R}[\alpha+\theta] - \frac{c^2\lambda_0(S)\,\vec{P}\cdot\vec{R}[\alpha+\theta]}{H(Mc^2+H)}\vec{P},$$

where $\vec{R}[\alpha]$ is a definite 3-vector, function of $\vec{S}$ and α, such that $\vec{R}[\alpha]\cdot\vec{S} = 0$, $|\vec{R}| = 1$, $\lambda_0(S)$ is the Poincarè invariant

$$\lambda_0(S) = \sqrt{S^2 - I}/M(S)c, \quad (I \gtrless 0, \text{ integration constant}),$$

and θ is an angle dynamically generated by the boost from the center-of-momentum frame Σ_o ($\vec{P} = 0$) to the laboratory Σ , which is implicitly defined by

$$\theta + \frac{\omega_o \lambda_o}{H} \vec{P} \cdot \vec{R}\,[\alpha + \theta] = 0\,, \quad \left(\omega_o \equiv c^2 \frac{dM(s)}{ds}\right).$$

4) The kinematics of $\vec{x}(t)$ can be worked out explicitly in the form

$$\vec{x}(t) = \vec{X} + \frac{c^2 t}{H}\vec{P} + \cos[\eta(t) + \tilde{\alpha}]\,\frac{\vec{S}}{S}\wedge \lambda_o(s)\,\vec{R}[\alpha] -$$
$$- \sin[\eta(t) + \tilde{\alpha}]\,\lambda_o(s)\vec{R}[\alpha] + \sin\eta(t)\,\frac{c^2|\vec{S}\wedge\vec{P}|\,\lambda_o(s)}{SH(Mc^2+H)}\,\vec{P}$$

where $\tilde{\alpha} = \tan^{-1}\left[S(\vec{S}\wedge\vec{P})_z / [(\vec{S}\cdot\vec{P})S_z - S^2 P_z]\right]$ and $\eta \equiv \alpha + \theta - \tilde{\alpha} - \frac{\pi}{2}$ is a sort of "eccentric anomaly" satisfying the Kepler-like equation

$$\eta(t) - \varepsilon \frac{V_o}{c} \sin\eta(t) = \omega t \qquad (1)$$

where $V_o(s) = \lambda_o \omega_o$, $\varepsilon = \frac{V_P}{c}\frac{\vec{S}\wedge\vec{P}}{SP}$, $V_P = \frac{c^2 P}{H}$, $\omega = \frac{Mc^2}{H}\omega_o$.

The motion of $\vec{x}(t)$ results from the superposition of the inertial drift of the center-of-inertia $\vec{X}$ along the total linear momentum $\vec{P}$ (constant of the motion) and an inner elliptical revolution taking place on a uniformly translating plane orthogonal to the space part of the Pauli-Lubansky 4-vector W^μ ,with center at $\vec{X}$, major semi-axis of lenght $\lambda_o(s)$ and direction of $\vec{S}\wedge\vec{P}$, and eccentricity ε ;(note that $\vec{\rho}$ lies at a focus if $I = 0$). The oscillating radius of this classical zitterbewegung is

$$\lambda = |\vec{x} - \vec{X}| = \sqrt{\lambda_o^2(s) - c^2\theta^2/\omega_o^2(s)}$$

and reduces to $\lambda_o(s)$ in the frame Σ_o , where the velocity 3-vector

$$\vec{v} \equiv \{\vec{x}, H\} = \frac{c^2}{H}\vec{P} + \frac{\partial\theta}{\partial\alpha}\frac{Mc^4}{H(Mc^2+H)}\vec{P} + \left(1 + \frac{\partial\theta}{\partial\alpha}\right)\frac{V_o M c^2}{H}\frac{\vec{S}}{S}\wedge\vec{R}[\alpha+\theta]$$

is simply $\vec{V}_o = V_o \frac{\vec{S}}{S}\wedge\vec{R}[\alpha]$,

corresponding to a steady circular motion with linear velocity V_o.

Putting

$$\gamma \equiv \left(1 - \frac{v^2}{c^2}\right)^{-1/2}, \quad \gamma_o \equiv \left(1 - \frac{V_o^2}{c^2}\right)^{-1/2}, \quad \gamma_P \equiv \left(1 - \frac{V_P^2}{c^2}\right)^{-1/2},$$

it follows the "addition law"

$$\gamma = \gamma_o \gamma_P / \left(1 + \frac{\partial\theta}{\partial\alpha}\right).$$

5) Defining the covariant spin

$$\vec{\sigma} = \vec{J} - \vec{x}\wedge\vec{P}\,, \qquad \left(\vec{\sigma}\big|_{\vec{P}=0} = \vec{S}\right)$$

and the corresponding time part

$$\vec{\tau} = c\vec{K} + \frac{H}{c}\vec{x}\,, \qquad \left(\vec{\tau}\big|_{\vec{P}=0} = Mc\lambda_o\vec{R}[\alpha]\right),$$

it follows

$$\frac{d\vec{\sigma}}{dt} \equiv \{\vec{\sigma}, H\} = -\vec{v}\wedge\vec{P}\,, \quad \frac{d\vec{\tau}}{dt} \equiv \{\vec{\tau}, H\} = \frac{H}{c}\vec{v} - c\vec{P}$$

with $h \equiv \vec{\sigma}\cdot\vec{P}/P = \vec{S}\cdot\vec{P}/P$ <u>constant of the motion</u>. As for $\vec{x}(t)$, the complicated precessional motion of $\vec{\sigma}(t), \vec{\tau}(t)$ can be worked out explicitly. The relevant algebraic relations of the spin D.V. are

$$\{\vec{x}, \sigma^k\} = \{\vec{x}, \tau^k\} = \{\vec{P}, \tau^k\} = \{\vec{P}, \sigma^k\} = 0,$$
$$\{\sigma^i, \sigma^j\} = \varepsilon^{ijk}\sigma^k, \{\sigma^i, \tau^j\} = \varepsilon^{ijk}\tau^k, \{\tau^i, \tau^j\} = -\varepsilon^{ijk}\sigma^k,$$
$$i, j, k = 1, 2, 3.$$

Therefore the (instant form) Poincarè tensor $\sigma^{\mu\nu} \equiv [\sigma^{ok} \equiv \tau^k, \sigma^{ij} \equiv \varepsilon^{ijk}\sigma^k]$ generates a C.R. of the Homogeneous Lorentz Group [1] with invariant foliation $\vec{x}$ = const., $\vec{P}$ = const. The values of the Casimir invariants are

$$I_1 \equiv \frac{1}{2}\sigma_{\mu\nu}\sigma^{\mu\nu} = \vec{\sigma}^2 - \vec{\tau}^2 = S^2 - M^2c^2\lambda_o^2 = I$$
$$I_2 \equiv \frac{1}{2}\tilde{\sigma}_{\mu\nu}\sigma^{\mu\nu} = \vec{\sigma}\cdot\vec{\tau} = Mc\lambda_o\vec{S}\cdot\vec{R}[\alpha] = 0,$$

therefore, for given $\vec{x}$, $\vec{P}$, the C.R. is irreducible. There exists a canonical transformation connecting the symplectic chart defined by $\vec{P}$ $\vec{x}$ and suitable four independent functions of $\vec{\sigma}, \vec{\tau}$ with the symplectic chart defined by $\vec{P}, \vec{Q}$ and suitable four independent functions of $\vec{S}, \alpha$ which is the best classical counterpart of the (generalized) Foldy-Wouthuysen transformation [6].

6) A manifest covariantization of the theory can be better achieved by defining the 4-vectors

- intrinsic spin $\quad h^\mu \equiv \frac{1}{c}\tilde{\sigma}^{\mu\nu}u_\nu, \quad h^\mu h_\mu \equiv -b^2(s)$
- intrinsic dipole $\quad \varepsilon^\mu \equiv \frac{1}{c}\sigma^{\mu\nu}u_\nu, \quad \varepsilon^\mu\varepsilon_\mu \equiv -d^2(s)$
- transverse momentum $\quad \zeta^\mu \equiv \frac{1}{c}\varepsilon^{\mu\nu\rho\sigma}h_\nu\varepsilon_\rho u_\sigma,$

where u^μ is the velocity 4-vector which turns out to be

$$u^\mu \equiv [\gamma c, \gamma\vec{v}] = \frac{b}{MS}P^\mu - \frac{\gamma_o\omega_o}{SM^2c^2}\sigma^{\mu\lambda}\sigma_{\lambda\rho}P^\rho \equiv \Lambda^\mu{}_\rho P^\rho. \tag{2}$$

The intrinsic spin and dipole magnitudes are related to S by

$$b = S\,\mathrm{Cosh}\,\delta_o + \sqrt{S^2-I}\,\mathrm{Sinh}\,\delta_o, \quad d = \sqrt{S^2-I}\,\mathrm{Cosh}\,\delta_o + S\,\mathrm{Sinh}\,\delta_o$$

where $\delta_o = \mathrm{Cosh}^{-1}\gamma_o$ is the Lorentz angle of the boost $\Sigma_o \rightarrow \Sigma_*$ being Σ_* the frame in which $\vec{x}$ is instantaneously at rest ($\vec{v} = 0$). Clearly

$$b^2(s) - d^2(s) = I.$$

It is remarkable that eq.(2) can be inverted: since

$$\mathrm{Det}|\Lambda^\mu{}_\rho| = (b/Ms)^2(d/M\sqrt{s^2-I})^2,$$

provided that
$$b \neq 0, \quad d \neq 0,$$

putting
$$\mu(s) \equiv \gamma_o(s)M(s), \tag{3}$$

it follows

$$P^\mu = \mu u^\mu + \frac{\mu c}{bd}\sqrt{\gamma_o^2-1}\,\zeta^\mu, \tag{4}$$

which gives in particular
$$u_\mu P^\mu = \mu(s)c^2.$$

Being $(H = cP_0)|_{\vec{V}=0} = \mu(s)\, c^2$, μ is essentially the intrinsic energy mass. The invariant spin leaves are best parametrized by b instead of S. Assuming that the function $\mu(b)$ is assigned, S itself, the inertial mass and the other relevant parameters are easily derived, for example

$$M(b) = \left[\mu^2 - \left(\frac{d\mu}{db}\right)^2 (b^2 - I)\right]^{1/2}.$$

A more transparent physical insight is achieved by defining the intrinsic mass

$$m = M/\gamma_0 . \qquad (5)$$

Actually a Lorentz rotation in the plane ε^μ, h^μ carries the family of moving (normalized) proper tetrads

$$\hat{u}^\mu = u^\mu/c, \quad \hat{\varepsilon}^\mu = \varepsilon^\mu/d, \quad \hat{\zeta}^\mu = \zeta^\mu/bd, \quad \hat{h}^\mu = h^\mu/b \qquad (6)$$

into a new family of tetrads given by

$$\hat{\Gamma}^\mu = \gamma_0 \hat{u}^\mu + \sqrt{\gamma_0^2 - 1}\, \hat{\zeta}^\mu \equiv \frac{P^\mu}{Mc}, \quad \hat{\Lambda}^\mu = \sqrt{\gamma_0^2-1}\, \hat{u}^\mu + \gamma_0 \hat{\zeta}^\mu$$
$$\hat{\varepsilon}^\mu, \; \hat{h}^\mu \; ; \quad (\hat{\Gamma}_\mu \hat{\Gamma}^\mu = 1 = -\hat{\Lambda}_\mu \hat{\Lambda}^\mu), \qquad (7)$$

which provide a further decomposition of the total linear momentum into "current" and "spin" momentum

$$P^\mu = m u^\mu + mc\sqrt{\gamma_0^2 - 1}\, \hat{\Lambda}^\mu . \qquad (8)$$

From eqs.(3),(4') it follows the relation among intrinsic energy, intrinsic mass and intrinsic spin energy

$$\mu c^2 = mc^2 + \frac{\omega_0^2}{mc}(S^2 - I) = mc^2 + \frac{c^2}{\mu}\left(\frac{d\mu}{db}\right)^2 (b^2 - I),$$

and the Dirac-like expression of the canonical Hamiltonian

$$H = \vec{V}\cdot\vec{P} + \frac{\gamma_0^2}{\gamma} mc^2 = \vec{V}\cdot\vec{P} + \frac{\mu}{\gamma} c^2 = \vec{V}\cdot\vec{P} + \frac{\gamma_0}{\gamma} Mc^2.$$

7) The manifestly covariant equations of motion can be written in either of the following equivalent forms (dot ≡ derivation with resp. to the proper time $d/d\tau = \gamma\{\cdots, H\}$)

$$\dot{P}^\mu = 0$$
$$\ddot{x}^\mu = \frac{1}{c}\omega_0^2 \gamma_0^2 \frac{\lambda_0}{d} \sigma^{\mu\lambda} u_\lambda = \frac{\gamma_0^2 \omega_0^2}{\mu^2 c^2} \sigma^{\mu\lambda} P_\lambda \equiv a^\mu$$
$$\dot{\sigma}^{\mu\nu} = \frac{M}{bd}\sqrt{\gamma_0^2 - 1}\left[(\sigma^{\mu\rho}\sigma_{\rho\lambda} u^\lambda) u^\nu - (\sigma^{\nu\rho}\sigma_{\rho\lambda} u^\lambda) u^\mu\right] =$$
$$= P^\mu u^\nu - P^\nu u^\mu . \qquad (9)$$

In terms of the tetrads family $N^\mu_{(\pi)} \equiv [\hat{u}^\mu, \hat{\varepsilon}^\mu, \hat{\zeta}^\mu, \hat{h}^\mu]$ we have the linear equations

$$\dot{N}^\mu_{(\pi)} = \Omega^{\mu\nu} N_{(\pi)\nu} \qquad , (\Omega^{\mu\nu} = -\Omega^{\nu\mu})$$

where the antisymmetrical tensor operates both a Fermi-Walker transport and a spatial rotation

$$\Omega^{\mu\nu} = \omega_0 s\{h^\mu, h^\nu\} = \frac{1}{c}(a^\mu u^\nu - a^\nu u^\mu) + \frac{\omega_0 \gamma_0^2}{bc}\varepsilon^{\mu\nu\rho\sigma} h_\rho u_\sigma .$$

Explicitly we have

$$\begin{aligned}
\dot{\hat{u}}^\mu &= \omega_0\gamma_0\sqrt{\gamma_0^2-1}\;\hat{\varepsilon}^\mu && \equiv \frac{1}{r_1(b)}\hat{\varepsilon}^\mu \\
\dot{\hat{\varepsilon}}^\mu &= \omega_0\gamma_0\sqrt{\gamma_0^2-1}\;\hat{u}^\mu + \omega_0\gamma_0^2\hat{\zeta}^\mu && \equiv \frac{1}{r_1(b)}\hat{u}^\mu + \frac{1}{r_2(b)}\hat{\zeta}^\mu \\
\dot{\hat{\zeta}}^\mu &= -\omega_0\gamma_0^2\hat{\varepsilon}^\mu && \equiv -\frac{1}{r_2(b)}\hat{\varepsilon}^\mu \\
\dot{\hat{h}}^\mu &= 0 .
\end{aligned} \tag{10}$$

Alternatively, in terms of the rotated tetrads,

$$\begin{aligned}
\dot{\hat{\Gamma}}^\mu &= 0 \qquad\qquad \dot{\hat{h}}^\mu = 0 \\
\left.\begin{aligned}\dot{\hat{\varepsilon}}^\mu &= \omega_0\gamma_0\hat{\Lambda}^\mu \\ \dot{\hat{\Lambda}}^\mu &= -\omega_0\gamma_0\hat{\varepsilon}^\mu\end{aligned}\right\} &\longrightarrow \ddot{\hat{\varepsilon}}^\mu + \omega_0^2\gamma_0^2\hat{\varepsilon}^\mu = 0 .
\end{aligned} \tag{11}$$

For any given spin leaf b = const., eqs.(10) can be viewed as evolution equations for the principal Frenet tetrad of the w.l. $x^\mu(\tau)$, being $\hat{u}^\mu$ $\hat{\varepsilon}^\mu$, $\hat{\zeta}^\mu$, $\hat{h}^\mu$ and $1/r_1(b)$, $1/r_2(b)$ the tangent, normal, binormal, trinormal and the first and second curvature, respect.(zero 3rd curv.). Then, following Gürsey [7] one can describe each tetrad by a "wave matrix" which parametrizes the Lorentz transformation boosting it into the local Cartesian axes or, equivalently, by the associated 4-spinor φ. Then, putting $\psi \equiv \exp[\frac{i}{2}\omega_0\gamma_0\tau]\varphi$, one obtains the spinor equation (γ^λ = Dirac matrices)

$$\left[\gamma^\lambda P_\lambda - M(b)c\right]\psi_b = \frac{2Mc}{\omega_0\gamma_0}\, i\,\frac{d\psi_b}{d\tau} .$$

It is worth stressing the formal simplicity of the manifestly covariant description in contrast with the complexity of the underlying canonical formalism.

8) The mass-spin relation is in principle arbitrary at the present level of analysis. A first natural restriction on the possible mass formulae $\mu(b)$ or $M(b)$ is obtained, however, by requiring that the zitterbewegung be confined for all values of the spin. Assuming the existence of an asymptotic expansion for $\mu(b)$, it is easily seen that this condition delimitates a set of functions which includes the class determined by imposing that the <u>intrinsic mass</u> m be a constant. This class, which I call provisionally the <u>physical class</u>, is characterized by

$$M^2(b) = \frac{m^2}{2} + \frac{m}{4l_0 c}\left[b + \sqrt{b^2 - I} + \frac{m^2c^2 l_0}{b + \sqrt{b^2-I}}\right]$$

where the lenght l_0 is the asymptotic value of the radius λ_0 . Then for the particular subclass $m^2c^2l_0^2 = I \longrightarrow l_0 = \frac{\sqrt{I}}{mc}$, ($|I|$ if $I<0$), $M(b)$ has a strict Regge behaviour

$$M^2(b) = \frac{I}{2l_0^2c^2} + \frac{\sqrt{I}}{2l_0^2c^2}\, b ,$$

with $\omega_0 = \frac{c}{2\ell_0}$, $\lambda_0 = \ell_0 [b\sqrt{I} - I]^{1/2} [b\sqrt{I} + I]^{-1/2}$.

An interesting limiting case is obtained from the physical class using the S parametrization and letting $m \to 0$. In this limit $V_0 \to c$, $V \to c$, however the whole canonical formalism remains well defined. The inversion of eq.(2) and manifestly covariant equations of motion like eqs.(9) or (11) can still be maintained replacing the proper time τ by the invariant parameter $\xi = \gamma_0 \tau$ (proper time of Σ_0). On the other hand the 4-vector

$$z^\mu \equiv dx^\mu / d\xi$$

becomes a null vector and the proper Frenet tetrads degenerate into the triads

$$z^\mu \quad , \quad E^\mu \equiv \frac{1}{c} \sigma^{\mu\nu} z_\nu \quad , \quad H^\mu \equiv \frac{1}{c} \tilde{\sigma}^{\mu\nu} z_\nu$$

$$E_\mu E^\mu = H_\mu H^\mu \equiv -\chi^2(s) \; ; \; \chi \equiv s + \sqrt{s^2 - I} \; .$$

The relevant parameters are now

$$M = \frac{1}{2\ell_0 c} \chi \; , \; \omega_0 = \frac{c}{\ell_0} \frac{\chi^2}{\chi^2 - I} \; , \; \lambda_0 = \ell_0 \frac{\chi^2 - I}{\chi^2} \; .$$

Finally it is a remarkable result that the "eccentric anomaly" $\eta(t)$ satisfies in this limit the true Kepler equation (see eq.(1)),

$$\eta(t) - \varepsilon \sin \eta(t) = \omega t \; , \; \left(\varepsilon = \frac{c}{H} \frac{|\vec{S} \wedge \vec{P}|}{s} \right)$$

with $\vec{\varrho}$ lying at the focus if $I = 0$! The velocity 3-vector, as measured in the laboratory, is

$$\vec{V} = \frac{c^2}{H} \vec{P} + \vec{V}_{zitterb.}$$

with $|\vec{V}| = c$ but

$$\vec{V}(t)^2_{zitterb.} = \frac{c^2}{\gamma_P} \frac{1 + \varepsilon \cos \eta(t)}{1 - \varepsilon \cos \eta(t)} \; .$$

The equations of motion of practically all the models of classical relativistic spinning particles existing in the literature are contained in the present theory as particular cases, in the limit of zero external fields (see Refs.(8) for a bibliography). It is worth emphasizing moreover, that, unlike the majority of these models, the present formulation is based on a definite canonical formalism and derived quite generally from the structure of the Poincarè group, conditions a, b, c, being in fact the only assumptions needed for the goal. The spin foliation and the existence of mass-spin trajectories are then particular consequences. The existing models typically contain some constraint of the form

$$\sigma^{\mu\nu} u_\nu = 0 \; , \quad \text{(Bhabha-Corben class: pure gyroscope)}$$

$$\tilde{\sigma}^{\mu\nu} u_\nu = 0 \; , \quad \text{(Hönl-Papapetrou class: pure dipole)}$$

and from the present point of view ($d = 0$, $b = 0$, respect.) are singular cases in which any tetrad structure collapses and the inversion of eq. (2) is no longer feasible. However the underlying canonical formalism still survives in all cases. These constrained models are found to be subcases of the class which is derived by imposing the condition that the intrinsic energy $\mu(s)$ be a constant (instead of $m(s)$!). The consequent expression of the inertial mass is

$$M(s) = \Xi \frac{2(s+\sqrt{s^2-I})}{(\Xi/\mu c)^2+(s+\sqrt{s^2-I})^2}, \quad \Xi \text{ real constant} > 0$$

and in particular

Bhabha-Corben class: $(\Xi/\mu c)^2 = I > 0 \quad : \quad M(s) = \frac{\mu c\sqrt{I}}{s}$

Hönl-Papapetrou class: $(\Xi/\mu c)^2 = -I < 0 \quad : \quad M(s) = \frac{\mu c\sqrt{|I|}}{\sqrt{s^2+|I|}}$.

All these models have unphysical descending mass-spin trajectories and unlimited zitterbewegung radius. It is interesting to note that in the limit $\mu \to \infty$ one still has $V_0 \to c$ and for asymptotic values of s these P.C.R. approach the so called zero mass, infinite spin, realizations corresponding to the Casimir values

$$P_\mu P^\mu = 0 \quad , \quad W_\mu W^\mu = -\Xi^2 \neq 0 \quad (I=0)$$

and usually ruled out on simple physical grounds [9].

REFERENCES

1- M.Pauri and G.M.Prosperi, J.Math.Phys.,7,366,(1966); 8,2256, (1967); 9,1146,(1968); 16,1503,(1975); 17,1468,(1976).
2- T.F.Jordan and N.Mukunda, Phys.Rev.,132,1842,(1963).
3- C.Møller, Ann.Ist.H.Poincarè,11,251,(1949).
4- M.H.L.Pryce, Proc.Roy.Soc.(London),195A,6,(1948).
5- G.N.Fleming, Phys.Rev.,137B,188,(1965); 139B,963,(1965).
6- L.L.Foldy and S.A.Wouthuysen, Phys.Rev.,78,29,(1950).
7- F.Gürsey, Nuovo Cimento,5,784,(1957).
8- J.B.Hughes, Suppl.Nuovo Cimento, 20,89,(1961); 20,148,(1961); P.Nyborg, Nuovo Cimento, 23,47,(1962); W.G.Dixon, Nuovo Cimento,34, 317,(1964); 38,1616,(1965); J.R.Ellis, Math.Proc.Camb.Phil.Soc.,78, 145,(1975).
9- E.P.Wigner, Ann.Math.,40,149,(1939); "Invariant Quantum Mechanical Equations of Motion" in Theoretical Physics (I.A.E.A., Wien 1965).

SIMILARITY ANALYSIS OF WAVE PROPAGATION IN AXIALLY NONLINEAR NONHOMOGENEOUS SYSTEMS

Mansa C. Singh
Associate Professor
Department of Mechanical Engineering
The University of Calgary
Calgary, Alberta, Canada

Waclaw Frydrychowicz
Visiting Scientist
Department of Mechanical Engineering
The University of Calgary
Calgary, Alberta, Canada

Similarity analysis of hyperbolic type of partial differential equations arising in wave propagation problems in uniaxial nonlinear-nonhomogeneous systems is complicated by the presence of a moving boundary at the wave front. In order that the original system be totally represented by a similarity representation, location of similarity coordinate at the wave front is required. For this purpose, similarity coordinate is established at the moving front by making use of the property of its invariance and that of the characteristics of the original system under the same group of transformations. The similarity representation so obtained reduces the original hyperbolic system into a boundary value problem consisting of an ordinary differential equation with one boundary condition at the near end and the other at the wave front. This analysis is applied to wave propagation in axially nonlinear-nonhomogeneous systems. Even though, the problems of wave propagation in a nonlinear axial system on the one hand and a nonhomogeneous axial system on the other have been dealt with by various authors, hardly any literature seems to be available on wave propagation in a system which is simultaneously nonlinear and nonhomogeneous. Similarity analysis augmented by similarity-characteristic relationship, thus helps to solve problems which are difficult to track by the known similarity methods.

A common form of the equation representing wave propagation phenomena in a one-dimensional system is given by

$$M(u) \equiv \psi\,(x,t,u,u_x,u_t)\,u_{xx} - u_{tt} = o\ , \tag{1}$$

where ψ is a function of the arguments shown in the parenthesis, x is the Lagrangian coordinate, t is the time, u is the particle displacement and the variables in the subscript denote differentiation. Equation (1) is a quasilinear equation of the hyperbolic type.

The fixed auxiliary conditions are

$$B_{\bar\alpha}\,(u_x,\ u_t,\ u,\ x,\ t) = 0;\ \bar\alpha = 1,\ 2,\ \ldots,\ m\ , \tag{2}$$

and the conditions at the moving boundary or the wave front;

$$u(x = D\ (t);\ t) = 0\ , \tag{3}$$

as the displacement is zero ahead of the wave-front, $x = D(t)$.

Let the system of equations (1-3) be invariant under the one-parameter infinitesimal group of transformation

$$\begin{aligned} \bar{u} &= u + \varepsilon\, U^*\ (x,\ t,\ u) + 0(\varepsilon^2)\ , \\ \bar{x} &= x + \varepsilon\, X\ (x,\ t,\ u) + 0(\varepsilon^2)\ , \\ \bar{t} &= t + \varepsilon\, T\ (x,\ t,\ u) + 0(\varepsilon^2)\ , \end{aligned} \tag{4}$$

where ε is an infinitesimal parameter, U^*, X and T are known as infinitesimals.

Assuming that a unique solution $u = \theta(x,t)$ of $M(u) = 0$ exists,

$$u(x,\ t) = \bar{u}(\bar{x},\ \bar{t}) = u(x,\ t,\ \theta(x,\ t),\ \varepsilon)\ . \tag{5}$$

The differential equations for the invariant surface can be obtained from the Equation (5) as

$$X\ (x,\ t,\ \theta)\ \frac{\partial \theta}{\partial x} + T\ (x,\ t,\ \theta)\ \frac{\partial \theta}{\partial t} = U^*\ (x,\ t,\ \theta)\ . \tag{6}$$

The characteristic equation corresponding to the equation (6) obtained from invariance of equation (1) under the transformation (4) is

$$\frac{dx}{X\ (x,\ t,\ \theta)} = \frac{dt}{T\ (x,\ t,\ \theta)} = \frac{d\theta}{U^*\ (x,\ t,\ \theta)}\ . \tag{7}$$

Solution of equation (7) gives the similarity transformation [1].

In the case of wave propagation in one dimensional rods, there is no displacement ahead of the wavefront. Therefore equation (3) is satisfied.

Let the path of the wavefront be described by a characteristic

$$\phi\ (x = D(t);\ t) = 0\ . \tag{8}$$

In order for the wave propagation problem to be expressible in terms of a similarity representation, $\phi(x,\ t) = \phi(\bar{x},\ \bar{t})$ at the moving boundary, $x = D(t)$, giving the expression

$$X(x,\ t,\ u)\ \frac{\partial \phi}{\partial x} + T\ (x,\ t,\ u)\ \frac{\partial \phi}{\partial t} = 0\ . \tag{9}$$

The infinitesimals $X\ (x,\ t,\ u)$ and $T\ (x,\ t,\ u)$ are related by the equation

$$\frac{dx}{X\ (x,\ t,\ u)} = \frac{dt}{T\ (x,\ t,\ u)} = \frac{d\phi}{0}\ . \tag{10}$$

Thus the invariance of the characteristics leads to the same subsystem relating

X (x, t, u) and T (x, t, u) as equation (7).

Let a characteristic of the equation (4) be expressible in the form

$$\phi\ (x,\ t,\ u,\ u_x,\ u_t) = 0\ . \tag{11}$$

From the equivalence of equations (6) and (10), similarity characteristic relation is obtained as [2],

$$\phi\ (x,\ t,\ F(\eta_w)\ F'(\eta_w)) = \eta_w\ \eta(x,\ t)\ , \tag{12}$$

where η is the similarity variable and η_w is its value at the wave front.

The system of equations of a nonlinear nonhomogeneous rod subjected to a time dependent velocity impact is:

$$\frac{E_o}{\rho q} x^n \left(-\frac{\partial u}{\partial x}\right)^{\frac{1-q}{q}} \frac{\partial^2 u}{\partial x^2} - \frac{E_o n}{\rho} x^{n-1} \left(-\frac{\partial u}{\partial x}\right)^{\frac{1}{q}} = \frac{\partial^2 u}{\partial t^2} \tag{13}$$

$$\frac{\partial u}{\partial t} (x = o, t) = V_c t^{\delta}\ , \quad \text{boundary condition} \tag{14a}$$

$$u = (x = D(t),\ t) = 0\ , \quad \text{condition at the wave front, where} \tag{14b}$$

$$e = \left[\frac{\sigma}{E(x)}\right]^q\ , \quad E(x) = E_o x^n\ , \quad e = -\frac{\partial u}{\partial x}\ ; \quad q > 0\ . \tag{15a,b,c,d}$$

In the above equations the coordinate x is measured along the rod, t denotes time, σ is stress, u is displacement, e is strain, E is modulus of elasticity, ρ is the density E_o, n are material parameters and δ and V_c are parameters of the velocity impact.

Similarity transformations [3,4] are obtained as

$$u = v_c t^{\delta+1}\ F(\eta)\ , \quad \eta = \frac{K\ x^{\frac{1+q\ -\ nq}{1+q}}}{t^m}\ , \quad \text{where} \tag{16a,b}$$

$$K = \left(\frac{\rho q}{E_o}\right)^{\frac{q}{q+1}} \left[\frac{1}{V_c}\right]^{\frac{1-q}{1+q}}\ , \quad m = 1 + \delta\ \frac{1-q}{1+q}\ . \tag{17a,b}$$

Making use of the equations (16) in (13) and (15a), the transformed equations are obtained in the form

$$\left[\left(\frac{1+q\ -\ nq}{1+q}\right)^{\frac{1+q}{q}} (-F')^{\frac{1-q}{q}} - m^2\eta^2\right] F''(\eta) - \left(\frac{1+q\ -\ nq}{1+q}\right)^{\frac{1}{q}} \frac{nq^2}{1+q}\ \eta^{-1}\ [-F'(\eta)]^{\frac{1}{q}}$$

$$- m\ (m-2\delta-1)\ \eta\ F'(\eta) - \delta(\delta+1)\ F(\eta) = 0\ , \tag{18}$$

$$F(\eta = 0) = \frac{1}{1+\delta}\ . \tag{19a}$$

Similarity characteristic relationship is used at the wave front to express (2b) in the form [2]

$$F(\eta = \eta_w) = 0 \ , \quad \eta_w = \left(\frac{1+q - nq}{1+q}\right)^{\frac{1+q}{2q}} \frac{1}{m} [-F'(\eta_w)]^{\frac{1-q}{2q}} \ . \tag{19b,c}$$

For an almost nonlinear rod, namely for the values of q close to unity, the similarity representation, (6,7) assumes the form

$$[(1+q - nq)^{\frac{1+q}{q}} \eta - (1+q)^{\frac{1+q}{q}} m^2 \eta^3] F''(\eta) + [n q^2 (1+q - nq)^{\frac{1}{q}} - (1+q)^{\frac{1+q}{q}}$$

$$m(m-2\delta-1)\eta^2] F'(\eta) - (1+q)^{\frac{1+q}{q}} \delta(1+\delta)\eta F(\eta) = 0 \tag{20a}$$

$$F(\eta = 0) = \frac{1}{1+\delta} \ , \quad F(\eta = \eta_w) = 0 \tag{20b,c}$$

where

$$\eta_w = \frac{1}{m} \left(\frac{1+q - nq}{1+q}\right)^{\frac{1+q}{2q}} \tag{20d}$$

and for

$$q > 0 \ , \quad m > 0 \ , \quad n < \frac{1+q}{q} \ .$$

Equation (20a) is an ordinary differential equation with variable coefficients and varying parameters and with $\eta = 0$, and $\eta = \eta_w$ as its two regular singular points [5]. The solution of the boundary value problem (20) has been obtained in the form

$$F(\eta) = \frac{1}{1+\delta} [F_1(\eta) - \frac{F_1(\eta_w)}{F_2(\eta_w)} F_2(\eta)] \ , \text{ for} \tag{21}$$

$$0 \leq \eta \leq \left(\frac{1+q - nq}{1+q}\right)^{\frac{1+q}{2q}} \frac{1}{m} \ , \text{ and for} \tag{22a}$$

$$(1+q - nq)(1 + 2\delta) + m(1+q - nq + nq^2) > 0 \ ; \tag{22b}$$

$$n < \frac{1}{q} \ ; \quad 1 + \delta \frac{1-q}{1+q} > 0 \quad \text{and} \quad q > 0 \ . \tag{22c,d,e}$$

The functions $F_1(\eta)$ and $F_2(\eta)$ in (21) are obtained in the series form as solution of equation (20) as

$$F_1(\eta) = 1 + \sum_{s=1}^{\infty} \eta^{2s} \frac{(1+q)^{\frac{1+q}{q}s}}{2^s s! (1+q - nq)^{\frac{s}{q}}} \frac{\delta(\delta+1)}{(1+q - nq + nq^2)}$$

$$\frac{(\delta-2m+1)(\delta-2m)}{[3(1+q - nq) + nq^2]} \cdots \frac{[\delta-2m(s-1)+1][\delta-2m(s-1)]}{[(2s-1)(1+q - nq) + nq^2]} \tag{23a}$$

and

$$F_2(\eta) = \eta^{\frac{1+q-nq-nq^2}{1+q-nq}} \left\{ 1 + \sum_{s=1}^{\infty} \eta^{2s} \frac{(1+q)^{\frac{1+q}{q}s}}{2^s s! (1+q - nq)^{\frac{s}{q}}} \right.$$

$$\frac{(\delta-m+\frac{mnq^2}{1+q-nq}+1)(\delta-m+\frac{mnq^2}{1+q-nq})}{[3(1+q-nq)-nq^2]} \cdots\cdots$$

$$\cdots\cdots \left.\frac{[\delta-2m(s-1)+\frac{mnq^2}{1+q-nq}+1-m]\,[\delta-m(2s-1)+\frac{mnq^2}{1+q-nq}]}{[(2s+1)(1+q-nq)-nq^2]}\right\} \qquad (23b)$$

The above solution has been specialized to a linear nonhomogeneous rod [6], a nonlinear homogeneous rod and a linear homogeneous rod. All the results agree with those available. The nonlinear system (18,19) is solved numerically for different values of q and behaviour of the rod studied for compatible values of the parameters q, n and δ. The numerical solutions obtained for values of q close to unity compare well with those given by (21,23) for almost nonlinear rod.

REFERENCES

1. Bluman, G.W., Cole, J.D., Similarity Methods for Differential Equations, Springer-Verlag, New York, 1974.
2. Seshadri, R., Singh, M.C., Archives of Mechanics 6, 1980.
3. Singh, M.C., Brar, G.S., J. Acoustical Society Am. 63 (4), 1978.
4. Seshadri, R., Singh, M.C., Archives of Mechanics 28, 1976.
5. Kaplan, W., Ordinary Differential Equations, Addison-Wesley, Reading, 1958.
6. Singh, M.C., Frydrychowicz, W., "Wave Propagation in Nonhomogeneous Thin Elastic Rods Subjected to Time Dependent Velocity Impact", Department of Mech. Eng., Report #169, May 1980.

Author index

Selected Issues from

Lecture Notes in Mathematics

Vol. 684: E. E. Rosinger, Distributions and Nonlinear Partial Differential Equations. XI, 146 pages. 1978.

Vol. 690: W. J. J. Rey, Robust Statistical Methods. VI, 128 pages. 1978.

Vol. 691: G. Viennot, Algèbres de Lie Libres et Monoïdes Libres. III, 124 pages. 1978.

Vol. 693: Hilbert Space Operators, Proceedings, 1977. Edited by J. M. Bachar Jr. and D. W. Hadwin. VIII, 184 pages. 1978.

Vol. 696: P. J. Feinsilver, Special Functions, Probability Semigroups, and Hamiltonian Flows. VI, 112 pages. 1978.

Vol. 702: Yuri N. Bibikov, Local Theory of Nonlinear Analytic Ordinary Differential Equations. IX, 147 pages. 1979.

Vol. 704: Computing Methods in Applied Sciences and Engineering, 1977, I. Proceedings, 1977. Edited by R. Glowinski and J. L. Lions. VI, 391 pages. 1979.

Vol. 710: Séminaire Bourbaki vol. 1977/78, Exposés 507–524. IV, 328 pages. 1979.

Vol. 711: Asymptotic Analysis. Edited by F. Verhulst. V, 240 pages. 1979.

Vol. 712: Equations Différentielles et Systèmes de Pfaff dans le Champ Complexe. Edité par R. Gérard et J.-P. Ramis. V, 364 pages. 1979.

Vol. 716: M. A. Scheunert, The Theory of Lie Superalgebras. X, 271 pages. 1979.

Vol. 720: E. Dubinsky, The Structure of Nuclear Fréchet Spaces. V, 187 pages. 1979.

Vol. 724: D. Griffeath, Additive and Cancellative Interacting Particle Systems. V, 108 pages. 1979.

Vol. 725: Algèbres d'Opérateurs. Proceedings, 1978. Edité par P. de la Harpe. VII, 309 pages. 1979.

Vol. 726: Y.-C. Wong, Schwartz Spaces, Nuclear Spaces and Tensor Products. VI, 418 pages. 1979.

Vol. 727: Y. Saito, Spectral Representations for Schrödinger Operators With Long-Range Potentials. V, 149 pages. 1979.

Vol. 728: Non-Commutative Harmonic Analysis. Proceedings, 1978. Edited by J. Carmona and M. Vergne. V, 244 pages. 1979.

Vol. 729: Ergodic Theory. Proceedings 1978. Edited by M. Denker and K. Jacobs. XII, 209 pages. 1979.

Vol. 730: Functional Differential Equations and Approximation of Fixed Points. Proceedings, 1978. Edited by H.-O. Peitgen and H.-O. Walther. XV, 503 pages. 1979.

Vol. 731: Y. Nakagami and M. Takesaki, Duality for Crossed Products of von Neumann Algebras. IX, 139 pages. 1979.

Vol. 733: F. Bloom, Modern Differential Geometric Techniques in the Theory of Continuous Distributions of Dislocations. XII, 206 pages. 1979.

Vol. 735: B. Aupetit, Propriétés Spectrales des Algèbres de Banach. XII, 192 pages. 1979.

Vol. 738: P. E. Conner, Differentiable Periodic Maps. 2nd edition, IV, 181 pages. 1979.

Vol. 742: K. Clancey, Seminormal Operators. VII, 125 pages. 1979.

Vol. 755: Global Analysis. Proceedings, 1978. Edited by M. Grmela and J. E. Marsden. VII, 377 pages. 1979.

Vol. 756: H. O. Cordes, Elliptic Pseudo-Differential Operators – An Abstract Theory. IX, 331 pages. 1979.

Vol. 760: H.-O. Georgii, Canonical Gibbs Measures. VIII, 190 pages. 1979.

Vol. 762: D. H. Sattinger, Group Theoretic Methods in Bifurcation Theory. V, 241 pages. 1979.

Vol. 765: Padé Approximation and its Applications. Proceedings, 1979. Edited by L. Wuytack. VI, 392 pages. 1979.

Vol. 766: T. tom Dieck, Transformation Groups and Representation Theory. VIII, 309 pages. 1979.

Vol. 771: Approximation Methods for Navier-Stokes Problems. Proceedings, 1979. Edited by R. Rautmann. XVI, 581 pages. 1980.

Vol. 773: Numerical Analysis. Proceedings, 1979. Edited by G. A. Watson. X, 184 pages. 1980.

Vol. 775: Geometric Methods in Mathematical Physics. Proceedings, 1979. Edited by G. Kaiser and J. E. Marsden. VII, 257 pages. 1980.

Vol. 779: Euclidean Harmonic Analysis. Proceedings, 1979. Edited by J. J. Benedetto. III, 177 pages. 1980.

Vol. 780: L. Schwartz, Semi-Martingales sur des Variétés, et Martingales Conformes sur des Variétés Analytiques Complexes. XV, 132 pages. 1980.

Vol. 782: Bifurcation and Nonlinear Eigenvalue Problems. Proceedings, 1978. Edited by C. Bardos, J. M. Lasry and M. Schatzman. VIII, 296 pages. 1980.

Vol. 783: A. Dinghas, Wertverteilung meromorpher Funktionen in ein- und mehrfach zusammenhängenden Gebieten. Edited by R. Nevanlinna and C. Andreian Cazacu. XIII, 145 pages. 1980.

Vol. 786: I. J. Maddox, Infinite Matrices of Operators. V, 122 pages. 1980.

Vol. 787: Potential Theory, Copenhagen 1979. Proceedings, 1979. Edited by C. Berg, G. Forst and B. Fuglede. VIII, 319 pages. 1980.

Vol. 791: K. W. Bauer and S. Ruscheweyh, Differential Operators for Partial Differential Equations and Function Theoretic Applications. V, 258 pages. 1980.

Vol. 792: Geometry and Differential Geometry. Proceedings, 1979. Edited by R. Artzy and I. Vaisman. VI, 443 pages. 1980.

Vol. 793: J. Renault, A Groupoid Approach to C*-Algebras. III, 160 pages. 1980.

Vol. 798: Analytic Functions, Kozubnik 1979. Proceedings. Edited by J. Ławrynowicz. X, 476 pages. 1980.

Vol. 799: Functional Differential Equations and Bifurcation. Proceedings 1979. Edited by A. F. Izé. XXII, 409 pages. 1980.

Vol. 801: K. Floret, Weakly Compact Sets. VII, 123 pages. 1980.

Vol. 802: J. Bair, R. Fourneau, Etude Géometrique des Espaces Vectoriels II. VII, 283 pages. 1980.

Vol. 804: M. Matsuda, First Order Algebraic Differential Equations. VII, 111 pages. 1980.

Vol. 805: O. Kowalski, Generalized Symmetric Spaces. XII, 187 pages. 1980.

Vol. 807: Fonctions de Plusieurs Variables Complexes IV. Proceedings, 1979. Edited by F. Norguet. IX, 198 pages. 1980.

Vol. 810: Geometrical Approaches to Differential Equations. Proceedings 1979. Edited by R. Martini. VII, 339 pages. 1980.

Vol. 816: L. Stoica, Local Operators and Markov Processes. VIII, 104 pages. 1980.

Vol. 819: Global Theory of Dynamical Systems. Proceedings, 1979. Edited by Z. Nitecki and C. Robinson. IX, 499 pages. 1980.